P9-BJT-352

PRINCIPLES OF MICROBIOLOGY

SEVENTH EDITION

Principles of MICROBIOLOGY

ALICE LORRAINE SMITH
A.B., M.D., F.C.A.P., F.A.C.P.

Associate Professor of Pathology, The University of Texas Health Science Center at Dallas, Texas; Assistant Professor of Microbiology, Department of Nursing, Dominican College and St. Joseph's Hospital, Houston, Texas

WITH 305 ILLUSTRATIONS

THE C. V. MOSBY COMPANY
Saint Louis 1973

Seventh edition

Printed in the United States of America

Distributed in Great Britain by Henry Kimpton, London

Library of Congress Cataloging in Publication Data

Smith, Alice Lorraine, 1920-
Principles of microbiology.

First ed. by C. F. Carter.
Includes bibliographies.
1. Medical microbiology. I. Carter, Charles Franklin, 1890-1957. Principles of microbiology.
II. Title. [DNLM: 1. Microbiology. QW 4 S642p 1973]
QR46.S635 1973 616.01 72-14372
ISBN 0-8016-4680-4

TS/CB/B 9 8 7 6 5 4

To my mother

Preface

> I keep six honest serving-men
> (They taught me all I knew);
> Their names are What and Why and When
> And How and Where and Who.
>
> RUDYARD KIPLING

The scientific answers to the What and When and How and Who are coming so thick and fast in these exciting days that to no one's surprise it is time for a revision of *Principles of Microbiology*. Here then is the book reworked to give contemporary coverage in microbiology and ordered to progress in a design readily assimilated by students in health science training programs.

The basic scheme of prior editions, a unit system in biologic orbit around microorganisms, continues in this one. In the first two of six units, the pattern begins with basic concepts in microbiology. In the rest, it develops the events of injury, indicts culprits, and emphasizes restraints.

What happens when microbes and their products contact living cells of the human body is a theme permeating this book. One full unit is devoted to it. This unit includes a chapter on immunology, drastically reorganized to coordinate the immune system, immunoglobulins, cell-mediated immunity, immunopathology, tissue transplantation, and tumor immunology. The largest unit of the book makes up the roster of significant pathogens and parasites; a smaller but more compact unit categorizes various agents found to destroy or impede microorganisms. Important additions were incorporated into both of them. An unusual unit, the last one, relates the student to the microbial life of our environment. A survey, yet a practical unit, it accommodates such items as the measures to safeguard food and swimming pool sanitation. Herein two chapters are paired to focus on the best available information on immunization from the United States Public Health Service, the American Academy of Pediatrics, and the United States Armed Forces. The companion chapters sort out modern biologic products, outline technics in passive immunization, tabulate the latest schedules for active immunization, and give the crucial guidelines for administration of biologic products.

I have examined all parts of this book, editing and updating to keep frame and substance relevant to the contemporary scene in health careers. Examples of topics in microbiology expanded are the nosocomial infections, the vulnerability of the compromised host, organ transplantation, immunologic reactions, venereal diseases, viral oncogenesis, and viral teratogenesis. Discussions of pertinent subjects such as the Australia antigen, the Epstein-Barr (EB) virus, cytomegalic inclusion disease, pneumocystosis, *Echinococcus granulosus*, bacteriology of dog and cat bites, the mast cell in allergy, T-strain mycoplasmas, and molluscum contagiosum represent new material. The historical survey, "Milestones of Progress," augmented over that of the past, appears as the second chapter in the book.

I chose the projects for the laboratory survey at the end of each unit for their adaptability to the needs of students and suitability for use with varied, sometimes limited, facilities. To support the laboratory sections, a sequence of four chapters, spaced within the first three units, indicates both conventional and newly emerging methods for studying microbes. Standard immunologic procedures are collected into a new chapter. For maximum laboratory safety, at the outset of the laboratory sessions, please sound the warnings!

The illustrations were carefully selected to enrich the meaning of the prose. For instance, when the subject of tuberculin testing is presented, the photograph of the tine test skin reactions makes the impact. Every teacher knows that tables dramatize related facts and give quick access to significant information. Immunization schedules, sterilization technics, incubation periods, comparative sizes, metric equivalents, differential characteristics, and biologic properties have been thus arranged more effectively. Current references are assembled at the end of a unit except where they relate directly to a given group of organisms. In addition to thought-provoking questions at the end of a chapter, evaluation of the student's progress may be made with a cluster of exercises arranged at the end of each unit. The classification of microbes used throughout is that found in *Bergey's Manual of Determinative Bacteriology* (1957) with certain exceptions. The enteric bacteria are classified according to the scheme of W. H. Ewing, and the modern classification of viruses, that based on their properties, is given. Bedsoniae (chlamydiae) are treated separately. Sources for the glossary are found in the text, standard medical dictionaries, and Webster's unabridged dictionary.

This revision would not have been possible without the counsel, technical know-how, and cooperation of certain talented persons at The University of Texas Health Science Center in Dallas. In the Department of Pathology, I gratefully acknowledge the kindness of Dr. V. A. Stembridge, Chairman, Dr. R. C. Reynolds and Dr. B. D. Fallis, professors, Mrs. Phyllis Kitterman, Secretary, and Mrs. Linda Bolding, Laboratory Technical Assistant; in the Department of Medical Art, Mr. W. A. Osborne, Chairman, Miss Jean Gionas, medical photographer, Miss Karen Waldo and Mr. Ron Brandon, medical illustrators; and in the Library, Mrs. Elinor Reinmiller, Reference Librarian, and other able members of Dr. Donald Hendricks' staff. I am indebted deeply to Mrs. Earline Kutscher, Chief Technologist, and her staff at the Bacteriology Laboratory of the Dallas County Hospital District for invaluable assistance during preparation of this text.

Now a special word of appreciation to teachers and students whose ever-welcome criticisms and comments have guided me: may I voice a heartfelt thanks to the many of you who have used this text and who carefully consider this edition.

ALICE LORRAINE SMITH

Contents

UNIT ONE **MICROBIOLOGY: prelude and primer**

1 Microbiology: definition and dimension, 2
2 Milestones of progress, 12
3 The basic unit, 26
4 The bacterial cell, 40
5 Methods for microbes: visualization, 53

UNIT TWO **MICROBIOLOGY: principles and procedures**

6 Bacteria: biologic needs, 78
7 Methods for microbes: cultivation, 84
8 Bacteria: biologic activities, 97
9 Methods for microbes: biochemical reactions, animal inoculation, 107
10 Specimen collection, 114

UNIT THREE **MICROBES: production of infection**

11 Microbes and disease, 140
12 The body's defense, 155
13 Immunology, 165
14 Methods for microbes: immunologic reactions, 189
15 Allergy, 202

UNIT FOUR **MICROBES: preclusion of disease**

16 Physical agents in sterilization, 230
17 Chemical agents in sterilization, 239
18 Practical technics, 264

UNIT FIVE **MICROBES: pathogens and parasites**

19 The gram-positive cocci: pyogenic cocci, 288
20 The gram-negative cocci: neisseriae, 310
21 The aerobic gram-positive bacilli: diphtheria bacilli, anthrax bacilli, 324
22 The anaerobic gram-positive bacilli: clostridia, 332
23 The gram-negative enteric bacilli, 340

24 The gram-negative small bacilli, 357
25 Miscellaneous microbes, 371
26 The spiral microorganisms, 379
27 The viruses, 396
28 Viral diseases, 420
29 The rickettsiae and bedsoniae, 456
30 The acid-fast bacteria, 470
31 Fungi: medical mycology, 484
32 Protozoa: medical parasitology, 508
33 Metazoa: medical parasitology, 527

UNIT SIX MICROBES: public welfare

34 The microbe everywhere, 568
35 The microbiology of water, 577
36 The microbiology of food, 586
37 Biologic products for immunization, 596
38 Recommended immunizations, 617

GLOSSARY, 643

UNIT ONE

MICROBIOLOGY: prelude and primer

1 Microbiology: definition and dimension

2 Milestones of progress

3 The basic unit

4 The bacterial cell

5 Methods for microbes: visualization

Laboratory survey of unit one

Evaluation for unit one

1 Microbiology: definition and dimension

Take interest, I implore you, in those sacred dwellings which one designates by the expressive term: laboratories. Demand that they be multiplied, that they be adorned. These are the temples of the future—temples of well-being and of happiness. There it is that humanity grows greater, stronger, better.

LOUIS PASTEUR

DEFINITION

Biology is the science that treats of living organisms. The branch of biology dealing with *microbes*—organisms structured as one cell and studied with the microscope—is *microbiology* ("microbe-biology"). Microbiology considers the occurrence in nature of the microscopic forms of life, their reproduction and physiology, their participation in the processes of nature, their helpful or harmful relationships with other living things, and their significance in science and industry.

Within the province of microbiology lies the study of bacteria, rickettsiae, viruses, fungi, and protozoa. The subordinate sciences are *bacteriology*, the study of bacteria and rickettsiae (both unicellular plants); *virology*, the study of viruses; *mycology*, the study of fungi (unicellular and multicellular plants); and *protozoology*, the study of protozoa (unicellular animals). Many microbes are parasitic. The science dealing with organisms dependent upon living things for their sustenance is *parasitology*. So closely associated with microbiology as to be considered a part of it is the science of *immunology*, the study of those mechanisms whereby one organism deals with the harmful effects of another.

Although man has lived with microorganisms from time immemorial and has used certain of their activities, such as fermentation, to his advantage, the science of microbiology is a product of only the last 100 years. The studies of Antonj van Leeuwenhoek in the seventeenth century had shown the existence of microscopic forms of life, but it was not until the work of Louis Pasteur toward the end of the nineteenth century (some 200 years later) that the science of microbiology really took shape. The new science stated the germ theory of disease, demonstrated patterns of communicable disease, and gave man a measure of protection he had not known in his struggle against the injurious forces in the biologic environment.* In its time this very young science has influenced practically every phase of human endeavor.

*For scientific knowledge to bring results, such as in the organization of public health programs, it must be disseminated. Such is the aim of *health education.* To the individual it explains the mechanisms by which he can protect himself against microbial hazards. To the social group it designates the available community resources.

DIMENSION

Biologic classification

All living things are placed in either the plant or animal kingdom, and their overall separation is indicated in Table 1-1. To classify plants and animals, the following terms are basic units (in ascending order):

1. Species—organisms sharing a set of biologic characteristics and reproducing only their exact kind
   a. Strain—organisms within the species varying in a given quality
   b. Type—organisms within the species varying immunologically
2. Genus (*plural,* genera)—closely related species
3. Family—closely related genera
4. Order—closely related families
5. Class—closely related orders
6. Phylum (*plural,* phyla)*—related classes

TABLE 1-1. MAJOR FEATURES DISTINGUISHING PLANTS FROM ANIMALS

	Structure		Function			
	Cell wall	Growth	Energy source	Chlorophyl	Main reserve food	Mobility
Plants	Well defined	Continuous	Photosynthesis	Present	Starch	Absent
Animals	None	Limited	Organic materials	Absent	Glycogen, fat	Present

Naming of microbes

The scientific name of an animal or plant is usually made up of two words that are Latin or Greek in form. The first name begins with a capital letter and denotes the genus. The second name begins with a small letter and denotes the species. Either the genus or the species name may be derived from the proper name of a person or place or from a term describing some feature of the organism. The proper name may be that of the scientific investigator or that of the related geographic area. Biologic characteristics indicated include the color of the organism, the location in nature, the disease produced, and the presence of certain enzymes. For examples, *Staphylococcus aureus* is the scientific name for bacteria of genus *Staphylococcus* (Greek *staphylē,* 'bunch of grapes' + *kokkos,* 'berry') and species *aureus* (Latin *aureus,* 'golden'). It indicates that the bacteria grow in typical clusters and produce a golden pigment. Honoring the German bacteriologist G. T. A. Gaffky, *Gaffkya tetragena* names bacteria occurring characteristically in tetrads. Whether a term indicates a class, order, or family may be determined from its ending. Classes end in *-etes,* orders end in *-ales,* and families end in *-aceae.* For instance, *Schizomycetes* is the name of a class, *Eubacteriales* is the name of an order, and *Bacillaceae* is the name of a family.

*In plant biology, the term *division* is used instead of phylum.

Scope of microbiology

The lower forms of life incorporate variably the features of both plants and animals and do not show the dramatic differences of the higher forms. It is often difficult to define many microorganisms sharply as being either plant or animal. As bacteria and related forms classified as plants have been more closely studied, the differences between these one-celled units and the more complex plants appear even greater.

Although the differences do exist, most of the microorganisms—fungi, bacteria, and related forms—surveyed in a treatise on microbiology remain classified as plants and fall into two divisions of the plant kingdom: (1) *Protophyta,* primitive plants, and (2) *Thallophyta,* thallus plants (Greek *thallos,* young shoot or branch). A separate division *Protophyta* is advocated to contain bacteria, related forms, rickettsiae, and viruses in order to emphasize their special features. *Thallophytes* are defined as simple forms of plant life that do not differentiate into true roots, stems, or leaves, and include the "fungi" (a term ordinarily used to refer to molds, yeasts, and certain related forms). Table 1-2 gives the subdivisions of these two.

Since some of the most important disease-producing agents known to man are lower forms of animal life, the science of microbiology extends its coverage to the unicellular protozoa, the simplest forms of animal life, and to a restricted number of the more complex multicellular or metazoan animals. (These forms of life are discussed in Chapters 33 and 34.)

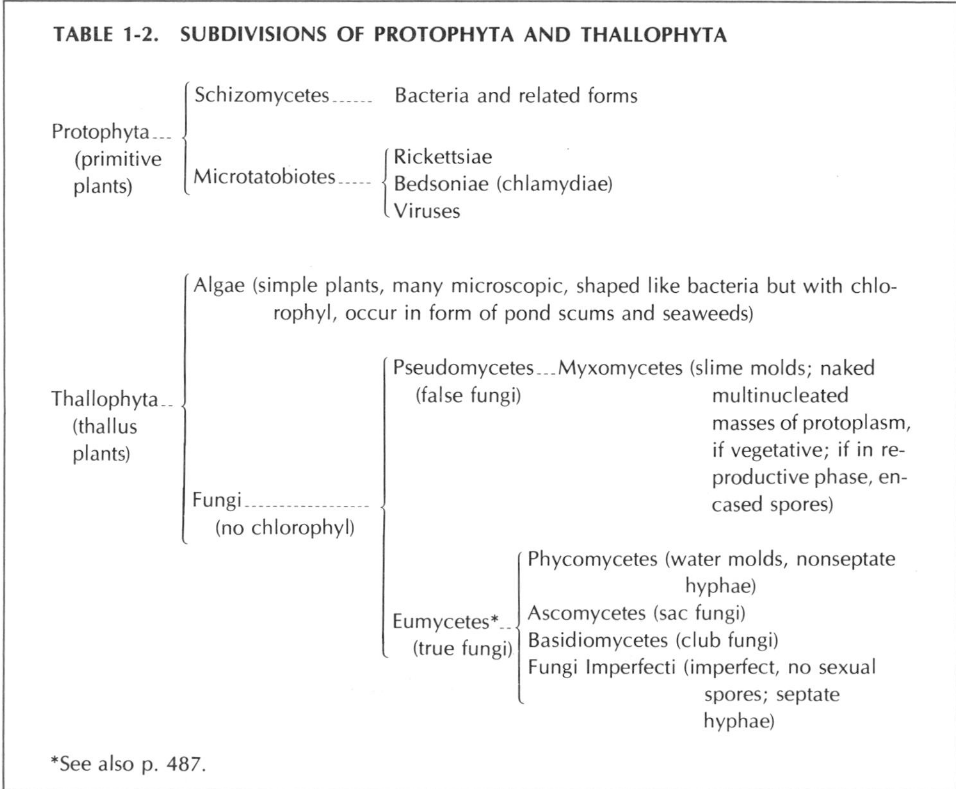

TABLE 1-2. SUBDIVISIONS OF PROTOPHYTA AND THALLOPHYTA

Division	Subdivision	Group	Members
Protophyta (primitive plants)	Schizomycetes		Bacteria and related forms
	Microtatobiotes		Rickettsiae
			Bedsoniae (chlamydiae)
			Viruses
Thallophyta (thallus plants)	Algae (simple plants, many microscopic, shaped like bacteria but with chlorophyl, occur in form of pond scums and seaweeds)		
	Fungi (no chlorophyl)	Pseudomycetes (false fungi)	Myxomycetes (slime molds; naked multinucleated masses of protoplasm, if vegetative; if in reproductive phase, encased spores)
		Eumycetes* (true fungi)	Phycomycetes (water molds, nonseptate hyphae)
			Ascomycetes (sac fungi)
			Basidiomycetes (club fungi)
			Fungi Imperfecti (imperfect, no sexual spores; septate hyphae)

*See also p. 487.

Classification of bacteria

The classification of bacteria is difficult. This applies both to the separation of bacteria into groups and to the placing of certain organisms into the proper group. Biologic classification is based largely upon morphology, but the morphology of bacteria as a whole is so uniform that it is useful only in dividing bacteria into comparatively large groups. Shape has been an important factor in general classification, but for more exact identification such criteria as staining reactions, cultural characteristics, biochemical behavior, and immunologic differences must be used.

In this book we adhere to the scientific classification embodied in *Bergey's Manual of Determinative Bacteriology*, with certain important exceptions. This is the one most generally accepted in the United States. Table 1-3 presents in abbreviated form an overall survey of Bergey's classification of the microorganisms designated as bacteria, related forms, rickettsiae, and viruses.*

Bacteria that live in or about the human body have been much more completely studied than have those from animals or any other source in nature. Of these, the ones producing disease are most familiar. Discussions in subsequent chapters will necessarily be limited largely to those microorganisms significantly related to human disease.

---

*The American Type Culture Collection, located outside Washington, D.C., in Rockville, Maryland, 20852, maintains and distributes for research or other purposes authentic cultures of practically all known microorganisms. This nonprofit, independent agency grew out of the Bacteriological Collection and Bureau for the Distribution of Bacterial Cultures, established in 1911, at the American Museum of Natural History in New York City. There are over 12,000 strains of bacteria and fungi, 356 classified viruses and rickettsiae, and 58 certified cell lines in this collection. Each year it supplies over 14,000 cultures of bacteria and fungi and 1400 of viruses to industrial and educational institutions, such as medical schools, breweries, wineries, oil companies, pharmaceutical houses, and food processors.

TABLE 1-3. ABBREVIATED CLASSIFICATION OF MICROBES FROM *Bergey's Manual of Determinative Bacteriology* (1957)*

Division I. *Protophyta* (primitive plants)
- Class I. *Schizophyceae* (blue-green algae)
- Class II. *Schizomycetes* (bacteria and related forms)
  - Order I. *Pseudomonadales* (soil and water bacteria; some forms pigmented)
    - Suborder I. *Rhodobacteriineae* (red, purple, brown, or green bacteria containing chlorophyl-like pigments, or bacteriochlorophyl; freshwater or marine habitat)
    - Suborder II. *Pseudomonadineae* (soil and water bacteria; some forms pathogenic for man)
      - Family I. *Nitrobacteraceae* (soil and freshwater bacteria converting ammonia to nitrite, nitrite to nitrate)
        - *Nitrosomonas* species (ellipsoidal bacteria oxidizing ammonia to nitrite)
        - *Nitrobacter* species (rod-shaped bacteria oxidizing nitrite to nitrate)
      - Family II. *Methanomonadaceae* (soil and water bacteria using methane, hydrogen, or carbon monoxide as source of energy)
      - Family III. *Thiobacteriaceae* (sulfur bacteria of soil and sulfurous water)

Continued.

*Bracketed listings reflect current trends in nomenclature.

TABLE 1-3. ABBREVIATED CLASSIFICATION OF MICROBES FROM *Bergey's Manual of Determinative Bacteriology* (1957)*—cont'd

Family IV. *Pseudomonadaceae* (soil and water bacteria, including seawater or even heavy brine; many water-soluble, fluorescent pigments notable; many plant, a few animal pathogens)

Pseudomonas aeruginosa (causes many different disease conditions in man and animals; found in polluted water and sewage; the organism of blue pus)

Pseudomonas pseudomallei (causes melioidosis)

[*Pseudomonas mallei* (causes glanders)]

Acetobacter species (vinegar bacteria participating in carbon cycle)

Aeromonas hydrophila (nonluminescent water bacteria, sometimes in human feces; pathogens of cold-blooded animals; cause of red leg, a bacteremic disease of frogs; rare pathogen of "compromised host")

Photobacterium species (luminescent saltwater bacteria on dead fish and crustaceans)

Family V. *Caulobacteraceae* (salt- or freshwater bacteria; characteristic stalk for attachment to substrate; grouping in clusters or rosettes)

Family VI. *Siderocapsaceae* (iron bacteria of fresh water)

Family VII. *Spirillaceae* (curved, comma-shaped, or spiral bacteria of water; some forms pathogenic)

Vibrio comma [*Vibrio cholerae*] (causes cholera)

[Vibrio Eltor (also causes cholera)]

Spirillum minus (found in the blood of rats and mice; causes one kind of rat-bite fever)

Order II. *Chlamydobacteriales* (colorless, algalike bacteria of fresh water and seawater; may possess a sheath)

Order III. *Hyphomicrobiales* (budding bacteria)

Order IV. *Eubacteriales* (true bacteria)

Family I. *Azotobacteraceae* (large soil and water bacteria; important for nitrogen fixation)

Azotobacter species (bacteria of single genus within family)

Family II. *Rhizobiaceae* (important for nitrogen fixation in nodules on roots of leguminous plants)

Rhizobium species (bacteria of the nitrogen-fixing genus within the family)

Family III. *Achromobacteraceae* (soil and fresh- and saltwater bacteria; some plant pathogens)

Alcaligenes faecalis (nonpathogenic inhabitant of intestinal tract of man and animals)

Achromobacter species (genus of nonpigmented soil and water bacteria)

Flavobacterium species (proteolytic bacteria producing characteristic yellow, orange, or red pigments; some pathogenic)

Family IV. *Enterobacteriaceae*† (pathogenic and saprophytic species; many are normal inhabitants of intestinal tract of man and animals; coliform-dysentery-typhoid group of bacteria; some are plant parasites and pathogens)

Tribe I. *Escherichieae* (members ferment lactose; have no affinity for plants)

Escherichia coli (colon bacillus of man and other vertebrates; widespread in nature; important pathogen out of its normal setting)

Aerobacter aerogenes (coliform organism widely distributed in nature, an important pathogen in urinary tract of man)

Klebsiella pneumoniae (pathogenic in the respiratory tract)

†Note modification of Bergey's classification of this family in Chapter 23.

Paracolobactrum aerogenoides (normal occupant of intestinal tract of man and animals; sometime pathogen)

Tribe II. *Erwinieae* (plant pathogens)

Tribe III. *Serratieae* (members of single genus variably produce characteristic red pigment, prodigiosin)

Serratia marcescens (saprophyte noted for its red pigment; opportunist pathogen)

Tribe IV. *Proteeae* (members found in feces, decompose urea, can be formidable pathogens)

Proteus vulgaris (found in decomposing and putrefying animal and vegetable matter; causes urinary tract infections, wound infections, and rarely peritonitis and meningitis)

Proteus morganii (causes summer diarrhea in infants)

Proteus rettgeri (causes gastrointestinal disease)

Tribe V. *Salmonelleae* (members are important pathogens; do not decompose urea nor ferment lactose)

Salmonella choleraesuis (causes salmonellosis)

Salmonella typhimurium (causes food poisoning)

Salmonella enteritidis (causes salmonellosis)

Salmonella typhosa [*Salmonella typhi*] (causes typhoid fever)

Salmonella hirschfeldii (causes paratyphoid fever and other forms of enteritis)

Salmonella paratyphi (causes paratyphoid fever and other forms of enteritis)

Salmonella schottmuelleri (causes paratyphoid fever and other forms of enteritis)

Salmonella gallinarum (causes fowl typhoid)

Shigella dysenteriae (Shiga's bacillus; causes bacillary dysentery complicated by effects of a diffusible neurotoxin)

Shigella boydii (Boyd's bacillus; causes bacillary dysentery)

Shigella flexneri (Flexner's bacillus; causes bacillary dysentery)

Shigella sonnei (one cause of summer diarrhea of young children; causes milder form of bacillary dysentery in adults)

Family V. *Brucellaceae* (virulent pathogens in man; said to penetrate unbroken skin or mucous membrane)

Pasteurella multocida (causes chicken cholera and hemorrhagic septicemia in warm-blooded animals)

Pasteurella pestis (causes plague)

Pasteurella pseudotuberculosis (causes pseudotuberculosis in animals)

Pasteurella tularensis [*Francisella tularensis*] (causes tularemia)

Bordetella pertussis (causes whooping cough)

Bordetella parapertussis (causes pertussis-like disease)

Bordetella bronchiseptica (causes bronchopneumonia in rodents)

Brucella melitensis (chief host is milch goat; a cause of undulant fever in man)

Brucella abortus (chief host is milch cow; a cause of undulant fever in man)

Brucella suis (chief host is hog; causes undulant fever in man, abortion in swine)

Haemophilus influenzae (occupant of respiratory tract; can cause acute respiratory infection, acute conjunctivitis, and purulent meningitis in young children)

Haemophilus aegyptius (causes acute conjunctivitis)

Haemophilus suis (acting with virus causes swine influenza)

Continued.

TABLE 1-3. ABBREVIATED CLASSIFICATION OF MICROBES FROM *Bergey's Manual of Determinative Bacteriology* (1957)*—cont'd

Haemophilus parainfluenzae (causes rarely subacute endocarditis, usually nonpathogenic in upper respiratory tract)
Haemophilus ducreyi (causes chancroid)
Actinobacillus lignieresii (causes actinobacillosis of cattle)
Actinobacillus mallei [*Pseudomonas mallei*] (causes glanders)
Calymmatobacterium granulomatis [*Donovania granulomatis*] (causes granuloma inguinale)
Moraxella lacunata (causes conjunctivitis)

Family VI. *Bacteroidaceae* (strict anaerobes, a few microaerophiles; found in intestinal tract; may demonstrate marked pleomorphism and are sometimes pathogenic)
Bacteroides fragilis (opportunist pathogen; causes pulmonary gangrene and other forms of bacteroidosis)
Bacteroides melaninogenicus (opportunist pathogen; causes bacteroidosis)
Fusobacterium fusiforme [*Bacteroides fusiformis*] (found in Vincent's infection and about the gums in normal mouths)
Streptobacillus moniliformis (causes one type of rat-bite fever)

Family VII. *Micrococcaceae* (gram-positive cocci, including saprophytes and important pathogens; normal occupants of skin and associated structures in man)
Staphylococcus aureus (coagulase-positive; type lesion is the abscess; pus formation is characteristic; important diseases are boils, wound infections, tissue abscesses, pyelonephritis, osteomyelitis, and food poisoning)
Staphylococcus epidermidis (coagulase-negative; normal occupant of skin and mucosae of man; implicated as a pathogen in certain skin abscesses and in wound infections)
Gaffkya tetragena (occupant of respiratory tract; pathogenic for small animals)
Sarcina lutea (saprophyte of air, soil, and water; produces characteristic yellow pigment)
Methanococcus species (anaerobic saprophytes producing methane)
Peptococcus species (anaerobic cocci; human habitat)

Family VIII. *Neisseriaceae* (gram-negative parasitic cocci)
Neisseria gonorrhoeae (causes gonorrhea)
Neisseria meningitidis (causes epidemic cerebrospinal fever or meningitis)
Neisseria catarrhalis (occupant of lining of respiratory tract in man)
Neisseria sicca (inhabitant of respiratory tract of man)
Neisseria flava (inhabitant of respiratory tract of man)
Veillonella species (anaerobic gram-negative cocci; harmless parasites of natural cavities of man and animals; opportunist pathogens)

Family IX. *Brevibacteriaceae* (gram-positive rods, sometimes coccobacilli, widely distributed in nature—in soil, water, decomposing organic matter, in dairy products, and in a variety of plants)

Family X. *Lactobacillaceae* (lactic acid bacteria in mouth and intestinal tract of man and in food and dairy products; some forms highly pathogenic)
Tribe I. *Streptococceae* (gram-positive cocci occurring in chains; pathogens and saprophytes)
Diplococcus pneumoniae (inhabitant of respiratory tract in man; causes lobar pneumonia, otitis, meningitis, and peritonitis)

Streptococcus pyogenes (causes many types of infections, usually of a diffuse or spreading nature—erysipelas, puerperal sepsis, scarlet fever, tonsillitis, and pharyngitis)
Streptococcus mitis (occupant of mouth and throat of man; found about carious teeth; low-grade pathogen ordinarily; causes subacute bacterial endocarditis)
Streptococcus faecalis (enterococcus in intestine of man; sometimes causes urinary tract infections)
Streptococcus lactis (important in dairy industry; some strains used as "starter" cultures in making cheeses and cultured milk drinks)
Streptococcus MG (associated with primary atypical pneumonia; found in respiratory tract of man)
Pediococcus species (microaerophilic saprophytes in fermenting vegetable juices; produce lactic acid from carbohydrates)
Leuconostoc mesenteroides (found in slimy sugar solutions, in fermenting plant materials, and in prepared meat products)

Tribe II. *Lactobacilleae* (gram-positive rods, microaerophilic to anaerobic, occurring singly or in chains)
Lactobacillus lactis (causes milk to sour)
Lactobacillus acidophilus (found in feces of children and adults on high milk diets)
Lactobacillus casei (most common lactic acid rod in milk and milk products)

Family XI. *Propionibacteriaceae* (irregularly shaped rods occupying intestinal tract of animals and found in dairy products)

Family XII. *Corynebacteriaceae* (animal and plant parasites and pathogens; found in dairy products and in soil)
Corynebacterium diphtheriae (causes diphtheria; toxigenic)
Corynebacterium pseudotuberculosis (causes pseudotuberculosis in sheep, horses, and cattle; toxigenic)
Corynebacterium xerosis (harmless occupant of conjunctiva; nontoxigenic)
Corynebacterium pseudodiphtheriticum (found in normal throats; nontoxigenic)
Corynebacterium equi (found in pneumonia and other infections of horses; nontoxigenic)
Listeria monocytogenes (pathogen of listeriosis in man and animals; causes widespread lesions in tissues, meningitis, and granulomatosis infantiseptica; induces marked increase in circulating monocytes in bloodstream)
Erysipelothrix insidiosa [*Erysipelothrix rhusiopathiae*] (causes erysipelas of swine and turkeys, erysipeloid of man)

Family XIII. *Bacillaceae* (spore-forming saprophytes of soil)
Bacillus subtilis (widespread in soil and decomposing organic matter; common laboratory contaminant; some strains produce antibiotics)
Bacillus anthracis (causes anthrax)
Bacillus polymyxa (source of antibiotic polymyxin)
Clostridium tetani (its potent exotoxin produces tetanus)
Clostridium botulinum (its potent exotoxin produces botulism)
Clostridium perfringens, novyi, and *septicum* (cause gas gangrene)

Order V. *Actinomycetales* (elongated organisms of soil sometimes branching; single or simply branched forms are often acid-fast)

Family I. *Mycobacteriaceae* (important acid-fast organisms, both parasitic and saprophytic; branching inconspicuous)

Continued.

TABLE 1-3. ABBREVIATED CLASSIFICATION OF MICROBES FROM *Bergey's Manual of Determinative Bacteriology* (1957)*—cont'd

Mycobacterium phlei (found in soil and on plants)
Mycobacterium smegmatis (found about genitals; widespread in soil)
Mycobacterium fortuitum (found in soil and in infections of man, cattle, and cold-blooded animals)
Mycobacterium marinum (causes tuberculosis in saltwater fish)
Mycobacterium ulcerans (causes skin ulcers in man)
Mycobacterium tuberculosis (causes tuberculosis in man)
Mycobacterium bovis (causes tuberculosis in cattle and in man)
Mycobacterium avium (causes tuberculosis in birds)
Mycobacterium leprae (causes leprosy)

Family II. *Actinomycetaceae* (characteristic branching mycelium, special spores for reproduction)
Nocardia asteroides (causes nocardiosis in man, often a tuberculosis-like disease of lungs)
Actinomyces bovis (causes lumpy jaw in cattle)
Actinomyces israelii (causes human actinomycosis)

Family III. *Streptomycetaceae* (primarily soil forms with branching mycelium and special spores for multiplication; many species important in production of antibiotics)
Streptomyces griseus (source of streptomycin)
Streptomyces rimosus (source of oxytetracycline)
Streptomyces aureofaciens (source of chlortetracycline)
Streptomyces erythraeus (source of erythromycin)
Streptomyces noursei (source of nystatin)
Streptomyces venezuelae (source of chloramphenicol)
Streptomyces halstedii (source of carbomycin)
Streptomyces fradiae (source of neomycin)
Streptomyces niveus (source of novobiocin)
Streptomyces kanamyceticus (source of kanamycin)

Order VI. *Caryophanales* (saprophytic bacteria with visible nucleus)
Order VII. *Beggiatoales* (filamentous gliding bacteria)
Order VIII. *Myxobacterales* (slime bacteria)
Order IX. *Spirochaetales* (slender spiral bacteria)

Family I. *Spirochaetaceae* (free-living, coarse spiral organisms in stagnant, fresh, or salt water)
Spirochaeta species (undulating rods; protoplast spirals about axial filament; free living in fresh- and sea-water slime; common in sewage)

Family II. *Treponemataceae* (parasitic in man and animals)
Borrelia recurrentis (causes epidemic relapsing fever)
Borrelia vincentii (associated with a fusiform bacillus in cases of Vincent's angina)
Treponema pallidum (causes syphilis)
Treponema pertenue (causes yaws)
Treponema microdentium (found in normal mouth)
Leptospira icterohaemorrhagiae (causes infectious jaundice—Weil's disease)

Order X. *Mycoplasmatales* (pleuropneumonia group—tiny, very pleomorphic, delicate saprophytes and pathogens)

Family I. *Mycoplasmataceae* (gram-negative organisms with distinctive colonial features; recovered from human and animal sources)
Mycoplasma mycoides (causes contagious pleuropneumonia in cattle)

Mycoplasma hominis (implicated in nongonococcal urethritis)
[*Mycoplasma pneumoniae* (causes primary atypical pneumonia)]

Class III. *Microtatobiotes* (smallest living things; all parasitic, mostly within cells of host)

Order I. *Rickettsiales* (small pleomorphic microorganisms, usually intracellular, parasitic for reticuloendothelial and vascular endothelial cells in man and animals)

Family I. *Rickettsiaceae* (parasites of tissue cells; carried by arthropod vectors; important pathogens for man)

Rickettsia prowazekii (causes epidemic typhus fever)
Rickettsia typhi (causes murine typhus fever)
Rickettsia rickettsii (causes Rocky Mountain spotted fever)
Rickettsia tsutsugamushi (causes scrub typhus)
Rickettsia conorii (causes boutonneuse fever)
Rickettsia akari (causes human rickettsialpox)
Rickettsia quintana (causes trench fever)
Coxiella burnetii (causes Q fever)

Family II. *Chlamydiaceae* [*Bedsonia*] (intracellular parasites; arthropod vectors not implicated; cause ophthalmic and urogenital diseases in man)

Chlamydia trachomatis (causes trachoma)
Chlamydia oculogenitalis (causes inclusion conjunctivitis)
Miyagawanella lymphogranulomatosis (causes lymphopathia venereum)
Miyagawanella [*Chlamydia*] *psittaci* (causes psittacosis)
Miyagawanella ornithosis (causes ornithosis)

Family III. *Bartonellaceae* (parasitic for red blood cells in man and animals)

Bartonella bacilliformis (causes Oroya fever and verruga peruana, both forms of bartonellosis in man)

Order II. *Virales* (ultramicroscopic, filtrable parasites and pathogens of bacteria, plants, and animals)*

*See p. 402 for classification of viruses.

QUESTIONS FOR REVIEW

1. What is biology? Microbiology?
2. How are plants distinguished from animals?
3. What is the difference between thallophytes and other plants? Do all forms of life contain chlorophyl?
4. State the purpose of health education.
5. Give the divisions of microbiology. Briefly define each.
6. List basic units used to classify both animals and plants.
7. How is the scientific name of an animal or plant derived?
8. Give the word ending that indicates reference to a class, order, or family. Carefully spell out three examples.
9. Make pertinent comments regarding classification of bacteria. What classification is most generally used at present?

REFERENCES. See at end of Chapter 5.

2 Milestones of progress*

Microbes were probably the first living things to appear on the earth. The study of fossil remains indicates that microbial infections and infectious diseases existed thousands of years ago. Epidemics and pandemics have been witnessed at all periods of written history, but an exact diagnosis of many outbreaks incompletely described in their time is not possible today.

Microbes were not seen until some 3 centuries ago because lenses of sufficient power to render them visible were not perfected until that time. Even after microbes were discovered, almost 200 years elapsed without any great progress in their study being made.

1546: *Girolamo Fracastoro wrote a treatise (De Contagione et Contagiosis Morbis), in which he said disease was caused by minute "seed" and was spread from person to person.* The idea of communicability of disease by contact is an ancient one. Fracastoro (1483-1553), a sixteenth century scholar of northern Italy, published a work, which contained his theory of contagion, accounts of contagious diseases, and comments regarding their cure. Using the word "contagion" in a sense different from its accepted meaning today, Fracastoro defined it as an "infection" passing from one individual to another by contact alone, by fomites, or at a distance. He spoke of "seeds" or "germs" of disease, but he compared contagion to exhalation of an onion causing shedding of tears. He did realize from his observations, however, that all contagions do not behave alike, some attacking one organ and some another. For example, he believed in the infectivity of consumption (tuberculosis) and knew that the lungs were mainly involved. His work represents a great, although isolated, landmark in the doctrine of infectious diseases and was the result of wide and practical study of infectious diseases prevalent in his day.

1658: *Athanasius Kircher, a Jesuit priest, published a treatise on microscopy in which he recorded seven experiments on the nature of putrefaction.* The magnifying glass was known to the ancients. Not quite 100 years after the birth of Christ, Seneca, the Roman philosopher-statesman, recorded the magnifying effects of even a glass bulb filled with water. In 1590 father and son spectacle makers in Holland by the names of Johannes and Zacharias discovered accidentally the magnification produced by two convex

*To facilitate review and permit a general study of the history of microbiology, the important contributions (milestones) are briefly set up in italics. In most cases the milestones not necessarily in exact chronologic order are followed by a short discussion. Also, in several parts of the book, historical items in footnotes are related to the text.

lenses in sequence. If the lenses were arranged in a metal tube with sliding barrels, they found the magnified image could be focused. In 1624 Galileo, the Italian astronomer who made several telescopes, also made a microscope. The first person to engage in what might be called medical microscopy, however, was Kircher (1602-1680), who was well informed in mathematics, physics, medicine, and music and who first drew attention to the Egyptian hieroglyphics.

1675: *Antonj van Leeuwenhoek of Holland first described microbes (bacteria and protozoa) under the microscope. He made his own microscopes, which were superior to any of that time. He is known as the father of bacteriology because it was he who first accurately described the different shapes of bacteria (coccal, bacillary, and spiral) and pictured their arrangement in infected material (1683).* Leeuwenhoek (1632-1723), the bedellus of Delft, was a wealthy man who devoted the major portion of his time to scientific studies. He fashioned about 250 microscopes, grinding most of the lenses himself. Leeuwenhoek was obsessed with his microscopic findings and examined with his scopes everything he could think of. He is also known as the father of scientific microscopy. In addition to his work in microbiology, Leeuwenhoek made other contributions to medicine: He gave the first complete account of the red blood cell, he discovered spermatozoa, he demonstrated the capillary connections between arteries and veins, and he made other important anatomic observations.

Kircher's microscope magnified only 32 times. The better scopes fashioned by Leeuwenhoek magnified about 270 times. The compound light microscope of today magnifies 1000 times, the electron microscope 100,000 times.

1762: *Marc Antony von Plenciz, Viennese physician, published a thesis stating his belief that disease was caused by living organisms and that each disease had its own organism.* Not only did von Plenciz (1705-1786) believe that there was an organism for each disease, but he also thought that the organisms multiplied within the body and suggested that they might be transferred from person to person through the air. This is reminiscent of the treatise *(De Contagione)* written by Fracastoro in 1546.

1748: *John Needham, probably the greatest supporter of the theory of spontaneous generation, published material advocating that theory.*

1765: *Abbé Lazzaro Spallanzani introduced experiments to disprove the doctrine of spontaneous generation.* One of the factors that accelerated the development of microbiology was the argument that had been going on for years concerning spontaneous generation. The proponents believed that living forms sprang from nonliving matter, and the opponents believed that every living thing descended from parents like itself. The older proponents of the theory believed that eels originated from mud, and they gave formulas for producing mice from decaying rags and cheese. As time went on, both sides conceded that spontaneous generation did not occur in the higher forms of life. Much experimentation was done on both sides, and the argument continued until Pasteur completely and finally disproved the theory of spontaneous generation and showed the beliefs of its opponents to be correct.

Spallanzani (1729-1799), spoken of as the great Italian master of experiment, performed several experiments to disprove the doctrine of spontaneous generation. He concluded that what he called "animalcules" carried by air into infusions of organic material were the explanation for spontaneous generation. Much of Spallanzani's work was criticized by a man of Welsh origin, John Needham (1713-1781), whose observations on infusions of organic material in bottles closed with cork stoppers led him to a firm belief in spontaneous generation. To answer Needham's criticisms,

Spallanzani performed a series of experiments whereby he boiled the infusion, removed the air from the flask, and hermetically sealed the container. He observed that, after a flask of infusion had remained barren a long time, only a small crack in the neck was sufficient to allow for development of animalcules in the infusion. This carefully done work might have ended the discussion on spontaneous generation had the precautions taken by Spallanzani been observed by subsequent workers.

1796: *Edward Jenner, an English physician, introduced the modern method of vaccination to prevent smallpox.* One of the world's epochal contributions to preventive medicine was made on May 14, 1796, when Jenner (1749-1823), having been impressed by the countryside tradition in Gloucestershire, England, that milkmaids who contracted cowpox while milking were subsequently immune to smallpox, performed a vaccination against smallpox by transferring material from a cowpox pustule on the hand of a milkmaid (Sarah Nelmes) to the arm of a small boy named James Phipps. Six weeks later the boy was inoculated with smallpox and failed to develop the disease. In 1798 Jenner published his results in 23 cases. By 1800 about 6000 persons had been inoculated with cowpox to prevent smallpox, and the scientific basis of the method, which in a more refined form is used today, was firmly established. The word "vaccination" was first used by Pasteur out of deference to Jenner.

1837: *Theodor Schwann proved that yeasts were living things.* Schwann (1810-1882) may be regarded as the founder of the germ theory of putrefaction and fermentation. In 1837 he concluded that the processes of putrefaction and fermentation were related to living agents that obtained their sustenance from fermentable or putrescible material. In the course of these experiments Schwann discovered and gave an accurate account of yeasts and their mode of reproduction by budding. He anticipated Pasteur's work when he asserted that fermentation of sugar was a chemical decomposition brought about by yeast attacking sugar.

1838-1839: *The cell theory was advanced.* In 1665 the Englishman Robert Hooke (1635-1703) first referred to the cell in an early work on microscopy. With his simple and imperfect microscope, he studied thin slices of a cork bottle stopper, noting the "little empty rooms," or cells, each with its own distinct walls. The word "cell," derived from Latin *cella,* meaning 'storeroom,' has persisted to designate the basic unit for all living things. The acceptance of the biologic theory that the structural unit of all plant and animal life is the cell came after the conclusions of Matthias Jacob Schleiden (1804-1881), German sea lawyer and botanist, and Theodor Schwann. Their conclusions were voiced in 1838 and 1839 and were based on their scientific studies of plants and animals.

1850: *Casimir Joseph Davaine observed minute infusoria (anthrax bacilli) in the blood of sheep dead of anthrax and transmitted the disease by inoculating this blood into other animals.* Davaine (1812-1882) was a French pathologist, parasitologist, and experimenter who, with Pierre Rayer (1793-1867), first saw the bacillus that causes anthrax. He was a doctor in general practice who had to keep his experimental animals in the garden of a friend. Yet he was one of the early writers on infectious diseases and published classical works on anthrax and septicemia.

1843: *Oliver W. Holmes, physician-poet, published an important article on the contagiousness of puerperal sepsis.*

1861: *Ignaz Philipp Semmelweis published his views on the cause of puerperal fever in a book and stated that the cause of this disease could be found in "blood poisoning."* Oliver Wendell Holmes (1809-1894) in 1843 published an article "On the Contagiousness of Puerperal Fever," which appeared in an early New England medical journal. In this article he indicated that this disease process might be spread by the examining hand of the nurse or doctor. Some time later Semmelweis (1818-1865), a Hungarian obstetrician in Vienna, noted that the death rate seemed to be higher in certain clinics where the medical students and physicians came directly to the wards from the morgue or autopsy room. He witnessed the postmortem examination of a pathologist who had died of an infectious process that developed after a dissection wound, and he observed that the changes brought about by the infection in this man were similar to those present in women dead of puerperal fever. Semmelweis then rightly concluded as to the infectious nature of puerperal fever and instituted certain practices intended to prevent the spread of the infection. On the wards that he supervised, all hands had to be cleansed carefully before a patient was examined, and the rooms were scrupulously cleaned. The mortality on his service dropped immediately. He made a statement in 1847 to the Vienna Medical Society that the cause of puerperal fever could be found in "blood poisoning" and published a book expressing his views in 1861. However, these views as to the infectious nature of the disease met with strong opposition (as did those of Holmes earlier). They were not accepted for 20 years, long after the death of Semmelweis.

1868: *Without knowing its cause, Jean Antoine Villemin showed that tuberculosis could be transmitted from man to animal and from animal to animal by inoculation of infectious material.* The early physicians from the time of Hippocrates believed that tuberculosis was infectious or contagious. An Alsatian physician of the nineteenth century (1827-1892), Villemin was the first to prove experimentally (before the bacteriologic era) that tuberculosis could be transmitted by the injection of tuberculous material from man into animals.

1861: *Louis Pasteur disproved the theory of spontaneous generation.*

1863-1865: *Pasteur devised the process of destroying bacteria known as pasteurization.*

1881: *Pasteur developed anthrax vaccine.*

1885: *The first preventive treatment for rabies (the Pasteur treatment) was used by Louis Pasteur.* The development of modern bacteriology began with the work of the great French scientist, Pasteur, who probably has had more influence on future generations than any other man who ever lived. He was born the son of a tanner at Dôle, France, in 1822, and was not a physician but a chemist. At about 30 years of age, having already distinguished himself as a chemist, Pasteur became interested in the process of fermentation, which, among other things, led to his disproving the theory of spontaneous generation.

By 1865 Pasteur had proved that the "diseases of wine" could be prevented without altering the flavor by heating the wine for a short time to a temperature (55° to 60° C.) a little more than halfway between its freezing and boiling points. This process known as *pasteurization* is employed throughout the civilized world today to preserve milk and certain other perishable foods.

Pasteur proved that pébrine, a disease of silkworms that threatened to destroy the silk industry of France, was caused by an animal parasite and devised methods for its prevention. Later he confirmed Koch's studies on anthrax, which was the first animal disease proved to be caused by bacteria. He discovered the staphylococcus, described the streptococcus, and was codiscoverer of one of the organisms causing gas gangrene.

Pasteur's next contribution, the one for which he is most widely known, was the development of the Pasteur treatment for rabies. He gave the first treatment in 1885, and his method, modified slightly, is used throughout the world today. Pasteur died in 1895, and his body lies in the Pasteur Institute of France.

1865: *Joseph, Lord Lister, English surgeon, applied antiseptic treatment to the prevention and care of wound infections.* The famous English surgeon Lister (1827-1912) was so impressed with the similarity between wound infections and certain fermentative and putrefactive changes, which Pasteur had already proved to be caused by microorganisms, that he concluded that wound infections, too, were caused by microbes. With this idea in mind, he protected wounds with dressings saturated with a solution of carbolic acid (phenol) and devised operating room procedures calculated to destroy microorganisms. The establishment of these methods was so far reaching in effect that Joseph, Lord Lister will always be known as the father of antiseptic surgery.

1873: *William Budd recognized the role of milk and water in the transmission of typhoid fever.* His very accurate study of typhoid fever outbreaks in 1856 led Budd (1811-1880) to believe that the disease was infectious and that the causative agent was excreted in the feces of the patient. He suspected that contaminated milk and water played an important part in the spread of the disease. In 1872 he noted that an outbreak in Lausen, Switzerland, was related to contaminated water. In a treatise on typhoid fever in 1873 Budd summarized his views. The discovery of the causative organism of typhoid fever did not come until 1880 (Karl Joseph Eberth, 1835-1926); Georg Gaffky (1850-1918), one of Koch's co-workers, was the first to cultivate it (1884).

1876: *Robert Koch reported the isolation of the anthrax bacillus in pure culture and the proof of its infectiousness.*

1881: *Koch had perfected his technic of isolating bacteria in pure culture. He introduced the use of solid culture media.*

1882: *Koch discovered Mycobacterium tuberculosis, the organism causing tuberculosis.*

1890: *Koch discovered tuberculin.* The first contribution of the German bacteriologist, Robert Koch, (1843-1910) was the isolation of the anthrax bacillus. (That this agent alone caused anthrax had been confirmed by Pasteur.) Koch is remembered for both his discovery of important disease-producing organisms and his fundamental contributions to bacteriologic technic. He was the first to stain bacterial smears as is done today. He devised a liquefiable solid culture medium (gelatin) and worked out pure culture technics (triggered by chance observations on a boiled potato left out in his laboratory one day).

1880: *Alphonse Laveran discovered the parasite of malaria.*

1895-1898: *Sir Ronald Ross discovered that mosquitoes transmitted malaria.* In 1880 Charles Louis Alphonse Laveran (1845-1922), a French Army surgeon, discovered the parasite of malaria, and within the next 20 years its transmission by the mosquito was worked out by Ross (1857-1932), an English pathologist and parasitologist. Ross also devised suitable methods for the elimination of mosquitoes over wide areas.

1888: *Roux and Yersin discovered diphtheria toxin.* Émile Roux (1853-1933), research associate of Pasteur, was considered the greatest French bacteriologist after Pasteur. Roux and Élie Metchnikoff were two of Pasteur's most noted pupils. Roux, who adored Pasteur, said that in the hands of his master the scientific approach was one "which resolves each difficulty by an easily interpreted experiment, delightful to the mind, and at the same time so decisive that it is as satisfying as a geometrical demonstration."

Together with Alexandre Yersin (1863-1943), a Swiss bacteriologist, Roux made a number of important contributions to the bacteriology of diphtheria, one of which was the demonstration of a toxic substance in filtrates of broth cultures of the diphtheria organism.

1889: *Hans Buchner discovered the bactericidal effect of blood serum.* Buchner (1850-1902), a German bacteriologist and immunologist, by his experimental work demonstrated the bactericidal effect of blood serum. He showed that bactericidal substances occur in blood and also in cell-free serum.

1889: *Kitasato cultivated the tetanus bacillus and proved it to be the cause of lockjaw.* In 1889 Shibasaburo Kitasato (1852-1931), a pupil and colleague of Koch, cultivated the tetanus bacillus and proved it to be the cause of tetanus or lockjaw. Since he was unable to find this organism in animals that had died of the disease, he felt the disease process to be the result of chemical intoxication rather than of bacterial invasion.

1888: *Ludwig Brieger discovered tetanus toxin.*

1890: *Kitasato and von Behring discovered tetanus antitoxin (passive immunization). Von Behring discovered diphtheria antitoxin.*

1894: *Roux and Martin produced antitoxin by immunizing horses.*

1913: *Von Behring first used toxin-antitoxin to produce a permanent active immunity against diphtheria in human beings.*

1914: *Park used toxin-antitoxin mixtures for immunization in New York City.*

1923: *Ramon prepared diphtheria toxoid, which has supplanted toxin-antitoxin.*

In 1888 Brieger (1849-1909), Berlin physician, discovered the toxin of tetanus and studied its action. In 1890 Kitasato and Emil von Behring (1854-1917), also an assistant in Koch's laboratory, found that experimental animals treated with tetanus toxin developed a transferable immunity. The cell-free serum from such animals rendered harmless the toxin elaborated by the bacillus. This protective serum could be given to other laboratory animals. Normal serum did not affect tetanus toxin. Kitasato and von Behring published their results, and in a footnote in their article the word "antitoxic" was used. From this word, *antitoxin* became firmly established. Only a week after the discovery of tetanus antitoxin was announced, von Behring published another article, this one on immunization against diphtheria. In 1894 Émile Roux and his co-worker Louis Martin began to immunize horses to produce antitoxin. Martin (1864-1946) also was noted for his studies on diphtheria and antidiphtheritic serum.

Before the end of the nineteenth century, William Hallock Park (1863-1939) organized the first municipal bacteriologic laboratory in the United States. As dynamic head of this laboratory in New York City, he was the first to apply the best known methods of the time to protect the children of the city against diphtheria. For his widespread immunization program he used the toxin-antitoxin combination (first used by von Behring) and deserves the credit for practical application of von Behring's work.

By 1923 Gaston Leon Ramon (1885-1963), French bacteriologist and veterinarian, had prepared diphtheria toxoid, an agent devoid of the dangers of toxin-antitoxin and with superior immunizing capacity.

1892: *George M. Sternberg, called by Koch the "father of American bacteriology," published A Manual of Bacteriology.* Among Americans who added to the science of microbiology in the nineteenth century, Sternberg (1838-1915), military surgeon, pioneer bacteriologist, epidemiologist, and Surgeon General of the United States Army for nearly a decade, contributed much to the epidemiology of cholera and yellow fever. In 1881 he announced simultaneously with Pasteur the discovery of the pneumococcus. His treatise on bacteriology, published the year before he became Surgeon General, was an early landmark, being the first extensive presentation on the subject of microbiology published in the United States. During his term of office as Surgeon General, the Army Medical School was established (in 1893), probably the first school of modern hygiene in America; the Nurse and Dental Corps were organized; and in 1900, Major Walter Reed (1851-1902) was detailed to lead the famous Yellow Fever Commission, which incriminated the mosquito in the spread of yellow fever.

1896: *Gruber and Durham described the phenomenon of agglutination.* In 1896 Max von Gruber (1853-1927) and Herbert Edward Durham (1866-1945) discovered the phenomenon of agglutination. Practical application of this has resulted in agglutination tests for such diseases as typhoid fever, typhus fever, undulant fever, and tularemia.

1884: *Élie Metchnikoff announced the phagocytic theory of immunity.*

1898: *Paul Ehrlich expounded his "side-chain" theory of immunity.* In 1884 Metchnikoff (1845-1916), zoologist and native of the Ukraine, proposed the phagocytic theory of immunity. (In some of his earliest experiments on phagocytosis Metchnikoff studied the ingestion of thorns from a rosebush by the larval form of a starfish.) With the development of the science of immunology in the 1800's two schools arose. The first, the French school, headed by Metchnikoff, believed the whole process of immunity to be centered in the ingestion and destruction of invading disease-producing agents by the leukocytes and certain of the body cells. This is the phagocytic or cellular theory of immunity.

The second school of immunologists, the German school, was led by Paul Ehrlich (1854-1915), an outstanding German scientist who did important research in immunology from 1890 to 1900. Although he had drawn the outlines of his side-chain theory of immunity more than 10 years before, it was not until the turn of the century that he made his first exposition of it. In his studies Ehrlich conceived the structure of protein molecules to be similar to a benzene ring with unstable side chains. These could act as chemoreceptors that would combine with harmful materials to neutralize them and be released in excess after the need had been met. The German school believed that the establishment of immunity stemmed from the development in body fluids, especially in the blood, of certain substances known as antibodies or immune bodies with the capacity to destroy invading disease-producing agents. This is the humoral or chemical theory of immunity.

The two views on immunity were once held as opposing ones, and the controversy that raged between their adherents did much to put immunology on a scientific basis. Today we know the two concepts are interrelated.

1909: *Ehrlich introduced salvarsan ("the great sterilizing dose") as a treatment for syphilis.* Paul Ehrlich was a genius of extraordinary activity who added a great mass to our knowledge of medical science, and a large number of technical laboratory methods he proposed are now in daily use. To him we are indebted for the basic procedures in staining, for he was one of the first to apply aniline dye solutions to cells and tissues. He founded and was the greatest exponent of the science of hematology and created the new branch of medicine known as *chemotherapy.* He began to synthesize many new drugs of slightly varying chemical formulas and to test each experimentally until a maximum therapeutic effect was obtained. Collaborating with a Japanese physician,

Sahachiro Hata (1873-1938), Ehrlich found in the six hundred and sixth compound of a series tested, salvarsan (arsphenamine), a drug capable of controlling the manifestations of syphilis over a long period of time.

1898: *Beijerinck recognized the first virus (virus of tobacco mosaic disease).* In 1892 Dmitri Iwanowski reported the first experiments that pointed to the presence of an infectious agent much smaller than any known up to that time. The subject of these experiments was the virus of tobacco mosaic disease. This work was repeated and extended by Martinus Willem Beijerinck (1851-1931), Dutch botanist, in 1898. Iwanowski was doubtful of his own findings, but Beijerinck realized that a new type of infectious agent had been discovered. In addition to being the first virus discovered, the virus of tobacco mosaic was the first to be purified in a crystalline form. (About 1 ton of infected tobacco must be processed for 1 tablespoon of virus crystals.) This task was accomplished in 1935 by Wendell Meredith Stanley (1904-1971), biochemist and virologist. In 1955, in Stanley's laboratory at the University of California, the poliomyelitis virus was crystallized, the first virus affecting animals or man to be so purified.

1901: *The process of complement fixation was developed.* In 1901 Jules Jean Baptiste Bordet (1870-1961) and Octave Gengou (1875-1957) described the phenomenon of complement fixation, and in 1906 August von Wassermann (1866-1925) applied it to the diagnosis of syphilis, giving the medical profession the highly important Wassermann test, the first reliable serologic test for syphilis. In 1907, Leonor Michaelis (1875-1949), German physical chemist, described the phenomenon of precipitation, which is now used in such tests as the Kline, Kahn, Mazzini, VDRL, and Hinton tests for syphilis.

1901: *Karl Landsteiner discovered the basic human blood groups.*

1940: *The Rh factor was discovered by Landsteiner and Wiener.* Karl Landsteiner (1868-1943), American pathologist, had been associated with Max von Gruber in Vienna. When he came to America, he held an important post with the Rockefeller University of New York. He found that normal human blood falls into four main groups, the blood of some of these groups destroying blood belonging to other groups. This discovery laid the foundation for modern blood transfusion. In 1940 Landsteiner and Alexander S. Wiener (1907-   ) identified the Rh factor of the blood. Landsteiner thought blood to be as unique a characteristic of an individual as his fingerprints.

1906: *Novy and Knapp discovered the spirochete of the American variety of relapsing fever.* Frederick George Novy (1864-1957), American bacteriologist, together with R. E. Knapp, discovered the spirochete causing the American variety of relapsing fever. The first teaching course in bacteriology in the United States was established by Novy at the University of Michigan in 1889. Three years before, in 1886, the first department of bacteriology in any United States medical school had been instituted at Columbia University.

1894 to present: *Women microbiologists have made important contributions:*

1913: *Edna Steinhardt first used tissue culture technics to grow virus.*

1917-1923: *Alice C. Evans investigated undulant fever.*

1921-1927: *Gladys Dick assisted her husband George Dick in work on the bacteriology and serology of scarlet fever.*

1931: *Alice Woodruff grew a virus in a fertile egg.*

1928-1933: *Rebecca Lancefield published methods of classification of streptococci.*

1953 to present: *Sarah Stewart has carried on significant research in tumors in animals induced by virus.*

These milestones are by way of special mention of the distaff side who have played no inconsiderable part in the development of modern microbiology. One of the first such women was Anna Wessels Williams (1863-1954). In 1894 she was appointed assistant bacteriologist of a New York Board of Health laboratory just organized by W. H. Park. For 40 years she was closely associated with this outstanding health officer during a period of remarkable achievement in the field of preventive medicine in New York City. During the first part of this century the Chicago bacteriologist Ruth Tunnicliff (1876-1946) studied streptococci, anaerobic organisms, measles, and meningitis. Pearl L. Kendrick (1890-   ) has added much to our knowledge of whooping cough. Eleanor A. Bliss (1899-   ) is known for her work on streptococci and the sulfonamide drugs.

Senior Bacteriologist at the National Institutes of Health for 27 years, Alice Catherine Evans (1881-   ) was a notable predecessor of women working in investigative laboratories. As a result of the work done by the Dicks, a husband-wife team in Chicago, important information was gained about the bacteriology of scarlet fever. In 1923 they causally linked the disease to hemolytic streptococci and in 1924 devised the Dick test. Rebecca C. Lancefield (1895-   ) developed the serologic classification of human and other groups of streptococci in wide use today.

Edna Steinhardt and her associates at Columbia University, who propagated cowpox virus in bits of cornea from the eye of a guinea pig, have been credited the first to cultivate virus by technics of tissue culture. The news in 1931 that viruses could be grown in embryonated hen's eggs made a major impact on the entire microbiologic world. A physiologist, another member of a husband-wife team, Alice Miles Woodruff, working with fowlpox virus, was first able to do this. The work was reported from the Pathology Laboratory at Vanderbilt University School of Medicine under the direction of E. W. Goodpasture. Since virus cultivation in fertile eggs and in tissue cultures produces an ample supply of virus for study or for vaccine production, the introduction and standardization of these two technics have in large measure contributed to the explosive surge of knowledge that has occurred in the field of virology in recent years. In fact, this period of the last several decades has been called the "golden age" of virology.

Feminine scientists figure prominently in the field of cancer research. A noted one with a background as a microbiologist is Sarah Stewart of the National Institutes of Health, who is working on tumors induced by viruses. In 1953 her first report came out on the salivary gland tumors in mice induced by the polyoma virus, a fairly common virus readily infecting many laboratory animals. With the help of Bernice Eddy and co-workers, she has subsequently propagated the virus and shown that it can produce a great variety of cancers in the animals studied.

1928: *Penicillin was discovered by Sir Alexander Fleming of England.*

1940: *The value of penicillin as a therapeutic agent was determined by Florey and Chain in England.* In 1928, upon return from a week's vacation, Fleming (1881-1955), Scottish bacteriologist, discovered a patch of green mold that had fallen haphazardly upon one of his cultures, and he became interested in its antibacterial action. In 1929 he reported that this mold, *Penicillium notatum,* elaborated during its growth a substance that inhibited the development of certain bacteria. He called this substance penicillin. Fleming's own life at one time was saved from an overwhelming pneumonia by the very agent he had discovered. To his great delight, he made a speedy and dramatic recovery.

Fleming's monumental work went practically unnoticed until 1938, when Sir Howard Florey (1898-1968) and Ernst Boris Chain (1906-   ) at Oxford University began a systematic study of Fleming's findings. Not only did they develop suitable methods for producing penicillin and for demonstrating its usefulness, but they were also responsible for placing it before the American medical world and for gaining the interest of American industry. American industrial organizations have made possible the production and availability of penicillin on a wide scale.

1935-1936: *Domagk, Colebrook, and Kenny established the remarkable value of Prontosil (the forerunner of the sulfonamide compounds) in streptococcal infections.*

1938 to present: *Many important sulfonamide compounds have been developed.* Ehrlich, from his work in chemotherapy, had the idea that dyes might act as selective poisons, and other workers had found that certain azo dyes had some bactericidal power. In 1927 the pathologist Gerhard Domagk (1895-1964) began an eventful collaboration with two organic chemists, Mietzsch and Klarer, at the Elberfeld Laboratories of the German firm of I. G. Farbenindustrie. These two chemists prepared a number of azo dyes, and in 1932 Domagk found that one of them, when injected into mice infected with streptococci, had a definite curative power. The molecule of this effective agent contained a grouping of atoms that chemists call the *sulfonamide* group. This discovery was a considerable stimulus to further research, and subsequently more than 1000 new dyes were made. In 1932 Mietzsch and Klarer prepared a red dye that proved to be the most effective of the compounds, and it was given the trade name of Prontosil. In, 1935 Domagk published his experimental data (in German), demonstrating the antibacterial powers of Prontosil, and about the same time several clinical reports confirming the therapeutic value of this compound in streptococcal infections appeared in the German literature. Shortly after, a group of French investigators showed that the compound Prontosil was broken down within the body to form sulfanilamide and demonstrated the antibacterial properties of this compound (first prepared by Paul Gelmo in 1908). Leonard Colebrook and Méave Kenny were responsible for arousing the interest of the English-speaking world in bacterial chemotherapy and in these new sulfonamide compounds. They demonstrated the clinical effectiveness of sulfanilamide in puerperal fever, a streptococcal infection.

Since the introduction of Prontosil, more than 5000 sulfonamides have been prepared. Of these, only a few show antibacterial activity comparable to or greater than that of the parent drug.

1939: *René Jules Dubos discovered tyrothricin.*

1943: *Streptomycin was discovered by Waksman.*

1947 to present: *Many new antibiotic agents are constantly being developed and tested.* In 1939 Dubos (1901-   ), now of the Rockefeller University, discovered tyrothricin and demonstrated that microorganisms isolated from the soil can produce, under certain conditions, substances possessing antibacterial properties. Tyrothricin was the first member of the group of agents known as antibiotics to be given extensive study and was the first to be produced in a pure form. This information is not so new as it might seem because Pasteur in 1877 had demonstrated that the growth of anthrax bacilli was retarded by the presence of common soil organisms, and several antibiotic-like substances (example, pyocyanase) were prepared before 1900.*

---

*The ancient Egyptians listed painstakingly on papyri concoctions with activity corresponding to that of the modern day antibiotics. One item they felt to be most effective was "moldy wheaten loaf."

The development of streptomycin came as the result of a planned, careful search for an antibiotic agent effective for gram-negative bacteria and of minimum toxicity. Soils, composts, and other natural substrates harboring large numbers of organisms capable of producing antibiotic substances were carefully studied. Within a period of 5 years nearly 10,000 cultures were examined before streptomycin was obtained by Selman Abraham Waksman (1888-   ) and his co-workers at Rutgers University. This antibiotic agent was isolated from two strains of *Streptomyces* identified as *Streptomyces griseus.* Less than a year after the isolation of streptomycin, its effectiveness against the tubercle bacillus was demonstrated. For the first time in history an effective drug for the treatment of tuberculosis had been found.

Practical application of the observations of the soil microbiologist, the plant pathologist, and the medical bacteriologist has resulted in the development of many important antibiotics since World War II. Many diseases that formerly progressed unchecked are now harmless and easily controlled.

1941: *Albert Coons developed the fluorescent antibody test.* The fluorescent antibody technic represents an important example of the modern advances being made in the field of immunology. It was developed by an immunologist, Albert H. Coons (1912-   ), and his associates at Harvard Medical School, who demonstrated that fluorescein, a fluorescent dye, could be bound to an antibody without changing the behavior of the antibody. The attachment of the fluorescent dye would, however, cause the antibody to become luminous if it were viewed under ultraviolet light. These workers were the first to mark antibodies for future identification. Greer Williams refers to this test as one "equipped with neon lights."

1944: *Avery, MacLeod, and McCarty discovered the genetic role of DNA.*

1953: *Watson and Crick demonstrated its structure.* The story of deoxyribonucleic acid (DNA) goes back nearly a hundred years, but the dramatic chapters in the elucidation of its biologic significance did not unfold until well into the twentieth century. An acid substance referred to as nucleic acid, something as to its chemical composition, its relation to a carbohydrate (deoxyribose-deoxyribonucleic acid), and its association with chromosomes of the nucleus were known before Frederick Griffith, a British pathologist, made a significant discovery in 1928. When he inoculated mice (animals peculiarly susceptible to pneumococci) with a mixture of pneumococci, alive but *weakened,* and pneumococci, heat-killed but *virulent,* the mice developed a pneumonia produced by *virulent* organisms. A mysterious transformation had occurred whereby harmless microorganisms had changed their character. This work constituted a prologue to what has been called a revolution in biology.

In the years following, researchers at the Rockefeller University, because of their interest in the disease-producing properties of pneumococci, turned their attention to Griffith's heat-killed pneumococci and repeated his work. Over a period of 10 years they slowly eliminated, one by one, all genetic factors possibly involved in the transformation of the pneumococci, including protein of the nucleus (long thought to be the carrier of genetic information), the capsule of the microorganisms, and so on, until nothing was left except deoxyribonucleic acid (DNA), the nucleic acid of the nucleus. When DNA was removed from the heat-killed virulent pneumococci, it became apparent that it was indeed the transforming principle. By 1944, three workers, Oswald T. Avery, Colin MacLeod, and Maclyn McCarty, reported that they had for the first time the proof that DNA is the stuff of which heredity is made. The work of three biophysicists, one American, two British, is responsible for the elucidation of the molecular structure of DNA. In 1953 J. D. Watson and F. H. C. Crick reported their construction of the double helix model of the DNA molecule for which M. H. F. Wilkins had laid the groundwork by his x-ray diffraction studies.

1949-1963: *The dramatic development of poliomyelitis vaccine took place.* By the early 1930's research on poliomyelitis was well under way, and a vaccine had been prepared from the brain and spinal cord of monkeys. This first experimental vaccine was crude, only partially effective, and very impractical to make. Therefore, when the National Foundation for Infantile Paralysis (the National Foundation of today) began its support of scientific research in 1938, it could be rightly said that not a great deal was known about poliomyelitis. Before a successful program of immunization could be launched, more basic facts had to be obtained about the disease and its causative agent.

One of the most significant events in the history of the poliomyelitis vaccine was the report of John F. Enders and associates (Thomas H. Weller and Frederick C. Robbins) at Harvard that poliomyelitis virus could be successfully grown in cultures of living cells of nonnervous tissue. To this point poliomyelitis research had been slow and costly because of its dependence upon monkeys as experimental laboratory animals, animals often hard to obtain and expensive to keep. The introduction of the tissue culture method marked the end of the monkey era in poliomyelitis research. Through tissue cultures extremely large amounts of virus, free of the presence of injurious impurities in nervous tissue, could be produced quite inexpensively and hundreds of preliminary and pertinent investigations as to the nature of the virus could be easily carried out.

In 1954 a field trial using the Salk vaccine was launched by the National Foundation for Infantile Paralysis. The vaccine was an experimental one that Jonas E. Salk (1914-) and associates at the University of Pittsburgh had prepared by formaldehyde inactivation of poliovirus. The events of this field trial were unique in medical history. It was a project of universal interest that reached out to involve the talents of research teams, the methods of commerce and industry, the facilities of universities and government, and the resources of the publicly subscribed National Foundation, meanwhile engrossing the press of the entire world. The success of the field trial was formally announced by the Poliomyelitis Evaluation Center of the University of Michigan on April 12, 1955, the anniversary of the death of Franklin D. Roosevelt.

The development of a live poliovirus vaccine started with a number of experimental, clinical, and epidemiologic studies on a series of small well-controlled trial runs in the United States. It then jumped to a mass immunization program of global extent and astronomic proportions. The trial vaccines for the final large-scale tests were experimental vaccines (comprising the three strains of poliomyelitis virus) developed by the major workers in the field: Hilary Koprowski, Herald A. Cox, and Albert B. Sabin (1906-). Koprowski was the first to make an attenuated live virus vaccine. At the time that the Salk vaccine was introduced Sabin was busy searching for nonvirulent strains among the polioviruses growing in monkey tissue cultures in his laboratory. In 1956 he developed his first poliovirus for live virus (oral) vaccine.

From 1957 to 1959, beginning in the Belgian Congo, a program of mass immunization was carried out all over the world. In 1960 a committee set up by the United States Public Health Service in 1958 to follow developments in the laboratory and in the field trials reported that more than 60 million persons had been fed attenuated live poliovirus vaccines with no ill effects and that of the experimental vaccines available, the Sabin strains seemed to be more highly attenuated and therefore safer but were as immunogenic as the Cox and Koprowski vaccines. On the recommendation of its committee, the U.S. Public Health Service gave its approval in 1960 for the commercial manufacture of the Sabin vaccine under specified regulations. Although it had been used extensively abroad, the Sabin oral vaccine was first used in the United States in 1962. After the three strains of the Sabin oral vaccine had been licensed, immunization of the American public was carried out through community programs set up to dispense oral vaccine on a mass scale.

In the history of infectious disease, the conquest of poliomyelitis is probably the most dramatic chapter. Once a major threat, this crippling disease has almost disappeared.

1954-1963: *Another viral vaccine, the measles vaccine, was developed.* In 1954 John F. Enders of Harvard University and Thomas Peebles, an associate, adapted measles virus to tissue cultures of human cells. The virus used was isolated from the blood of an 11-year-old boy named David Edmonston in Bethesda, Maryland. Designated the Edmonston strain, it was weakened by successive growth in cell cultures of human and animal tissues and in the chick embryo. The Edmonston strain has provided the source for the different measles vaccines subsequently used for research purposes and field trials. In initial tests, approximately 25,000 children in the United States were vaccinated with live virus vaccine.

Recently safe and effective viral vaccines have been developed for rubella and mumps.

1964: *Baruch S. Blumberg discovered the Australia antigen.* In 1964 at the Institute for Cancer Research in Philadelphia, Blumberg, a geneticist, noted a foreign substance in the blood of an Australian aborigine, that he labeled the Australia antigen. In 1968 Alfred M. Prince, a virologist, of New York City Blood Center found the same substance in the blood of patients with serum hepatitis (hepatitis B), and subsequently the Australia antigen was specifically linked to the disease. The isolation of this substance has stimulated a considerable amount of research in relation to hepatitis and the Australia antigen conceivably may represent an entirely new class of infectious agents.

1910: *Peyton Rous (1879-1970) demonstrated experimentally a significant relation between cancer and viruses in chickens.*

The last several years in biologic sciences have been remarkable ones as to the many and diverse discoveries of far-reaching significance. It is not possible in this chapter to mention the many implications and ramifications. We choose to conclude therefore with only a hint as to coming events in an amazing story—that of viruses and cancer.

At the turn of the century the suggestion was made that viruses might cause cancer. In 1910 Peyton Rous, pathologist and medical researcher of the Rockefeller University, used material from a large tumor on the breast of a Plymouth Rock hen to show that cell-free filtrates of it induced the same tumor in hens, that the induced tumor behaved as did the original one, and that the properties of the agent in his experiments were those of a filterable virus. This important landmark in cancer research was years ahead of the times. To the early fifties, nearly half a century, the viral etiology of cancer made slow headway although substantiated by a number of significant observations (see Table 27-6), and for many years it took considerable courage to hold to such a theory.

1958: *Denis Burkitt defined Burkitt's lymphoma epidemiologically and postulated a transmissible viral agent.*

In 1958 Burkitt defined the 'jaw tumor in central African children' that now bears his name, and mapped out its geographic distribution. He described it as a cancer but much about it impressed him as an infectious disease. Partly because it is confined sharply to the yellow fever and mosquito belt of central Africa, he thought at first that it might be caused by a virus carried in a mosquito; later he dismissed the idea of an insect vector. This time the climate of the scientific world is right. The disease is a dramatic one, and the tools and technics for study are highly developed. It is being studied intensively. Burkitt's lymphoma is thought by all to be caused by a virus, but the precise one and the mechanisms are not established.

The relation of human neoplasms to viruses is one of the most challenging of research problems today. Scientifically we are living in exciting times!

3 The basic unit

THE CELL

Every living thing, plant or animal, is made up of one or many cells. In the lowest forms of life the whole organism is only a single cell, one in which are carried on all the processes associated with life. Such is a *unicellular* organism. In the higher forms of life, multiplied millions of cells arrange themselves into groups of different type cells in order to meet the varied physiologic needs of that organism. Such an organism with a division of labor is a *multicellular* one.

Regardless of how simple or complex the organism, the unit of structure is the *cell.* Cells are the bricks with which living organisms are built. Not only is the cell the structural unit, but it is also the operational one. One living cell, be it ever so small, is a dynamic biologic unit, and the performance of the organism or of any of its parts is the sum total of the functions of the constituent cells, acting singly or in groups. The science of the cell is *cytology.*

Anatomy of the cell

■ **Protoplasm.** A cell is composed of a colorless, translucent viscid substance of a colloidal nature known as protoplasm (Greek, 'first formed'), which is the physicochemical basis of life. Protoplasm is a complex but dynamic system incorporating units of varying size, physical nature, and chemical composition. Although adapted to cells of specific types, protoplasm presents basic features in all cells. It is made up principally of water (up to 90%). Contained therein are inorganic mineral ions and organic chemical compounds of varying complexity and physiologic significance—proteins (including enzymes), carbohydrates, fats, nucleic acids (see Table 3-1), hormones (usually either protein or lipid or derivatives), and vitamins (chemically heterogeneous compounds).

Certain organic compounds in the protoplasm of the cell have special biochemical significance in determining the form and biologic properties of the cell. Designated *macromolecules,* they are composed of organic chemical building blocks linked together in a characteristic fashion. Important examples of protoplasmic macromolecules of the cell are select carbohydrates (polysaccharides), proteins, enzymes, and nucleic acids. Most of the unique qualities of protoplasm stem from its contained macromolecules.

■ **Shape.** In keeping with their location inside the body of an organism and the work that they do, cells vary in shape, being spherical, oblong, columnar, six sided, or spindle shaped.

TABLE 3-1. OUTLINE OF CHIEF PROTOPLASMIC COMPOUNDS

	Biochemical compound			
	Carbohydrates	Lipids	Proteins	Nucleic acids
Composition (building blocks)	Simple sugars, some with amino groups	Triglycerides of fatty acids; phosphorus in phospholipids	Amino acids (at least five elements here); most complex; can be very large	Pentose nucleotides; complex combinations with protein are nucleoproteins
Role in cell	Synthesis of cell wall Essential fuel for all living organisms Mostly found as glycogen (storage form of glucose)	Phospholipids important to cell membrane Large part of energy needs met by these fuels	Most are enzymes Synthesis of organelles and other parts of cell	DNA—primary function in heredity; the genetic code for cell structure and function RNA—protein synthesis

■ **Size.** Cells are of microscopic size, except eggs of various animals (an ostrich egg is several inches around), striated muscle cells under certain conditions, and cells of the pith, stem, and flesh of some pulpy fruits. Some cells are so small as to be barely visible with the highest magnification of the ordinary compound microscope.

For a living object of such small dimensions as the cell, a special unit of measurement is needed, the micromillimeter, or *micron,* designated by the Greek letter *mu* (μ). A micron is 1/1000 of a millimeter or 1/25,000 of an inch in length. (The paper on which this book is printed is about 100μ in thickness. See Table 3-2 for metric equivalents.) The average diameter of the red blood cell is 7.5μ. Cells show considerable variation in size, but most range in diameter from 10μ to 100μ. Individual varia-

TABLE 3-2. EQUIVALENTS IN THE METRIC SYSTEM

UNIT	Meter (m.)	Centimeter (cm.)	Millimeter (mm.)	Micron (μ)	Millimicron (mμ)	Angstrom (Å)
Meter (m.)	1.0	100.0 10^2	1,000.0 10^3	1,000,000.0 10^6	1,000,000,000.0 10^9	10,000,000,000.0 10^{10}
Centimeter (cm.)	0.01 10^{-2}	1.0	10.0	10,000.0 10^4	10,000,000.0 10^7	100,000,000.0 10^8
Millimeter (mm.)	0.001 10^{-3}	0.1 10^{-1}	1.0	1,000.0 10^3	1,000,000.0 10^6	10,000,000.0 10^7
Micrometer (μm.)	0.000001 10^{-6}	0.0001 10^{-4}	0.001 10^{-3}	1.0	1,000.0 10^3	10,000.0 10^4
Nanometer (nm.)	0.000000001 10^{-9}	0.0000001 10^{-7}	0.000001 10^{-6}	0.001 10^{-3}	1.0	10.0
Angstrom (Å)	0.0000000001 10^{-10}	0.00000001 10^{-8}	0.0000001 10^{-7}	0.0001 10^{-4}	0.1 10^{-1}	1.0

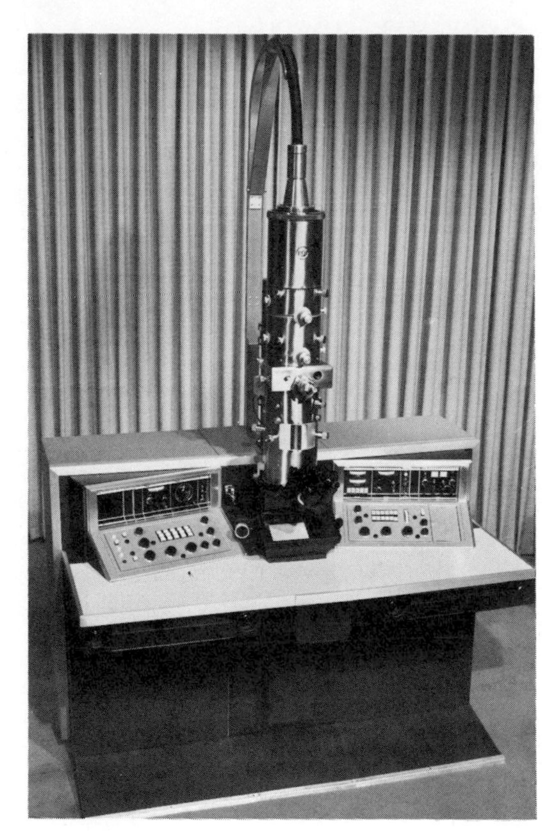

FIG. 3-1. Electron microscope. In this instrument particles $1/10$ millionth inch in diameter are examined. (Courtesy Forgflo Corporation, Sunbury, Pa.)

tions in the size of multicellular organisms of the same species depend upon the number, not the size, of their cells.

The information gained through the study of cells by the compound light microscope in a little over 100 years has contributed significantly to the development of modern medicine. However, the nature of light is such that it is not possible to get a clear image in the light microscope when the magnification of the object for study is greater than 2000 times. Hence many minute (submicroscopic) structures within the cell were not visualized until the electron microscope was directed to cytologic study.

ELECTRON MICROSCOPE. The electron microscope (Fig. 3-1), one of our most powerful research tools, differs from the compound microscope. In the latter, rays of ordinary light pass from the object being examined to the eye to be focused there, whereas in the electron microscope electrons pass through the specimen to be focused on a viewing screen from which a photograph is made—an *electron micrograph.* With the most modern electron microscopes the dimensions of the object under observation may be magnified 200,000 times. If careful photographic technics are used, a further enlargement of 10 times can be accomplished so that today a potential magnification of 2 *million* times is possible. (With this magnification, a human hair viewed in its entirety in the electron microscope would appear twice the size of a California redwood.) A unit even smaller than the micron must be used in designating the size of objects too small to be visualized in the light microscope. This unit is the *angstrom* (Å).

In Table 3-3 the sizes of microscopic and *ultramicroscopic* objects are compared. An ostrich egg is added to illustrate the magnitude of the size discrepancy.

■ **Structure.** If we magnify and examine a typical cell (Fig. 3-2), we see that there are two main parts. The central portion is occupied by a more densely arranged and usually compact structure, known as the *nucleus*. The *nuclear membrane* separates the *nucleoplasm* (protoplasm of the nucleus) from a zone of less dense spongy protoplasm surrounding and suspending the nucleus, known as the cytoplasm. Generally, the cytoplasm of the cell (1) aids in cell multiplication, (2) assimilates and stores food, (3) eliminates waste products, and (4) secretes enzymes* and other products of physiologic significance. The cytoplasm varies in different cells, for enzymes and enzyme systems may be possessed by only one specific kind of cell.

The nucleus (1) rules over the growth and development of the cell, (2) controls the

*Enzymes enable metabolites in the cell to be transformed at a rate and temperature unattainable in man-made laboratories.

TABLE 3-3. COMPARATIVE SIZES OF BIOLOGIC OBJECTS

Limit of resolution	Biologic object	Diameter		
		Microns (μ)	Millimicrons (mμ)	Angstrom units (Å)
Human eye	Ostrich egg	200,000	200,000,000	2,000,000,000
	Mature human ovum	120	120,000	1,200,000
	†			
Light microscope	Erythrocyte (red blood cell)	7.5	7,500	75,000
	Serratia marcescens (bacterium)	0.75	750	7,500
	Rickettsia	0.475	475	4,750
	Psittacosis bedsonia	0.27	270	2,700
	‡			
Electron microscope	Mycoplasma	0.15	150	1,500
	Influenza virus	0.085	85	850
	Genetic unit (Muller's estimation of largest size of a gene)	0.02 × 0.125*	20 × 125*	200 × 1,250*
	Poliomyelitis virus	0.027	27	270
	Tobacco necrosis (plant virus)	0.016	16	160
	Egg albumin molecule (protein molecule)	0.0025 × 0.01*	2.5 × 10*	25 × 100*
	Hydrogen molecule	0.0001	0.1	1

*Width × length.
‡Limit of resolution of compound light microscope, 0.2μ.
†Limit of resolution of human eye, 0.1 mm. or 100μ.

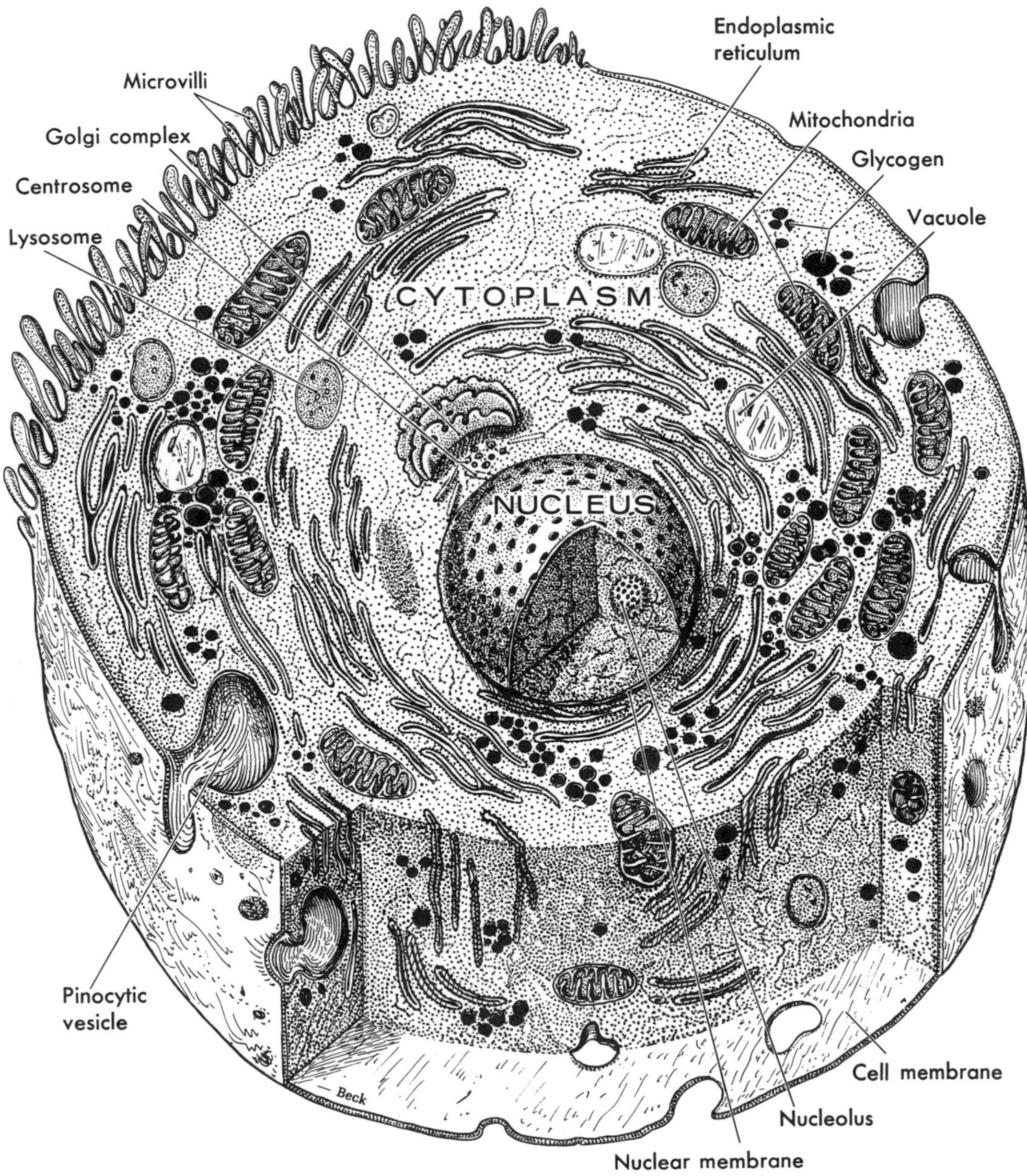

FIG. 3-2. The "cell." Diagram of structure observed in electron micrograph (discussion in text). Note pinocytic vesicle representing a droplet of watery fluid going into cytoplasm. *Pinocytosis* (cell drinking) indicates uptake of certain substances in solution by the cell. (From Anthony, C. P., and Kolthoff, N. J.: Textbook of anatomy and physiology, ed. 8, St. Louis, 1971, The C. V. Mosby Co.)

metabolic processes that go on inside the cell, (3) transfers the hereditary characteristics of the cell, and (4) controls reproduction, In short, the nucleus is the center or initiating point for all the vital activities of the cell, and a cell without a nucleus is dead.*

The pattern of intracellular organization is not the same in all cells, and many cells do not show all the typical structures to be given.

CYTOPLASM. Within the ground substance of the cytoplasm are (1) numerous small purposeful configurations of living substance called *organelles* and (2) stores of lifeless materials, sometimes transient, referred to as *inclusions.* Organelles include the plasma membrane (plasmalemma), centrosome, centrioles, mitochondria, endoplasmic reticulum, ribosomes, Golgi apparatus, lysosomes, microtubules, and several kinds of filaments and fibrils. Inclusions may be starch granules, glycogen, fat globules, proteins, pigment granules, secretory products, and crystals. Certain organelles and inclusions were well known to the cytologists, but with the electron microscope to reveal the complexities of cell architecture others were seen for the first time.

Encasing the cytoplasm is the *plasma membrane* (Fig. 3-3). This is a delicate film limiting the cytoplasm and presiding as a physiologic gatekeeper over the exchange

*The human red blood cell, with no nucleus, cannot divide and is limited in its metabolic behavior.

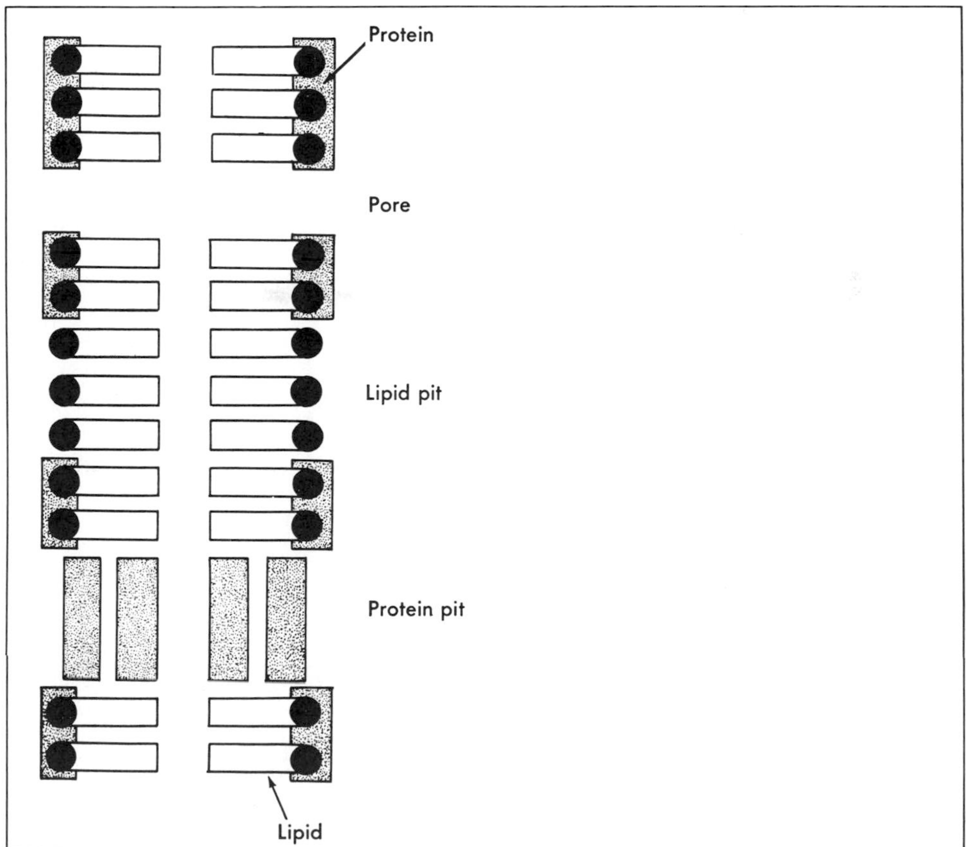

FIG. 3-3. Plasma membrane. Diagram to show precision of structure. Note clothespin appearance to molecules. (From Trumbore, R. H.: The cell: Chemistry and function, St. Louis, 1966, The C. V. Mosby Co.)

of food materials, metabolic waste products, and various other chemicals. Between the cell and its surroundings, numerous biologic events take place on or in the plasma membrane. A three-layered membrane, 80 to 100 Å thick, it is formed chemically by a precise association of protein, fat, and carbohydrate molecules. About the plasma membrane in certain vegetable cells and adherent to it is the *cell wall,* made of cellulose; this wall is absent or indistinct in animal cells. The many thin, filamentous projections of cytoplasm seen along the free border of certain cells are referred to as *microvilli.*

The *cell center* or *centrosome* is a cytoplasmic region of altered texture usually adjacent to the nucleus. It is occupied by a pair or more of rod-shaped *centrioles.* In electron micrographs centrioles are seen as hollow cylinders 300 to 500 mμ long and 150 mμ wide, placed at right angles to each other. Although the function of the centrosome is poorly understood, it is thought to play an important role in cell division.

The powerhouse of the cell is the *mitochondrion* (plural, *mitochondria*), the site of many important biochemical reactions. A cell may contain a thousand or more. Barely visible through the light microscope as a threadlike form (0.2μ to 3μ), the mitochondrion through the electron microscope is an intricate oval or elongated structure consisting of a double membrane, the inner layer of which is a system of folds called *cristae* (Fig. 3-4). The mitochondria house the bulk of chemicals (enzymes) that provide energy for cell activities; they are the principal sites at which the energy in food materials is released to the cell. (Biochemically, mitochondria can generate this energy for the cell because they contain cytochrome chains that are linked by the Krebs cycle.) With special technics in isolated mitochondria, the necessary enzymes and coenzymes required for the complex chemical reactions can be located on the membranes. A mitochondrion may contain anywhere form 50 to 50,000 such enzymes, depending upon the type of cell and its functional state.

The *endoplasmic reticulum* (the "cytoskeleton"), seen only by electron microscopy, is an elaborate pattern of channels with a secretory function. Its membrane-bound tubules measure 400 to 700 Å in diameter. The endoplasmic reticulum may be *granular*

FIG. 3-4. Mitochondrion. Cutaway sketch to show inner and outer membranes and cristae. (From Schottelius, B. A., and Schottelius, D. D.: Textbook of physiology, ed. 17, St. Louis, 1973, The C. V. Mosby Co.)

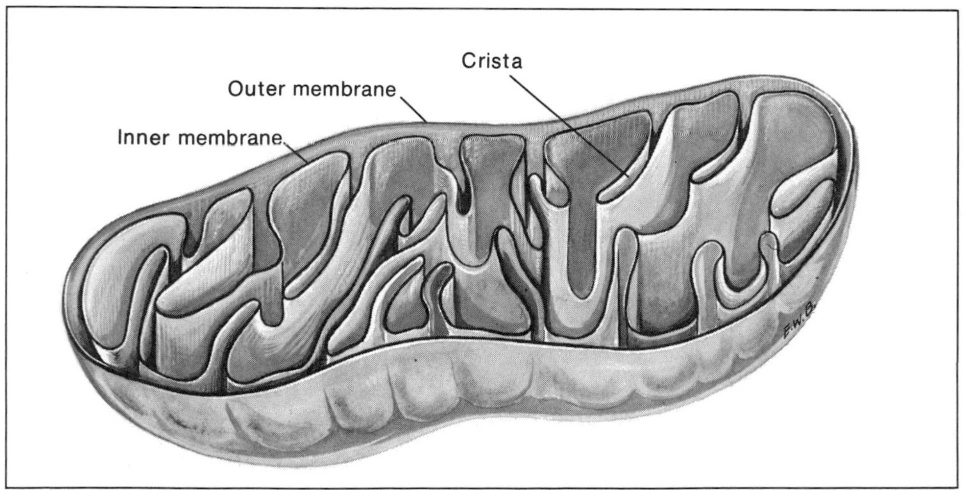

(rough surfaced) or *agranular (smooth surfaced).* Many tiny, uniform, beadlike granules called *ribosomes,* which are also found free in the cytoplasmic matrix, are adherent to the outer surface of the membranes of the granular reticulum. Ribosomes are rich in ribonucleic acid (RNA) and represent the places in the cell where proteins are formed. A single cell may contain billions of them. Granular reticulum is highly developed in glandular cells that elaborate protein secretions. If the secretory product is an export from the cell, the ribosomes must be closely associated with the endoplasmic reticulum, for it is through the pathways of the reticulum that the product is transported to the Golgi region for packaging. The secretory product stored as droplets or granules for a while in the cytoplasm is released in time from the cell surface. Agranular endoplasmic reticulum is related to other metabolic functions of cells; in the liver cell it is involved in lipid and cholesterol metabolism, and in the cells of the gonads it is related to the production of steroid hormones.

Placed near one pole of the nucleus, the *Golgi apparatus* is made up of aggregates of flat saccules in parallel fashion with vesicles and vacuoles clustered about. Although it still remains somewhat a mystery, it does play an important role in cell secretion. The polysaccharide fraction of certain complex secretions is known to be formed in the Golgi complex and there to be bound to protein supplied from other organelles of the cell.

Greatly variable in different cells are small dense particles, *lysosomes,* limited by a membrane. Lysosomes are rich in lytic enzymes, including deoxyribonuclease and ribonuclease. They are also known as "suicide bags," for their potent enzymes could digest the cell containing them if they were released from the organelle. A prime function of lysosomes is tied in with the engulfment of foreign material in the body, and they are abundant in the phagocytic cells (see p. 157). Close by and related to the centrioles are *microtubules,* cytoskeletal elements helping to maintain cell shape. These are straight tubules 200 to 270 Å in diameter with a dense filamentous wall 50 to 70 Å thick. *Filaments* and *fibrils* are commonly seen in the cytoplasmic matrix of many cells. Filaments enable muscle cells to contract, but for many cytoplasmic filaments the biologic significance is undetermined. A newly described cytoplasmic organelle is the *peroxisome.* It is thought to play a role in carbohydrate metabolism.

Cellular inclusions may serve as stores of energy for the cell, for example, deposits of glycogen, the polysaccharide storage form of carbohydrate, and fat (triglycerides of fatty acids). The enzymatic breakdown of the glucose derived from glycogen supplies energy to the cell and short-chain carbon skeletons to be reused in cell protoplasm. *Melanin,* the brown pigment of the skin and hair, is an example of an intracellular pigment. Certain *crystalline* inclusions visible by light and electron microscopy are assumed to be protein but are poorly understood. Tears, hydrochloric acid, mucus, milk, and digestive enzymes are well-known secretory products of cells. Mucus is one such product that may be readily visible microscopically as a droplet in the cytoplasm of the cell elaborating it.

NUCLEUS. The executive nucleus, usually found near the center of the cell, presides over the functional activities of that cell. Commonly a spherical body, it is made up of a framework of fibrils or threads on which are located the *chromatin granules,* which unite to form the *chromosomes* (Fig. 3-5) during cell division. For the time that the cell is not dividing, the chromosomes are in an extended state and cannot be counted. Early in cell division the chromosomes begin to contract and by metaphase the small rod-shaped bodies appear distinctly. The number and size of the chromosomes depend upon the species and are constant for each species.

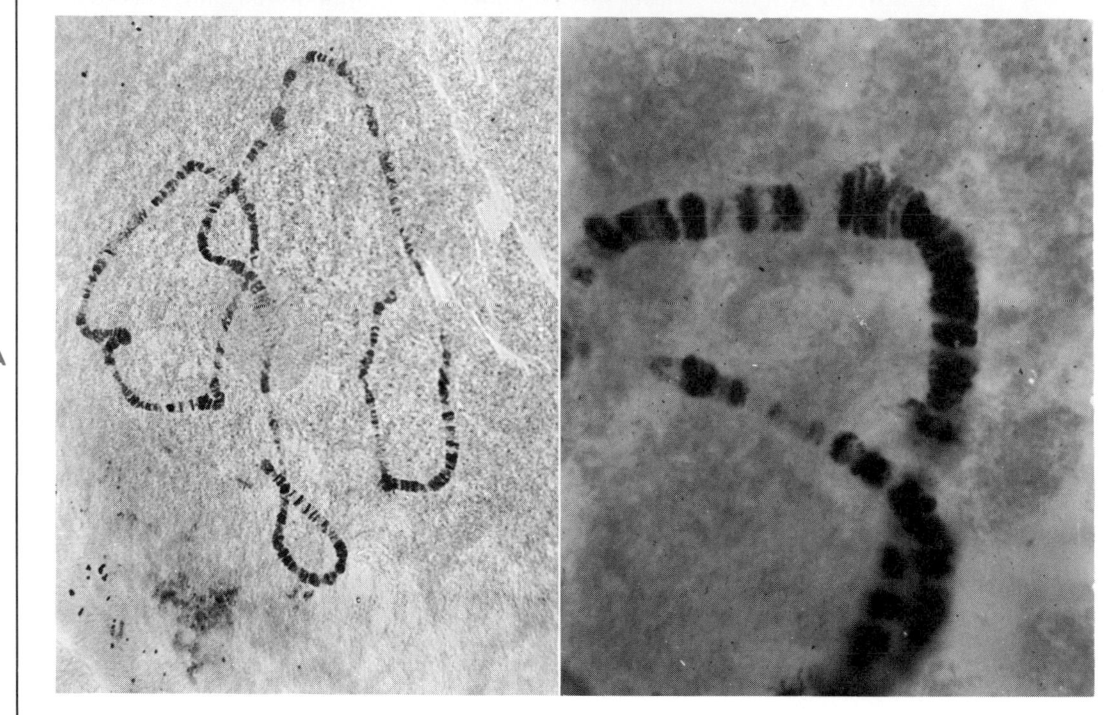

FIG. 3-5. Chromosomes from salivary gland of fly (species unidentified); prepared by "squash" technic. Note structure in photomicrographs **A,** low-power, and **B,** higher magnification. (From Brown, W. V., and Bertke, E. M.: Textbook of cytology, St. Louis, 1969, The C. V. Mosby Co.)

Structurally, a chromosome consists of coiled threads, each made up of a large number of strands of the giant molecule deoxyribonucleic acid, DNA. A chromosome is the unit of organization for DNA. In a way not fully understood, it represents a binding together of DNA, ribonucleic acid, histone (low molecular weight basic protein), and a more complex acidic residual protein.

In DNA three smaller chemical molecules—organic nucleotide bases, a pentose sugar (deoxyribose), and phosphoric acid—are fastened together in a characteristic spiral pattern. Two sugar-phosphate ladders with ever so many paired nucleotide rungs—that is, two identical, unbranched, rigid, intertwining, spiral chains—are thus formed. These are at least 1,500 times as long as they are wide and are twisted in opposite directions about a central shaft (Fig. 3-6 and 3-7).

Every living organism must possess somewhere in its makeup a master plan that formulates all aspects of its appearance and behavior. For even simple organisms such a master plan encompasses vast amounts of biologic information that to be stored efficiently must be converted to some sort of code. Because of its vantage point in the nucleus of the cell deoxyribonucleic acid is just right to do this biochemically. Its building blocks are linked in such a way that the four nucleotide bases (adenine, thymine, cytosine, and guanine) can serve as letters of a four-character code alphabet. Words can be formed in a biochemical and genetic language.* In the language of life

*The Morse or International telegraphic code, composed of the dot and dash, is a two-character code.

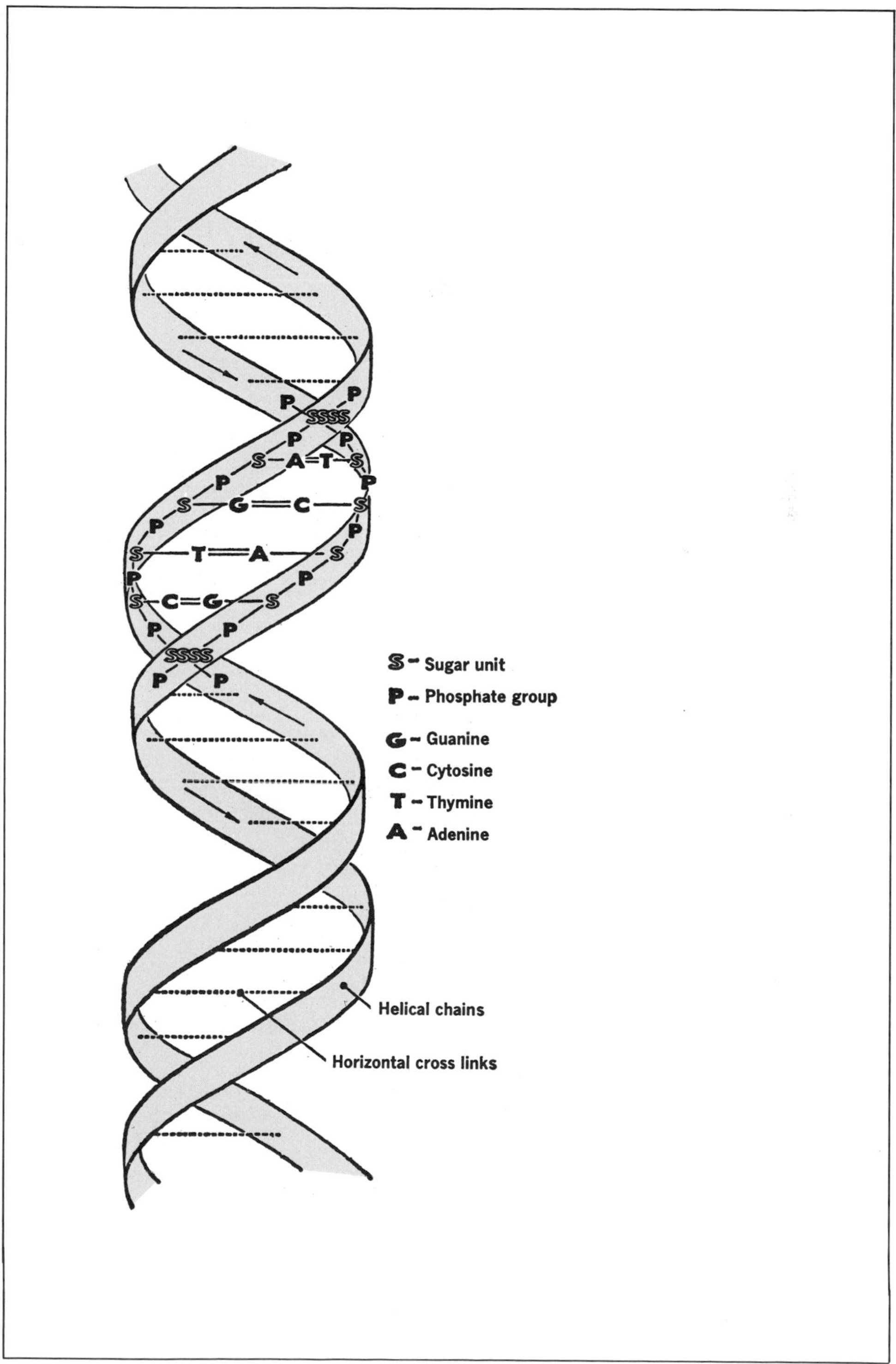

FIG. 3-6. DNA molecule, sketched. (From Arnett, R. H., Jr., and Braungart, D. C.: An introduction to plant biology, ed. 3, St. Louis, 1970, The C. V. Mosby Co.)

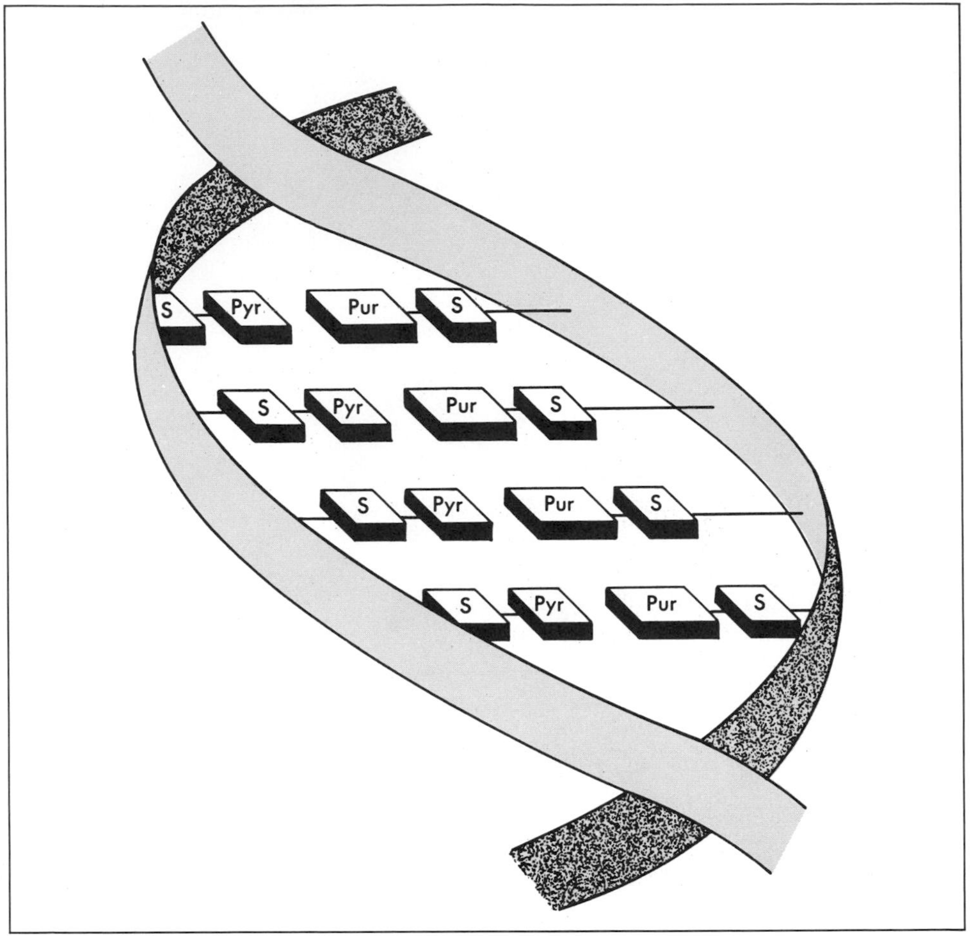

FIG. 3-7. Purine and pyrimidine bases, sketched in relation to helical structure of DNA. Part of sugar moiety also projects into center of spiral. *Pur,* purine; *Pyr,* pyrimidine; *S,* sugar. (From Trumbore, R. H.: The cell: Chemistry and function, St. Louis, 1966, The C. V. Mosby Co.)

the four characters are expressed in linkages of three, called triplet codes. There are 64 such triplet codes with so many possible arrangements that they can describe easily the 3.7 billion human beings populating the world. For a given organism the specifications for its every structure and life process are contained in specific chemical sequences in its DNA.

The chromosome, then, is a biologic document. From it can be transcribed biologic messages in code, over and over again, even the same message. The chromosomes carry the particular hereditary pattern for the given organism. The hereditary pattern for each of the 100 million million tiny cells of the human body is carried in each cell in its chromosomes.

An unusual property of the DNA molecule is its ability to reproduce itself, that is, to replicate into two exact copies (Fig. 3-8). (Replication occurs before cell division.) Because of this, the hereditary biochemical pattern, or *genetic code,* may be passed from one generation to the next. (Think of the thousands of billions of times the DNA of

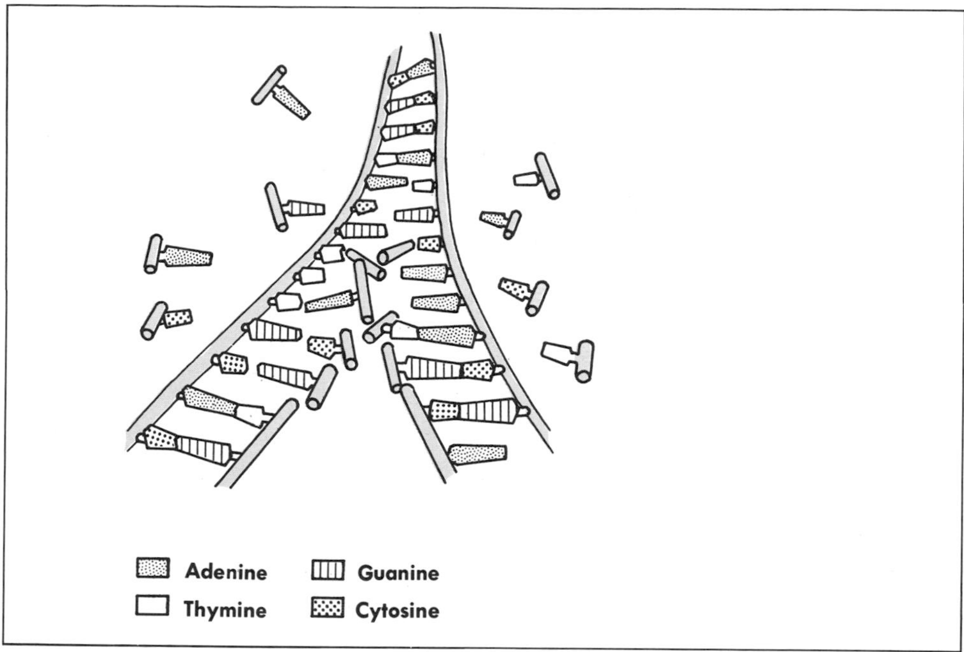

FIG. 3-8. Replication of DNA molecule, diagram. Note that molecule splits down middle, like a zipper. The two bases of each rung of ladder snap apart. Free nucleotides within nucleus move to halves of splitting ladder and reconstitute it, according to precise specifications of genetic code. Shortly thereafter, there are two exact copies of original molecule.

the fertilized human ovum* replicates itself to form the myriads of cells of the mature human being.)

If the chromosomes of the nucleus are compared to two measuring tapes side by side when the spirals are straightened, each point or locus marks the position of a *gene,* which is paired with a matching gene on the corresponding tape. The majority of the life processes within the cell are carried on by enzymes, which are mostly protein. Since a gene physiologically is responsible for the development of a protein,† collectively the influence of genes on the sum total of enzymatic action makes unicellular and multicellular organisms what they are biologically.

The other nucleic acid of physiologic significance, *ribonucleic acid,* or *RNA,* is similar to DNA chemically except that its sugar is a different pentose, ribose, and its chemical pattern is laid out as a *single* coil, not a double one. About 10% of the cell's quota of RNA is found in the nucleolus, the rest in the cytoplasm. Between cell divisions the nucleus transfers information coded in the chromosomes into specific sequences of amino acids in *messenger* RNA, which then passes out of the nucleus to transmit instructions to the ribosomes. These structures can decipher a sequence of several

---

*The DNA of the ova resulting in the entire earth's human population would fit into a ⅛-inch cube. One-tenth of a trillionth of an ounce of DNA is present in one fertilized human egg, although the number of DNA molecules it contains is astronomically large and virtually incalculable.

†A given protein (such as an enzyme) is made up of hundreds or thousands of amino acid building blocks. The biologic property of the protein molecule is completely dependent upon genes acting to align the amino acids precisely in the protein molecule.

thousand words (in the biochemical language) to map out complex protein patterns. (Cells of bacteria, plants, and animals contain both of the nucleic acids; viruses contain either, but only one of the two.)

Within the nucleus is usually found one spherical *nucleolus* (little nucleus), sometimes more, which is incorporated with the chromosomes into the background material of the nucleus, the *nuclear sap*. There is no membrane about it. The nucleolus disappears during cell division but reforms in the daughter cells. It plays a key role in nucleic acid metabolism and protein synthesis and is itself composed largely of RNA. It is prominent in rapidly growing embryonic cells.

The outer boundary of the nucleus, the *nuclear membrane*, is seen in the electron micrograph to be made up of two membranes, each 75 Å thick, separated by a space 400 to 700 Å wide. The nuclear membrane is similar in its makeup to the endoplasmic reticulum and is thought to be derived from it. The canals of the endoplasmic reticulum appear to open into small pores along the course of the nuclear envelope, and there may be numerous ribosomes on its cytoplasmic surface.

Interesting lines of communication are set up between the inner part of the nucleus and the outer portions of the cytoplasm, extending through to the outside of the cell. These pathways go from the pores of the nuclear membrane along the channels of the endoplasmic reticulum, incorporate the membranes of the Golgi apparatus, and end with the plasma membrane.

FIG. 3-9. Mitosis, sketch of various stages. Note nuclear changes, appearance of chromosomes, and loss of nuclear membrane. Chromosomes take shape in cell III, are divided in cell IV, and are aligned on the mitotic spindle in cell V. By cell VI they are separated along the spindle in the elongated cell.

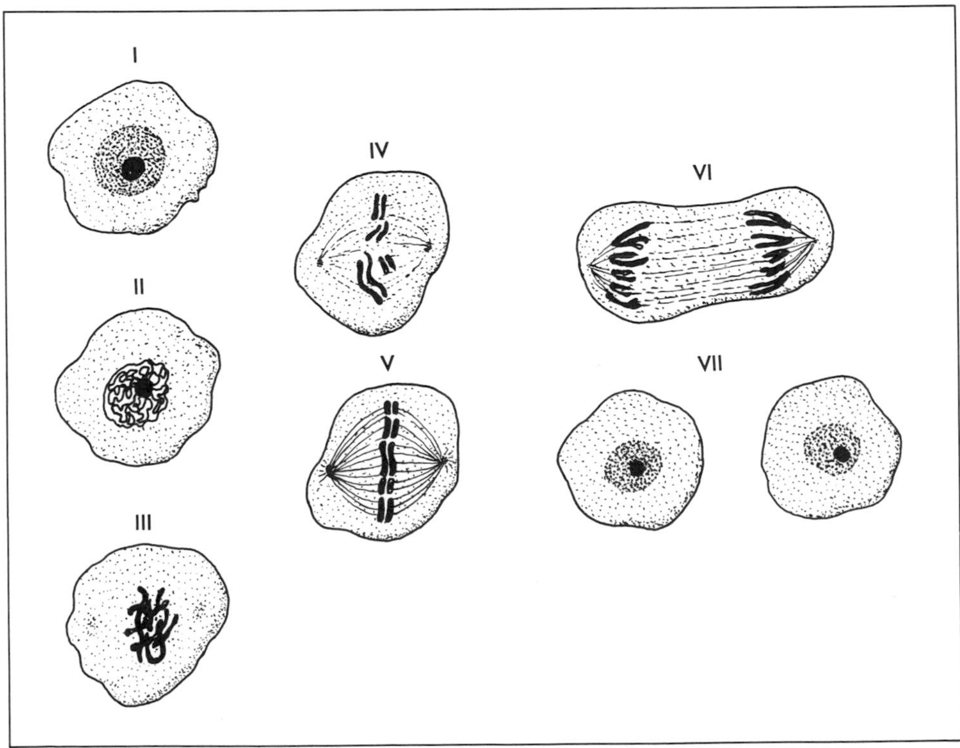

Division of the cell

Every cell owes its existence to the division of a preexisting one. Cells divide by the process of *mitosis* (Fig. 3-9), or indirect cell division, in which cleavage of the cell is preceded by a series of complicated nuclear and cytoplasmic changes. Mitotic division is characteristic of the higher animals and plants and may be seen in the lower forms as well.

QUESTIONS FOR REVIEW

1. Define protoplasm, cytology, mitosis, nanometer, macromolecule.
2. Outline the parts of a typical cell and give the function of each.
3. In your own words define the cell.
4. What structure controls and regulates vital activities in the living cell?
5. What instruments are used to study cells and unicellular organisms?
6. Give the unit of measurement for cells and unicellular organisms. State why it is used.
7. How are the hereditary characters of a cell transmitted?
8. What cells may be seen with the unaided eye?
9. What is DNA? Why is it important? Where does it occur?
10. State the function of a gene.
11. What is RNA? Where is it found in the cell? What is its physiologic significance?
12. Briefly characterize the important biologic molecules in protoplasm.

REFERENCES. See at end of Chapter 5.

4 The bacterial cell

Definition. *Bacteria* (singular, *bacterium*) are minute unicellular microorganisms that ordinarily do not contain chlorophyl and that may be able to move independently in their environment.

KINSHIP TO PLANTS. In the early days of microbiology it was thought that bacteria belonged to the animal kingdom, but for the following reasons it seemed more suitable to classify them with plants: Like plants generally, bacteria start with rather simple substances and build them into complicated organic chemical molecules. In bacteria, division of the cell occurs along the short axis of the cell, whereas in animal cells the division occurs along the long dimension of the cell. Like most plants, bacteria utilize food materials only if they are in solution. Waste products must be excreted in fluids that are diffused through the cell wall. There is no provision, as in animals, for intake of solid particles with digestion to occur inside the organism. There are no specialized structures for release of solid particles from within the bacterial cell. Bacteria possess a rigid cell wall, a feature of plants. Among plants, bacteria represent one of the lowest forms of life, being simpler in structure and mode of development than are the fungi. Nevertheless, they have an elaborate and complicated life history.

■ **Distribution.** Bacteria are widely distributed in nature.* They have adapted to every conceivable habitat. They are found within and upon our bodies, in the food we eat, in the water we drink, and in the air we breathe. They are plentiful in the upper layers of the soil, and no place on earth, except possibly the peaks of snowcapped mountains, is free of them. They are found in frozen Antarctica and in the hot water of the geysers in Yellowstone. Our skin has a large bacterial population, and bacteria make up a generous portion of the contents of the alimentary tract. There are several thousand species of bacteria; of this number about 100 species produce disease in man. Some of the bacteria that produce disease in man also produce disease in the lower animals. Others produce disease in the lower animals only, and still others attack only plants. The majority, however, do not attack man, lower animals, or living plants, and either do not affect animals and plants at all or are actually helpful to them. In fact, if the activities of bacteria were to cease, all plant and animal life would soon become extinct. Bacteria that cause disease are spoken of as *pathogenic;* those that do not cause disease are spoken of as *nonpathogenic.*

---

*The total living mass of microbes is about 20 times that of animal life.

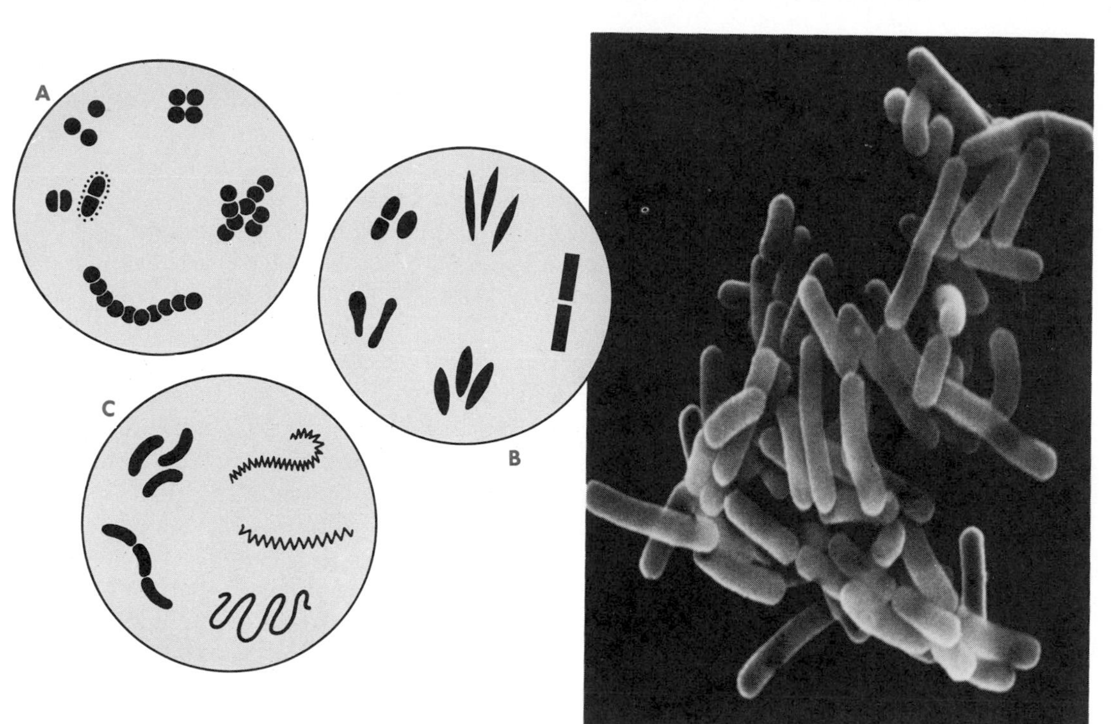

FIG. 4-1. Basic shapes of bacteria. **A,** Spherical—coccus (pair, diplococci; four, tetrad; chain, streptococci; cluster, staphylococci). **B,** Rod shaped—bacillus (including coccobacillus). **C,** Spiral shaped—vibrio, spirillum, and spirochete.

FIG. 4-2. Micrograph of bacilli (*Lactobacillus* species) taken with scanning electron microscope. Note three-dimensional effect obtained from this instrument. (Courtesy Dr. R. C. Reynolds, University of Texas Health Science Center at Dallas.)

Morphology

■ **Shape.** Bacteria occur in three basic shapes (Figs. 4-1 and 4-2):

1. Spherical—coccus*
2. Rod-shaped—Bacterium or bacillus
3. Spiral-shaped—vibrio, spirillum, spirochete

A *vibrio* is a curved organism shaped like a comma. A *spirillum* is a spiral organism whose long axis remains rigid when the organism is in motion. A *spirochete* is a spiral organism whose long axis bends when the organism is in motion.

Cocci are not necessarily round but may be elongated, oval, or flattened on one

---

*The following are the singular and plural forms:

Singular	*Plural*
Coccus	Cocci
Bacillus	Bacilli
Spirillum	Spirilla
Bacterium	Bacteria
Medium	Media

side. Some bacilli are long and slender, whereas others are so short and plump that they may be mistaken for cocci. These short, thick, oval bacilli are known as *coccobacilli*. The ends of bacilli are usually round but may be square or concave.

When bacteria, especially cocci, divide, the manner in which they do so and their tendency to cling together often give them a distinct arrangement. Cocci that divide so as to form pairs are known as *diplococci*. The opposing sides of diplococci may be flattened (examples, gonococci and meningococci). Cocci that divide and cling end to end to form chains are known as *streptococci*. Those that divide in an irregular manner to form grapelike clusters or broad sheets are known as *staphylococci*. Other characteristic arrangements of cocci are in groups of four *(tetrads)* and packets of eight *(sarcinae)*. No pathogenic cocci are found in the latter group. Bacilli that occur in pairs are known as *diplobacilli* and those that occur in chains *streptobacilli*. The diplobacillus and streptobacillus arrangements are not common. When some bacilli divide, they bend at the point of division to give two organisms arranged in the form of a V. This is known as *snapping*. In other cases they tend to arrange themselves side by side. This is known as *slipping*.

■ **Size.** Bacteria are so small (no larger than 1/50,000 of an inch*) that the highest magnification of the ordinary compound microscope must be used to study them. Cocci range from 0.4μ to 2μ in diameter. The smallest bacillus is about 0.5μ in length and 0.2μ in diameter. The largest pathogenic bacilli are seldom greater than 1μ in diameter and 3μ in length; the average diameter and length of pathogenic bacilli are 0.5μ and 2μ, respectively. Nonpathogenic bacilli may be much larger, reaching a diameter of 4μ and a length of 20μ. The spirilla are usually narrow organisms and are from 1μ to 14μ in length. Different species of bacteria show marked variation in size, and there is some variation within a species, but as a rule, the size of each species is fairly constant.

■ **Structure.** Bacteria, always unicellular, are so tiny and transparent (about the density of water) and so slightly refractile that, unless stained with dyes, they are difficult to see even with the compound light microscope. When stained they appear homogeneous or slightly granular. With the electron microscope, however, microbiologists can visualize minute details of bacterial structure (Fig. 4-3).

The shape of the bacterial cell is maintained by a rigid *cell wall*. The protoplasmic substance of bacteria exerts such a high osmotic pressure, equivalent to that of a 10% to 20% solution of sucrose, that in ordinary environments the cell wall is necessary to prevent the cell from bursting. If the bacterial cell is placed in a suitable hypertonic medium and the cell wall dissolved, the remainder of the bacterium is converted into a spherical *protoplast*. In an isotonic environment protoplasts remain viable and grow. The stability of the cell wall is derived from its chemical makeup; this varies in the two major groups of bacteria (see p. 60). In gram-positive bacteria, the chief component is mucopeptide—a polymer of the amino sugars *N*-acetylglucosoamine and *N*-acetylmuramic acid and short peptide linkages of amino acids. Sometimes teichoic acids or a mucopolysaccharide complex of amino sugars and simple monosaccharides may be present. In gram-negative bacteria, mucopeptide inner layer is chemically bound to two outer layers of lipopolysaccharide and lipoprotein. There are no teichoic acids.

---

*A cubic inch would hold 10 trillion medium-sized bacteria, or as many as there are stars in 100,000 galaxies.

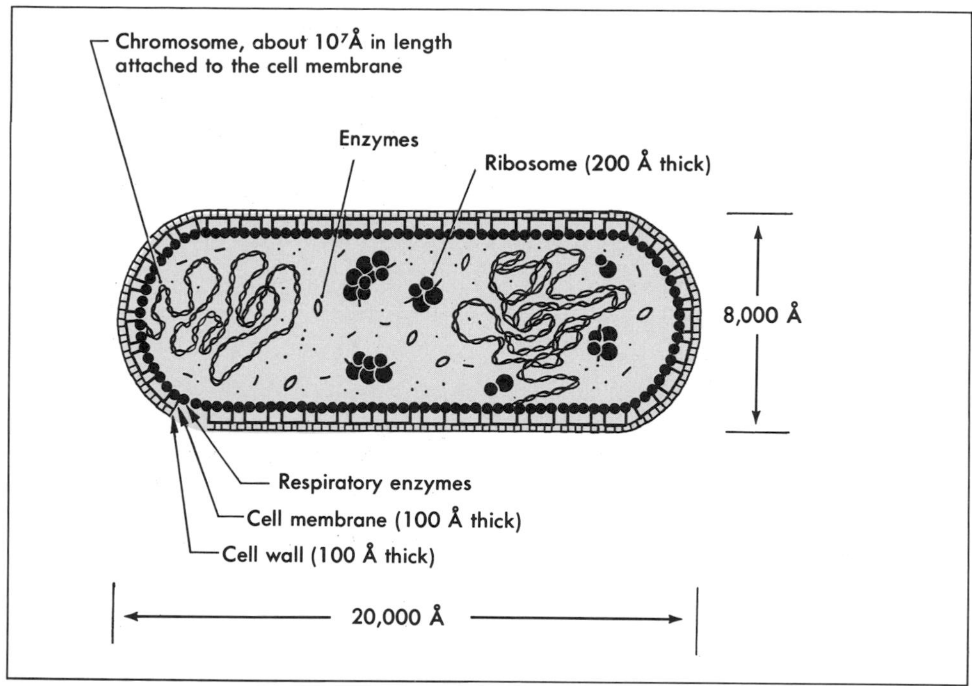

FIG. 4-3. Ultrastructure of young bacterial cell *(Escherichia coli)*, diagram. (Adapted from Watson, J. D.: Molecular biology of the gene, New York, 1970, W. A. Benjamin, Inc.)

The cell wall is so narrow that it cannot be seen with the ordinary compound light microscope. In ultrathin sections it is revealed by the electron microscope as a well-defined structure surrounding a distinct layer, the cell membrane or *plasma membrane* (Fig. 4-3), which separates it from the cytoplasm of the bacterial cell. The plasma membrane is the site of important enzyme systems, including the respiratory enzyme system (cytochrome enzymes). In fact, in bacteria it corresponds to the mitochrondria of higher organisms.* In regulating the passage of food materials and metabolic by-products between the interior of the cell (where metabolic activities are carried on) and the surroundings, it functions osmotically both as a barrier and as a link. It blocks the entry of certain substances and catalyzes the transport of others into the cell.

Surrounding many bacteria is a mucilaginous envelope or *capsule* (Fig. 4-4, *A*). Indistinct in most bacteria, it is well developed in a few (examples, *Diplococcus pneumoniae, Clostridium perfringens,* and *Klebsiella pneumoniae*). The capsule is formed by an accumulation of slime excreted by the bacterium. This material is usually a complex polysaccharide. If it is present about the cell in only small amounts, a distinct capsule does not appear.

Capsule formation is most prominent in organisms taken directly from the animal body, for when grown on artificial media, the same organisms often lose their ability to form capsules. A capsule does not stain with the ordinary bacteriologic dyes but may appear as a clear halo around the bacterium even two or three times as broad as

---

*Bacteria do not contain mitochondria.

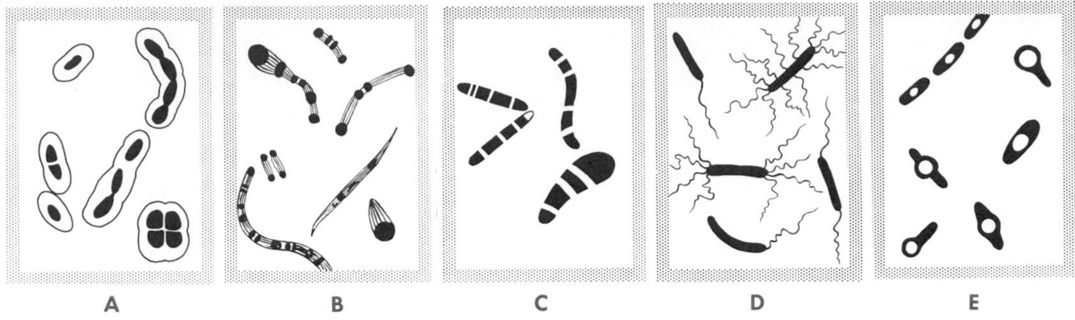

FIG. 4-4. Special features of bacteria. **A,** Capsules. **B,** Metachromatic granules. **C,** Barred forms (cross-hatching from uneven staining). (Note pleomorphism in **B** and **C**.) **D,** Flagella. **E,** Spores.

the bacterium. It is stained by special methods. The presence of a capsule appears to increase the virulence of an organism by protecting it against phagocytosis, and in some cases the capsule gives the organism its specific immunologic nature. For instance, relative to the nature of their capsules, pneumococci are divided into at least 79 types. The specific antigenic nature of a capsule depends upon its carbohydrate content.

Within some bacteria (example, *Corynebacterium diphtheriae*) at the ordinary magnification of the light microscope are seen granules that stain more deeply than the remainder of the cell. They are known as *metachromatic granules* (Fig. 4-4, *B* and C) and are enzymatically active. They are thought to be reserves of inorganic phosphate stored as polymerized metaphosphate (volutin). Sulfur-oxidizing bacteria convert excess hydrogen sulfide from the environment into intracellular granules of elemental sulfur. In some species granules are arranged irregularly within the cell, whereas in others they are located in one or both ends of the cell, where they are known as *polar bodies*.

Electron microscopy reveals a dense packing of ribosomes in bacterial cytoplasm and the presence of spherules and various submicroscopic granules. The ribosomes, as would be expected, play an important role in protein synthesis. The submicroscopic granules are known to be biochemically complex and active; they may represent stores of food materials, even elimination products. There does seem to be no relation of these granules to the ability of the bacteria containing them to produce disease.

As special staining methods designed to visualize the chemical substances making up the nucleus in the higher forms of life were applied to the study of bacteria, it became evident that bacteria contain a true nucleus, consisting essentially of deoxyribonucleic acid (DNA). Nuclear material is seen as a distinct but relatively transparent structure of rounded proportions in electron micrographs (Fig. 4-8). No nuclear membrane is detectable. The DNA present is a single chromosome (Fig. 4-3); if unfolded, it would stretch approximately 1 mm. It is hooked at one point to an infolding of the plasma membrane known as a *mesosome*. Many bacteria possess one nucleus, but in young, actively dividing cells, two or more nuclei may be seen. As it is for all cells, the nucleus is precisely the governing force for the bacterial cell, and vital activities cannot be carried on in a bacterial cell without it.

■ **Chemical composition.** Even in a unit such as the bacterial cell, 500 times smaller than the average plant or animal cell, the chemical composition is exceedingly complex. To gain an idea as to the chemical makeup of any bacterial cell, let us look at one that, since it is easily grown and manipulated in the laboratory, is undergoing almost as intensive study today as man. This is the colon bacillus or *Escherichia coli* (see Figs. 4-3 and 4-8).

Biochemically this microbe contains perhaps 3000 to 6000 different types of molecules (Table 4-1). Of these, specific proteins account for 2000 to 3000 kinds. The amount of DNA present is postulated to be that required to code for the necessary amino acid sequences in these proteins. About the DNA are 20,000 to 30,000 spherical ribosomes composed of protein (40%) and RNA (60%). Water, water-soluble enzymes, and a large number of various small and less complex molecules are associated with these nucleic acids.

TABLE 4-1. CHEMICAL MAKEUP OF A SINGLE, YOUNG, ACTIVELY DIVIDING, COLON BACILLUS (ESCHERICHIA COLI)*

Component	Average molecular weight	Estimated number of molecules	Number of different kinds of molecules	Percent total cell weight
Proteins	40,000	1,000,000	2000 to 3000	15
Carbohydrates and precursors	150	200,000,000	200	3
Lipids and precursors	750	25,000,000	50	2
Nucleic acids				
DNA	2,500,000,000	4	1	1
RNA	25,000 to 1,000,000	1000 to 400,000	More than 1000	6
Amino acids and precursors	120	30,000,000	100	0.4
Nucleotides and precursors	300	12,000,000	200	0.4
Other small molecules (breakdown products of food molecules)	150	15,000,000	200	0.2
Inorganic ions (Na^+, K^+, Mg^{++}, Ca^{++}, Fe^{++}, Cl^-, $PO_4^{\equiv}$, SO_4^{--})	40	250,000,000	20	1
Water	18	40,000,000,000	1	70 to 75

*From Watson, J. D.: Molecular biology of the gene, New York, 1970, W. A. Benjamin, Inc.

■ **Motility.** Many bacilli and all spirilla are motile when suspended in a suitable liquid at the proper temperature.* True motility is seldom observed in cocci. The organs of bacterial locomotion are fine hairlike appendages (Fig. 4-4, *D*) known as *flagella* (little whips), that spring from the bacterial cell and cause it to move along by their wavelike, rhythmic contractions. Some spirochetes aid the action of their flagella with a sinuous motion of the entire cell body. A bacterium may have one flagellum (monotrichous), a few, or many flagella in a tuft (lophotrichous), and the flagella may be attached to one end, both ends, or all around the organism (peritrichous) (Fig. 4-5). Flagella, which chemically are elastic proteins, do not take the ordinary bacteriologic dyes but have to be stained by special methods.

Bacteria may be motile when grown on one medium and nonmotile when grown

*True motility, in which the organism changes its position in relation to its neighbors, should not be confused with *brownian motion,* a peculiar dancing motion possessed by all finely divided particles suspended in a liquid.

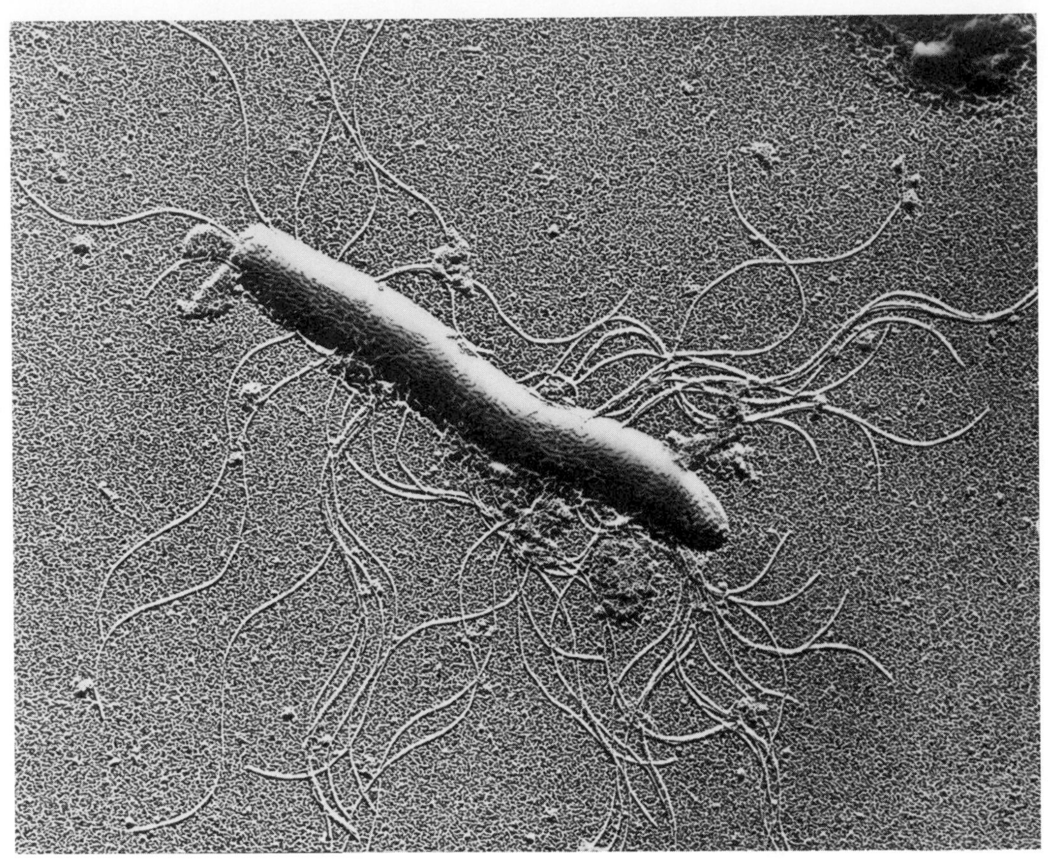

FIG. 4-5. Peritrichous bacillus from mouth, electron micrograph. Note flagella—number and arrangement. In preparation specimen shadowed with gold. (×28,000.) (Courtesy Dr. J. Swafford, Arizona State University, Tempe, Arizona; from Brown, W. V., and Bertke, E. M.: Textbook of cytology, St. Louis, 1969, The C. V. Mosby Co.)

on another. Also they may be motile at one temperature and nonmotile at another. Different organisms travel at different rates of speed. The rod-shaped organism that causes typhoid fever *(Salmonella typhi)* is able to progress at a rate of 2000 times its own length per hour.

Pili (Latin, 'hairs') are surface projections like flagella found in gram-negative bacteria (Fig. 4-6). However they are shorter and finer and do not propel the cell. Also called *fimbriae,* they may be part of the attachment of cells in conjugation; but otherwise their purpose is unknown.

■ **Spores.** Under certain poorly understood conditions, some species of bacteria (for example, species in genera *Bacillus* and *Clostridium*) form within their cytoplasm bodies that are resistant to influences adverse to bacterial growth. These bodies are known as *spores (endospores)* (Fig. 4-4, *E*). Spore formation seems to be a characteristic of bacilli, being exceedingly rare in cocci and spirilla. Among bacilli about 150 species, most of which are nonpathogenic, form spores. The important pathogenic, spore-forming bacteria are those that cause tetanus, gas gangrene, botulism, and anthrax.

Since spores seem to form when conditions for bacterial growth are unfavorable,

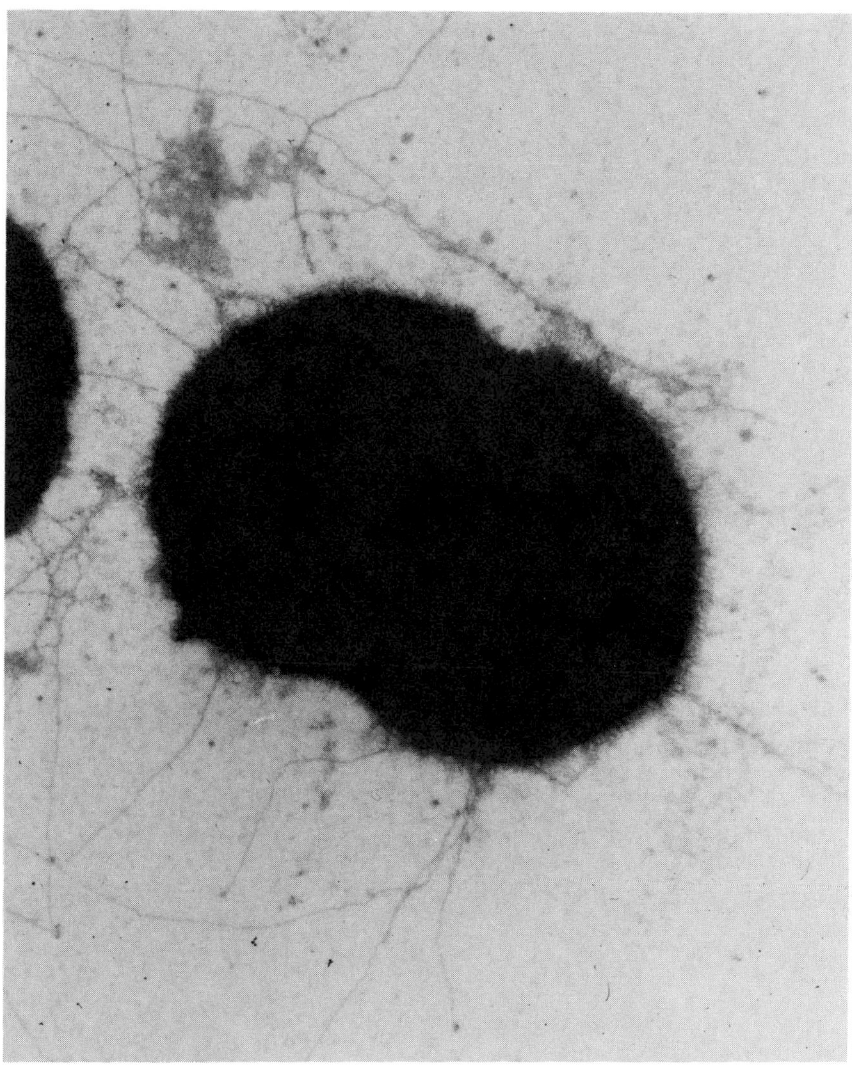

FIG. 4-6. Pili in electron micrograph of gram-negative gonococcus *(Neisseria gonorrhoeae)*. Negative staining with 2% uranyl acetate (see p. 63). (×88,020.) (Courtesy Janice C. Bullard and Dr. Stephen J. Kraus of the Venereal Disease Research Unit, Center for Disease Control, Health Services and Mental Health Administration, Department of Health, Education, and Welfare, Atlanta, Ga.)

spore formation may be a protective mechanism, but in the life of certain species of bacteria it seems to be a normal phase. Bacteria that do not form spores and spore-bearing bacteria in which spores are not forming are known as *vegetative* bacteria. Those in which spores are forming are *sporulating* bacteria. When conditions suitable for bacterial growth are established, the spore converts itself back to the actively multiplying form (germinates). The germinating spore becomes the vegetative form of the bacterium.

There is a temperature for each species at which spore formation is most active, and spore formation is preceded by a period of active vegetative reproduction. Some species form spores only in the presence of oxygen; others form spores only in its absence. Some observers regard the spore phase as a period of rest or hibernation for the organism. Spore formation is not a reproductive phenomenon because a spore forms only one bacterium and a bacterium forms only one spore.

Sporulation occurs in the following manner: The bacterial cell forms within its substance a round or oval, highly refractile body surrounded by a capsule. This body increases in size until it is broad or broader than the cell. The portion of the cell that remains gradually disintegrates, leaving only the spore. Spores from which the remainder of the cell has disappeared are *free* spores. In the electron microscope a spore has a structural pattern: a coat, a cortex, and a nuclear core of chromatinic material.

Spores do not take the ordinary bacteriologic dyes but may be stained by special methods. Ordinary stains of sporulating organisms may show the spore as a clear, unstained area situated in the end of the bacterium (terminal), near its center (central), or in an intermediate position (subterminal). In shape, spores may be spherical, ellipsoidal, or cylindrical. In anaerobic bacteria the spore is broader than the remainder of the bacterial cell, causing the bacterium to assume a spindle shape if the spore is central and a club shape if it is terminal or subterminal. The shape and position of the spores help to identify certain bacteria.

Although spores are especially resistant to heat, chemicals, and drying, the vegetative forms are no more resistant to these or other adverse influences than nonsporebearing bacteria. An idea of the protective value of spores can be drawn from the observation that some spores withstand boiling for hours, whereas the temperature of boiling water kills vegetative bacteria within 5 minutes. On the one hand, spores resist tremendous pressures, but on the other hand, they can persist in a vacuum approaching the emptiness of space. Spores have also been known to resist the temperature of liquid air ($-190°$ C.) for 6 months. Their resistance is probably the result of the membrane about them and of the concentrated, water-free nature of their substance. Although only a few species of bacteria produce spores, they are of great significance because their spores are everywhere. This means that spore-killing methods in bacteriologic and surgical sterilization and in the canning industry are absolutely necessary.

Reproduction

■ **Cell division.** The typical mode of bacterial reproduction, an asexual process, is by simple transverse division (binary fission) (Figs. 4-7 and 4-8). Bacteria do not divide by mitosis—there is no mitotic spindle. The cell elongates, reaches its maximum size, and then becomes constricted in the middle, after which the constriction deepens and separates the organism into halves. In preparation for cell division the nuclear chromosome of the dividing cell replicates with equal division of nuclear material into sister chromosomes. Replication initiates active membrane synthesis at the periphery, and a transverse membrane is formed that moves into the bacterium. The membrane with a new cell wall constricts it along the short axis, and partly because of the presence of the mesosome, pushes the two chromosomes apart and into each of two newly formed daughter cells. Each newly formed cell soon reaches its full size and in turn divides. A newborn bacterium requires from 15 to 30 minutes to reach adult size and divide. Reproduction is always specific; for example, staphylococci always reproduce staphylococci.

■ **Production of L-forms.** In certain species of bacteria (examples, *Proteus, Bacteroides,* and some of the coliforms) a normal cell may swell to a large entity only to disintegrate into numerous particles, some 0.2μ in diameter, known as *L-forms.* This kind of change may occur without known stimulus or with a well-defined one, such as a drug. Al-

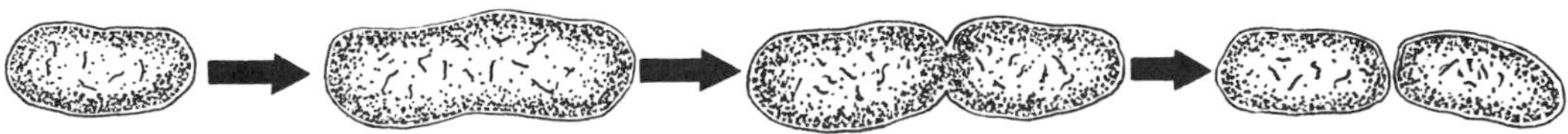

FIG. 4-7. Simple transverse division (binary fission). Note that young bacterium on left becomes full grown and divides into two young ones, as seen on the right.

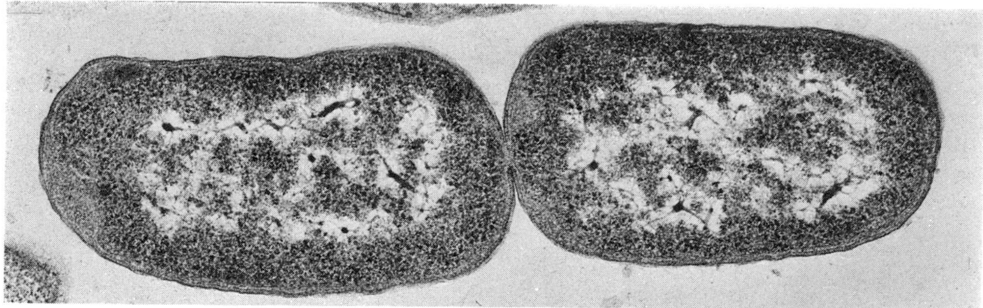

FIG. 4-8. Transverse fission of normal bacterium *(Escherichia coli)*, electron micrograph. Two microorganisms are still attached along line of division. (From Morgan, C., Rose, H. M., Rosenkranz, H. S., and Carr, H. S.: Electron microscopy of chloramphenicol-treated Escherichia coli, J. Bact. **93:**1987, 1967.)

though these forms possess distinguishing characteristics, they are not only akin to the parent cell but can revert to it.

Variation

Living organisms, or the aggregate of living organisms, are seldom if ever exactly the same. Microbes are no exception. The deviation from the parent form in bacteria of the same species growing under different or identical conditions is known as *variation.*

Variation may be caused by external or internal influences. Environmental factors include the kind of culture medium, the temperature of growth, the length of time grown artificially, and the event of exposure to chemicals or to radiant energy (X rays, for example). Variation may also result from factors inherent in the bacteria themselves.

■ **Observed variations.** Variation may affect all the biologic properties—size, shape, biochemical nature, colonial characteristics, and physiologic activities—of bacteria and may be temporary or permanent.

DISSOCIATION. A change in the kind of colony formed on a semisolid culture medium as an example of bacterial variation is termed bacterial (or microbial) *dissociation.* Although a pure culture of an organism is used to streak the surface of a culture medium, the resultant colonies may present contrasting appearances. There may be *smooth* or *mucoid* colonies, regular in outline, round, moist, and glistening—the *S-type* colonies. On the other hand there may be *rough* colonies, larger, irregular in outline, indented, and wrinkled—the *R-type* colonies. Between the two are the intermediate forms. By proper laboratory manipulation organisms forming S-type colonies can be made to form R-type colonies and vice versa. Usually organisms forming S-type colonies are more vigorous in their disease-producing capacities than those forming R-type colonies. This variation offers some explanation as to the rise and fall of epidemics.

MORPHOLOGIC VARIANTS. Bacteria of the same species, growing under the most favorable conditions, may show considerable variation in size, shape, and appearance. This is *pleomorphism.* Growing under unfavorable conditions (for example, in old cultures), the members of some species assume irregular, bizarre shapes, and stain irregularly. These swollen, shrunken, or granular, aberrant (or abnormal) variants are known as *involution* or *degenerative forms.* In this instance variation in morphology probably reflects the presence of many dying as well as dead cells in the culture medium and the injurious effect of the accumulation therein of metabolic by-products.

Organisms that are surrounded by a capsule when grown in the animal body often lose that capsule when grown artificially. In some cases capsule formation may be restored by their return to a susceptible animal. Capsule formation is pronounced in anthrax bacilli infecting a susceptible animal, but from the artificial media of the laboratory there is hardly a sign of a capsule for the very same organism. Bacteria with capsules tend to form smooth colonies and those with none tend to produce rough colonies.

■ **Adaptation.** One of the attributes of living cells is their power to adapt themselves to their surroundings. This is probably truer of bacterial cells than of many other types of cells. For instance, certain bacteria that require specially prepared media to sustain their continued growth, when first isolated from the animal body, gradually acquire the ability to grow on media devoid of the growth-promoting and enriching materials necessary for their early growth. They may also be grown artificially in a gradually changing environment until at last they are able to grow under conditions of food supply, temperature, moisture, and oxygen supply far different from those in which they originally grew best.

Those variations (changes in bacterial makeup) that represent the physiologic adjustment to the environment are designated by the term "*adaptation.*"

ATTENUATION. An important form of adaptation is *attenuation* (an important concept in immunology), which indicates a loss in disease-producing ability of a given organism. An organism whose virulence is decreased is attenuated. A highly pathogenic organism may be rendered temporarily or permanently nonpathogenic if repeatedly subcultured on artificial laboratory media. For example, by cultivation of a strain of bovine tubercle bacilli on media containing bile until they had lost their ability to cause disease, a suitable preparation (BCG vaccine) was developed for vaccination against tuberculosis. Virulence, although artificially eliminated, may often be restored by animal passage, that is, the serial injection of microorganisms into and their recovery from susceptible animals.

■ **Genetic factors.** Certain internal factors operative in bacterial variations pertain to changes in the genetic makeup of the bacteria.

MUTATIONS. Mutations in bacteria are analogous to those in the higher forms of life. Since the specific intracellular enzyme (protein) is regulated biochemically by a specific gene positioned on the chromosome of the cell, the related structure (and function) of the cell must depend upon the integrity of that gene. Within the gene the significant component is the sequence of the nucleotide bases with any change in this pattern projecting an effect on the cell. Such a change constitutes a mutation and is inheritable.

Rearrangement of the nucleotide sequence of a gene can result from an error in replication or from a breakage of the sugar-phosphate backbone of the DNA molecule. Mutations occur spontaneously or are induced by certain mutagenic agents that can

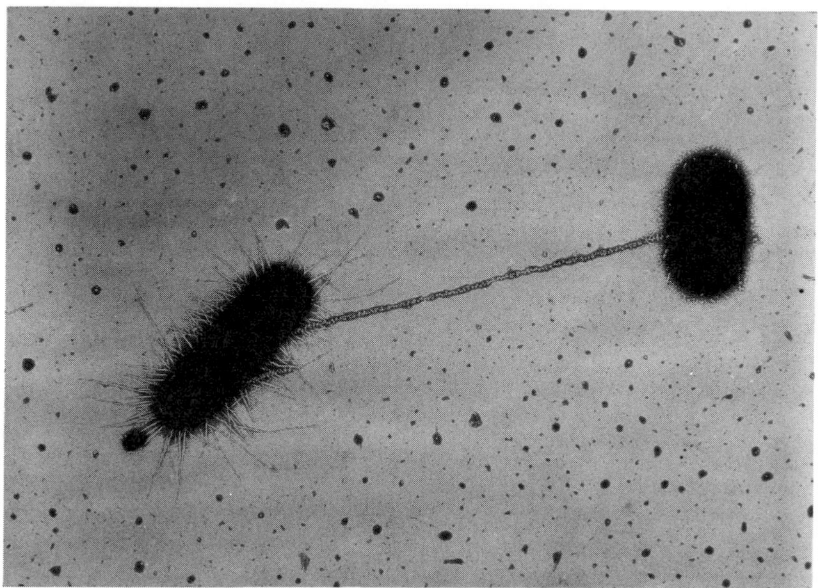

FIG. 4-9. Conjugation in bacteria. Passage of genetic material by means of a long thin filament between bacterium of one species and that of another but both in family Enterobacteriaceae. A factor inducing resistance to certain antibiotics is known to be so transferred (see also p. 256). Note pili on surface of microbe on left. (Courtesy Dr. C. C. Brinton, Pittsburgh, Pa.)

alter the nucleotide bases in such a way as to promote replication errors. The most effective mutagens are certain alkylating agents and forms of radiant energy.

INTERMICROBIAL TRANSFER. Variations resulting from the passage of nuclear material from one bacterium to another are easily demonstrated. If two bacterial strains, variants in a single species, are incubated together under special circumstances, such a transfer is borne out by the appearance in the culture of new organisms displaying qualities of both original strains. When genetic material (nuclear DNA) is transferred from one cell to another, the receiving cell does not gain the full complement of chromosome but only a portion. Immediately after the event of transfer this portion must be matched to the corresponding segments of the chromosome in the cell and genetic material exchanged and eliminated. From this rearrangement a newly formed chromosome emerges containing DNA from two bacterial cells. This is the *recombinant chromosome,* shaped by the process of *recombination.*

Conjugation is a process effecting transfer of genetic material in a special way. Recent studies indicate that on occasions a type of sexual reproduction occurs in bacteria wherein hereditary material is passed from one organism to the other on a transient physical contact (see also p. 256). In fact, in electron micrographs (Fig. 4-9) two individual bacterial cells are seen to unite by means of a cytoplasmic bridge extending between them. Even male and female mating types have been defined for the closely related groups of enteric (gram-negative) bacteria in the family Enterobacteriaceae that have been studied. Conjugation can occur between any two members of the family, not just within a given genus. However, there is no new cell as a consequence. The progeny of the cell receiving the nuclear material are produced in the usual way, by binary fission of the parent.

Transformation is a process of direct transfer of nucleic acid. Certain species of bacteria release DNA into the medium in which they are growing, such as certain species of *Neisseria* that release it in their extracellular slime. A small number of bacteria in a given population, referred to as *transformable* bacteria, can pick up this DNA. How they do it is not known. Within the bacterial cell the DNA is integrated with the DNA already present, and what is left over is degraded. The net effect is the replacement of a *short* region of the chromosome with a new portion of genetic material.

Transduction is a special form of indirect transfer of genetic material. A bacterial virus or bacteriophage carries a small fragment of the chromosome from the bacterial cell in which it was produced to the bacterium that it invades. Transduction occurs in coliforms, enteric pathogens, and staphylococci.

These above mentioned internal factors may pertain to changes in the genetic substance of bacteria. When nuclear material (DNA) is transferred from one bacterium to another (transformation), there is alteration of some property in the recipient.

■ **Drug fastness.** A microbial variation with considerable therapeutic importance is one that arises from a mutation or other genetic mechanism giving the organism increased tolerance for an antimicrobial drug. This tolerance for the drug is termed *drug resistance* or *drug fastness.* An organism that becomes resistant to a drug used in the treatment of disease is said to be drug fast. Strains of bacteria are always emerging resistant to one or more antibiotics, a resistance usually acquired when clinical infectious disease has been inadequately treated with the given antibiotic.

Bacteria may become tolerant of more than one antimicrobial drug. This *cross resistance* is noted especially in the case of closely related antibiotics.

QUESTIONS FOR REVIEW

1. Define bacteria. Indicate their distribution.
2. Discuss the function of the cell wall. Of what is it composed?
3. Name and describe the three basic forms of bacteria.
4. What is the function of the plasma membrane?
5. Give two characteristics of bacteria at least partly dependent upon the capsule.
6. How do bacteria move about in a fluid medium?
7. By what processes do bacteria reproduce?
8. What are spores? Comment as to their formation and purpose.
9. Give examples of bacterial variation.
10. Discuss adaptation.
11. State the consequences of drug fastness and cross resistance.
12. Briefly discuss the bacterial nucleus.
13. List significant contributions of the electron microscope to our knowledge of the bacterial cell.
14. Briefly define L-forms, conjugation, mutation, transformation, transduction, attenuation, recombination.

REFERENCES. See at end of Chapter 5.

5 Methods for microbes: visualization*

PLAN OF ACTION

The microbiologist studies microbes in many ways. To visualize them, he must use an instrument of precision, the microscope. The test specimen from a contaminated site or diseased area that he inspects with the microscope may be unstained, or, as is often the case, it may be stained by one of several methods. He may make a culture from the suspicious material. After the microbes have multiplied sufficiently to form visible growth, he may note the physical pattern of this growth, study the microorganisms in stained or unstained microscopic preparations made from the growth, and use some of it to determine their biochemical and biologic properties. He may inject organisms recovered from the test specimen into a suitable laboratory animal and observe the effect on the animal. Finally, he may apply well-known immunologic tests.

Note: In summary, the methods used to study microbes include (1) direct (microscopic) examination, (2) culture, (3) biochemical tests, (4) animal inoculation, and (5) immunologic reactions. They will be presented in a sequence of five chapters distributed through three units.*

Microbes must be handled with extreme caution and in accord with well-known principles of conduct in microbiologic laboratories. *Bacteria or other disease-producing agents can be very dangerous, and accidental laboratory infections can be fatal.* In a recent survey of 1300 infections among laboratory workers, 39 ended fatally. In many cases the infections occurred in research workers and highly trained technologists.

TOOLS FOR STUDY

The microbiologist has many instruments of precision at his command; some are in constant use; others are needed only in special investigations.

■ **The microscope.** The instrument most often used by the microbiologist and one to be handled with the greatest of care is the compound light microscope (Fig. 5-1). Needless to say, its workmanship should be of the highest quality.

*Discussions of the laboratory methods for identification and diagnosis of microbes are not confined to the five-chapter sequence entitled "Methods for microbes" but are included in material specifically related to the different microbes throughout the book.

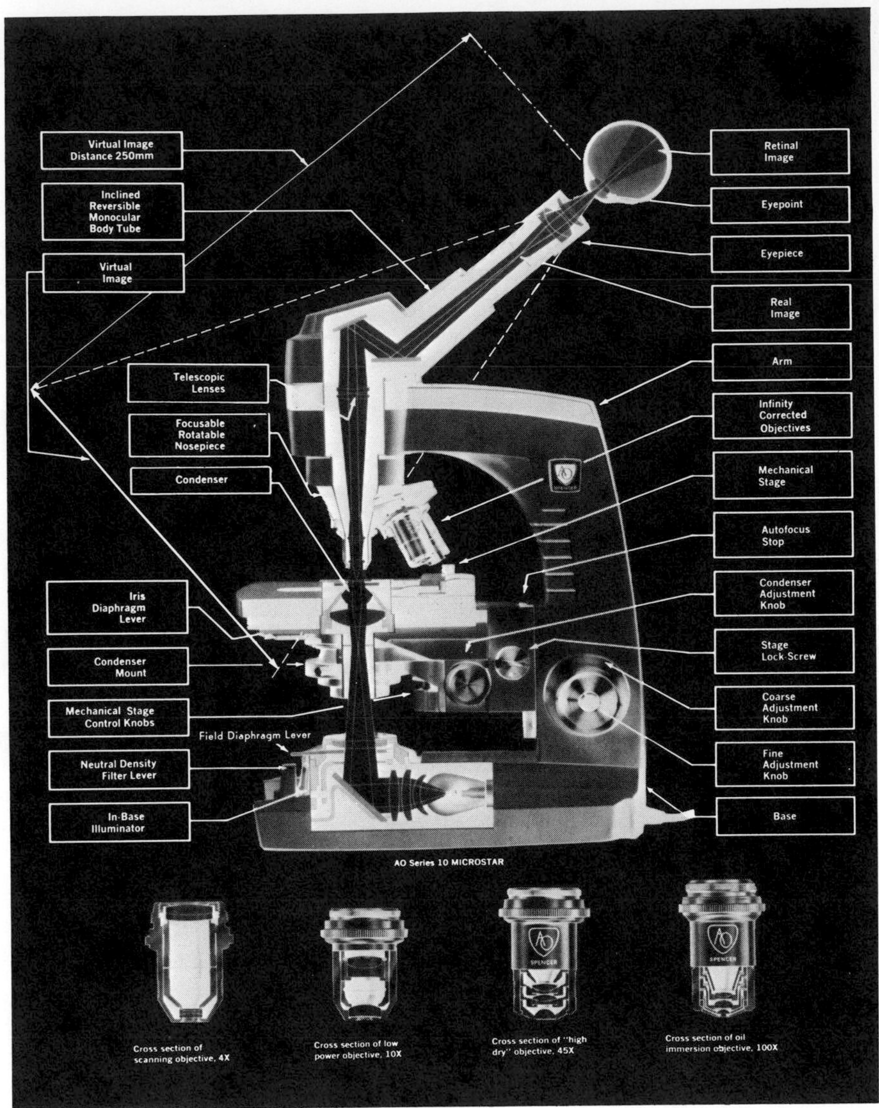

FIG. 5-1. Dissection of microscope—parts and optical features of a monocular microscope. Note how light rays are reflected from the mirror through the microscope to eye of observer. This modern microscope is fitted with a focusable stage whereby specimen may be focused to objective, making it unnecessary for observer to change level of his eye. (Courtesy American Optical Corp., Scientific Instrument Division, Buffalo, N. Y.)

GENERAL DESCRIPTION. Microscopes are of two kinds, simple and compound. A *simple* microscope is little more than a magnifying lens. A *compound* microscope incorporates two or more lens systems so that the magnification of one system is increased by the other. Practically, it consists of two parts, the supporting stand and the optical system. The supporting stand includes (1) a base and pillar, (2) an arm to support the optical system and house the fine adjustment, (3) a platform (stage) on which the object to be examined rests, and (4) a condenser and mirror fitted beneath the stage. The condenser and mirror focus the light through a central opening in the stage on the object to be examined.

The optical system consists of a body tube, which supports the *ocular* lenses (eyepiece) at the top end and the *objective* lenses attached to a revolving *nosepiece* at the other end. The optical system is connected to the arm of the supporting stand by an *intermediate slide,* which moves up and down on the arm in response to movement of the *fine adjustment.* The intermediate slide contains the rack and pinion for the *coarse adjustment,* which acts directly on the tube of the optical system. The platform of the microscope is usually equipped with a mechanical device to hold firmly the microslide, upon which the object is mounted, so that it can be moved from place to place by set screws. The advantages of this device are that the specimen can be examined systematically and, unless moved, the specimen remains in a fixed position.

The magnification of an objective is usually designated by its equivalent focal distance in inches or millimeters. By *equivalent focal distance* is meant the focal distance of a lens having the same magnification as the objective. The higher the number of the objective, the less is its magnification. American microscopes are usually fitted with 16 mm., 4 mm., and 1.8 mm. objectives. The last is known as an *immersion objective* because for best results there must be a liquid (oil or water) between the objective and the object being examined. Usually this is immersion oil. The 16 mm. objective magnifies 10 times; most 4 mm. objectives magnify 43 times, and most 1.8 mm. objectives magnify 97 times.

The oculars of a microscope are given 6×, 10×, and similar designations to indicate that they increase the magnification of the objective 6, 10, or more times, respectively. To obtain the magnification of any combination of ocular and objective lenses, multiply the magnification of objective by that of the ocular (see Table 5-1). Remember that magnification refers to both the length and the width of an object; that is, a magnification of 100 means that the object is made to appear 100 times as long and 100 times as wide.

TABLE 5-1. MAGNIFICATION WITH LENS COMBINATIONS OF THE COMPOUND MICROSCOPE

	Objectives		
Oculars	16 mm. 10×	4 mm. 43×	1.8 mm. 97×
6×	60×	258×	582×
10×	100×	430×	970×
15×	150×	645×	1455×

■ **Ocular micrometer.** Microbes can be measured microscopically by means of a device known as a micrometer (Fig. 5-2). The simplest type is the ocular micrometer, which consists of a scale on a glass disk that fits (scale side down) between the lenses of the eyepiece. The spaces between the lines on this scale do not represent true measurements. Real values are obtained by calibrating the ocular micrometer against a stage micrometer. This is a glass slide with a true measurement scale on it—the lines on the

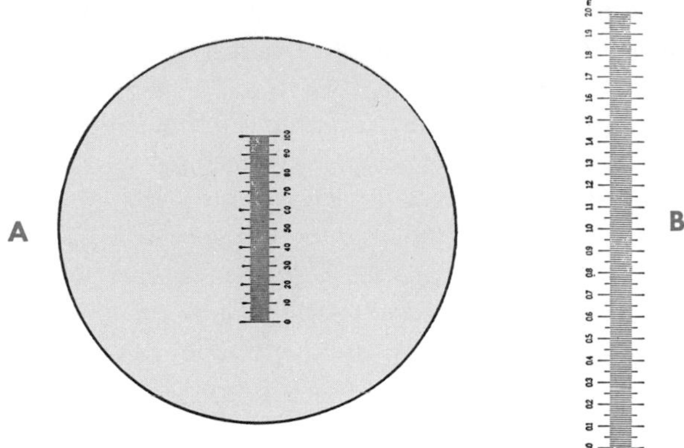

FIG. 5-2. A, Ocular micrometer. **B,** Stage micrometer. (Courtesy American Optical Corp., Scientific Instrument Division, Buffalo, N. Y.)

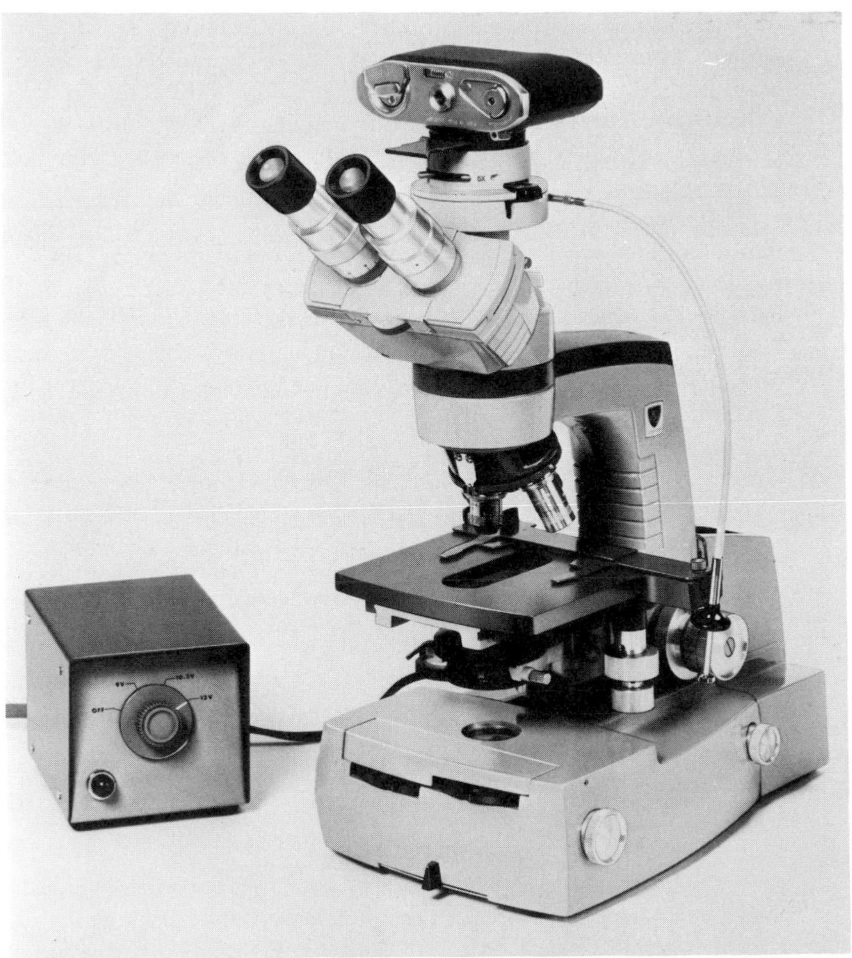

FIG. 5-3. Photomicrographic camera. Note binocular compound microscope fitted with camera for making photomicrographs (photographs of microscopic objects). Camera is fitted onto an adaptor. (Courtesy American Optical Corp., Scientific Instrument Division, Buffalo, N. Y.)

scale are either exactly 10 μ or 100 μ apart. By placing the stage micrometer in the position occupied by a smear of cells and by looking through the ocular of the microscope, one may superimpose the scale of the ocular on that of the stage micrometer. It is then easy to determine the actual unit length that the distance between the lines of the ocular micrometer represents.

■ **Photomicrographic camera.** Another instrument of great value to microbiologists is the photomicrographic camera, with which pictures are made of objects seen under the microscope (Fig. 5-3). This gives an easily preserved visual record of microscopic findings and renders a wealth of material available for future study.

EXAMINATION OF UNSTAINED BACTERIA

It may be desirable to examine unstained bacteria to determine their biologic grouping, their motility, and their reaction to chemicals or specific serums. These features may be determined in a hanging-drop preparation. A few species of bacteria that cannot be stained by the methods to be discussed are often examined by dark-field illumination.

■ **Hanging-drop preparations.** To examine bacteria microscopically in a hanging drop, one must use (1) a platinum loop for transferring the material to be examined to (2) a cover glass and (3) a hanging-drop slide.

The platinum loop is a piece of fine platinum wire about 3 inches in length. One end is fastened in a handle and the other end into a loop about 1/16 inch in diameter (Fig. 5-4). Platinum is used for making wire loops because heating this metal in a flame repeatedly to sterilize it does not destory it; the wire cools quickly after being heated.

A hanging-drop slide is a thick glass slide with a circular concavity or depression at its center. A cover glass is a piece of very thin glass about 7/8 inch square.

Make the preparation as follows: spread a small amount of petroleum jelly around the concavity of the slide. If the specimen to be examined is a culture growing on a solid

FIG. 5-4. Platinum wire loops and inoculating needles with holders. About 1½ inches of platinum wire makes a microbiologic loop. Gauge 26 or 27 wire is satisfactory for most routine inoculations; gauge 24 wire is better for stabbing.

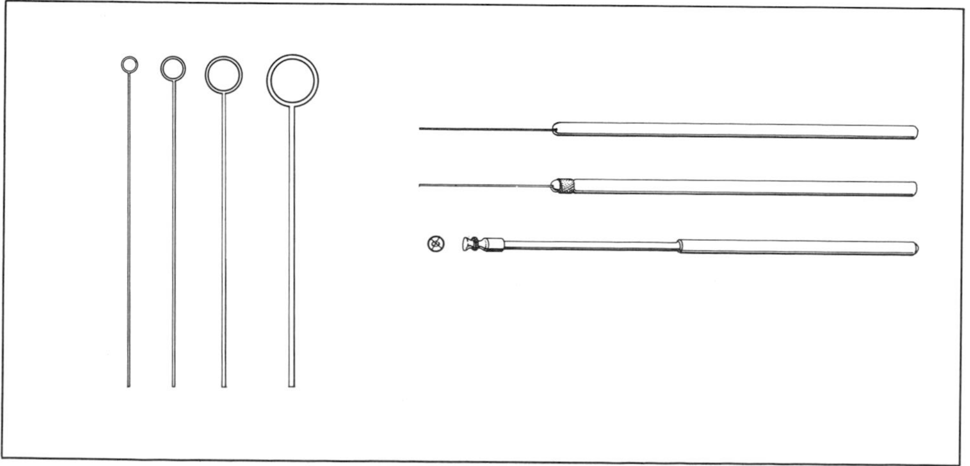

medium or material such as thick pus, take up a loopful of specimen with the platinum loop and mix thoroughly with a drop of sterile isotonic salt solution placed in the center of the cover glass. If bacteria growing in a liquid medium are to be examined, transfer a drop of the fluid to the cover glass by means of the wire loop. Place the hanging-drop slide over the cover glass in such a way that the center of the depression lies over the drop. The petroleum jelly seals the cover glass to the slide, holds it in place, and prevents evaporation. Invert the slide now so that the drop to be examined hangs from the bottom of the cover glass but does not touch the surface of the concavity at any point. The preparation is ready for microscopic examination. Examine with the 4 mm. (high dry) lens, and reduce the amount of light passing through it by partly closing the diaphragm of the substage condenser of the microscope. When hanging-drop preparations are observed, brownian motion and flowing of organisms in currents should not be mistaken for true motility.

The platinum loop should be sterilized immediately before and after each transfer of material containing bacteria (Fig. 5-7, *D*). Since hanging-drop preparations contain living bacteria, discard the slide and cover glass into a suitable container of disinfectant after the examination is finished.

The *wet mount* is similar to the hanging-drop preparation except that an ordinary microslide is used instead of the thick hanging-drop slide with its central depression (Fig. 5-5). Many of the applications are the same. (See also p. 488.)

■ **Dark-field illumination.** Dark-field illumination is used to examine certain delicate bacteria that are invisible in the living state in the light microscope, that cannot be stained by standard methods, or that are so distorted by staining as to lose their identifying characteristics. Its greatest usefulness is in the demonstration of *Treponema pallidum* in chancres and other syphilitic lesions, but it is of value in the examination of many other organisms as well.

FIG. 5-5. Wet mount preparation: Rim a clean cover glass with petroleum jelly (or make a ring of it on a clean microslide). First flame the platinum loop. Then, **A,** use it to take a drop of fluid containing test specimen (either a fluid or test material in sterile isotonic saline solution). **B,** Place the drop on the cover glass. Invert cover slip onto microslide as in **C**. This simple procedure is easily carried out and readily adaptable.

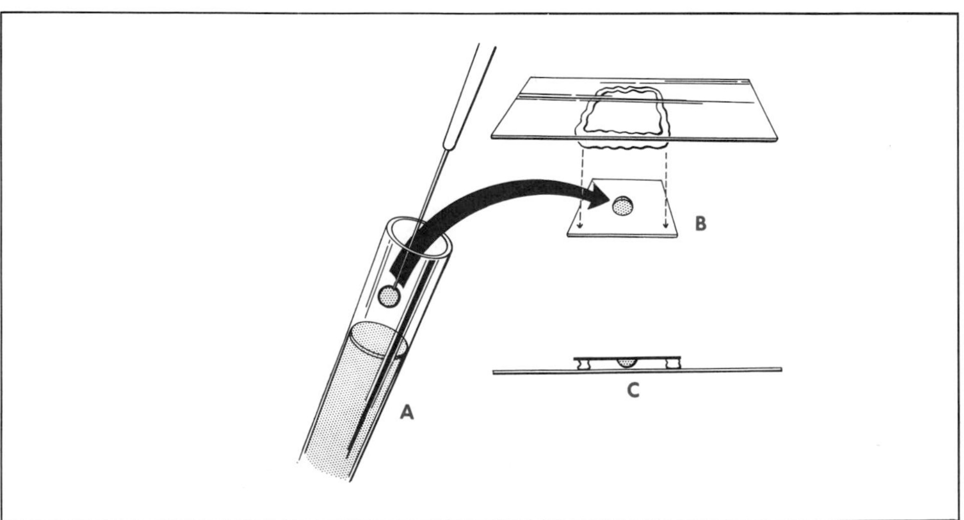

The material to be examined is placed on an ordinary slide and covered with a cover glass. Sealing the cover glass to the slide with a ring of melted paraffin prevents the cover glass from slipping and accidental infection of the fingers. Dark-field illumination depends on the use of a substage condenser so constructed that the light rays do not pass directly through the object being examined, as is the case with an ordinary condenser, but strike it from the sides at almost a right angle to the objective of the microscope. The microscopic field becomes a dark background against which bacteria or other particles appear as bright silvery objects (Fig. 5-6). A similar effect is seen when a beam of light enters a darkened room and renders visible particles of dust that cannot be seen in a better-lighted room.

EXAMINATION OF STAINED BACTERIA

■ **Staining.** The bodies of bacteria are so small that, when examined in hanging-drop preparations, little of their finer structure can be made out; to be studied more closely, they must be colored with some dye. This process is called *staining*. The dyes most often used are aniline dyes, derivatives of the coal tar product, aniline.

For a stained preparation, place a small amount of the material to be examined on a perfectly clean glass slide and spread out into a thin film by means of a platinum loop or swab (Fig. 5-7, *A*). The film is known as a *smear*. After it is allowed to dry in the air, slowly pass the slide, smear side up, through a flame two or three times (Fig. 5-7, *B*). *Flaming* kills the bacteria in the smear and causes them to stick to the slide. This is known as *fixing*. Other methods of fixing, such as immersion in methyl alcohol or in Zenker's solution, are sometimes used, but heat is the most suitable for routine work. Apply the stain to the fixed smear and wash off with water (Fig. 5-7, *C*); blot the slide

FIG. 5-6. Dark-field microscopy. Note large spiral organism is white against black background. The dark-field microscope is an important tool in the identification of various kinds of spiral microbes. (Courtesy American Optical Corp., Scientific Instrument Division, Buffalo, N. Y.)

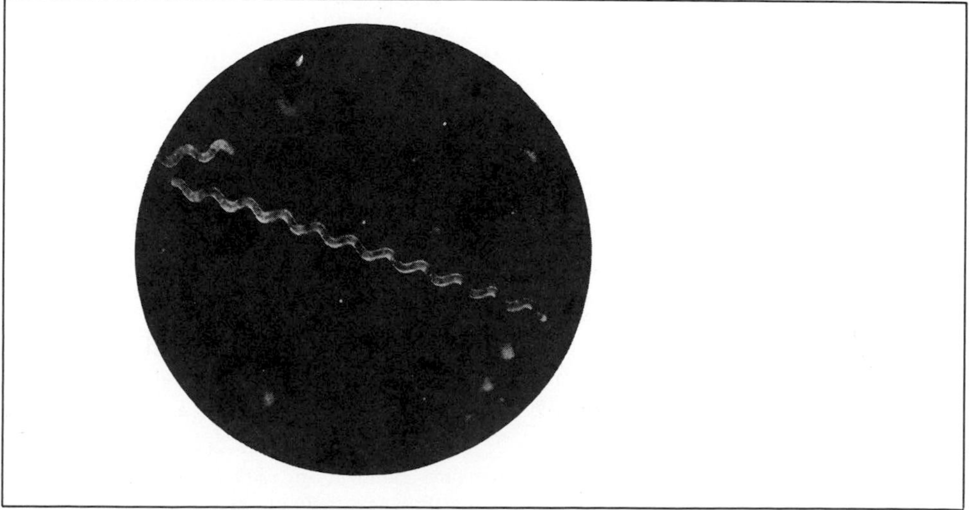

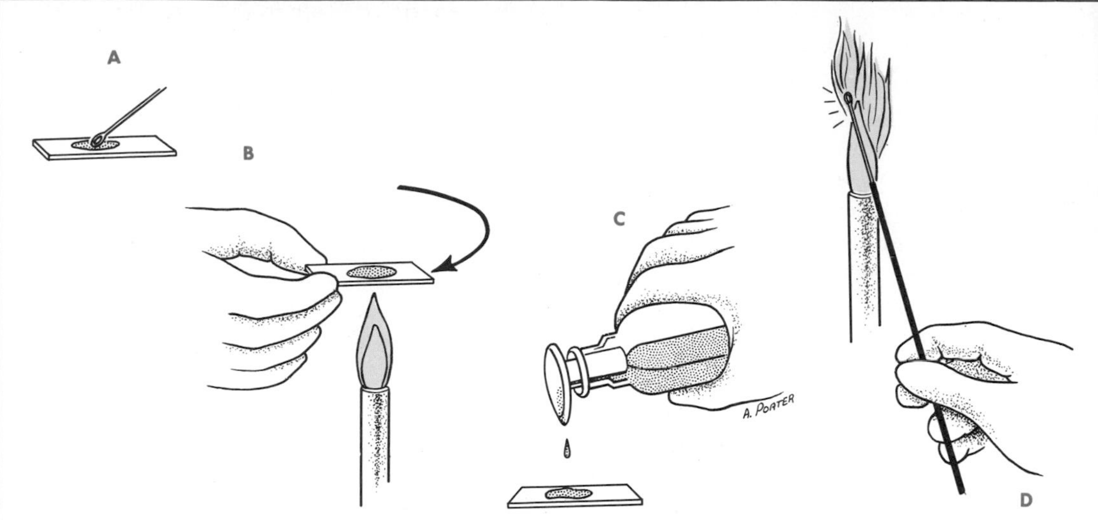

FIG. 5-7. Preparation of stained bacterial smear. **A,** Spread loopful of test specimen thinly on clean glass microslide. Allow to dry. **B,** Fix specimen onto glass slide (held between fingers) by passing it through flame. **C,** Apply drops of stain from dropper bottle to slide (according to technic indicated). **D,** Sterilize inoculating loop (or needle) before and after it is used. Hold the length of wire in flame until it glows. *This is a very important step!*

dry between sheets of absorbent paper. Flame sterilization of the platinum loop is sketched in Fig. 5-7, *D*.

Three classes of stains are used in bacteriology: (1) simple stains, (2) differential stains, and (3) special stains. In addition, there is the process of negative staining.

SIMPLE STAINS. A simple stain is usually an aqueous or alcoholic solution of a dye. It is applied to the fixed smear from 1 to 5 minutes and washed off. The stained preparation is then ready for microscopic examination. The widely used simple stains are Löffler's alkaline methylene blue, carbolfuchsin, gentian violet, and safranine. The length of time that the stain remains on the smear depends upon the avidity with which it acts. Sometimes added to the solution is a chemical that makes it stain more intensely. Such chemicals are called *mordants.*

Most bacteria stain easily and quickly with simple stains, some do not stain so readily, and a few do not stain at all. Capsules and spores are not stained with these simple stains but may give contrast as clear, *unstained* structures. Flagella cannot be stained in this way and are not seen as contrasting structures.

DIFFERENTIAL STAINS. More complex staining methods divide bacteria into groups, depending upon their reaction to the chemicals used for staining. Of these, the *Gram stain* and the *acid-fast stain* are most often used.

The method of staining introduced by Hans Christian Gram* divides bacteria into two great groups: those that are *gram-positive* and those that are *gram-negative.* This method depends upon the fact that when bacteria are stained with either crystal violet or gentian violet and the smear is then treated with a weak solution of iodine (mordant), the bodies of some bacteria combine with the dye and iodine to produce

*Hans Christian Gram (1853-1938), a Danish physician working in Berlin, introduced this important method of differential staining in 1884. It remains essentially unaltered in use today.

a color that cannot be removed by alcohol, acetone, or aniline, whereas the color is readily removed from certain other bacteria by these solvents. Bacteria from which the color cannot be removed are spoken of as being *gram-positive*, and those from which it can be removed are spoken of as being *gram-negative*. A few bacteria that sometimes keep the stain and that at other times do not are *gram-variable*. Certain physiologic differences are generally correlated with Gram staining. Gram-positive bacteria tend to be more resistant to the action of oxidizing agents, alkalis, and proteolytic enzymes than are gram-negative ones. They are more susceptible to the acids, detergents, sulfonamides, and antibiotics, such as penicillin, than are gram-negative ones. Many modifications have been devised for the original Gram's method. The method outlined below is often used. The explanations appended will apply to any technic used.

1. Make a thin smear of material for study and fix in a flame.
2. Stain with crystal violet or gentian violet (Gram I*) for 1 minute.
3. Blot thoroughly to take up excess stain.
4. Cover the smear with Gram's iodine solution (Gram II†) for 1 minute.
5. Drain and blot dry. Both gram-positive and gram-negative bacteria are now stained a dark violet or purple color.
6. Drop acetone on the smear (Gram III) until no more color flows from the smear (about 5 to 10 seconds). Blot dry. All gram-negative bacteria are completely decolorized. The gram-positive ones are not affected.
7. Cover the smear with a stain (Gram IV‡) that gives a contrast in color (counterstain) for 1 minute.
8. Wash with water, blot dry, and examine.

Stains used to give contrast in color are known as *counterstains*. The ones most often used in the Gram stain are safranine and dilute carbolfuchsin, both of which give a red color, and Bismarck brown, which, as its name implies, gives a brown color. Gram-negative bacteria are stained with the counterstain. Gram-positive bacteria do not stain with the counterstain because they are completely stained with the stain-iodine-bacterial cell combination that gives them their gram-positive microscopic appearance.

Table 5-2 gives the reaction to the Gram stain of important pathogenic bacteria.

A method of staining known as the acid-fast or Ziehl-Neelsen stain§, is discussed next. When most bacteria and related forms are stained with carbolfuchsin, they stain easily, but when the smear is treated with acid, they are completely decolorized. It is difficult to stain certain other microbes with carbolfuchsin, but once stained, they retain the dye, even when treated with an acid. Those that retain the stain are spoken of as

---

*Gram I is usually a mixture of a 10% solution of crystal violet in 95% ethyl alcohol (solution A) with a 1% solution of ammonium oxalate in distilled water (solution B) in the ratio of one part of solution A to four parts of solution B (Hepler's ratio 1:8).

†Gram II is prepared by dissolving 2 gm. of potassium iodide and 1 gm. of iodine crystals in 300 ml. of distilled water; it should be stored in a brown glass bottle.

‡Gram IV is usually made by mixing 10 ml. of a 2.5% alcoholic solution of safranine with 90 ml. of distilled water.

§The credit for staining tubercle bacilli belongs to Paul Ehrlich (1854-1915), whose method slightly modified is used today. The modifications were made by Franz Ziehl (1857-1926) and Friederich Neelsen (1854-1894), whose names are attached to the staining method.

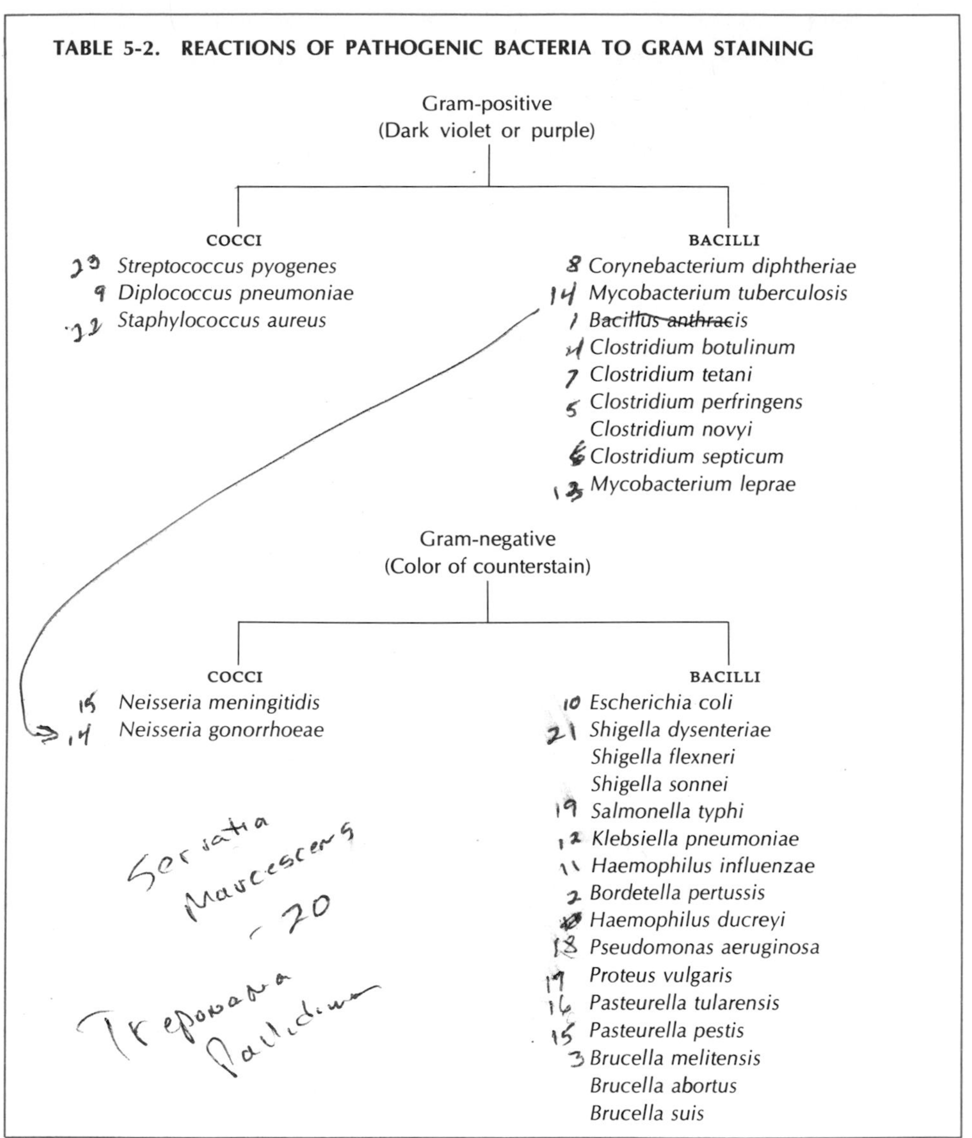

Gram-positive
(Dark violet or purple)
COCCI
Streptococcus pyogenes
Diplococcus pneumoniae
Staphylococcus aureus
BACILLI
Corynebacterium diphtheriae
Mycobacterium tuberculosis
Bacillus anthracis
Clostridium botulinum
Clostridium tetani
Clostridium perfringens
Clostridium novyi
Clostridium septicum
Mycobacterium leprae
Gram-negative
(Color of counterstain)
COCCI
Neisseria meningitidis
Neisseria gonorrhoeae
BACILLI
Escherichia coli
Shigella dysenteriae
Shigella flexneri
Shigella sonnei
Salmonella typhi
Klebsiella pneumoniae
Haemophilus influenzae
Bordetella pertussis
Haemophilus ducreyi
Pseudomonas aeruginosa
Proteus vulgaris
Pasteurella tularensis
Pasteurella pestis
Brucella melitensis
Brucella abortus
Brucella suis

TABLE 5-2. REACTIONS OF PATHOGENIC BACTERIA TO GRAM STAINING

being *acid-fast*. The property of being acid-fast probably derives from the presence of complex fatty substances within the bacterial cell.

The technic of acid-fast staining is as follows*:

1. Make a smear on the slide and fix.
2. Flood the slide with carbolfuchsin† (red) and gently steam over a flame (do not boil) for 3 to 5 minutes. Keep the slide covered with dye.
3. Allow the slide to cool. Wash off the excess stain with water (all bacteria are now red).

*There are several standard modifications of this method.

†The carbolfuchsin solution for the acid-fast stain is made by combining one part of a saturated solution of basic fuchsin in 95% ethyl alcohol with nine parts of a 5% aqueous solution of phenol.

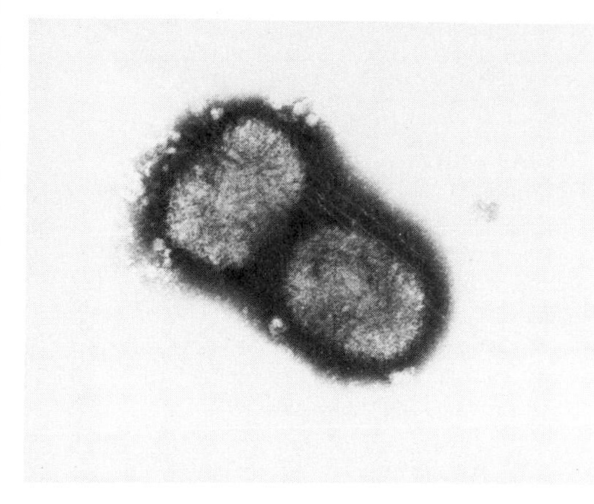

FIG. 5-8. Negative staining applied to study of viruses. Example here is virus of smallpox. Phosphotungstic acid provides a background opaque to electrons when specimen is examined in the electron microscope and one against which surface structures of a viral particle are defined by contrast. (×120,000.) (Courtesy Dr. Derrick Baxby, University of Liverpool, England.)

4. Dip slide repeatedly in acid-alcohol* until all the red color is removed from smear. Wash with water. (At this step the acid has removed the red color from all bacteria that are not acid-fast. The acid-fast organisms are unaffected and remain stained a bright red color.)
5. Apply a counterstain for 1 minute to give contrast. Löffler's alkaline methylene blue† is frequently used. (The acid-fast organisms are completely saturated with the red carbolfuchsin and therefore will not take any of the counterstain. The nonacid-fast organisms, having had all their stain removed by the acid, stain a deep blue.)

Important examples of acid-fast organisms encountered in medicine are *Mycobacterium tuberculosis, Mycobacterium leprae,* the anonymous mycobacteria, and the actinomycete, *Nocardia asteroides.*

SPECIAL STAINS. Important special stains are those for capsules, spores, flagella, and metachromatic granules. Stains primarily designed to demonstrate metachromatic granules are especially valuable in identifying *Corynebacterium diphtheriae* and in differentiating it from related organisms. Important stains of this type are Albert's stain, which uses toluidine blue and malachite green, and Neisser's stain, which uses methylene blue.

NEGATIVE (RELIEF) STAINING. Microorganisms such as *Treponema pallidum,* not stained by ordinary dyes, may be made visible by the process known as negative or relief staining, in which the background, but not the microorganisms, is stained. The microorganisms are mixed with india ink or 10% nigrosin (both of which are black); the mixture is spread out into a thin smear and allowed to dry. The microbes appear as colorless objects against a black background. Negative staining may be done with the dye Congo red; this has been used to display spiral organisms and bedsoniae (chlamydiae). Viruses are prepared by negative staining for visualization in the electron microscope (Fig. 5-8).

---

*Acid-alcohol for decolorizing is 3% hydrochloric acid (concentrated) in 95% ethyl alcohol.

†Löffler's alkaline methylene blue is made by mixing 30 ml. of a saturated solution of the dye methylene blue chloride in 95% alcohol (1.5% filtered) with 100 ml. of a weak (0.01%) aqueous solution of potassium hydroxide. Modern samples of the dye have been considerably purified, and the addition of alkali to 100 ml. of distilled water may not be necessary.

QUESTIONS FOR REVIEW

1. Name the procedures used in the study of microbes.
2. What are the purposes of a hanging-drop preparation? How is one made?
3. What is fixing?
4. What are the three classes of stains used in microbiology? Describe each one and give examples.
5. How is a simple stain made? Do many bacteria stain with simple stains?
6. Name two very important differential stains. Give the underlying principles of each.
7. Name (a) two important gram-positive cocci, (b) three important gram-positive bacilli, (c) two important gram-negative cocci, (d) three important gram-negative bacilli.
8. Name two important acid-fast bacilli and one acid-fast actinomycete.
9. What is negative or relief staining?
10. Briefly describe the microscope.
11. Give the steps in the preparation of a bacterial smear.
12. When is dark-field illumination used to study microbes?

REFERENCES FOR UNIT ONE

Athanasius Kircher (1602-1680) clerical scholar, editorial, J.A.M.A. **199:**336, 1967.

August von Wassermann (1866-1925), Wassermann reaction, editorial, J.A.M.A. **204:**1000, 1968.

Baker, F. J.: Handbook of bacteriological technique, New York, 1967, Appleton-Century-Crofts.

Bartlett, R. C., and Carrington, G. O.: How to avoid hazards in microbiology, Med. Lab. Observer **1:**46, (July) 1969.

Bauer, J. D., Ackermann, P. G., and Toro, G.: Bray's clinical laboratory methods, ed. 7, St. Louis, 1968, The C. V. Mosby Co.

Beaver, W. C., and Noland, G. B.: General biology—the science of biology, ed. 8, St. Louis, 1970, The C. V. Mosby Co.

Bodily, H. L., Updyke, E. L., and Mason, J. O., editors: Diagnostic procedures for bacterial, mycotic and parastic infections, Washington, D.C., 1970, American Public Health Association, Inc.

Bonner, J. T.: The size of life, Natural History **78:**40, (Jan.) 1969.

Bradbury, S.: The evolution of the microscope, New York, 1967, Pergamon Press, Inc.

Breed, R. S., Murray, E. G. D., and Smith, N. R., editors: Bergey's manual of determinative bacteriology, Baltimore, 1957, The Williams & Wilkins Co.

Brown, W. V., and Bertke, E. M.: Textbook of cytology, St. Louis, 1969, The C. V. Mosby Co.

Bryson, V., editor: Microbiology; yesterday and today, New Brunswick, N. J., 1959, Institute of Microbiology, Rutgers, The State University.

Burrows, W.: Textbook of microbiology, Philadelphia, 1968, W. B. Saunders Co.

Chedd, G.: The new biology, New York, 1972, Basic Books, Inc.

Corrington, J. D.: Getting acquainted with the microscope, Rochester, N.Y., 1968, Bausch & Lomb, Inc.

Culliton, B. J.: Cell membranes: a new look at how they work, Science **175:**1348, 1972.

Dalton, A. J., and Haguenau, F., editors: The nucleus, New York, 1968, Academic Press Inc.

Davis, B. D., Dulbecco, R., Eisen, H. N., et al.: Microbiology, New York, 1967, Hoeber Medical Division, Harper & Row.

Davis, J. G.: Aspects of laboratory hygiene, Lab. Pract. **21:**101, 1972.

Dickerson, R. E.: The structure and history of an ancient protein, Sci. Amer. **226:**58, (April) 1972.

Doctor Koch and the boiled potato, Today's Health **41:**77, (Dec.) 1963.

Dowben, R. M., editor: Biological membranes, Boston, 1969, Little, Brown & Co.

Elie Metchnikoff (1845-1916) advocate of phagocytosis, editorial, J.A.M.A. **203:**139, 1968.

Fawcett, D. W.: The cell: its organelles and inclusions, an atlas of fine structure, Philadelphia, 1966, W. B. Saunders Co.

Fisher manual of laboratory safety, Pittsburgh, 1965, Fisher Scientific Co.

Focus on the microscope, Today's Health **46:**44, (Nov.) 1968.

Fox, C. F.: The structure of cell membranes, Sci. Amer. **226:**30, (Feb.) 1972.
Freese, A. S.: Pasteur: bold genius of science, Today's Health **46:**58, (March) 1968.
Garrison, W.: He discovered healing in the soil, Today's Health **48:**42, (March) 1970.
Gibson, W. C.: Some Canadian contributions to medicine, J.A.M.A. **200:**860, 1967.
Gillies, R. R., and Dodds, T. C.: Bacteriology illustrated, Baltimore, 1968, The Williams & Wilkins Co.
Goldblith, S. A.: A condensed history of the science and technology of thermal processing, Part I, Food Technology **25:**44, (Dec.) 1971; Part II, **26:**64, (Jan.) 1972.
Gray, P.: The dictionary of the biological sciences, New York, 1967, Reinhold Publishing Corp.
Hamburger, M.: Wall-defective bacteria, Arch. Intern. Med. **122:**175, 1968.
Hammon, W. McD.: Human infection acquired in the laboratory, J.A.M.A. **203:**647, 1968.
Haskins, C. P.: Science advances in 1967, Amer. Scientist **56:**165, (Summer) 1968.
Holmes, O. W.: The contagiousness of puerperal fever, New Eng. Quart. J. Med. Surg. **1:**503, 1843.
Kit, S.: Basic research and the health needs of the nation, Texas Med. **66:**98, (Sept.) 1970.
Kornberg, A.: The synthesis of DNA, Sci. Amer. **219:**64, (Oct.) 1968.
Leikind, M. C.: Stanhope Bayne-Jones: Physician, teacher, soldier, scientist-administrator, friend of medical libraries, Bull. N.Y. Acad. Med. **48:**584, 1972.
Lentz, J.: The age of the enzyme, Today's Health **48:**32, (April) 1970.
Lillie, R. D., editor: H. J. Conn's biological stains, Baltimore, 1969, The Williams & Wilkins Co.
Lipman, C. B.: Discovery of living microorganisms in ancient rocks, Science **68:**272, 1928.
Lovejoy, E. P.: Women doctors of the world, New York, 1957, The Macmillan Co.
MacFate, R. P.: Introduction to the clinical laboratory, Chicago, 1971, Year Book Medical Publishers, Inc.
Mayr, E.: Theory of biological classification, Nature **220:**545, 1968.
Metric medicine for a metric America, (editorial), Ann. Intern. Med. **76:**138, 1972.
Minckler, J., Anstall, H. B., and Minckler, T. M., editors: Pathobiology: An introduction, St. Louis, 1971, The C. V. Mosby Co.
Mirsky, A. E.: The discovery of DNA, Sci. Amer. **218:**78, (June) 1968.
Needham, G. H.: The microscope, a practical guide, Springfield, Ill, 1968, Charles C Thomas, Publisher.
Neutra, M., and Leblond, C. P.: The Golgi apparatus, Sci. Amer. **220:**100, (Feb.) 1969.
Nyka, W.: New technics for staining tubercle bacilli, Bull. Path. **9:**48, 1968.
Nyka, W.: New techniques used to study tubercle bacilli in lungs of treated patients. Lab. Med. **3:**42, (March) 1972.
Paul, J. R.: A history of poliomyelitis, New Haven, Conn., 1971, Yale University Press.
Pike, R. M., Sulkin, S. E., and Schulze, M. L.: Continuing importance of laboratory-acquired infections, Amer. J. Public Health **55:**190, 1965.
Robert Hooke (1635-1703)—master inventor, editorial, J.A.M.A. **194:**453, 1965.
Russell, R. L., and Reynolds, J. W.: Quality control of microbiology, Bull. Path. **10:**65, (March) 1969.
Ryter, A.: Association of the nucleus and the membrane of bacteria: a morphological study, Bact. Rev. **32:**39, (March) 1968.
Schmidt, J. E.: Medical discoveries—who and when, Springfield, Ill., 1959, Charles C Thomas, Publisher.
Shapiro, L., Agabian-Keshishian, N., and Bendis, I.: Bacterial differentiation, Science **173:**884, 1971.
Smith, D. T., et al.: Zinsser microbiology, New York, 1968, Appleton-Century-Crofts.
Snyder, B.: Pitfalls in the gram stain, Lab. Med. **1:**41, (July) 1970.
Symposium on membrane structure and function, editorial, J.A.M.A. **219:**1756, 1972.
The National Library of Medicine, editorial, J.A.M.A. **218:**1937, 1971.
Toner, P. G., and Carr, K. E.: Cell structure. An introduction to biological electron microscopy, Edinburgh, 1971, E. & S. Livingstone, Ltd.
Watson, J. D.: Molecular biology of the gene, Menlo Park, Calif., 1970, W. A. Benjamin, Inc.
Weissmann, G.: The many-faceted lysosome, Hosp. Pract. **3:**30, (Feb.) 1968.
Whittaker, R. H.: New concepts of kingdoms of organisms, Science **163:**150, 1969.
Widmann, F. K.: Notes: on the early history of blood transfusion, Lab. Med. **1:**9, (Feb.) 1970.
Winchester, J. H.: Colonel Gorgas' Panama victory, Today's Health **46:**58, (May) 1968.
Wischnitzer, S.: Introduction to electron microscopy, Elmsford, New York, 1970, Pergamon Press.
Zugibe, F. T.: Diagnostic histochemistry, St. Louis, 1970, The C. V. Mosby Co.

LABORATORY SURVEY OF UNIT ONE

Design

The purposes of a laboratory course in microbiology are (1) to present the student with the scientific method, (2) to impress the fact of microbes as living agents, and (3) to demonstrate characteristics of microbes and their effects upon the human body.

The following exercises may be varied to meet the needs of the individual student and to fit the teaching facilities of the school. Details of technic have been omitted in this type of abbreviated survey because it is felt that the student can be taught technics to which the instructor is accustomed. In all instances the exercises have been designed to be carried out with a minimum of equipment and a maximum of safety for the student.

The student should obtain a notebook in which to keep an accurate record of each experiment. Included in the notes should be (1) the purpose of the experiment, (2) how it is performed, (3) the results obtained, and (4) the conclusions reached. The notebook should be neatly kept and be corrected at frequent intervals by the instructor.

SAFETY FIRST!

Rules for your own personal protection

Certain general laboratory rules are based on experience and good judgment. Avoid exposing yourself or others at all times to unnecessary danger. Observe the following precautions for safety in the laboratory:

1. Wash your hands thoroughly with soap and water at the beginning and at the end of each laboratory period. Disinfect your hands if necessary.
2. Wear a laboratory coat or jacket made of material easily cleaned and sterilized. Coat or jacket should be laundered at least once a week.
3. Scrub with a cleansing agent and disinfect the working surface of your desk at the beginning and at the end of each laboratory period.
4. Do not put anything in your mouth during the laboratory period except as specifically indicated in the laboratory exercise (pipetting, etc.). Do not under any circumstances lick gummed labels or put pencils in your mouth. Do not eat, drink, or smoke in the laboratory during laboratory time.
5. At all times, remember to keep your hands away from your face and mouth. Fingers and the areas under the nails can become contaminated quite easily.
6. Dispose of infectious material, cultures, and contaminated materials carefully in the way prescribed by the instructor. Note the location of the special containers of disinfectant and use regularly.
7. Always flame-sterilize platinum loops and needles carefully, faithfully, and thoroughly before laying them aside.
8. Be very careful to avoid spilling any kind of contaminated material. If infectious material contacts the desk, hands, clothing, or floor, notify the instructor at once.
9. Report even a minor accident to the laboratory instructor *immediately!*

10. Arrange cultures and infectious material in a secure place on the desk. Do not allow unused equipment to accumulate in your work area.
11. Keep glassware and other equipment clean and in its proper place. Be as clean and orderly as possible at all times.
12. Keep personal items in the places designated and away from the work area.

The compound light microscope

Care of the microscope

The following instructions are given to the student:

1. Keep both eyes open. Very little practice will enable you to do this.
2. Avoid direct sunlight. North light is advantageous. Good results are obtained with daylight.
3. When a slide is placed on the stage, see that it lies flat against the platform. Adjust the light so that the object is evenly illuminated.
4. By means of the coarse adjustment, move the objective to be used until it nearly (but *not quite*) touches the cover glass or upper surface of the mounted specimen. Then focus *up* until the object comes plainly into view. Complete focusing with the fine adjustment. Beginners should learn to focus with the low-power objectives (such as the 16 mm. objective). When the immersion objective is used, a drop of immersion oil must first be placed on the object before it can be clearly brought into view.
5. Keep the microscope clean and handle all parts with care. Do not touch the glass parts of the microscope with the fingers. Do not allow chemicals to contact the microscope, as they may injure it. The mechanical parts may be cleaned by application of olive oil on gauze, which can be wiped off with chamois or lens paper. To remove oil from the optical glass parts, wipe with lens paper moistened with xylol. Do this as rapidly as possible to prevent injury to the optical settings.
6. Clean the microscope thoroughly when you are finished. Leave objectives with the lowest power in the working position. This precaution ensures that the least expensive objective will be injured should the optical system be jammed down accidentally. Keep the microscope covered when not in use.

PROJECT

Use of the compound light microscope

1. *Parts of the microscope*
   a. Place the microscope upon a table in the proper working position at a convenient height. Be sure you are comfortably seated and that you do not have to stretch or that you are not cramped.
   b. Locate all parts shown in Fig. 5-1. Refer to the discussion of the microscope on pp. 53 to 55.
   c. By means of their numbers, locate the low-power, high dry, and oil immersion objectives. Explain the use of each in your notebook.
2. *Focusing*
   a. Place a prepared microslide on the stage of the microscope. (A section of tissue may be used.) Be sure that the specimen slide lies flat on the stage.

b. Elevate the top of the condenser so that it is flush with the surface of the stage.
c. With the aid of the instructor, focus low-power objective on specimen.
d. Observe the changes when the coarse and fine adjustments are manipulated.

3. *Illumination*
   a. Observe the changes in lighting brought about when the position of the mirror is changed. (Note: a microscope with a built-in lamp base will not have a mirror.) When you can fill the field of observation with light, have the instructor check to see if the lighting can be improved.
   b. Adjust the illumination so that it is even throughout the microscopic field of view.
4. *Attaining contrast*
   a. Observe the changes in the specimen when the iris diaphragm is opened or closed.
   b. Note changes in contrast of specimen image in the microscope.
5. *Comparison of three objectives*
   a. Focus on a specimen image with the low-power objective. (Complete focusing with the fine adjustment.) Center specimen image.
   b. Move high-power objective into position. Complete focusing with fine adjustment.

   Note: The constant movement of the fine adjustment is essential in producing a three-dimensional effect on the specimen image in microscopy. The fine adjustment is rotated slightly clockwise and then slightly counterclockwise.

   c. Under the direct supervision of the instructor, carefully focus with the oil immersion objective: First place a drop of immersion oil on the cover glass of the microslide (or on the upper surface of the specimen preparation). Lower the oil immersion objective with the coarse adjustment into the drop of oil until the objective engages the drop and spreads it somewhat. The objective does not quite touch the surface of the microslide. Complete focus with the fine adjustment.
   d. Under the supervision of the instructor, clean the microscope and return it to the locker.
6. *Problem*
   a. Calculate the magnification of the different objectives of your microscope with each eyepiece and complete the following table:

Focal length	Magnification objective lens	Final magnification with 8× ocular	Final magnification with 12.5× ocular
16 mm.	10×		
4 mm.	43×		
1.8 mm.	97×		

PROJECT

The study of cells

1. *Cells from inside of cheek*—unstained
   a. Scrape inner surface of the cheek with a wooden tongue blade.
   b. Mount the scrapings in a drop of water on a glass microslide and apply cover glass.
   c. Examine with low power of the microscope.
   d. Using the high dry objective for observation, sketch a few epithelial cells. You can recognize these by their flattened, pavement appearance. They may occur singly or in clusters. Locate any blood cells or bacteria that may be present.
2. *Cells from inside of cheek*—stained
   a. Mount scrapings from the inside of the cheek as outlined above, spreading the material out into a thin even smear with a toothpick applied to the microslide. Do not cover with a cover glass. Let dry.
   b. Pass the slide rapidly through a flame two or three times to fix the smear.
   c. Cover the preparation with methylene blue stain and allow to remain for 3 minutes.
   d. Wash the stain off with distilled water. Blot between sheets of blotting or bibulous paper.
   e. Examine with the oil immersion objective. Sketch three or four cells.
3. *Cells in the urine*
   a. Place a 10 ml. sample of urine in a conical centrifuge tube and centrifuge for 5 minutes at low speed (about 1500 rpm). Decant supernatant.
   b. Place a drop of the urinary sediment on a microslide and cover with a cover glass.
   c. Examine with low-power and high-power objectives of microscope. Look for epithelial cells and leukocytes (white blood cells). Make a sketch showing the different objects seen. With the aid of the instructor, identify them.

   Note: Close down the iris diaphragm of the scope to enhance contrast in the cellular specimen.
4. *Cells of the blood*
   a. Obtaining the specimen of blood
      (1) Cleanse side of the tip of the fourth finger of either hand with 70% alcohol, removing alcohol with dry sterile cotton.
      (2) With sterile lancet or needle, prick the finger with enough force to get a large drop of blood. Discard the *first* drop.
      (3) Gently squeeze finger until a second drop of blood forms.

   Note: Students may work in pairs to obtain finger puncture specimens of blood.

   b. Making the blood smear
      (1) Touch a clean slide to a drop of blood on the finger. The drop of blood should be about 1.5 cm. from one end of the microslide.
      (2) Place the slide on a level surface with the drop of blood up. With one hand, steady the slide, and with the other hand, place the end of another slide flat against the surface of the blood slide so that an angle of about 45 degrees is formed. The second slide acts as the spreader or pusher.

(3) Draw the spreader slide back until its edge contacts the drop of blood. Let the blood spread along junction of the two slides. Push spreader slide forward, keeping the 45 degrees of contact to form a film of blood.

Note: The thickness or thinness of the blood smear depends on the acuteness of the angle and the rapidity with which the slide is pushed along. A wider angle and increased speed of spreading tend to thicken the blood film greatly. Since thin films are more easily studied, the angle of contact should be kept small and the speed of spread fairly slow.

c. Staining the blood smear

(1) Allow the blood film to dry in air.

(2) Cover with Wright's stain* and let stand 4 to 6 minutes.

(3) Do not remove any stain but add an equal volume of distilled water or preferably phosphate buffer.

(4) Let remain until a metallic scum forms (usually 2 to 3 minutes).

(5) Wash with distilled water and dry between sheets of blotting paper.

Note: A phosphate buffer† used for 5 to 10 minutes after the application of Wright's stain (instead of distilled water) gives nicely stained smears. The staining times for Wright's stain and phosphate buffer are variable and the actual time for a given stain must be determined by trial intervals of time.

d. Examining the smear

(1) Use the low-power objective of the microscope first to examine the blood film.

(2) Place a drop of immersion oil on a thinner part of the smear.

(3) Use the oil immersion lens to study the different kinds of cells present.

Note: Red cells appear reddish orange in a well-stained smear. Those from normal adult human blood do not have nuclei. White cells have dark purple nuclei with light blue or lavender cytoplasm. Some white blood cells have granules in the cytoplasm.

(4) Sketch a few red cells and a few white cells. Note the proportion of red blood cells to white blood cells.

(5) From your study of this slide, do you think that there are different types of red blood cells or are all essentially alike? How would you answer this question when applied to white blood cells?

5. *Cells in pus*

a. Examine microscopically a prepared and stained smear of pus. Use the oil immersion objective.

b. Sketch a few pus cells.

c. What seems to be the relation between pus cells and certain white blood cells that you have just visualized?

---

*Wright's stain working solution is made by grinding 0.3 gm. of commercial powder with a pestle in 3 ml. of glycerin and successive small portions of 100 ml. of acetone-free methyl alcohol in a mortar. The solution so made is allowed to stand, with occasional shaking, for at least one week and must be filtered before use.

†Potassium phosphate, monobasic, 6.63 gm., and sodium phosphate, dibasic, 2.56 gm., in 1 liter of distilled water give a buffer with a final pH of 6.4.

PROJECT

Microscopic appearance of bacteria

1. *Study of prepared slides*
   a. Using the oil immersion objective of the microscope, study the prepared stained slides carefully. Make drawings.
   b. Note the different shapes of bacteria. Make drawings of cocci, bacilli, and spirilla.
   c. Note the arrangements of bacteria. Make drawings to show bacteria arranged as staphylococci, diplococci, streptococci, and streptobacilli.
   d. Note the following structures related to bacteria and make drawings: (1) spores, (2) capsules, (3) metachromatic granules, and (4) flagella.
2. *Demonstration of motility of bacteria*
   a. Observe motility in hanging-drop preparations set up as demonstrations.
   b. Contrast true motility with brownian motion.
   c. Note the steps in making a hanging-drop preparation (demonstration and discussion by instructor).
3. *The hanging-drop preparation*—motile organisms
   a. Consult p. 57 for the method.
   b. Using a broth culture of a motile organism, make a hanging-drop preparation.
   Note: Be careful to sterilize the platinum loop in the flame each time *before* and *after* it is used.
   c. Examine the preparation with the high dry objective of the microscope. The amount of light passing through the substage condenser must be somewhat reduced by partly closing the diaphragm. A properly made preparation shows the cells standing out distinctly against a dimly lighted background.
   d. Discard the preparation according to the method given by your instructor. *Why is this important?*
4. *The hanging-drop preparation*—nonmotile organisms
   a. Make hanging-drop preparations of the following:
      (1) A suspension of carmine—note brownian motion.
      (2) Nonmotile organisms—note that these organisms do not change their position in relation to each other. Note brownian motion.
   b. Define brownian motion.
5. *Demonstration of the use of the dark-field microscope*
   The instructor should demonstrate the use of the dark-field microscope and give a brief discussion of its application.
6. *Negative or relief staining*
   a. Place a loopful or two of a broth culture or watery suspension of bacteria on a slide.
   b. Add an equal amount of commercial india ink that has been diluted with two parts of water.
   c. Mix thoroughly and spread out slightly.

d. Allow to dry and examine with the oil immersion objective. In satisfactory portions of the smear the bacteria stand out as colorless bodies against a gray-brown or black background. This is known as *relief* staining; that is, the background is stained but the bacteria are not.

PROJECT

Staining of bacteria

1. *Simple stains*
   a. Make three smears of organisms furnished by the instructor according to directions given previously (p. 59). *Do not neglect proper sterilization of the platinum loop.*
   b. After the smears have been fixed, pour a few drops of methylene blue on the first, a few drops of gentian violet on the second, and a few drops of carbolfuchsin on the third. Let stains act for 1 minute.
   c. Wash the stains off with distilled water and drain the slides. Blot with blotting paper. Examine with the oil immersion lens. Make drawings. What is the color given to the organisms by the dye used in each smear?
2. *Gram stain*
   a. Fixed smears already prepared by the instructor may be used, or the students may prepare such smears from suitable organisms furnished by the instructor.
   b. Refer to the technic of staining outlined previously (p. 61).
   c. Examine smears with the oil immersion objective of the microscope. Are the organisms furnished you gram-positive or gram-negative? Why?
3. *Acid-fast stain*
   a. Prepared and fixed slides of sputum are furnished by the instructor.
   b. Refer to the technic of acid-fast staining outlined previously (p. 62).
   c. Examine smears with the oil immersion objective. What color is *Mycobacterium tuberculosis* when stained by this method? What is the color of nonacid-fast organisms? Sketch acid-fast bacilli.
4. *Special stains*
   a. Make a methylene blue stain of a sporeforming organism. Note spores, which appear as unstained areas in the cell body. Make a drawing.
   b. Make a Gram stain of pneumococci. Note the capsule, which stands out as an unstained halo around the cell. Make a drawing.

EVALUATION FOR UNIT ONE

Part I

In the following statements or questions, encircle the number in the column on the right that correctly completes the statement or answers the question.

1. The sciences that include the study of animals are:

(a) Botany	1. a and e
(b) Biology	2. b only
(c) Microbiology	3. b, c, and d
(d) Zoology	4. c only
(e) Bacteriology	5. e only

2. Sciences that include *only* unicellular organisms are:

(a) Bacteriology	1. a, b, and c
(b) Botany	2. a only
(c) Zoology	3. b and c
(d) Protozoology	4. a and d
(e) Microbiology	5. none of the above

3. In what order do we arrange the following groups of living organisms beginning with the largest or most general classification?

(a) Species	1. c, d, f, b, e, a
(b) Family	2. b, c, a, f, e, d
(c) Phylum	3. d, a, f, c, b, e
(d) Order	4. c, f, d, b, e, a
(e) Genus	5. none of above is correct
(f) Class	

4. Which of the following possess chlorophyl?

(a) Bacteria	1. a only
(b) Fungi	2. b only
(c) Pond scums	3. a and b
(d) Seaweeds	4. c and d
(e) Slime molds	5. c, d, and e

5. Which of the following are associated with the discovery of the genetic role of DNA?

(a) MacLeod	1. b only
(b) Watson	2. a, b, c, d, e
(c) Avery	3. a, b, c, e
(d) Coons	4. b, c, d, e
(e) Crick	5. e only

6. Which of the following are associated with the development of the poliomyelitis vaccine?

(a) Salk	1. a and b
(b) Edmonston	2. a, c, and e
(c) Enders	3. b, c, and e
(d) Dubos	4. e only
(e) Sabin	5. a only

7. In which of the following parts of the cell are the hereditary characteristics transmitted?
   (a) Cytoplasm 1. a
   (b) Nucleolus 2. b
   (c) Chromosomes 3. c
   (d) Mitochondrion 4. d
   (e) Endoplasmic reticulum 5. e
8. Which of the following functions belong to the nucleus of a cell?
   (a) Storage of food 1. a only
   (b) Control of metabolism 2. a, b, and c
   (c) Elimination of waste products 3. b only
   (d) Control of reproduction 4. b, d, and e
   (e) Transfer of hereditary characteristics 5. a, b, d, and e
9. Which of the following functions belong to the cytoplasm of a cell?
   (a) Storage of food 1. a only
   (b) Control of metabolism 2. a and b
   (c) Enzyme secretion 3. a, b, and c
   (d) Assimilation of food 4. a, c, d, and e
   (e) Transfer of hereditary characteristics 5. a, c, d
10. Which of the following bacteria are spiral in shape?
   (a) Diplococci 1. a and e
   (b) Bacilli 2. a
   (c) Vibrios 3. b
   (d) Spirochetes 4. c and d
   (e) Streptococci 5. e
11. Which of the structures listed below are resistant to conditions adverse to bacterial growth?
   (a) Capsules 1. a only
   (b) Flagella 2. d only
   (c) Metachromatic granules 3. a and d
   (d) Spores 4. a, b, c, and d
   (e) Polar bodies 5. none of these
12. The method by which bacteria reproduce is:
   (a) Budding 1. a only
   (b) Transverse fission 2. b and d
   (c) Spore formation 3. c only
   (d) Binary fission 4. d only
   (e) Capsule formation 5. c and e
13. Which of the following maintains the shape of the bacterial cell?
   (a) Spore 1. e only
   (b) Flagella 2. a and e
   (c) Plasma membrane 3. c and e
   (d) Capsule 4. d and e
   (e) Cell wall 5. c only
14. A structure in the bacterial cell rich in enzymatic activity is the:
   (a) Plasma membrane 1. a and b
   (b) Cell wall 2. d and e
   (c) Metachromatic granule 3. b and c
   (d) Capsule 4. b only
   (e) Flagellum 5. a only
15. A protoplast is not:
   (a) A bacterial cell with a spore 1. a and e
   (b) A bacterial cell without a spore 2. a, c, and e
   (c) A bacterial nucleus 3. a, b, and c
   (d) A bacterial spore without a coat 4. d only
   (e) A bacterial cell without a cell wall 5. a, b, c, and d

16. The unit of measurement for most purposes in the study of cells with the compound light microscope is the:

(a) Angstrom
(b) Micron
(c) Millimicron
(d) Nanometer
(e) Meter

1. a
2. b
3. c
4. d
5. e

Part II

1. Match the discovery or contribution to the science of microbiology listed in column B to the name of the person responsible in column A. Give the *one* best answer.

COLUMN A

_______ von Behring
_______ Beijerinck
_______ Budd
_______ Eberth
_______ Ehrlich
_______ Fleming
_______ Holmes
_______ Jenner
_______ Koch
_______ Landsteiner
_______ Leeuwenhoek
_______ Lister
_______ Metchnikoff
_______ Pasteur
_______ Ross
_______ Schwann
_______ Semmelweis
_______ Snow
_______ Spallanzani
_______ Waksman
_______ von Wassermann

COLUMN B

1. Discovery of penicillin
2. Phagocytic theory of immunity
3. Humoral theory of immunity
4. Technic of pure culture
5. Preventive treatment for rabies
6. Observations on epidemiology of typhoid fever
7. Observations on epidemiology of cholera
8. Discovery of diphtheria antitoxin
9. Proof that yeasts are living things
10. Recognized first virus
11. Transmission of malaria
12. Discovery of basic human blood groups
13. Early experiments to disprove spontaneous generation
14. Laid foundation for aseptic surgery
15. Observations on transmission of puerperal sepsis
16. Discovery of streptomycin
17. Established serological diagnosis of syphilis
18. Discovery of typhoid bacillus
19. Introduced smallpox vaccination
20. First drawing of bacteria

2. Consider the characteristics as listed on the left. Indicate by a check mark in the appropriate column the relation to plants or animals.

	PLANTS	ANIMALS
Possess a well-defined cell wall	_______	_______
Show continuous growth	_______	_______
Show limited growth	_______	_______
Are nonmotile	_______	_______
Can move independently	_______	_______
Carry on photosynthesis	_______	_______
Possess no chlorophyl	_______	_______
Use starch as main reserve food	_______	_______

3. Match the constituent of the cell as given in column B to the item in column A that best indicates its structure or function. Give the *one* best answer.

COLUMN A	COLUMN B
______ Hydrolytic enzymes	1. Golgi apparatus
______ Protein synthesis	2. Microtubules
______ Protein excretion	3. Chromosomes
______ Cell shape	4. Cell wall
______ Glycogen	5. Mitochondria
______ Cellulose	6. Endoplasmic reticulum
______ Pigment granules	7. Ribosomes
______ Cytochrome enzymes	8. Lysosomes
______ Cytoskeleton	9. Plasma membrane
______ Exchange of food material	10. Inclusions
______ Crystals	11. Gene
______ Powerhouse of cell	12. Nucleolus
______ "Suicide bags"	13. Peroxisome
______ Nucleic acid metabolism	14. Nuclear sap
______ Role in carbohydrate metabolism	
______ Unit of organization for DNA	
______ Genetic code	

4. Identify the organisms from the following list as being either gram-positive or gram-negative by placing the number before the name in the appropriate column.

GRAM-POSITIVE	GRAM-NEGATIVE	GENUS OF ORGANISMS TO BE IDENTIFIED:
______	______	1. *Neisseria*
______	______	2. *Staphylococcus*
______	______	3. *Streptococcus*
______	______	4. *Shigella*
______	______	5. *Escherichia*
______	______	6. *Salmonella*
______	______	7. *Diplococcus*
______	______	8. *Clostridium*
______	______	9. *Haemophilus*
______	______	10. *Brucella*
______	______	11. *Pasteurella*
______	______	12. *Bacillus*
______	______	13. *Corynebacterium*
______	______	14. *Pseudomonas*
______	______	15. *Proteus*

5. Identify the organisms from the following list as being either acid-fast or nonacid-fast by placing the number before the genus name in the appropriate column.

ACID-FAST	NONACID-FAST	GENUS OF ORGANISMS TO BE IDENTIFIED:
______	______	1. *Escherichia*
______	______	2. *Proteus*
______	______	3. *Mycobacterium*
______	______	4. *Nocardia*
______	______	5. *Actinomyces*
______	______	6. *Staphylococcus*

UNIT TWO

MICROBIOLOGY: principles and procedures

6 **Bacteria: biologic needs**

7 **Methods for microbes: cultivation**

8 **Bacteria: biologic activities**

9 **Methods for microbes: biochemical reactions, animal inoculation**

10 **Specimen collection**

Laboratory survey of unit two

Evaluation for unit two

6 Bacteria: biologic needs

ENVIRONMENTAL FACTORS

For bacteria to grow and to multiply most rapidly, certain requirements must be met: (1) Sufficient food of the proper kind must be present, (2) moisture must be available, (3) the temperature must be most suitable for the species, (4) the proper degree of alkalinity or acidity must be present, (5) the oxygen requirements of the species must be met, (6) light must be partially or completely excluded, and (7) by-products of bacterial growth must not accumulate in great amounts. Significant departure from any of these will modify bacterial growth although bacteria generally possess a greater degree of resistance to unfavorable conditions in their environment than do plants and animals.

Food materials prepared for the growth of bacteria in the laboratory are known as *culture media*. Some bacteria will grow on practically any properly prepared culture medium. Others grow only on specially nutritious ones, and a few will not grow on any artificial medium.

■ **Nutrition.** The protoplasm of the bacterial cell is composed of numerous organic compounds, including proteins, fats, and carbohydrates, as well as various inorganic components containing sulfur, phosphorus, calcium, magnesium, potassium, and iron. Proteins comprise about 50% of the dry weight of the cell (each species has a type of protein peculiar to itself), and bacterial nitrogen makes up 10%. In some species carbohydrates are plentiful, and some of the important traits of the species depend upon these compounds.

Nutrition is the provision of food materials (that is, chemical substances) to bacteria so that they can grow, maintain their constituents, and multiply. For their nourishment bacteria require sources of carbon and nitrogen, growth factors, certain mineral salts, and sources of energy. With the exception of some saprophytic species, bacteria derive their carbon and nitrogen from organic matter. A number of minerals are required, the most important of the salts being those of calcium, phosphorus, iron, magnesium, potassium, and sodium. Certain minerals are important in the activation of enzymes.

Many microorganisms can synthesize all the organic compounds of their complex makeup if supplied with the basic nutrients. Many cannot, however, since they require vitamins and certain organic growth factors (a growth factor is utilized as the intact substance) for their activities. In this respect they resemble rather closely higher forms of life. In fact, such vitamins as nicotinic acid, pantothenic acid, para-aminobenzoic acid, biotin, and folic acid—requirements for animal nutrition—were first studied and identified as substances necessary for the growth of microorganisms.

KINDS OF ORGANISMS. Organisms that obtain their nourishment from nonliving organic material are known as *saprophytes.* Those that depend on living matter for their nourishment are *parasites. Facultative saprophytes* usually obtain nourishment from living matter but may obtain it from dead organic matter. *Facultative parasites* usually obtain nourishment from dead organic matter but may obtain it from living matter. Some pathogenic bacteria can exist only on living material (example, the spirochete of syphilis cannot be grown outside a living organism). Most, however, are capable of leading either a parasitic or a saprophytic existence. A few pathogenic bacteria, usually saprophytic, may adapt themselves to a parasitic existence (example, bacteria that cause gas gangrene). The organism on which a parasite lives is known as a *host.*

Organisms that obtain their nourishment by breaking down organic matter into simpler chemical substances are *heterotrophic.* Those that obtain their food by building the organic compounds in protoplasm from the simpler inorganic substances are *autotrophic.* All pathogenic bacteria and many nonpathogenic ones are heterotrophic.

■ **Moisture.** Water is necessary for the growth and multiplication of bacteria. Not only is water a major component of the cytoplasm (on an average, 75% to 80% of the bacterial cell is made up of water), but it also dissolves the food materials in the environment of the bacterial cell so that they can be absorbed.

Drying is highly detrimental to bacterial growth. Delicate bacteria such as the gonococcus resist drying for only a few hours, and even highly resistant bacteria such as the tubercle bacillus succumb within a few days. Spores, however, may resist drying for years. As a rule, bacteria with capsules are more resistant to drying than are those with none.

■ **Temperature.** For each species of bacteria there is a minimum, optimum, and maximum temperature, meaning, respectively, the lowest temperature at which the species will grow, the temperature at which it grows best, and the highest temperature at which growth is possible (Fig. 6-1). The optimum temperature for a species corresponds to the average temperature of its usual habitat. For instance, bacteria that naturally live in or attack the human body live best at 37° C.* (the normal temperature of the body); these are *mesophiles.* The lowest temperature at which any of these species will continue to multiply is around 20° C. (Celsius) and the highest is from 42° to 45° C.

Many bacteria will not grow at a temperature more than a few degrees above or below their optimum. Some pathogenic bacteria die off rapidly at only 38° C. The majority of saprophytic bacteria (mesophiles) grow best between 25° and 40° C., but some *thermophiles* or heat-loving species grow at a temperature above 45° C. and as high as 65° C. A few *psychrophiles* or cold-loving species grow at temperatures just above the freezing point (20° C. or less). They proliferate slowly in the refrigerator.

Cold retards or stops bacterial growth, but when the bacteria are later exposed to a temperature favorable for their growth, multiplication is resumed. Refrigeration

*The temperature scale called centigrade in America has been known in many other countries as Celsius after Anders Celsius (1701-1744), the Swede who originated it in 1742. According to the Eleventh General Conference on Weights and Measures (1960), the term "centigrade" is inexact and the name Celsius replaces it, with the capital C retained.

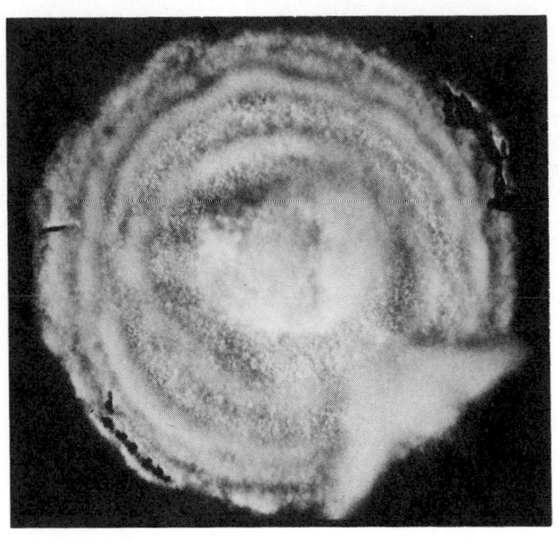

FIG. 6-1. Variation in growth rate produced by changes in temperature. Note rings of rapid and slow growth in giant colony of fungus *(Histoplasma capsulatum).* One sector of mutant growth is seen in lower right hand corner. (Courtesy Dr. R. H. Musgnug, Haddonfield, N. J.)

(4° to 6° C.) is one of the best methods of preserving bacterial cultures since bacteria are generally resistant to low temperatures and even to freezing. Prolonged freezing, however, destroys them.

High temperatures are much more injurious to bacteria then are low ones and are used effectively in the practical situations where bacteria and their spores must be destroyed (see Chapter 16).

■ **Reaction.** For each species of bacteria there is a certain degree of alkalinity or acidity, (a certain pH) at which growth is most rapid. The reaction of culture media must be carefully adjusted to the desired hydrogen ion concentration. The best growth of most microorganisms is found in a narrow pH range of not less than 6 nor more than 8. Most pathogens grow best in a neutral or slightly alkaline medium. Regardless of the influence of the environment, the reaction of the interior of the cell is just at the neutral point (pH 7).

■ **Oxygen.** Organisms that grow in the presence of free atmospheric oxygen are known as *aerobes.* If an organism cannot develop at all in the absence of free oxygen, it is an *obligate aerobe.* Those that cannot grow in the presence of free oxygen but must obtain it from oxygen-containing compounds (inorganic sulfates, nitrates, and carbonates, or certain organic compounds) are *anaerobes.* Few of the pathogenic organisms are anaerobic. Of these, the organisms of tetanus and gas gangrene are significant. Organisms adaptable either to the presence of atmospheric oxygen or to its absence are *facultative.* Organisms growing best in an amount of oxygen less than that contained in the air are known as *microaerophiles,* whereas organisms vulnerable to free oxygen are *obligate anaerobes.* The enzyme systems of obligate anaerobes are inactivated by the oxygen of the atmosphere.

Organisms designated *capnophiles* need a 3% to 10% increase in carbon dioxide in the environment to initiate growth.

■ **Light.** Violet, ultraviolet,* and blue lights are highly destructive to bacteria, green light is much less so, and red and yellow lights have little bactericidal action. Because of its content of ultraviolet light, direct sunlight kills most bacteria within a few hours. Bright daylight has an effect similar to that of sunlight but one of less potency.

A few species of saprophytic bacteria containing chlorophyl can utilze sunlight to build up the compounds of which they are composed. This bacterial chlorophyl is not contained as in the chloroplasts of the cells of higher plants but is scattered throughout the cytoplasm.

■ **By-products of bacterial growth.** If it were not for inhibitory influences, bacteria could completely submerge the whole world. It is estimated that the progeny from the unrestricted growth of a single bacterium would be 280 trillion at the end of only 24 hours. In cultures, bacterial reproduction is so rapid that bacteria soon exhaust their food supply and release products that inhibit further bacterial growth. Notable are organic acids from carbohydrate metabolism that inhibit growth by changing the reaction of the medium. The practical application of this is found in the pickling industry, where the acid medium is used to prevent bacterial contamination.

■ **Electricity and radiant energy.** By itself electricity does not destroy bacteria, but it causes heat and changes that may be lethal in the medium in which the bacteria are growing. Electric light inhibits bacterial growth, but the inhibition is an effect of light, not of electricity.

The effects of roentgen rays are generally harmful to most bacteria, being about 100 times more effective in their action against bacteria than is ultraviolet light.

■ **Chemicals.** Certain chemicals destroy bacteria and others inhibit their growth (p. 239). Some substances attract bacteria *(positive chemotaxis)*, whereas others repel them *(negative chemotaxis)*.

■ **Osmotic pressure.** The bacterial cell is encased in a membrane said to be *semipermeable* because it allows water to pass freely in and out of the cell but gives a varying degree of resistance to dissolved substances in the fluid medium in which the cell is suspended. This makes the cell a small osmotic unit responsive to changes in its fluid environment (Fig. 6-2).

Under normal conditions there is a higher concentration of dissolved substances within the cell than without it. The greater osmotic pressure on the inside of the cell keeps the protoplasm of the cell firmly against the cell wall, and the cell is said to be *turgid*. If a bacterial cell is placed in solutions having varying concentrations of dissolved substances, changes take place. In a solution with a high concentration of dissolved substances (*hypertonic* solution) water leaves the interior of the cell and the cell begins to shrink. If the difference in concentration between the interior and the exterior of the cell is not too great, the cell may be able to adjust to the hypertonic solution, regain its turgor, and continue its growth. If not, the cell continues to shrink and finally dies *(plasmolysis)*.

If a cell is placed in a solution with a low concentration of dissolved substances or in distilled water (*hypotonic* solution), water passes into the cell, the cell swells, and it may burst *(plasmoptysis)*. A solution containing that concentration of dissolved substances in which the cell neither swells nor shrinks is said to be *isotonic*.

Most bacteria resist small changes in osmotic pressure but are killed or inhibited by high concentrations either of salt (as used in brines) or of sugar. This fact is utilized

---

*The use of ultraviolet light in sterilization is discussed on p. 237.

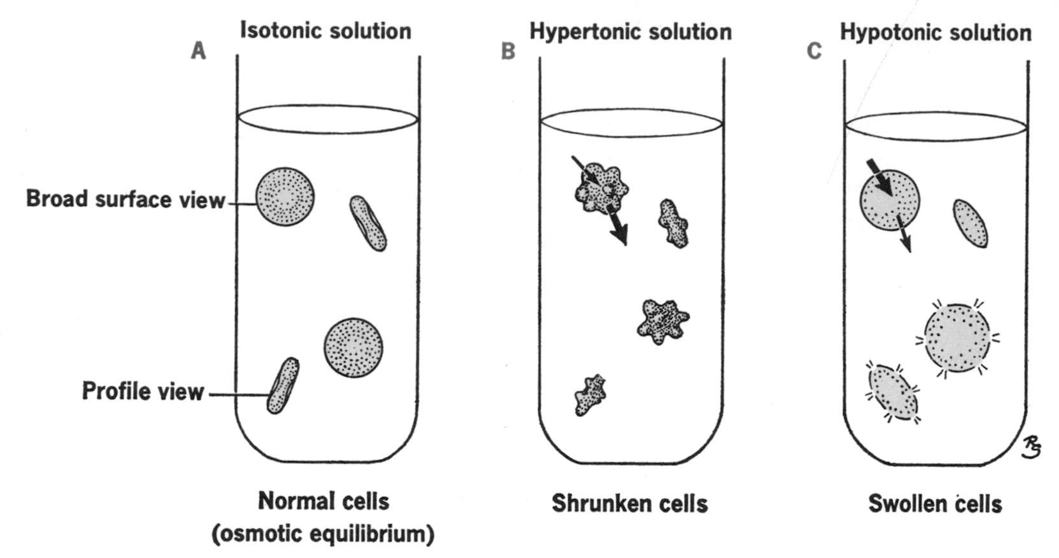

FIG. 6-2. Effect of osmotic pressure on cells. Note red blood cell suspended in liquid containing **A,** the same amount of dissolved substance as the cell protoplasm (no transfer of water; no change in cell size or shape); **B,** more dissolved substance than in cell protoplasm (water passes from cell to surrounding fluid; cell shrinks); and **C,** less dissolved substance than cell protoplasm (water moves into cell; cell swells and bursts).

in the preservation of foods such as syrups and jellies and in the preservation of meats in brine. That some microorganisms can adjust to a high concentration of sugar is seen in molds on jellies. To preserve foods safely, higher concentrations of sugar must be be used than of salt.

That a few species of bacteria *(osmophiles)* do well in a hypertonic solution is seen from the bacterial life of the oceans. Even the Dead Sea with its high salt content supports a bacterial population *(halophiles)*. A small amount of various salts in the fluid medium for bacteria is beneficial to bacterial growth. Traces of such salts are furnished by natural foods such as meat extracts.

INTERRELATIONS

■ **Symbiosis.** Certain species of bacteria grow well together, and the associated species accomplish harmful or beneficial results that neither does alone. For instance, staphylococci and influenza bacilli multiply more rapidly when grown together than either does when grown alone. This is known as *synergism*.

Symbiosis refers to the relation of mutual benefit existing between two organisms. For example, there is the beneficial relation between the leguminous plants and the nitrogen-fixing bacteria living in the root nodules of these plants (p. 570). *Commensalism* (mutual tolerance) is the term applied when two organisms live together without benefit or injury to each other.

■ **Antagonism.** Sometimes the presence of certain species of bacteria inhibits the growth of others. For instance, growth of the gonococcus is inhibited by the presence of almost any other species of bacteria. This is *antagonism*. Theories put forth to explain antagonism are (1) that one organism secretes a substance toxic to the growth of the other and (2) that one organism promotes a defense mechanism of the animal body against the other. The appearance of certain infections after the administration of antibiotics may be explained in terms of an antagonistic relationship between two organisms, only one of which succumbs to the action of the antibiotic. Released from the restraining influence and not affected by the antibiotic, the other microorganisms can now become quite aggressive.

QUESTIONS FOR REVIEW

1. Name at least five requirements for bacteria to grow and multiply.
2. Classify bacteria from the standpoint of food requirements. Name elements required for the nourishment of bacteria.
3. Define host, autotrophic, heterotrophic, facultative, obligate, hypertonic, hypotonic, and isotonic.
4. How are bacteria affected by heat and cold?
5. Classify bacteria from the standpoint of oxygen requirements.
6. What are the effects of light on bacteria?
7. What is the difference between negative and positive chemotaxis? Plasmoptysis and plasmolysis?
8. Consult outside sources and briefly discuss semipermeable membranes.
9. What is commensalism? Synergism? Symbiosis? Antagonism?
10. What is the preferred designation of the scientific temperature scale?
11. What are psychrophiles, halophiles, thermophiles, and osmophiles?

REFERENCES. See at end of Chapter 10.

7 Methods for microbes: cultivation

The small size and the similarity in appearance and staining reactions of microbes often make the identification by microscopic methods alone impossible. One of the most important ways to identify them is to observe their growth on artificial food substances prepared in the laboratory. This is cultivation of the microorganisms or *culturing*. The food material on which they are grown is known as a *culture medium* (plural, *media*), and the growth itself is the *culture*. Some 10,000 different kinds of culture media have been prepared.

Cultural methods help to identify microbes and also to pick up organisms in test material of low microbial content. When such material is placed on a suitable culture medium, each organism present multiplies many times. Practically speaking, the smear may contain few or no organisms, whereas the culture soon produces hundreds of them.

CULTURE MEDIA

■ **General considerations.** For the most satisfactory growth of bacteria and related forms on artificial culture media, the proper temperature, the right amount of moisture, the required oxygen tension, and the proper degree of alkalinity or acidity (pH) must be provided. The culture medium itself must contain the necessary nutrients and growth-promoting factors and, of course, must be free from contaminating microorganisms; that is, *it must be sterile.*

Most disease-producing bacteria require complex foods similar in composition to the fluids of the animal body. For this reason the basis of many culture media is an infusion of meat of neutral or slightly alkaline pH containing meat extractives, salts, and peptone to which various other ingredients may be added.*

Culture media are of three types: (1) a liquid, (2) a solid that can be liquefied by heating and that upon cooling returns to the solid state, and (3) a solid that cannot be liquefied. It is many times desirable that the culture medium be a solid, since bacteria and related forms can be viewed on the surface. Agar (Japanese seaweed) is a solidify-

*The basic infusion can be prepared by soaking 500 grams of fresh lean ground beef in 1 liter of distilled water in the refrigerator. After 24 hours, the surface layer of fat is removed with absorbent cotton, the mixture is passed through muslin, or gauze, and the meat discarded. After ingredients related to the organism under study are added and the reaction of the medium is adjusted to neutral or slightly alkaline, the infusion is heated to 100° C. for 20 minutes to remove coagulated tissue proteins and is filtered through coarse paper. At each step the volume is reconstituted to 1000 ml.

ing agent widely used. It melts completely at the temperature of boiling water and solidifies when cooled to about 40° C. With a few minor exceptions, it has no effect on bacterial growth and is not attacked by bacteria growing on it. The low temperature at which agar solidifies is very important if test material must be inoculated directly into melted media before it solidifies. If the agar began to solidify at a temperature high enough to kill bacteria or fungi, this could not be done. Gelatin is a well-known but less frequently used solidifying agent.

Numerous enriching materials are found in different culture media—carbohydrates, serum, whole blood, bile, ascitic fluid, and hydrocele fluid. Carbohydrates are added (1) to increase the nutritive value of the medium and (2) to indicate the fermentation reactions of the microbes being studied. Serum, whole blood, and ascitic fluid are added to promote the growth of the less hardy organisms. Dyes added to culture media act as indicators to detect the formation of acid when fermentation reactions are being tested or act as inhibitors of the growth of certain bacteria but not that of others. An example of an indicator dye is phenol red, which is red in an alkaline or neutral medium but yellow in an acid one. An example of an inhibitory dye is gentian violet, which inhibits the growth of most gram-positive bacteria.

In actual practice, media making is greatly simplified by the availability of commercially prepared mixtures containing the necessary and any special ingredients for growth of bacteria or fungi in just the right proportions. Such preparations are sold in the dehydrated state as a powder or tablet, either of which may be reconstituted in water.*

Before sterilization, hydrated culture media are poured into suitable test tubes or flasks that are closed with cotton plugs. After sterilization, some of the tubes of hot liquid media containing agar remain vertical for the agar to solidify, but some are laid on a flat surface with their mouths raised so that when the medium cools and solidifies there is a slanting surface, an *agar slant* (Fig. 7-8). This gives a larger surface area for growth. Cotton plugs allow the access of moisture and oxygen but block the entrance of contaminating microorganisms; therefore, the medium remains sterile until used, and microbes with which it is inoculated are not contaminated by those from the outside.

A large surface area for bacterial growth is provided when a solid culture medium partly fills a Petri dish. A *Petri dish* is a circular glass (or plastic) dish about 3 inches in diameter with perpendicular sides about ½ inch high. Inverted over it is a glass, metal, or plastic cover exactly like it except for a slightly greater diameter. The edges of the dishes are smooth, and a container is formed that prevents either the entrance or exit of bacteria. The Petri dish may be filled with a sterile agar medium still hot from the sterilization process, or a tube of solidified agar may be melted and poured into the dish. Until the agar has completely cooled, the lid is slightly raised on the Petri dish. When the agar is solid, the lid is lowered.

Most media are sterilized by autoclaving (pp. 232 and 235). Those that contain carbohydrates may have to be sterilized by the fractional method because many carbohydrates will not withstand the high temperature of the autoclave. Enrichment materials, such as serum and whole blood, that are injured by even moderate heat are collected in such a manner as to be kept sterile and are mixed with the medium only

---

*Dehydrated culture medium is weighed out and dispensed into a given volume of distilled water. The mixture is heated carefully over a water bath to effect solution.

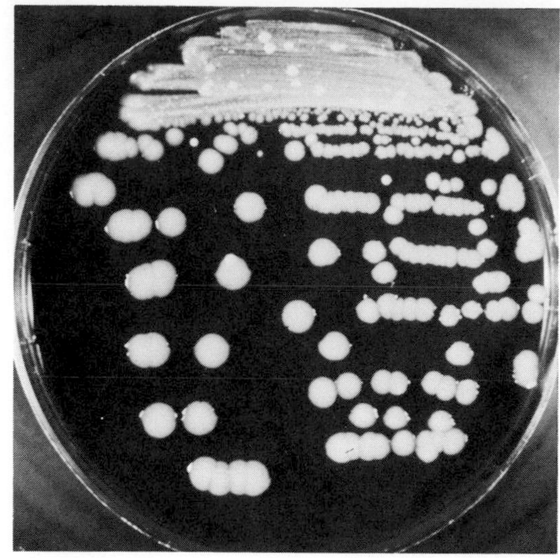

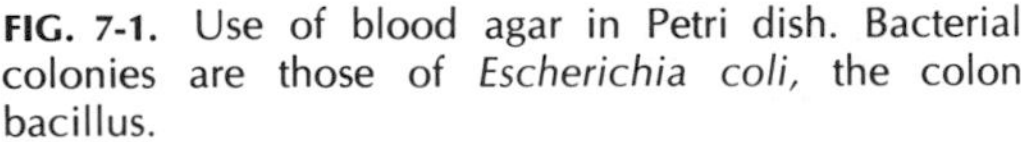

FIG. 7-1. Use of blood agar in Petri dish. Bacterial colonies are those of *Escherichia coli,* the colon bacillus.

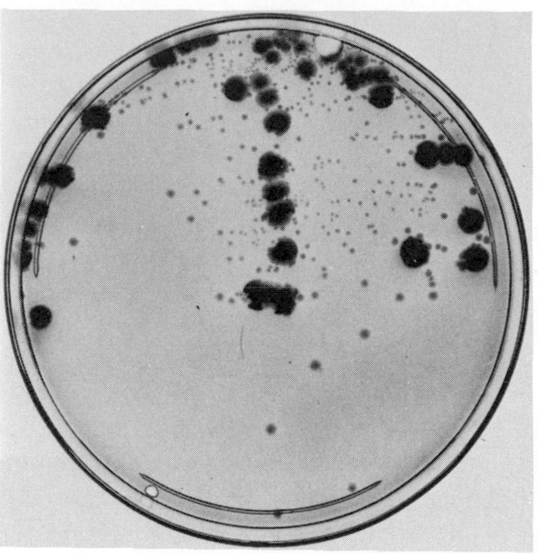

FIG. 7-2. Use of differential culture medium. On MacConkey agar the dark (red purple) colonies of lactose-fermenting *Escherichia coli* are separated from and contrasted with the colorless colonies of non—lactose-fermenting *Shigella* species.

after it has been sterilized and allowed to cool. In the case of agar media, these substances are added when the medium reaches 40° to 42° C. because when the temperature falls slightly lower than this, the agar begins to solidify.

Almost every species of microbe has some medium upon which it grows best. A few will grow only on media specially prepared for them, but most of the pathogenic ones will grow more or less luxuriantly on certain routine or standard media (Fig. 7-1). Table 7-1 lists media suitable for the isolation and growth of important pathogenic microorganisms.

■ **Selective and differential media.** Media that promote the growth of one organism and retard the growth of other organisms are *selective* media. Examples are bismuth sulfite agar and Petragnani medium. Bismuth sulfite agar, used to isolate typhoid bacilli from feces, promotes the growth of typhoid bacilli but retards the growth of bacteria normally resident in the feces. Petragnani medium promotes the growth of tubercle bacilli but retards the growth of any other organism present.

Media that differentiate organisms growing together are *differential* media. Examples are eosin-methylene blue (EMB) agar and MacConkey agar used in the differentiation of the gram-negative bacteria of the intestinal tract. The incorporation of lactose into such differential media makes possible a sharp differentiation between the organisms that ferment this sugar and those that do not. The colonies of lactose fermenters are deeply colored; colonies of nonfermenters are colorless. This point is important, since generally the pathogens in the intestinal tract do not ferment lactose whereas the normal inhabitants, the coliforms, do (Fig. 7-2).

Selective and differential media are of great help in the diagnosis of infections in such areas of the body as the respiratory and intestinal tracts, where normally a variety of organisms resides. The presence of pathogenic bacteria in the area may not disrupt

TABLE 7-1. MICROBES RELATED TO CULTURE MEDIA

Organism	Media most suitable for growth
Actinomyces bovis	Brain-heart infusion glucose broth and agar; thioglycollate broth medium (Brewer-modified) (anaerobic)
Bacillus anthracis	Growth on almost all media
Bedsonia (Chlamydia)	Chick embryo; tissue culture
Blastomyces dermatitidis	Blood agar; beef infusion glucose agar; Sabouraud glucose agar (antibiotics added to media used)
Bordetella pertussis	Glycerin-potato-blood agar (Bordet-Gengou agar)
Brucella abortus, melitensis, and *suis*	Trypticase soy broth and agar
Candida albicans	Sabouraud glucose agar; brain-heart infusion blood agar; corn meal agar; rice infusion agar
Clostridia of gas gangrene	Thioglycollate broth medium (Brewer-modified); anaerobic blood agar; cooked meat medium under petroleum; Clostrisel agar
Clostridium tetani	Same as for clostridia of gas gangrene
Coccidioides immitis	Sabouraud glucose agar
Corynebacterium diphtheriae	Löffler blood serum medium; cystine-tellurite-blood agar
Cryptococcus neoformans	Blood agar; beef infusion glucose agar; Sabouraud glucose agar
Diplococcus pneumoniae	Blood infusion agar; tryptose phosphate broth; trypticase soy broth; brain-heart infusion media
Entamoeba histolytica	*Entamoeba* medium (Difco) with dilute horse serum and rice powder
Escherichia coli and coliforms	Eosin-methylene blue (EMB) agar; Endo agar; MacConkey agar; desoxycholate agar; blood infusion agar (growth on almost any medium)
Fungi, including dermatophytes	Sabouraud glucose agar with antibiotics; Littman oxgall agar
Haemophilus influenzae	Chocolate agar with yeast extract; rabbit blood agar with yeast extract
Histoplasma capsulatum	Brain-heart infusion glucose blood agar with antibiotics; Sabouraud glucose agar with antibiotics; brain-heart infusion glucose broth
Mycobacterium tuberculosis	Slow growth on special media such as Petragnani medium, Löwenstein-Jensen medium, or Dubos oleic agar
Neisseria gonorrhoeae	Chocolate agar
Neisseria meningitidis	Chocolate agar
Nocardia asteroides	Beef infusion blood agar; beef infusion glucose agar; Sabouraud glucose agar; Czapek agar
Pasteurella tularensis	Cystine glucose blood agar

Continued.

TABLE 7-1. MICROBES RELATED TO CULTURE MEDIA—cont'd

Organism	Media most suitable for growth
Proteus vulgaris	EMB agar; Endo agar; MacConkey agar; desoxycholate agar; blood infusion agar (growth on almost any medium)
Pseudomonas aeruginosa	Blood agar; EMB agar; Endo agar; MacConkey agar; desoxycholate agar (growth on almost any medium)
Rickettsia	Chick embryo; tissue culture
Salmonella	Desoxycholate agar; desoxycholate-citrate agar; *Salmonella-Shigella* (SS) agar; MacConkey agar; tetrathionate broth; selenite-F enrichment medium
Salmonella typhi	Bismuth sulfite agar; SS agar; desoxycholate agar; desoxycholate-citrate agar; MacConkey agar; tetrathionate broth; selenite-F enrichment medium
Shigella	SS agar; desoxycholate agar; desoxycholate-citrate agar; MacConkey agar; tetrathionate broth; selenite-F enrichment medium
Staphylococcus	Growth on almost any medium
Streptococcus	Blood infusion agar; trypticase soy broth; tryptose phosphate broth; brain-heart infusion media
Viruses	Chick embryo; tissue culture

the normal bacterial pattern for the area (the normal flora); such pathogens tend to be mixed with other organisms in material taken from the site.

■ **Synthetic media.** The culture media just discussed are made up of components of variable composition, such as meat infusions, serum, or other body fluids. Such media are *nonsynthetic media. Synthetic media,* on the other hand, contain components of definite chemical composition. Used primarily in research work, such media have the advantage that whenever prepared their composition is the same. Therefore, the results of microbial action on these media in one laboratory are strictly comparable with those in other laboratories even thousands of miles away.

CULTURAL METHODS*

■ **Inoculation.** A culture is made or inoculated (Fig. 7-3) when some of the material to be cultured, such as sputum, urine, or pus, is placed into a fluid medium or is rubbed gently over the surface of a solid medium with either a sterile swab or a flame-sterilized platinum wire loop. The inoculated medium is then incubated for a period of time, routinely 24 to 48 hours.

■ **Bacteriologic incubator.** The bacteriologic incubator consists of an insulated cabinet fitted with an electrical heating element and a thermoregulator (Fig. 7-4). When the temperature reaches the point for which the thermoregulator is set, the regulating device cuts the heat off but turns it on again when the temperature falls slightly below

*Standard aseptic bacteriologic technics are implied in the discussion to follow.

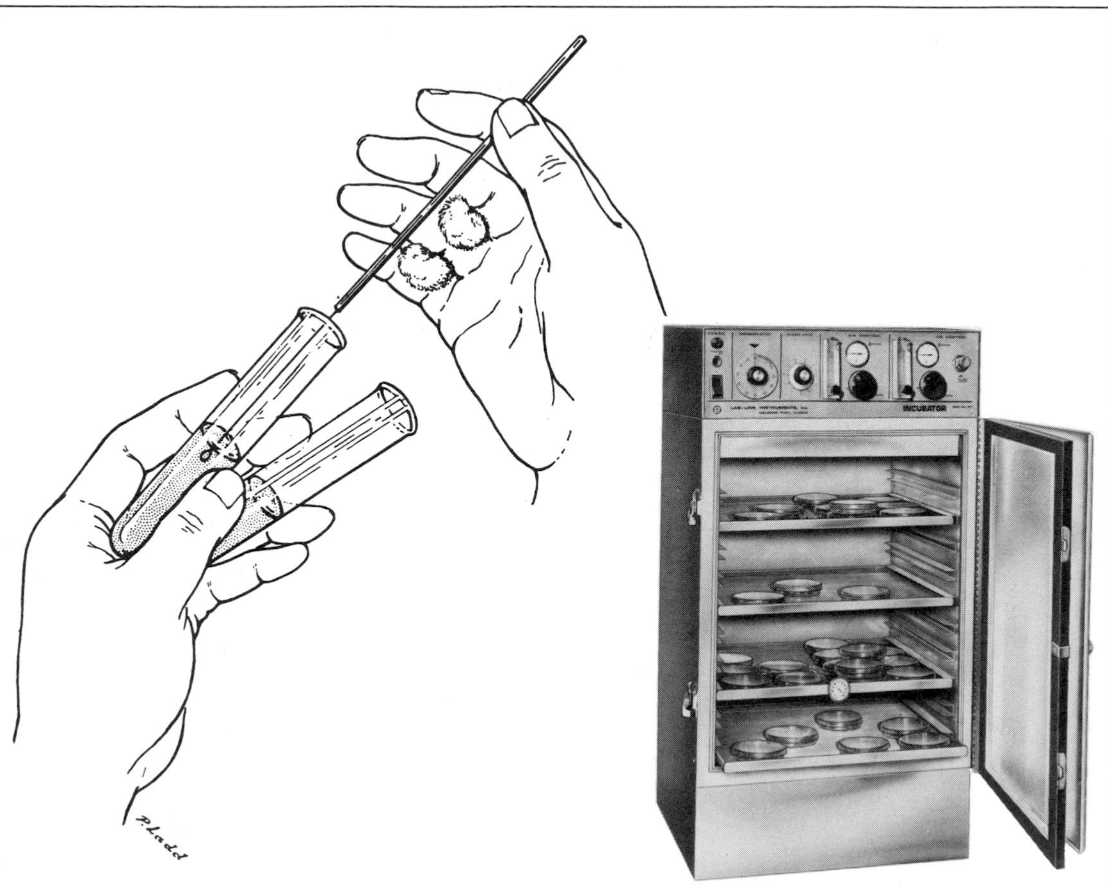

FIG. 7-3. Inoculating a tube of liquid culture medium. Note slanting position of test tubes during maneuver. Contaminating organisms from air are less likely to drop into the open tubes. The fingers of the right hand are lifted away from the handles of the platinum loop to show how cotton plugs are held.

FIG. 7-4. Bacteriologic incubator. It is electrically heated, and the temperature is kept constant by a delicate thermoregulator. This one also provides a measured carbon dioxide atmosphere, a feature promoting the growth of certain microbes. (Courtesy Lab-Line Instruments, Inc., Melrose Park, Ill.)

that point. A good thermoregulator placed in a properly constructed incubator will maintain the temperature constant from day to day to within 0.5° C. Electrically heated incubators maintain a more nearly constant temperature than do those heated otherwise.

The incubator must be properly ventilated. It may be constructed so that circulation of air is brought about by the combined effects of gravity and the difference in weight of warm and cold air. Incubators are also ventilated with blowers or fans. All incubators are fitted with perforated shelves. Most are fitted with sets of double doors; the inner ones of glass allow the contents of the incubator to be viewed, and they keep out cold air.

■ **End result.** Each microbe in the medium inoculated multiples rapidly, and within a few hours there are many more microorganisms in the culture than there were in an equal amount of the material from which the culture was made. Consequently bacteria

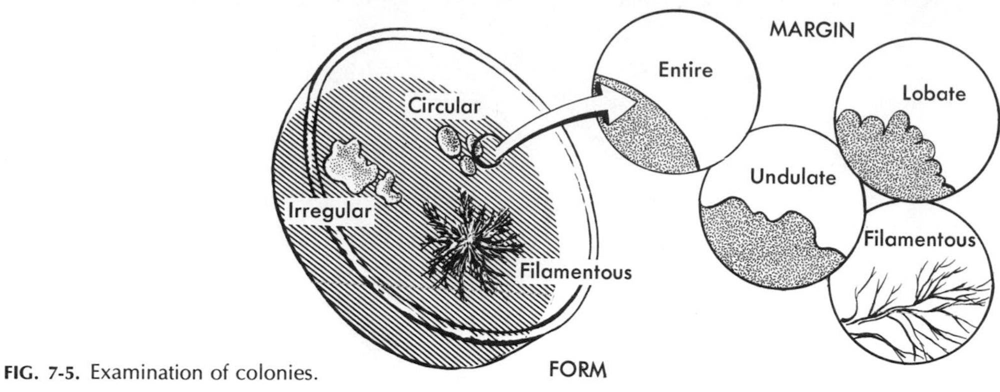

FIG. 7-5. Examination of colonies.

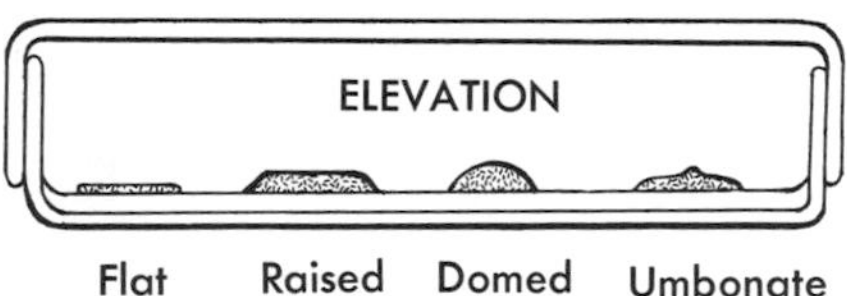

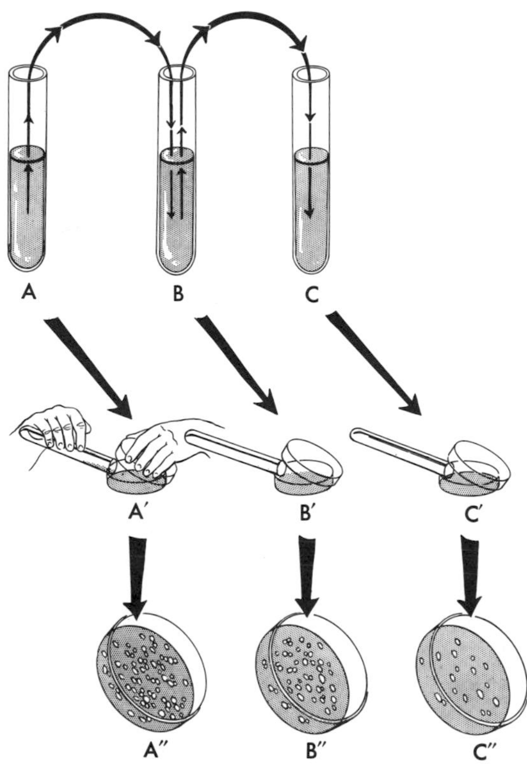

FIG. 7-6. Pure culture, pour plate method (see also text). *Step I:* Inoculate tubes **A, B,** and **C,** as indicated by arrows. *Step II.* Transfer the contents of each tube to Petri dishes **A′, B′,** and **C′,** respectively. Incubate plates. *Step III.* Inspect plates (**A″, B″,** and **C″**) for number and distribution of colonies. Fish a representative colony for further study.

or fungi are found in cultures when they are found with difficulty or not at all in smears ot the test material. Diphtheria bacilli can be found in throat cultures twice as often as in throat smears.

When a culture is made on a solid medium, all of the bacteria that grow from each bacterium deposited on the medium cling together to form a mass visible to the naked eye. This is called a *colony,* The colony has characteristics such as texture, size, shape, color, and adherence to the medium (Fig. 7-5). These are fairly constant for each species and are valuable in differentiating one species from another. Theoretically each organism should give rise to one colony, but two or more may cling together and, when planted on a medium, give rise to only one colony. If organisms that cling together are different, the colony will contain the different kinds. As a rule, with *good laboratory technic* a colony contains only one kind of microorganism.

Petri dish cultures permit good observation of colonies. The large surface area favors separation of individual microoganisms. Petri dish cultures are usually spoken of as plates and the process of making them as *plating* or *streaking.*

■ **Pure cultures.** A *pure culture* contains only one kind of bacteria; a *mixed culture* contains two or more kinds. As a rule, infectious material contains more than one kind of bacteria. So that one kind alone can be studied, it must be separated from all other kinds and grown alone as a pure culture. Pure cultures are usually obtained by the pour plate or streak plate method. The pour plate method is as follows (Fig. 7-6):

1. Melt three tubes of agar in boiling water and then allow to cool to 40° or 42° C. If an enriching material that is injured by heat, such as serum or whole blood, is needed, add it to the medium at this time.
2. Transfer to one of the tubes a loopful of the test material from which one expects to obtain a pure culture. Replace the cotton plug, and roll the tube between the palms of the hands to distribute the bacteria throughout the medium.
3. Flame sterilize the loop thoroughly. Transfer three loopfuls of the contents of the inoculated tube to the second tube.
4. Mix the contents of the second tube with the inoculum. Sterilize the platinum loop. Transfer five loopfuls of the mixture to the third tube and mix.
5. Pour the contents of each tube of medium into a Petri dish and allow to solidify.
6. Incubate the Petri dishes for 24 to 48 hours, and observe the colonies carefully.

Successive dilution of the specimen in the three tubes above reduces the number of bacteria and disperses them in the medium so that the colonies in the Petri dish cultures are more likely to be distinctly separated from each other. Those that are to be studied further are removed from the Petri dish with a straight platinum wire and rubbed over the surface of one or more slants of suitable media. This is known as *fishing* (Figs. 7-7 and 7-8). If the colonies are not separated from each other, it is impossible to fish one colony without touching other colonies.

The inoculated slants are allowed to incubate for 24 to 48 hours, and these cultures are studied further (Fig. 7-9). In the majority of cases all bacteria growing on a slant will be alike because they all grew from the members of a single colony on the plate, and these in turn grew from a single bacterium in the original material. If, as may sometimes happen, the original colony on the plate contains two or more kinds of bacteria, the same two or more kinds of bacteria will grow on the slant. Separation is made by suspending some of the bacteria from the slant in sterile salt solution and replating.

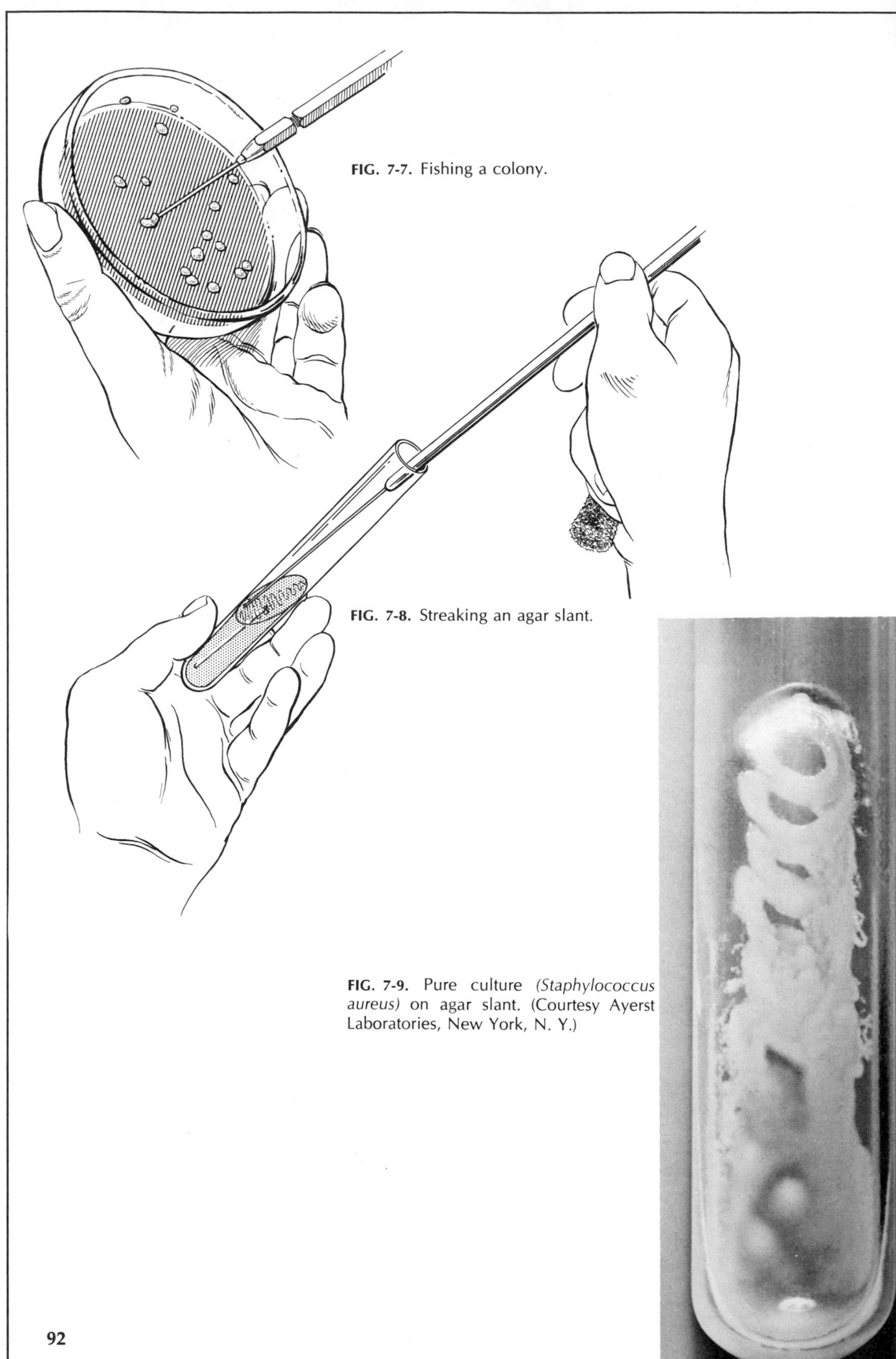

FIG. 7-7. Fishing a colony.

FIG. 7-8. Streaking an agar slant.

FIG. 7-9. Pure culture *(Staphylococcus aureus)* on agar slant. (Courtesy Ayerst Laboratories, New York, N. Y.)

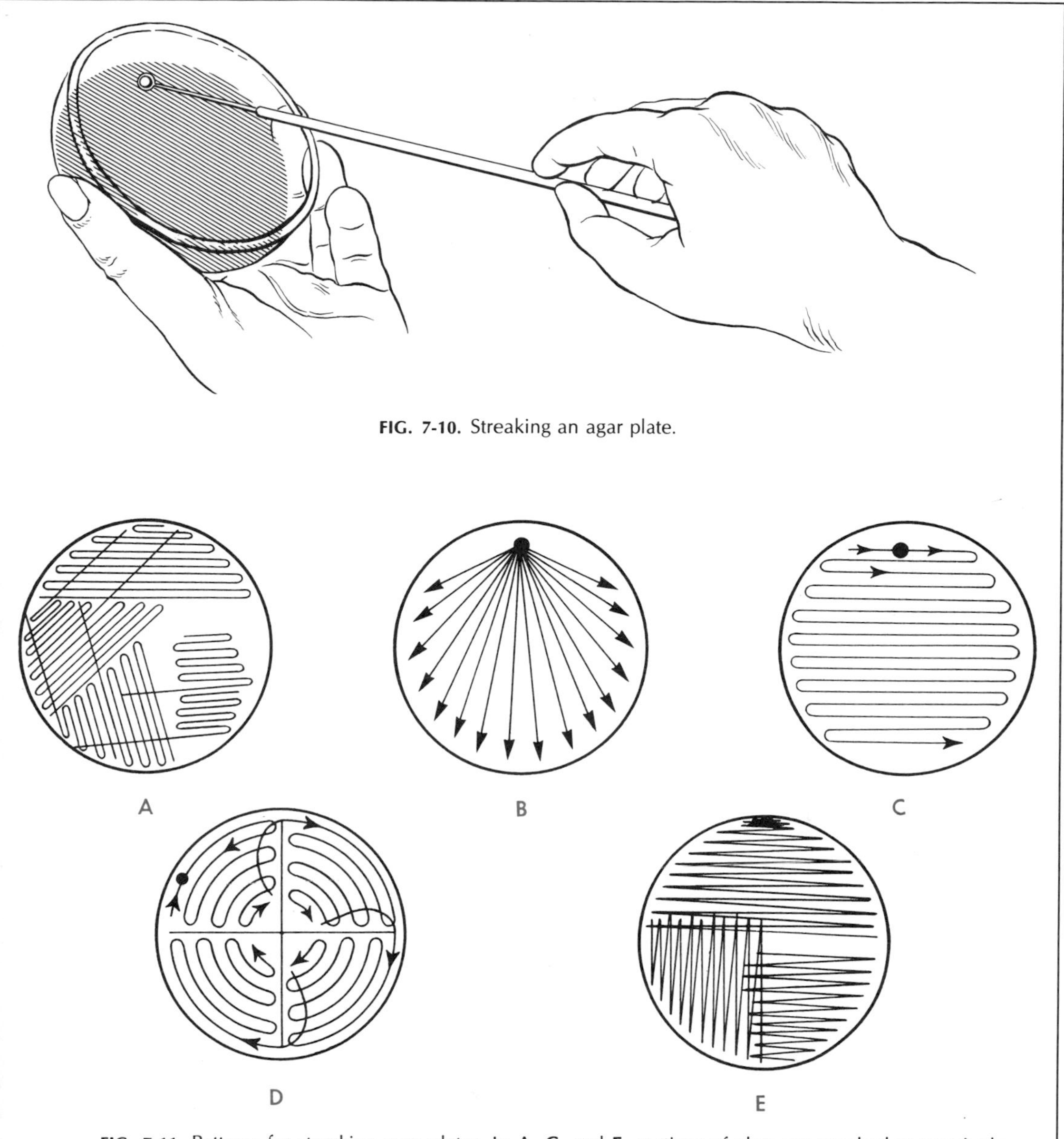

FIG. 7-10. Streaking an agar plate.

FIG. 7-11. Patterns for streaking agar plates. In **A, C,** and **E,** sections of plate are streaked successively. After each section is streaked, the platinum loop is flame sterilized. A bit of inoculum for next section in turn is obtained when loop is passed back into or through section just inoculated. (From Smith, A. L.: Microbiology: Laboratory manual and workbook, ed. 2, St. Louis, 1969, The C. V. Mosby Co.)

■ **Streak plates.** To prepare a streak plate, a single loopful of infectious material is streaked over the surface of the solid medium (agar) in a Petri dish. There are several ways to streak an agar plate. The method illustrated in Figs. 7-10 and 7-11 is widely used and gives good separation of colonies.

■ **Bacterial plate count.** The accuracy of the bacterial plate count (Fig. 7-12) depends upon the theory that when material containing bacteria is cultured every bacterium present develops into a colony. This statement is not strictly true because

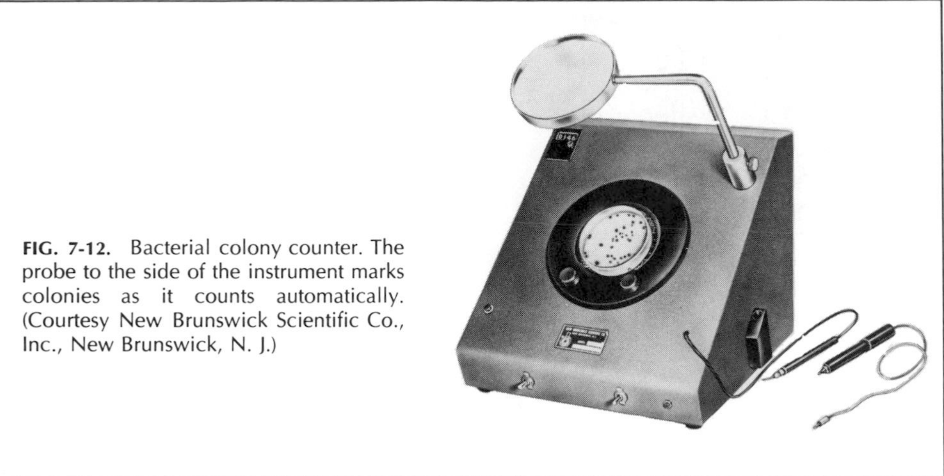

FIG. 7-12. Bacterial colony counter. The probe to the side of the instrument marks colonies as it counts automatically. (Courtesy New Brunswick Scientific Co., Inc., New Brunswick, N. J.)

some bacteria may fail to multiply, and two or more bacteria may cling together to form a single colony. In spite of this fallacy, the method is of decided value in the examination of water and milk (pp. 581 and 589).

A bacterial plate count is carried out as follows: Four tubes containing 9 ml. of sterile distilled water are needed. To the first tube, add 1 ml. of the material to be examined and mix. From this tube, transfer 1 ml. to the second tube and mix. From the second tube transfer 1 ml. to the third tube and mix, and from the third tube, transfer 1 ml. to the fourth tube and mix. This gives 1:10, 1:100, 1:1000, and 1:10,000 dilutions of the original material. With a sterile pipette transfer exactly 1 ml. from each tube to a Petri dish, and add sufficient melted agar (the agar cooled to 42° C.) Mix the contents of each dish by rotating and allow to solidify. After incubation for 24 to 48 hours, count the colonies that have developed. Let us say that the second plate shows 200 colonies. This means that at least 200 bacteria were present in the 1 ml. of material placed in the tube, and since this was a 1:100 dilution, the original test material must have contained at least 20,000 bacteria per milliliter. More than this number of bacteria may have been introduced into the dish because some may have failed to grow and some may have clung together in groups of two or more that developed into single colonies. In other words, the number of bacteria indicated by the count is certainly present, and more may be also.

The bacterial plate count is an important part of the microbiologic evaluation of urine from a patient with possible urinary tract infection. A measured amount of urine is handled as indicated above. A colony count is made from the bacterial growth on solid media. Quantitation helps to indicate whether bacteria recovered from urine are disease producers or contaminants. Generally with less than 1000 to 10,000 colonies from 1 ml. of specimen the microorganisms isolated represent normal flora of the urethra or contamination.

■ **Culture of anaerobic bacteria.** To grow anaerobic bacteria, a tube of suitable culture medium about 8 inches in length and about half full of medium may be heated in a boiling water bath for several minutes to expel any oxygen present in the medium. The tube is quickly cooled and inoculated; then sterile melted petroleum jelly to make a layer about ¾ inch thick is poured onto the top of the medium, thus sealing off the culture from the air. If agar is used, the inoculation is made when the heated medium cools to a temperature of 42° C.

Thioglycollate broth, a special medium containing thioglycollic acid, supports the growth of anaerobes within the depths of a partly filled tube without special seal. Methylene blue indicator colors the upper layers of the fluid medium with oxidation.

Cultures may be made in the ordinary manner and placed in a specially constructed glass or metal chamber from which the oxygen either is removed by some chemical reaction that is made to take place in the chamber (see also Fig. 20-3) or is replaced by a gas, such as hydrogen, that does not affect the growth of the bacteria.

■ **Slide cultures.** Slide cultures (see p. 488), used to study fungi, may be placed on the stage of the microscope and actual growth observed from time to time. The depression of a sterile culture slide may be filled with suitable medium, inoculated, and covered with a sterile cover glass. Liquid cultures may be made by inoculating a drop of medium on a sterile cover glass and proceeding as for a hanging-drop preparation.

■ **Cultures in embryonated hen's egg.** Important sites of growth for viruses, bedsoniae (chlamydiae), and rickettsiae are the yolk sac and the embryonic membranes of the developing chick embryo (p. 405). Bacteria are occasionally grown in this way.

■ **Cultivation of microbes in tissue culture.** Tissue cultures are cultures in which animal cells are growing. Cells used are from mammalian or fowl embryos or selected adult tissues such as the cornea, the kidney, and the lung. Minced tissue is placed in a suitable substrate in which cells multiply and grow. The growing cells may be supported on a solid or semisolid substrate such as fibrin, agar, and cellulose, or they may be suspended in a liquid. Certain pathogenic agents such as rickettsiae, bedsoniae, and viruses, which do not grow in lifeless media, multiply with ease when incorporated into tissue cultures.

ANTIBIOTIC SENSITIVITY TESTING

The susceptibility (or resistance) of bacteria to different antibiotics may be determined by suitable laboratory tests, usually designated as *sensitivity tests.* Two methods are used: the disk sensitivity test (Fig. 7-13) and the tube-dilution method. The first employs disks of filter paper impregnated with known amounts of different antibiotics. A Petri dish is heavily inoculated with the test organisms. The disks are dropped onto the freshly inoculated surface and the plate incubated for 24 hours. As the organisms grow, the antibiotic is diffusing out in the culture medium. If the organism is sensitive to the antibiotic, there is a zone of inhibition (no growth) around the disk. If the microorganism is unaffected by the drug, growth will cover the area around the disk. Several preparations can be tested on one plate. The method is simple and rapid, and the results correlate well with the effectiveness of the clinical treatment.

The second method utilizes a tube of culture medium containing both the test organism and a specified amount of a given antibiotic. The tube is incubated and the amount of growth evaluated. Since only one drug can be tested at a time, this method is laborious.

With this sort of testing, the microorganisms are reported as being sensitive or resistant to a given antibiotic. There are technics whereby the amount of certain antibiotics can be determined in the blood.

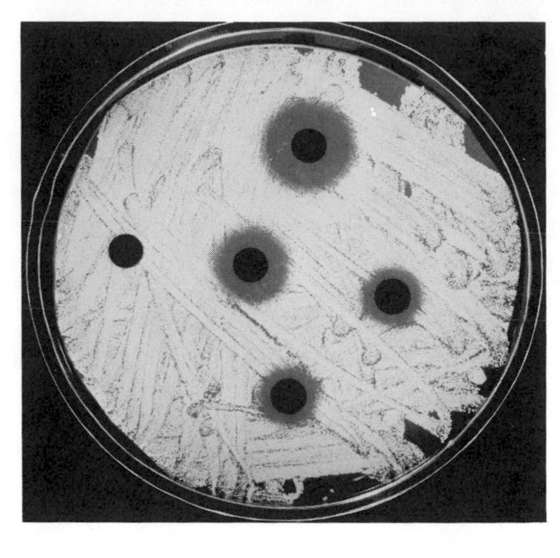

FIG. 7-13. Antibiotic sensitivity testing: single paper disk diffusion method. A Petri dish of agar was heavily inoculated with test organisms. Disks of filter paper impregnated with different antibiotics were dropped onto freshly seeded surface. Note heavy white bacterial growth with zones of inhibition about some of the disks and none about one. A zone of inhibition indicates that growth of organisms would probably be limited in a patient receiving that antibiotic.

QUESTIONS FOR REVIEW

1. Name important ingredients of culture media.
2. Why is agar the most commonly used solidifying agent?
3. What purposes do dyes serve in culture media?
4. What advantages are there to culture by comparison with direct visualization on smears?
5. Explain differential media and selective media. Give examples.
6. Give the practicality of the Petri dish in microbiologic culture.
7. Outline the steps in making a pure culture. What is meant by a pure culture? What purpose does it serve?
8. Briefly describe how a bacterial count is made. Under what circumstances are bacterial counts made?
9. Briefly explain: tissue culture, slide culture, culture on synthetic media, and culture of anaerobic bacteria.
10. How is antibiotic sensitivity testing carried out? What is its purpose?

REFERENCES. See at end of Chapter 10.

8 Bacteria: biologic activities

Bacteria engage in a complex of biologic activities of varying importance. Of prime consideration are those relating to the growth and integrity of the microorganism. Those resulting in the elaboration of toxins and substances harmful to other living cells have special significance in the production or aggravation of disease. Other events in the life of the bacterial cell, although striking in their manifestations, are not vital and are of secondary importance.

MAJOR METABOLIC EVENTS

Bacteria first drew attention to their very existence by the dramatic changes resulting from their metabolic activities. Today their biochemical capabilities are obvious, when one considers the phenomena of fermentation (p. 101), putrefaction (p. 109), decay (p. 109), soil fertility (p. 571), and infectious disease (p. 140).

Much of the general body of information concerning metabolism in creatures more complex than bacteria has come from studies of comparable processes in bacteria and other microorganisms. Because bacteria multiply rapidly, we can learn much about their activities in 2 days. To obtain comparable information about man would require 200 years or more.

The bacterial cell makes two biologic demands of its environment: It must obtain therein the chemical ingredients with which it can build and maintain itself, and at the same time it must derive therefrom the energy necessary to do its work. *Metabolism* encompasses all biochemical reactions occurring within the bacterial cell by which these two requirements are met.

■ **Enzymes.** Enzymes are essential, very efficient ingredients in the complex maze of metabolic activity. Life is not possible without them.* The term "enzyme" is of Greek origin and means 'leavened' (*en*, 'in'; *zymē*, 'leaven').

CHARACTERISTICS. Enzymes are catalysts† and, being protein in nature, are therefore organic catalysts. At the relatively low temperature of the cell they bring about chemical changes and are not consumed themselves. They contribute nothing to the end product and do not furnish energy. Although produced by a living cell, an enzyme

*In space exploration, the enzyme phosphatase, common to most forms of life on this planet, is considered a good indicator for the presence of life or prelife if detected in even trace amounts in outer space.

†A catalyst is a substance that accelerates a chemical reaction that otherwise would occur either very slowly and ineffectively or not at all. In a reaction the catalyst is not chemically altered.

operates independently of that cell. Its activity is not lessened when it is separated from the cell that produced it.

Over a thousand enzymes have been identified among living cells. Well-known enzymes in animals include ptyalin of the salivary glands, pepsin from the glands of the stomach, and trypsin elaborated in the pancreas. Bacteria produce enzymes that are remarkably like those produced by the organs of higher forms of life.

In the cell, a protein is a key component and functional unit that is made up of a characteristic chain of some 200 amino acids. Twenty distinct amino acids compose proteins, and the identity of given proteins is largely the result of different patterns of these amino acids, the number present, and the sequence in which they are placed. There is an almost infinite number of proteins belonging to different animal and plant species. A given organism, however, builds only a certain number.

Of the multitude of proteins built by living cells, the majority function as enzymes. A bacterium contains, on an average, about 2000 enzymes. Elaboration of proteins and therefore enzymes within a cell is under the direction of the DNA of the bacterial nuclear apparatus. For 2000 enzymes there would be corresponding 2000 control positions (genes) in the bacterial nucleus (the one gene–one enzyme hypothesis).

Enzymatic action (Fig. 8-1) is specific; that is, each enzyme causes only its own peculiar type of chemical change on its specific substrate.* Enzymes represent large protein molecules of special chemical configuration. The specific activity of the enzyme comes from its particular shape and the shape of the molecules reacting with it. The enzyme molecule is a kind of template into which the reacting compound can fit. Once the chemical reaction is complete, a new compound is released. However, the enzyme by definition remains intact. Enzymatic action can result in separation as well as combination of molecules. For the activation of an enzyme, a small molecule, usually an inorganic metallic ion such as magnesium, manganese, zinc, or iron, may be needed. The *activator* molecule does not participate in the process.

Very small amounts of enzyme can induce most extensive chemical changes. One molecule of enzyme can catalyze the reaction for 10,000 to 1,000,000 moles of substrate per minute. Although the enzyme is not destroyed during the chemical reaction, its activity may be inhibited by the end products formed. The accumulation of end products may also block the formation of more enzyme. This is spoken of as a *negative feedback control.*

Enzymes are most active at temperatures from 35° to 40° C. and within a limited pH range. They are delicate structures; a temperature of 60° C. destroys them within 10 to 30 minutes (heat denaturation). Freezing retards their action but does not destroy them. A medium that is too acid or too alkaline inhibits their action or destroys them. Since the biochemical reactions occurring in bacteria are complicated and interrelated ones, enzymes may function in a group or as an *enzyme system,* with several enzymes being responsible for closely related chemical changes.

A *coenzyme* is a nonprotein organic compound that may be necessary for enzymatic activity. It serves to position the substrate to the enzyme. Some coenzymes are tightly bound to their enzymes; others are less so. Although it is changed during the catalytic reaction, the coenzyme reappears at the end of the reaction in its original form. Many coenzymes come from dietary vitamins.

It is possible for enzymes performing the same chemical function to exist with

---

*"Substrate" is the term designating the material acted on by the enzyme.

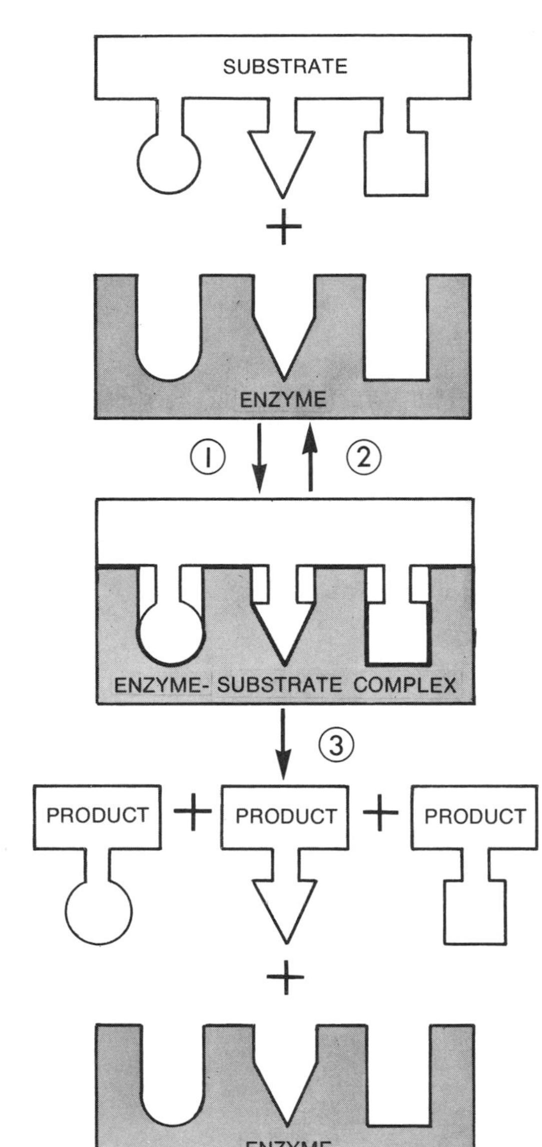

FIG. 8-1. Action of enzyme, in outline. Note tight fit between substrate and enzyme molecule to form enzyme-substrate complex. (From Tuttle, W. W., and Schottelius, B. A.: Textbook of physiology, ed. 16, St. Louis, 1969, The C. V. Mosby Co.)

different molecular structures. The physiologic equivalent for a given enzyme is an *isoenzyme*. An enzyme may have one or more isoenzymes. The enzyme lactic dehydrogenase has five.

CLASSIFICATION. From the standpoint of a given cell, enzymes may be *exoenzymes* or *endoenzymes*. The former are elaborated within the cell and are diffused through the plasma membrane into the surrounding medium, where they are active. Endoenzymes are liberated only by the disintegration of the cell that produced them. Exoenzymes are more important; some have been separated and crystallized. Enzymes are also related to the type of chemical reaction catalyzed or to the particular substrate affected. They are named by adding the suffix *-ase* to the name of their substrate or to the name of the reaction influenced.

In 1964, the Commission on Enzymes of the International Union of Biochemists adopted a classification separating enzymes into six major categories according to their type reaction:

1. *Oxidoreductases* catalyze oxidation-reduction reactions (includes oxidases and dehydrogenases).
2. *Transferases* catalyze the transfer of a specific group from one substrate to another (as in transamination or transmethylation). Examples are transaminases, transacetylases, and kinases.
3. *Hydrolases* catalyze hydrolytic reactions. Here are found digestive enzymes (proteinases, lipases, amylases) that break down large molecules to smaller, readily assimilated ones. Others include peptidases, esterases, and phosphatases.
4. *Lyases* (nonhydrolytic enzymes) catalyze the removal of specific groups from given substrates, such as when bonds are split between carbon atoms (decarboxylases) and between carbon and oxygen atoms (carbonic anhydrase). Others are deaminases and aldolases.
5. *Isomerases* catalyze the interconversion of two compounds of the same atomic composition but with different structure and properties. An isomerase is an important component of the glycolytic pathway.
6. *Ligases* catalyze the joining together of two molecules, coupled with the breakdown of ATP or a similar energy-yielding triphosphate. Ligases are also known as synthetases and as "activating" enzymes.

■ **Chemosynthesis.** The plant cell, possessing chlorophyl, obtains energy from the environment in the form of light (photosynthesis). Microorganisms with no chlorophyl must gain their energy from the chemical alteration of the substances at hand, a process termed *chemosynthesis.* Because of the complexity of their metabolic enzymes most bacteria can break down the proteins, fats, carbohydrates, and other organic compounds that they contact.* A series of enzymatic steps characterizes the involved chemical changes. The starting point is bacterial digestion; the end point is biologic oxidation.

DIGESTION. The principal sources of energy for bacteria in their environment are mostly complex molecules, too large for the bacterial cell to deal with directly. They must be reduced to workable particle size. To accomplish this, bacteria rely upon certain enzymes (hydrolases) that they release into the medium to split the large molecules chemically by a process that includes the addition of water. This is *hydrolysis.* The hydrolytic enzymes are examples of exoenzymes. Briefly, digestion is the exoenzyme-catalyzed hydrolysis of a large molecule to secure fragments small enough for passage into the bacterial cell.

ABSORPTION. After the bacterial cell has reduced the large molecules into smaller particles, it must next absorb them. There are two ways of doing this: Because of the osmotic pressure gradient, the cell can passively allow the molecules to move in by diffusion. This is slow and inefficient at best. On the other hand, the cell can actively participate to speed the movement. The manner in which this is brought about is referred to as *active transport.* As would be expected, the mechanisms of active trans-

---

*Organisms utilizing inorganic matter as their source of energy are *lithotrophs.* Those utilizing organic matter are *organotrophs.* Most bacteria are organotrophs.

port including the enzymes required are located on the cell membrane, and their activity constitutes a "pump."

OXIDATION. Some of the molecules that the cell pulls into itself from without are ready for the sequences of biologic oxidation. Most are not, and these must be prepared, that is, changed into a form that can be oxidized. If a phosphate group is attached to the molecule (phosphorylation), a very significant chemical bonding is made. When this phosphorylated compound is oxidized, the chemical bond of the phosphate group not only traps energy in a form suitable for later use but also in large amounts. Phosphorylation is the most important preliminary sequence to oxidation, but other chemical reactions may be encountered at this step.

Biologic oxidation is the chemical setup whereby a cell is provided with energy that it can use biologically. The complicated reactions contributing to it, many of them chain reactions, are carried out notably by enzymes (oxidases, dehydrogenases, or oxidoreductases), coenzymes, and hydrogen acceptors or carriers. *Oxidases* act on a substrate to cause it to undergo oxidation, that is, a reaction wherein electrons are removed from a substrate with accompanying hydrogen ions. Enzymes making possible the removal of hydrogen (or electrons) are called *dehydrogenases.* As the hydrogen ions are removed, there must be a suitable compound to receive them, *a hydrogen acceptor* or *carrier,* which in so accepting becomes reduced. Hydrogen acceptors or carriers, which can be oxidized and reduced, function in the transport of hydrogen (or electrons) from tissue metabolites to oxygen or some other end product. Transfer of electrons constitutes oxidation and reduction, and the end results must include an oxidized product, a reduced one, and energy liberated or trapped.

STORAGE OF ENERGY. There is no dissipation of heat in biologic oxidation. The energy released is held in the chemical bond between an organic molecule and a phosphate group. Bacteria store this energy-rich bond solely on the organic molecule adenosinediphosphate, ADP. When a phosphate group is transferred to ADP, there is formed adenosinetriphosphate, ATP, from which energy-rich reserves are available.

■ **Classes of biologic oxidation.** If the ultimate and final hydrogen acceptor is molecular oxygen, the events (aerobic) of biologic oxidation to generate energy are those of *respiration.* If it is an inorganic nitrate, sulfate, or carbonate, the metabolic processes are those of *anaerobic respiration.* The sulfate- and carbonate-reducing organisms are obligate anaerobes relying exclusively on these compounds for energy. The denitrifying bacteria are facultative aerobes using nitrate only in the absence of oxygen. If the final hydrogen acceptor is an organic compound, the processes of biologic oxidation are those of *fermentation.*

Fermentation then is metabolism utilizing organic compounds as both electron donors and electron acceptors. Many organic compounds can be fermented, and many microorganisms including all disease producers can live under the anaerobic conditions wherein they must satisfy their energy needs through fermentation reactions. Seven different biochemical types have been elucidated.

Using microbial metabolism of glucose, let us compare respiration (aerobic) with fermentation (anaerobic). In the presence of molecular oxygen, respiration takes place in all animal cells and in microbes, and glucose is changed in a biochemical sequence to carbon dioxide and water. In the absence of oxygen, microbes ferment the glucose but with different end products. Among bacteria generally, incomplete oxidation of this nature is the rule, and a wide variety of end products accumulate to a considerable extent. The most commonly occurring sequence for glucose degradation is the Embden-Meyerhof glycolytic pathway in which glucose is changed to pyruvate. Respiration is a much more efficient process than is fermentation. Many more intermediate and intervening steps mean much more energy generated. The potential from complete oxidation of glucose is far greater than that from its conversion to alcohol and carbon dioxide; for instance, respiration yields over 30 molecules of ATP (the energy source) as compared to the net of only two for each mole of sugar fermented.

Microbial fermentation reactions however are of great value to man because of the many by-products of practical worth to home and industry. They are also useful in the laboratory study of microbes.

■ **Macromolecular synthesis.** In a microbial cell basic physiologic phenomena are geared primarily for growth, that is, an orderly increase in mass or number of its constituents, most of which are macromolecules. Macromolecules are found in the protein-synthesizing groups of the cytoplasm, on membranes of organelles, in chromosomes, in the cell wall, and in the various enzyme systems. The metabolic activities of the cell revolve around these large molecules, and one of the crucial functions of the cell is their biosynthesis. To put together a macromolecule requires two things: (1) the subunits for linkage and (2) the energy to do the work. The energy to induce linkage comes from the cellular reserves of ATP. The subunits must come from either the environment or the metabolism of the cell and depend upon the chemical nature of the unit complex. For proteins these are amino acids; for carbohydrates, simple sugars; for lipids, glycerol (or other alcohols) and fatty acids; for nucleic acids, nucleotides; and for phospholipids, such substances as choline. The alignment of subunits in the macromolecule is determined in one of two ways. In nucleic acids and proteins, it is template-directed: DNA serves as template for its own synthesis and for the synthesis of the types of RNA. Messenger RNA serves as template for the synthesis of proteins. In carbohydrates and lipids the arrangement of subunits is entirely enzymatic.

Although many metabolic activities are universally the same, some are unique to bacteria, for example the biochemical pathway by which the mucopeptide of the cell wall is structured. This provides a chemical basis for the design of an antimicrobial compound with a selective effect on one of the steps in the biosynthesis of the cell wall. Since bacterial protoplasm is under greatly increased osmotic pressure in relation to the isotonicity of the host cells, weakening the cell wall jacketing the bacterium can easily explode it.

MEDICALLY RELATED ACTIVITIES

■ **Toxins.** Practically all species of bacteria produce poisonous substances known as *toxins.* The capacity to elaborate toxins is *toxigenicity.* In some the toxin is almost entirely responsible for the specific action of the bacteria. Toxins are of two types: *exotoxins,* diffused by the bacterial cell into the surrounding medium, and *endotoxins,* liberated only when the bacterial cell is destroyed. Some bacterial exotoxins are more deadly than any mineral or vegetable poison. Among the comparatively few bacteria (mostly gram-positive) noted for their exotoxin production are *Corynebacterium diphtheriae, Clostridium tetani,* and *Clostridium botulinum.* In disease caused by *Corynebacterium diphtheriae* (diphtheria) and *Clostridium tetani* (tetanus), the bacteria grow restricted to a superficial area in the body. Within themselves these organisms produce little effect, but the soluble exotoxins elaborated at the site of growth are absorbed into the body to cause serious and often fatal illness. Such illness is better thought of as chemical poisoning rather than as disease induced by growth of bacteria.

Exotoxins are protein in composition, antigenic, and specific. When a small amount of a toxin is injected into an animal, it stimulates the production of *antitoxin.* The *specificity* of exotoxins refers to the fact that diphtheria bacilli elaborate a toxin that causes diphtheria and nothing else and tetanus bacilli elaborate a toxin that causes tetanus and nothing else. When an exotoxin is inactivated with formaldehyde, it no longer causes the disease but still can produce an immunity to the disease. Such modified toxins are known as *anatoxins* or *toxoids.* Toxoids are used to produce permanent immunity to diphtheria and tetanus.

Substances contained in the seeds of certain plants and some of the secretions of animals resemble exotoxins. Among these are crotin derived from the seed of the croton plant, ricin from the castor bean, and the venoms of snakes, spiders, and scorpions.

Endotoxins are lipopolysaccharides and are intimately related to the cell wall of most gram-negative bacteria. Endotoxins are not found in the surrounding medium when bacteria are grown in a liquid. They do not effectively promote the formation of antitoxins, do not possess the specificity of exotoxins, and cannot be converted into toxoids (see Table 8-1). Noteworthy for their production of endotoxins are *Salmonella typhi,* the cause of typhoid fever, and *Shigella flexneri,* one of the causes of bacillary dysentery.

The biochemical mechanisms by which toxins injure cells are poorly understood.

■ **Harmful metabolic products.** Bacteria produce a number of enzymatic substances not directly toxic but related significantly to disease. *Hemolysins* are substances that

TABLE 8-1. DISTINGUISHING FEATURES OF TOXINS

Differential point	Exotoxins	Endotoxins
Location	Gram-positive bacteria	Gram-negative bacteria
Relation to cell	Extracellular	Closely bound to cell wall (release with destruction of cell)
Toxicity	Great	Weak
Tissue affinity	Specific (examples, nerves—tetanus toxin; adrenals, heart muscle—diphtheria toxin)	Nonspecific (local reaction—injection site; systemic reaction—fever, shock)
Chemical nature	Protein	Lipopolysaccharide complex
Stability	Unstable (denatured with heat, ultraviolet light)	Stable
Antigenicity	High (neutralizing antibodies)	Weak
Conversion to toxoid	Yes	No
Diseases where action important	Diphtheria Tetanus Gas gangrene Staphylococcal food poisoning Botulism	Typhoid fever and certain salmonelloses Asiatic cholera Brucellosis

cause the lysis (dissolution) of red blood cells. There are many types, among which are immune hemolysins, hemolysins of certain vegetables, hemolysins contained in venoms, and bacterial hemolysins. The bacterial hemolysins are of two types: (1) those that may be separated from the bacterial cells by filtration (filtrable hemolysins) and (2) those that are demonstrated about the bacterial colony on a culture medium containing red blood cells. The filtrable hemolysins are named after the bacteria that give rise to them (examples, *staphylolysin* derived from staphylococci and *streptolysin* derived from streptococci). The filtrable hemolysins have a certain degree of specificity; that is, a given hemolysin may act on the red blood cells of one species of animal but not on those of another species. They are protein in nature and inactivated by exposure to a temperature of 55° C. for 30 minutes. In the body of an animal they induce the formation of antibodies against themselves.

Staphylococci, streptococci, pneumococci, and *Clostridium perfringens* are important producers of these hemolysins. An organism may have more than one filtrable hemolysin; for instance, streptococci produce two hemolysins known respectively as streptolysin O and streptolysin S.

The hemolysins detected when bacteria are grown on a culture medium containing blood are of two types, *alpha* and *beta*. Beta hemolysins give rise to a clear, colorless zone of hemolysis around the bacterial colony. Alpha hemolysins give rise to a greenish zone of partial hemolysis. The relation existing between filtrable hemolysins and those detected by growing the bacteria on a culture medium containing blood is not known. The relation between hemolysin production and virulence is obscure. As a rule, the hemolytic strains of pathogenic bacteria have greater capacities to cause disease than do the nonhemolytic strains. On the other hand, certain nonpathogenic bacteria produce hemolysins. Hemolysins probably contribute to the invasive capacity of some bacteria.

Leukocidins destroy polymorphonuclear neutrophilic leukocytes. The leukocytes, or white blood cells, take an active part in the battle of the body against infection. Leukocidins are formed by pneumococci, streptococci, and staphylococci. Their relation to immunity is not known. It is likely that they enhance virulence. Certain hemolysins and leukocidins appear to be identical.

Coagulase is elaborated by bacteria to accelerate the coagulation of blood and, under suitable laboratory conditions, causes oxalated or citrated blood to clot. Coagulase formation appears to be confined to the staphylococci. The *coagulase test,* performed by mixing a culture of staphylococci with a suitable amount of oxalated or citrated blood, under specified conditions, is used to differentiate pathogenic and nonpathogenic staphylococci. Because the former cause the blood to coagulate, they are said to be *coagulase-positive.* The role of coagulase in immunity is not known. Possibly it protects organisms against the destructive action of leukocytes (phagocytosis). The coagulum from plasma that has leaked into the tissues forms a barrier between the bacteria and the white blood cells.

The *bacterial kinases* act on certain components of the blood to liquefy fibrin. Kinases interfere with coagulation of blood since a blood clot is made up of red blood cells enmeshed in interlacing strands of fibrin. Kinases liquefy clots already formed. These seems to be a relation between the kinases and virulence because kinases destroy blood or fibrin clots that form around a site of infection to wall it off. The most important kinase is *streptokinase,* also known as *fibrinolysin,* elaborated by many hemolytic streptococci. Staphylococci and certain other bacteria also produce kinases. Strepotokinase has been used in certain locations in the body to dissolve clots and to prevent the formation of adhesions that would be laid down on the fibrin precipitated in the body cavities.

Hyaluronidase is secreted by certain bacteria to make the tissues more permeable to the bacteria elaborating it. It hydrolyzes hyaluronic acid, a constituent of the intercellular ground substance of many tissues that helps to hold the cells of the tissue together. Hyaluronidase was formerly spoken of as the "spreading factor." It is produced by pneumococci and streptococci.

Bacteriocins are bacterial protein or polypeptide substances produced by strains of a family of microbes and active only on certain other strains in the given family. *Colicins* are those produced by strains in the family Enterobacteriaceae. They are highly specific and seem to be directed chiefly against the bacterial membrane. Nearly 20 different ones have been described and grouped by their bactericidal specificity.

SECONDARY EFFECTS

■ **Pigment production.** So far as is known, pigment production has no relation to disease and no significance other than in identification of pigment-producing organisms. Yellow pigments are most prevalent, but pigments of almost any color may occur. The red and yellow pigments probably belong to the same chemical group as those in turnips, egg yolk, and fruits. Pigments are produced by both parasitic and saprophytic bacteria; but most species of bacteria do not produce any pigment. Common pigment producers are *Staphylococcus aureus,* with a golden pigment, and *Pseudomonas aeruginosa,* with a bluish-green pigment. *Serratia marcescens,* produces a red pigment. Pigment-producing or chromogenic bacteria lose that property when grown under unfavorable conditions.

■ **Heat.** During the growth of all bacteria heat is produced, but in such a small amount that it can be detected in ordinary cultures only by most delicate methods. The heating of damp hay is in part caused by bacterial action.

■ **Light.** A few species of bacteria have the ability to produce light (Fig. 8-2). Most of them live in salt water, and some live a parasitic existence on the bodies of saltwater fish. Fox fire, seen on decaying organic material, especially in the woods, is from luminescent bacteria. None of the light-producing bacteria is pathogenic.

■ **Odors.** Certain odors are characteristic of some species of bacteria; some arise from the decomposition of the material on which the bacteria are growing.

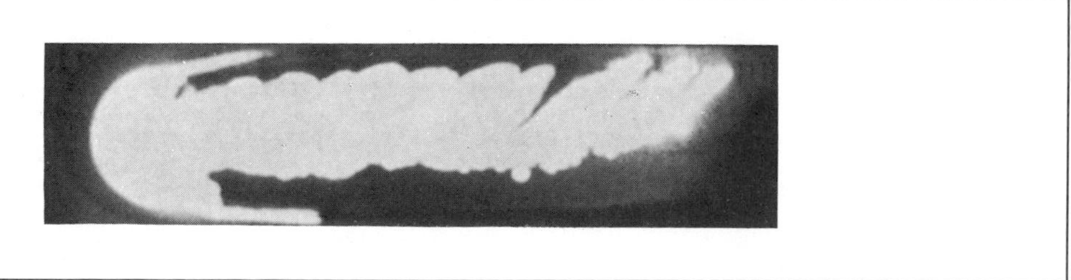

FIG. 8-2. Luminescent bacteria *(Photobacterium fischeri).* Photograph made in total darkness to show light produced by culture. (Courtesy Carolina Biological Supply Co., Burlington, N. C.; from Beaver, W. C., and Noland, G. B.: General biology, ed. 8, St. Louis, 1970, The C. V. Mosby Co.)

QUESTIONS FOR REVIEW

1. Briefly discuss bacterial metabolism. What are the two main functions of metabolism?
2. Define enzymes. Give salient characteristics.
3. Differentiate an exoenzyme from an endoenzyme. Give an important example of an exoenzyme.
4. Classify enzymes. Give examples.
5. What are toxins? Classify them and define each class.
6. Explain chemosynthesis, digestion, phosphorylation, oxidation-reduction, fermentation.
7. Name three species of bacteria important as exotoxin producers and two producers of endotoxins.
8. What are toxoids? For what are they used?
9. Define hemolysin, leukocidin, coagulase, and streptokinase. Name an organism that produces each.
10. Classify bacterial hemolysins. What is meant by alpha and beta hemolysis?
11. What is hyaluronidase? Why is it sometimes called the "spreading factor"? In infections with what organisms is it important?
12. Discuss pigment production by bacteria.
13. Give an example of heat production by bacteria.
14. What is the importance of the coagulase test?
15. Define catalyst, substrate, macromolecule, colicins.

REFERENCES. See at end of Chapter 10.

9 Methods for microbes: biochemical reactions, animal inoculation

BIOCHEMICAL REACTIONS

Biochemical tests demonstrate the presence of enzyme systems within the microbial cell, such as those responsible for the fermentation of carbohydrates and the decomposition of proteins.

Miniaturization of microbiologic technics refers to the use of the commercially prepackaged test units wherein basic reagents for a given biochemical reaction are premeasured, standardized, and compacted into a tablet or onto a paper disk or filter paper strip. The test unit is applied to cultures in liquid or solid media, and results are quick and easy. In the design of the test unit, enzymatic action of the test organism plus the reagents results in a color change as end point. There are many varieties of such tests available.

■ **Fermentation of sugars.** Microbes ferment many organic compounds, including carbohydrates, generating in the process energy-rich chemical bonds. Simple sugars through fermentation can serve as the main source of energy for many different kinds of microbes, and in the laboratory it is practical to deal with these relatively simple test solutions. The end products of fermentation depend upon the substrate, the enzymes present, and the conditions under which the reaction proceeds. For instance, the yeast enzymes that break down glucose into carbon dioxide and alcohol are without effect on sucrose (cane sugar), and certain enzymes produced by pneumococci and streptococci change glucose to lactic acid. Common products of bacterial fermentation are lactic acid, formic acid, acetic acid, butyric acid, butyl alcohol, acetone, ethyl alcohol, and the gases carbon dioxide and hydrogen.

Fermentation reactions, though varying among species of microorganisms, are quite constant and therefore of great value in differentiating the species. From their specific action on a given sugar in the laboratory, bacteria (and other microbes) are classified as (1) those that do not ferment it, (2) those that do ferment it, with the production of acid only, and (3) those that ferment it, with the production of both acid and gas. Media containing different sugars are inoculated with the bacteria (or other microbes) and observed for gas production and acid formation. Gas production in liquid media is detected by an accumulation of gas that displaces fluid medium contained in the closed arm of a special tube. It may also be detected in a small tube (one end of which is sealed off) placed in an inverted position within a larger tube of liquid medium, the

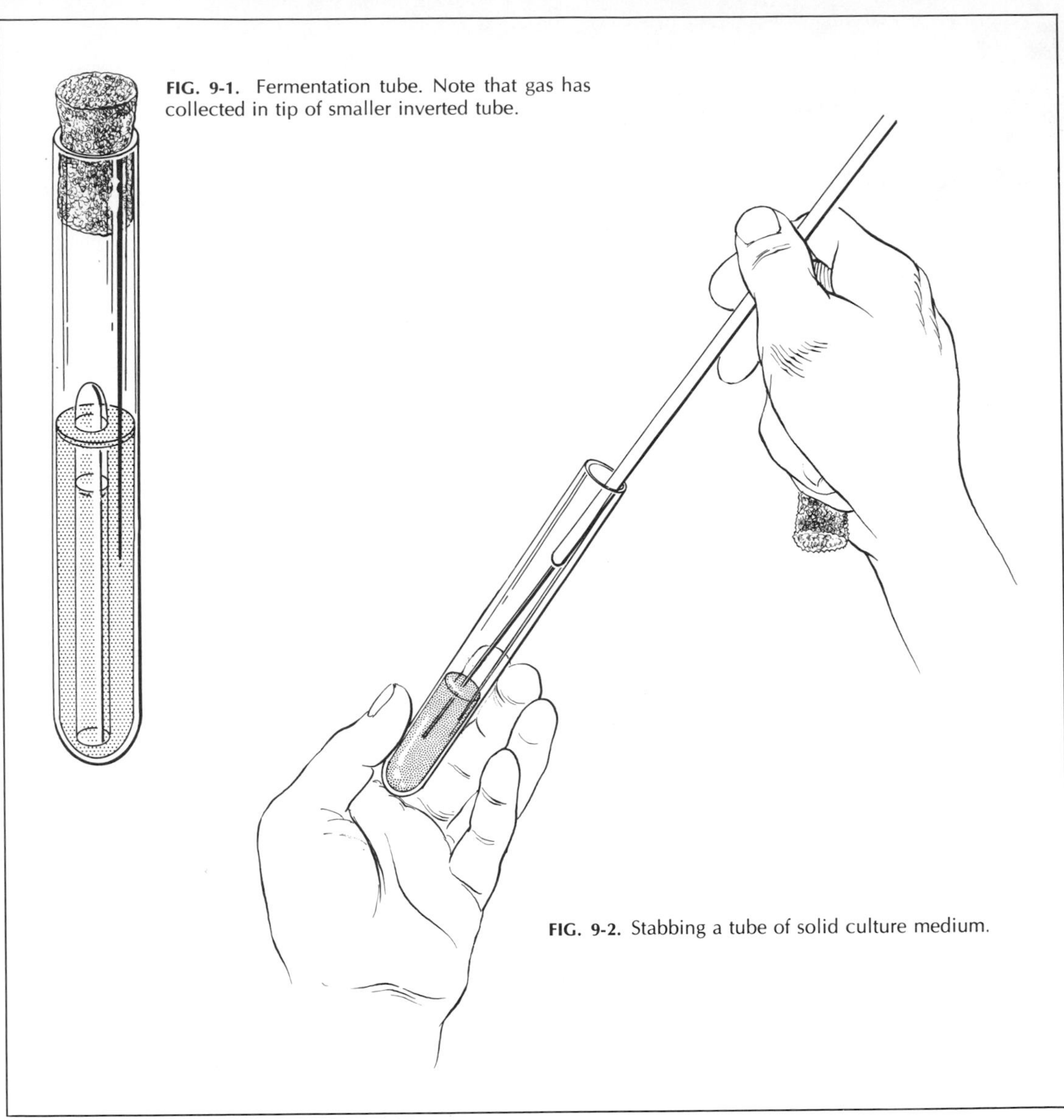

FIG. 9-1. Fermentation tube. Note that gas has collected in tip of smaller inverted tube.

FIG. 9-2. Stabbing a tube of solid culture medium.

sealed end projecting slightly above the level of the fluid. As gas is formed in the inoculated liquid, it collects in the small tube within the depths of the culture and rises toward the sealed end (Fig. 9-1).

Solid media are inoculated for fermentation studies by plunging a straight needle carrying some of the bacteria deep into the medium. (This plunging maneuver is known as "stabbing"—Fig. 9-2.) The solid medium may be melted, cooled to 40° C., and then inoculated. Gas production in solid media is indicated by disruption of the medium by gas bubbles. Acid formation in both liquid and solid media is indicated by color changes in indicators incorporated therein.

Many carbohydrates are used in even routine fermentation studies. Important sugars are glucose, lactose, maltose, sucrose, xylose, mannitol, and salicin. The fermentation of lactose is crucial in the identification of bacteria recovered from the intestinal tract since it makes a division between the nonpathogenic coliforms and the pathogenic *Salmonella* and *Shigella*, those organisms that cannot ferment lactose (with a few exceptions) being the pathogens.

■ **Hydrolysis of starch.** When microbes are grown on starch agar, a clear zone develops around the colonies of those that digest starch. If the agar is covered with a weak solution of iodine, the *undigested* starch assumes a blue color.

■ **Liquefaction of gelatin.*** A gelatin medium is stabbed with a wire having some of the bacteria on it, incubated at 20° C., and observed for liquefaction. It may be incubated at 37° C. (gelatin remains liquefied at this temperature) and then placed in the refrigerator, where the unliquefied gelatin will solidify. Incubation should be continued for at least 2 weeks unless liquefaction occurs before that time, and a tube of uninoculated gelatin should be incubated as a control.

■ **Indole production.** Production of indole (from the amino acid tryptophane) is determined by culture of the organisms in a medium containing tryptophane. A strip of filter paper soaked with a saturated solution of oxalic acid may be hung over the culture and held in place either by the cotton plug or by the screw cap of the culture tube. A pink color on the paper indicates indole production. Important indole producers are *Escherichia coli* and *Proteus*, which avidly decompose proteins.

■ **Nitrate reduction.** Nitrate reduction means the removal of oxygen from the nitrate radical (NO_3) to convert it into nitrite (NO_2). The organisms are incubated in broth containing 0.1% potassium nitrate, and the broth is tested for nitrite with sulfanilic acid and alpha-naphthylamine reagents. A red color means a positive test.

■ **Deoxyribonuclease elaboration.** Formation of DNAse (deoxyribonuclease) is demonstrated by culture of test microbes on the surface of an agar plate into which deoxyribonucleic acid has been incorporated. After 24 hours incubation, the surface is flooded with either N/1 hydrochloric acid to highlight the clear zones about the DNAse-positive growth or with 0.1% toluidine blue, in which case, a bright rose pink color is the end point. Staphylococci produce this enzyme.

■ **Hydrogen sulfide production.** A stab culture in an agar that contains basic lead acetate is incubated for 1 to 4 days. The production of hydrogen sulfide from the sulfur-containing amino acids of the medium is indicated by the appearance of a black compound, lead sulfide, formed from the combination of the hydrogen sulfide with the lead acetate. Instead of being incorporated into the medium, the basic lead acetate may be impregnated on sterile filter paper that is suspended over the medium as in the test for indole production. Iron salts also may be used to detect hydrogen sulfide formation. The production of hydrogen sulfide facilitates the identification of *Brucella* species and enteric bacilli.

■ **Splitting of urea.** Certain microorganisms when cultivated on media containing urea convert it to ammonia. With phenol red as the indicator, the presence of urease is seen by the appearance of a red color. The enzyme *urease*, responsible for the conversion, is produced by species of the genus *Proteus*. This test separates these organisms from the urease-negative *Salmonella* and *Shigella*.

---

*The proteolytic enzymes (protein-splitting enzymes, proteinases), produced by bacteria, split complex proteins into proteases, peptones, polypeptides, amino acids, ammonia, and free nitrogen. Protein decomposition is known as *putrefaction*. Some authorities restrict the term "putrefaction" to the decomposition of proteins by anaerobic bacteria, which results in the formation of hydrogen sulfide and other foul-smelling decomposition products, and they use the term "decay" for the decomposition of proteins by aerobic bacteria. The latter does not result in the malodorous decomposition products.

■ **Digestion of milk.** Bacterial growth in sterile milk may be alkaline or acid and with or without curdling. Curdling may or may not be followed by liquefaction of the curd. Excessive gas production in milk produced by *Clostridium perfringens* is "stormy fermentation."

■ **Oxidase reaction.** The enzyme *oxidase* is produced by *Neisseria* species and its detection is of great value in the identification of *Neisseria gonorrhoeae*. The oxidase reagent colors the positive colonies pink to red to black.

■ **Niacin test.** The niacin test is useful in distinguishing *Mycobacterium tuberculosis* from the atypical mycobacteria. A 4% alcoholic aniline solution and a watery solution of cyanogen bromide are added to 1 or 2 ml. of emulsified bacterial growth. A complex yellow compound is formed when niacin or nicotinic acid, formed by the tubercle bacillus but not by atypical mycobacteria, reacts with the cyanogen bromide and a primary amine.

■ **Demonstration of specific enzymes.** Reagent systems for the identification of diagnostic enzymes are commercially impregnated onto easy-to-use strips of paper suitable for application to bacterial cultures. Three enzymes of practical importance are so packaged: phenylalanine deaminase, cytochrome oxidase, and lysine decarboxylase.

Phenylalanine deaminase catalyzes the metabolism of phenylalanine to phenylpyruvic acid, which in turn reacts with ferric ions present to give a brownish or gray-black color. The presence of this enzyme in cultures of *Proteus* species is responsible for the positive reaction obtained with these organisms.

$$\text{Phenylalanine} \xrightarrow{\text{enzyme}} \text{Phenylpyruvic acid} + \text{Ferric ions} \longrightarrow \text{Color}$$

Cytochrome oxidase is produced by *Pseudomonas, Alcaligenes, Neisseria, Vibrio, Brucella,* and some *Flavobacterium* species. This enzyme catalyzes a coupling reaction to give a blue color (the positive reaction) as follows:

$$\text{Dimethyl-}p\text{-phenylenediamine} + \text{Alpha naphthol} + \text{Oxygen} \xrightarrow{\text{enzyme}} \text{Indophenol blue}$$

Lysine decarboxylase, which is formed by most of the *Salmonella* species, catalyzes the conversion of lysine to cadaverine, a more alkaline compound than lysine. In one commercial prepackaged test kit, the Prussian blue reaction is used to indicate the positive test.

$$\text{Lysine} \xrightarrow{\text{enzyme}} \text{Cadaverine} + CO_2 \longrightarrow \text{Bromthymol blue (yellow) changing to blue as pH rises}$$

ANIMAL INOCULATION

The inoculation of suitable laboratory animals is an essential part of the study of many microbes. After an animal has been inoculated, it is observed for effects produced by the microbes. In some cases it is killed after a certain length of time and examined for evidence of disease. Smears and cultures are made, and gross and microscopic changes in the organs observed. In other cases the animal is not killed, but blood and body fluids are removed at intervals for examination.

The well-known laboratory animals are guinea pigs, white mice, white rats, hamsters, and rabbits (Fig. 9-3 and 9-4). The inoculations may be given with syringe and needle either subcutaneously (beneath the skin), intradermally (between the layers

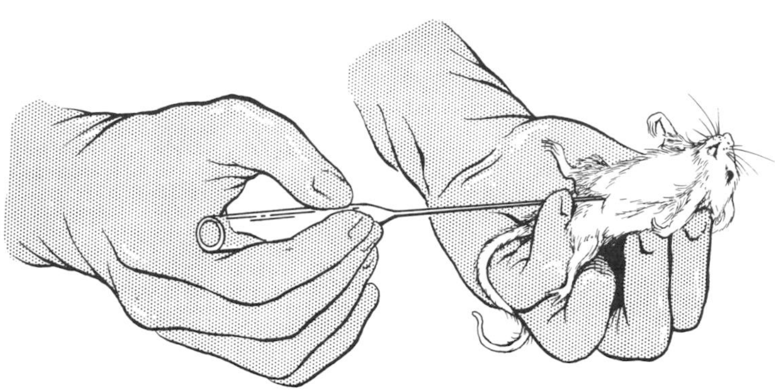

FIG. 9-3. Intraperitoneal inoculation of a white mouse. Note point of capillary pipette directed into peritoneal cavity as mouse is held firmly by nape of neck.

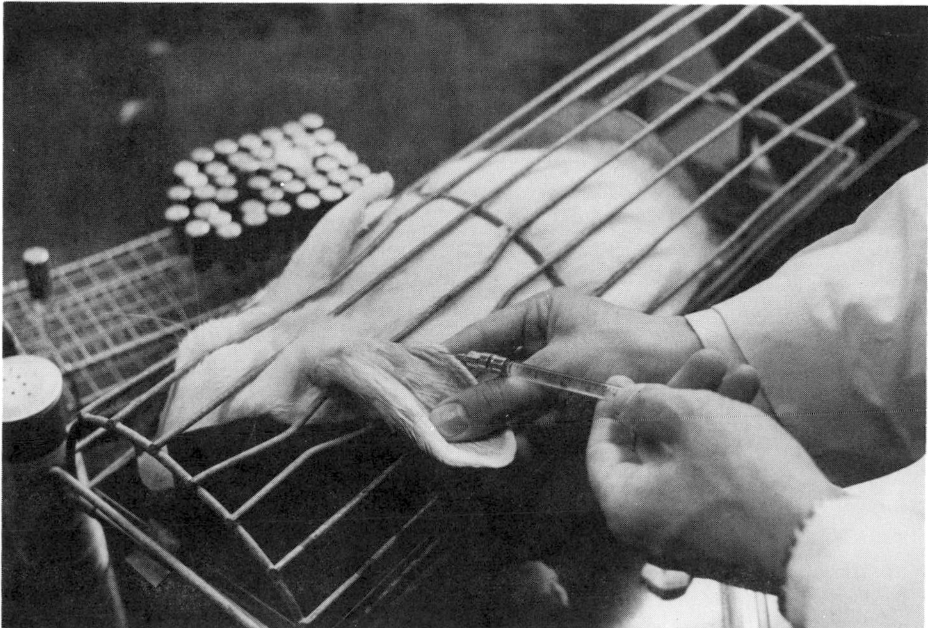

FIG. 9-4. Intravenous inoculation of a rabbit. Note needle in marginal vein of ear. Rabbit is held in a restraining cage. (Courtesy Joseph Nettis Productions, Philadelphia, Pa.)

of the skin), intravenously (into a vein), intraperitoneally (into the peritoneal cavity), subdurally (beneath the dura of the brain), or intracerebrally (into the brain).

■ **Advantages.** The advantages of animal inoculation in recovery and identification of microbes are as follows:

1. Some microbes are most easily detected by animal inoculation (examples, *Pasteurella tularensis* and, under some conditions, *Mycobacterium tuberculosis*).
2. The virulence of microbes can be exhibited (such as *Corynebacterium diphtheriae*).
3. It is sometimes the easiest way of obtaining the pure culture (*Diplococcus pneumoniae*).

4. It is often necessary in order to determine the action of drugs on microbes.
5. Microorganisms that cannot be cultured at all on artificial media can be readily recovered from suitable laboratory animals (spiral bacteria, rickettsiae, and most viruses). A very few microbes can be propagated in no other way.
6. The action of pathogenic and nonpathogenic agents on the animal body can be determined experimentally.
7. Animal inoculation is basic to the manufacture of antitoxins and other antiserums.

A good example of animal inoculation as the best method of detecting the microbes is the ease with which tuberculosis of the kidney is proved in a patient whose urine shows no bacteria on direct microscopic examination but whose urine produces classic disease in a guinea pig. A good example of animal inoculation being used to obtain a pure culture is seen in pneumococcal typing. Some of the sputum in which the type of pneumococcus is to be determined is injected into the peritoneal cavity of a white mouse, and within only a few hours the growth of pneumococci has outstripped that of all other organisms to such an extent that the peritoneal contents consist practically of a pure culture of pneumococci ready to be typed.

SUMMARY: IDENTIFICATION OF BACTERIA

When the various methods are applied to the study of bacteria, it will be seen that classification into comparatively large groups may be made from several standpoints, as follows:

- I. Direct microscopic examination
  - A. Staining reactions
    - 1. Reaction to Gram stain
      - a. Gram-positive
      - b. Gram-negative
    - 2. Reaction to acid-fast stain
      - a. Acid-fast
      - b. Nonacid-fast
  - B. Size
  - C. Shape
    - 1. Coccus—spherical
    - 2. Bacillus—rod shaped
    - 3. Vibrio, spirillum, spirochete—spiral shaped
  - D. Presence of spores
    - 1. Sporeformers
    - 2. Nonsporeformers
  - E. Capsule formation
    - 1. Encapsulated
    - 2. Nonencapsulated
  - F. Motility
    - 1. Motile
    - 2. Nonmotile
- II. Culture
  - A. Food requirements
    - 1. Saprophytes—growth on dead organic matter
    - 2. Parasites—growth on living matter
  - B. Media most suitable for growth
    - 1. Growth on simple media
    - 2. Growth only on special media
    - 3. No growth on any media
  - C. Appearance of growth on different media
  - D. Oxygen requirements
    - 1. Aerobes—growth only in the presence of free oxygen
    - 2. Anaerobes—growth only in the absence of free oxygen
    - 3. Microaerophiles—growth in the presence of small amounts of free oxygen
  - E. Optimum temperature for growth
  - F. Characteristics of the pure culture (Fig. 7-5)
    - 1. Size, shape, texture of colonies
    - 2. Pigment production
  - G. Production of hemolysins on blood agar
    - 1. Beta-hemolytic—complete hemolysis of red blood cells
    - 2. Alpha-hemolytic—partial hemolysis of red blood cells
    - 3. Nonhemolytic

III. Biochemical reactions
   A. Fermentation of sugars
      1. Fermentation with acid and gas
      2. Fermentation with acid only; no gas
      3. No fermentation
   B. Fermentation of lactose
      1. Lactose fermentors
      2. Non–lactose fermentors
   C. Splitting of urea
      1. Urease-positive
      2. Urease-negative

IV. Animal inoculation
   A. Disease production (virulence tests)
      1. Pathogenic—disease produced
      2. Nonpathogenic
   B. Toxin production
      1. Exotoxin producers
      2. Endotoxin producers

QUESTIONS FOR REVIEW

1. How are fermentation reactions used in the study of microbes?
2. List five sugars important in routine fermentation studies.
3. What two observable changes may occur in the culture medium when carbohydrates are fermented?
4. What is the importance of the fermentation of lactose?
5. List the routine laboratory animals.
6. State the advantages of animal inoculation.
7. What is meant by putrefaction? Decay?
8. Name two organisms that produce indole.
9. What genus produces urease?
10. Briefly outline the niacin test.
11. How may specific enzymes be demonstrated? Name three of diagnostic value.
12. Summarize methods used to classify and categorize bacteria.

REFERENCES. See at end of Chapter 10.

10 Specimen collection

To the extent that microbiologic investigation defines the diagnosis and nature of disease, it determines the mode of treatment and outlook for the patient. In the practice of scientific medicine, specimen collection must be a crucial first step upon which rests the validity of each succeeding step.

GROUND RULES

In the collection of any specimen for microbiologic examination remember the following:

1. Collect the specimen from the actual site of disease and do not contaminate it with microbes from nearby areas (example, in making smears and cultures from ulcers in the throat be very careful to take the material from the actual site of ulceration and not to contaminate it unduly with the secretions of the mouth).
2. Always use sterile paraphernalia in collecting the specimen. Place the specimen in a sterile container.
3. Sterilize material used in collecting the specimen as soon as possible after proper disposal of the specimen.
4. Collect adequate amounts of material (for instance, when pus is to be examined, collect several milliliters, if possible. If it is necessary to collect the material on swabs, use more than one swab).
5. Insofar as possible, do not send specimens to the laboratory on swabs (a swab can take up a small amount of material, dry quickly, and enmesh in its fibers important cells and organisms that will fail to be transferred to smears or culture media).
6. Collect the specimen in such a manner as not to endanger others (example, sputum or other excreta must not soil the outside of the container).
7. Take great care in handling specimens collected in cotton-plugged tubes. Cotton plugs can soak up a small specimen. Microbes from the environment pass through wet plugs and contaminate the specimen.
8. Whenever possible, make smears from the original material.
9. Do not add any preservative or antiseptic to the specimen. If possible, take the specimen before the patient has received any antimicrobial drug or before his wound has had local treatment. If the patient has already received a sulfonamide or antibiotic, notify the laboratory.

10. Label the specimen properly with the patient's name, the source of specimen, and the tentative diagnosis.
11. Deliver the specimen to the laboratory immediately.

The person making the delivery of the specimen may have to care for the specimen in the laboratory if at the time laboratory personnel are off duty. In this event specimens already inoculated onto culture media are placed in the incubator and those not on culture media are placed in the refrigerator. When cultures must be made without delay, the medium chosen should be that most likely to grow the suspected organisms, as well as any other that might be present. Since it supports the growth of organisms that may be present in the specimen better than any other culture medium found in the usual microbiologic laboratory, blood agar is best to use. Blood agar plates are superior to the tubed media, except in those instances in which the cultures must be shipped a distance into the laboratory. Plates are not easily transported.

CLINICAL SPECIMENS

Urine for microbiologic examination may be carefully collected as a clean-voided specimen. The collection of a urine specimen with a sterile catheter has long been recommended, especially in the female, but many physicians now feel that catheterization is unnecessary. No matter how carefully done, the maneuver extends infectious organisms up into the urinary tract. For most purposes a distinction can be made between infectious and contaminating microorganisms in a *carefully collected,* clean-voided specimen. The meatus (opening) of the urethra is gently cleansed with soap (or detergent) and water. In both voided and catheterized specimens the first portion should be discarded, and the last portion should be received into a sterile container. A "clean catch" urine specimen is one obtained during the mid-part or toward the end of the act of voiding; it is received into a sterile receptacle. The sample should be promptly refrigerated or a suitable preservative added if laboratory examination is delayed. The urine in health is free of microorganisms.

Blood for cultures must be collected with special care because microbes, especially staphylococci, are present on the surface of the skin and within its superficial layers. A blood culture is taken by venipuncture (Fig. 10-1) as follows:

1. Paint the skin over the veins in the bend of the elbow with tincture of iodine or other suitable disinfectant (70% alcohol may be used). NOTE: Some persons are allergic to iodine.
2. Remove disinfectant with 70% alcohol (or make a second application of 70% alcohol).
3. Place a 70% alcohol compress on the area.
4. Secure a tourniquet, not too tightly, around the arm just above the elbow, and ask the patient to close and open his hand three or four times.
5. Puncture a prominent vein with a 20- or 21-gauge needle to which may be attached a 20 ml. syringe or Vacutainer, a specially designed holder for a vacuum tube (Fig. 10-1). Remove 10 to 15 ml. of blood.
6. Add the blood directly to culture media or place it in a sterile bottle containing sodium citrate to prevent coagulation and carry to the laboratory, where it is distributed to suitable culture media. To prevent contamination, remove the needle from the syringe before expelling the blood.

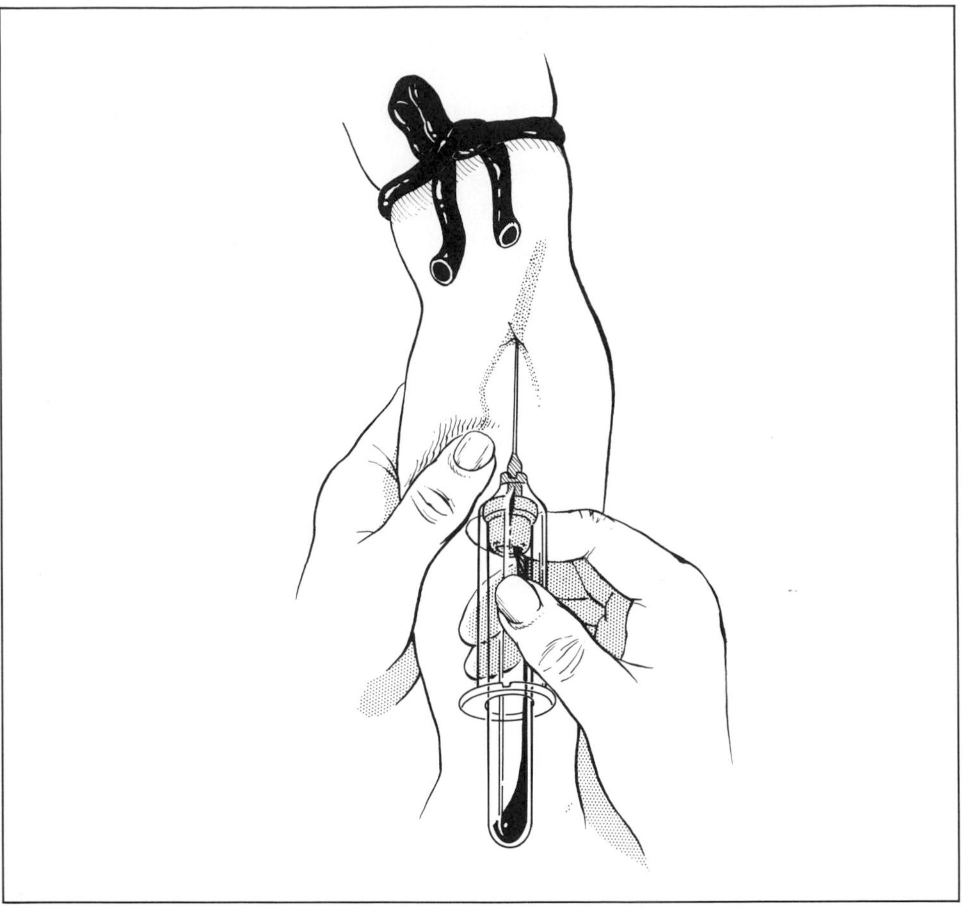

FIG. 10-1. Collection of blood sample by venipuncture. Note use of evacuated blood collection tube in plastic holder. (Syringe can also be used.)

Bacteremia is often of short duration. If the blood culture is not collected at the proper time, the causative organism of the disease may not be found. Since the advent of the antimicrobial drugs, positive blood cultures are fewer. Even one dose of an antimicrobial drug before the culture is taken may mask the infection. The bloodstream in health is free of microorganisms.

Blood for serologic examinations (such as the precipitation tests for syphilis and various agglutination tests for the continued fevers) may be collected from a vein in the bend of the elbow. The technic is the same as that for collecting blood for culture, except that (1) the preliminary sterilization consists of scrubbing the area with 70% alcohol only and (2) the blood is placed in a chemically clean test tube and allowed to clot. Five milliliters are usually sufficient.

The syringe as well as the needle used should be sterilized by boiling or autoclaving. Hepatitis can be conveyed by a clean but *unsterile* syringe to the person from whom the blood is being removed. Sterile *disposable* needles and syringes are used today to prevent hepatitis.

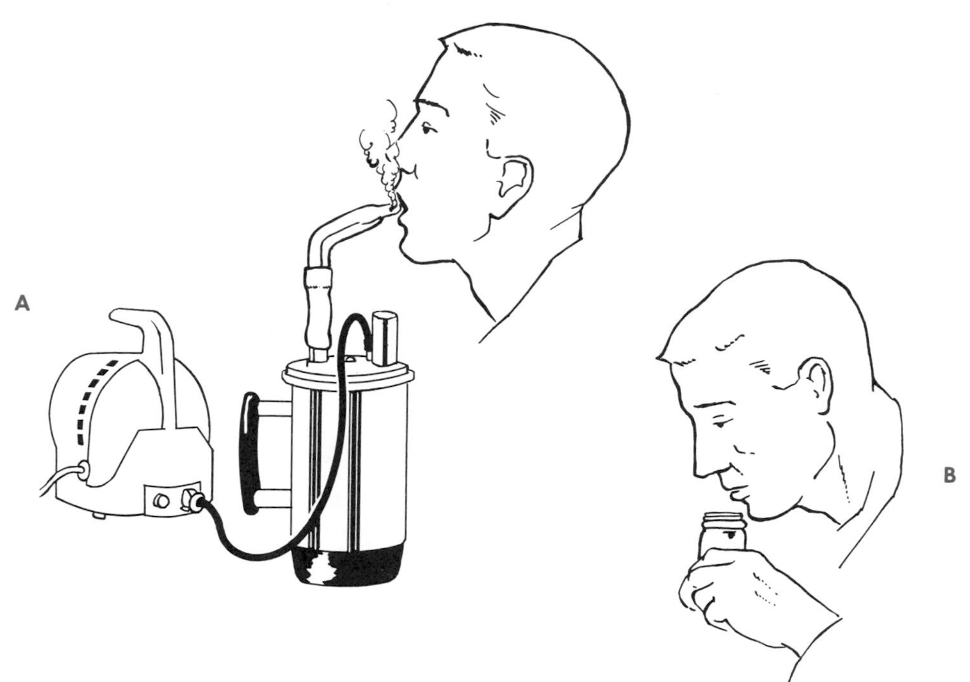

FIG. 10-2. Collection of sputum. Note aerosol method of promoting sputum in **A.** In **B,** sputum is received into widemouthed container. (From O'Donnell, W. E.; Day, E., and Venet, L.: Early detection and diagnosis of cancer, St. Louis, 1962, The C. V. Mosby Co.)

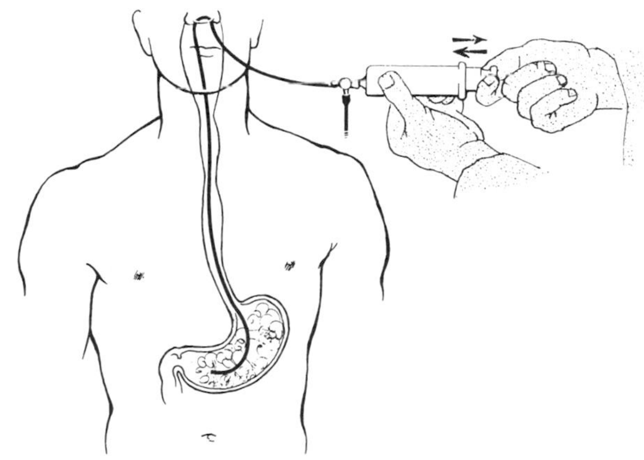

FIG. 10-3. Aspiration of gastric contents. Note nasal tube passed into stomach. This is an important way of collecting sputum (swallowed) in children, and adults as well. (From O'Donnell, W. E.; Day, E.; and Venet, L.: Early detection and diagnosis of cancer, St. Louis, 1962, The C. V. Mosby Co.)

Sputum is often collected improperly, and samples tend to be too small. Many specimens consist of secretions from the nose, mouth, and throat and contain no sputum at all. The sputum should represent a true pulmonary secretion and should be expelled after deep coughing. It is worthwhile to instruct the patient as to how to bring the specimen up from the lower part of the respiratory tract. If necessary nebulized aerosols can be used (Fig. 10-2). The teeth should be thoroughly scrubbed with toothpaste and the mouth rinsed with sterile water before the specimen is taken. Usually more sputum is raised in the morning, and this sputum may contain *Mycobacterium tuberculosis*, whereas specimens taken later in the day do not. Generally specimens for bacteriologic study are collected daily for at least 3 separate days. The ideal container for sputum is a 6-ounce widemouthed bottle (or plastic receptacle) with a tight-fitting stopper or cover.

Children swallow their sputum, and adults do also in their sleep. In children the only way to collect sputum is to aspirate stomach contents (Fig. 10-3). When it is impossible to obtain a satisfactory cough specimen of sputum in adults, the stomach contents may be examined. The stomach should be aspirated early in the morning before any food or water is taken.

Bronchial secretions are secured by means of the bronchoscope and may be subjected to microbiologic study. After bronchoscopy, sputum specimens are collected again for a 3-day period.

Cultures from the nose and throat are made many times since sore throat and nasal discharge are common disorders and frequently part of systemic illness. To take the culture just after an antiseptic has been used is to defeat the purpose of the examination because the antiseptic retards the growth of bacteria present. (This applies to other cultures as well as to throat cultures.) When a throat culture is taken, use a good light, and insofar as possible allow the swab to contact only the diseased area. At least two swabs are required. When cultures are to be taken from the nose, pass a small tightly wound swab directly back through the nose. Avoid large loose swabs because they may slip off the applicator and lodge in the nose.

In cases of suspected diphtheria, cultures from the nose and throat are of special importance. More cases of diphtheria would be detected if cultures were made from both nose and throat instead of from the throat alone. The organisms causing diphtheria are often found in abundance in cultures after they have not been found in direct smears. Although this is true, smears should always be made when diphtheria-like lesions of the throat are encountered, because the exudate in lesions of Vincent's angina, the causative organisms of which are detected only in smears, often bears a definite resemblance to the membrane of diphtheria.

To detect meningococcal carriers, *cultures from the nasopharynx* (the upper portion of the throat back of the soft palate) are made. To make a swab, wrap cotton around the end of a piece of wire bent at a right angle about 1 inch from the end. When specimens from the throat and nasopharynx are to be obtained, take care to avoid contaminating the swabs with saliva.

Feces for ordinary microbiologic examination may be collected directly by the patient into a sterilized cardboard carton or other suitable container. At times it may be desirable for the patient to pass a stool into a larger, previously sterilized container, with a sterile tongue blade or spoon being used to remove a small amount to a sterile widemouthed bottle. The stool in the latter case should be examined superficially. Portions of the fecal material with mucus are preferable for microbiologic study.

Rectal swabs are very satisfactory for most microbiologic purposes and are easily obtained from both adults and children. If a disease process involves the lining of the terminal portion of the rectum, such swabs are likely to obtain material from within the focus of disease and are therefore more likely to contain disease-producing agents than is the stool of that person. Swabs are of value in testing for carriers of bacteria causing typhoid fever, dysentery, or cholera. To obtain the rectal swab, cleanse the skin about the anus with soap, water, and 70% alcohol. Introduce a sterile cotton swab moistened with either a sterile isotonic solution or sterile broth through the anus and gently rotate it about the circumference of the lower rectum to contact a large portion of the lining mucous membrane. A fecal specimen may also be obtained from the physician's glove after digital examination of the rectum has been made.

In all cases the stool specimen should be sent to the laboratory at once because nonpathogenic intestinal bacteria may quickly overgrow the pathogenic organisms, and pathogenic amebas rapidly lose their motility if the fecal material containing them cools. In cases of suspected acute amebiasis, the warm feces should be examined immediately for pathogenic amebas showing their typical and diagnostic movement. If some delay is anticipated, receive the specimen into a vessel that has been warmed. A portion may be then placed in a small tightly corked bottle. This small bottle may be placed in a fruit jar that has been filled with water just a few degrees above body temperature. This will keep the specimen warm for a considerable time so that it can be speeded to the nearest laboratory. In the hospital the specimen may be sent to the laboratory at once, without this special preparation. Examination for typhoid bacilli is facilitated by collecting the feces in a special brilliant green bile medium because bile facilitates the growth of typhoid bacilli, and brilliant green retards the growth of many other intestinal organisms.

Pus from abscesses and boils may be obtained after drainage. The abscess or boil is painted with suitable disinfectant, the area is allowed to dry, and an incision is made with a sterile scalpel. Some of the contents are obtained by means of sterile swabs. If the lesion is opened widely, as much of the superficial portion should be removed as is possible and the specimen taken from the deeper part. When this is not done, the specimen is usually contaminated with surface microorganisms. Protect all specimens taken on swabs from drying. This is best done by placing the swabs in a sterile test tube containing either a drop or two of sterile physiologic salt solution or a nutrient broth that may be directly inoculated with the swabs. The swabs, which are longer than the test tube, should be placed in the test tube in such a manner that the cotton pledget is just above the salt solution but does not touch it. The cotton plug is inserted to hold the swabs in place.

Smears for gonococci (see also p. 313) in the female should be taken from the meatus of the urethra, the cervix uteri, and the rectum. In acute gonorrhea in the male the smears are obtained from the urethra. In chronic cases in the male a specimen may be obtained from the physician's massage of the prostate gland and seminal vesicles. In little girls suspected of having gonorrheal vulvovaginitis, the smears and cultures are made from the vagina because it is the vagina that is primarily attacked.

Cerebrospinal fluid is obtained by a procedure known as *lumbar puncture* (Fig. 10-4). It is carried out by the physician as follows: After the overlying skin of the lower back has been disinfected and anesthetized, a special needle with stylet is introduced slightly to one side of the midline and between the third and fourth lumbar vertebrae and passed into the spinal subarachnoid space. The strictest asepsis must be observed and a sterile dressing placed over the site of puncture. If an infant or child is seated with head and arms resting on a folded sheet, blanket, or pillow set against the abdomen, he may be held securely while the physician performs the lumbar puncture.

Cerebrospinal fluid may be obtained from the cisterna magna at the base of the brain by the skilled neurosurgeon. However, *cisternal puncture* is infrequent because

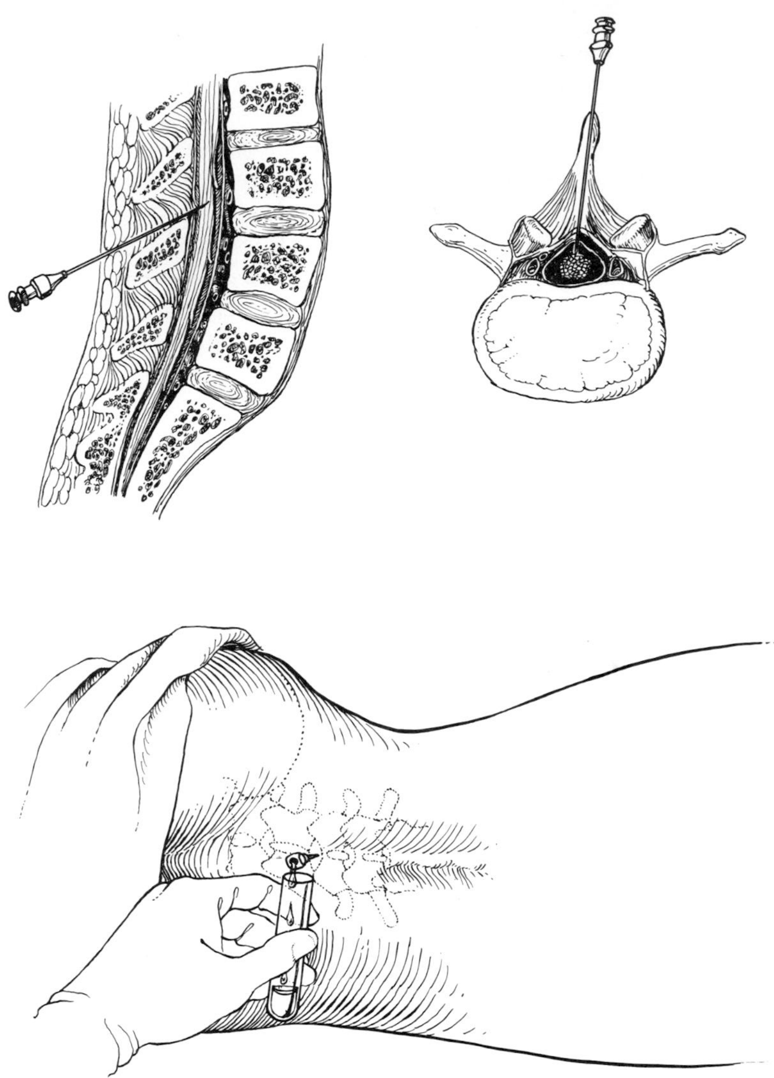

FIG. 10-4. Collection of cerebrospinal fluid by lumbar puncture. Note spinal needle in place in lower back. Anatomic inserts indicate position of needle in subarachnoid space.

of the danger at this location that the penetrating needle might impale the brainstem or cervical cord. The cerebrospinal fluid is normally sterile.

Pleural and peritoneal fluids are obtained through the lumen of a trochar or specially designed needle inserted through the wall of the chest or abdomen. With the trochar in place, fluid is aspirated with a syringe or freely drained into a sterile container. The strictest asepsis should be observed during the procedure, which is done by the physician.

Smears and cultures from the conjunctiva are made with swabs wet with sterile physiologic solution. The swabs should be handled carefully to prevent injury and spread of infection to adjacent parts of the eye or to the other eye.

TABLE 10-1. THE SPECIMEN AND ITS PATHOGENS

Specimen	Important pathogens
Blood	*Staphylococcus aureus* *Streptococcus pyogenes* and other streptococci *Diplococcus pneumoniae* *Neisseria meningitidis* *Neisseria gonorrhoeae* *Haemophilus influenzae* *Brucella* *Pasteurella tularensis* *Spirillum minus* *Streptobacillus moniliformis* *Leptospira* *Escherichia coli* and other coliforms *Salmonella* species *Proteus* *Pseudomonas* Hepatitis viruses Rubeola and rubella viruses Psittacosis bedsonia *Coxiella burnetii* Rickettsiae of spotted fevers
Urine	*Staphylococcus aureus* *Streptococcus pyogenes* and other streptococci *Escherichia coli* and other coliforms *Proteus* *Pseudomonas aeruginosa* *Candida albicans* *Mycobacterium tuberculosis* *Salmonella* and *Shigella* Measles viruses Mumps virus
Nose and throat secretions	*Streptococcus pyogenes* and other streptococci *Neisseria meningitidis* *Corynebacterium diphtheriae* *Staphylococcus aureus* *Bordetella pertussis* Coliforms Predominance of *Haemophilus influenzae, Staphylococcus aureus,* or *Candida albicans* Adenoviruses Enteroviruses Myxoviruses Mumps virus (saliva) Respiratory viruses Psittacosis bedsonia Measles viruses *Mycoplasma pneumoniae*

Continued.

TABLE 10-1. THE SPECIMEN AND ITS PATHOGENS—cont'd

Specimen	Important pathogens
Sputum	*Mycobacterium tuberculosis* *Coccidioides immitis* *Cryptococcus neoformans* *Diplococcus pneumoniae* *Streptococcus pyogenes* *Staphylococcus aureus* *Klebsiella pneumoniae* Anaerobic streptococci *Nocardia asteroides* *Actinomyces israelii* *Candida albicans* *Blastomyces dermatitidis* *Histoplasma capsulatum*
Stool (feces)	*Staphylococcus aureus* *Salmonella typhi* *Salmonella paratyphi* *Salmonella* species *Shigella dysenteriae* *Shigella sonnei* *Shigella flexneri* *Shigella* species *Candida albicans* *Vibrio cholerae* *Proteus* *Pseudomonas* Enteroviruses Hepatitis virus Protozoa (*Entamoeba histolytica*, etc.) Metazoa (*Taenia*, etc.)
Cerebrospinal fluid	*Neisseria meningitidis* *Mycobacterium tuberculosis* *Diplococcus pneumoniae* *Streptococcus pyogenes* *Staphylococcus aureus* *Haemophilus influenzae*, type b *Cryptococcus neoformans* *Listeria monocytogenes* Coliforms Enteroviruses Arboviruses
Fluid from conjunctiva of eye	*Staphylococcus aureus* *Streptococcus pyogenes* and other streptococci *Diplococcus pneumoniae* *Haemophilus* *Neisseria gonorrhoeae* *Moraxella lacunata* and *liquefaciens* Adenoviruses Trachoma bedsonia

TABLE 10-1. THE SPECIMEN AND ITS PATHOGENS—cont'd

Specimen	Important pathogens
Pleural fluid	*Mycobacterium tuberculosis* *Diplococcus pneumoniae* *Staphylococcus aureus* *Streptococcus pyogenes* and other streptococci *Haemophilus influenzae*
Peritoneal fluid	*Mycobacterium tuberculosis* Coliforms Enterococci
Pus, exudates, wound drainages, etc.	*Staphylococcus aureus* *Streptococcus pyogenes* and other streptococci Coliforms Enterococci *Proteus* species *Pseudomonas* species *Mycobacterium tuberculosis* *Pasteurella tularensis* *Corynebacterium diphtheriae* *Bacillus anthracis* *Clostridium tetani* *Clostridium* species *Actinomyces* *Nocardia* *Blastomyces* and other systemic fungi

Specimens for the viral and rickettsial laboratory, such as blood, cerebrospinal fluid, respiratory secretions, feces, and urine, should be collected with the strictest aseptic technic and placed in sterile containers. Further contamination of throat swabs, nasal washings, stools, and specimens already containing bacteria is to be avoided. Swabs may be received into tubes holding a small amount of sterile nutrient broth.

Rush all specimens to the diagnostic laboratory immediately after collection. Often this means that they must be transported a distance to a medical center, a large hospital, or the laboratory of a state health department. When this is the case, freeze the specimen (with the exception of blood) immediately and keep it so until shipment can be made. Refrigerate blood samples and allow to clot. It is desirable to separate serum from the clot. Transit is best made with specimen container packed in dry ice and posted airmail special delivery. A frozen shipment is likely to be a satisfactory one provided that the dry or wet ice has not thawed out by the time the destination is reached. (Small dry-ice packages are likely to thaw in 8 to 12 hours, or less.) To send blood samples on dry ice, place the clot and serum in separate tubes. Although cold shipment of fresh material is preferable most of the time, specimens may be preserved in sterile 50% glycerin solution (but *never* in formalin).

The paired samples of blood for serologic examinations in viral and rickettsial diseases (p. 407) are to be collected aseptically, the blood allowed to clot, and the serum separated from the clot. The serum is sent at once to the out-of-town laboratory by airmail.

Remember that proper labels on the outside of the package help to prevent accidental infection of laboratory personnel receiving the specimen.

RELATED PATHOGENS

In this chapter we have considered body fluids and anatomic sites in the body from which specimens for microbiologic examination are most commonly taken. Table 10-1 lists certain routine specimens with the important pathogenic microorganisms encountered therein.

QUESTIONS FOR REVIEW

1. Why is careful specimen collection stressed?
2. State the ground rules for the collection of any specimen for microbiologic examination.
3. What precautions should be taken in collecting urine specimens for microbiologic examination? Is it always necessary to catheterize the female?
4. How should the skin be prepared for the collection of a blood culture? Why is this necessary?
5. What vein is usually used as a collection site for most blood specimens?
6. What are the precautions to be considered in collecting sputum specimens? In collecting stool specimens?
7. How would you handle a specimen of pus that is to be cultured?
8. Briefly discuss the collection of cerebrospinal fluid.
9. How are specimens handled that must be shipped to an out-of-town laboratory for diagnosis of viral disease?

REFERENCES FOR UNIT TWO

Alexander, E., Jr.: Lumbar puncture, J.A.M.A. **201:**316, 1967.

Andriole, V. T., Kunin, C. M., Stamy, T. A., and Martin, C. M.: Preventing catheter-induced urinary tract infections, Hosp. Pract. **3:**61, (Feb.) 1968.

Bailey, W. R., and Scott, E. G.: Diagnostic microbiology, St. Louis, 1970, The C. V. Mosby Co.

Bartlett, D. I.: The survival of bacteria on swabs, Lab. Med. **1:**32, (July) 1970.

Blair, J. E., Lennette, E. H., and Truant, J. P., editors: Manual of clinical microbiology, Bethesda, Md., 1970, American Society for Microbiology.

Bologna, C. V.: Understanding laboratory medicine, St. Louis, 1971, The C. V. Mosby Co.

Chitwood, L. A.: Isolation and identification of Group A streptococci with commercially available agents, Lab. Med. **2:**26, (Jan.) 1971.

Cole, M.: Pitfalls in cerebrospinal fluid examination, Hosp. Pract. **4:**47, (July) 1969.

Duncan, I. B. R.: Virology in a general hospital, Canad. Med. Ass. J. **98:**1050, 1968.

Finegold, S. M., White, M. L., and Ziment, I., et al.: Rapid diagnosis of bacteremia, Appl. Microbiol. **18:**458, (Sept.) 1969.

Frankel, S., Reitman, S., and Sonnenwirth, A. C., editors: Gradwohl's clinical laboratory methods and diagnosis, St. Louis, 1970, The C. V. Mosby Co.

Graber, C. D.: Rapid diagnostic methods in medical microbiology, Baltimore, Md., 1970, The Williams & Wilkins Co.

Harris, P. C. F.: Show and tell in microbiology, Med. Lab. Observer **4:**48, (Jan.-Feb.) 1972.

Hartman, P. A.: Miniaturized microbiological methods, In Umbreit, W. W., editor: Advances in applied microbiology, supp. 1, New York, 1968, Academic Press Inc.

Heinzl, D.: Evaluation of the use of Clostrisel agar, Lab. Med. **2:**36, (Oct.) 1971.

Herrmann, E. C., Jr.: Experience in providing a viral diagnostic laboratory compatible with medical practice, Mayo Clin. Proc. **42:**112, 1967.

Horowitz, N. H., Cameron, R. E., and Hubbard, J. S.: Microbiology of the dry valleys of Antarctica, Science **176:**242, 1972.

Jawetz, E., Melnick, J. L., and Adelberg, E. A.: Review of medical microbiology, Los Altos, Calif., 1972, Lange Medical Publications.

Johnson, N. E.: Coping with complications of IV's, Nursing '72 **2:**5, (Feb.) 1972.

Knights, E. M.: Commercially available products for screening tests, Bull. Pathol. **10:**411, 1969.

Langford, T. L.: Nursing problem: Bacteriuria and the indwelling catheter, Amer. J. Nurs. **72:**113, 1972.

Lee, L. W.: Elementary principles of laboratory instruments, St. Louis, 1970, The C. V. Mosby Co.

Lee, S., Schoen, I., and Malkin, A.: Comparison of use of alcohol with that of iodine for skin antisepsis in obtaining blood cultures, Amer. J. Clin. Path. **47:**646, 1967.

Lowis, M. J.: An identification key for some aerobic bacteria, Lab. Pract. **20:**331, 1971.

Mason, K. N.: How to isolate *Nocardia asteroides,* Lab. Med. **1:**37, (Jan.) 1970.

McMennamin, A. M., and Wenk, R. E.: Use of selective media in the dip-slide method of estimating colony counts, Lab. Med. **2:**51, (Jan.) 1971.

Motzkin, D.: Office urinary tract bacteriology, Springfield, Ill, 1970, Charles C Thomas, Publisher.

Payne, J. E., and Kaplan, H. M.: Alternative techniques for venipuncture, Amer. J. Nurs. **72:**702, 1972.

Pike, R. M., and Sulkin, S. E.: The laboratory as a source of infection, Bull. Coll. Amer. Path. **22:**229, 1968.

Portnoy, B.: Diagnosis of viral disease in the clinical laboratory, Lab. Med. **1:**38, (Sept.) 1970.

Rosenzweig, A. L.: Venipuncture technique—a reappraisal, Lab. Med. **1:**36, (Sept.) 1970.

Rosner, R. F.: Identification of enteric bacteria directly from primary isolation media using reagent-impregnated paper strips, Amer. J. Clin. Path. **54:**587, 1970.

Rowan, D. F., Helgeson, N. G. P., Race, G. J., and McCraw, M. F.: The role of a virus laboratory as an aid in the management of rubella infection and vaccination, Dallas Med. J. **55:**452, 1969.

Russell, R. L., and Reynolds, J. W.: Quality control of microbiology, Bull. Path. **10:**65, (March) 1969.

Smith, H.: Biochemical challenge of microbial pathogenicity, Bact. Rev. **32:**164, 1968.

Stier, A. R., and Miller, L. K.: Reagent—water specifications and methods of quality control, Pathologist **26:**41, 1972.

Swartz, J. H., and Medrek, T. F.: Rapid contrast stain as diagnostic aid for fungous infections, Arch. Derm. **99:**494, 1969.

Taylor, W. I.: Rapid detection of bacteria in urine, Lab. Med. **2:**29, (Sept.) 1971.

Thompson, C.: Blood drawing and your patient's emotions, Med. Lab. Observer **3:**51, (March-April) 1971.

Tilton, R. C.: Methodologic advances in clinical microbiology, Lab. Med. **1:**54, (Feb.) 1970.

Ungvarski, P.: Mechanical stimulation of coughing, Amer. J. Nurs. **71:**2358, 1971.

Zabransky, R. J.: Isolation of anaerobic bacteria from clinical specimens, Mayo Clinic Proc. **45:**256, 1970.

Zager, J., Kalmanson, G. M., and Guze, L. B.: Dip stick determination of bacteriuria, Amer. J. Med. Sci. **258:**214, 1969.

LABORATORY SURVEY OF UNIT TWO

PROJECT
Distribution of bacteria in our environment

Part A — Demonstration by the instructor with discussion

1. *The Petri dish*—what it is, how it is used, how it is manipulated with sterile technic, how a culture medium is poured into the dish, why it is inverted in the incubator
2. *The microbiologic incubator*—its mechanism, how it is used

Part B — Student participation — sampling the bacterial population

Note: It is suggested that for the following exercise student work in groups of three or six. Each group is furnished six Petri dishes containing sterile nutrient agar for the manipulations of the exercise.

1. *Bacteria in the air*
   a. Open one Petri dish and expose the surface of the nutrient agar to the air in the laboratory for 5 minutes. (Exposure is made by removing the lid from the dish, but the lid is not inverted.)
   b. Replace the lid and put the dish, bottom up, in the incubator for 24 to 48 hours.
2. *Bacteria at specific sites*
   a. Select other areas of the laboratory building, for example, table tops, door knobs, cabinet handles, the floor, water from the faucet, the walls, or the window sill. Expose the surface of the nutrient agar in a given Petri dish to a given site for 5 minutes.
   b. Cover the dish and incubate.
3. *Study of colonial morphology*
   a. After incubation, observe the growth of bacterial colonies on the Petri dishes inoculated as indicated above. Wherever a bacterium comes in contact with the surface of the medium, a colony will develop.
   b. Note the number of colonies on each plate. Do they differ in size, shape, and color? Are the edges smooth or irregular? What does this difference indicate?

PROJECT
Conditions affecting growth of bacteria

1. *Influence of temperature on bacterial growth*
   a. The instructor will furnish you three broth cultures just inoculated with bacteria. Note that all are clear because bacterial growth has not occurred yet.

b. Place one in the refrigerator, another at room temperature, and another in the microbiologic incubator.
c. Observe at end of 24 hours.
d. Record results. In which tube is the growth, as indicated by cloudiness, most abundant? What does this prove?

	Refrigerator temperature 4° C. (24 hr.)	*Room temperature 20° C. (24 hr.)*	*Incubator temperature 37° C. (24 hr.)*
Bacterial growth			

2. *Effect of reaction of medium (pH) on bacterial growth*
   a. The instructor will furnish you five recently inoculated tubes of broth. Make additions of alkali and of acid to four of them as follows: (a) 0.2 ml. N/1 NaOH, (b) 0.5 ml. N/1 NaOH, (c) 0.2 ml. N/1 HCl, and (d) 0.5 ml. N/1 HCl. The fifth serves as a control.
   b. Incubate at 37° C. for 24 hours. Compare the multiplication of bacteria in the different tubes as indicated by clouding of the medium.
   c. Record results.

	Alkaline pH		*Acid pH*		
	0.2 ml. N/1 NaOH	*0.5 ml. N/1 NaOH*	*0.2 ml. N/1 HCl*	*0.5 ml. N/1 HCl*	*Control*
Bacterial growth					

3. *Effects of sunlight on bacterial growth*
   a. The instructor will furnish you a recently inoculated Petri dish. Invert and cover one half of the Petri dish in such a manner that direct sunlight is not allowed to strike the medium. A piece of black paper may be used.
   b. Expose to sunlight for several hours.
   c. Incubate at room temperature until colonies appear. On which half of the plate are the colonies more numerous? What does this prove?

PROJECT
Special activities of bacteria
Part A—Fermentation

1. *Fermentation of sugars*
   a. The instructor will furnish each group of students three Smith fermentation tubes containing sterile sugar solutions. The first contains 10% dextrose (grape sugar) solution; the second contains 10% lactose (milk sugar) solution; and the third contains a 10% solution of sucrose (cane sugar).
   b. Rub up a small piece of yeast cake in water and add a portion to each tube.
   c. Set in a warm place and observe at the end of 24 hours.

d. Chart. What differences are noted? Why?

	Dextrose 10% (grape sugar)	Lactose 10% (milk sugar)	Sucrose 10% (cane sugar)
Changes observed after inoculation with yeast			

2. *Proof that gas formed in no. 1 is carbon dioxide*
   a. Measure the length of the gas column in the arm of the fermentation tube.
   b. Remove a small amount of material from the bulb of the tube with a pipette and reserve for no. 3.
   c. Fill bulb of tube with 10% potassium hydroxide (KOH).
   d. Place thumb over mouth of tube and mix by inverting so that gas in arm of tube comes in contact with KOH.
   e. Collect all remaining gas in arm of tube. Reduction in amount of gas proves CO_2 was present and was absorbed according to the following chemical equation: $2KOH + CO_2 = K_2CO_3 + H_2O$.
3. *Proof that the fermentation in no. 1 produced alcohol*
   a. Filter the liquid removed in no. 2.
   b. To the filtrate add a few drops of weak iodine solution.
   c. Add enough 10% sodium hydroxide solution to change the color from a brown to a distinct yellow.
   d. Warm slightly. A distinct odor of iodoform will be obtained, and a yellow precipitate, appearing under the microscope as small hexagonal crystals, will be formed. If the amount of alcohol is small, the crystals may not appear until the solution has cooled or stood some time.
4. *Fermentation with acid and gas formation*
   a. The instructor will furnish you a phenol red dextrose broth fermentation tube recently inoculated with *Escherichia coli.*
   b. Incubate 24 to 48 hours.
   c. Note multiplication of bacteria as indicated by turbidity. What other two changes have taken place? What does each indicate? What is the purpose of the phenol red?
5. *Fermentation causing reduction of pigment*
   a. Nearly fill a long narrow test tube with milk.
   b. Add enough weak aqueous solution of methylene blue to give the milk a distinct blue color.
   c. Add a little milk that has soured naturally.
   d. Mix and incubate at 37° C. for several days.
   e. Observe. What change has taken place?

Part B—Pigment production by bacteria

1. The instructor will give you three agar slants. One has been inoculated with *Pseudomonas aeruginosa,* one with *Serratia marcescens,* and one with *Staphylococcus aureus.*

a. Incubate at 37° C. until pigment production is well developed.
b. Observe. What color is produced by each? Is the color confined to the growth of bacteria or does it diffuse into the medium?

2. Examine cultures of *Staphylococcus albus, Staphylococcus citreus,* and *Staphylococcus aureus.*
   a. Observe. What color is produced by each?
   b. Record colors.

	Pseudomonas aeruginosa	*Serratia marcescens*	*Staphylococcus aureus*	*Staphylococcus citreus*	*Staphylococcus albus*
Color produced					

PROJECT

Preparation of culture media

Part A—Demonstration of bacteriologic technics by the instructor with discussion

1. Flaming the mouth of a container to preserve sterility of contents
2. Holding tubes of culture media in a slanting position during inoculations
3. Removing and inserting cotton plugs or other stoppers during inoculations
4. Transferring bacterial growth
   a. Broth-to-agar slant cultures
   b. Agar slant-to-broth cultures
5. Pouring melted agar into a Petri dish

Part B—Preparation of nutrient broth

The composition of 1 liter of nutrient broth is as follows:

Meat extract	3 grams
Peptone	5 grams
Sodium chloride (may be omitted)	5 grams
Distilled water	1 liter

Commercially, one can obtain Difco (or equivalent product)—a dehydrated medium that contains the meat extract, peptone, and buffer substance in such proportions that the finished product has a pH of 6.8. It is not necessary to adjust the pH of this product. Using the commercial dehydrated medium to make nutrient broth, proceed as follows:

a. Dissolve 4 gm. of dehydrated nutrient broth (containing 1.5 gm. of meat extract and 2.5 gm. peptone) in 500 ml. of distilled water contained in a liter Erlenmeyer flask. Add 2.5 gm. of sodium chloride if desired. (For the subsequent experiments, 500 ml. amounts of nutrient broth will be sufficient.)
b. After solution is complete, pour 10 to 15 ml. amounts in each of 25 culture tubes.
c. Plug tubes with cotton plugs and autoclave at 121° C. for 15 minutes. During the process of plugging and when handling the culture medium, do not let the liquid come in contact with the cotton plugs. Under no condition can the pressure of the autoclave be rapidly released when steam pressure

sterilization is finished because rapid release of pressure causes the liquid in the tubes to boil over and wet the plugs. What are some of the objections to wetting of cotton plugs?

Part C — Preparation of nutrient agar

The composition of 1 liter of nutrient agar is as follows:

Meat extract	3 grams
Peptone	5 grams
Sodium chloride (may be omitted)	5 grams
Agar	15 grams
Distilled water	1 liter

The Difco dehydrated medium (or equivalent commercial product) contains the meat extract, peptone, and agar in a dehydrated condition buffered so that the final pH of the medium is 6.8. No further adjustment of pH is necessary.

To make nutrient agar, proceed as follows:

a. To 500 ml. of cold distilled water contained in a liter flask, add 11.5 grams dehydrated nutrient agar (Difco).

b. Place in pan of water or on boiling water bath and heat until solution is effected. The material may be boiled over a free flame, but there is great danger of scorching.

c. In order to have sufficient culture medium for the experiments to follow in this and the next unit, put up the following:

(1) Tubes containing 5 ml. of agar (these are slanted to solidify after sterilization)	6
(2) Test tubes containing 25 ml. agar	6
(3) Erlenmeyer flasks (150 ml.) containing about 89 ml. agar	4

d. Plug containers with cotton plugs and autoclave at 121° C. for 15 minutes.

e. After pressure in autoclave has returned to normal, slant agar tubes to solidify as previously indicated.

PROJECT
Cultivation of bacteria

1. *Taking a throat culture*

   a. Demonstration by the instructor with discussion

   The instructor will show how a sterile cotton swab can be used to take a throat culture. The swab, after it has been rubbed over the back part of the throat of a given individual, is then rubbed over the surface of a suitable culture medium. No attempt in this exercise is made to separate bacterial organisms. Media such as Löffler's serum medium, blood agar, or nutrient agar may be used.

   b. Student participation

   (1) Using a sterile swab, take a throat culture. (Students may work in pairs.) Apply swab to the surface of a suitable culture medium.

   (2) Incubate cultures and observe at the end of 24 hours.

   (3) Make gram-stained smears of the bacterial growth. Make drawings to show representative microorganisms.

2. *Cultural characteristics of bacteria*
   a. Demonstration by the instructor with discussion
      (1) Demonstration of bacterial growth in tubes of solid media and on plates (Petri dishes).
      (2) Demonstration of the differences in size, shape, and contour of bacterial colonies—enumeration of the various cultural characteristics. Use descriptive terms for the identification of bacterial colonies. (See Fig. 7-5, p. 90.)
      (3) Demonstration of the difference in ability of different organisms to grow on various laboratory media.
   b. Student participation
      (1) Obtain a culture (tube or plate) showing growth of a nonpathogenic organism.
      (2) Observe a demonstration by the instructor of the method of transferring microorganisms from one kind of culture medium to another.
      (3) Make the transfer of organisms from your culture to a different kind of culture medium.
      (4) Incubate the culture for 24 to 48 hours. Observe and describe growth.
3. *Anaerobic cultures*
   Demonstration by instructor with discussion—setting up anaerobic cultures

---

PROJECT
Isolation of pure cultures

1. *Pour plate method*

Note: The students, working in small groups, will use tubes of culture media that they have prepared, sterile Petri dishes, and a mixed culture furnished by the instructor.

   a. Refer to the pour plate method as outlined on p. 91.
   b. Use this method to obtain pure cultures of organisms in the mixed culture provided to you. Incubate the Petri dish cultures for 24 to 48 hours.

2. *Streak plate method*

Note: The students are each given three Petri dishes of sterile culture medium, and mixed cultures are laid out in the laboratory.

   a. Refer to the paragraph on streak plates on p. 93 of the text and to the illustration of the method in Fig. 7-11. Consult the instructor for assistance when you carry out this technic for the first time.
   b. Make a streak plate using one of the mixed cultures available in the laboratory. Incubate for 24 to 48 hours.
   c. Observe colonial growth as obtained by the two methods of isolation. How do the distributions of the colonies on the streak and pour plates differ?
   d. Note the importance of making gram-stained smears for microscopic study of different types of colonies obtained. How can a gram-stained smear be used to check a pure culture? Make gram-stained smears for representative colonies that you have obtained from your cultures.

3. *Fishing of colonies*

Demonstration by the instructor with discussion

The instructor should demonstrate how given colonies are transferred from Petri dishes to agar slants. Why is it often necessary to do this?

PROJECT
Bacterial plate count

1. Work in small groups of two to four.
2. Obtain necessary materials for this experiment as follows:
   a. A small amount of milk
   b. Four tubes containing 9 ml. sterile distilled water
   c. Five sterile 1 ml. pipettes
   d. Four sterile Petri dishes
3. Place a flask of agar in a pan of water and heat water until agar melts. Let cool to about 40° C. This may be determined roughly by considering a temperature of 40° C. as that temperature at which the flask may comfortably be held against the back of the hand. Keep at this temperature.
4. Proceed as follows:
   a. Transfer 1 ml. of milk to a tube of 9 ml. sterile distilled water and mix. With a fresh pipette transfer 1 ml. of the mixture to a second tube of 9 ml. sterile distilled water. Continue in this manner until the fourth tube is mixed, being careful to use a fresh sterile pipette to make each mixture. You now have 1:10, 1:100, 1:1000, and 1:10,000 dilutions of milk.
   b. With a sterile pipette transfer 1 ml. of each dilution to a Petri dish. Begin with the highest dilution and go to the lowest when making transfers. The material is transferred to the Petri dishes by slightly raising the lid and depositing the material in the middle of the plate. Gently raise the lid and add about 15 ml. of melted agar to each dish. Mix contents of each dish by gently rotating. Let agar solidify; invert and incubate 48 hours.
   c. Select a plate showing well-distributed, easily counted colonies and count the colonies present. Multiply the number of colonies counted by the times the milk in that plate was diluted. The result is the number of bacteria in 1 ml. of the milk used. Why was a new pipette used for making each dilution of the milk? Why can transfers from the tubes to Petri dishes be made with a single pipette by beginning with the highest dilution and proceeding to the lowest?

PROJECT
Animal inoculation

1. Demonstration by instructor with discussion
   a. Technic of animal inoculation
   b. Value of animal inoculation

      The organisms suspected and the results obtained should be explained.
2. Study of preserved organs from autopsies of guinea pigs inoculated with *Mycobacterium tuberculosis* and *Pasteurella tularensis*.

   Compare diseased animal organs with normal ones.

EVALUATION FOR UNIT TWO

Part I

In the following statements or questions, encircle the number in the column on the right that correctly completes the statement or answers the question.

1. Why is refrigeration a good method for preserving bacterial cultures?

(a) Cold is necessary for bacterial growth	1. a
(b) Cold retards bacterial growth	2. c
(c) Freezing may destory bacteria	3. a, c, and e
(d) Cold excludes oxygen	4. b
(e) Cold protects bacteria from sunlight	5. a, b, and d

2. Which of the following bacterial metabolic products are associated with disease production?

(a) Hemolysins	1. a and d
(b) Coagulases	2. a and b
(c) Leukocidins	3. a, b, and c
(d) Kinases	4. c, d, and e
(e) Spreading factor	5. a, b, c, d, and e

3. Bacteria are capable of which of the following variations?

(a) A change from coccus to bacillus	1. all are true
(b) A change from encapsulated to a nonencapsulated organism	2. all but d are true
	3. all but a and d
(c) A change from a nonvirulent to a virulent organism	4. all but a, b, and e
	5. none is true
(d) A change from a spirillum to a spirochete	
(e) A change from an antibiotic-susceptible to an antibiotic-resistant organism	

4. Which of the following organisms exist *only* on living material?

(a) Facultative saprophyte	1. b
(b) Strict parasite	2. b and d
(c) Facultative parasite	3. c
(d) Strict saprophyte	4. a

5. Which terms indicate that two organisms can live together?

(a) Commensalism	1. d
(b) Symbiosis	2. b and e
(c) Antagonism	3. a and b
(d) Positive chemotaxis	4. c
(e) Plasmoptysis	5. a, b, and c

6. A spirochete can best be studied by which of the following preparations?

(a) Gram stain	1. a
(b) Hanging-drop preparation	2. b
(c) Dark-field illumination	3. c
(d) Acid-fast stain	4. d
(e) Culture on an artificial medium	5. e

7. In which groups would you *not* expect to find concentrated the organisms of greatest medical importance?

(a) Autotrophs	1. a
(b) Lithotrophs	2. b
(c) Psychrophiles	3. all of these
(d) Obligate halophiles	4. a, b, c, and d
(e) Thermophiles	5. c and d

8. Strict anaerobes grow well:

(a) Under 5% to 10% increase of carbon dioxide tension	1. a, b, and c
	2. a, b, c, and d
(b) In atmospheric oxygen	3. b, c, d, and e
(c) In broth beneath layer of petroleum jelly	4. c, d, and e
	5. c and d
(d) In thioglycollate broth	
(e) On routine blood agar	

9. The process by which the red cell gives up water to a hypertonic plasma is called:

(a) Hemolysis	1. c
(b) Crenation	2. b
(c) Osmosis	3. b and c
(d) Metabolism	4. a and c
(e) None of the above	5. e

10. Testing the antibiotic susceptibility of a bacterial culture by the sensitivity disk method has:

(a) Become unnecessary since discovery of broad-spectrum drugs	1. a and b
	2. c and d
(b) Little application to subsequent treatment of patient	3. c and e
	4. a, b, c, and d
(c) Been used only with penicillin	5. none of these
(d) Been too time consuming to be practical	
(e) Interfered with routine blood cultures	

11. The bacterial plate count is important in the microbiologic evaluation of urine. Generally the microbes recovered per milliliter of specimen are not considered significant until the count exceeds:

(a) 100 per milliliter of urine	1. d
(b) 1000 per milliliter of urine	2. e
(c) 10,000 per milliliter of urine	3. a
(d) 100,000 per millilter of urine	4. b
(e) 1,000,000 per milliliter of urine	5. c

12. The usual complication that follows urethral catheterization is:

(a) Urinary tract infection	1. a and b
(b) Urinary tract closure	2. a
(c) Rupture of urethra	3. b
(d) Bleeding from the bladder	4. c and d
(e) Patient is unable to void naturally	5. e

Part II

A. Before each principle, place the letter from the column on the right indicating the procedure based on that principle.

PRINCIPLES

_______ 1. Bacteria can be partially classified by their characteristic action on media containing different carbohydrates.

_______ 2. A single species of bacteria can be isolated from infectious material by making a dilute mixture of the medium and the infectious material. Then, when the inoculated medium is allowed to harden in a Petri dish, the colonies will be widely separated.

_______ 3. We can assume that for each colony appearing on a pour plate, there was only one bacterium originally present in the medium.

_______ 4. Oxygen can be removed from an airtight container by bringing about some chemical reaction in the container that uses up the oxygen.

PROCEDURES

(a) Pure culture, pour plate method
(b) Culture of anaerobic bacteria
(c) Pure culture, streak plate method
(d) Bacterial count, plate method
(e) Determination of fermentation reactions

B. Consider the characteristics as listed in the column on the right. Indicate by a check mark in the appropriate column the relation to exotoxin or endotoxin.

Exotoxin	*Endotoxin*	
		Found in gram-positive bacteria
		Found in gram-negative bacteria
		Great toxicity
		Protein
		Lipopolysaccharide complex
		Important in Asiatic cholera
		Weakly antigenic
		Conversion to toxoid

C. Comparisons: Match the item in column B to the word or phrase in column A best describing it.

COLUMN A

Gram-positive with gram-negative bacteria:

_______ 1. Walls often contain teichoic acid
_______ 2. Walls contain mucopeptide
_______ 3. Walls rich in lipid
_______ 4. Usually more susceptible to antibiotics
_______ 5. Survive autoclaving
_______ 6. Usually more resistant to oxidizing agents

COLUMN B

(a) Gram-positive bacteria
(b) Gram-negative bacteria
(c) Both
(d) Neither

	COLUMN A	COLUMN B
	Capsules with flagella:	
________	7. Rare in cocci	(a) Capsules
________	8. Complex polysaccharide	(b) Flagella
________	9. Elastic protein	(c) Both
________	10. Stained by special stains	(d) Neither
________	11. Stained by routine Gram stain	
________	12. Quellung reaction	
________	13. Related to virulence	
	Respiration (aerobic oxidation) with fermentation:	
________	14. Yields more energy per mole of substrate	(a) Aerobic oxidation of glucose (b) Fermentation of glucose
________	15. End products carbon dioxide and water	(c) Both
________	16. End product ammonia	(d) Neither
________	17. End product pyruvate	
________	18. Carried out by facultative organisms	
________	19. Carried out by strict anaerobes	
________	20. Important laboratory application in identification of microbes	
	Differential culture media with selective media:	
________	21. Retard growth of microbes in certain instances	(a) Selective culture media (b) Differential media
________	22. Important in culture of feces	(c) Both
________	23. Petragnani for growth of tubercle bacilli, an example	(d) Neither
________	24. MacConkey agar for growth of gram-negative bacilli, an example	
________	25. Lactose may be incorporated	
________	26. Dye added to indicate fermentation in certain instances	
________	27. Dye added to suppress growth in certain instances	
________	28. Selected embryonic tissues incorporated in certain instances	

D. Match the item in column B to the phrase best defining it in column A.

	COLUMN A	COLUMN B
________	1. Enhanced at times by radiant energy	(a) Adaptation
________	2. Transfer of large segment of DNA	(b) Pleomorphism
________	3. Smooth and rough colonies grown from pure culture	(c) Dissociation (d) Commensalism
________	4. Variation in size, shape, and other features	(e) Attenuation (f) Recombination
________	5. Loss in disease-producing ability of microbe	(g) Transformation (h) Mutation
________	6. Direct transfer of DNA into culture medium	(i) Transduction (j) Conjugation
________	7. Indirect transfer of genetic material by virus	
________	8. Physiologic adjustment to environment	

E. Complete the following statements by filling in the blanks to the left. Be sure that the number of the answer corresponds with that in the statement. Note vocabulary for reference to the right.

1. ________________ Certain bacteria can form a (1) when surrounding conditions are unfavorable for growth.
2. ________________ A visible mass formed by rapid reproduction of an organism on culture media is called a (2).
3. ________________ When the organism sought is found in a culture, we designate the culture a (3) one.
4. ________________ Iodine is used as a (4) in the Gram-staining procedure.
5. ________________ When the presence of two species of microbes in the same environment favors the development of both, the condition is called (5).
6. ________________ A growth of one kind of bacteria is called a (6) culture.
7. ________________ Chromogenic bacteria produce (7).
8. ________________ Microbes without chlorophyl must obtain their energy by (8)
9. ________________ The (9) is an important structural constitutent of cells.
10. ________________ Excessive gas formation in milk produced by *Clostridium perfringens* is (10).
11. ________________ The enzyme (11) is produced by *Neisseria* species.
12. ________________ The (12) test is important in the differentiation of tubercle bacilli from atypical mycobacteria.
13. ________________ If a culture contains two or more kinds of microbes, it is termed a (13) culture.
14. ________________ Culture of media of known and definite chemical composition are (14) media.
15. ________________ The enzyme (15) is responsible for conversion of urea to ammonia.
16. ________________ A nonprotein compound necessary for enzyme action is (16).
17. ________________ The temperature most suitable for growth of a given microbe is termed the (17) one.
18. ________________ Organisms growing best in an amount of oxygen less than that of the air are (18).
19. ________________ The organic catalyst of the body is the (19).
20. ________________ A protein may be composed of several hundred (20) building blocks.
21. ________________ The material acted upon by a given enzyme is termed the (21).
22. ________________ An enzyme released from a given cell into the extracellular medium is called (22).
23. ________________ Microbes utilizing organic matter as source of energy are termed (23).
24. ________________ The process whereby a cell works actively to absorb molecules is (24).
25. ________________ (25) destroys polymorphonuclear neutrophilic leukocytes.
26. ________________ (26) accelerates clotting of blood.

VOCABULARY

Absorption
Active transport
Aerobic
Agar
Amino acid
Amylase
Anaerobic
Antagonism
Biologic oxidation
Chemosynthesis
Chlorophyl
Coagulase
Coenzyme
Colony
Commensalism
Contaminated
Decolorizer
Dye
Endoenzyme
Enzyme
Exoenzyme
Fatty acid
Fermentation
Kinase
Leukocidin
Light
Lipase
Macromolecule
Maximal
Microaerophilic
Mixed
Mordant
Negative
Niacin
Nicotinic acid
Optimal
Organotrophic
Oxidase
Penetration
Pigment
Pili
Positive
Pure
Routine
Spore
Stormy fermentation
Substrate
Symbiosis
Synthetic
Urease

UNIT THREE
MICROBES: production of infection

11 Microbes and disease

12 The body's defense

13 Immunology

14 Methods for microbes: immunologic reactions

15 Allergy

Laboratory survey of unit three

Evaluation for unit three

11 Microbes and disease

INFECTION

When microbes or certain other living agents enter the body of a human being or animal, multiply, and produce a reaction there, this is an *infection*. The host is *infected*, and the disease is an *infectious* one. The reaction of the body may or may not be accompanied by outward signs of disease. Infection is differentiated from *contamination*, which refers to the mere presence of infectious material (no reaction produced). Superficial wounds in skin and mucous membranes are often the site of a large microbial population. As long as these microorganisms do not invade the deeper tissues and induce a reaction there, they are contaminants instead of agents of infection.

THE RESIDENT POPULATION

■ **Microbes normally present.** *The mere presence of microbes in the body does not mean infection because microorganisms normally inhabit many parts of the body without invading the deeper tissues to cause disease.* Certain of the body fluids, such as blood and urine, are normally sterile, that is, free of the presence of microorganisms. In parts of the body other than the circulatory system and urinary tract, the secretions or excretions are normally in contact with a resident population of microorganisms. These microbes are consistently present in varying proportions. This pattern of growth, conspicuously bacterial, is associated with the well-being of the person and in an area such as the intestinal tract is even necessary to his health. These microorganisms constitute the *normal flora* of the body. Table 11-1 gives their normal habitats in the human body.

INVASION OF THE BODY

■ **Source of microbes causing infections.** The microbes that cause infections fall into two classes: (1) those that can cause disease in healthy persons and reach such persons directly or indirectly from animals, persons ill of the disease, or carriers (these are pathogens) and (2) those that enter the host at a time of injury or lowered resistance or when they themselves are increased in virulence (these are ordinarily saprophytes and can be opportunists).

Microbes belonging to the first group cause communicable diseases. To the second

TABLE 11-1. DISTRIBUTION OF THE RESIDENT POPULATION

Anatomic site	Important resident microbes
Skin	Staphylococci Streptococci Coliforms Enterococci Diphtheroids (aerobic and anaerobic) *Proteus* species *Pseudomonas* species *Bacillus* species Fungi (lipophilic)
Respiratory tract	Staphylococci Streptococci Diphtheroids *Neisseria* species *Haemophilus* species Pneumococci Spirochetes Certain amebas Actinomycetes *Candida albicans* and other fungi
Gastrointestinal tract*	Coliforms Enterococci *Clostridium* species *Proteus* species Yeasts *Penicillium* species Enteroviruses *Pseudomonas aeruginosa* Aerobic and anaerobic streptococci Staphylococci Enterococci *Alcaligenes faecalis* *Bacteroides* species *Lactobacillus* species†
Genital tract	*Mycobacterium smegmatis* Streptococci Spirochetes Coliforms Enterococci Lactobacilli Mycoplasmas *Proteus* species Staphylococci

*The stomach is normally free of microbial growth because of its acid content. Microorganisms are much more plentiful in the large than in the small intestine.

†A baby is born with a sterile intestinal tract, but before or with the first feeding, bacteria are introduced. If the child is breast fed, the predominating organism is *Lactobacillus bifidus* (bacteria of the genus *Lactobacillus* convert carbohydrates to lactic acid). If the child is bottle fed, *Lactobacillus acidophilus* predominates. In addition other bacteria are present.

Continued.

TABLE 11-1. DISTRIBUTION OF THE RESIDENT POPULATION—cont'd

Anatomic site	Important resident microbes
Genital tract—cont'd	Diphtheroids Saprophytic yeasts
Vagina	
Before puberty	Streptococci Coliforms Diphtheroids
During childbearing period	Döderlein's bacilli (*Lactobacillus* species)
After childbearing period	Streptococci Coliforms Diphtheroids
Organs of special sense	
Eye	Staphylococci Diphtheroids Pneumococci Alpha streptococci
External ear	Staphylococci and certain cocci Diphtheroids *Bacillus* species *Gaffkya tetragena* Nonpathogenic acid-fast organisms

group belong microorganisms that normally inhabit the body but produce disease only under given conditions and certain ones that are inadvertently introduced into the body by wounds, injuries, and such.

■ **How microbes reach the body.** According to the manner in which the causative agent reaches the body, infectious diseases may be classified as communicable and noncommunicable.

A *communicable disease* is one whose agent is directly or indirectly transmitted from host to host. Examples are typhoid fever and tuberculosis. The host from which the infection is spread is usually of the same species as the recipient, but not necessarily so. For instance, cattle may transmit tuberculosis and undulant fever to man. A *noncommunicable disease* is one whose agent either normally inhabits the body, only occasionally producing disease, or resides outside it, producing disease only when introduced into the body. For example, tetanus bacilli, inhabitants of the soil, produce disease only when introduced into abrasions or wounds. Although *not* communicable, tetanus is infectious. The term "contagious" is applied to diseases that are easily spread directly from person to person.

Infectious diseases are also classified as exogenous and endogenous. *Exogenous* infections are those in which the causative agent reaches the body from the outside and enters through one of the portals of entry. An *endogenous* infection is one from organisms normally present in the body. To repeat, endogenous infections occur when the defensive powers of the host are weakened or, for some reason, the virulence of the microorganism is increased.

■ **How microbes enter the body.** Microorganisms invade the body by several avenues, and each species has its own favored one. The way of access to the body is known as the *portal of entry,* as indicated in the following areas:

SKIN. Most pathogenic organisms do not penetrate the unbroken skin or mucous membrane, but some do. Staphylococci and some of the fungi can penetrate the hair follicles and cause disease in the deeper tissues of the skin. The organisms of tularemia pass through the unbroken skin, as do hookworm larvae. In their penetration of the intact skin, malarial parasites have the help of a biting insect. Many bacteria are found in the superficial layers of the skin but under normal conditions are not able to penetrate deeply. As soon as the superficial barrier is broken, infection is easily accomplished.

RESPIRATORY APPARATUS. Pulmonary tuberculosis, pneumonia, and influenza are contracted by way of the respiratory tract, and the viruses causing measles, smallpox, and German measles enter the body this way.

ALIMENTARY TRACT. Some of the most important pathogenic organisms gain entrance to the body by way of the alimentary tract. In many cases food and drink are the vehicles. Such organisms are typhoid and dysentery bacilli, cholera vibrios, and amebas of dysentery. The great majority of pathogenic microorganisms enter the body via the respiratory system or the digestive tract.

GENITOURINARY SYSTEM. Certain infections are acquired chiefly through the genitourinary system, notably the venereal diseases.

PLACENTA. Many microorganisms do not pass through the placenta, but the spirochete of syphilis and the smallpox virus may be so transmitted.

EVENT OF INFECTION

■ **Factors influencing the occurrence of infection.** The fact that microbes have entered the body in no manner indicates that infection has occurred. Whether infection supervenes is related to (1) the portal of entry, (2) the virulence of the organisms, (3) their number, and (4) the defensive powers of the host (see Chapter 13).

Most pathogenic bacteria have definite portals by which they enter the body, and they can fail to produce disease when introduced into the body by another route. For instance, typhoid bacilli produce typhoid fever when swallowed but only a slight local inflammation when rubbed on the abraded skin. Streptococci rubbed into the skin produce intense inflammation (cellulitis) but are generally without effect if swallowed. If streptococci are breathed into the lungs, they may cause pneumonia. A few bacteria, such as *Pasteurella tularensis,* the cause of tularemia, are able to enter the body by several different routes and produce disease in each area. Usually the route of entry determines the type of disease process induced for a given organism.

By *virulence* or *pathogenicity* one means the ability of microbes to induce disease by overcoming the defensive powers of the host. There is a difference in virulence not only among species but also among members of the same species. As a rule, microbes are most virulent when freshly discharged from the person ill with the disease that they cause. Organisms harbored by carriers are usually of comparatively low virulence. Virulence may be increased by rapid transfer of organisms through a series of susceptible animals. As each animal becomes ill, the organisms are isolated from its excreta and transferred to a well animal. The explanation of epidemics is that the causative organism of the disease has, by repeated passage from person to person, become so virulent that everyone whom it contacts is made ill.

An organism that is highly virulent for one species of animals and less virulent for another may, upon repeated passage through the animal for which it is less virulent, show a *transposal of virulence;* that is, it becomes less virulent for the animal for which it was originally highly virulent and highly virulent for the animal for which it was originally less virulent. Advantage is taken of this trait in the production of antirabic vaccine.

The *number* of microbes is crucial to infection. If only a few enter the body, they are likely to be overcome by the local defenses of the host even though they are highly virulent. Therefore, if infection is to occur, enough microorganisms to overcome the local defenses of the host must penetrate.

■ **How microbes cause disease.** When pathogenic microorganisms enter the body, two opposing forces are set in motion. The organisms strive to invade the tissues and colonize there. The body, utilizing its defensive powers, strives to block the invasion of the microbes, to destroy them, and to cast them off. If the body wins the contest, the microbes are destroyed, and the body suffers no ill effects. It the microbes win, infection occurs.

Microbes cause disease in different ways. In some cases the mechanical effects of microorganisms operate, for example, when the organisms of estivoautumnal malaria occlude the capillaries to the brain.

In most cases, however, biochemical effects are foremost. The ability of pathogenic microorganisms to produce disease is closely tied in with certain complex chemical substances that the organisms either elaborate and release or that are a vital part of their make-up. A soluble exotoxin (p. 102) is an example of a product released into body fluids and even into the bloodstream of the host. A constituent of the bacterial cell wall may be an injurious factor. A part of the microbial cell, such as its capsule, though not harmful in itself, may so protect the microbe as to enhance its virulence. Their carbohydrate (polysaccharide) capsules afford protection from the defensive cells of the lungs to pneumococci invading the lower respiratory tract, thereby promoting the development of a pneumonia.

Biochemical substances implicated in pathogenicity of bacteria and other disease-producing microbes are many, and their action is not always well understood. The major effects of these are summarized as follows:

1. They interfere with mechanical blocks to the spread of infection set up in the body (bacterial kinases).
2. They slow or stop the ingestion of microbes by the phagocytic white blood cells (leukocidins).
3. They destroy body tissues (hemolysins, necrotizing exotoxins, lethal factor).
4. They cause generalized unfavorable reactions in the host, resulting in fever, discomfort, and aching (endotoxins). Such features are usually collected together under the term "toxicity."

Most disease-producing microbes have a preference for certain parts of the body; they have their favored site for involvement. This is *elective localization.* For instance dysentery bacilli attack the intestine, pneumococci attack the lungs, and meningococci localize in the leptomeninges. Toxic products released by microbes also have an affinity for certain anatomic sites; the toxin of tetanus attacks the central nervous system, and the toxin of diphtheria affects not only the central nervous system but also the heart.

■ **Local effects of microbes.** By local effects of microbes one means the changes produced in the tissues in which they are multiplying, summed up in the process of *inflammation.* Inflammation is the body's answer to injury. Its design is to halt the invasion and destroy the invaders. The features of the inflammatory process vary greatly with the different causative organisms and are significantly related to the disease-producing capacity of the microbe.

■ **General effects.** There are certain host reactions found in nearly all infectious diseases. In addition, many diseases have their own peculiar features. Among the general effects are fever (p. 156), increased pulse rate (tachycardia), increased metabolic rate, and signs of toxicity. The degree of fever approximates the severity of the infection. Anemia is a result of prolonged and severe infections.

A very common effect of infection is a change in the total number of circulating leukocytes (white blood cells) and in the relative proportion of the different kinds. This is why total leukocyte and differential white cell counts are so important in the diagnosis of disease. In most infections the total number of leukocytes is increased *(leukocytosis).* In some the number is decreased *(leukopenia),* and in a few it is unchanged. If fever or leukocytosis fails to occur in infections where either ordinarily is present, a severe infection or decreased host resistance is indicated.

An important consequence of infection is the development of immunity (Chapter 13).

■ **How disease-producing agents leave the body.** Just as they have definite avenues of entry, pathogenic agents have definite routes of discharge from the body, known as *portals of exit,* which to a great extent depend upon the part of the body that is diseased. The following list gives the important ones with examples:

1. *Feces*—bacteria of typhoid fever, salmonellosis, bacillary dysentery, and cholera; protozoa of dysentery; and viruses of poliomyelitis and infectious hepatitis
2. *Urine*—bacteria of typhoid fever, tuberculosis (when affecting the genitourinary tract), and undulant fever
3. *Discharges from the mouth, nose, and respiratory passages*—bacteria of tuberculosis, whooping cough, pneumonia, scarlet fever, and epidemic meningitis; viruses of measles, smallpox, mumps, poliomyelitis, influenza, and epidemic encephalitis
4. *Saliva*—virus of rabies
5. *Blood (removed by biting insects)*—protozoa of malaria; bacteria of tularemia; rickettsiae of typhus fever and Rocky Mountain spotted fever; virus of yellow fever

PATTERN OF INFECTION

■ **Course of infectious disease.** The course of many infectious diseases extends over the following:

1. *Period of incubation*—interval between the time the infection is received and the appearance of disease

   In some diseases its length is constant. In others it varies greatly. The length of the period of incubation depends upon (a) the nature of the disease-producing agent (for example, the incubation period of diphtheria is less than that of rabies), (b) its virulence, (c) the resistance of the host, (d) the distance from the site of entrance to the focus of action (for instance, the incubation period of rabies, which affects the brain, is shorter when the site of inoculation is about the face), and (e) the number of infectious agents invading the body. Table 11-2 gives the incubation periods of different infections.

2. *Period of prodromal symptoms*—short interval (the prodrome) that sometimes follows the period of incubation, characterized by such symptoms as headache and malaise
3. *Period of invasion*—disease reaching its full development and maximum intensity

   Invasion may be rapid (a few hours, as in pneumonia) or insidious (a few days, as in typhoid fever). At the onset of acute infectious disease, rigors and chills often precede the rise in temperature. The sensation of cold is difficult to explain but is probably due to a difference between the temperatures of the deep and superficial tissues of the body. To conserve heat, the superficial vessels of the skin constrict and sweating stops. The skin is pale and dry. As heat loss is decreased, the temperature rises rapidly.
4. *Fastigium or acme*—disease is at its height
5. *Period of defervescence or decline*—stage during which the manifestations subside

   During this state profuse sweating occurs. Heat loss soon exceeds heat production. As the temperature falls, the normal hue of the body returns and sweating ceases. Defervescence may be by *crisis* (within 24 hours) or by *lysis* (within several days, with the temperature going down a little each day until it returns to normal). Fever that begins abruptly usually ends by crisis. During the stage of *convalescence* the patient regains his lost strength.

Many diseases are *self-limited,* which means that, under ordinary conditions of host resistance and microbial virulence, the disease will last a certain, rather definite length of time and recovery will take place.

■ **Types of infection.** In a *localized* infection, the microbes remain confined to a particular anatomic spot (example, boils and abscesses). In a *generalized* infection, microorganisms or their products are spread generally over the body by the bloodstream or lymphatics. A *mixed* infection is one caused by two or more organisms. If a person infected with a given organism becomes infected with still another, a *primary* infection is complicated by a *secondary* one. Secondary infections of the skin and respiratory tract are quite common, and in some cases the secondary infection is more dangerous than the primary, for example, the streptococcal bronchopneumonia that often follows measles, influenza, or whooping cough. Lowered body resistance resulting from the primary infection facilitates the development of the secondary infection.

A *focal* infection is one confined to a restricted area from which infectious material spreads to other parts of the body. Examples are infections of the teeth, sinuses, and prostate gland. An infection that does not cause any detectable manifestations is an *inapparent* or *subclinical* one. An infection held in check by the defensive forces of the body but which may spread when the body resistance is reduced is a *latent* infection. An infection from the accidental or surgical penetration of the skin or mucous membranes is sometimes spoken of as an *inoculation* infection.

When bacteria enter the bloodstream but do not multiply, the condition is spoken of as a *bacteremia.* If they enter the bloodstream and multiply, causing infection of the

TABLE 11-2. INCUBATION PERIODS OF IMPORTANT INFECTIOUS DISEASES*

Disease	Usual incubation period
Adenovirus infections	5 to 7 days
Amebiasis	2 weeks (varies)
Ascariasis	8 to 12 weeks
Brucellosis	5 to 30 days (varies)
Cat scratch fever	3 to 10 days (varies)
Chickenpox	14 to 16 days
Coccidioidomycosis, primary infection	1 to 3 weeks
Coxsackievirus infections	2 days to 2 weeks
Dengue fever	3 to 15 days
Diphtheria	2 to 6 days
Echovirus infections	3 to 5 days
Encephalitis	3 to 15 days (varies as to type)
Gas gangrene	1 to 5 days
Gonorrhea	3 to 5 days
Hepatitis, infectious	15 to 50 days
Hepatitis, serum	2 to 6 months
Herpesvirus infections	4 days
Histoplasmosis	5 to 18 days
Hookworm disease	6 weeks
Impetigo contagiosa	2 to 5 days
Infectious mononucleosis	4 to 14 days
Influenza	1 to 3 days
Leprosy	3 to 5 years (?)
Leptospirosis	6 to 15 days
Lymphopathia venereum	1 to 4 weeks
Measles	9 to 14 days
Meningitis, acute bacterial	1 to 7 days
Mumps	14 to 28 days
Mycoplasmal pneumonia (primary atypical)	7 to 14 days
Pertussis	5 to 21 days
Pinworm infection	3 to 6 weeks
Plague	3 to 8 days
Poliomyelitis	7 to 14 days
Psittacosis	4 to 15 days
Q fever	14 to 26 days
Rabies†	2 to 6 weeks (to 1 year)
Rocky Mountain spotted fever	2 to 12 days
Rubella	14 to 21 days
Salmonelloses	
Food poisoning	6 to 48 hours
Salmonellosis	1 to 10 days
Typhoid fever	7 to 21 days
Scabies	4 to 6 weeks
Shigellosis (bacillary dysentery)	1 to 7 days
Smallpox	12 days
Streptococcal infections (scarlet fever, etc.)	2 to 5 days
Syphilis, primary lesion	14 to 30 days

*In this table and throughout the book the length of the incubation period is usually that given in the "Red Book" of the American Academy of Pediatrics.
†Rabies incubation in dogs is 21 to 60 days.

Continued.

TABLE 11-2. INCUBATION PERIODS OF IMPORTANT INFECTIOUS DISEASES—cont'd

Disease	Usual incubation period
Tapeworm infections	8 to 10 weeks
Tetanus	3 days to 3 weeks
Trichinosis	2 to 28 days
Tuberculosis, primary lesion	2 to 10 weeks
Tularemia	1 to 10 days
Yellow fever	3 to 6 days

bloodstream itself, the condition is *septicemia.* Septicemia is the layman's "blood poisoning." When pyogenic bacteria (pus formers) in the bloodstream are spread to different parts of the body to lodge and set up new foci of disease, the condition is called *pyemia.* When toxins liberated by bacteria enter the bloodstream and cause disease, the condition is *toxemia.* Diphtheria is a good example of a toxemia. Saprophytic bacteria may grow on dead tissue, such as a retained placenta or a gangrenous limb, and produce poisons that cause disease when absorbed into the body. This condition is *sapremia.* Patients with chronic wasting diseases like cancer often die from the immediate effects of some bacterial infection, especially streptococcal and pneumococcal infections. These are *terminal* infections.

A *sporadic* disease occurs only as an occasional case in a community. An *endemic* disease is one that is constantly present to a greater or lesser degree in a community. When a disease attacks a larger number of persons in the community in a short time, an *epidemic* is said to exist. Endemic diseases may become epidemic. When a disease becomes epidemic in a great number of countries at the same time, it is said to be *pandemic. Epidemiology* presents the pattern of disease in a given community and is the study of those factors influencing its presence or absence.

SPREAD OF INFECTION

■ **Transmission of communicable diseases.** The etiologic agents of communicable diseases may be transmitted from the source of infection to the recipient by (1) direct contact, (2) indirect contact, or (3) insect carriers (Fig. 11-1).

"Direct contact" is the term applied when an infection is spread more or less directly from person to person or from a lower animal to a human being. It does not necessarily mean actual bodily contact but does indicate a rather close association. Actual bodily contact is necessary in some instances, for example, the venereal diseases (see p. 317). *Droplet infection* is infection by microbes cast off in the fine spray from the mouth and nose during coughing, talking, or laughing. It is a form of direct contact as is placental transmission. By direct contact, blood transfusion transmits malaria and viral hepatitis. Diseases spread directly from man to man include tuberculosis, diphtheria, measles, pneumonia, scarlet fever, the common cold, smallpox, syphilis, gonorrhea, and epidemic meningitis, and from lower animals to man, rabies and tularemia. (Some of these are transmitted also by indirect contact.)

Indirect contact refers to the spread of the etiologic agent of a disease by conveyers such as milk, other foods, water, air, contaminated hands, and inanimate objects.

Infections arising from indirect contact are usually more widely separated in space and in time than those arising by direct contact. The diseases ordinarily spread by indirect contact are those in which the infectious material enters the body via the mouth.

Food, including milk, may convey infection. Salmonellosis, bacillary dysentery, and cholera may be spread by contaminated food. Botulism is caused by the consumption of canned foods insufficiently heated. Trichinosis and tapeworm infection are contacted from improperly cooked pork.

Air is established in the spread of infection; particles of dried secretion from the mouth or nose *(droplet nuclei)* may float in the air for a considerable time and be carried a rather long distance. Naturally, airborne transfer is most significant in respiratory infections.

Contaminated fingers are important conveyors of infection. A person with contaminated fingers may pass infection to other people or to different parts of his own body. Fingers easily soil food and drink.

Fomites are inanimate objects that spread infection. The most important are items such as handkerchiefs, towels, blankets, bed sheets, diapers, pencils, and drinking cups. Money is grossly contaminated as indicated by recent studies, especially the small unit coins and paper bills that more frequently change hands. Potential pathogens were cultured from 13% of the coins and 42% of the bills collected.

Filth is responsible for disease, and certain diseases are primarily ones of filth. Of these, typhus fever, as it is seen in parts of the world, is an example. However, filth plays little or no part in the spread of typhus fever in the United States; it is rare in this country. Persons who are unclean in their personal habits are more likely to contract an infectious disease than are those who are clean, and disease is more difficult to control in an unsanitary community.

Insects convey disease mechanically or biologically. In the mechanical transfer of disease, insects merely hold the pathogenic organisms on their feet or other parts of their body. Flies can so transfer the agents of typhoid fever and bacillary dysentery from the feces of patients to food to infect the consumer of that food. Flies act as more or less important carriers of more than 20 other diseases. The distance that they may carry

FIG. 11-1. Summary outline of events in transmission of disease. (From Hanlon, J. J.: Principles of public health administration, ed. 5, St. Louis, 1969, The C. V. Mosby Co.)

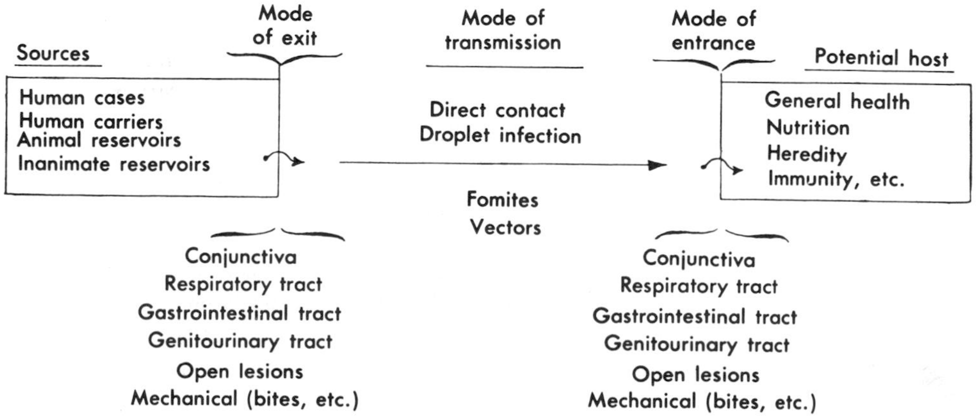

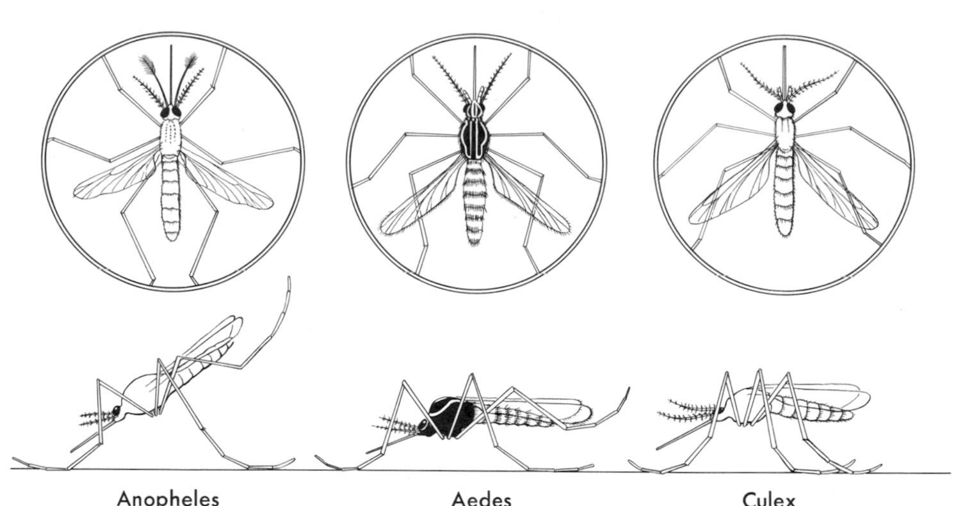

FIG. 11-2. Three mosquitoes important as biologic vectors in spread of disease. Note typical resting positions. Mosquitoes of genus *Anopheles* transmit malaria; members of genus *Aedes* transmit yellow fever and dengue fever. The common house mosquito (genus *Culex*) is the vector for arboviral encephalitis and filariasis.

infectious agents is surprising because they have been known to travel miles in search of food.

In the biologic transfer of disease, the insect in biting a person or animal ill of a disease, or a carrier, ingests some of the infected blood. The microbes therein undergo a cycle of development within the body of the insect within a specified time. The insect can then transfer the infection to a well person, usually by biting that person. Spread in this manner are malaria, yellow fever, and encephalitis by mosquitoes (Fig. 11-2); typhus fever by lice; Texas fever (a disease of cattle) by ticks; African sleeping sickness by tsetse flies; and plague by fleas. For diseases that are biologically transferred to be prevalent in a community, both the transmitting insect and hosts to harbor the infection must be present. As a rule, a particular infection is spread by only one species of insect, and a given insect is able to spread only one type of infection. There are, however, important exceptions.

■ **Sources of infection in communicable diseases.** In only a few instances does the infectious agent live in lifeless surroundings outside the body long enough to maintain the source of infection. For the continuous existence of a disease there must be a *reservoir of infection.* The most important reservoirs of infection and the sources of most communicable diseases are found in practically all communities as (1) human beings or animals with typical disease, (2) human beings or animals with unrecognized disease, and (3) human or animal carriers (Fig. 11-3). Plants may be the reservoir in some of the fungous infections.

Human carriers are persons who harbor pathogenic agents in their bodies but show no signs of illness. The carrier is infected but asymptomatic. *Convalescent carriers* harbor an organism during recovery from the related illness. *Active carriers* harbor an organism for a long time *after* recovery. *Passive carriers* shelter a pathogenic organism without having had its disease. *Intestinal* and *urinary carriers* discharge the infectious agent from the body via the feces and urine, respectively. *Oral carriers* discharge in-

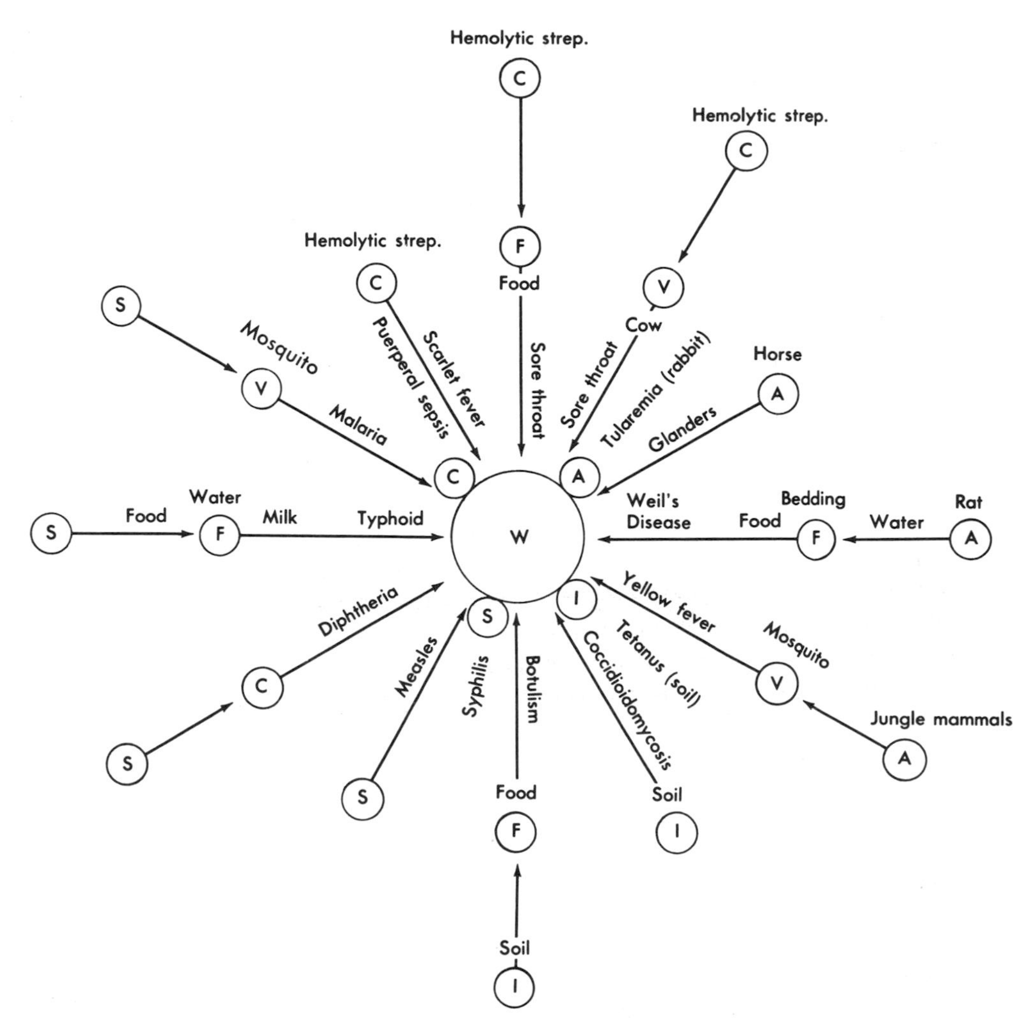

FIG. 11-3. Spread of communicable disease. **A,** Animal reservoir. **C,** Carrier (well but infected human). **F,** Fomite (inanimate transmitter). **I,** Inanimate reservoir. **S,** Sick human. **V,** Vector (animate transmitter). **W,** Well human. (From Hanlon, J. J.: Principles of public health administration ed. 5, St. Louis, 1969, The C. V. Mosby Co.)

fectious material from the mouth. *Intermittent carriers* discharge organisms only at intervals. (Intestinal carriers are often intermittent carriers.) Human carriers play an important part in the spread of diphtheria, epidemic meningitis, salmonellosis, amebic dysentery, bacillary dysentery, streptococcal infections, and pneumonia.

As a rule, an actual case of a disease is more likely to spread infection to others than is a carrier. Carriers and persons with unrecognized disease keep epidemic diseases in existence during interepidemic periods. It has been found that just before the epidemic (for example, diphtheria and acute bacterial meningitis) the number of carriers increases. The significance of carriers in the spread of a disease depends upon the frequency with which people become carriers and the length of time the carrier state persists.

Animals spread disease to man in a number of ways: Man may so acquire disease from (1) direct contact with an infected animal, (2) contamination of food by discharges of the animal, (3) insect or rodent vectors, (4) contaminated air or water, and (5) consumption of animal products, such as milk or eggs.

Rabies is acquired by the bite of rabid dogs, cats, or other animals. Children may contract tuberculosis by drinking the raw milk of infected cows. Bubonic plague is essentially a disease of rats that is transmitted from rat to man by the flea. Undulant fever may be transmitted to man by milk, or it may be acquired when man handles the meat of infected animals. Tularemia is contracted from rodents, especially wild rabbits. Psittacosis is an infectious disease of parrots transmitted from the sick parrot to man. Typhus fever is primarily a disease of rats that is transmitted from man to man by lice. Those handling the hides of anthrax-infected animals may contract anthrax. Tetanus may be indirectly contracted from horses because the tetanus bacillus is a normal inhabitant of the intestinal canal of the horse. This is why wounds contaminated by barnyard dirt are likely to be followed by tetanus. Shellfish, such as oysters, may transmit typhoid fever, hepatitis, and other enteric infections if human excreta contaminates the water in which the shellfish are grown. All told, there are about 150 different infections that are spread from animals to man.

Animals found in the home as pets may be important sources of infection (see Table 11-3) since many persons live closely associated with them. The two most common household pets are cats and dogs.* In North America the dog can transmit 24 diseases to man and the cat can transmit 12. Exotic animals that are becoming more popular as household pets may present real hazards. For example, monkeys in the home, pet shop, or even some zoos may spread such diseases to man as shigellosis, salmonellosis, and tuberculosis.

The most prevalent of the *zoonoses* (infections of animals secondarily transmissible to man) are cat scratch disease, salmonellosis, and fungal infections. There are about 100 different diseases known to be transmitted naturally from man to animals.

■ **Spread of disease in the jet age.** This is an age of travel at high speeds. More and more people are and will be traveling over the world aboard the jumbo jets or supersonic airliners and consequently exposing themselves to the infections of the geographic localities they visit. The problems in public health are enormous. Some are as follows: First, disease is so easily and quickly spread as to jeopardize public health controls. A traveler can bring back smallpox, for example, into a country that had virtually eliminated it. Second, since no two points on the earth are separated by more than 48 hours travel time, the tourist can return home during the incubation period (feeling well), to come down with the disease days or weeks later. By this time, the diagnosis is not necessarily obvious. Third, the world sightseer may fail to take adequate precautions against a disease he inevitably contacts. The best example of this is malaria, the major risk. It is a widespread disease but nowhere is chemoprophylaxis required. The traveler must find out for himself what to do for protection. Even then, he may fail to take the suppressive drugs as required and may discontinue them after leaving an infected area, feeling they are no longer needed. Fourth, still another problem has to do with the establishment of international standards and guidelines for sanitary facilities. Maintenance of these is required for the tourist in all areas, but especially aboard the big jet, in the airport terminal, and in the accommodations related to it.

---

*There are an estimated 26 million canine pets in the United States.

TABLE 11-3. SOURCES OF INFECTION IN HOUSEHOLD PETS

Disease	Dogs	Cats	Farm animals	Caged birds (pigeons, parrakeets, parrots, myna birds, canaries, etc.)	Poultry	Rodents (mice, rabbits, rats, hamsters, etc.)	Reptiles (snakes, turtles, lizards, etc.)
Viral							
Rabies	✓	✓	✓				
Encephalitides			✓	✓			
Bedsonial (chlamydial)							
Cat scratch disease		✓					
Psittacosis-ornithosis				✓	✓		
Rickettsial							
Q fever			✓				
Bacterial							
Salmonellosis	✓	✓	✓	✓	✓	✓	✓
Anthrax	✓		✓				
Brucellosis	✓	✓	✓		✓	✓	
Tularemia	✓	✓				✓	
Tuberculosis	✓	✓	✓				
Leptospirosis	✓		✓	✓		✓	
Fungal							
Ringworm	✓	✓	✓			✓	
Parasitic							
Roundworm infection	✓	✓					
Tapeworm infection	✓	✓	✓				
Scabies	✓	✓					
Toxoplasmosis	✓	✓					

Koch's postulates

Absolute proof that an organism is the cause of a given disease rests upon the fulfillment of certain requirements known as Koch's postulates.* These are as follows:

1. The organism must be observed in every case of the disease.
2. The organism must be isolated and grown in pure culture.
3. The organism must, when inoculated into a susceptible animal, cause the disease.
4. The organism must be recovered from the experimental animal and its identity confirmed.

In exceptional cases an organism has been accepted as the cause of a disease although all of these requirements have not been met.

---

*Robert Koch (1843-1910) stated certain principles related to the germ theory of disease with such clarity that they are known as Koch's postulates. They remain to this day the basis of the experimental investigation of infectious disease.

QUESTIONS FOR REVIEW

1. Explain infection and contamination.
2. Give the routes by which microbes enter the body. Name two diseases whose causative agent enters the body by each route.
3. What is virulence? How do microbes produce disease?
4. What is one important local effect of bacterial invasion?
5. List some of the general effects of bacterial invasion.
6. What is the difference between bacteremia and septicemia?
7. Give the routes by which disease-producing agents leave the body. Name two agents eliminated by each route.
8. List five diseases spread by animals to man.
9. Define carrier. List the different kinds.
10. Name three diseases spread by carriers.
11. List important insect vectors and diseases spread.
12. State Koch's postulates.

REFERENCES. See at end of Chapter 15.

12 The body's defense

PROTECTIVE MECHANISMS

If infection occurred each time infectious and injurious agents entered our bodies, we would be constantly ill. When microorganisms attempt to invade the body, they must first overcome certain mechanical, physiologic, and chemical barriers existing on the body surface or in the body cavity at the site of entry. The body not only reacts to the event of invasion but is also endowed with a measure of protection against that event.

■ **Anatomic barriers.** During his agelong struggle for existence man has developed certain mechanisms that enable him to overcome many agents of potential injury in his environment. In the first place, nature has provided him with special senses that act as watchdogs against danger. Man protects himself from major injuries by moving objects by batting his eyes and jumping aside. Although some of these acts are voluntary, they are executed as reflex movements so that they may be considered protective natural mechanisms.

The body's first line of defense is the epithelium, which covers the exterior of the body and lines its internal surfaces. A condensation of resistant cells along the most superficial part of the skin offers a barrier against physical and chemical injuries or microbial invasion. In many cases the epithelium may become quite thick at a site of irritation (example, a callus or corn). The lining cells of the mucous membranes opening upon body surfaces also afford a barrier, but these cells are softer and more vulnerable to injury than are the surface cells of the skin. The hairs in the anterior nares protect the respiratory tract by filtering bacteria and larger particles from the inspired air. The respiratory passages are lined with epithelial cells from the surface of which spring hairlike appendages, known as cilia, that sweep overlying material from the deeper portion of the tract to the upper portion from where it may be discharged to the environment.

Closure of the glottis during swallowing prevents food from entering the respiratory tract. Coughing and sneezing serve to expel mechanically any irritating materials from the respiratory tract. In a similar manner vomiting and diarrhea mechanically rid the intestinal tract of irritants.

■ **Chemical factors.** Body cavities opening on the surface are protected by secretions that wash away bacteria and foreign materials. Body fluids, such as saliva, tears, gastric juice, and bile exert an antiseptic action, which reduces microbial invasion. For instance, the gastric juice destroys bacteria and almost all important bacterial toxins except that of *Clostridium botulinum.* The antimicrobial action of mucus and tears is partly related to the presence of *lysozyme* (muramidase), a substance dissolving the cell walls of certain gram-positive bacteria. It is found in secretions coming from organs exposed to airborne bacteria. Lysozyme, not an antibody, was discovered by Fleming (in 1922) before he discovered pencillin. Because of their acid reaction and content of fatty acids, sweat and sebaceous secretions on the skin have antimicrobial properties.

Another internal defense is a naturally occurring substance *interferon* (not an antibody). Its target action is against viruses to block viral infection. Many cells form it quickly if so treatened.

■ **Physiologic reserves.** The blood supply of many parts of the body is protected by a series of *anastomoses* (connections) among the branches of the supply vessels. If a vessel is occluded, the blood is detoured around the obstruction by way of the anastomoses. This is known as the establishment of a *collateral* or *compensatory circulation.*

When the body becomes chilled, the superficial vessels of the skin contract to prevent heat from being dissipated at the surface, and sweating stops to slow cooling caused by evaporation. To aid in this conservation of heat, the involuntary muscles of the skin contract (gooseflesh). In animals having hair or feathers, the hair or feathers are raised to enclose in their meshes a thick layer of air, which is a poor conductor of heat. The muscular activity associated with rigor and shivering increases heat production. When the body becomes too hot, the superficial vessels dilate and sweating occurs. Heat is dissipated from the surface, and evaporation has a cooling effect.

The functional capacities of the vital organs (organs necessary for life) extend far beyond the demands of normal life. We have much more liver, pancreas, adrenal gland, and parathyroid gland tissue than we ordinarily use. That we are able to lead a normal life after one kidney has been removed or after one lung is collapsed illustrates the abundance of reserve possessed by our vital organs.

Vital organs (brain, heart, lungs) are enclosed within protective bony cases, and their surfaces are bathed with a watery fluid. The fluid surrounding the brain serves as a water bed to prevent undue jarring, and the fluid that bathes the surface of the heart and lungs acts as a lubricant.

■ **Fever.** Fever is a condition characterized by an increase in body temperature. It is usually thought to result because toxic substances arising from the disintegration of microbes and other cells in an injured (or inflamed) area gain access to the bloodstream. There are many causes, ranging from mechanical to microbial injury but bacterial infection is by far the most important. Although the manifestations of fever may be disagreeable, it is primarily beneficial. If the temperature is not too high, fever accelerates the destruction of injurious agents by increasing phagocytosis and the production of immune bodies (see Chapter 13).

The temperature of warm-blooded animals is rather closely controlled by the temperature-regulating mechanism of the brain and is dependent upon two factors: (a) heat production (thermogenesis) and (b) heat dissipation (thermolysis). When less heat is dissipated, the body temperature is elevated. In fever the temperature control mechanism in the brain is set to give a higher degree of heat in the body in much the same manner as the temperature of the incubator is raised by an adjustment of the thermoregulator.

In health the body temperature of man remains remarkably constant, and, as determined by a thermometer placed in the mouth, ranges from 96.7° F. (35.94° C.) to 99° F. (37.22° C.) with an average of about 98.6° F. (37° C.). The rectal temperature is about 1° F. higher and the temperature taken in the armpit is about 1° F. lower than the oral temperature. The rectal temperature is the most dependable because both oral and axillary readings are influenced by many outside factors. The temperature of the internal organs is 2° or 3° F. higher than that of the skin.

There is a close relation between pulse rate and temperature. Ordinarily an increase of around eight pulse beats per minute occurs for each 1.8° F. increase in temperature. Certain exceptions to this rule are diagnostically significant: For instance, the pulse in typhoid fever, malaria, miliary tuberculosis, and yellow fever is comparatively slow.

THE RETICULOENDOTHELIAL SYSTEM (THE PHAGOCYTIC SYSTEM)

The reticuloendothelial system with its widely dispersed cells constitutes one of the main organs in the body designed primarily for man's defense against both the living and the nonliving agents in his environment that harm him. The reticuloendothelial cells not only can directly fight the invader, which they do by phagocytosis, but can also clean up the debris and eliminate the cellular and metabolic breakdown products incident to the encounter. In short, they are also scavengers.

The production and regulation of the cells in the peripheral blood and bone marrow are crucial functions of the RE system. Formation of the different blood cells (hemopoiesis) is possible because of the presence within the system of primitive nonphagocytic cells. Being multipotential cells they give rise to the range of blood cell types found.

■ **Phagocytosis.** The ingestion of microbes or other particulate matter by cells is known as phagocytosis (Fig. 12-1), and the cells that ingest such materials are phagocytes. Phagocytosis bears a close resemblance to the feeding process of unicellular organisms and is a universal response on the part of the body to penetration by microbes, alien cells, or other foreign particles. It is an essential protective mechanism against infection and probably plays an important part in natural immunity. Phagocytosis is a general process because the few kinds of body cells having this power can ingest many different kinds of particulate matter (bacteria, dead body cells, mineral particles, dusts, and pigments).

Study of the process has revealed that phagocytosis takes place in three phases. (See Fig. 12-2.) First, the particle (for example, a microbe) to be engulfed is isolated and incorporated into an invagination of the cell membrane of the phagocyte. The result is a *phagosome,* a vacuole (within the cell) surrounded by cytoplasm. Afterwards there is a burst of metabolic activity within the cell. The phagosome is moved into the interior of the cell contacting its lysosomes. When hydrolytic enzymes are released into the phagosome, it becomes a *phagolysosome.* A special kind of microbicidal system is activated in preparation for the second phase. This is the killing or inactivation of the microbe, so that in the third phase it can be digested.

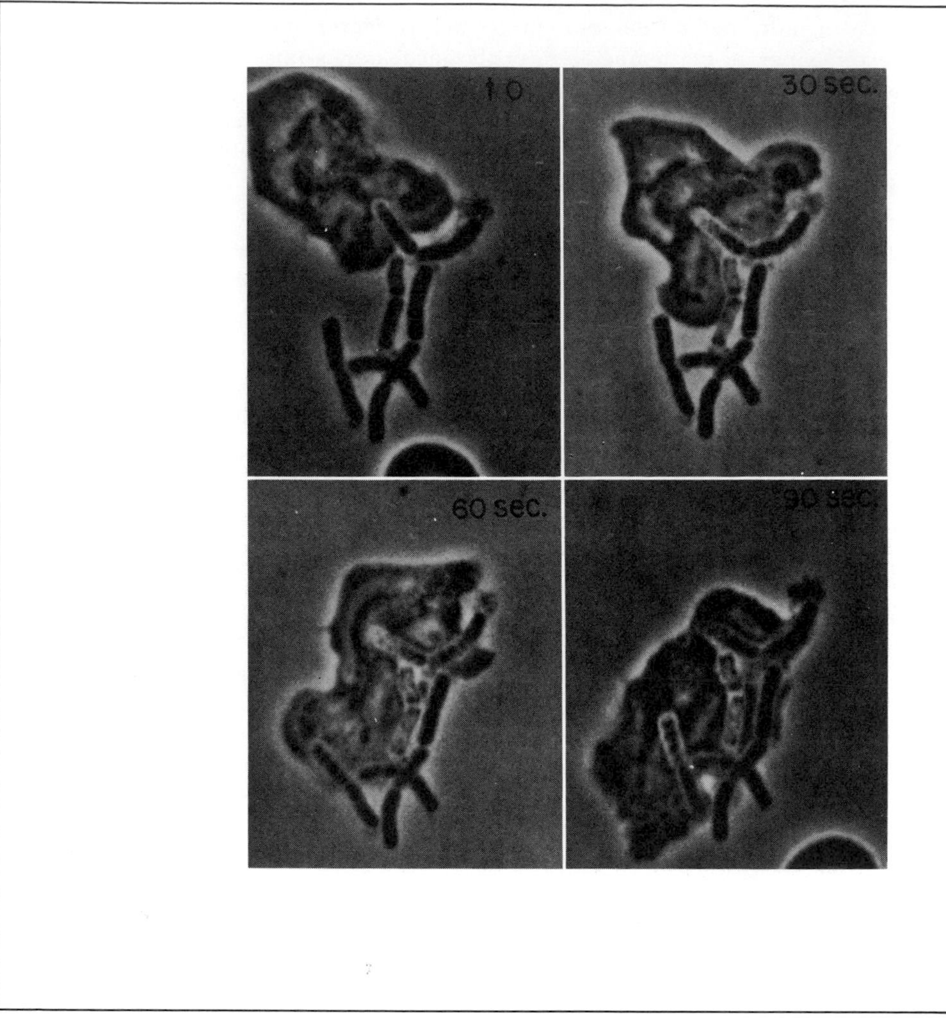

FIG. 12-1. Phagocytosis in phase contrast microscopy: successive stages in ingestion of bacteria by human white blood cell (polymorphonuclear neutrophilic leukocyte). Photographs taken with phase microscope initially, 30 seconds, 60 seconds, and 90 seconds later. (×1500.) (Courtesy Dr. J. G. Hirsch, Rockefeller University, New York; from Dubos, R. J.: The unseen world, New York, 1962, Rockefeller Institute Press.)

If after ingesting bacteria, phagocytes fail to destory them, the phagocytes themselves may be destroyed with liberation of the ingested organisms. Relatively harmless microbes are usually completely destroyed. Sometimes the microbes persist and even continue to multiply within the cytoplasm of the phagocyte. Virulent bacteria, especially those with capsules, are resistant to phagocytosis.

The student who wishes to see bacteria undergoing phagocytosis should stain a smear from an ordinary boil. Staphylococci can be found easily within the cytoplasm of the leukocytes. Smears from gonorrheal exudates and from the cerebrospinal fluid in epidemic meningitis usually show organisms undergoing phagocytosis inside the leukocytes.

Different bacteria show differences in their susceptibility to phagocytosis. For

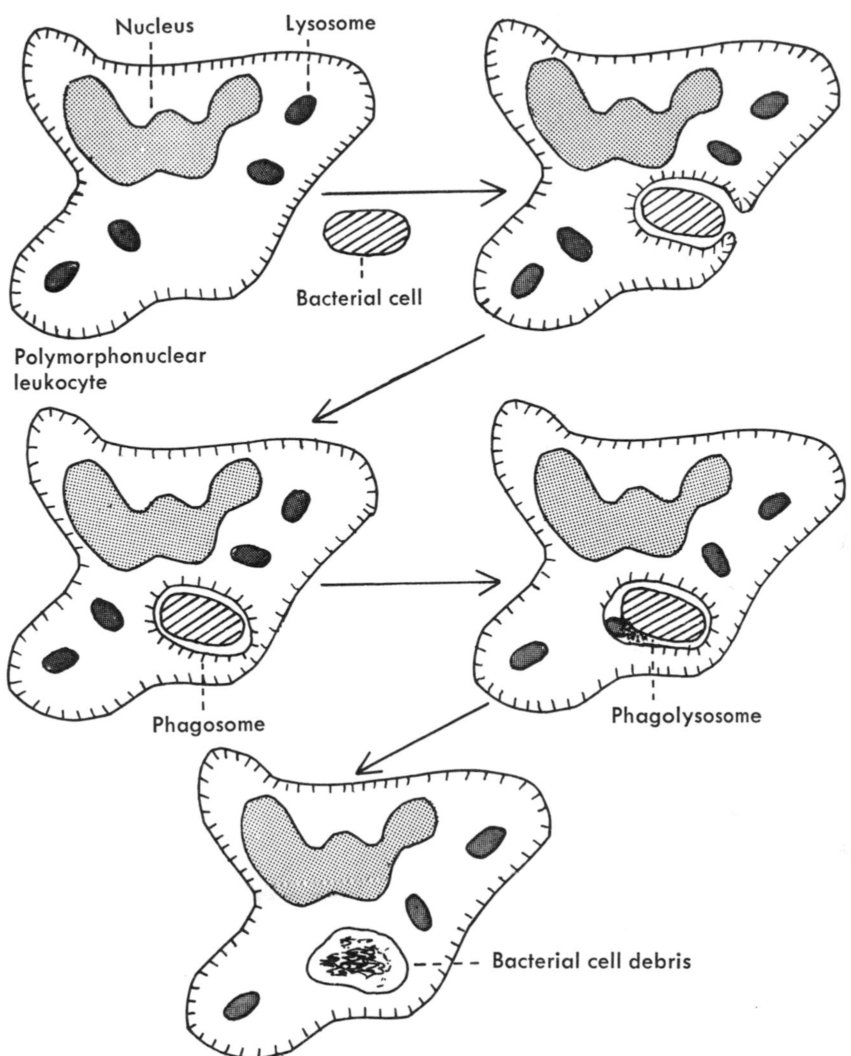

FIG. 12-2. Phagocytosis in sketch. Stages in digestion of bacterium by neutrophil. Chemotaxins attract the neutrophil. Surrounding the bacterium, the neutrophil engulfs and retains it in a phagosome. When phagosome contacts lysosomes, hydrolytic enzymes are released into vacuole, now termed a "phagolysosome" or "phagolyosome." (Undigested debris from bacterium is eventually egested.) (From Barrett, J. T.: Textbook of immunology, St. Louis, 1970, The C. V. Mosby Co.)

instance, gonococci undergo phagocytosis easily, whereas tubercle bacilli are quite resistant.

Phagocytosis is not so great in the first 3 years of human life as it is later on.

■ **Anatomy of the RE system.** The RE system is a group of cells in the body especially endowed with phagocytic powers—both the cells themselves and those to which they give rise. In short, it is a phagocytic system. Though widespread throughout the body (Fig. 12-3) these cells are concentrated in lymph nodes, spleen, bone marrow, and liver, where they are arranged as the lining cells for the peculiar kind of sinusoidal arrangement common to these organs. Though scattered, these uniform cells are spoken of collectively as the *reticuloendothelial system.* There are two main cell types: (1) the stationary or fixed (littoral) cells and (2) the free or wandering cells. The wandering cells in tissues are known as *macrophages;* in the bloodstream, as *monocytes.*

The components of the reticuloendothelial (RE) system are shown in the following schema:

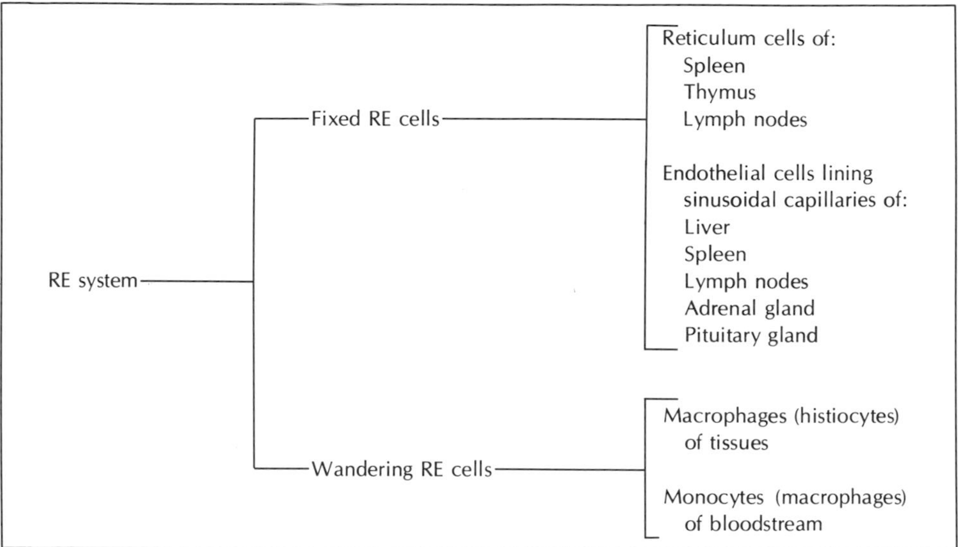

Because the macrophage moves about in body fluids and because its observed properties of phagocytosis are dramatic, it receives a lot of attention. The details of its life history, including even its exact origin, are not completely worked out. It is known by a variety of names—RE cell, histiocyte, epithelioid cell, clasmatocyte, multinucleated giant cell (if confluence of several cells)—relating it (the same cell) to different circumstances and reflecting an active participation in its diverse roles.

In *granulomatous* inflammation the disease-producing agent is dealt with directly by the reticuloendothelial system, and the main reacting cell is this wandering phagocyte or macrophage.

A most important derivative cell of the RE system and a phagocyte is the *polymorphonuclear neutrophilic leukocyte,* familiarly known as the "poly," a cell (microphage) especially designed to phagocytize bacteria and one of several kinds of white blood cells (leukocytes). With the majority of bacterial inflammations, there is an increase in the number of these cells at the site of injury and in the bloodstream *(leukocytosis).* They have been produced in the bone marrow in increased numbers in response to chemotaxins released from the inflammatory site and carried by the blood to the marrow. In conditions that produce a leukocytosis the phagocytic power of the leukocytes is increased in patients who are doing well. Neutrophils are rich in proteolytic enzymes contained within their lysosomes. These enzymes known as leukoproteases participate in intracellular digestion of phagocytized particles. The phagocytic cells especially related to processes of immunity (discussed in Chapter 13) are the polymorphonuclear neutrophils and the macrophages of the RE system.

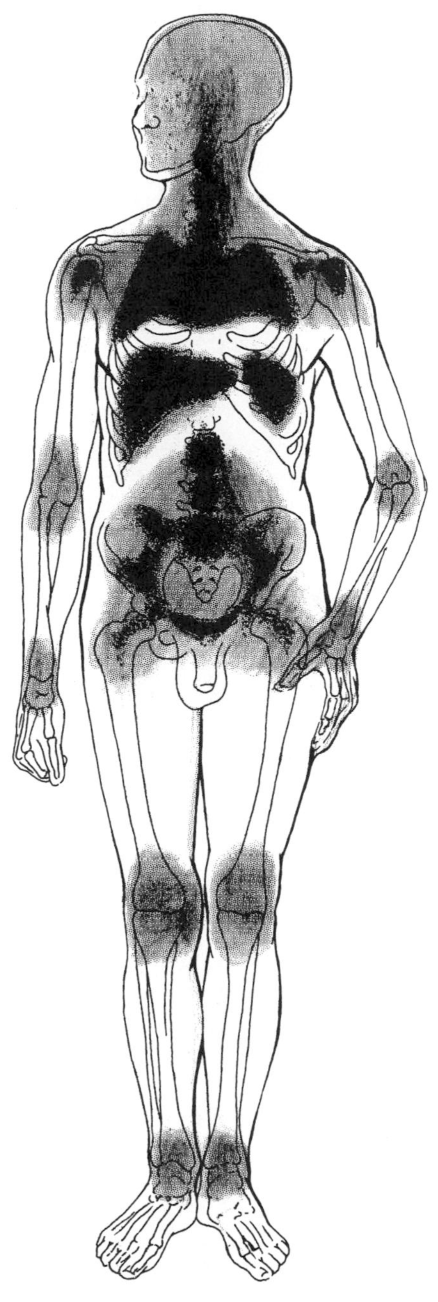

FIG. 12-3. The reticuloendothelial system. Note anatomic distribution of maximal activity in the system as indicated by black areas over the body. To produce such an image, certain radioactive colloidal particles are given to a subject and radiation detection technics delineate the tissue uptake. Note definition of liver, spleen, and active bone marrow in the axial skeleton and proximal parts of long bones. (Adapted from McIntyre, P. A.: Visualization of the reticuloendothelial system, Hosp. Pract. **6:**77, July 1971.)

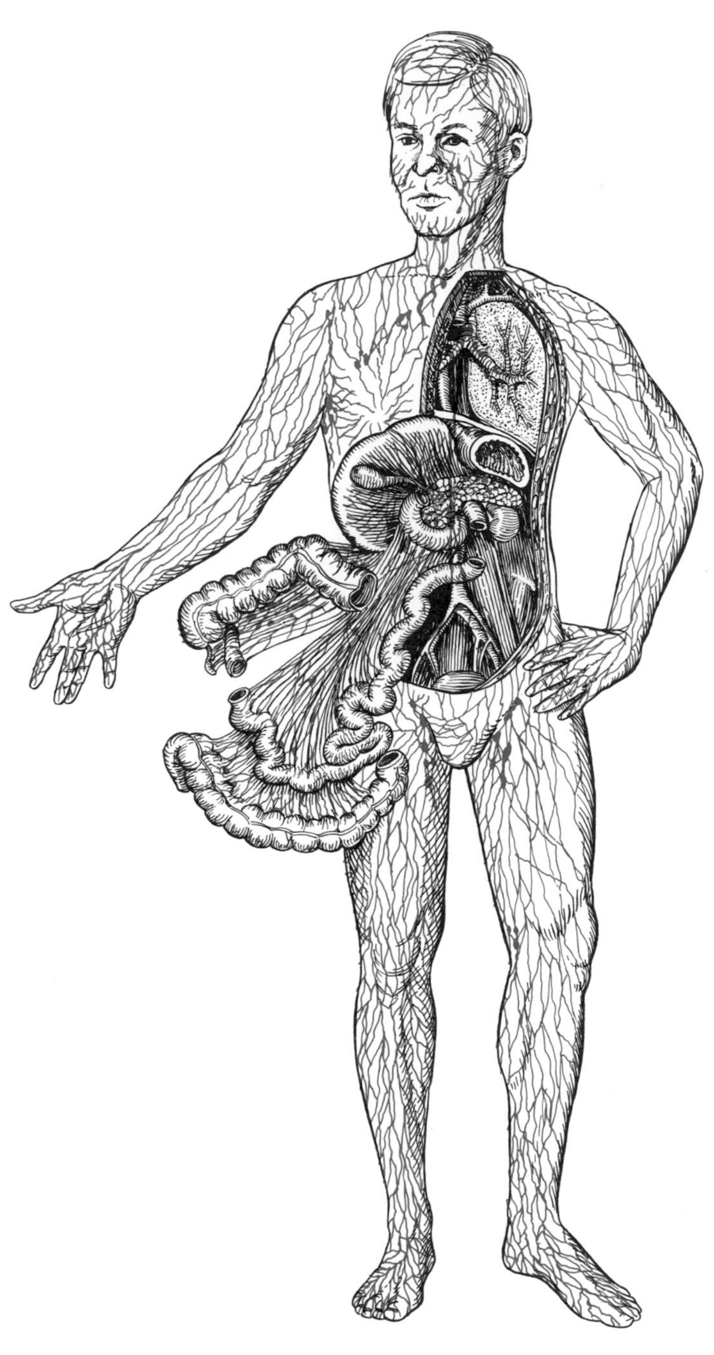

FIG. 12-4. The lymphoid system. Note pattern of lymphatic vessels and lymph nodes in all areas of the body (superficial and deep).

THE LYMPHOID SYSTEM (THE IMMUNOLOGIC SYSTEM*)

■ **Lymphoid tissue.** Closely interwoven with the reticuloendothelial system (the phagocytic system) and encompassed by it is the lymphoid system. This comprises the lymphoid tissues and organs of the body. The chief cells of the system are the lymphocytes, small round cells with a relatively large and darkly staining nucleus, and their constant companions, the plasma cells—slightly larger cells with an eccentric nucleus of distinct appearance. The lymphoid cells are packed, more or less densely, onto a spongelike support provided by a special type of cellular stroma, and collections of lymphocytes including their precursors and progeny in a well defined usually rounded aggregate constitute a lymphoid nodule. A nodule may be a solitary one, or several nodules may be grouped together, as in Peyer's patches of the small intestine. If lymphoid tissue is organized and encapsulated, a small bean-shaped organ is formed, the lymph node. (See p. 168.) Lymph nodes are found throughout the body, strategically placed in all major organ systems and body areas (Fig. 12-4).

■ **Lymphatic circulation.** In the exchange of nutrients for waste products that goes on continuously between the cells and the blood of the capillaries, excess fluid leaks into the tissue spaces. From here it moves into the lymphatic capillaries, the many small thin-walled tubes of the lymphatic circulation draining the vast intercellular region of the body. The fluid circulates as lymph into successively larger vessels and ultimately pours back into the bloodstream. Along the course of the lymphatic channels at definite intervals are placed the lymph nodes in chain formations (Fig. 12-4). All the lymph of the body is filtered and strained through these chains.

■ **Spleen.** The largest mass of lymphoid tissue in the body is the spleen, found in the upper left abdomen. Like the lymph nodes, it is a filter, but unlike them it strains the blood not the lymph stream. It functions as an important member of the reticuloendothelial system in the filtration of foreign particles from the blood including living organisms. The fixed cells of the reticuloendothelial system line the lumens of all lymphatic channels, lymph sinuses, and splenic sinusoids. There is an intimate relationship between elements designated as lymphoid and those as reticuloendothelial.

■ **Thymus.** The thymus, a key organ in immunologic processes, is situated in the front part of the chest just behind the breast bone. At or about the time of birth it processes certain lymphocytes, which then migrate out to colonize the spleen, lymph nodes, and other areas of the body. These in turn give rise to cells that maintain cell-mediated immunity. (See Chapter 13.)

■ **The lymphatics in infection.** When an agent gains access to the tissues, it may sometimes be carried away from the site of entry by the lymphatic circulation and be deposited in the nearest lymph nodes. (In the lungs of persons residing in coal-burning regions, coal dust is thus deposited in the drainage nodes.) Likewise bacterial invaders may be borne from the site of disease to the regional nodes where their presence stimulates phagocytic cells normally present to increased activity. Hopefully these phagocytes dispose of the invaders. Should they fail to do so, the bacteria flow with the lymph stream through successive nodes in the chains into the thoracic duct from

*Immunologic function is an important property of lymphoid tissue. Mechanisms are discussed in the next chapter.

there to be emptied into the bloodstream and disseminated throughout the body. Infection of the bloodstream (septicemia) before the days of antibiotic therapy was a rapidly fatal condition.

■ **Responses to infection.** If an infection is localized, the reticuloendothelial cells of the regional lymph nodes proliferate and the nodes enlarge as a consequence (lymphadenopathy). Regional lymph nodes are an important second line of defense. Many injurious agents carried there are destroyed. If an infection becomes generalized, the reticuloendothelial system throughout the body responds. One notable feature of this is the enlargement of the spleen (splenomegaly) seen with such acute infectious processes as typhoid fever, malaria, and septicemia.

QUESTIONS FOR REVIEW

1. List mechanisms that serve to protect the body from disease or injury and explain how each functions.
2. What is lysozyme? Who discovered it?
3. What are cilia? What is their function in the respiratory passages?
4. What is a vital organ? Name three. How are the vital organs protected in the body?
5. Define fever.
6. What is considered normal body temperature?
7. What is the relation between pulse rate and temperature?
8. Briefly describe the reticuloendothelial system. What is its chief function?
9. Briefly discuss phagocytosis. What are the three phases? Why is this an important process?
10. Outline the anatomic distribution of the RE system.
11. Give the component cells and organs of the lymphoid system.
12. What is meant by lymphadenopathy? By splenomegaly? With what are they associated?

REFERENCES. See at end of Chapter 15.

13 Immunology

MEANING OF IMMUNITY

Immunology is the division of biology concerned with the study of *immunity*. Today the concept of immunity is hard to define. In terms of the steady progression of knowledge in the field of immunology, the meaning of the word is rapidly expanding.

Basic to an understanding of the complexity of physiologic mechanisms encompassed by immunity are the concepts of "self" and "nonself" or "foreign." For a given individual to preserve his biologic integrity (self) he must be able to recognize and deal with factors in his environment threatening it. On the one hand there is self, which must remain intact. On the other there are the elements it contacts, not self, although perhaps of comparable makeup, and potentially harmful. Because of the many biologic parallels in nature, the distinction is a fine one. We require an arrangement whereby the cells and cell substances of our bodies are marked off as "ours" and the substances that appear on the scene are accosted as "intruders."

The scheme of things in the body that function in this discriminatory way is the machinery of immunity, and the elements making it up are collectively the *immune system*. The term *immunobiology* surveys the subject of immunity and its many ramifications in modern medicine.

IMMUNITY AND INFECTION

Infectious agents, such as bacteria and other microorganisms, are easily "foreign" to the body (nonself), their effect detrimental, and the cause-effect relationships clear cut. The applications of immunity with infectious processes have been dramatic. Our classic definition that immunity is a highly developed state of resistance refers directly to infectious disease. *Susceptibility* is the reverse of immunity and is the result of an absence or suppression of the factors that produce immunity.

■ **Kinds of immunity.** Traditionally immunity has been classified as follows:

1. NATURAL
   Species
   Race
   Individual

2. ACQUIRED

Naturally	
Active	Passive
By attack of disease (clinical or subclinical)	By placental transmission By colostrum

Artificially	
Active	Passive
By vaccination with: live organisms attenuated ones dead ones toxins toxoids toxin-antitoxin	By injection of antiserum (prophylactic, therapeutic)

A *natural* immunity is a more or less permanent one with which a person or lower animal is born; that is, it is a natural heritage. It may be the heritage of a species, race, or individual. It is also known as *innate* or *genetic* immunity. *Species* immunity is that peculiar to a species (for example, man does not have distemper, nor do dogs have measles). A *racial* immunity is one possessed by a race (for example, ordinary sheep are susceptible to anthrax, whereas Algerian sheep seldom contract the disease). Species and racial immunities are more highly developed in plants than in animals. *Individual* immunity is a rare condition, and most so-called cases are from unrecognized infections.

An *acquired* immunity is one in which protection *must* be obtained. It is never the heritage of a species, race, or individual. It may be *naturally* acquired when a mother transmits antibodies (protective substances) to her fetus by way of the placenta or when a person has an attack of a disease. When a person has a disease such as measles, the disease is rather severe for a time, after which recovery occurs. The person will not then contract the disease again though repeatedly exposed; he has developed a permanent immunity.

An immunity may be *artificially* acquired by vaccination or the administration of an immune serum.

Depending upon the part played by the body cells of the animal or person being immunized, acquired immunity (natural or artificial) is classified as *active* and *passive*. If a person has an attack of measles or is vaccinated against it, his body cells respond to the presence of the agent (or its products) by producing antibodies that destroy the causative agent of the disease should it again gain access to the body. When certain body cells react to the agent or its products in this way, the naturally acquired immunity is an *active* one. For example, if a horse is given frequent injections of diphtheria toxin, beginning with a small dosage and with a gradually increased amount at each subsequent injection, the horse will eventually be able to withstand thousands of times the amount of toxin required to kill an untreated animal; it has become immune to the action of diphtheria toxin.

Immunity transferred by way of the placenta to the child from his mother is a *passive* immunity, which the child has naturally acquired. When the serum of an actively immunized animal is injected into a nonimmune animal, the latter becomes temporarily immune. This artificially acquired immunity is *passive* because the immunity-producing principle is introduced in the serum, and the cells of the recipient animal take no part in the process. For the first 4 to 6 months of his life an infant may have a passive immunity to measles, smallpox, diphtheria, mumps, tetanus, influenza, and certain staphylococcal and streptococcal infections because of the transfer of immune bodies from the blood of the mother to the child via the placenta. The immunity that the child receives in his mother's womb may be enhanced by the content of protective substances that he receives from his mother's milk (especially from colostrum). Naturally, the child will not receive any immunity if the mother is not herself immune. Except in newborn babies, all passive immunities are established by the administration of an immune serum (the serum of an animal that contains antibodies because the animal has been actively immunized). The production of passive immunity shows its most brilliant application in the use of diphtheria antitoxin to prevent and cure diphtheria.

Active immunity artificially acquired does not last so long as when it is naturally acquired, but active immunity always lasts longer than passive immunity. One should bear in mind that an active immunity can be established only with an attack of a given disease or with vaccination against it and that the immunity is slowly established (days or weeks) but of long duration (months or years). On the other hand, a passive immunity is ordinarily established at once by the injection of an immune serum but is of short duration (1 or 2 weeks). Upon these facts is based the principle of producing an active immunity when there is no immediate danger from disease and establishing a passive immunity when the danger from disease is imminent or the patient is already ill.

■ **Level of immunity.** It seems that when an infectious disease attacks a population that has never been exposed, the attack rate is very high and many succumb to its ravages. On the other hand, when a disease endemic in a population for years becomes epidemic, the number of persons attacked and the percentage of deaths are not so great. People develop a degree of inherent resistance to a given disease through exposure for generation after generation. There is little doubt that syphilis was at one time a far move virulent disease and more often fatal that it is now. The susceptibility of aboriginal people to tuberculosis is much greater than that of people in communities in which the disease has existed for a long time.

In 1951 measles was introduced into Greenland by a recently arrived visitor who attended a dance during the onset of symptoms. Within 3 months there were more than 4000 cases, with seventy-two deaths. The attack rate reached the unprecedented figure of 999 cases per 1000 people. When measles was first introduced into the Fiji Islands in 1875, 30% of the population died. It has been said that the white man subjugated the American Indian more by the diseases he brought to him, to which the Indian possessed no immunity, than he did by superior knowledge or more effective weapons.

THE IMMUNE SYSTEM

Immunity is important as a defense against infection, but in light of current achievements it is more. In a broad way of speaking immunity comprises the things that help man to maintain his structural and functional integrity and to ward off certain kinds of injury. As such it can be separated into two major categories. One, that of *nonspecific* or *innate* immunity, includes built-in factors of resistance, sometimes hard to measure, basic to all living organisms. Such factors are the inflammatory process, the presence of anatomic barriers, and the effect of naturally occurring antimicrobial substances. The other category is that of *specific* or *adaptive* immunity. It represents more highly developed, more precisely specialized physiologic mechanisms, topics of great concern to immunologists, biochemists, microbiologists and geneticists.

There are two expressions of the fundamental processes in adaptive immunity. One, relating to the elaboration of certain chemicals,* is *humoral immunity*, and the other, relating to the activities of certain cells, is *cell-mediated immunity (cellular immunity)*. Reflecting the dual nature of the immune response, our discussion will take the twofold approach, with first a consideration of humoral immunity, and second a note of the cellular aspects.

■ **Role of lymphoid tissue.** In man the lymphoid tissue is responsible for the events of specific immunity. Widely distributed in the body, it is both concentrated (lymph nodes, spleen, thymus, Peyer's patches) and scattered (lining of alimentary tract, bone marrow). The major constituent cells are lymphocytes in a mixed population that takes in plasma cells, macrophages, and their precursors (forebears). Lymphocytes circulate as single cells in the bloodstream and account for one-fifth of the white blood cells present.

The cells (predominantly lymphocytes) in the different areas of lymphoid tissue look very much alike when viewed under the microscope, but there is reason to believe that in the immunologic setting they behave quite differently. At least two distinct cell populations have emerged. Both may have a common origin, but they diverged in development of function.

One is a so-called *thymus-dependent* (processed) *system* of cells, because its early development is influenced by the thymus gland. It is made up of small lymphocytes that circulate in the blood and lymph. These cells have a very long life expectancy—many years, a decade, or so—and regulate the cell-mediated immune responses (graft rejection, delayed hypersensitivity).

The other population of lymphoid cells is a *thymus-independent system* of cells. It owes its differentiation to a site not as yet identified in man. The small lymphocytes of this cell population are short-lived—only a week or two. The thymus-independent system of cells regulates the production of humoral immunity as is seen in the body's main defense against bacterial infection.

A lymph node (Fig. 13-1) shows a diffuse arrangement of lymphoid tissue well enscribed by a connective tissue capsule. In the outer part (cortex) of the node there is an orderly alignment of lymphoid aggregates, the *lymphoid follicles* (nodules). If lymph nodes are studied during the course of an immune response, definite changes may be seen within these follicles. With hormonally mediated responses, they become very active and numerous plasma cells appear within the greatly enlarged follicles. With cellularly mediated immune responses the follicles appear to be spared but pronounced changes are found in the lymphoid tissue adjoining the cortical nodules. Within these *paracortical* areas numerous mononuclear cells containing increased amounts of RNA accumulate and later become small lymphocytes ready for release into the circulation.

**Immunochemistry* is the study of the complex chemical reactions involved in immunity.

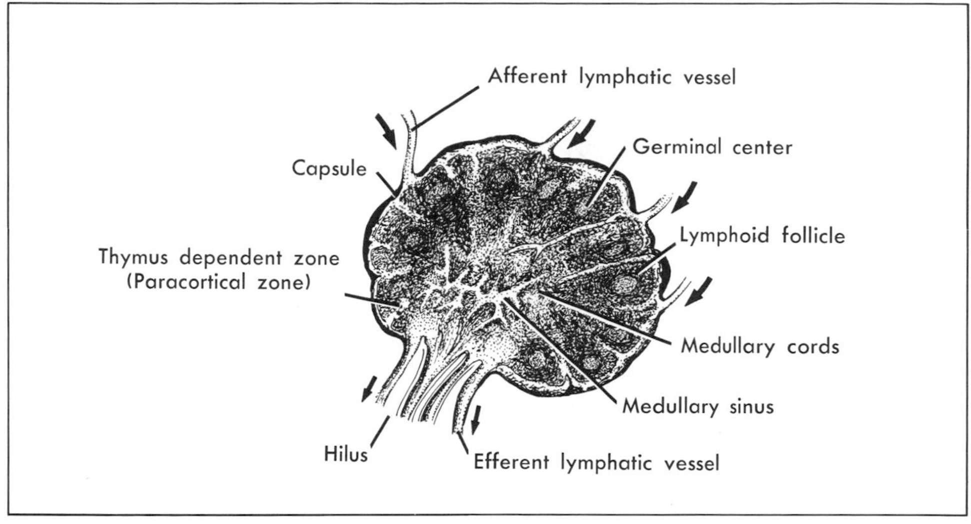

FIG. 13-1. Lymph node in sketch. Note lymphoid follicles in cortex and paracortical areas adjoining.

HUMORAL IMMUNITY

Humoral immunity is that related to antibodies.

■ **Antigens.** When a person or animal becomes immune to a disease, the immunity is largely the result of the development within the body of substances capable of destroying or inactivating the causative agent of the disease should it gain access to the body. These substances, known as *antibodies* or *immune bodies*, are produced by the body in response to a specific stimulus. Microbes and their products may stimulate the body cells to antibody production, as may also certain vegetable poisons, snake venoms, and, from an animal of a different species, red blood cells, serum, and other proteins. A substance, such as any of these, eliciting the production of antibodies against itself when introduced into the animal body, is known as an *antigen*. To act as an antigen, it *must be* introduced into the body; that is, it must be a substance foreign to the body of a given individual. Antigens are usually high molecular weight substances of a protein nature, but in some cases complex carbohydrates (polysaccharides) may act as antigens. Enzymes and many hormones are antigenic.

When certain body proteins (examples, thyroglobulin and lens protein from the eye) of one animal are injected into another animal of the same species, antibodies against the proteins are produced. These antigens are *isoantigens* (*iso*, from an individual of the *same* species), and the antibodies to them are *isoantibodies*. Naturally occurring isoantigens are those residing on the red blood cells and making up the blood groups (p. 182).

Comparatively simple chemical substances (certain low molecular weight lipids and carbohydrates) may be combined with the protein of an antigen to give it specificity. When separated from the protein of the antigen, these substances do not elicit the formation of antibodies but do combine with antibodies already formed against the antigen. These are *haptens* or *partial antigens*.

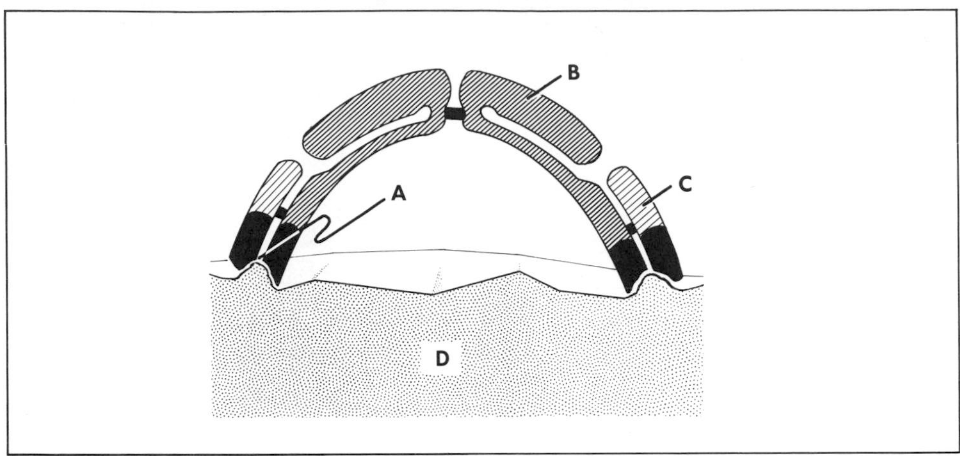

FIG. 13-2. Immunoglobulin molecule, IgG class, diagrammatic alignment of components. **A,** Combining site (with antigen). **B,** H (heavy-chain). **C,** L (light-chain). **D,** Antigen. Note two identical light and two identical heavy chains; they are united by disulfide bonds.

Certain chemicals that within themselves are not antigenic increase the potency of antigens. They are known as *adjuvants.* Among these are alum and aluminum hydroxide, used in preparation of toxoids. In addition to increasing the potency of the antigen, they concentrate it.

■ **Antibodies.** As you may recall, plasma is the liquid portion of the blood in which the red corpuscles and other formed elements are suspended. It is composed chiefly of water (90% to 92%), of proteins (6% to 7%), and various mineral constituents. The plasma proteins are albumin, globulin, and fibrinogen.* By electrophoresis it is shown that the globulin is made up of these main fractions—alpha 1, alpha 2, beta, and gamma globulins. Antibodies migrate electrophoretically in the gamma and beta globulin regions. They are largely associated with gamma globulin, representing 1% to 2% of total serum protein. This is why this fraction (immune globulin) is used so extensively in the prevention of diseases such as infectious hepatitis, poliomyelitis, and measles.

NATURE. Much is being learned of the true nature of antibodies, the specialized globulins that react precisely with the antigen inducing their formation. Antibodies are more precisely designated as *immunoglobulins* (Ig) and belong to a family of related proteins with distinctive properties when studied by special laboratory technics. They are large protein molecules and the most complex group known. Chemically, immunoglobulins adhere to a basic pattern upon which are superimposed significant variations. (See Fig. 13-2.) The basic subunit or building block is structured of four polypeptide chains in two pairs. Two of the paired chains are of greater molecular weight—heavy or H chains—and two of small molecular weight—light or L chains. The light chains are linked to the heavy ones by single disulfide bonds. Light chains are about 200 amino acids long; heavy chains, approximately 450 amino acids.

CLASSIFICATION. The major classes of immunoglobulins as defined by the World Health Organization Committee on Nomenclature of Immunoglobulins are given in Table 13-1 with their distinguishing features; Table 13-2 indicates their immunologic capacity.

FORMATION. The process of antibody (immunoglobulin) formation occurs in the lymphoid tissue of the body in response to the presence of a particular antigen. The principal cells involved are macrophages (mobile cells capable of ingesting fairly large particles of foreign matter), lymphocytes, and plasma cells. On the surface of an antigen

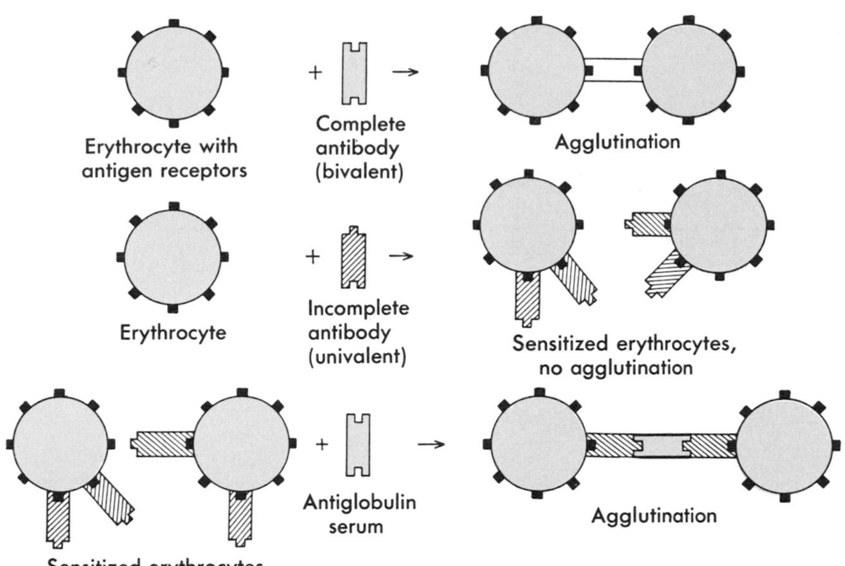

FIG. 13-3. Antigen-antibody reaction, graphic copy. Note three horizontal reactions between antigens depicted on red blood cells (erythrocytes) and antibody. Agglutination (visible clumping) results from bonding of antigens and antibody in topmost reaction. Antigen-antibody complex is indicated in middle diagram but without clumping. Red cells here are altered or sensitized. In contact with antiglobulin serum (see also p. 186), a special complete antibody, sensitized red cells are clumped in lowermost reaction. (From Miale, J. B.: Laboratory medicine: Hematology, ed. 4, St. Louis, 1972, The C. V. Mosby Co.)

there is a special pattern of atoms designated as the *antigenic determinant* or *marker.* The immunoglobulin, to be formed because of the stimulus of the given antigen, must be structured so as to make a neat fitting into the unique arrangement of the antigenic determinant. When an antigen gains access to the body, there is a poorly understood process whereby it is recognized and processed by the RE system. Authorities postulate that the macrophage contacts it, ingests it, and breaks it down into smaller units, preserving meanwhile its antigenic determinant. In some way at the cellular level the macrophage transfers the prescription for the corresponding antibody to the lymphocyte. This causes the lymphocyte to become transformed into an immunologically active cell. Many observers believe that some of the altered lymphocytes differentiate into plasma cells, the cells known to be directly concerned with the synthesis of antibodies. Immunoglobulin has been repeatedly demonstrated in their cytoplasm. However, other cells among the reacting lymphocytes are also thought to form antibodies. The main anatomic sites in which antibody production is found are the spleen, lymph nodes, and bone marrow, but antibody formation may be demonstrated wherever there is lymphoid tissue *with the exception of the thymus.*

Today we speak of the lymphocyte as an *immunologically competent* cell, meaning one that can undertake an immunologic response when engaged by an antigen. During a lifetime, because of the many and varied antigens of our environments, each of us produces tens of thousands of different kinds of antibodies.

■ **Antigen-antibody reaction.** The antigen-antibody reaction is specific: A given antigen promotes the production of antibodies only against itself, and a given antibody

TABLE 13-1. ANALYSIS OF MAJOR CLASSES OF IMMUNOGLOBULINS (ANTIBODIES)

WHO term*	Where found in body	Sedimentation coefficient (ultracentrifugal analysis)†	Chemical composition known	Molecular weight
IgG	Serum (40% intravascular)	7S	4 polypeptide chains— 2 light, 2 heavy	160,000
IgA	Serum (40% intravascular) External secretions: saliva, parotid gland, gastrointestinal, respiratory tract, colostrum (secretory IgA)	9-15S	4 polypeptide chains— 2 light, 2 heavy	160,000 (monomer) to 900,000 (secretory IgA: 390,000)
IgM	Serum (80% intravascular)	19S	20 polypeptide molecules of five basic 4-chain units —2 light, 2 heavy	1,000,000
IgD	Serum (75% intravascular)	7S	—	160,000
IgE	Serum	8S	4 polypeptide chains— 2 light, 2 heavy	196,000

*Designation recommended by the World Health Organization Committee on Nomenclature of
†The ultracentrifuge is an important tool in the identification of antibodies. By means of the exceed according to their molecular weights. Results are quantitated in Svedberg units (S).

acts only against that antigen promoting its development. A person vaccinated against smallpox is protected against smallpox only. The hemolysin in the serum of a rabbit that has been repeatedly injected with the red blood cells of a sheep dissolves the red blood cells of the sheep but not those of other animals.

Within the antibody molecule there is a pattern of small areas specially designed for a close-fitting union with the antigen (like a key fitted into a lock). Most immunoglobulins are bivalent; this is, their structure allows combination with the antigen molecule at two areas.

If the union of antigen and antibody occurs in such a way that the aggregation is manifest as a visible phenomenon, the antibody is said to be a *complete antibody* (Fig. 13-3). When union of an antibody occurs with a specific antigen but the antibody

Estimated percent Ig in population	Serum level (gm./ 100 ml.)	Crosses placenta	Binding of complement	Percent carbohydrate	Electrophoretic mobility (principal)
75%	1.2	Yes	+	2.5%	γ
21%	0.18	No	No	8%	Slow β
7%	0.12	No	+	10%	Between γ and β
0.2%	0.003	No	No	?	Between γ and β
0.5%	0.0001	?	No	11%	γ

Immunoglobulins.
ingly high gravitational forces possible in this instrument, serum protein fractions may be separated

lacks the capacity to change the surface properties of the antigen and therefore a visible reaction does not take place, that antibody is referred to as an *incomplete* or *blocking antibody*.

DETECTION OF ANTIBODIES AND THE ANTIGEN-ANTIBODY REACTION. Discussion of this subject is found in Chapter 14.

■ **Complement.** Although complement (alexin) is not an antibody, its presence is necessary for the complete action of certain antibodies, notably, hemolysins, bacteriolysins, and other cytolysins, and complement or a complement-like substance enhances the action of opsonins. Basically an enzyme system of proteins taking part in the antigen-antibody reaction, complement is a component of fresh serum (10% of globulin fraction) and is not increased by immunization. There are at least eleven complement

TABLE 13-2. IMMUNOLOGIC ACTIVITIES OF THE MAJOR CLASSES OF IMMUNOGLOBULINS

Immunoglobulin	Participation in well-known antigen-antibody reactions: Agglutination	Precipitation	Complement fixation	Lysis	Neutralization: Viruses	Neutralization: Toxins (and enzymes)	General remarks
IgG (gamma G) (γG)	Weak	Strong	Strong	Weak	+	+	1. Ones best studied 2. Responsible for passive immunity of newborn 3. Identified here: a. Certain Rh antibodies b. Bacterial agglutinins c. L. E. cell factor* d. Antitoxins e. Antiviral antibodies
IgA (gamma A) (γA)	+	±	−	−	+	?	1. Known to be made by plasma cells 2. Chief Ig in external secretions 3. Protective function on body surfaces exposed to environment 4. Identified here: a. Diphtheria antitoxin b. Blood group antibodies c. Antibodies against *Brucella* and *Escherichia coli* d. Antibodies against respiratory viruses e. Antinuclear factors in collagen disease* f. Certain other auto-antibodies* (against insulin in diabetes, thyroglobulin in chronic thyroiditis)
IgM (gamma M) (γM)	Strong	Variable	Weak	Strong	+	−	1. Rapid protection here —first antibodies noted after antigen injection

*For discussion of autoimmunity and autoimmune diseases, see pp. 178 and 179.

TABLE 13-2. IMMUNOLOGIC ACTIVITIES OF THE MAJOR CLASSES OF IMMUNOGLOBULINS—cont'd

	Participation in well-known antigen-antibody reactions						
					Neutralization		
Immuno-globulin	Agglu-tination	Precipi-tation	Comple-ment fixation	Lysis	Viruses	Toxins (and enzymes)	General remarks
IgM (gamma M) (γM) —cont'd							2. Powerful agglutinins and hemolysins here (700 to 1000 times stronger than those of IgG in agglutinating red cells or bacteria) 3. Identified here: a. Blood group antibodies (ABO) b. Antibodies against somatic O factors of gram-negative bacteria c. Human anti-A isoantibody d. Isohemagglutinins, cold agglutinins, rheumatoid factor
IgD							Functions not identified
IgE (reagin)							1. Role in allergy—governs responses of immediate-type hypersensitivity 2. Carries skin-sensitizing antibody

components with inhibitors in the complement system. Complement is destroyed by exposure to room temperature for a few hours or to a temperature of 56° C. for 30 minutes. It acts only after the antigen has been sensitized (acted on) by its antibody. During the process of antigen-antibody-complement union, the complement is destroyed or at least inactivated. This is known as *complement fixation.* The activity of complement directed to cell membranes is responsible for the cytolytic effect of complement fixation reactions. The fresh serum of a guinea pig is regularly used as a source of complement because its content shows little variation in different animals and is comparatively high. Complement may be preserved by various chemical methods or by drying the fresh frozen guinea pig serum to a powder in a vacuum.

CELL-MEDIATED IMMUNITY

In acute bacterial infections with gram-positive cocci, the mechanisms of humoral immunity in the normal individual are efficient. Persons with agammaglobulinemia, who cannot make measurable amounts of antibody globulin, do not fare well with the cocci but they do adequately resist infections caused by viruses, fungi, protozoa, and certain other bacteria. Such resistance must stem from an immunologic mechanism other than the classic antigen-antibody combination.

To the host, infection with these latter agents presents a peculiar problem in defense as these microbes colonize the interior of host cells. There they obtain shelter from antimicrobial elements of blood and tissue fluids. Some are in danger if they pass from cell to cell, but others are unaffacted in transit. The patterns of cell-mediated immunity seem designed to meet this situation.

Though much less is known of immunity related to the activities of certain cells than is known of that related to antibodies, *cell-mediated (cellular) immunity,* because of its practical implications for tissue transplantation and rejection, is a subject of considerable current interest. The notion that without demonstrable antibody cells can bring about immune responses was hard to accept for a long time.

■ **Comparisons.** Like humoral immunity, cell-mediated immunity is part of the body's defense. Unlike it, there is no demonstrable, clear-cut antibody. One has been postulated to remain bound in some way to certain cells producing it, but details are lacking. Some observers believe it does not exist. For this reason there is no equivalent of passive immunization with cellular immunity—it cannot be passed in serum. However, transfer can be made in a less predictable fashion with lymphoid cells from a sensitized host.

For the study of cellular immunity there have been few, if any, immunologic procedures to match the extended list of well-standardized tests used to describe antibodies. The traditional skin test, as important as it is, depends on the subject's prior contacts with antigens and is influenced by local tissue factors. As is illustrated by the skin test, the immune responses of cellular type emerge more slowly (several days) than those mediated by antibodies, which can evolve in a matter of hours.

In both kinds of immunity, contact with antigen triggers the sequence of events, but in cell-mediated immunity the antigen does not reach the lymphoid tissues; it seems blocked peripherally. With this type of immunity, concurrence with presence of antigen is crucial. If antigen is destroyed, the immune process subsides. In tuberculin testing, for example, the previously sensitized person has a reaction lasting until all injected tuberculoprotein is gone.

■ **Principal cells.** The principal cells involved are lymphocytes and macrophages. The mediating and primary reacting cells are small, *thymus-dependent* lymphocytes. Authorities think that these lymphocytes possess their unique specificity for reaction if in contact with a certain kind of antigen as a result of the influence of the thymus gland. When contact with that particular antigen is made, the cells become sensitized and respond immunologically.

The macrophages are secondary participants important for their property of phagocytosis. They make up 80% to 90% of cells mobilized to the reaction site.

■ **The antigen.** Antigens that induce the sequence of events in cellular immunity are usually protein or molecules conjugated with protein. (If antibody is formed, it remains fixed to its cell of origin, the sensitized lymphocyte.) The antigenic protein may be quite small or more complex, for example, the lipoprotein transplantation antigen in the mouse.

■ **The events.** The first step is the recognition of the antigen as "foreign" by the population of thymus-dependent lymphocytes. Then ensues a series of changes in lymphoid tissue associated with the release of biologically active substances intended to summon fully equipped macrophages to the scene. They assemble in large numbers, primed to exert their capabilities to attack, sequester, and destroy the nonself that does not belong. Although the activation of the macrophages by the lymphocytes is a poorly understood immunologic phenomenon, it does mean that all of their microbicidal faculties are enhanced.

■ **Manifestations.** Cell-mediated immunity is a major component of the body's resistance to viruses, fungi, and certain bacteria. Classic expressions of it are found in delayed-type hypersensitivity (once known as bacterial hypersensitivity), graft rejection, and graft-versus-host reactions. It may also be the capacity possessed by the body to resist some forms of cancer.

DISTURBANCES IN IMMUNITY

Considerable attention is given today to the elucidation of both normal and abnormal mechanisms in immunity. As a result immunologic disorders are better understood. Some have only recently been defined. The nature of the defect is indicated in the following listing:

1. Immunologic accidents—best example, reaction after mismatched blood transfusion
2. Immunologic depression (immunosuppression)—administration of immunosuppressive agents materially interfering with or inhibiting the immune system (humoral or cellular); such are the steroid hormones in huge doses, irradiation, cytotoxic chemicals (antimetabolite drugs and alkylating agents), and antilymphocytic serum (ALS)
3. Immunologic deficiency—failure or impairment of normal development of immunologically competent cells of the lymphoid system of the body, both antibody deficiencies and defects of cell-mediated immunity included

   An absence of lymphocytes and plasma cells and hence of all immunoglobulins is termed *lymphoid aplasia.* Abnormal development of the thymus gland *(thymic dysplasia)* or a congenital absence of the thymus will result in severe impairment of immunologic mechanisms. Death comes early in infancy.

   An absence of plasma cells with very low levels of gamma globulin is either *hypogammaglobulinemia* (antibodies in greatly decreased amounts) or *agammaglobulinemia* (no demonstrable gamma globulin in the blood). An absence of or a defect in some of the immunoglobulin components of serum is *dysgammaglobulinemia.*

   Certain patients with depressed cellular immunity are peculiarly vulnerable to infections caused by *Mycobacterium tuberculosis, Pseudomonas aeruginosa,* and *Pneumocystis carinii.* An infection such as vaccinia may progress even where there is a demonstrably high titer of circulating antibodies.
4. Immunologic effects of nonimmunologic diseases—infections with bacteria, viruses, and protozoa producing striking elevations in globulin levels of serum *(hyperglobulinemia)*
5. Allergic states—see Chapter 15
6. Autoimmunity and autoimmune diseases—defects of immunologic tolerance*

*Immunologic tolerance refers to the ability of the immune system to discriminate between what belongs to its host (which it tolerates) and what does not (which it attacks).

7. Immunologic aberrations secondary to malignancies of the lymphoid system—as would be expected, associated with striking abnormalities

   The malignant prototype of the lymphoid cell can yield an abnormal type of globulin, often in large quantities. Examples of frankly abnormal globulins synthesized by malignant cells are the myeloma proteins, the macroglobulins of Waldenström, and the Bence Jones proteins. The malignancies include the leukemias, lymphomas, or other neoplastic cell proliferations arising in the reticuloendothelial tissues.

■ **Autoimmunization.** Generally the immunologic mechanisms in the body are protective. Microbes or substances that would enter and injure are attacked. Because the body recognizes its own cells and tissues, it differentiates between the protein of the foreign invader and that which belongs to itself. "Unfortunately," as William Boyd says, "the immune system is a two-edged sword which can be turned against the body in that biological paradox nicknamed autoimmunity. The same forces that normally reject foreign material act in reverse and reject the cells and tissues of the body itself with unpleasant consequences."

Many potentially antigenic substances exist in or on the surface of a person's own cells, but before these can induce an immune response, it is thought that they must be altered in some way, possibly by the action of bacteria, viruses, chemicals, or drugs. Once that change has been made, these antigens, now *autoantigens*, stimulate the body of the individual in whom they occur in certain instances to form the corresponding antibodies, or *autoantibodies*. The process by which antibodies are made by the body against its own cellular constituents is referred to as *autoimmunization*. The importance of autoimmunization lies in the unfavorable and disease-producing effects of certain of the immune responses to the antigen. The category of human disease resulting is spoken of as the *autoimmune diseases*.

■ **Autoimmune diseases.** Autoantibodies can exist without disease. Under certain conditions, however, the reaction of autoantibodies (or sensitized immunocytes of the lymphoid series) with the specific tissues (the autoantigens) does result in disease, the clinical and pathologic picture of which is determined by the particular tissue attacked immunologically, its distribution in the body, and the extent of the damage done by the reaction. Autoimmunity (autoallergy) is seen in disorders of the thyroid gland, brain, eye, skin, joints, kidneys, and blood vessels.

Six features are emphasized in discussions of autoimmune disease: (1) the elevation of serum globulin, (2) the presence of autoantibodies in the serum, (3) the deposition of denatured protein in tissues, (4) the infiltration of lymphocytes and plasma cells in damaged tissues, (5) the benefit of treatment with corticosteroids, and (6) the coexistence of multiple autoimmune disorders.

In some autoimmune disorders the autoantigens involved are known, and circulating autoantibodies can be demonstrated in patient's serum, especially when there is an associated elevation of gamma globulin. In lupus erythematosus, for example, factors are present in the patient's serum that are active against nuclear material and certain cytoplasmic substances of cells (Table 13-3). As a group these diseases often give a history of preceding infection, tend to run in families, and are relieved by steroid hormones (ACTH and cortisone); the patient's serum gives a false positive test for syphilis. Virus particles have been associated with autoimmune diseases, but the relationship is unclear at this time.

Examples of autoimmune diseases are allergic (demyelinating) encephalomyelitis,

TABLE 13-3. EXAMPLES OF DISEASE AND ANTIGEN IN AUTOIMMUNITY

Autoimmune disease	Autoantigen
Acquired hemolytic anemia	Antigens on surface of red blood cells
Idiopathic thrombocytopenic purpura	Platelet antigens
Rheumatoid arthritis	Denatured gamma globulin
Systemic lupus erythematosus	Nuclear material including DNA; certain cytoplasmic substances

chronic thyroiditis, rheumatoid arthritis, acquired hemolytic anemia (both warm and cold antibody types), and idiopathic thrombocytopenic purpura.

DISSEMINATED (SYSTEMIC) LUPUS ERYTHEMATOSUS. One of the most interesting of the autoimmune diseases is disseminated lupus erythematosus. (Its name is derived from the Latin word for "wolf.") A disease primarily of women, it runs an acute or subacute course interspersed with periods of improvement. Although the mechanisms are obscure, immunologic injury is implicated. It is postulated that disease-producing complexes are formed when the autoantibodies, the antinuclear globulins, react with the breakdown products of nuclei in the normal wear and tear of cells (Fig. 13-4). Anatomically, connective tissue which is widely distributed over the body, is destroyed. Therefore lesions can occur anywhere but select the kidney, heart, spleen, and skin. The characteristic butterfly rash is found on the skin of the face. With the changes in the connective tissue are joint pains, fever, malaise, and anemia.

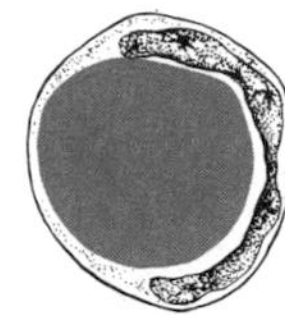

FIG. 13-4. Lupus erythematosus (L.E.) cell. Drawing of distinctive cell found in blood and bone marrow of patient with lupus. It is a mature neutrophil that has phagocytized a homogeneous mass of nuclear material (degraded DNA) in the presence of antinuclear autoantibody in serum.

TISSUE TRANSPLANTATION

There are many times in medicine when it would be highly desirable to take a normal tissue (or an organ) from a healthy person (donor) and transplant it into the body of a patient (recipient) whose corresponding tissue (or organ) is severely diseased. It is true that a piece of skin can be relocated as a graft from one site on the same body to another *(autograft)*; that a transplant of living tissue from one identical, *not* fraternal, twin will take in the other (*isograft, isogeneric* transplant*); and that in plastic surgery pieces of nonviable bone, cartilage, and blood vessel from one individual can be used in another as a framework for new growth in certain areas *(homostatic graft)*. But immunologically we are rugged individuals! In the natural course of things what would be lifesaving grafts of viable skin, bone marrow, kidney, and so on are cast off, rejected,

**Heterografts* or *xenografts* are those between members of different species.

because very subtle antigenic differences exist among even closely related members of the human race.

■ **Graft rejection.** A graft from one member to another within a noninbred species is a *homograft* or *allograft.* Without intervention of any kind this graft is characteristically rejected. For about the first week there is every indication that the graft has taken. After that time the situation changes. The circulation of blood established in the graft is cut down; the graft is infiltrated by mononuclear cells, loses its viability, and is soon cast off.

What has happened immunologically is an expression of cell-mediated immunity. (Although antibodies may also play a part, they are not primarily implicated.) Small thymus-dependent lymphocytes detected the foreign antigen, became transformed, were sensitized in the regional lymph nodes, and were released back into the circulation as "activated" lymphocytes. They returned to the graft site to set the rejection process in motion. Although they initiated it, the actual demolition of the graft was carried out by macrophages, with the help of a few polymorphonuclear leukocytes, that the lymphocytes secured to the site.

■ **Transplantation antigens.** *Histocompatibility* refers to the degree of compatibility, or lack of it, between two given tissues. Of some eight to ten genetically independent systems in the body affecting graft survival or rejection, there are two that figure significantly and that are of practical importance.

The first, a very important histocompatibility factor in man, is the ABO blood group system, which is present on red blood cells and practically all tissues.

The second is determined by a complex genetic region known as the HL-A locus (histocompatibility locus-A) because it controls homograft immunity. It contains a group of antigens associated with cell membranes that are found in the majority of cells and tissues of the body, and especially in lymphoid and blood-forming tissues (not on red blood cells, however). This complex system of at least eleven antigens, of which a person may have as many as four, is most easily identified in leukocytes. Therefore tissue typing is commonly done with white blood cells. Practically speaking, the antigens usually referred to as *transplantation antigens* are those of the HL-A system. Part of the graft, they are the ones that mark the graft as "trespasser" to the recipient host.

■ **Testing for histocompatibility.** When tissue transplantation is anticipated, typing and cross matching of blood specimens from the postulated donor and patient-recipient are done as though for blood transfusion. Any incompatibility in the ABO blood groups is an *absolute* contraindication to the use of the donor's tissue.

Tissue typing for transplantation, done with white blood cells or platelets, is partly analogous to red blood cell typing and cross matching for clinical blood transfusion. Of the various technics for tissue typing, lymphoagglutination and lymphocytotoxicity are widely used. *Lymphoagglutination* compares the clumping reactions of would-be donor and recipient lymphocytes when tested with a panel of selected antiserums. *Lymphocytotoxicity* measures the lethal effect of plasma membrane injury to presumed donor and recipient lymphocytes under similar conditions. Recipient serum can be mixed with the would-be donor's lymphocytes as a "cross match." Reaction in this indicates probability of immediate graft rejection. In evaluating the degrees of donor-recipient incompatibility for HL-A types determined from tissue typing, there are no absolute matches (except with identical twins). A "good" match still means differences, although small.

■ **Immunosuppression.** Various clinical and experimental technics have been designed to suppress or manipulate artificially the body's normal immune responses so that tissue transplantation would be possible. *Immunosuppression* (induction of tolerance) is the term used for the sum of approaches. Several are available. A potent antiproliferative drug such as 6-mercaptopurine (6-MP) that selectively interferes with either DNA or RNA synthesis may be used singly or combined with steroid hormones and irradiation. Depletion of lymphocytes has been attempted by drainage of the thoracic duct. Antilymphocyte globulin (ALG) has been administered to selectively eliminate the cellular population mediating the immune responses and perhaps to spare the sorely needed humoral ones. The disadvantage of immunosuppressive agents is that they concurrently impair normal host resistance. Such an individual with either subnormal or practically nonexistent immune responses is spoken of as the *compromised host*. An effort is being made experimentally to induce tolerance to the graft under conditions that do not compromise the body's normal defenses when they are needed elsewhere in the body.

■ **Graft-versus-host reaction.** It has been observed in animals and a few humans that at times instead of the host rejecting the graft, the graft has rejected the host. This has been found especially with transplantation of bone marrow and lymphoid tissue. In the immunologically depressed host, the graft survives. The immunologically competent lymphoid cells of the graft recognize the "foreignness" of the environment, and the stage is set for rejection. The end of it—runt disease in animals, secondary disease in man—is characterized by wasting, diarrhea, dermatitis, infection, and death.

■ **Present status.** Much has been accomplished in tissue transplantation. Transplantation of the kidney is established; worldwide more than 3000 recipients of a kidney transplant have survived. Transplantation of other organs is still experimental. Fallout has been considerable in terms of new knowledge in basic disciplines, control of infection, and a breakthrough in cancer immunology. It was discovered that with immunosuppression there is also a greater likelihood for the development of certain cancers. Problems are formidable still but optimism and enthusiasm prevail.

IMMUNITY IN CANCER

In man and animals there are many reasons for thinking that a relationship exists between the individual's immunologic system and his cancer. Some stem from the abundance of experimental work and a number of interesting observations made on patients with organ transplants. (Immune mechanisms must be tampered with dramatically for organ transplantation.) Malignancies complicate immunosuppression, and in congenital immunologic deficiencies as well, there is a high incidence of cancer.

The immunology of cancer is complex, and a lot of work is being done on it. It is safe to say that most and maybe all neoplasms contain tumor-associated antigens. Tumor antigens are tumor specific and provoke an antitumor response either in the host animal or upon transplantation of the tumor into a comparable animal. Typically the immune response is mediated by cells of the lymphoid series. Tumor antigens arise because the cancer cell differs from the normal cell (not because of differences between donor and recipient animal). They probably result from the initial transformation of the normal cell into a cancerous one. At this early time a neoplasm is most vulnerable to the immune response. Probably many incipient ones are completely and quietly destroyed.

As to be expected there are different kinds and degrees of tumor immunity. A special kind is found with tumors induced by viruses. Surface antigens appear as a direct effect of viral activity, and these are consistent for the same viral oncogen in other circumstances. Not so if the oncogenic mechanism is nonviral. That neoplasm has its own distinct antigen, it is true, but even though the same carcinogen is applied in exactly the same way to other anatomic areas in the same animal, that tumor antigen is not repeated. As a result of this effect, many antigenic types become possible with varying immunologic potential. If tumor cells are highly antigenic, they may immunize against themselves so efficiently that tumor growth is blocked.

In view of the existence of an immune mechanism, how does one explain the prevalence of clinical cancer? One can only surmise. It may be that the bulk of experimental cancers and cancers in human beings arise at a time when immunologic defenses are down. It may be that a small fraction of starter growths can make the clinical scene because they combine the least in antigenicity with the highest in proliferative activity.

IMMUNOLOGY OF THE RED BLOOD CELL

The red blood cell (Fig. 13-5) is composed of a compact stromal protein that supports molecules of lipid and hemoglobin. Antigens of the blood groups are contained in both the stroma and on the surface of the red cells. On the surface, exposed antigens are ready to react with specific antibodies. The spatial features of the relatively small antigenic groups out on the cell surface are such that each red cell can have a large number and variety of them.

■ **Agglutinins of the blood.** The blood of every person falls into one of four ABO blood groups, depending upon the ability of his serum to agglutinate the cells of persons whose blood is in another group (Fig. 13-6). This is the result of the distribution of two agglutinins in the serum and two agglutinogens (antigens) on the red blood cells. The agglutinins are known as *anti-A* and *anti-B*. The agglutinogens are designated *A* and *B*. Agglutinin anti-A causes cells containing agglutinogen A to clump. Agglutinin anti-B causes cells containing agglutinogen B to clump. The blood groups and their agglutinin and agglutinogen content are as follows:

Group AB: Serum contains no agglutinin; cells contain agglutinogens A and B
Group A: Serum contains agglutinin anti-B; cells contain agglutinogen A
Group B: Serum contains agglutinin anti-A; cells contain agglutinogen B
Group O: Serum contains agglutinins anti-A and anti-B; cells contain no agglutinogen (See Table 13-4.)

TABLE 13-4. DETERMINATION OF BLOOD TYPE FROM AGGLUTINATION OF RED BLOOD CELLS IN SPECIFIC TYPING SERUM

If agglutination of red cells occurs in		Then blood type is	How many persons in the United States with it?	
Anti-A serum	Anti-B serum		If it is Rh positive	If it is Rh negative
−	−	O	1 in 3	1 in 15
+	−	A	1 in 3	1 in 16
−	+	B	1 in 12	1 in 67
+	+	AB	1 in 29	1 in 167

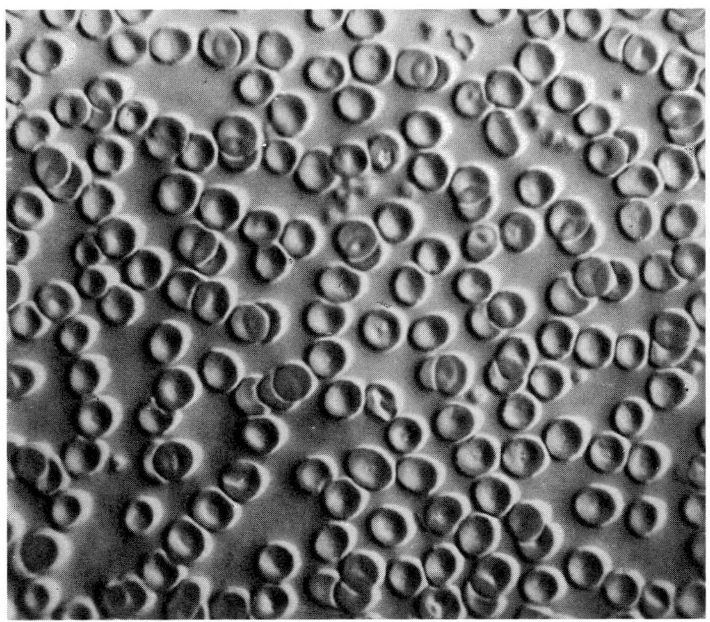

FIG. 13-5. Red blood cells in thin film upon which vaporized chromium was shadow-cast at fixed angle. Note biconcave shape. (From Mountcastle, V. B., editor: Medical physiology, ed. 12, St. Louis, 1968, The C. V. Mosby Co..

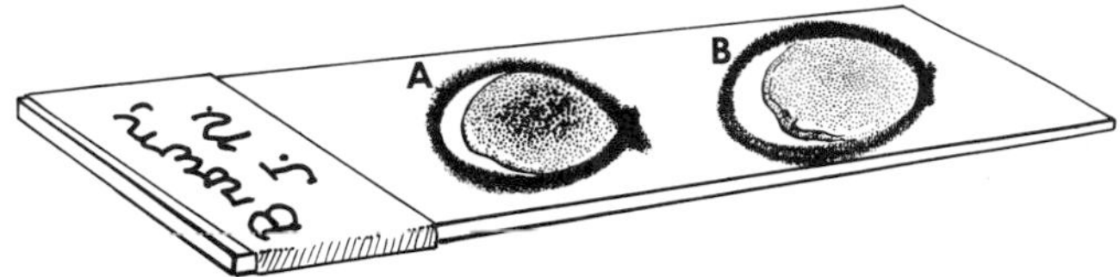

FIG. 13-6. Technic of blood grouping. Make a suspension of red blood cells in isotonic saline from person whose blood is to be typed. Mark out two circles (wells) on a microslide; label **A** and **B.** Place a drop of test suspension of red cells in each well. Mix a drop of serum from a person belonging to blood group A (anti-B serum) with the drop in the well labeled **B,** and a drop of serum from a person belonging to blood group B (anti-A serum) in the well labeled **A.** Macroscopic agglutination of red blood cells in the drop of typing serum is seen in **A.** In **B** there is no agglutination of red cells. If **A** identifies the typing serum as anti-A serum and **B** as anti-B serum, what type of blood is here represented?

If a person whose blood belongs to one group receives a transfusion of blood from a donor whose blood belongs to another group, a hemolytic transfusion reaction is likely to occur. The reason is that the serum of the recipient may agglutinate the cells of the donor, or the serum of the donor may agglutinate the cells of the recipient. Reactions to transfusion of mismatched blood are associated with agglutination of red cells in the circulation, hemolysis of red cells, and liberation of free hemoglobin into the plasma; this leads to hemoglobinuria, fever, prostration, failure of kidney function, and in some cases death. A reaction is much more likely to occur if the recipient's serum agglutinates the donor's cells than if the donor's serum agglutinates the recipient's cells because the donor's serum is considerably diluted in the circulation of the recipient. This effect tends to minimize the agglutinating capacity of the donor's serum. For kinds of transfusion reactions, see Table 13-5.

TABLE 13-5. OUTLINE OF TRANSFUSION REACTIONS

Kind	Cause	Clinical features
Hemolytic	Mismatched transfusion; example, blood type A given to patient with blood type O	Severe chill; lumbar pain; nausea and vomiting; fever; suppression of urine (urine may be reddish brown); jaundice; death
Pyrogenic	Fever-producing substance in blood (sterile chemical contaminants, bacterial toxic products, antibodies to white blood cells)	Occurs 30-60 min. after transfusion; flushing; nausea and vomiting; headache; muscular aches and pains; chills and fever
Contaminated blood	Break in aseptic technic: bacteria (usually gram-negative rods) introduced at time blood collected or with use of unsterile equipment	Chills, fever; generalized aching and pain; marked redness of skin; drop in blood pressure; shock
Allergic	Occurs in patients with history of allergy; cause unknown	Occurs within 1 or 2 hours of transfusion; itching of skin; hives or definite rash; swelling of face and lips; sometimes asthma
Sensitivity to donor white blood cells, platelets, or plasma	Multiple transfusions favor development of leukoagglutinins (those to white cells) and similar substances	Chill-fever; headache; malaise. Serum may agglutinate white blood cell suspension from donor
Circulatory overload	Blood transfused too rapidly into person with failing circulation; likely to occur in elderly, debilitated, or cardiac patients, also in children	Difficulty in breathing; cyanosis; cough with frothy, blood-tinged sputum; heart failure
Embolic	Infusion of air: (1) transfusion under pressure, (2) tubing not completely filled before venipuncture	Sudden onset of cough, cyanosis, syncope, convulsions

Although not a transfusion reaction, a serious complication of blood transfusion is posttransfusion hepatitis.* It is a serious concern to blood banks. The link between the disease and the Australia antigen (AuAg) has led in many blood banks to the routine screening of all would-be donors for its presence. The high-risk carriers of the agent (see p. 449) are most likely to be the "commercial" donors from the skid-row, drug-addict, and prison-inmate populations.

THE RH FACTOR. An agglutinogen with no relation to the ABO blood groups just listed is the *Rh factor*. When a guinea pig is given repeated injections of the red blood cells of a rhesus monkey, the serum of the animal acquires the ability to agglutinate the red cells of rhesus monkeys. It also agglutinates the red blood cells of a majority of human beings (85% of the Caucasian race; more in others). The red blood cells that are agglutinated contain the Rh factor, and the person from whom the cells were re-

*Synonyms are serum hepatitis, long-incubation hepatitis, virus B hepatitis, tattoo hepatitis, postvaccinal hepatitis, HAA-positive hepatitis, homologous serum jaundice, syringe hepatitis.

moved is said to be *Rh positive*. Persons who do not have the Rh factor on their red blood cells are said to be *Rh negative*. The plasma of Rh-negative persons does not contain agglutinins against the Rh factor, but such agglutinins may develop if blood transfusions of Rh-postive blood are given to these people. Such incompatible blood transfusions if continued may lead to serious reactions.

If an Rh-negative mother and Rh-positive father have children, one or more will probably be Rh positive. The Rh-positive cells of her fetus can sometimes get into the bloodstream of the mother during gestation but usually in such small amounts that they do not induce antibody formation. It is thought that the Rh-negative mother has antibodies against the Rh factor during a given pregnancy because she was immunized in a previous one. When the Rh-positive cells of the baby enter the mother's circulation at time of delivery, they can be most effective in stimulating antibody formation. Antibodies formed against the Rh-positive cells begin to appear about 6 weeks later.

When antibodies against the Rh factor are present in the plasma of the Rh-negative mother, they freely pass through the placenta to attack the red blood cells of the fetus and destroy them. As a consequence, the child may be born with *erythroblastosis fetalis* or *hemolytic disease of the newborn*. The disease attacks the fetus before it is born or within the first few days of postnatal life. Its features are anemia, jaundice, edema, and enlargement of the infant's spleen and liver. As to be expected, erythroblastosis seldom occurs during the first pregnancy, but when once it happens, the condition will recur in about 80% of subsequent pregnancies.

Modern treatment of hemolytic disease of the newborn is by means of *replacement* or *exchange transfusion*. Small amounts of the infant's blood are alternately removed and replaced by corresponding amounts of compatible Rh-*negative* blood (the cells of which are not affected by the antibodies) until about 1 pint of blood has been transfused. By this procedure about 85% of the infant's vulnerable cells may be removed, with considerable improvement in the blood picture and in the infant's clinical condition.

Designated the Rh "vaccine," an immunizing agent* has become available recently to prevent a susceptible Rh-negative mother from developing anti-Rh antibodies. This agent is human gamma globulin containing anti-Rh antibodies and obtained from sensitized Rh-negative mothers who have given birth to erythroblastotic babies. Injected as a single dose within 72 hours of delivery, it cancels out the antigenic effect of the Rh-positive cells that have entered the circulation. Immune mechanisms for the production of Rh antibodies are blocked. In the event of future pregnancy with an Rh-positive fetus, the immunologic basis for hemolytic disease of the newborn has been eliminated.

■ **Blood grouping (typing).** Human blood is routinely grouped (or typed) in modern blood bank laboratories into one of the four major blood groups, and the presence or absence of the Rh factor is determined. This means that the blood is typed as O, A, B, or AB and as either Rh positive or Rh negative. However, the red blood cell is a very complex structure and contains many antigens or factors (at least 60), in addition to the A and B agglutinogens and the Rh factor, and new blood factors are constantly being discovered. As these antigens are studied, an attempt is made to classify them into categories containing related antigens, designated *systems*. The ABO system is made up of four blood groups relating to the presence or absence of the agglutinogens A and B as just explained. There are today at least 15 well-known systems of blood groups:

---

*Trade name RhoGAM.

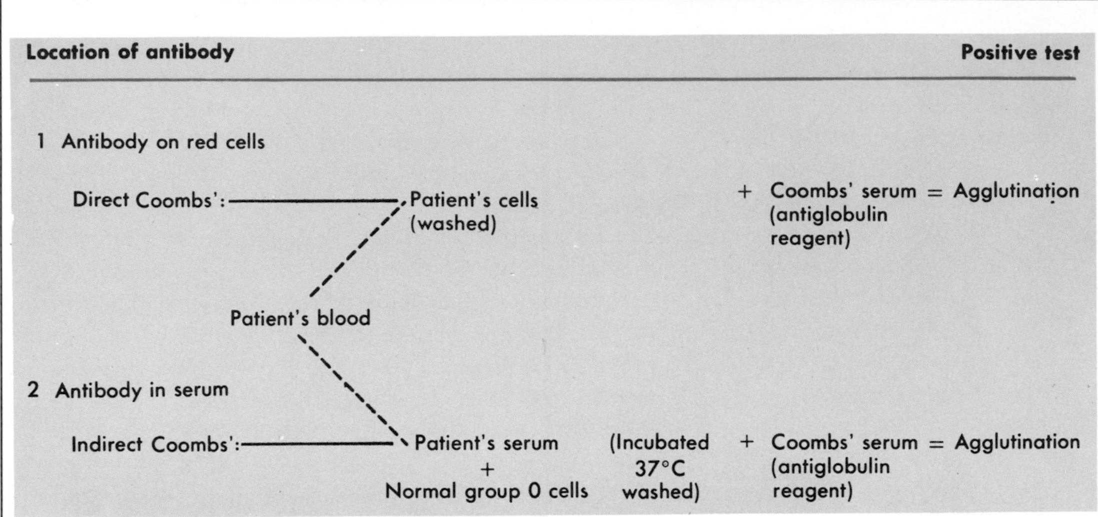

FIG. 13-7. Coombs' antiglobulin test (direct and indirect), outline. The antiglobulin test (patient's cells plus Coombs' serum) indicating presence of antibodies on red blood cells is the *direct* Coombs' test. The *indirect* Coombs' test (patient's serum plus 0 cells plus Coombs' serum) identifies presence of antibodies in patient's serum. No agglutination is a negative test. (Modified from Bove, J. R.: Coombs' [antiglobulin] test, J.A.M.A. **200:**459, 1967.)

(1) ABO, (2) MNS, (3) Rh, (4) P, (5) Lewis, (6) Kell, (7) Lutheran, (8) Duffy, (9) Kidd, (10) Sutter, (11) Vel, (12) Diego, (13) I, (14) Auberger, and (15) Xg (the only sex-linked blood group).

CROSS MATCHING. In routine blood grouping or typing for purposes of blood transfusion, the presence or absence of antigens in all of these systems is not determined. But it is important to determine whether blood from a donor will be safe to give to a recipient, most often a patient with some medical or surgical condition. Certain technics have been devised to determine the relative safety of such a procedure and to prevent a blood transfusion reaction. The first step is called the *major cross match,* and in this the red blood cells of the donor are mixed with the serum of the person to receive the blood. The mixture is carefully observed for any sign of clumping of the red blood cells, and any agglutination means that the donor blood is incompatible with the blood of the recipient. If the recipient were to receive this blood even in small amounts, a transfusion reaction could easily occur.

The second step is the *minor cross match,* in which red blood cells, this time from the recipient, are admixed with serum from the donor. Again the mixture is carefully observed for any sign of clumping of the red blood cells, which would mean an incompatibility of the two bloods. Agglutination at this step indicates the possibility of a transfusion reaction should the donor blood be transfused into the recipient, but the reaction would probably not be so severe as an incompatibility picked up in the major cross match.

A third step is further indicated to detect unusual or uncommon antibodies that for some reason are not demonstrated in the major or minor cross match. In many blood bank laboratories this is accomplished by the use of *Coombs' test* (also known as the antiglobulin test). If we remember that antibodies are primarily related to the globulin fraction of the plasma proteins, and if we know that the protein globulin itself can function as an antigen to stimulate the production of antibodies, then it should be easy to understand that an antihuman globulin serum can be prepared by successively injecting human serum into a suitable animal, such as the rabbit or goat. The serum of this immunized animal will then contain an antibody against the globulin of human serum. Since this fraction contains the antibodies, it means that this Coombs' serum will also have an action against antibodies, an action, as one can see, of a general or nonspecific nature. The Coombs' test does not in any way identify antibodies; its chief use lies in the fact that it can detect antibodies or indicate that they are present and attached to red blood cells (see Fig. 13-7). It helps to indicate the presence of antibodies even when an antigen-antibody combination has not resulted in an observable reaction. Such a test is of immeasurable value in blood typing and, on the whole, is easily carried out. If this test is negative when donor red blood cells and recipient serum are used in the procedure and if the major and minor cross matches show no incompatibility, it is then considered safe for the donor blood to be given as a transfusion to the recipient in question.

QUESTIONS FOR REVIEW

1. Fill out the following table, indicating whether each example of acquired immunity (a) is obtained naturally or artificially and (b) is active or passive.

	Acquired			
Cause of immunity	*Naturally*	*Artificially*	*Active*	*Passive*
Attack of measles				
Attack of smallpox				
Vaccination against poliomyelitis				
Vaccination against smallpox				
Pasteur treatment				
Diphtheria antitoxin				
Diphtheria toxoid				
Tetanus antitoxin				
Tetanus toxoid				
Immune globulin				
Bacterial vaccine				
Placental transmission of antibodies				

2. Classify immunity. Give an example of each kind.
3. What type of immunity is of the longest duration? The shortest? Explain.
4. What are immunoglobulins? How may they be classified?
5. Compare humoral immunity with cellular immunity.
6. What is the role of the lymphoid system in immunity? Role of thymus gland?
7. Comment on the immunologically competent cell.
8. Categorize in general terms the disturbances in immunity.
9. In the following diagram, place arrows pointing from the serum of each blood group to the cells agglutinated by that serum.

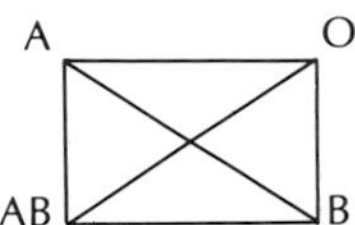

10. Why is a transfusion reaction more likely to occur when the donor's cells are agglutinated by the recipient's serum than when the recipient's cells are agglutinated by the donor's serum?
11. List the different kinds of transfusion reactions. Briefly explain.
12. Cite examples of immunosuppressive therapy.
13. What is meant by autoimmune disease? How is it thought to come about?
14. State the importance of the transplantation antigens.
15. Explain the difference between autograft and heterograft; between isograft and homostatic graft.
16. Discuss briefly immunity in cancer.
17. Define or explain: immunity, immunology, immunopathology, immunoglobulin, immunochemistry, immunobiology, immunosuppression, agglutination, antigenic determinant, isoantigen, adjuvant, immune system, and susceptibility.
18. What practical application is made of the Coombs' test?

REFERENCES. See at end of Chapter 15.

14 Methods for microbes: immunologic reactions

Because of the specificity of the antigen-antibody reaction, if either the antigen or antibody is known, it is possible to identify the other. In many instances, by diluting the one known it is possible to obtain reliable information as to how much of the other is present. Since antigen-antibody reactions are usually measured in serum,* they are called *serologic reactions. Serology* is the term that means study of the antigen-antibody combination. With a positive serologic test, the unit that measures the number of antibodies present is the *titer.* There is a variety of technics available. Some of the important ones will be discussed here.

DETECTION OF ANTIBODIES

The type of antigen-antibody reaction that applies in any given situation depends largely on the physical state of available antigen. Practically speaking, antibodies (immunoglobulins) are detected by what they do to produce a *visible* reaction with a specific antigen under specified conditions (Table 14-1). In the hospital and clinical laboratory antibodies are commonly classified as (1) complement-fixing antibodies, (2) antitoxins (neutralizing), (3) precipitins, (4) cytolysins, (5) agglutinins, (6) opsonins, (7) virus-neutralizing antibodies, and (8) fluorescent antibodies. In Table 14-1 this list of antibodies is tied in with antigens involved, and the nature of the antigen-antibody reaction is briefly indicated.

KINDS OF REACTIONS

■ **Complement fixation.** The complement-fixation test is designed to detect complement-fixing antibodies and is based upon this fact: When serum containing the complement-fixing antibodies against the etiologic agent of a disease, the microbe itself, and complement are mixed in suitable proportions and incubated, the three enter into a kind of combination whereby the complement is destroyed or inactivated (fixed).

*To collect serum: Allow a sample of blood to clot. The fibrin formed in the process entraps the blood cells. When the clot shrinks, the fluid expressed is the serum, an important source of antibodies. Practically in the clinical laboratory, serology is the study of antigen-antibody reactions in a specimen of serum.

TABLE 14-1. DETECTION OF ANTIGEN-ANTIBODY COMBINATIONS

The positive reaction	Antigen	Antibody (in serum)	Nature of reaction
Complement fixation	Microbes	Complement-fixing antibodies	Inactivation of complement detected by hemolytic indicator system; *absence* of hemolysis a positive test
Flocculation	Exotoxin	Antitoxin	Flocculent precipitate
Precipitation	Microbes Animal proteins	Precipitins	Fine precipitate from clear solutions
Cytolysis	Intact cells	Cytolysins	Cells dissolved
Bacteriolysis	Bacteria	Bacteriolysins	
Hemolysis	Red blood cells	Hemolysins	
Agglutination	Agglutinogens	Agglutinins	Gross clumping of cells
	Bacteria		
	Proteus bacilli	Heterophil antibodies	
	Human red blood cells	Isoantibodies	
	Sheep red blood cells	Heterophil antibodies	
Inhibition of viral hemagglutination	Human type O red cells coated with viral antigen	Inhibiting antibodies	Clumping of red cells *blocked*
Opsonization	Bacteria	Opsonins	Bacteria phagocytosed in greater numbers by leukocytes
Neutralization or protection	Viruses	Virus-neutralizing antibodies	Protection of an animal or tissue culture from harmful effects of agent
	Toxins	Antitoxins	
Fluorescence under ultraviolet light	Microbes	Fluorescein-tagged antibodies	Reaction tagged with marker

Since this combination is not accompanied by any visible change, some kind of indicator system must be used to detect it. Because hemolysis is dramatic, a hemolytic system is a good indicator. In this system the combination of complement, *hemolysin (amboceptor)*, and sensitized red cells (of the species of animal against which hemolysin was prepared) effects *hemolysis* or dissolution of the red cells.

If a sample of serum from a patient is mixed with complement and a bacterial suspension and incubated, complement will be fixed when complement-fixing antibodies to the given bacteria are present in that serum. If suitable proportions of hemolysin and red blood cells are then added and the mixture incubated a second time, hemolysis *cannot* occur. This is a positive test, graphically illustrated in the schema (Table 14-2). If, on the other hand, the patient's serum does not contain the specific antibodies, complement is *not* fixed during the first incubation and hemolysis can occur when hemolysin and red blood cells are combined with complement. This is a negative test.

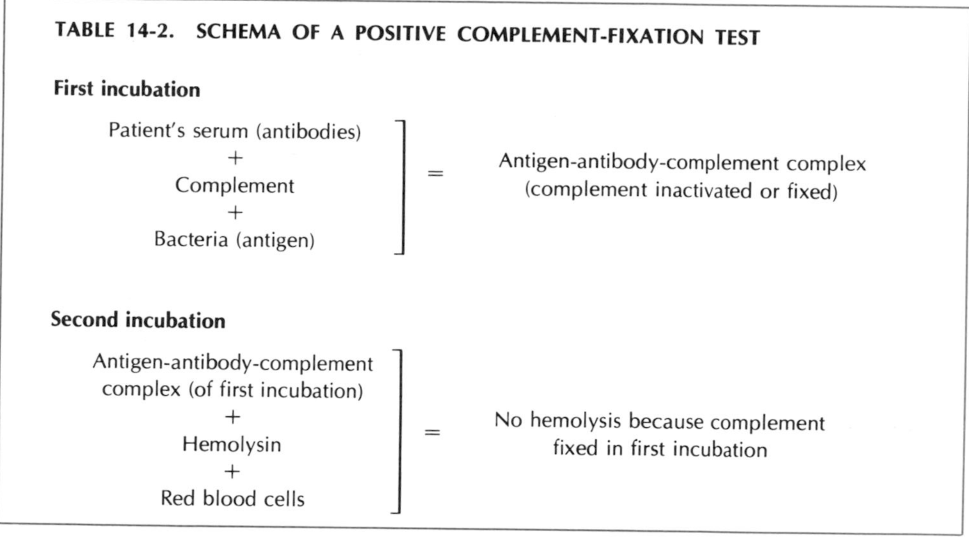

TABLE 14-2. SCHEMA OF A POSITIVE COMPLEMENT-FIXATION TEST

First incubation

Patient's serum (antibodies) + Complement + Bacteria (antigen)	=	Antigen-antibody-complement complex (complement inactivated or fixed)

Second incubation

Antigen-antibody-complement complex (of first incubation) + Hemolysin + Red blood cells	=	No hemolysis because complement fixed in first incubation

WASSERMANN TEST. The Wassermann test for syphilis depends upon the principle of complement fixation. It has always been impossible to grow the causative organisms of syphilis by artificial laboratory methods. When the Wassermann test first came into use, extracts of the livers of syphilitic fetuses were used as antigens. These extracts were used because the liver of a syphilitic fetus was known to contain myriads of the spirochetes of syphilis and, as such, were the nearest approach to extracts of the organisms themselves. Today certain lipid extracts of normal organs are efficient antigens. A lipid extract of beef heart gives clinically reliable and consistent results. Of course, the term "*antigen*" applied to extracts of normal tissue used in the Wassermann test is scientifically incorrect. In other diseases the antigens used in complement-fixation tests are derived directly from the causative organisms.

■ **Toxin-antitoxin flocculation.** Antitoxins are antibodies that neutralize toxins, the combination demonstrable by flocculation. *Flocculation* refers to the presence of aggregates visible microscopically or macroscopically with certain antigen-antibody reactions. Toxin-antitoxin flocculations are similar to precipitin reactions (described below) but show a very sharp end point, with the reaction being blocked by either too much antigen or too much antibody.

■ **Precipitation.** When a suitable animal is immunized with certain bacteria, the animal's serum acquires the ability to cause a fine, powdery precipitate from the clear filtrate of the bacterial growth (*soluble* antigen of the bacterial extract is present). The antibodies formed are *precipitins*. The same phenomenon is noted when animals are immunized with soluble antigens of certain animal and vegetable proteins. Precipitins are specific; bacterial precipitins react only with the filtrates of the bacteria that produced them, and precipitins formed by immunizing an animal with the serum of an animal of a different species react only with the proteins of the species used for immunization. Precipitins do not form regularly in infectious disease, but where they are present they are nicely demonstrated by technics of immunodiffusion.

The precipitin reaction has a wide medicolegal application in determining whether bloodstains are of human origin and in detecting the adulteration of one kind of meat with another. It is the test upon which Lancefield's classification of streptococci is based. Flocculation tests for syphilis, such as the Kline, Kahn, and VDRL, are modified precipitation tests.

C-REACTIVE PROTEIN. A biologic coincidence occurs in a number of inflammatory conditions of either infectious or noninfectious nature. A peculiar protein (a globulin) that can form a precipitate when in contact with the somatic C-polysaccharide of pneumococci appears in the blood of the affected person. Because of this precipitation, the protein is spoken of as C-reactive protein; it is *not* an antibody. The serum of an animal immunized to this protein is used in a precipitation test to detect C-reactive protein in serum of persons suspected of having one of the diseases in which the protein appears. Among such diseases are acute rheumatic fever, subacute bacterial endocarditis, staphylococcal infections, infections with enteric bacteria, and neoplasms.

■ **Cytolysis.** Antibodies that effect the dissolution or lysis of cells are known as *cytolysins* (or *amboceptors*). *Bacteriolysins* are cytolysins that lyse bacteria and are produced or at least increased by bacterial infection. Cytolysis is dependent upon the presence of both cytolytic antibodies and complement; if only one is present, cytolysis does not occur. A cytolysin of great importance in immunologic procedures is hemolysin, which effects dissolution of red blood cells. It is prepared by giving an animal (most often a rabbit) a series of injections of the washed red blood cells from an animal of another species (most often man or sheep). Hemolysin develops in the serum of the immunized animal. It is specific for the red blood cells of the species used, it does *not* affect the red cells of another species.

■ **Agglutination.** If the serum of a person who has had a certain disease, such as salmonellosis, is mixed with a suspension of the causative bacteria, the bacteria will adhere to one another and in the test tube form easily visible clumps that sink to the bottom. Likewise, if an animal receives several injections of a given microbe, its serum will acquire the ability to clump or *agglutinate* the microorganisms used. This kind of clumping is *agglutination* (Figs. 14-1 and 14-2), and the antibodies that bring it about are called *agglutinins.* Particulate substances (antigens) that, upon injection into an animal, induce the formation of agglutinins are known as *agglutinogens.* Agglutinogens may be microorganisms, red blood cells, or latex spheres coated with adsorbed antigen. In each instance these must be particles for the characteristic clumping to occur.

Motile bacteria lost their motility before agglutination occurs. Bacteria are not necessarily killed by agglutination, and dead bacteria are agglutinated as easily as are live ones. Agglutinated microorganisms are more easily taken up by phagocytic cells.

KINDS OF AGGLUTININS. Normally occurring agglutinins may exist in the blood of certain persons. For instance, human serum sometimes shows a weak agglutinin content against typhoid bacilli although the person has never been knowingly infected with the organisms. The agglutinin content is so low, however, that, when the serum is diluted 20 or 40 times and then mixed with a suspension of organisms, agglutination does not occur.

Immune agglutinins are agglutinins brought about by infection or artificial immunization. Such agglutinins can occur plentifully in serum, and such serum will cause agglutination though diluted 500 to 1000 times. Agglutinin formation is usually specific; that is, when an organism is introduced into the body, the body forms agglutinins against that organism only, although, in some cases, against very closely related organisms. Agglutinins for closely related organisms are known as *group* agglutinins. However, the agglutinin content of the serum is much higher for the organism directly inducing the production of the agglutinins than it is for the closely

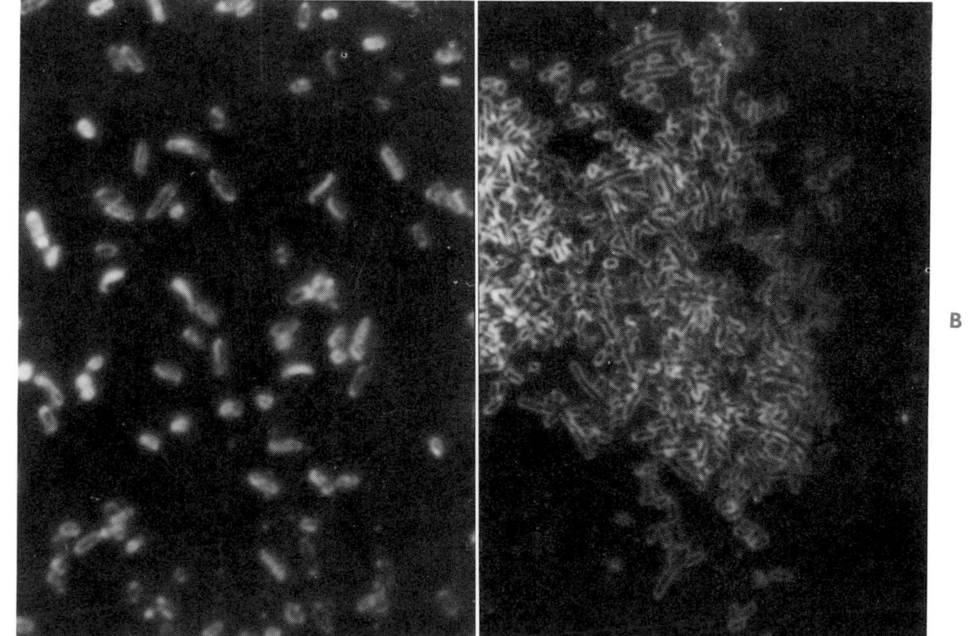

FIG. 14-1. Bacterial agglutination, dark-field illumination. **A,** Saline suspension of killed *Salmonella,* bacteria evenly spread with no clumping. **B,** Addition of positive antiserum; note large clump of organisms. (Courtesy Gwyn Hopkins and Dr. Gerard Noteboom, Dallas, Texas.)

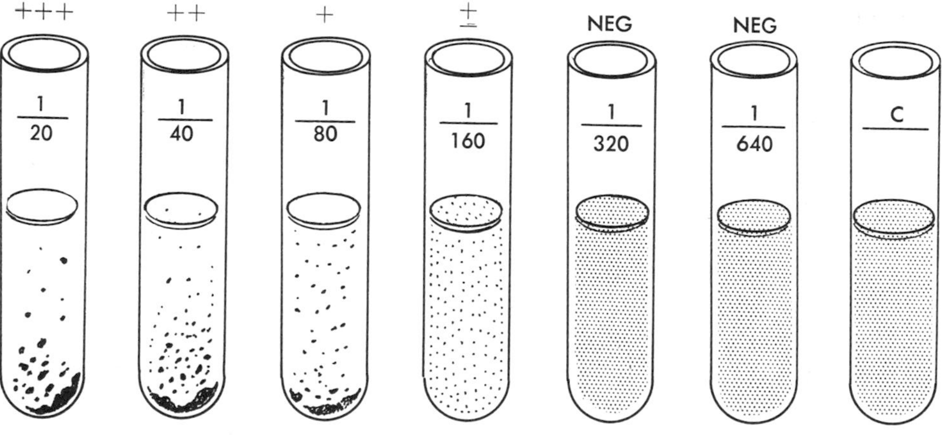

FIG. 14-2. Bacterial agglutination, test tube method. Two-tenths milliliter of patient's serum is serially diluted in isotonic saline through first six tubes (left to right). One milliliter of a suspension of bacteria is added to each tube through all seven. Final dilutions indicated by numbers on each tube (as 1 part serum in 20 parts suspension). Control tube on far right contains only bacterial suspension and saline. Note maximal agglutination of bacteria (large masses) in first three tubes on left, some in the fourth, and none in fifth and sixth. Agglutination test positive; titer of serum 1:160.

related organisms.* Usually in an infection, the human (or animal) body has not manufactured enough agglutinins for positive identification until the infectious disease has lasted a week or 10 days.

Comparatively few organisms induce appreciable agglutinin formation with infection or artificial immunization. Most important of these are *Salmonella typhi*, other *Salmonella* species, *Brucella abortus*, *Brucella melitensis*, and *Pasteurella tularensis*. In infections with these organisms (salmonellosis, undulant fever, and tularemia) the agglutination test (Fig. 14-2) becomes an important diagnostic procedure. For example, serum from a patient is mixed with a suspension of *Salmonella typhi* in the *Widal test* for typhoid fever. If agglutination occurs when the serum is diluted enough to exclude the action of natural and group agglutinins, we know that the patient's serum contains immune agglutinins against *Salmonella typhi*. We may not know, however, whether the agglutinins result from a current attack of typhoid fever, a previous attack, or the recent administration of typhoid vaccine. (The diagnosis of typhoid fever is usually resolved with the case history and clinical manifestations of the disease.) In the lower animals the agglutination test is used to detect Bang's disease (contagious abortion of cattle) and bacillary white dysentery in chickens caused by *Salmonella gallinarum*.

Sometimes *nonspecific* agglutinins are produced against organisms with apparently no relation to the disease. Nonspecific agglutination is of special diagnostic value in typhus fever and other rickettsial diseases. In typhus fever nonspecific agglutinins develop against certain members of the *Proteus* group of bacteria; typhus fever is caused by rickettsiae, not by *Proteus*. It is thought that many such agglutinins are formed as the result of the introduction into the body of *heterophil* antigens* (*hetero*, from an individual of a *different* species). Heterophil antigens cause heterophil antibodies to be formed not only against themselves but against other antigens as well. Such antigens are common in the lower orders of life. In typhus fever the causative agent of the disease, a rickettsia, seems to act as a heterophil antigen, and the antibodies formed react not only to the rickettsiae but to certain types of *Proteus* bacilli as well. The agglutination of certain *Proteus* organisms by the serum of a patient with typhus fever constitutes the *Weil-Felix reaction*.

The serum of human patients with infectious mononucleosis agglutinates the red blood cells of the sheep. The antibody that agglutinates the red cells is a *heterophil antibody* since it agglutinates an antigen (in the red cells) of a different species. This important diagnostic test is the *heterophil-antibody test* or the *Paul-Bunnell test*.

Agglutination tests are utilized in the laboratory for the identification of bacteria. Agglutinating serums are prepared by artificially immunizing separately several individual animals of the same species with known species or types of bacteria so that a battery of typing serums is available. After an immunized animal is bled, its serum is separated and preserved for use. Such typing serums retain their potency for months. A suspension of unknown bacteria may be mixed with the battery of typing serums. If one of the serums shows agglutination, the unknown bacteria must be the same as those against which the serum was prepared.

In the application of the agglutination test to the identification of bacteria, the antibody is known and the antigen is unknown; in its application to the diagnosis of disease, the antigen is known and the antibody (in serum from a patient) is unknown.

---

*The heterophil antigen is an exception to the tenet of antigen specificity.

Agglutination forms the basis of many useful laboratory tests and is thought to represent a body mechanism for clearing the bloodstream of would-be invaders.

HEMAGGLUTINATION. Many types of antigens can be adsorbed onto the surface of a red blood cell. A red cell properly prepared with a test antigen can serve as a reagent for the detection of antibodies in the serum of a patient. This technic is *hemagglutination.* The titer of antibody is the dilution of serum in which no further agglutination of red cells takes place.

HEMAGGLUTINATION-INHIBITION. Red cell agglutination is an important procedure in the identification of certain viruses (p. 408). The action of a virus to clump red cells (viral hemagglutination) can be blocked by specific antibody to the virus. This is *hemagglutination-inhibition.* It is a very important test in a number of viral infections.

■ **Opsonocytophagic test.** Opsonins are substances that act on bacteria or other cells in such a manner (opsonization) as to render them more easily ingested by phagocytes and to favor their breakdown within the cytoplasm of the phagocyte. When the white blood cells of a person or animal suspected of having a disease are mixed with a suspension of the bacteria causing the disease, the degree of phagocytosis taking place is of diagnostic significance and can be evaluated in the *opsonocytophagic* test. This test has been used in the diagnosis of undulant fever and tularemia.

Phagocytosis (see p. 157) is not an activity of the phagocytes alone, but is dependent upon the action of opsonins in the serum. Opsonins are to some degree present in normal serum but are present in an increased amount in immune serum, and their action is enhanced by complement. This test is not in wide use today.

NITROBLUE TETRAZOLIUM (NBT) TEST. When normal neutrophils (microphages) in an individual with a normally functioning immune system are confronted by bacterial pathogens, certain metabolic changes consequent to their phagocytic activity take place within the neutrophils. (These changes probably relate to rearrangements in the cell membrane; opsonins are not specifically pinpointed in this phenomenon.) Because of these, the neutrophils can be shown to reduce in considerable amounts a supravital dye, pale yellow nitroblue tetrazolium to blue black formazan crystals. An in vitro test, the nitroblue tetrazolium (NBT) test, has been devised to display precisely in human blood the ability of the neutrophils to do so. This test is applied to the separation of certain bacterial infections from nonbacterial illness in patients where clinical manifestations are confusing and to the monitor of patients with increased vulnerability to infection (because of tissue transplant, immunosuppressive therapy, and the like).

■ **Neutralization or protection tests.** Neutralization tests are important in a number of infections caused by viruses and a few caused by bacteria. Antibodies that give immunity to viruses are known as *virus-neutralizing antibodies.* They can block the infectivity of a given viral agent under test conditions in a susceptible host. (See p. 407.)

■ **Immunofluorescence.** In recent years the phenomenon has been demonstrated that antibodies can be chemically tied to the fluorescent dye fluorescein, without changing their basic properties as antibodies. Such tagged antibodies *(fluorescent antibodies)* react with their specific antigens in the usual way. When fluorescent antibodies are allowed to react with their specific antigens in cultures or smears containing certain microorganisms or in the tissue cells of the human or animal body (the *direct test*), the result is a precipitate formed from the combination of antigen and antibody, which, because of the accompanying fluorescein, can be seen as a luminous area if viewed under ultraviolet light. A special type of microscope (Fig. 14-3) has been devised with

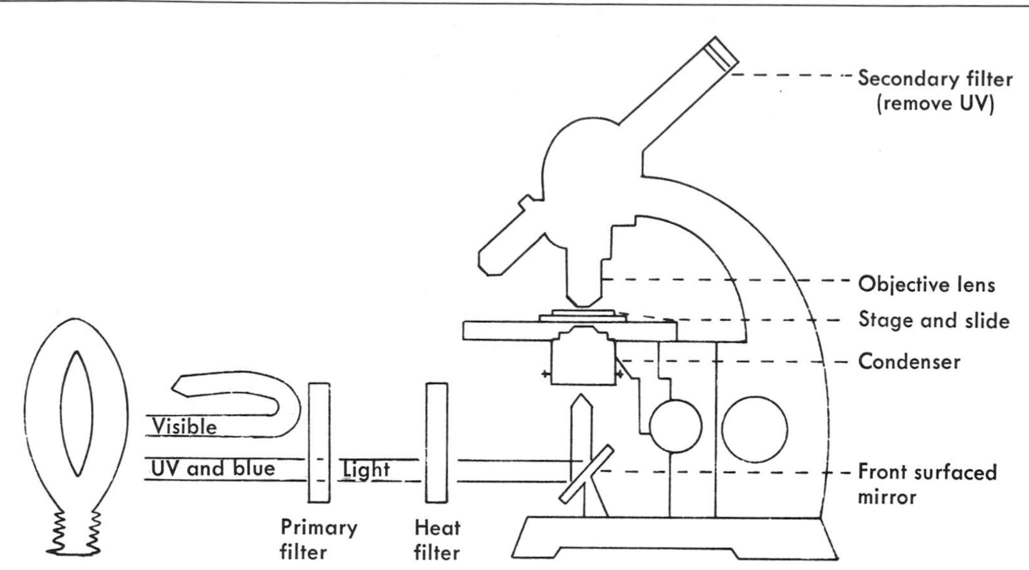

FIG. 14-3. Fluorescent antibody microscopy, physical arrangement. Note path from source of light (ultraviolet and short blue) on left through primary filter (transmits only ultraviolet and blue light) to remove longer wavelengths, through heat filter to remove heat rays, and on to front surface mirror on right. From here light is diverted into dark-field condensor (used to give black background) to strike specimen. Fluorescent-labeled material emits visible light to observer through secondary filter in microscope tube to remove stray ultraviolet. (From Barrett, J. T.: Textbook of immunology, St. Louis, 1970, The C. V. Mosby Co.)

which one can do just this. Specimens for testing with fluorescent antisera may be taken from the various body fluids and sites of disease. The direct test is applicable to infections with the following agents:

- Group A streptococci
- Enteropathogenic *Escherichia coli*
- *Neisseria meningitidis*
- *Neisseria gonorrhoeae*
- *Haemophilus influenzae*
- *Diplococcus pneumoniae*
- *Staphylococcus aureus*
- *Bordetella pertussis*
- *Shigella*
- *Brucella*
- *Listeria monocytogenes*
- Herpesvirus
- Enteroviruses
- Rabies virus
- Influenza virus
- Parainfluenza virus
- *Actinomyces*
- *Blastomyces dermatitidis*
- *Candida albicans*
- *Cryptococcus neoformans*
- *Histoplasma capsulatum*

In the *indirect test,* nonfluorescent antibody is bound to its antigen. The combination is made visible after application of a second antibody (antihuman serum), this time a fluorescent one. The test is set up so that the unconjugated serum acts as antibody in the first part and as antigen when exposed to fluorescent serum in the second part. The indirect test has been used with infections caused by the following:

- *Haemophilus influenzae*
- *Brucella*
- *Treponema pallidum*
- *Leptospira* species
- Rubella virus
- Cytomegalovirus
- Herpesvirus
- Agent of infectious mononucleosis
- Trachoma bedsonia
- *Rickettsia prowazekii*
- *Mycoplasma pneumoniae*
- *Toxoplasma gondii*
- *Plasmodium* of malaria
- *Cryptococcus neoformans*

Because of the rapid results with this technic, there is the wide variety of applications in identification of antigens related to bacteria, rickettsiae, viruses, fungi, and protozoa. Pathogenic amebas can be quickly distinguished from nonpathogenic ones, streptococci may be detected quickly in throat swabs, and rabies virus can be tagged in animals. In tissues fluorescent staining helps to localize the precise anatomic sites of antigen-antibody combination in diseases such as lupus erythematosus, rheumatoid arthritis, and certain hypersensitivity reactions.

■ **Other technics.** There are several other laboratory tests important to the study of the structure and nature of antibodies (immunoglobulins) and to the elucidation of the immunologic disorders.

ELECTROPHORESIS. When proteins are placed in an electrical field at a given pH, they will move in a distinctive path according to their own specific electrical charges. The pattern of migration of different proteins can be visualized under prescribed test conditions on a strip of paper stained with dye. (See Figs. 14-4 and 14-5.)

Zone electrophoresis refers to the technic of using a stabilizing medium to trap migrating proteins into more or less separate areas or zones that can then be stained and identified later. Commonly used solid support media include paper, starch, agar, and cellulose acetate.

RADIAL DIFFUSION IMMUNOASSAY. *Gel diffusion* is a term used for the antigen-antibody precipitin reaction detected in semisolid media. If protein (or antigen) is placed in a specially designed hole or well in an agar gel medium (no electrical charge), the protein diffuses concentrically out of the well. Just as specific antibody precipitates antigen in gels, this radial diffusion can be shown nicely in the clear gel if specific antibody has been impregnated into it. As antigen diffuses out of the well, it forms a ring of precipitate at the points of contact with specific antibody. This reaction can be quantitated easily for amounts of protein (or antigen). Known also as *single gel diffusion,* this is an important method for analysis of immunoglobulins.

Electroimmunodiffusion is single gel diffusion in an electric field.

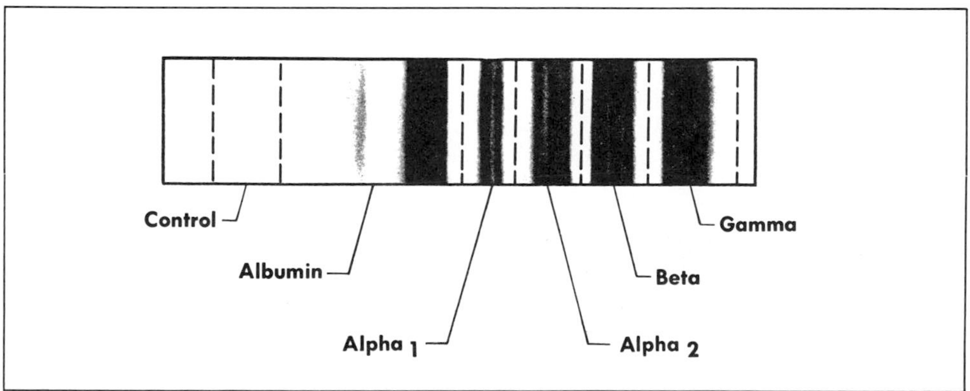

FIG. 14-4. Paper electrophoresis of normal human serum. Serum sample is placed near one end of paper strip moistened with buffer. An electric potential is applied. Serum proteins migrate along strip at different speeds. Afterward, strip is dried and proteins are fixed and stained (bromphenol blue). Black bands correspond to separated and stained proteins, with the depth of color in each band proportional to amount of that protein present. Paper strip may be cut crosswise as indicated and protein fractions analyzed further. (From Bauer, J. D., Ackermann, P. G., and Toro, G.: Bray's clinical laboratory methods, ed. 7, St. Louis, 1968, The C. V. Mosby Co.)

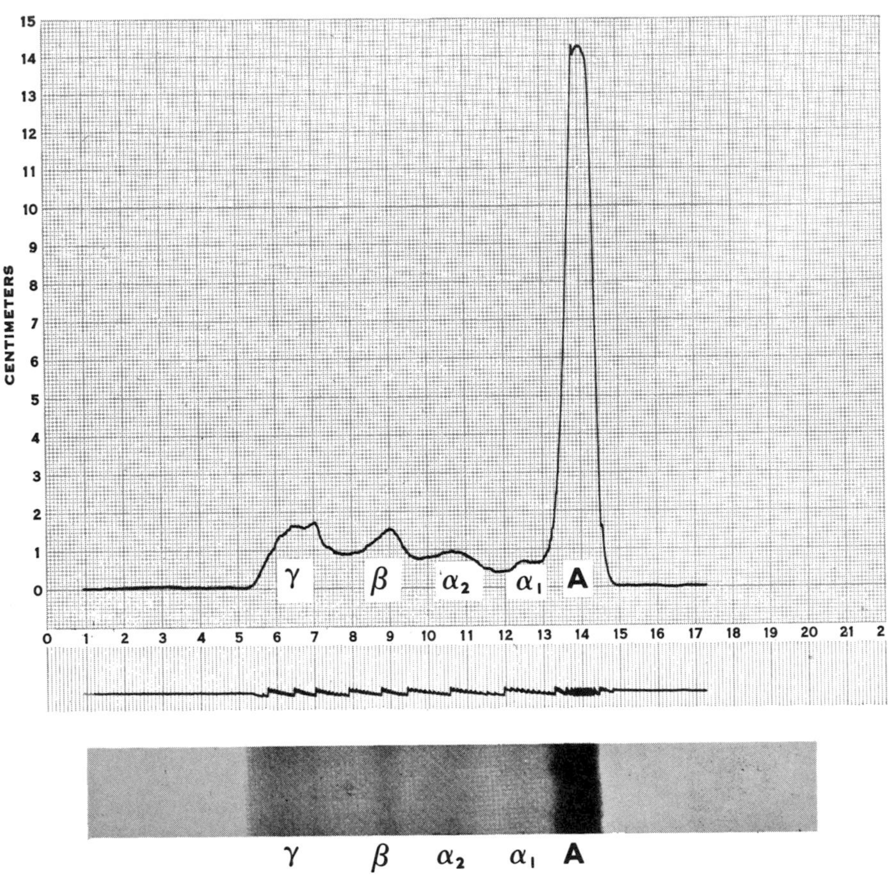

FIG. 14-5. Paper electrophoresis of normal human serum. Color of individual protein fractions may be measured directly on paper strip with a recording densitometer and a curve obtained as above. (From Bauer, J. D., Ackermann, P. G., and Toro, G.: Bray's clinical laboratory methods, ed. 7, St. Louis, 1968, The C. V. Mosby Co.)

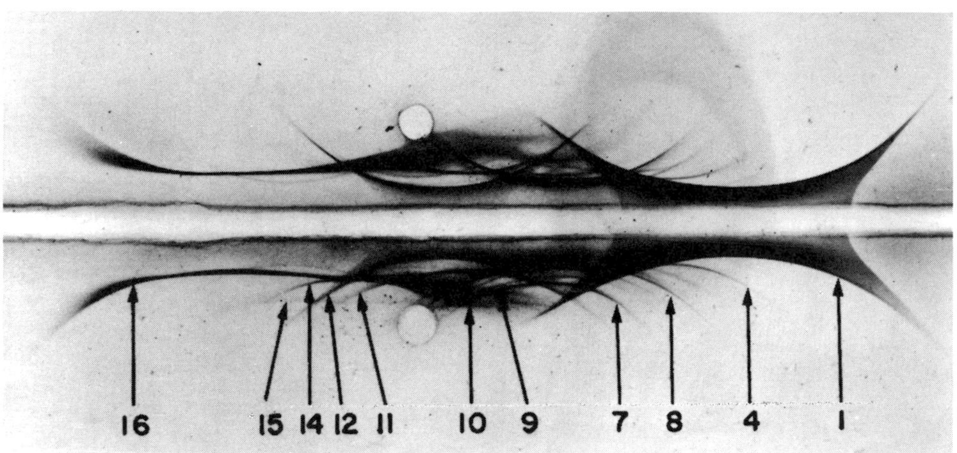

FIG. 14-6. Immunoelectrophoresis. Demonstration of proteins of normal serum when tested against polyvalent antiserum. Note pattern of arcs: **1,** albumin; **14,** IgA globulin; **15,** IgM globulin; **16,** IgG globulin; **4** and **7** to **12,** other proteins. (From Miale, J. B.: Laboratory medicine: Hematology, ed. 4, St. Louis, 1972, The C. V. Mosby Co.)

DOUBLE GEL DIFFUSION. In this method antibody is not incorporated throughout the agar as in single gel diffusion but is placed in a trench (or other reservoir) cut in the medium. The arrangement is such that soluble protein (antigen) and antibody diffuse out into the agar to form a band or line of precipitate at points of contact. If a mixture of proteins (or antigens) is analyzed, each specific antigen-antibody combination tends to present as a separate band at a definite position in the medium between the reservoir for the antigen and that for the antibody. This reaction is also a quantitative one.

FIG. 14-7. Protein abnormalities, electrophoretic study. In several technics demonstrated, an example of a malignant proliferation of lymphoid system (multiple myeloma) is used, a disease associated with large quantities of abnormal globulins. **1,** Paper electrophoresis showing abnormal peak (*arrow*). **2,** Starch gel electrophoresis showing abnormal gamma globulin component. **3** to **6,** Immunoelectrophoresis (in each instance normal control sample is at top, patient specimen at bottom): **3,** Testing done with polyvalent antiserum; **4,** testing with anti-IgA; **5,** with anti-IgG (which picks up, identifies, and locates the IgG) there is an increase of IgG; **6,** with anti-IgM there is a decrease of IgM. (From Miale, J. B.: Laboratory medicine: Hematology, ed. 4, St. Louis, 1972, The C. V. Mosby Co.)

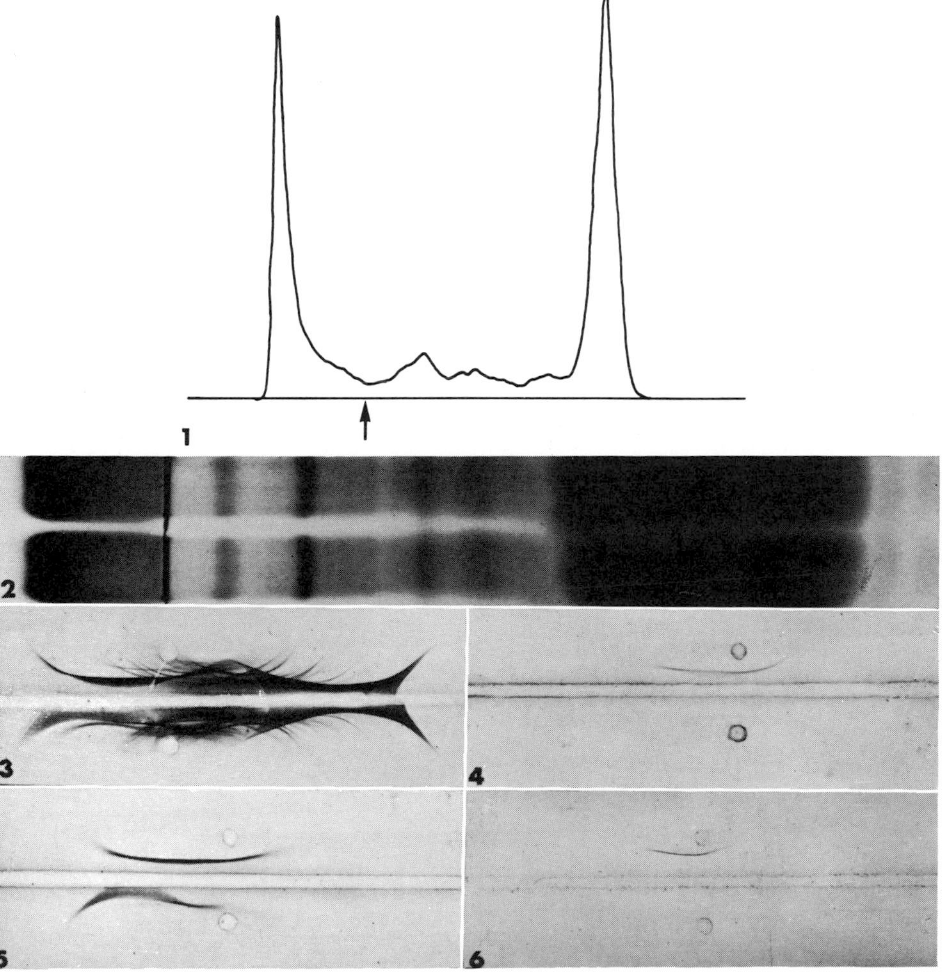

APPLICATION OF IMMUNOLOGIC METHODS

Some of the more frequently used immunologic methods for identifying and studying microbes and for defining their role in disease production are given in Table 14-3.

TABLE 14-3. IMMUNOLOGIC APPROACH TO DISEASE

Test(s)	Application
Serologic:	
1. Complement fixation	Diagnosis of syphilis, viral infections, and certain bacterial infections
2. Flocculation	Diagnosis of syphilis
3. Precipitation	Diagnosis of syphilis
Protein precipitation	Detection of origin of blood stains
	Detection of meat adulteration
4. Agglutination	Identification (serologic typing) of bacteria; diagnosis of undulant fever, tularemia, and salmonelloses
a. Widal	Diagnosis of typhoid fever
b. Weil-Felix	Diagnosis of typhus fever and Rocky Mountain spotted fever
c. Paul-Bunnell (heterophil antibody)	Diagnosis of infectious mononucleosis
d. Viral hemagglutination	Diagnosis of viral infections—rubella, rubeola, influenza, parainfluenzal infection
5. Virus neutralization	Diagnosis of viral infections
6. Fluorescent antibody	Identification of protozoa, fungi, bacteria, rickettsiae, bedsoniae, and viruses; diagnosis of rabies, syphilis, amebiasis
Intradermal (skin):	Diagnosis of infection, past or present, caused by microorganisms of:
1. Allergy to infection with:	
a. *Bacteria*	
Brucellergen	Brucellosis
Lepromin	Leprosy
Mallein	Glanders
Tuberculin	Tuberculosis
b. *Fungi*	
Blastomycin	Blastomycosis
Coccidioidin	Coccidioidomycosis
Trichophytin	Epidermophytosis
Oidiomycin	Candidiasis
Histoplasmin	Histoplasmosis
Sporotrichin	Sporotrichosis
c. *Bedsoniae*	
Frei	Lymphopathia venereum
2. Susceptibility	
Schick	Susceptibility to diphtheria
Dick	Susceptibility to scarlet fever

IMMUNOELECTROPHORESIS. If the technics of electrophoresis and double gel diffusion are combined, we have the process of *immunoelectrophoresis*, a valuable method for separating complex mixtures of proteins (antigens). In the test specimen in an electrical field, the unknown proteins are spread out in a series of differently charged masses. Specific antibody is placed in a trough parallel to the line of migration of the proteins in the gel. After electrophoresis proteins diffuse outward; antibody diffuses inward in an incoming wave. As specific antibody meets antigen, precipitation occurs in a series of arcs. (See Fig. 14-6.) The immunoprecipitin bands are readily seen, and the position and shape are consistent and stable for known proteins. Deviations represent abnormal constituents. (See Fig. 14-7.)

QUESTIONS FOR REVIEW

1. What is an agglutination test? Name diseases in which it is used as a diagnostic procedure.
2. Briefly define: antitoxin, cytolysin, agglutinin, precipitation, precipitin, serology, heterophil antibody, opsonin, hemolysis, hemolysin, amboceptor, serum, hemagglutination, virus-neutralization, flocculation.
3. How may antibodies be detected?
4. What is meant by the titer of a serum?
5. What are the advantages of the immunofluorescent technics?
6. Explain the difference between the direct and indirect fluorescent antibody test.
7. Briefly discuss immunoelectrophoresis and its applications.
8. Diagram a negative complement-fixation test.

REFERENCES. See at end of Chapter 15.

15 Allergy

ALLERGIC DISORDERS

■ **General characteristics.** Some persons exhibit unusual and heightened manifestations upon contact with certain substances, that are often, but not necessarily, protein and that do not affect the average or another person. For example, an individual may have asthma upon contact with horse dander,* and many persons have hay fever upon contact with plant pollens.

We have learned that immunity is a state wherein by prior contact a person obtains protection against an agent that would otherwise harm him. The reaction is specific for the particular agent, and a benefit comes from the immune setup. If we think of another situation, comparable immunologically but one in which the immune mechanisms injure or damage instead of protect, then we are referring to the state of *allergy* or *hypersensitivity*.† Allergy or hypersensitivity is a state of altered reactivity directly related to the operation of the immune system. The study of the effects in body tissues of the allergic reaction is included in *immunopathology*.

An antigenic substance that can trigger the allergic state is an *allergen*. It may be a protein or frequently a nonprotein of low molecular weight. Allergens may reach the body by way of the respiratory tract or the digestive tract, by direct contact with skin or mucous membranes, and by mechanical injection. Of these routes, the respiratory pathway is most regular. It is thought that unchanged protein must pass through the intestinal wall to the bloodstream for sensitization via the intestinal tract to occur. This can occur in children and in adults with digestive disturbances. A passive sensitization may be transmitted from the mother to her child in utero via the placenta. This type of sensitization is short lived, as is true of passive immunity in the newborn infant.

An allergic person is often sensitive to several different allergens. At least 10% to 20% of persons in the United States develop allergic disorders under the natural conditions of life.

---

*Dander is animal dandruff—the composite of cast-off epithelial cells of the skin, sebum, and other body oils.

†Both terms are ordinarily used to mean conditions such as emotional upsets, nonallergic food reactions, toxic responses to drugs, and others that are nonimmunogenic. Since *hypersensitivity* is more often so used than *allergy*, some authorities feel that the term *allergy* has priority. Allergy is a term first used by von Pirquet.

■ **Kinds of allergy.** As would be expected from our discussion of immunity, the reactions of allergy are separated into two main divisions (Table 15-1): (1) those in which the allergic responses result from the presence of humoral antibody, reactions of the *immediate* type, and (2) those in which the responses are not mediated by circulating antibodies but by specifically sensitized cells (that is, they are cell-mediated), reactions of the *delayed* type.

MECHANISMS. The mechanisms of allergy are enormously complex and poorly understood. Only a sketch is given of some of the background factors.

In immediate-type allergy, it is postulated that as a consequence of injury to a certain kind of cell, allergen combining with humoral antibody triggers a release of a number of highly active substances. The cell believed to be sensitized by circulating antibody and vulnerable is the mast cell, a cell widely distributed in the connective tissues of the body. The substances released include histamine, bradykinin, heparin, and slow-reacting substance. Each of these has the chemical potential to induce responses associated with immediate-type allergy. The spectrum of effects in immediate-type allergy includes contraction of smooth muscle, increase of gastric secretion, increase of nasal and lacrimal secretions, increase of vascular permeability, and vascular changes that can lead to circulatory collapse (shock). Of the chemical mediators of allergy, histamine has been the most extensively studied, and the allergic reaction has even been said to constitute a response to too much histamine.

TABLE 15-1. KINDS OF ALLERGIC DISORDERS

	Immediate-type reactions (anaphylactic type)	Delayed-type reactions (bacterial or tuberculin type)
Clinical state	Hay fever Asthma Urticaria Allergic skin conditions Serum sickness Anaphylactic shock	Drug allergies Infectious allergies Tuberculosis Tularemia Brucellosis Rheumatic fever Smallpox Histoplasmosis Coccidioidomycosis Blastomycosis Trichinosis Contact dermatitis
Onset	Immediate	Delayed
Duration	Short—hours	Prolonged—days or longer
Allergens	Pollen Molds House dust Danders Drugs Antibiotics Horse serum Soluble proteins and carbohydrates Foods	Drugs Antibiotics Microbes—bacteria, viruses, fungi, animal parasites Poison ivy and plant oils Plastics and other chemicals Fabrics, furs Cosmetics
Passive transfer of sensitivity	With serum	With cells or cell fractions of lymphoid series

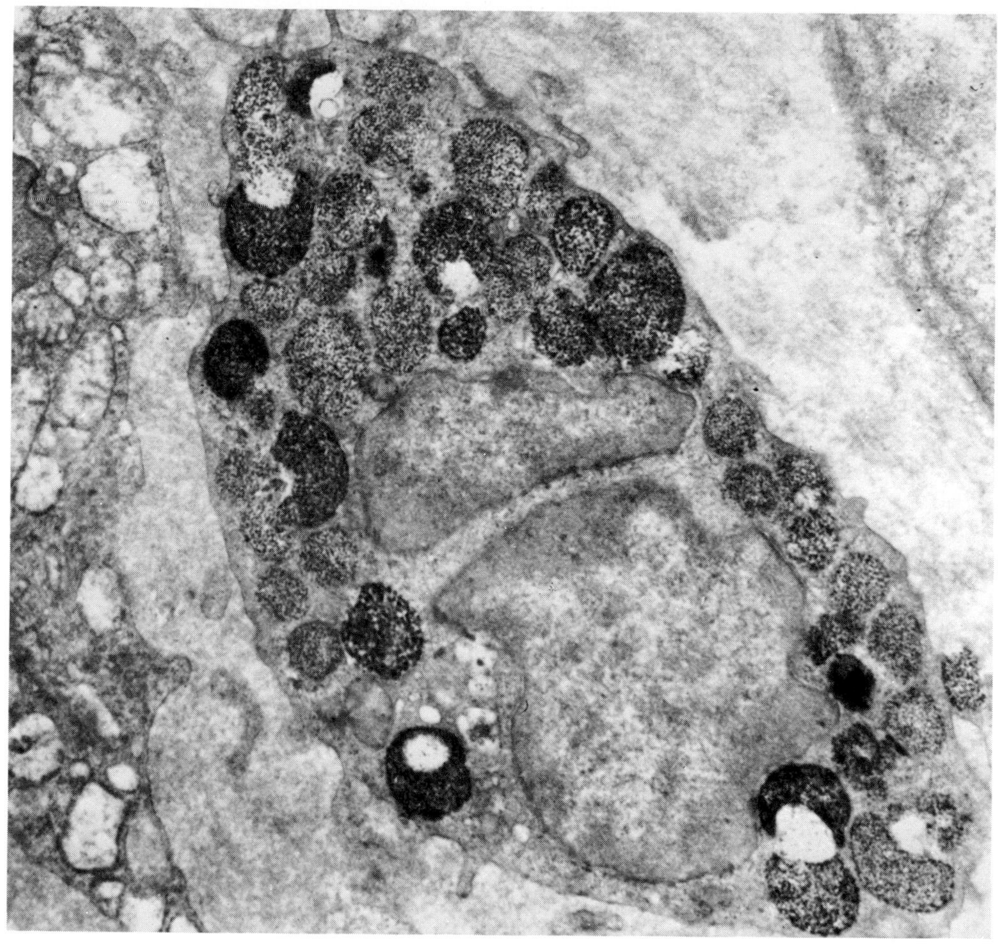

FIG. 15-1. Human mast cell (basophil of peripheral blood), electron micrograph. Note large dark granular masses (the large dark blue cytoplasmic granules of Wright-stained blood smears). They store histamine and heparin. (From Barrett, J. T.: Textbook of immunology, St. Louis, 1970, The C. V. Mosby Co.)

The mast cell (the basophil of peripheral blood) is a cell about 15μ wide and is known for its distinctive appearance (Fig. 15-1). Large, densely placed, metachromatically staining granules found in its cytoplasm are notable for their high concentration of pharmacologically potent compounds such as histamine, heparin, and proteolytic enzymes. Injury to the cell means degranulation. As the cell loses its unique granules, the substances contained therein are released into tissue fluids. Nonimmune as well as immune factors can cause degranulation.

Of 5000 kinds of immunoglobulin molecules found in normal persons, only one seems to be an immunoglobulin E (IgE). However, allergic individuals seem to produce considerably more of their antibodies from this class. (It is formed in lymph nodes.) Immunoglobulin E has been identified on the mast cell (basophil) and is involved with the release of histamine and the other substances. It is believed to play an important role in governing the allergic response.

In immediate-type allergy the consequences of the antigen-antibody combination follow shortly thereafter, within a matter of minutes, and the reaction reaches a maximum within a few hours. The responses occur in vascularized tissues and relate primarily to smooth muscle, blood vessels, and connective tissues. A specific shock tissue is important. Manifestations are acute and short lived. Epinephrine, antihistaminic drugs, and related compounds are of benefit. This type of reactivity can be transferred passively to a nonsensitized person by antibody-containing serum.

In delayed-type allergy, there is no relation to conventional antibody. The phenomenon is closely associated with cells. The responses are set in motion by lymphocytes specifically modified so that they can respond to the specific allergen deposited at a given site. The main feature of the reaction is the direct destruction of target tissue containing the antigen passively transferred to a nonsensitized person if immunologically competent cells (lymphoid cells) are used.

In allergic reactions of this type a much longer period of time elapses between the presentation of the antigen and the manifestations of the disorder. The period of incubation ranges from 1 to several days. Reactions do not favor any particular tissue but may occur anywhere. There is no specific shock organ or tissue as with immediate-type allergy. Many cells in the body are vulnerable. Manifestations tend to be slow in onset and prolonged. No histamine effect is demonstrated as in immediate-type reactions, and antihistamines are without effect. The steroid hormones seem to help in the treatment of some allergic disorders of this kind and in some of those of the immediate type as well. The exact mode of their action is not known.

Immediate-type allergic reactions

■ **Anaphylaxis.** In immediate-type allergy, if there is sudden release of the chemical mediators, the response is acute. The acute reaction is *anaphylaxis,* the prototype of immediate-type hypersensitivity. If a small amount of foreign protein (such as serum or egg white), within itself not poisonous, is injected into a suitable animal, the first dose will be without noticeable effect; but if a second injection is given after an interval of from 10 to 14 days, severe manifestations promptly appear within *minutes* and even death may occur. This condition is known as *anaphylaxis.* The first dose is known as the *sensitizing dose;* the second is the *provocative dose.*

Systemic anaphylaxis or anaphylactic shock is a generalized reaction brought about in a sensitized animal upon contact with adequate amounts of antigen administered so that rapid dissemination occurs (as intravascularly).

Anaphylaxis or *systemic shock* may be best demonstrated in a guinea pig: A small sensitizing injection of horse serum is followed at the end of 10 to 14 days by a second but larger injection of horse serum. With *1 or 2 minutes* after the provocative dose, the guinea pig becomes restless and breathes with difficulty. Frantic activity with extreme breathlessness is followed by death in respiratory failure.

Anaphylaxis is specific; that is, for anaphylaxis to occur, the sensitizing and provocative doses must be of the same substance. If the serum of a sensitized animal is injected into a normal animal, and if after an interval of 6 to 8 hours the normal animal is injected with some of the material to which the first animal is sensitive, signs of anaphylaxis appear in the normal animal. This is *passive anaphylaxis.* The passive transfer of the hypersensitive state assumes considerable importance when it is realized that the use of an allergic donor for blood transfusion may render the recipient hypersensitive. One interesting case is reported in which, after receiving blood from a donor sensitive to horse dander, the recipient had an attack of asthma on contact with a horse. If an actively sensitized animal survives an anaphylactic attack, it is desensitized for a few days but eventually becomes sensitive again.

THE ROLE OF SMOOTH MUSCLE. The dramatic manifestations of anaphylaxis result from the contraction of smooth muscle fibers, and the part of the body primarily attacked in a given animal depends upon the distribution of smooth muscle fibers in the species. In the guinea pig, smooth muscle fibers are plentiful in the lungs; contraction of these closes bronchioles and bronchi and leads to death from respiratory tract obstruction. In the rabbit, smooth muscle is plentiful in the pulmonary arteries. If the rabbit is used as test animal, these arteries contract, and a burden that leads to failure is thrown on the right side of the heart.

That anaphylactic shock causes contraction of smooth muscle may be proved in the following manner: Take a strip of fresh tissue containing an abundance of smooth muscle (for instance, the uterine muscle of the guinea pig) from an animal that has been sensitized and place it in a bath of Ringer's solution. Make one end of the strip stationary and attach the other end to a recording instrument (kymograph). Upon the addition of only a small amount of the sensitizing substance to the solution, vigorous contraction of the smooth muscle will occur and be registered on the kymograph. This is the *Schultz-Dale reaction.*

ANAPHYLAXIS IN MAN. In man anaphylaxis can be an immediate, severe, and fatal reaction, running its course within seconds or minutes (see Table 15-2). If the person is sensitive to the given allergen, a very small amount can precipitate the reaction.

When anaphylaxis was first described at the turn of the century, one of the chief causes then was a sting by a member of the insect order Hymenoptera. Included in

TABLE 15-2. MAKE-UP OF THE ANAPHYLACTIC SET*

Item	Number needed	Item	Number needed
1 ml. ampules of 1:1000 solution of epinephrine	2	2 ml. syringes	2
Hypodermic needles	2	Needles (1 and 4 in. long)	2
Ampules of aminophylline (0.24 gm. each)	2	1000 ml. bottle of 5% solution of dextrose in distilled water	1
Intravenous set	1	10 ml. ampule of diphenhydramine hydrochloride (10 mg. per ml.)	1
Bottle of hydrocortisone (dilution to 2 ml. gives 50 mg. per ml.)	1	Ampule of sterile water	1
Scalpel	1	Hemostat	1
Ampule of absorbable surgical suture (catgut) with needle	1	20 ml. syringe	1
Tongue depressors		Alcohol, 70% (disinfectant)	
Gauze sponges			

*This kind of emergency must be anticipated. At the Mayo Clinic, Rochester, Minnesota, "anaphylactic sets" are placed in strategic areas. They contain the above items, quickly available, to be used in the event of such a reaction.

this order are bees, yellow jackets, wasps, and hornets. The fact is still true. It is said today that far more deaths occur from insect stings than from snakebites.*

In all areas of our complex society the number of different substances to which man may become sensitized is astonishing. This is especially true in the field of medicine, where a variety of chemicals are used for many reasons so that most of the anaphylactic reactions today are the result of sensitivity to drugs that have been used in one way or another for diagnosis and treatment of disease. Reactions of this type are referred to as *iatrogenic* reactions. Table 15-3 lists chemical agents known to have caused systemic anaphylaxis in man.

■ **Atopic allergy.** Atopic allergy is the designation for a group of well-known human allergies to naturally occurring antigens, chronic manifestations of immediate-type allergy. Here are found hay fever and asthma. By contrast with systemic anaphylaxis, where large doses of antigen are given intravenously, low doses of antigen repeatedly contact mucous membranes.

Allergic diseases are not inherited, but the special tendency for the individual to develop the state of altered reactivity in tissues is. Atopic allergy seems to be present more in some families than in others; however, all the members of an affected family do not have the same condition. One may have hay fever, another asthma, and still another some type of skin eruption. Seldom will all the members of a family manifest the hypersensitive state.

Atopic allergy is associated with the type of antibody called *reagin,* with a special ability to bind to skin or other tissues. Recent observations indicate that reaginic antibody mostly belongs in immunoglobulin class IgE.

■ **Asthma.** Asthma is a recurrent type of breathlessness, coming in acute episodes and described by certain types of respiratory movements and wheezing. It is an allergic condition most often caused by animal hair, feathers or dander, house dust, foods, microbes, and certain cosmetics. Important asthma-producing foods are milk, milk products, eggs, meat, fish, and cereals. A person may become sensitive to the bacteria normally inhabiting the upper respiratory tract and thereby become a victim of asthmatic attacks. This type of asthma is spoken of as *endogenous* asthma. An estimated 4% of the population has asthma.

■ **Hay fever.** Hay fever, with its well-known nasal distress, results from sensitivity to pollens, and the period of attack corresponds to the time of pollination of the offending plant or plants. To be important as a cause of hay fever, a plant must produce a light dry pollen easily carried a long distance by the wind. This excuses both goldenrod and roses, long accorded an unearned distinction as causes of hay fever. Early spring hay fever is usually caused by the pollen of trees. Late spring and early summer hay fever is most often from grass pollens, and more than 80% of the cases of fall hay fever result from ragweed pollen. Ragweed is found only in the United States, parts of Canada, and Mexico. The importance of individual trees, grasses, or other plants depends on geographic location. Ten percent of the population has hay fever.

■ **Urticaria and allergic skin eruptions.** Allergic skin conditions may come from foods, drugs, chemicals, and a wide variety of other allergens. Urticaria, known some-

---

*A person with no history of allergy can have a fatal reaction to an insect bite. If one has had a reaction, he should take every precaution in areas of contact, wear the necessary protective clothing, and, if need be, carry an emergency kit of appropriate medication. Insect sting kits are available commercially.

TABLE 15-3. SYSTEMIC ANAPHYLAXIS IN MAN—DOCUMENTED AGENTS*

Chemical class	Trade name of drug	Route of inoculation	Deaths
Proteins			
Antiserum of horse (in passive immunization)		Parenteral	Yes
Antirabies serum			
Tetanus antitoxin			
Bivalent botulism antitoxin			
Hormones			
Insulin		Subcutaneous	
Corticotropin	Acthar	Intravenous	
Enzymes			
Chymotrypsin		Intramuscular	
Trypsin	Parenzyme	Intramuscular	
Penicillinase	Neutrapen	Intramuscular	
Sting of Hymenoptera (bees, wasps, hornets, and yellow jackets)		Subcutaneous (insect bite)	Yes
Pollen			
Bermuda grass		Intradermal	Yes
Ragweed		Intradermal	Yes
Food			
Egg white		Intradermal	Yes
Buckwheat		Intradermal	Yes
Cotton seed		Intradermal	Yes
Glue		Intradermal	Yes
Polysaccharides			
Acacia (emulsifier)		Intravenous	
Dextran (plasma expander)	Expandex, Gentran	Intravenous	
Others: action as haptens			
Antibiotics			
Penicillin		Oral and parenteral	Yes
Demethylchlortetracycline hydrochloride	Declomycin	Oral	
Nitrofurantoin	Furadantin	Oral	
Streptomycin		Intramuscular	
Other medications			
Sodium dehydrocholate	Decholin	Intravenous	Yes
Thiamine		Subcutaneous and intravenous	Yes
Salicylates (aspirin)		Oral	
Procaine	Novocaine	Parenteral	Yes
Diagnostic agents			
Sulfobromophthalein	Bromsulphalein (BSP)	Intravenous	Yes
Iodinated organic contrast agents		Oral and intravenous	Yes

*After Austen, K. F.: Systemic anaphylaxis in man, J.A.M.A. **192:**108, 1965.

times as nettle rash, is an allergic disorder of the skin depicted by the presence of wheals (whitish swellings) or hives (lesions of a strikingly transitory nature at times) (Fig. 15-2).

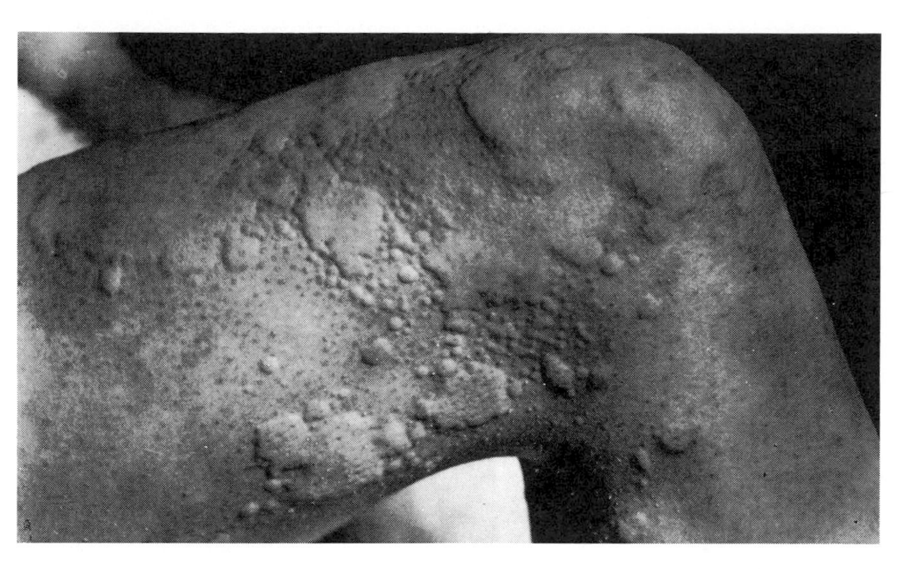

FIG. 15-2. Urticaria. Sensitivity to penicillin. (From Bergersen, B. S.: Pharmacology in nursing, ed. 12, St. Louis, 1972, The C. V. Mosby Co.)

■ **Serum sickness.** After the administration of an immune serum a reaction known as serum sickness may appear in persons who have never had a previous injection of horse serum and, so far as is known, are not sensitive to horse proteins.

Serum sickness is common, and its manifestations are unpleasant but seldom life threatening. It usually begins 8 to 12 days after the injection of an immune serum and is typified by skin eruption; swollen, painful, and stiff joints; enlargement of the lymph nodes; leukopenia; and decreased coagulability of the blood. A reaction may occur only around the site of injection (local serum disease). Serum sickness is thought to be caused by the combination of antibodies formed right after the injection with an excess of the serum. (Serum sickness is here classified as an immediate-type response although it is a disorder combining features of both major categories of hypersensitivity.) An immune serum should be administered with extreme caution to asthmatic patients and to those who have had a previous injection of horse serum. Before an immune serum is given, tests (see p. 614) to detect sensitivity to horse serum should always be done because the factor responsible for untoward reactions is the horse protein, the medium for many antitoxins and immune serums. The fall in blood pressure, the drop in body temperature, and respiratory difficulty resulting from the injection of an immune serum are most likely to appear when the second injection is given 2 or 3 weeks after the first, but severe reactions may occur when a second injection is given months or even years after the first. Anaphylaxis is most likely to occur in man after intravenous injection. Rarely, manifestations may lead to immediate collapse and death.

■ **Arthus phenomenon.** When material to which an individual is sensitized is reintroduced into his body, an allergic reaction may develop at the skin site. Although considered an immediate type of allergy, this reaction appears after a delay of 6 to 12 hours at least. There is localized change in the skin described as an area of swelling,

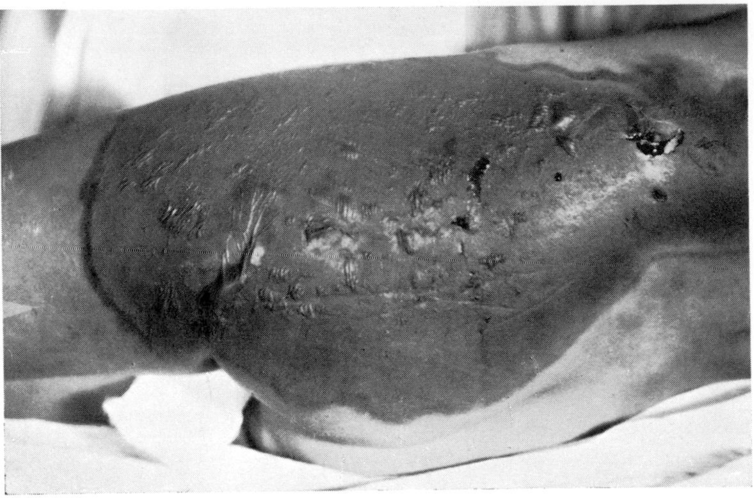

FIG. 15-3. Arthus phenomenon in sharply set-off area in skin of hip and thigh. Note discoloration associated with hemorrhage and death of tissue. (From Top, F. H., Sr.: Communicable and infectious diseases, ed. 6, St. Louis, 1968, The C. V. Mosby Co.)

reddening, hemorrhage, and even necrosis (Fig. 15-3). The explanation lies in an intricate sequence of events cumulating in tissue injury primarily to small blood vessels in the area. There are a number of interrelated factors, including the precipitation of antigen-antibody complexes in blood vessel walls, fixation of complement by these, and infiltration of neutrophilic leukocytes with release of their proteolytic enzymes. When carefully studied, the Arthus phenomenon gives insight as to how immunologic mechanisms damage tissues.

Delayed-type allergic reactions

■ **Sensitivity to drugs.*** It is not hard to find persons who manifest drug allergy, that is, an untoward reaction to a certain drug or drugs. It is believed that when drugs are regularly taken into the body, a chemical combination may occur between the drug and certain body proteins, forming a complete antigen in a protein compound foreign to the body and one against which the mechanisms responsible for allergic manifestations are directed. True allergic drug reactions may be associated either with humoral antibodies or cell-mediated responses.

It has been estimated that about 500 drugs are capable of bringing about the allergic state. The ones most commonly responsible for a hypersensitive state in the person using them are opiates (morphine, codeine), salicylates (aspirin), barbiturates, iodides, bromides, arsenicals, sulfonamides, and antibiotics. Among the drugs inducing allergy, penicillin is the worst offender; an allergy to the drug occurs in 10% to 15% of the instances where it is given. For other drugs, fortunately the incidence is much lower.

**Drug intolerance* is the state in which a person receiving the drug reacts in a characteristic way to unusually small doses of the drug, doses with no physiologic effect. *Drug idiosyncrasy* refers to the situation in which the recipient of the drug reacts in an unusual way, such as when a dose of a barbiturate (a sedative) produces excitement.

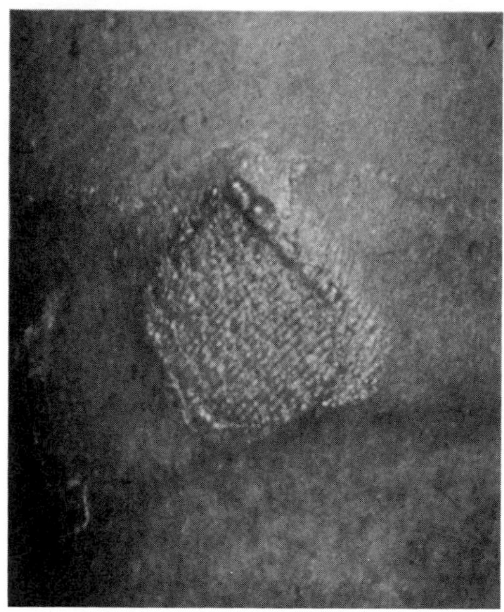

FIG. 15-4. Contact dermatitis—positive patch test to rubber from bathing cap. Note blisters and texture of skin. Similar reaction occurs if formalin, mercury, or other simple chemicals are used as test antigens in sensitized person. (From Leider, M.: Practical pediatric dermatology, ed. 2, St. Louis, 1961, The C. V. Mosby Co.)

Sensitivity to the sulfonamide drugs is more likely when the drugs are used in the form of topical ointments. Because of the likelihood of sensitivity after the local use of the sulfonamide drugs and penicillin as well, this type of therapy has almost completely disappeared. Sensitivity to the antibiotics regularly appears in the form of a skin eruption. In a general way, a reaction to penicillin and to certain other antibiotics often closely resembles serum sickness. The same is true of reactions to insulin and liver extract. If a previous reaction to penicillin has taken place or if there is reason to think that the patient is hypersensitive to the drug, it should *not* be given. Penicillin is said to have caused more deaths than any other drug.* No reliance should be placed on desensitization.

■ **Infectious allergies (hypersensitivity to infection).** Repeated or chronic infections may sensitize a patient to the microbes causing the infection. This is well illustrated in the allergy to *Mycobacterium tuberculosis* and its products. When an extract of *Mycobacterium tuberculosis* is applied to the skin of the person who has or who has had tuberculosis, there is a reaction, and the tuberculin test is said to be positive. Patients may likewise be hypersensitive to other organisms such as *Brucella* (the causative agents of undulant fever). In fact the manifestations of acute rheumatic fever are considered as an allergic reaction to streptococci. Other infections in which hypersensitivity to the causative agent are measured are leprosy, chancroid, lymphopathia venereum, coccidioidomycosis, blastomycosis, and histoplasmosis. In veterinary practice, hypersensitivity of infection is encountered in glanders, Bang's disease (undulant fever), and Johne's disease. In some of these, diagnostic skin tests are used to detect sensitivity (see Table 14-5).

*Penicillin as such does not induce hypersensitivity; it is the degradation products that can become haptens. A skin test can be done with one of these, penicilloyl-polylysine, to indicate the probable risk of an allergic reaction.

■ **Contact dermatitis.** Contact dermatitis localizes in the skin, especially in areas exposed to direct physical contact with irritant substances of many different kinds. As would be expected, this inflammatory condition is often seen on the hands. The first time that the skin contacts the offending substances there is no visible effect, but sensitization means that subsequent contacts will result in skin changes. Contact dermatitis is sometimes referred to as eczema or eczematous dermatitis. The list of offenders is a long one. It includes the well-known poisons of poison ivy and poison oak* and a vast host of substances related to trade and industry, such as dyes, soaps, lacquers, plastics, woods, fabrics, furs, formalin, cosmetics, drugs, chemicals, metals, and explosives (Fig. 15-4).

■ **Autoallergies.** (See p. 178.)

LABORATORY TESTS TO DETECT ALLERGY†

If some of the antigen to which a person is sensitive is rubbed into a scratch on the skin or if some of the dilute antigen is injected between the layers of the skin, a wheal and flare 0.5 to 2 cm. in diameter will occur at the site of contact within 20 to 30 minutes (cutaneous anaphylaxis). If the person is not sensitive to the antigen, no reaction will occur. This is the *skin test.*

The *ophthalmic test* helps to detect sensitization to serum. A drop of the *diluted* serum is instilled into the conjunctival sac; if the patient is sensitive to horse serum, redness of the conjunctiva and a watery discharge will appear within 10 to 20 minutes. The reaction may be controlled by epinephrine.

The *patch test* is useful in detecting the cause of *contact dermatitis.* In this test the suspect material is applied directly to the skin and held in place by means of adhesive tape from 1 to 4 days. A positive reaction reproduces the lesion from which the patient is suffering, with blister and papule formation.

DESENSITIZATION

In some cases desensitization may be accomplished by giving repeated injections of very small amounts of the antigen to which the patient is sensitive. A patient hypersensitive to horse serum may sometimes be given an immune serum if the administration is preceded by several injections of very small amounts at 30-minute intervals. The desensitizing doses are graded by the reaction of the patient, and the administration of an immune serum to a hypersensitive person should be undertaken only by those with experience in immune therapy. A similar but much prolonged method of desensitization is used in the treatment of hay fever and selected patients with asthma and urticaria. Desensitization, unlike sensitization, is of relatively short duration. In some forms of allergy it is accomplished with great difficulty.

---

*Poison ivy, oak, and sumac belong to the *Rhus* genus of trees and shrubs; the skin lesions may be called rhus dermatitis. The *Rhus* plants are responsible for more allergic contact dermatitis than all other allergens combined.

†See also p. 614.

QUESTIONS FOR REVIEW

1. Compare the two major categories of allergic disorders.
2. Briefly discuss anaphylaxis.
3. Describe briefly the most important allergic diseases.
4. What is serum sickness? Why is it less common than previously?
5. What effect has the development of sensitivity to the sulfonamides and antibiotics had upon the topical use of these drugs?
6. Outline the laboratory diagnosis of allergy.
7. Explain the role in allergy postulated for immunoglobulin E.
8. Explain the relation of the basophil to allergic disorders.
9. How is desensitization accomplished?
10. List five agents known to have caused anaphylaxis in man.
11. What are the infectious allergies? How can they be detected?
12. Define or briefly explain: hypersensitivity, allergy, sensitizing dose, provocative dose, allergen, immunopathology, Arthus phenomenon, Schultz-Dale reaction, reagin, iatrogenic, endogenous, insect sting kit, drug intolerance, drug idiosyncrasy, cutaneous anaphylaxis.

REFERENCES FOR UNIT THREE

Abrams, B. L., and Waterman, N. G.: Dirty money, J.A.M.A. **219:**1202, 1972

Anderson, C. L.: Community health, St. Louis, 1969, The C. V. Mosby Co.

Barnako, D.: The environment: Rats, J.A.M.A. **218:**663, 1971.

Barnett, S. A.: Rats, Sci. Amer. **216:**78, (Jan.) 1967.

Barrett, J. T.: Textbook of immunology, St. Louis, 1970, The C. V. Mosby Co.

Bazaral, M., Orgel, H. A., and Hamburger, R. N.: IgE levels in normal infants and mothers and an inheritance hypothesis, J. Immunol. **107:**794, 1971.

Bernton, H. S., and Brown, H.: Insect allergy: allergenicity of excrement of the cockroach, Ann. Allerg. **28:**543, 1970.

Billingham, R., and Silvers, W.: The immunobiology of transplantation, Englewood Cliffs, N. J., 1971, Prentice-Hall, Inc.

Block, V.: Wild pets could turn them off, Today's Health **46:**56, (Dec.) 1968.

Blumberg, B. S., London, W. T., and Sutnick, A. I.: Practical applications of the Australia antigen test, Postgrad. Med. **50:**70, (Dec.) 1971.

Boyd, W.: Textbook of pathology, Philadelphia, 1970, Lea & Febiger.

Bruninga, G. L.: Complement—a review of the chemistry and reaction mechanisms, Amer. J. Clin. Path. **55:**273, 1971.

Buckley, C. E., III: Immunologic evaluation of older patients, Postgrad. Med. **51:**235, 1972.

Burnet, F. M.: Immunological surveillance, New York, 1970, Pergamon Press.

Burnet, F. M.: Autoimmunity and autommune disease, Philadelphia, 1972, F. A. Davis Co.

Cancer antigens, editorial, J.A.M.A. **221:**66, 1972.

Castillo, P.: Methods of warming blood, A.O.R.N. J. **14:**82, (Aug.) 1971.

Cawley, L. P.: Electrophoresis and immunoelectrophoresis, Boston, 1969, Little, Brown & Co.

Chapman, J. S.: The environment: animal-to-human infection, J.A.M.A. **213:**1559, 1970.

Craven, R. F.: Anaphylactic shock, Amer. J. Nurs. **72:**718, 1972.

Feigin, R. D., Shackelford, P. G., and Choi, S. C.: Prospective use of nitroblue tetrazolium dye test in febrile disorders, J. Pediat. **79:**943, 1971.

Fenner, F.: Infectious disease and social change, Part 2, Med. J. Aust. **1:**1099, 1971.

Forrester, R. H.: Army research in military blood banking, Southern Med. Bull. **57:**17, (June) 1969.

Fort, A. T., and Walker, R. H.: Passive transfer of maternal antibody to the fetus, Lab. Med. **1:**40, (June) 1970.

Fotino, M.: Tissue typing for organ transplantation, Bull. S. Cent. Ass. Blood Banks **14:**4, (Nov.-Dec.) 1970.

Frazier, C. A.: Those deadly insects, RN **34:**49, (April) 1971.

Goldman, L.: Prevention and treatment of eczema, Amer. J. Nurs. **64:**114, (March) 1964.
Good, R. A.: On threshold of biologic engineering, Amer. J. Med. Tech. **38:**153, (May) 1972.
Good, R. A., and Fisher, D. W., editors: Immunobiology, Stamford, Conn., 1971, Sinauer Associates, Inc., Publishers.
Gotoff, S. P.: Cell mediated immune deficiency, editorial, J. Pediat. **78:**379, 1971.
Grothaus, R. H., and Adams, J. F.: An innovation in mosquito-borne disease protection, Milit. Med. **137:**181, 1972.
Halliday, W. J.: Glossary of immunological terms, New York, 1971, Appleton-Century-Crofts.
Hanlon, J. J.: Principles of public health administration, St. Louis, 1969, The C. V. Mosby Co.
Harmount, J. G.: How Waukegan won its war against rats, Today's Health **46:**16, (May) 1968.
Heinz, W. C.: The man who said, "They don't have to die," Today's Health **49:**26, (Jan.) 1971.
Henderson, L. L.: Acute reaction to insect sting, Postgrad. Med. **49:**191, (May) 1971.
Herbert, W. J., and Wilkinson, P. C.: A dictionary of immunology, Philadelphia, 1971, F. A. Davis Co.
Higdon, R. S.: Common insect, mite, and parasite problems in the United States, GP **37:**84, (May) 1968.
Holborow, E. J.: An ABC of modern immunology, Boston, 1968, Little, Brown & Co.
Immunological power of the lymphocyte, editorial, J.A.M.A. **206:**2514, 1968.
Irwin, T.: Fever: How to play it cool, Today's Health **46:**52, (Dec.) 1968.
Jet set diseases, Med. World News **11:**30, (Feb. 27) 1970.
Levine, P.: Suppression of Rh sensitization by Rh immunoglobulin, Bull. Path. **9:**46, (March) 1968.
Limax, A.: What immunity to disease means to you, Today's Health **43:**54, (Nov.) 1965.
Marples, M. J.: Life on the human skin, Sci. Amer. **220:**108, (Jan.) 1969.
Medical news: Cancer vaccine outlook; hopeful, but not soon, J.A.M.A. **219:**1555, 1972.
Miller, M. E.: WHO's immunodeficiencies, editorial, Ann. Intern. Med. **77:**149, 1972.
Minkin, W., and Lynch, P. J.: Incidence of immediate systemic penicillin reactions, Milit. Med. **133:**557, 1968.
Montes, L. F., Pittillo, R. F., Hunt, D., et al.: Microbial flora of infant's skin, Arch. Derm. **103:**400, 1971.
Murillo, G. J.: Synthesis of secretory IgA by human colostral cells, Southern Med. J. **64:**1333, 1971.
Murray, J. E., and Barnes, B. A.: Organ transplant registry, editorial, J.A.M.A. **217:**1546, 1971.
Nichols, G. A., and Kucha, D. H.: Taking oral temperatures, Amer. J. Nurs. **72:**1090, 1972.
Oxborrow, G. S., and Puleo, J. R.: Microbiological studies of spacecraft, Lab. Med. **1:**17, (Oct.) 1970.
Palmer R. L.: Diagnostic aids for immunologic disorders, Texas Med. **67:**79, (March) 1971.
Park, B. H.: The use and limitations of the nitroblue tetrazolium test as a diagnostic aid, J. Pediat. **78:**376, 1971.
Patterson, R., Head, L. R., Suszko, I. M., and Zeiss, C. R., Jr.: Mast cells from human respiratory tissue and their in vitro reactivity, Science **175:**1012, 1972.
Pent, J. H.: Testing for the presence of the hepatitis associated antigen, Bull. S. Cent. Ass. Blood Banks **14:**3, (July-Aug.) 1971.
Petersdorf, R. G.: The physiology, pathogenesis and diagnosis of fever, Res. Phys. **14:**67, (March) 1968.
Pierce, J. C., Madge, G. E., Lee, H. M., and Hume, D. M.: Lymphoma, a complication of renal allotransplantation in man, J.A.M.A. **219:**1593, 1972.
Porter, R. R.: The structure of antibodies, Sci. Amer. **217:**81, (Oct.) 1967.
Questions and answers about blood and blood banking, Chicago, 1970, American Association of Blood Banks.
Reeves, W. C.: Can the war to contain infectious diseases be lost? Amer. J. Trop. Med. **21:**251, 1972.
Reisfeld, R. A., and Kahan, B. D.: Markers of biological individuality, Sci. Amer. **226:**28 (June) 1972.
Remington, J. S.: The compromised host, Hosp. Pract. **7:**59, (April) 1972.
Reported incidence of notifiable diseases in the United States, 1971. (Summary), Morbid. Mortal. Wk. Rep. **20**(53) (ann. supp.): 1, 1971.
Reticuloendothelial function, editorial, J.A.M.A. **199:**419, 1967.
Ritzmann, S. E., and Daniels, J. C.: Immunology of transplantation, Texas Med. **66:**48, (May) 1970.
Rogers, R. S., III, and Callaway, J. L.: Contact dermatitis, Part 1. Plants (Rhus) and chemicals as causative agents, Hosp. Med. **7:**6, (Jan.) 1971.
Ross, W. S.: 4000 kidneys later. . . ., Today's Health **48:**35, (Dec.) 1970.
Roueché, B.: Poison ivy: "Its purpose has yet to be established," Today's Health **49:**38, (May) 1971.
Routine testing in blood banks for the hepatitis-associated antigen (current trends), Morbid. Mortal. Wk. Rep. **20** (14):123, 1971.

Samter, M., editor: Immunological diseases, Boston, 1971, Little, Brown & Co.
Scott, B.: Asthma—the demon that thrives on myths, Today's Health **48:**42, (June) 1970.
Shaffer, J. H., and Sweet, L. C.: Allergic reactions to drugs, Amer. J. Nurs. **65:**100, (Oct.) 1965.
Shelokov, A.: Australia antigen and hepatitis, Bull. S. Cent. Ass. Blood Banks **14:**5, (May-June) 1971.
Sindermann, C. J.: Diseases of marine animals transmissible to man, Lab. Med. **1:**50, (Jan.) 1970.
Sokol, A. B., and Houser, R. G.: Dog bites: prevention and treatment, Clin. Pediat. **10:**336, 1971.
Steele, J. H.: What is the current state of the zoonoses? Lab. Med. **1:**28, (Dec.) 1970.
Stiehm, E. R., moderator: UCLA Conference—Diseases of cellular immunity, Ann. Intern. Med. **77:**101, 1972.
Sussman, L. N.: Au antigen test—a simplified technique, Lab. Med. **2:**11, (Nov.) 1971.
Transplantation, editorial, J.A.M.A. **206:**2514, 1968.
Transplantation in a low key, editorial, J.A.M.A. **204:**999, 1968.
Transplantation of organs, editorial, Southern Med. Bull. **58:**7, (Feb.) 1970.
Ward, F. A.: A primer of immunology, New York, 1970, Appleton-Century-Crofts.
Yates, U. H.: Transplantation: today and tomorrow, Today's Health **46:**33, (April) 1968.

LABORATORY SURVEY OF UNIT THREE*

PROJECT

Sources of infection

Part A—Bacteria on the human body

1. *Hands*
   a. Melt a tube of nutrient agar and allow it to cool until it feels just comfortable to the back of the hand.
   b. Wash your hands in a sterile dish containing 300 ml. of sterile water. Do not use soap.
   c. With a sterile pipette place 1 ml. of the wash water in a sterile Petri dish.
   d. Pour the cooled agar into the same Petri dish, replace cover, and set the dish on your desk.
   e. Slide the dish *gently* over the desk with a rotary motion until the melted medium and water are well mixed.

   Note: Do not allow the agar to come up the sides of the Petri dish and touch the cover.

   f. Let the medium harden, invert the dish, and incubate at 37° C. for 24 to 48 hours.
   g. Observe the growth of colonies. Note the number and different kinds.
2. *Fingertips*
   a. Gently touch the ends of your fingers over the surface of a sterile agar plate.
   b. Incubate plate at 37° C. for 24 hours and examine.
3. *Breath*
   a. Normal breathing
      (1) Remove the lid from a sterile nutrient agar plate.
      (2) Hold the plate about 6 inches from your mouth and breathe normally but directly upon the medium for about 1 minute.
      (3) Replace the cover, invert, and incubate at 37° C. for 24 to 48 hours.
      (4) Observe the growth of colonies. Note different kinds.

*There is no end to the laboratory projects that may be carried out in connection with this unit. Some of them require such highly skilled technic and are so time consuming that it would be unwise to include them. Others yield little information. Our selection is therefore limited to a few —simple to perform and yet applicable to the content of this unit.

b. Violent coughing
    (1) Remove the lid from a sterile nutrient agar plate.
    (2) Hold the plate about 1 foot from your mouth and cough violently onto the surface of the agar.
    (3) Replace the cover, invert, and incubate at 37° C. for 24 to 48 hours.
c. Violent coughing—plate at arm's length from the mouth
    (1) Proceed as above.
    (2) Observe the growth of colonies on this plate. Note the different kinds.
    (3) Which plate of the three prepared in a, b, and c of this exercise has the most colonies? Why?

4. *Other body sites*
    a. Select body sites such as the nose, throat, or ears. You can pass a sterile cotton swab moistened with sterile isotonic saline solution over the given area. Make cultures of the areas chosen.
    b. Use tubes or plates of media that would be likely to grow all organisms present. Blood agar is suitable.
    c. Incubate cultures taken at 37° C. for 24 to 48 hours.
    d. Observe the growth. Note the appearance of colonies.
    e. Make smears from colonies and stain by Gram's method. Note the shape of bacteria and whether gram-positive or gram-negative.
    f. Compare the bacterial growth obtained from the body sites selected for culture.

Part B—Bacteria on insects

1. Prepare a Petri dish of nutrient agar.
2. Allow an insect, such as a fly, to crawl for a time on the surface of the hardened agar under the cover of the dish.
3. Invert the plate after the fly has escaped and incubate at 37° C. for 24 to 48 hours.
4. Note the colonies.
5. Make gram-stained smears and examine.

PROJECT
Measure of immunity

Part A—Demonstration of serologic technic by the instructor with discussion

1. Technic of pipetting solutions for serologic test
2. Technic of making serial dilutions for serologic testing
3. Technic of manipulating serologic tubes

Part B—Demonstration of complement fixation by the instructor with brief discussion

1. Principles of complement fixation
2. Explanation of results in positive and negative tests.
3. Demonstration of positive and negative tests, with ample opportunity for student observation of the exhibits

Part C — Detection of antigen-antibody combinations

1. *Demonstration of the phenomenon of agglutination*
   a. Set up a row of seven serologic tubes in a test tube rack.
   b. To the first tube add 1.8 ml. isotonic saline solution, and to each of the remaining six tubes add 1 ml. isotonic saline solution.
   c. To the first tube add 0.2 ml. agglutinating serum against *Salmonella typhi* (or other salmonellae) and mix.
   d. From tube 1 remove 1 ml. and transfer to tube 2 and mix. From tube 2 remove 1 ml., transfer to tube 3, and mix. Continue transferring and mixing until tube 6 has been mixed. Discard 1 ml. Tube 7 acts as a control.
   e. To all tubes add 1 ml. of a suspension of *killed* typhoid bacilli. Calculate the dilution of agglutinating serum in each tube.
   f. Incubate tubes in the test tube rack at 37° C. for 2 hours.
   g. Note clumping of bacteria (agglutination) in the tubes containing the lower dilutions of agglutinating serum.
   h. Relative to this experiment fill out the following table:

Tube	1	2	3	4	5	6	7
Dilution of serum							
Agglutination*							

*Complete, partial, or none.

Note: Suppose the serum of a person had been used in this test instead of an artificial agglutinating serum and the results had been as indicated in the table. What would be the significance of the test? What would agglutination in tube 7 indicate? What is the name of the agglutination test for typhoid fever?

2. *Agglutination in blood typing (grouping)*

   Note: Students should work in groups of two. The blood type of each student will be determined.

   a. Mark a microslide down the middle with a wax pencil. Mark the left end *A* and the right end *B*. Make two wax rings, one on each end of the slide (see Fig. 13-6).
   b. Place one drop of anti-A serum in the wax ring on the *A* side of the slide and a drop of anti-B serum on the *B* side.

   Note: These serums may be obtained commercially. One drop of each is sufficient for the test.

   c. Puncture the finger or the earlobe of the person to be tested. Discard the first drop of blood. Transfer a minute drop of blood, by means of a clean applicator, to the drop of anti-A serum, mixing to make a smooth suspension of the cells. Discard applicator. With a fresh applicator, transfer a like drop to the anti-B serum and mix thoroughly. Do *not* use the same applicator for both serums. Why?
   d. Allow the slide to stand for 5 minutes, occasionally rolling or tilting it to ensure thorough mixing. Examine under the low power of the microscope.

Note: It may be difficult to distinguish between true agglutination and formation of *rouleaux.* This may be determined by stirring with an applicator. Rouleaux will be broken up, but true agglutination will be unaffected.

e. If there is no agglutination at the end of 5 minutes, cover each mixture with a cover glass and examine at 5- to 10-minute intervals, making a final reading at the end of 30 minutes.

Note: Sometimes a final reading must be deferred for 60 minutes, especially in the case of group A or AB blood; in this event it is advisable to ring the preparation with petroleum jelly to prevent evaporation.

f. Determine the blood type as follows:
  (1) If the cells are not agglutinated on either end of the slide, the patient belongs in group O.
  (2) If the cells on both ends are agglutinated, the patient belongs in group AB.
  (3) If the cells on the A end are agglutinated but those on the B end are not, the patient belongs in group A.
  (4) If the cells on the B end are agglutinated but those on the A end are not, the patient belongs in group B.

3. *Demonstration of the phenomenon of hemolysis*
  a. Prepare serial dilutions of antisheep hemolysin.
    (1) Prepare a 1:100 dilution of antisheep hemolysin (having a titer of about 1:3000) by adding 0.1 ml. of hemolysin to 9.9 ml. of isotonic sodium chloride solution and mixing.
    (2) Set up a row of 10 clean serologic tubes in a test tube rack.
    (3) To the first tube add 1 ml. of the 1:100 hemolysin and 0.5 ml. of the 1:100 hemolysin to the ninth tube.
    (4) To the remaining tubes, except the ninth, add 0.5 ml. isotonic sodium chloride solution.
    (5) To make serial dilutions of hemolysin, remove 0.5 ml. from the first tube, add to the second tube, and mix. From the second tube remove 0.5 ml., transfer to the third, and mix. Continue in this manner until the contents of tube 7 have been mixed. Discard 0.5 ml. of contents of this tube.
    (6) Calculate the dilution of hemolysin so produced in each tube and chart the results in Table A, p. 220.
  b. Add complement.
    (1) To the first eight tubes add 0.3 ml. of a 1:30 dilution of fresh guinea pig serum (complement).
    (2) Note that complement is not added to tubes 9 and 10.
  c. Add isotonic saline solution.
    (1) Add 1.7 ml. to the first eight tubes.
    (2) Add 2 ml. isotonic sodium chloride to tube 9 and 2.5 ml. to tube 10.
    (3) Note that by this part of the exercise all 10 tubes contain isotonic saline solution.
  d. Add to all tubes 0.5 ml. of a 2% suspension of sheep red blood cells.
  e. Incubate tubes in the water bath at 37° C. for 1 hour.
  f. Observe. Note the solution of cells (hemolysis).
  g. Chart the results using Table B, p. 220.

TABLE A. SERIAL DILUTION OF HEMOLYSIN

Tube no.	1	2	3	4	5	6	7	8	9	10
Isotonic NaCl	—	0.5 ml.	0.5 ml.	0.5 ml.	0.5 ml.	0.5 ml.	0.5 ml.	0.5 ml.	—	0.5 ml.
Hemolysin	1.0 ml. of 1:100 dilution	0.5 ml. of 1:100 dilution No. 1	0.5 ml. from mixture of No. 2	0.5 ml from mixture of No. 3	0.5 ml. from mixture of No. 4	0.5 ml. from mixture of No. 5	0.5 ml. from mixture of No. 6 discard 0.5 ml.	—	0.5 ml. of 1:100 dilution	—
Dilution of hemolysin	1:100							—	1:100	—
Total volume	0.5 ml.	0.5 ml.	0.5 ml.	0.5 ml.	0.5 ml.	0.5 ml.	0.5 ml.	0.5 ml.	0.5 ml.	0.5 ml.

TABLE B. HEMOLYSIS OF RED BLOOD CELLS

Tube no.	1	2	3	4	5	6	7	8*	9*	10*
Isotonic NaCl	1.7 ml.	1.7 ml.	1.7 ml.	1.7 ml.	1.7 ml.	1.7 ml.	1.7 ml.	1.7 ml.	2.0 ml.	2.5 ml.
Hemolysin serial dilution	0.5 ml. 1:100	0.5 ml.	0.5 ml.	0.5 ml.	0.5 ml.	0.5 ml.	0.5 ml.	—	0.5 ml. 1:100	—
Complement 1:30 dilution	0.3 ml.	0.3 ml.	0.3 ml.	0.3 ml.	0.3 ml.	0.3 ml.	0.3 ml.	0.3 ml.	—	
Sheep red cell suspension 2%	0.5 ml.	0.5 ml.	0.5 ml.	0.5 ml.	0.5 ml.	0.5 ml.	0.5 ml.	0.5 ml.	0.5 ml.	0.5 ml.
Presence of hemolysis?										
Final volume	3.0 ml.	3.0 ml.	3.0 ml.	3.0 ml.	3.0 ml.	3.0 ml.	3.0 ml.	3.0 ml.	3.0 ml.	3.0 ml.

*Controls

h. Answer the following questions:
   (1) What is the highest dilution of hemolysin showing complete hemolysis?
   (2) In what dilutions is hemolysis partial?
   (3) In what dilution does hemolysis cease?
   (4) Is hemolysis present or absent in the eighth tube? Why?
   (5) Is it present in the ninth tube? Why?
   (6) What would hemolysis in the tenth tube indicate?

Part D—Phagocytosis

1. Examine stained smears of pus showing phagocytosis by pus cells.
2. Compare with smears of pus in which phagocytosis is absent.
3. What is the pus cell? What is its usual function?

PROJECT

Preparation of a bacterial vaccine

Part A—Taking the culture

Note: Students may work in groups of three or four. In the preparation of this vaccine no attempt will be made to procure one organism in a pure culture, but a mixed vaccine composed of all of the organisms in the throat will be made.

1. Prepare the following supplies and sterilize in autoclave at 121° F. for 15 minutes:

Isotonic saline solution containing 0.5% phenol (in 200 ml. flask plugged with cotton)	100 ml.
Cotton swabs (swab ends down in 60 ml. widemouthed bottle)	5
Small funnel with pledget of absorbent cotton to use as filter (set in mouth of flask to facilitate handling)	1
Vaccine bottle with diaphragm stopper (plug bottle with cotton and wrap stopper in paper)	1

2. Allow materials in the autoclave to cool.
3. Use a sterile swab to obtain material from the throat and inoculate four large tubes of nutrient agar.
4. Incubate cultures at 37° C. for 24 hours.
5. Examine cultures by gross inspection and by making gram-stained smears. Are spores present?

Note: If spores are found, discard the particular culture from which the gram-stained smear was made. Why?

Part B—Processing the culture

Note: During the preparation of the bacterial suspension you must use strict aseptic technic. Otherwise the vaccine will be contaminated by outside bacteria.

1. Use a sterile pipette to transfer 5 to 10 ml. of sterile isotonic saline solution to each of the four tubes previously cultured.
2. Using a fresh swab for each tube, rub bacteria off the culture medium into the isotonic saline solution. Discard the swab and replace the cotton plug in the tube.

Note: Swabs and contaminated material should be placed in a pan containing a small amount of water or disinfectant. After the preparation of the vaccine is completed, the pan can be boiled or its contents disinfected.

3. Remove the cotton plug from the vaccine bottle and place the funnel in the mouth of bottle. Filter contents of each culture tube into the vaccine bottle. Wash a small amount of isotonic sodium chloride through the funnel.
4. Remove the funnel and place in a waste pan.
5. Place the stopper in the vaccine bottle and secure in place with rubber bands.

Note: The vaccine bottle must not become unstoppered during the next step.

6. Place the vaccine bottle in a water bath at 60° C. Let remain 1 hour.

Part C—Testing the sterility of the vaccine

1. Remove the bottle from the water bath and sterilize the stopper with iodine.
2. Puncture the stopper with a sterile syringe and remove a small amount of the vaccine.
3. Plant the vaccine on a tube of nutrient agar. Incubate at 37° C.
4. Observe growth at the end of 24 and 48 hours. If the vaccine is satisfactory to give to a person, will bacterial growth be present or absent? What is a vaccine made from the patient's own organisms called?

Note: In the foregoing technic, anaerobic cultures and determination of the number of bacteria per milliliter have been purposely avoided because these procedures are too complicated to be attempted at this time.

PROJECT

Measure of allergy

Part A—Demonstration of anaphylaxis by the instructor with discussion

Note: Prepare test guinea pigs 2 weeks beforehand.

1. Administration of sensitizing dose—intraperitoneal injection of 0.1 ml. of horse serum (or suitable foreign protein) in 2 ml. of sterile isotonic saline solution
2. Injection of provocative dose—1.0 ml. of horse serum or suitable foreign protein *into the heart* of the test animal.
3. Observation and description of results by instructor and students

Part B—Demonstration of skin testing for allergy by the instructor with discussion

EVALUATION FOR UNIT THREE

Part I

Select the numbers on the right that represent completions of the statements or answers to the questions.

1. Which of these are infectious but *not* communicable diseases?

(a) Smallpox	1. c
(b) Tetanus	2. b and c
(c) Boils	3. c and d
(d) Typhoid fever	4. a, d, and e
(e) Common cold	5. e

2. Which of the following articles from an active case of typhoid fever most likely carry the organisms?

(a) Bedpan	1. c and d
(b) Urinal	2. e
(c) Mouth wipes with nasal secretion	3. c
(d) Eating utensils	4. a and b
(e) Bloody bandage from arm wound	5. none

3. In which of the following diseases should the nurse be careful to use a mask?

(a) Smallpox	1. a, c, and e
(b) Gonorrhea	2. b
(c) Epidemic meningitis	3. a and d
(d) Malaria	4. all but a
(e) Diphtheria	5. all but b

4. The skin is the common portal of entry for the causative agents of all of the following diseases of man *except:*

(a) Malaria	1. a, b, and c
(b) Tularemia	2. a, b, c, and d
(c) Hookworm disease	3. a and c
(d) Measles	4. d and e
(e) Cholera	5. e

5. In which of the following tests is the serum from a patient mixed with a suspension of bacteria to test its ability to clump bacterial cells?

(a) Agglutination test	1. a
(b) Complement fixation test	2. b and c
(c) Tuberculin test	3. a and d
(d) Precipitin test	4. c, d, and e
(e) Schick test	5. all but e

6. A patient has type B blood. A person of which of the following blood types could be used in unusual circumstances as a donor?

(a) Type A	1. AB and B
(b) Type AB	2. B and O
(c) Type B	3. O
(d) Type O	4. B
	5. all but A and B

Which of these types would be the best type to give to this patient under ideal conditions?

7. A nurse beginning her communicable disease service is given an injection consisting of an antigen that has been modified. Which of the following might she be receiving?
   (a) Gamma globulin — 1. a
   (b) Antitoxin — 2. b and d
   (c) Toxoid — 3. c
   (d) Immune serum — 4. all but e
   (e) Antivenin — 5. none of above

(a) Gamma globulin	1. a
(b) Antitoxin	2. b and d
(c) Toxoid	3. c
(d) Immune serum	4. all but e
(e) Antivenin	5. none of above

8. Which of the following most nearly states your chances of developing an allergy during your lifetime?

(a) A tendency to become hypersensitive runs in families to a noticeable degree.	1. a 2. b
(b) Allergy is always inherited.	3. c
(c) There is no connection between allergy and heredity.	4. d 5. e
(d) Hypersensitivity may be inherited as a constitutional abnormality, or it may develop later in life.	
(e) If hypersensitivity runs in a family, all members will suffer the same type of allergic condition.	

9. A grown person has just received a stab wound. What facts should be ascertained before tetanus antitoxin is injected?

(a) The patient's temperature, pulse, and respiration are normal.	1. a and e 2. b
(b) The patient has no allergic condition.	3. a, c, and d
(c) The patient has never reacted to horse serum and has not received a first dose within the last few months.	4. b, c, and d 5. b, c, d, and e
(d) The patient gives no history of active immunization with toxoid.	
(e) The results of a skin or ophthalmic test for sensitivity to horse serum are negative.	

10. The cell directly related to antibody formation is:

(a) Polymorphonuclear neutrophilic leukocyte	1. a 2. b
(b) Red blood cell	3. c
(c) Plasma cell	4. d
(d) Squamous cell	5. a and c
(e) Eosinophil	

11. Normal body temperature is usually given as:

(a) 37° C.	1. a
(b) 35° C.	2. a and d
(c) 99° F.	3. a and e
(d) 97.6° F.	4. d
(e) 98.6° F.	5. e

12. Koch's postulates:

(a) Can always be satisfied	1. a and e
(b) Are important in establishing that a given organism causes a given disease	2. b, c, and d 3. e
(c) Refer to Koch's procedure for isolating tubercle bacilli	4. b and e 5. b and d

(d) Are the basis of experimental investigation of infectious disease
(e) Are only of historical interest

13. All the following diseases are commonly transmitted to man from animals except:

(a) Rabies	1. a, b, and c
(b) Bubonic plague	2. b, c, and e
(c) Typhus fever	3. e
(d) Psittacosis	4. d
(e) Candidiasis	5. b

14. Hypersensitivity to horse serum protein may be induced by:

(a) Injection of toxoid	1. d
(b) Injection of immune serum human	2. d and e
(c) Injection of bacterial vaccine	3. e
(d) Injection of diphtheria antitoxin	4. c
(e) Specific passive prophylaxis against tetanus	5. a and b

15. Interferon is described as:

(a) An antibody to a virus	1. b
(b) A substance in the complement system	2. b and c
(c) A substance formed by host cells	3. c
(d) A substance with action against viruses	4. c and d
(e) A substance comparable to an antigen	5. a, b, c, d, and e

16. A tissue transplant from a person to his identical twin is designated:

(a) Autograft	1. a
(b) Homostatic graft	2. b
(c) Isograft	3. c
(d) Xenograft	4. d
(e) Allograft	5. e

17. Complement:

(a) Is found normally in circulation of man	1. a
(b) Is an antibody	2. b
(c) Is increased by immunization	3. a and b
(d) Is made up of at least 11 components	4. a, d, and e
(e) Is inactivated in combination with antigen and antibody	5. a, c, and e

18. Phagocytic cells:

(a) Are active producers of antibody	1. a and e
(b) Are found only in the bloodstream	2. b and c
(c) Can ingest bacteria in absence of antibody	3. a and d
	4. d
(d) Are not related to immune mechanisms	5. c
(e) Always digest ingested bacteria	

Part II

1. After each type of acquired immunity listed in Column A, place the number of each method by which it may be conferred as listed in Column B.

COLUMN A	COLUMN B
______________ Active	1. Injection of a vaccine
______________ Passive	2. Recovery from a given disease
	3. Injection of convalescent serum
	4. Injection of an antiserum
	5. Injection of an antitoxin
	6. Injection of a toxoid

2. Using the letter in front of the appropriate term from column *A*, designate the kind of immunity resulting from the situation indicated in column *B*.

COLUMN A

(a) Natural immunity
(b) Naturally acquired, passive immunity
(c) Naturally acquired, active immunity
(d) Artificially acquired, passive immunity
(e) Artificially acquired, active immunity

COLUMN B

_________ 1. Administration of tetanus antitoxin
_________ 2. Administration of Salk poliomyelitis vaccine
_________ 3. Administration of influenza vaccine
_________ 4. An attack of measles
_________ 5. Transfer of antibodies from the mother to her newborn child
_________ 6. An attack of whooping cough
_________ 7. Administration of antivenin
_________ 8. Administration of gamma globulin
_________ 9. An attack of German measles
_________ 10. Administration of duck embryo rabies vaccine
_________ 11. Administration of Sabin live poliomyelitis virus vaccine
_________ 12. Parenteral injection of convalescent serum
_________ 13. The Pasteur treatment
_________ 14. Vaccination with tetanus toxoid
_________ 15. Vaccination against smallpox
_________ 16. Injection of diphtheria toxoid
_________ 17. Resistance of Algerian sheep to anthrax

3. On the line at the right side of the page, place the number of the term from the left side that does *not* bear a significant relationship to the other terms in the aspect mentioned.

(a) 1. Sporadic, 2. endemic, 3. pandemic, 4. epidemic, 5. congenital (occurrence) _________
(b) 1. Phagocytes, 2. precipitins, 3. agglutinins, 4. lysins, 5. antitoxins (action) _________
(c) 1. Folic acid, 2. polysaccharide, 3. biotin, 4. nicotinic acid, 5. pantothenic acid (vitamins) _________
(d) 1. Scavenger cells, 2. polys, 3. macrophages, 4. lymphocytes, 5. microphages (phagocytes) _________
(e) 1. Antitoxin, 2. agglutinin, 3. bacteriolysin, 4. opsonin, 5. antigen (do not prepare bacteria for phagocytosis) _________
(f) 1. Hives, 2. hay fever, 3. asthma, 4. serum sickness, 5. contact dermatitis immediate-type allergic reactions) _________

4. Match the item in column *B* to the one in column *A* with which it is most closely associated.

COLUMN A		COLUMN B
(a) Contact dermatitis	________	1. Widal test
(b) Antiglobulin test	________	2. Paul Bunnell test
(c) Agglutination test	________	3. Fluorescent antibody test
(d) Heterophil antibody test	________	4. Coombs' test
(e) Skin test for fungous disease	________	5. Weil-Felix reaction
(f) Skin test for bacterial allergy	________	6. Wassermann test
(g) Skin test for bedsonial infection	________	7. Histoplasmin test
(h) Skin test for susceptibility to scarlet fever	________	8. Frei test
	________	9. Dick test
(i) Immunologic reaction requiring special microscope	________	10. Schick test
	________	11. Coccidioidin test
(j) Precipitin reaction	________	12. Tuberculin test
(k) Serologic test for viruses	________	13. Brucellergen
(l) Complement fixation test	________	14. Gel diffusion
(m) Skin test for susceptibility to diphtheria	________	15. Patch test
(n) Serologic test for syphilis	________	16. Flocculation test
	________	17. Neutralization test

5. Comparisons: Match the item in column *B* to the word or phrase in column *A* best describing it.

HUMORAL IMMUNITY WITH CELL-MEDIATED IMMUNITY:

	COLUMN A	COLUMN B
________	1. Graft rejection an example	(a) Humoral immunity
________	2. Major role in resistance to viruses	(b) Cell-mediated immunity
________	3. Major role in resistance to bacteria	(c) Both
________	4. Presence of antigen not necessary	(d) Neither
________	5. Passive immunization possible	
________	6. Immune response within hours	
________	7. Immune response within days	
________	8. Contact with antigen the trigger mechanism	
________	9. Antigen reaches lymphoid tissues	
________	10. Related to cancer immunology	
________	11. Mediating cells thymus-dependent lymphocytes	
________	12. Mediating cells thymus-independent lymphoid cells	
________	13. Part of body's defense against infection	
________	14. Presence of demonstrable antibody	
________	15. Runt disease	

IMMEDIATE-TYPE ALLERGIC REACTION WITH DELAYED-TYPE ALLERGIC REACTION:

________	16. Hay fever	(a) Immediate-type allergy
________	17. Tuberculin hypersensitivity	(b) Delayed-type allergy
________	18. Not associated with circulating antibody	(c) Both
		(d) Neither
________	19. Similar symptoms induced by histamine	
________	20. Drug allergies	
________	21. Passive transfer with cells or cell fractions of lymphoid series	
________	22. Short duration	
________	23. Prolonged duration	
________	24. Contact dermatitis	
________	25. Anaphylactic shock	
________	26. Arthus phenomenon	
________	27. Drug idiosyncrasy	
________	28. Triggered by well-defined allergen usually.	
________	29. Emotional upsets	
________	30. Asthma	

UNIT FOUR

MICROBES: preclusion of disease

16 Physical agents in sterilization

17 Chemical agents in sterilization

18 Practical technics

Laboratory survey of unit four

Evaluation for unit four

16 Physical agents in sterilization

Sterilization is the process of killing or removing all forms of life, especially microorganisms, associated with a given object or present in a given area. This includes bacteria and their spores, fungi, (molds, yeasts), rickettsiae, and viruses (which must be either destroyed or inactivated). An object upon and within which all microbes are killed or removed is said to be *sterile.* The length of time an object remains sterile depends upon how well it is protected from microorganisms after it is sterilized. For instance, the outside of a tube of culture medium soon becomes contaminated because it is in direct contact with the microorganisms of the air, whereas inside the tube the culture medium, protected from microorganisms by the cotton plug, remains sterile indefinitely. Bacteria that have been killed are unable to multiply, but their bodies are not necessarily completely destroyed. The dead bodies of certain bacteria may retain their shape and staining qualities and even promote the production of antibodies when introduced into the bodies of man or lower animals.

Sterilization may be accomplished by mechanical means, by heat, or by chemicals.

REMOVAL OF MICROBES BY MECHANICAL MEANS

Three chief mechanical methods for removing microbes are (1) scrubbing, (2) filtration, and (3) sedimentation.

■ **Scrubbing.** Scrubbing is usually done with water to which some chemical agent, such as soap, detergent, or sodium carbonate, has been added. The process is both mechanical and chemical. Scrubbing, within itself, removes many microorganisms mechanically while the incorporated compound acts on them chemically. Scrubbing with soap (or detergent) and water is a very important process basic to any discussion of sterilization because the removal of dirt, debris, and extraneous matter from an area or object is a preliminary step to the effective removal of microbes therefrom by any method. Hands and person, floors, walls, woodwork, furniture, utensils of all kinds, glassware, linens, clothing, instruments, thermometers—all must be clean.

■ **Filtration.** Bacterial filtration is the process of passing a liquid containing bacteria through a material, whose pores are so small that the bacteria are held back. The mechanical removal or separation of the bacteria from the fluid renders it sterile. In the laboratory this process is used for sterilizing liquids and culture media that cannot be heated and for separating toxins, enzymes, and proteins from the bacteria that produced them. Certain pharmaceutical preparations are sterilized in this way. The materials most often used for bacterial filtration are unglazed porcelain, diatomaceous earth, asbestos, sintered glass, and cellulose membranes. (The finest mesh filter paper of the best quality will not retain bacteria, but the pore size of a cellulose membrane filter may be reduced so that even certain viruses are retained.) Bacterial filters are constructed so that the material to be filtered is made to pass through a disk or the wall of a hollow tube made of the filtering material. Rickettsiae or viruses passing through bacteria-retaining filters are said to be *filtrable.*

Filtration is an important step in the purification of a city water supply (p. 582). Bacterial filtration by a plastic membrane technic is widely used as a laboratory procedure in sanitary microbiology (p. 581).

■ **Sedimentation.** The process by which suspended particles settle to the bottom of a liquid is sedimentation. It finds practical application in the purification of water by natural or artificial means. In nature, large particles and suspended bacteria sink to the bottom of lakes, ponds, and flowing streams. Sedimentation plays a significant role in the artificial purification of water (p. 582).

STERILIZATION BY MOIST HEAT

The most widely applicable and effective sterilizing agent is heat. It is also the most economical and easily controlled. The temperature that kills a 24-hour liquid culture of a certain species of bacteria at a pH of 7 (neutral reaction) in 10 minutes is known as the *thermal death point* of that species. Since this represents the temperature at which all bacteria are killed, it is obvious that many are destroyed before this temperature is reached. In fact, the majority are destroyed within the first few minutes. For standardization, bacteria must be in a neutral medium when their thermal death point is determined because in either a highly acid or a highly alkaline medium they are more susceptible to heat. The *thermal death time* is the time required to kill all bacteria in a given suspension at a given temperature.

Aside from burning (really a chemical process), heat may be applied as *moist heat* or *dry heat.* Moist heat may be applied as hot water or steam and is the method of choice in sterilization, except for those things altered or damaged by it.

■ **Boiling.** A commonly employed although an incompletely effective method of sterilizing by moist heat is boiling. Boiling kills vegetative forms of pathogenic bacteria, fungi, and viruses in a matter of a few minutes. Hepatitis viruses are probably destroyed at the end of 30 minutes. (For elimination of hepatitis viruses, boiling is *not* recommended.) Spores are less readily destroyed. Although most of the spores of pathogenic bacteria can be destroyed in a boiling time of 30 minutes, boiling is not a reliable method when materials are likely to contain spores. Certain heat-loving saprophytes can survive at high temperatures, and their spores resist *prolonged* boiling for many hours.

The addition of sodium carbonate to make a 2% solution in boiling water hastens the destruction of spores and helps to prevent the rusting of instruments. Surgical instruments, needles, and syringes that are boiled must be clean and free of organic material. Remember that microbes will not be eliminated in the interior of the materials boiled until heat has penetrated there. Boiling must be continued long enough to ensure even distribution of heat through the object being sterilized. Such objects must be completely immersed in the boiling water. Boiling time should be prolonged 5 minutes for each 1000 feet above sea level.

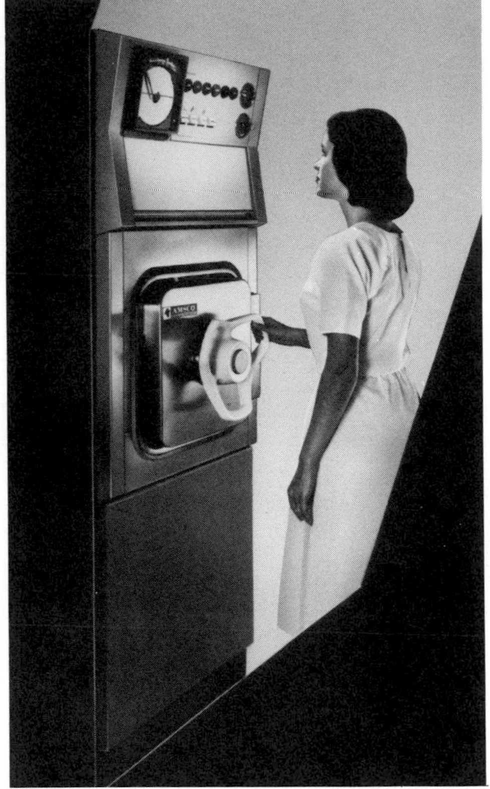

FIG. 16-1. Pressure steam sterilizer, or autoclave, cabinet model. (Courtesy American Sterilizer Co., Erie, Pa.)

■ **Sterilization by steam.** Steam gives up heat by condensing back into water. For instance, when a bundle containing fabrics, such as pads or sponges, is sterilized by steam, the steam contacts the outer layer, where a portion of it condenses into water and gives up heat. The steam then penetrates to a second layer, where another portion condenses into water and gives up heat. The steam thus approaches the center of the package, layer after layer, until the whole package is sterilized.

Steam may be applied as free-flowing steam or as steam under pressure. Free-flowing steam has about the same sterilizing action as boiling water. Steam under pressure is the most powerful method of sterilizing that we possess and is the preferred method unless the material being sterilized is injured by heat or moisture. The process is carried out in the *pressure steam sterilizer,* familiarly known as the *autoclave* (Figs. 16-1 and 16-2). It consists of a square sterilizing chamber surrounded by a steam jacket. The outside of the steam jacket is insulated and covered. The chamber is loaded with supplies to be sterilized (the load) through a door that closes the front end of the sterilizing chamber. This is a safety steam-locked door made tight against a flexible heat-resistant gasket. The design is such that steam can be admitted to the closed

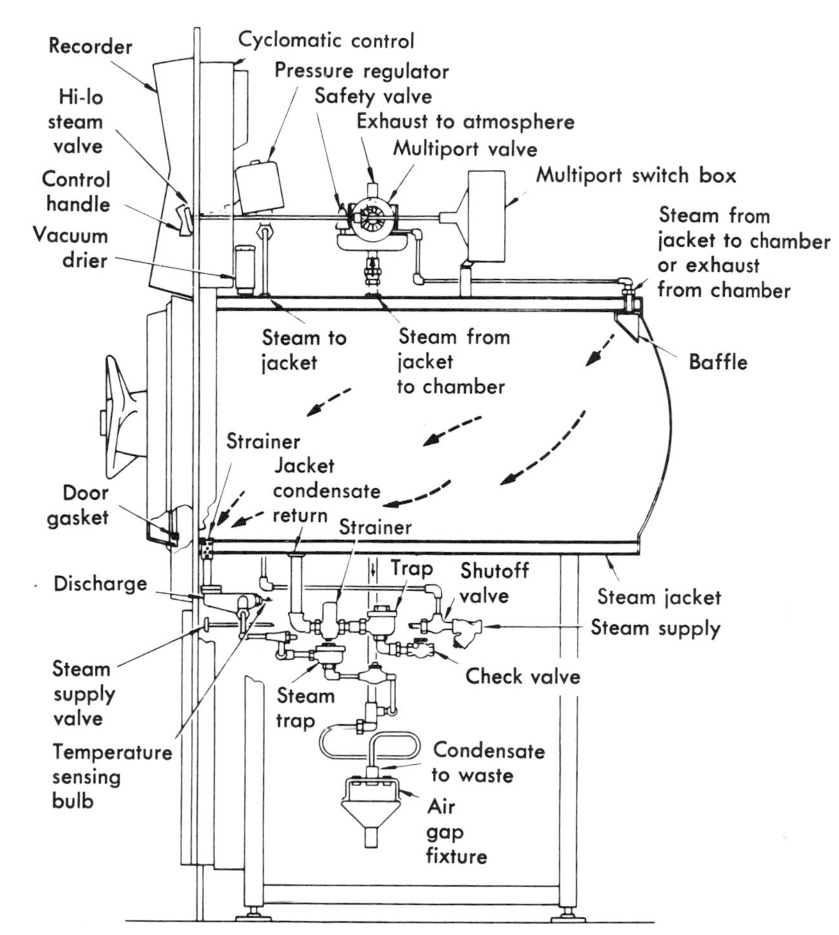

FIG. 16-2. The autoclave (pressure steam sterilizer, gravity air-removal type), diagram of longitudinal cross section. (Courtesy American Sterilizer Co., Erie, Pa.)

chamber under pressure. The source of the steam varies; it may come from the central boiler supply of a large hospital, or it may come from an electrically heated boiler on the instrument itself. Valves on the autoclave control the flow and exhaust of steam. On some of the pressure steam sterilizers, these valves are operated by hand, but in the modern versions, the entire operation of the instrument is automatically designed.

Steam under pressure is hotter than free-flowing steam, and the higher the pressure, the higher the temperature. The temperature of free-flowing steam (atmospheric pressure at sea level) is 100° C. At 15 pounds' pressure the temperature of steam is 121° C., and at 20 pounds' pressure it is 126° C. *Sterilization by steam under pressure is the result of the heat of the moist steam under pressure and not of the pressure itself.* Steam under a pressure of 15 or 20 pounds will kill all organisms and spores in 15 to 45 minutes (depending upon the materials involved). To maintain these temperatures at higher altitudes, the pressure must be increased 1 pound for each 2000 feet of increase in altitude.

OPERATION OF AN AUTOCLAVE. In the method of steam sterilization carried out in an autoclave (see Fig. 16-2) the long-used principle has been the "downward displacement gravity system." (Steam is admitted to the sterilizing chamber in such a way as to drive air down and out. Steam being lighter than air displaces air downward.) In the operation of an autoclave, pressure is first generated in the steam jacket. The connection to the sterilizing chamber is kept closed until jacket pressure is constant at 15 to 17 pounds. (The pressure within the steam jacket is kept constant during the procedure to keep the walls of the chamber heated and dry.) The load is placed in the chamber and the door secured. Then steam is admitted to the sterilizing chamber and the load is heated to the temperature of the steam in the chamber. At the time steam enters the chamber, the load and the chamber are both filled with air, which, if not evacuated, would reduce the moisture content of the autoclave and lessen its sterilizing capacity. Pressure steam sterilizers are provided with a vent arrangement whereby the air can escape to the atmosphere as the temperature is raised. When all of the air has been evacuated, steam will contact the thermostatic valve, and it closes. The moisture that condenses on the door or back part of the sterilizing chamber drains downward from behind a steam deflector plate to the bottom part of the chamber and is then discharged to the waste line. After the end of the sterilizing cycle, the steam is exhausted from the chamber but not from the jacket. At this point a drying cycle is effected before the door is opened. This is done by creating a partial vacuum in the chamber with steam from the jacket through an ejector tube on the autoclave and at the same time admitting air through a presterilized filter.

Since sterilization by steam under pressure is primarily a matter of temperature and the increase in pressure plays its part in the sterilizing process by increasing the temperature, the height to which the thermometer rises, rather than the reading on the pressure gauge, should be the guiding factor in the operation of the autoclave. This is particularly true because many pressure gauges are inaccurate. (Autoclaves usually have two pressure gauges, one to indicate the pressure in the steam jacket and the other to indicate the pressure in the sterilizing chamber.) As a whole, inaccurate pressure gauges read too high. A thermometer or the sensing element of a thermometer placed at the bottom of the sterilizer is a better indicator of the efficiency of the sterilizing process than one placed at the top, because, if any part of the autoclave fails to receive the full benefit of the steam, it is the bottom part. If the thermometer is placed in the discharge path of air and moisture coming from the sterilizing chamber, it can never indicate less than the lowest temperature in the system.

Modern autoclaves are equipped with a number of controls to increase the efficiency of sterilization and to remove insofar as possible the human factor. The *recording thermometer* is a clock-thermometer mechanism which indicates (1) the time at which the material being sterilized reaches the desired temperature, (2) whether the temperature remains stable, (3) how long the exposure lasts, and (4) how many times the autoclave is in operation during the day. The *indicating potentiometer* is an instrument for actually measuring the temperature of material in the autoclave. The *automatic time-temperature control* (1) operates the autoclave at the time and temperature for which it is set, (2) exhausts the steam from the chamber, (3) governs the process of drying, and (4) sounds an alarm indicating that the operation is complete.

Indicators are placed in with the load to be sterilized. An indicator changes in a characteristic and predictable way its physiochemical properties or biologic nature when the prescribed temperature for sterilization has been reached. Indicators used are strips of paper impregnated with biologic material such as the dried spores of *Bacillus stearothermophilus*, the thermal death time of which is known.

FIG. 16-3. High-prevacuum sterilizer (Medium Rectangular Vacamatic). Panel of controls for manual operation is behind white panel in upper right of instrument. (Courtesy American Sterilizer Co., Erie, Pa.)

HIGH-PREVACUUM STERILIZER.* An improved pressure steam sterilizer, the high-prevacuum sterilizer, is in wide use (Fig. 16-3). With a vacuum system incorporated into the sterilizer unit, a precisely controlled vacuum is pulled at the beginning and end of the sterilizing cycle. Saturated steam at a temperature of 275° F. (under a pressure of 28 to 30 pounds) enters the preevacuated chamber and instantly penetrates the load to be sterilized. Microbes present are killed within a few minutes. Under these high-temperature, high-pressure conditions of steam sterilization, there is a considerable shortening of the sterilizing (exposure) time (sometimes only 3 minutes). The vacuum pulled at the end of the sterilizing cycle dries the load. There is less damage to fabrics and to such items as rubber gloves because of reduced exposure.

■ **Fractional sterilization (intermittent sterilization).** When something that cannot withstand the temperature of an autoclave has to be sterilized, the procedure known as fractional or intermittent sterilization is used. This consists of exposing the material to free-flowing steam at atmospheric pressure for 30 minutes on 3 successive days;

*Basic principles for sterilization are the same in both the gravity air-removal type of sterilizer (just described) and the high-prevacuum one.

between times it is stored under conditions suitable for bacterial growth. With the first application of heat, all vegetative bacteria are killed, but the spores are not affected. Under conditions suitable for bacterial growth, the spores develop into vegetative bacteria, and the second application of heat kills them. The second incubation and third application of heat are added to ensure complete sterilization. This method is valueless unless the material being sterilized is of such a nature as to promote the germination of spores. Therefore, it is most useful in the sterilization of culture media. Sometimes referred to as *tyndallization,* it is infrequently used.

The low-temperature method of sterilizing vaccines (p. 276) may be applied in a fractional manner to materials such as serums that cannot withstand a temperature of 100° C. Such materials may be sterilized by heating to a temperature of 55° to 60° C. for 1 hour on 5 or 6 successive days.

■ **Pasteurization.** All nonspore-bearing disease-producing bacteria and most nonspore-bearing nonpathogenic bacteria are killed when exposed in a watery liquid to a temperature of 60°C. for 30 minutes. This is the basis of *pasteurization* (p. 587), a special method of heating milk or other liquids for a short time to destroy undesirable microorganisms without changing the compositon and food value of the material itself.

STERILIZATION BY DRY HEAT

Dry heat (hot air) sterilization consists of baking the material to be sterilized in a suitable oven. Dry heat at a given temperature is not nearly so effective a sterilizing agent as is moist heat of the same temperature. Under controlled conditions a dry temperature of 120° to 130° C. kills all vegetative bacteria within 1½ hours, and a dry temperature of 160° C. kills all spores within 1 hour; but a moist temperature of 120° C. kills all vegetative bacteria and most spores within 15 to 20 minutes. Whereas moist heat sterilization is primarily a process of protein coagulation, dry heat sterilization is a process of protein oxidation, and that oxidation goes on more slowly than coagulation. Moist heat also has greater penetrating power than dry heat.

For temperatures and times in practical dry heat sterilization, see Table 16-1. A temperature of more than 200° C. causes cotton and cloth to turn brown. Even a moderate degree of dry heat is injurious to most fabrics. Hot air is used mostly to sterilize glassware, metal objects, and articles injured by moisture, or items such as petrolatum (Vaseline), oils, and fats, that resist penetration by steam or water. An advantage is that dry heat does not dull cutting edges. In this form of sterilization the temperature should be slowly raised, and after sterilization is complete, the oven should be allowed to cool slowly to prevent breakage of glassware.

There are two causes of ineffective dry heat sterilization: (1) The materials to be sterilized are too closely packed, and (2) temperature is not uniform within the sterilizing oven. An attempt to overcome the uneven distribution of heat has been made in sterilizers and sterilizing ovens constructed in such a manner that either gravity aids in the circulation of hot air through the sterilizer (gravity convection) or circulation is carried on by means of blowers (mechanical convection). Mechanical convection is more satisfactory than gravity convection. Instruments to be sterilized must be clean and free of oil or grease films.

■ **Burning (incineration).** Burning is a form of intense dry heat very effective in removing infectious materials of various kinds. The platinum wire loop used to inocu-

TABLE 16-1. PHYSICAL STERILIZATION OF REUSABLE INSTRUMENTS AND SUPPLIES

	Administration		
Method	**Temperature**	**Time***	**Application**
Autoclave	121°-123° C. (250°-254° F.) 15-17 lb. pressure†	30 min.	Gloves, drapes, towels, gauze pads, instruments, glassware, and metalware
Dry heat	170° C. (340° F.)	1 hr.	Glassware, metalware, and dull instruments (any temperature listed)
	160° C.	2 hr.	Small quantities of powders, petrolatum (Vaseline), oils, and petrolatum gauze
	150° C.	3 hr.	Sharp instruments and metal-tip syringes
	121° C. (250° F.)	6 hr. or longer	
Boiling	100° C. (212° F.)	30 min.†	Method not recommended when dry heat and autoclave sterilization available

*With a high-prevacuum sterilizer, sterilizing (exposure) times are shorter; at a temperature of 132.8° to 135.5° C. (271° to 276° F.) sterilizing time is 4 minutes.
†Atmospheric pressure, sea level.

late cultures is repeatedly and quickly sterilized by *flaming,* that is, by heating the wire in an open flame until it glows. This form of sterilization is most important when materials and supplies are disposable or expendable.

All contaminated objects that are of no value or cannot be used again are preferably burned.

STERILIZATION BY OTHER PHYSICAL MEANS

■ **Natural methods of eliminating microbes.** If a culture of bacteria is dried, the majority of the bacteria are quickly killed but some live for quite a while. Spores and encysted protozoa resist drying for a long time. Although drying is an important natural method of removing or destroying microbes, it has little practical application to "artificial" sterilization, except that sterile dressings and similar objects should be kept dry.

Sunlight has a significant inhibitory and destructive action on microbes. It will kill *Mycobacterium tuberculosis* within a few hours and will kill many other bacteria in a shorter time. Sunlight is nature's great sterilizing agent, but it is so irregular in its presence that one cannot depend upon its action. The antimicrobial action of both drying and sunlight is applied to advantage in the home drying of food.

■ **Ultraviolet radiation.** The purity of the air in the "wide-open spaces" has long been recognized, and it is well known that the sterilizing effect of sunlight there is from the presence of ultraviolet rays. This fact has been applied to the construction of ultraviolet lamps, which are being widely used to prevent the airborne spread of disease-producing agents, especially in public places, in hospitals (operating rooms, treatment rooms, and nurseries), in microbiologic laboratories, and in quarters used to house animals.

Ultraviolet radiation is especially effective in killing organisms contained in the minute dried respiratory droplets that tend to disperse rapidly through the atmosphere of a building or hospital. It is not so active against dust-borne agents and microorganisms on surfaces. It inactivates certain viruses. Its bactericidal effect drops sharply when the humidity of the area exceeds 55% to 60%. In appropriate amounts it may damage the skin or conjunctivae.

■ **X rays and other ionizing radiations.** X rays and other ionizing radiations are known to be lethal to microbes and to living cells as well, but there is no practical application for their use in routine sterilization. In industry, beta rays or electrons sterilize prepackaged materials, such as sutures and plastic tubing.

One interesting application comes in the combination of heat and gamma radiation for the sterilization of spacecraft. If heat alone is used the spaceship is subjected to a temperature of 257° F. for 60 hours. The temperature has to be controlled carefully to prevent heat damage to and failure of components in such items as silver-zinc batteries and tantalum capacitors. If the spaceship with its equipment is sprayed with 150,000 rads of gamma radiation, the time at the temperature of 257° F. can be cut down to 2 hours.

■ **Ultrasonics.** Sound waves are mechanical vibrations. In the range of vibration where they are no longer heard as sound (supersonic or ultrasonic), these waves have been demonstrated to coagulate protein solutions and to destroy bacteria. *Cold boiling* results from the passage of ultrasonic pressure waves through a cleaning solution. Very tiny empty spaces in the liquid form and collapse thousands of times a second. This type of scrubbing action can blast material from the surface of objects made of metal, glass, or plastic. The use of such vibrations (pitched too high to be heard) is not widely practical, but the principle has been incorporated into a commercially available dishwasher. Cleaning medical instruments is a common application.

■ **Action of fluorescent dyes.** Certain dyes with the property of fluorescence, such as methylene blue, rose bengal, and eosin, are lethal to bacteria and viruses if in contact with these microbes in strong visible light.

QUESTIONS FOR REVIEW

1. Name three ways in which sterilization may be accomplished.
2. Define sterilization, bacterial filtration, sedimentation, thermal death point, pasteurization.
3. Name and describe briefly three mechanical means of removing or destroying microbes.
4. Why is moist heat more effective than dry heat as a sterilizing agent?
5. What is the effect of pressure on steam sterilization?
6. Explain intermittent or fractional sterilization.
7. What is an autoclave? Indicate the basic principles of its operation.
8. Briefly describe the high-prevacuum sterilizer. State its advantage.
9. How is dry heat applied for sterilization? Cite examples of items that must be sterilized in this way.
10. Tabulate all physical agents used for sterilization.

REFERENCES. See at end of Chapter 18.

17 Chemical agents in sterilization

EFFECTS OF CHEMICAL AGENTS ON MICROBES

There are certain definitions with which the student must be familiar to understand this and the following chapter. As we have already learned, "sterilization" is an absolute term referring to the destruction or removal of all microorganisms present under given conditions. *Disinfection* means death to disease-producing organisms and the destruction of their products. Disinfection does not consider the saprophytic organisms present in a given area that may or may not be killed. Disinfection is usually accomplished by means of chemical agents known as *disinfectants. Antiseptics* are agents that prevent the multiplication of bacteria but do not necessarily kill them. The terms "disinfection" and "disinfectant" are applied to procedures and chemical agents that are used to destroy microorganisms associated with inanimate objects. The term "antiseptic" is usually applied to an agent that acts on microorganisms associated with the living body. For instance, we disinfect the excretions from a sick person but apply an antiseptic to his skin or wounds. However, the terms are often used interchangeably.

Germicides are chemicals that kill microbes (not necessarily their spores). *Bactericides* are chemicals that kill bacteria. *Viricides* are agents that destroy or inactivate viruses. *Fungicides* destroy fungi. *Amebicides* destroy amebas, especially the protozoan *Entamoeba histolytica. Asepsis* means the absence of pathogenic microorganisms from a given object or a given area. In aseptic surgery the field of operation, the instruments, and the dressings are rendered free of microorganisms by sterilization. The operation is conducted in such a manner that it is kept as free of microbes as possible. The purpose of all of this is to avoid infecting the patient. *Fumigation* is the liberation of fumes or gases to destroy insects or small animals. *Deodorants* are substances that destroy or mask offensive odors. They may have neither disinfectant nor antiseptic action and may generally tend to obscure infectious material rather than to destroy it.

Bacteriostasis is that condition in which bacteria are prevented from multiplying (but in no other manner affected) by such agents as low temperature, weak antiseptics, and dyes. *Agents causing bacteriostasis are known as bacteriostatic agents.* Antiseptics and chemical bacteriostatic agents are synonymous terms. *Preservatives* are antiseptics or bacteriostatic agents used to prevent the deterioration of foods, serums, and vaccines.

Two terms with increasing applications are *degerm* and *sanitize.* To degerm is to remove bacteria from the skin by mechanical cleansing or application of antiseptics. To sanitize means to reduce the number of bacteria to a safe level as judged by public health requirements. It refers to the day-by-day control of the microbial population of utensils and equipment used in dairies and establishments where food and drink are served. In short, sanitization refers to a "good cleaning" and is basic to technics of sterilization. As used the term *sterilization* implies a mechanically clean item. The word *decontamination* applies to the process of killing all microbes from an item known to be mechanically dirty and containing a heavy growth of microorganisms.

■ **Qualities of a good disinfectant.** There are certain qualities that a chemical should have in order to be an ideal disinfectant for general use. Unfortunately at the present time none possesses all of them, and the selection of a disinfectant must depend upon the conditions under which it is to be used. The more of the following qualities that a disinfectant has, the more nearly it qualifies as an ideal general disinfectant. It should:

1. Attack all types of microorganisms.
2. Be rapid in its action.
3. Not be destructive to body tissues or act as a poison if taken into the body.
4. Not be retarded in its action by organic matter.
5. Penetrate the material being disinfected.
6. Dissolve easily in or mix with water to form a permanent solution or emulsion.
7. Not decompose when exposed to heat, light rays, or unfavorable weather conditions.
8. Not have a deleterious effect upon the materials being disinfected, such as instruments or fabrics.
9. Not have an unpleasant odor or discolor the material being disinfected.
10. Be easily obtained at a comparatively low cost and readily transported.

The most important feature of a disinfectant is its ability to form lethal combinations with microbial cells. *Remember that different species of microbes, especially bacteria, show much greater variation in their susceptibility to disinfectants than they do to sterilization by physical agents.*

■ **Action of antiseptics and disinfectants.** Antiseptics and disinfectants act by (1) oxidation of the microbial cell, (2) hydrolysis, (3) combination with microbial proteins to form salts, (4) coagulation of proteins, (5) modification of the permeability of the microbial plasma membrane, (6) inactivation of vital enzymes of microorganisms, and (7) disruption of the cell.

FACTORS INFLUENCING THE ACTION OF DISINFECTANTS. The factors influencing the action of disinfectants may be classified as to (1) the qualities of the disinfectant, (2) the nature of the material to be disinfected, (3) the concentration of the disinfectant, and (4) the manner of application. A chemical in a solution of one strength may be a disinfectant, whereas in a weaker solution it may act only as an antiseptic, and in certain very weak solutions it may actually stimulate microbial growth. The relative germicidal properties of the salts of a heavy metal are in proportion to their ionization.

The material being disinfected should be considered from the following standpoints: (1) kind and number of microbes present, (2) presence of vegetative forms or spores (3) distribution of microbes in clumps or in an even suspension, and (4) presence of organic compounds or other chemicals that inactivate the disinfectant.

Most chemical agents in common use as disinfectants are *germicidal but are not absolutely sporicidal;* that is, they do not kill all spores present. As a rule, the process of disinfection is a gradual one, and a few microorganisms survive longer than the majority do. To be effective, the disinfectant must be applied for a length of time sufficient to destroy all microorganisms. Many chemical disinfectants must be used for a long time to obtain the maximal effect; this often means 18 to 24 hours. An important factor relating to the disinfectant is the temperature at which it is applied. The higher the temperature, the more active the disinfectant. Remember that it must penetrate all parts of the material being disinfected because it must actually contact microbes to destroy them. An article being disinfected must be *completely submerged* in the disinfectant solution. Also, a disinfectant should be properly chosen in accordance with the physicochemical nature of the material to be disinfected.

SURFACE TENSION IN DISINFECTION. The molecules lying below the surface of a liquid are acted upon in all directions by the cohesive forces of neighboring molecules. Those at the surface are pulled downward and sideways by adjacent molecules but are not pulled upward because there are no molecules above the surface to attract them. This phenomenon is known as surface tension. Surface tension may be nontechnically defined as *that property due to molecular forces by which the surface film of all liquids tends to bring the contained volume into a shape having the least superficial area.* Since a sphere has the least area for a given volume, surface tension causes drops of liquid to assume the spherical form.

If a drop of mercury is placed on the skin or on a flat surface of metal or glass, it remains in the form of a distinct globule rolling from place to place. If, on the other hand, a drop of alcohol is placed on a surface, it does not form a globule but spreads out into a very thin layer over a large area. The surface tension of mercury is high; that of alcohol is low.

Surface tension is important in disinfection because liquids of low surface tension spread over a greater area and come into more intimate contact with cells than do liquids of high surface tension. Such low surface tension liquids are often spoken of as *wetting agents.* A good wetting agent spreads over a surface rapidly and remains in a thin film. Chemicals that reduce the surface tension of water when dissolved in it accumulate on the surface of cells in a more concentrated form than that in which they exist throughout the solution. Some wetting agents are also effective antiseptics. Wetting agents throught of primarily as surface-cleansing agents are *detergents.* Many detergents act as both cleansing agents and inhibitors of bacterial growth. The classic example of a wetting agent or detergent is soap. However, soap does not have the antiseptic action of certain other detergents. Many new synthetic organic detergents (liquids, granules, and such) that have been placed on the market are sometimes spoken of as soapless soaps or nonsoap cleaners. Some detergents, such as Tween 80, do not have an antiseptic action but promote bacterial growth.

■ **Standardization.** Many types of procedures have been designed to evaluate the antimicrobial activity of a given chemical agent as well as indicate its toxicity for tissues since Rideal and Walker in 1903 devised the original phenol coefficient test, which compared the disinfectant or antiseptic activity of a given compound with that of phenol under identical conditions. A phenol coefficient of greater than 1 indicated a stronger agent than phenol; a coefficient less than 1 indicated a weaker one. Despite the variety and number of methods existing today, there still remain inadequacies and the tests fail to give all the information needed.

COMMON DISINFECTANTS AND ANTISEPTICS

There are myriads of cleansing and disinfecting chemicals and combinations.* Only a few of the better-known ones are considered in this chapter.

Surface-active compounds

Soap is our most important cleansing agent. Although its utility as a disinfectant is limited, it is generally used before a given disinfectant is applied. The major action of soap is to aid the mechanical removal of microbes, primarily through scrubbing. In cleansing, soap and water separate particulate contamination of whatever kind from a given area, for example, the skin surface of the human body. This is an area constantly accumulating dead cells, oily secretions, dust, dried sweat, dirt, soot, and various microorganisms. Soap breaks up the grease film into tiny droplets; water and soap acting together lift up the emulsified oily materials and dirt particles, floating them away as the lather is washed off.

Although the term "detergent" means any cleansing agent, even water, it is used to distinguish the synthetic compounds from soap, both of which lower the surface tension of water. Soap is made from fats and lye; detergents are made from fats and oils by complicated chemical processes. All contain the one basic ingredient biodegradable linear alkylate sulfonate. Soaps depend for their cleansing action upon their content of alkali, which suspends the grime from the surface of an object in water washed off. The detergent ionizes in water; its electrically charged ions attach themselves to the dirt. The washing action releases the ions, which carry the dirt away. Detergents dissolve readily in cold water and completely in even the hardest water. Soap combines with the calcium and magnesium salts in the water, which make water hard, to form an insoluble scum.

Enzyme detergent refers to laundry presoak products in which certain proteolytic enzymes from bacteria are incorporated (in amounts up to 1.0% active enzyme). Enzymes are obtained through a fermentation process from the widely distributed nonpathogenic soil organism, *Bacillus subtilis.* Enzyme detergent dissolves organic (protein) stains such as blood, feces, and meat juices without harming fabric or user. Enzyme activity is quickly dissipated during the washing process and is inactivated by chlorine bleaches.

Soap is mildly antiseptic because of its sodium and alkali content, but to remove most bacteria effectively, scrubbing must be followed by the application of a suitable disinfectant. Organisms susceptible to the germicidal action of soap are pneumococci, streptococci, gonococci, meningococci, *Treponema pallidum,* and influenza viruses.†

If soap is to be followed by some germicide, it should be thoroughly washed off with 70% alcohol before the germicide is applied, because soap and germicide might combine

---

*With a few exceptions, there seems to be no general uniformity of opinion today as to which of these is best for application in any given situation, and the use of such chemical agents varies considerably, even in the same community.

†Most pathogenic bacteria and viruses are removed or chemically killed by the soaps and detergents ordinarily used in the commercial self-service laundry machines. The temperature of the dryer usually is high enough to eliminate any remaining bacteria. If the clothes are then ironed, the heat of the hot iron destroys any microbes that possibly could have survived.

to form an inert compound. The so-called germicidal soaps generally have little or no advantage over ordinary soaps. If soap is not properly handled and dispensed, it may become a source of infection within itself.

Benzalkonium chloride (Zephiran Chloride) is a mixture of high molecular alkyl dimethylbenzylammonium chlorides and is one of the most important members of the surface active chemical disinfectants known as quaternary ammonium disinfectants, or quats.* It is widely used in hospitals for the disinfection of hands and preparation of the field of operation. A 1:1000 aqueous solution may be used for the disinfection of instruments, especially endoscopes and sharp-edged cutting instruments, the blades of which would be dulled by autoclaving. Such a solution kills vegetative bacteria (except tubercle bacilli) in 30 minutes. *It has no effect on tubercle bacilli. It is not effective against spores.* The presence of serum or alkali retards its action. The same is true of soap; when Zephiran Chloride is used as a skin antiseptic, soap should be removed by thorough rinsing with 70% alcohol for 1 minute or more before the Zephiran Chloride solution is applied. (Water usually does not remove all soap, since soap and water constitute a colloidal solution.) Soap is completely soluble in 70% alcohol. When metal instruments are to be stored in a solution of Zephiran Chloride for a period of time, it is well to add an antirust agent or antirust tablets. One antirust tablet available commercially is a combination of sodium carbonate and sodium nitrite.

Diaparene Chloride is a quaternary ammonium compound particularly bacteriostatic toward *Brevibacterium ammoniagenes,* the intestinal saprophyte chiefly responsible for the production of ammonia in decomposed urine. It therefore is useful in the prevention of ammonium dermatitis in infants when used for the disinfection of diapers.

Ceepryn Chloride bears a close resemblance to the quaternary ammonium compounds just described. A commercially available 1:1000 solution is recommended as a mouthwash or gargle. It combines a foaming detergent cleansing action with an antibacterial activity against certain pathogenic organisms found in the mouth and throat.

Heavy metal compounds

Merbromin (Mercurochrome)† is a combination of mercury and a derivative of fluorescein. A 2% solution is used as a disinfectant of wounds. Stronger solutions are used as skin antiseptics. A 1% solution is tolerated by the bladder and kidney pelvis. A 2% solution of water (35%), alcohol (55%), and acetone (10%), is known as *surgical merbromin solution.* This antiseptic has probably received more publicity than its effectiveness warrants.

Nitromersol (Metaphen) is an organic mercury compound for the sterilization of instruments, for skin antisepsis, and for irrigation of the urethra. It is used in strengths ranging from 1:10,000 to 1:1000. It is said to be comparatively nontoxic, nonirritating, and nondestructive to metallic instruments and rubber goods. It kills

---

*A quaternary ammonium compound is built around a nitrogen atom of five valence bonds. Four of these bonds are attached to adjacent carbon atoms of organic radicals, and one is attached to an inorganic or organic radical. Soap reduces the germicidal action of the quats, as would hard water were it used to dilute a stock solution to make an aqueous preparation.

†In order to avoid the toxicity of the inorganic mercurials and still retain the disinfecting qualities of mercury, a number of organic mercury compounds have been developed. Among these are Mercurochrome, Metaphen, Merthiolate, and Mercresin (a mixture instead of a definite compound).

pathogenic bacteria that do not form spores other than tubercle bacilli. *Organic mercurials have no action against spores.*

Thimerosal (Merthiolate) is another organic mercury preparation with the advantages of its solubility in water and body fluids and low toxicity for tissues. It is used for disinfecting instruments, the skin, and mucous membranes and as a biologic preservative for vaccines, serums, and blood cells. This agent in aqueous solutions is considered to be fungistatic and bacteriostatic for nonspore-forming bacteria. The tincture or alcoholic solution is rather rapidly germicidal.

Mercresin combines the germicidal action of the mercurials with that of the phenolic derivatives, giving a maximum disinfectant action with a minimum of tissue injury. Its action is not inhibited by the presence of serum. It is widely used for local antisepsis of the skin.

Silver nitrate pencils are used to cauterize wounds. A 1:10,000 solution inhibits the growth of bacteria, and increasing the strength of solution, even up to 10% in certain instances, enhances germicidal activity. Silver nitrate has a selective action for gonococci. A 1% solution instilled into the eyes of newborn babies prevents ophthalmia neonatorum. Silver nitrate 0.5% solution is used in the continuously soaked dressings applied to burns. It reduces infection but *Pseudomonas* can still grow beneath the dry black eschar that results. The action of silver nitrate is retarded by chlorides, iodides, bromides, sulfates, and organic matter. It is reduced on exposure to light and, because of its many incompatibilities, should be used only with distilled water.

There are various unofficial salts of silver on the market, such as colloidal iodides, albuminates, and proteins, with fewer incompatibilities than silver nitrate. They have been used in various strengths, with varying results.

Protargol, Argyrol, Silvol, and Neo-Silvol are colloidal solutions of silver salts used because of their nonirritating qualities.

Zinc salts are mild antiseptics and also astringents.

Medicinal Zinc Peroxide, developed by the du Pont Laboratories in cooperation with Frank L. Meleney, is a mixture of zinc peroxide, zinc carbonate, and zinc hydroxide. The commercial powder should be sterilized at 140° C. dry heat for 4 hours. It is used in watery suspensions and ointments and is of special value in controlling infections caused by anaerobic and microaerophilic organisms in injuries such as gunshot wounds, bites, and deep puncture wounds.

Calamine Lotion is mostly zinc oxide with some ferric oxide.

Copper sulfate is valuable chiefly for its destructive action on the green algae that often grow in reservoirs and render water obnoxious. One part of copper sulfate per million parts of water kills algae if the water does not contain an excess of organic matter. One part of copper sulfate added to 400,000 parts of water destroys typhoid bacilli in 24 hours. It is not harmful to drink water for a short time that contains this amount of copper sulfate. Copper sulfate is an important ingredient of sprays used to combat fungous diseases of plants.

Alcohols and aldehydes

Ethyl alcohol is one of the most widely used disinfectants and one of the best. For alcohol to coagulate proteins (its disinfectant action), water must be present. Because of this fact, 70% has long been considered the critical dilution of ethyl alcohol. However, there is good reason to think that against microorganisms in a moist environment alcohol acts over a range including 70% and up to 95%. Alcohol in a 70%

dilution may be used for the disinfection of certain delicate surgical instruments although it does tend to rust instruments, and it dissolves the cement from around the lights of endoscopes. It does not kill spores. Alcohol kills tubercle bacilli rapidly and is a tuberculocidal disinfectant of choice.

Isopropyl alcohol is slightly superior to ethyl alcohol as a disinfectant. It is also cheaper, and its sale is not subject to legal regulations. Like ethyl alcohol, it acts against vegetative bacilli (not spores) and tubercle bacilli. Recent evidence indicates that this agent is an effective disinfectant in dilutions stronger than the conventional 70% and that the most effective one may well be full strength (99%).

Formaldehyde, a gas, occurs in commerce in a watery 37% solution known as *formalin.* In addition to its disinfecting properties, formaldehyde acts in some cases as a deodorizer and as a preservative of tissues. It is used to convert toxins into toxoids. Its action depends upon the presence of moisture, the concentration of the gas, the temperature, and the condition of the object to be sterilized. The presence of 1% of the gas in a room or chamber destroys all nonsporulating pathogenic bacteria.

Cytoscopes and certain specialized instruments that would be damaged by heat are sometimes disinfected in airtight sterilizing cabinets equipped with electric terminals so that formaldehyde pastils can be vaporized in them. The instruments are left in the cabinet exposed to the formaldehyde fumes for 24 hours.

Mixtures of alcohol, formalin, and hexachlorophene are frequently used for sterilizing surgical instruments. This combination makes one of the most active germicidal solutions commercially available; bacteria, spores, tubercle bacilli, and most viruses are speedily killed thereby. An example of such a mixture is *Bard-Parker Germicide,* a commercial mixture of formaldehyde, isopropanol, methanol, and hexachlorophene. It is especially useful in the sterilization of knife blades and suture needles.

Glutaraldehyde (Cidex) is one of the latest additions to any list and one of the most effective. *Activated glutaraldehyde solution* is bactericidal, viricidal, and sporocidal in short exposure times. It is useful for sterilization of anesthetic equipment and the intermittent positive-pressure breathing apparatus. It is recommended for use with instruments containing optical lenses.

Phenol and derivatives

Phenol (carbolic acid) is a corrosive poison. A 1:500 solution inhibits the growth of bacteria. A 5% solution kills all vegetative bacteria and the less resistant spores in a short time. Contact with alcohol or ether decreases its action, which is also inhibited by soap. The addition of 5% to 10% hydrochloric acid increases its efficiency.

Since the action of phenol is not greatly retarded by the presence of organic matter, it is an excellent disinfectant for feces, blood, pus, sputum, and proteinaceous materials. It does not injure metals, fabrics, or painted surfaces. The crude acid may be used for woodwork because it is cheaper and more effective than the pure substance.

The crystals or strong solutions should not be allowed to touch the skin. If this happens, alcohol should be applied at once. Dilute solutions should not be left in contact with the skin or mucous membrane for more than 30 minutes or 1 hour because they injure tissues.

Cresol has a higher germicidal power than does phenol and is less poisonous. *Saponated cresol solution* is an alkaline solution of cresol in soap. (Similar but generally more expensive proprietary preparations are on the market.) A 2.5% solution of saponated cresol makes a good disinfectant for feces and sputum.

Lysol, a widely used and popular proprietary disinfectant, is essentially a solution of cresol with soap sold under a trade name. In recent years improvements have been made in the mixture. Lysol is most important in the disinfection of inanimate objects, including instruments, furniture, table surfaces, floors, walls, rubber goods, rectal thermometers, and contaminated objects of varied description, especially when these have been contaminated by *Mycobacterium tuberculosis.*

In general, phenol derivatives such as cresol and Lysol can be used to disinfect excreta or contaminated secretions from patients with infectious diseases, but have special value in the disinfection of tuberculosis. They are of little value in antisepsis of the skin because in concentrations that would not injure the skin they possess little bactericidal activity. They do not destroy all spores present.

Amphyl is one of the modern and greatly improved phenolic disinfectants with wide and varied applications as a germicide. (Chemically it is a mixture of orthophenylphenol and paratertiary amylphenol with potassium ricinoleate in propylene glycol and alcohol.) Nontoxic, noncorrosive to metals, and nonirritating to skin and mucous membranes, it can be mixed with soap and certain other antiseptics. Because it is an agent with low surface tension, it spreads and penetrates materials. Surfaces treated with it tend to retain an antimicrobial action for several days. It is effective in dilutions of 0.25% to 3% for a range of routine disinfecting procedures involving skin, mucous membranes, floors, walls, furniture, dishes, utensils, and surgical instruments. Fungi, bacteria (including the tubercle bacillus), and viruses, but *not* spores, are destroyed by its action; heating increases its germicidal properties.

Staphene is a phenolic disinfectant related to Amphyl. It is a mixture of four synthetic phenols* (paratertiary amylphenol, ortho-benzyl-para-chlorophenol, orthophenylphenol, and 2,2′-methylene-bis[3,4,6-trichlorophenol]).

O-syl, an antiseptic, germicide, and fungicide, is considered to be especially effective against the causative agent of tuberculosis. Chemically it is one synthetic phenol (orthophenylphenol).

Hexachlorophene (G-11), a diphenol, is one of the few antiseptics and disinfectants not affected by soap. Therefore, in concentrations from 1% to 3%, it has been incorporated into soaps and combined with detergents that have been used for the preoperative hand scrubs of the surgical team and for the preoperative and postoperative preparation of the skin of the patient. Hexachlorophene is long retained on the skin, from where it can be recovered more than 48 hours after use. Its benefit is attributed to the bactericidal film it leaves on the skin after repeated application; the concentration on the skin is cumulative up to 2 or 4 days. The skin cannot be completely sterilized. Mechanical cleansing plus the use of germicidal agents removes the superficial growth of bacteria from the skin surface, which fortunately includes most of the pathogens. Hexachlorophene has been so widely used because of its bacteriostatic action against gram-positive microbes, notably the staphylococcus. Bathing of infants with a detergent containing hexachlorophene has been shown to reduce considerably the incidence of staphylococcal infection in a population of newborns.

Generally hexachlorophene is not irritating to the skin although it is sometimes associated with allergic manifestations. However, it is readily absorbed through the normal skin and more easily so through abraded or burned skin. Its use has been

---

*Some of the phenolic detergent germicides widely used in cleaning solutions can cause depigmentation of the skin of the user.

blamed for brain seizures developing in young burn victims who had been washed with it. Animal experiments relate the uptake through the skin over a period of days to subsequent brain damage and even paralysis.

The United States Food and Drug Administration warns against total body bathing of infants and adults with products containing 2% and 3% concentration of hexachlorophene and restricts the over-the-counter sale of products containing it.* Available only by prescription, it is still used generally in hospitals and similar institutions because of its effective antibacterial action.

pHisoderm, a detergent cream, is not a single compound but a proprietary mixture of wool fat, cholesterol, lactic acid, and sulfonated petrolatum. It cleanses faster and more effectively than does soap and is used for degerming the hands and skin. It may be used alone, but it is most often used combined with hexachlorophene.

pHisoHex is the name given to the mixture of the detergent base pHisoderm and 3% hexachlorophene. It is still a popular preparation for surgical hand scrubs. Regular use is recommended for maximal bactericidal effect. It is nonirritating, only small amounts need to be used, and the time required for the surgical scrubbing is much shorter than with soap preparations. Its emulsifying action on oily material and its ability to form suds are desirable characteristics.

Hexagerm is another proprietary antiseptic skin detergent containing 3% hexachlorophene.

Halogen compounds

Iodine is one of the best known and most generally used disinfectants. It is a potent amebicide, a bactericide in a wide spectrum, a good tuberculocide, fungicide, and viricide. The commonly used *tincture of iodine* containing 2% iodine is one of the best disinfectants for small areas of skin, minor cuts, abrasions, and wounds. The strong tincture of iodine (7%) is too toxic for most purposes. Very strong and very old solutions of iodine burn the skin. It should not be applied under a bandage. Although in some persons iodine compounds produce allergic skin rashes, iodine is less toxic than are other routinely used germicides. Within the last few years iodine has been used for disinfection of water in swimming pools.

Iodophors are compounds in which iodine is carried by a surface-active solvent. The germ-killing action results from the release of free iodine when the compound is diluted with water; only the free iodine has any appreciable disinfectant action. An iodophor enhances the bactericidal action of iodine and reduces odor. It does not stain the skin or materials on which it is placed as does the tincture. One such is the proprietary preparation *Wescodyne,* an iodine detergent germicide (also a good tuberculocide). Other examples are *Hi-Sine, Iosan,* and *Betadine* (povidone-iodine complex).

Chlorine is one of the most effective and widely used of all chemical disinfectants. It is used in the sterilization of drinking water, in the purification of swimming pools, and in the treatment of sewage. It is applied by releasing the gas from cylinders or by the use of compounds that release free chlorine. The chlorine liberated by chlorine-liberating compounds is spoken of as "available" chlorine. For sterilization to be effective the chlorine content of water must reach a concentration of 0.5 to 1 part per million

---

*Hexachlorophene has been perhaps the most universal of the antibacterial agents, having been incorporated into a wide variety and number of consumer products—soaps, shampoos, toothpastes, deodorants, lotions, powders, ointments, cosmetics, and medicinal cleansers.

(ppm) parts of water. If no organic material is present, 0.1 part of chlorine to 1 million parts of water destroys the poliovirus. In a strength of 1 ppm water, chlorine has no effect on the cercariae that represent one stage in the development of flukes (p. 527). (Nor are these minute worms removed by sand filtration or aluminum sulfate clarification of the water.) Cysts of *Entamoeba histolytica* (p. 511), the cause of amebic dysentery, can live as long as 1 month in water. The usual chlorination treatment for drinking water does not kill them; water containing an adequate amount of chlorine to destroy them would not be fit to drink. The World Health Organization recommends that commercial airlines filter their water supply and overchlorinate it, 8 to 10 p.p.m. of free available chlorine, in order to kill cysts of *Entamoeba histolytica* and viruses of infectious hepatitis. The water must then stand for at least 30 minutes. It is dechlorinated to eliminate the disagreeable taste and smell.

Sodium hypochlorite (NaOCl), made by the reaction of chlorine on sodium hydroxide, is a powerful oxidizing agent. It cannot be prepared in the form of a powder but is manufactured in solutions of varying strength. The stronger solutions are used as bleaches by laundries and other establishments. The weaker solutions (example, Clorox) are used as household bleaches and for the bactericidal treatment of food-handling equipment. *Dakin's solution* and modifications are weak neutral solutions of sodium hypochlorite, liberating from 0.5% to 5% available chlorine.

Chloramines are organic chlorine compounds that decompose slowly and liberate chlorine. They are used to sanitize glassware and eating utensils and to treat dairy and food-manufacturing equipment. They are inferior when rapid action is required. Their value lies in their prolonged action. The ones in most common use are chloramine-T, chloroazodin, and halazone. Halazone is used for the sterilization of relatively small amounts of drinking water.

Acids

Boric acid is a weak antiseptic most often used as an eyewash. However, when taken internally, boric acid is highly toxic. Because of the risk of accidental poisoning, the use of boric acid in hospitals has been condemned. It has no place in the pediatric division or in any nursery. If even a dilute solution of boric acid mistaken for distilled water should be used in an infant formula or in a parenteral fluid, the results could be disastrous. Dusting the powder repeatedly over the diaper area of a baby to remedy diaper rash can cause death. A so-called bland ointment applied to abraded areas of the skin can produce an unfavorable reaction.

Fuming nitric acid is the best agent for cauterizing the wounds inflicted by rabid animals.

Benzoic acid and *salicylic acid* are fungistatic agents. A mixture of benzoic acid 6% and salicylic acid 3% *(Whitfield's ointment)* is used to treat fungal infections of the feet.

Undecylenic acid (Desenex) is used for athlete's foot and other fungal infections of the skin.

Oxidizing agents

Hydrogen peroxide owes its disinfecting qualities to the free oxygen that it liberates. It is a spectacular but not overly reliable antiseptic that deteriorates rapidly.

Sodium perborate is an oxidizing agent used in the treatment of trench mouth (p. 389). It is a common ingredient of tooth powders.

Potassium permanganate owes its effectiveness to strong oxidizing qualities. However, its action is weakened by the presence of organic matter. It was once used extensively in the treatment of infections of the genitourinary tract.

Dyes

Dyes such as gentian violet and crystal violet in high dilutions inhibit the growth of gram-positive bacteria but have little effect on gram-negative bacteria, whereas dyes such as acriflavine and proflavine inhibit the action of gram-negative bacteria but have little effect on gram-positive bacteria.

Dyes are used to obtain pure cultures and to treat certain infectious processes. If it had not been for the advent of the sulfonamide drugs, the use of dyes as therapeutic agents would be presently more advanced.

Miscellaneous chemicals

Ethylene oxide is a gas with a broad spectrum of antimicrobial activity that has been used in industry for years for such purposes as the disinfection of the furnishings of railway cars. It is an odorless, poisonous, and explosive gas that can be easily kept as a liquid. Since its vapor is highly inflammable in air, it is transported commercially in a noninflammable mixture with carbon dioxide. This gas is used especially to treat items that are damaged by heat, water, or chemical solutions, and is a potent agent properly used. The article to be disinfected must be exposed to the gas in an appropriately designed sterilizing chamber in which the temperature can be raised and from which the air must be evacuated. Polyethylene and paper are suitable for packaging. A prolonged period of exposure is required, in one large model—at least 8 hours. Bacteria (including tubercle bacilli and staphylococci) and spores are killed. The hepatitis viruses are destroyed. Ethylene oxide properly used has an irreversible chemical reaction with all organisms including their spores. The reaction is speeded by heat. Ethylene oxide sterilization is safe and effective provided adequate aeration time is allowed for the gas to elude from items so sterilized. This is important since the gas and its by-products are highly irritating to skin and mucous membranes (including those of the eye). This form of *cold* sterilization is being widely used for delicate surgical instruments with optical lenses and for the tubing and heat-susceptible plastic parts of the heart-lung machine. Many prepackaged commercial items such as rubber goods and plastic tubes are sterilized with ethylene oxide. Because of the penetration of the gas, this form of sterilization is effective for blankets, pillows, mattresses, and bulky objects.

Lime is one of the most common and, when properly used, one of the most effective germicidal agents. Limestone, which occurs plentifully in nature, is converted into *quicklime* by heating as shown in the following equation:

$CaCO_3$ (Limestone or calcium carbonate)	+	Heat	=	CaO (Quicklime or calcium oxide)	+	CO_2 (Carbon dioxide)

When quicklime is treated with one half its weight of water, *slaked lime* is formed according to the following equation:

CaO (Quicklime)	+	H_2O (Water)	=	$Ca(OH)_2$ (Calcium hydroxide or slaked lime)

Slaked lime, a powerful disinfectant, is used in the form of *milk of lime,* which is prepared by mixing one part of freshly slaked lime with four parts of water. Milk of lime is especially useful as a disinfectant for feces. A more dilute suspension of slaked lime is known as *whitewash.*

Lime preparations must not be exposed to the air because they combine with the carbon dioxide of the air to form calcium carbonate, which is without any antiseptic action.

Chlorinated lime (chloride of lime, bleaching powder, calcium hypochlorite), made by passing chlorine through freshly slaked lime, is one of the most important of the chlorine-liberating compounds used as disinfectants. An unstable compound, its antiseptic action results from its release of hypochlorous acid, which is toxic to the bacterial cell. A 0.5% to 1% solution kills most bacteria in 1 to 5 minutes. A 1:100,000 solution destroys typhoid bacilli in 24 hours. Chlorinated lime bleaches and destroys fabrics and decomposes when exposed to the air. As a means of disinfecting excreta, chlorinated lime is probably without a rival.

Sulfur dioxide is a gas formed by burning sulfur. It was formerly used extensively for house and room fumigation. Its germicidal action depends upon the presence of moisture because it combines with water to form sulfurous acid ($H_2O + SO_2 = H_2SO_3$), which is the destructive agent. It injures fabrics and metals.

Ferrous sulfate (copperas) is commonly used in the form of the impure commercial salt. In addition to being a good disinfectant, it is a good deodorant because it combines with both ammonia and hydrogen sulfide. It is an ideal disinfectant for use in damp musty places and around houses.

Chlorophyl in the form of its water-soluble components has been used in the local treatment of wounds. It cleanses the wound, stops pus formation, destroys odors, and promotes healing.

Table 17-1 summarizes the antimicrobial effects of some of the more important chemical agents.

THERAPEUTIC CHEMICALS IN MICROBIAL DISEASES

Chemotherapeutic agents

Although it is possible to treat many local infections with chemicals, the finding of a chemical that, taken orally or parenterally, destroys microorganisms without injuring the body cells is not easy. Examples of such chemicals are quinine, against the parasites of malaria, and emetine (the active principle of ipecac), against *Entamoeba histolytica,* the organism causing amebic dysentery. Agents such as these are known as *chemotherapeutic agents.* The treatment of disease with chemotherapeutic agents is *chemotherapy.* The *chemotherapeutic index* compares the toxicity of a drug for the body with its toxicity for a disease-producing agent. It is obtained by dividing the maximal tolerated dose per kilogram of body weight by the minimal curative dose per kilogram of body weight.

The term "chemotherapy" was introduced by Paul Ehrlich (1854-1915), who was a pioneer in the development of chemotherapeutic agents. He introduced salvarsan, a drug used rather effectively in the treatment of syphilis for more than a quarter of a century.

Within recent years there have been phenomenal developments in the field of

TABLE 17-1. ANTIMICROBIAL ACTIVITY OF COMMONLY USED COLD "STERILANTS"*

Agent	Destructive action against				
	Bacteria	Tubercle bacilli	Spores	Fungi	Viruses
Alcohol—ethyl (70% to 90%)	+	+	0	+	±
Alcohol—isopropyl (70% to 90%)	++	+	0	+	±
Alcohol-formaldehyde (Bard-Parker Germicide)	++	+	+	+	+
Alcohol-iodine	++	+	±	+	+
Formalin (37%)	+	+	+	+	+
Glutaraldehyde (buffered, 2%) (Cidex)	++	+	++	+	+
Iodine (2% to 5% aqueous)	++	+	±	+	+
Iodophors (1%) (povidone-iodine)	+	+	±	±	+
Mercurials (Merthiolate)	±	0	0	+	±
Phenolic derivatives (O-syl 1% to 3%)	+	+	0	+	±
Quats (benzalkonium chloride 1:750 to 1:1000)	++	0	0	+	0

*After DiPalma, J. R., editor: Drill's pharmacology in medicine, New York, 1971, McGraw-Hill Book Co. ++, Very good; +, good; ±, fair (greater concentration or more time needed); 0, no activity.

chemotherapy, beginning with the discovery of the sulfonamide compounds. A brief discussion of these and certain specifically acting chemotherapeutic drugs follows.

Sulfonamide compounds contain the group $SO_2N<$. Many of them are derived from the compound known as sulfanilamide, which was among the first of the sulfonamides to be developed as a chemotherapeutic agent. Sulfanilamide was derived from a dye, known by the trade name of Prontosil, that had been found to be effective in treating streptococcal infections. Prontosil was used for only a short time, and sulfanilamide itself has been replaced by less toxic drugs. They are many sulfonamide compounds with varying actions on bacteria and varying toxic effects on the body

A patient receiving sulfonamide therapy should be continuously observed for early toxic symptoms. If such appear, the drug must be discontinued. Weight for weight, sulfathiazole is the most destructive to bacteria of all the sulfonamide compounds, but it has, to a great extent, been replaced by the less toxic and almost as effective sulfadiazine. The sulfonamide compounds most readily available today are sulfadiazine, sulfamerazine, sulfamethazine, sulfacetamide (Sulamyd), sulfadimethoxine (Madribon), sulfisoxazole (Gantrisin), sulfamethoxypyridazine (Kynex), sulfachloropyridazine (Sonilyn), sulfisomidine (Elkosin), sulfamethizole (Thiosulfil), and sulfaethidole, all for systemic use. Other compounds include sulfaguanidine, succinylsulfathiazole (Sulfasuxidine), and phthalylsulfathiazole (Sulfathalidine) for intestinal use.

The sulfonamide compounds are bacteriostatic in their mode of action; they interfere with certain enzymes in the affected bacterial cells. The sulfonamide compounds are not self-sterilizing. Their action is inhibited in the presence of pus. Sulfonamides are effective against gram-positive bacteria, some gram-negative diplococci and bacilli, certain large viruses, and *Actinomyces*. Generally the sulfonamides have been replaced by antibiotics that are not so toxic and that are more certain and rapid in their action. Where an antibiotic and a sulfonamide are of equal value, it is customary to give the antibiotic. The sulfonamides have been used for treatment of epidemic meningitis because the sulfonamides pass into the cerebrospinal fluid more easily than do the antibiotics. The highly soluble sulfonamides are important in the treatment of some infections of the urinary tract as well. Sulfonamides are given to protect an individual with a history of rheumatic fever against streptococcal infection, and certain sulfonamides for intestinal use suppress the growth of the intestinal flora, an important measure in the preparation of the patient for surgery of the large bowel.

Isoniazid (isonicotinylhydrazine) (INH) is a remarkably potent bacteriostatic agent against *Mycobacterium tuberculosis*, but it has little effect on most other microbes. Its mode of action is not known at the present time.

Isoniazid is an effective drug not only in treating tuberculosis but also in preventing tuberculous infection from becoming active disease. The American Thoracic Society recommends ideally that all tuberculin-positive individuals should receive a year of isoniazid prophylaxis (see also p. 447). Generally there have been few untoward reactions, but recently instances of drug toxicity have appeared. Some 10 out of 1000 individuals receiving isoniazid prophylactically will develop evidence of liver dysfunction, rarely a severe hepatitis. Most authorities feel this is no reason to discontinue the regimen but rather an indication to screen the recipients initially and that once on the drug to check them a monthly intervals for any sign of liver disorder.

Para-aminosalicylic acid (PAS) is another chemotherapeutic agent effective against the tubercle bacillus; it is often used in combination with isoniazid and streptomycin in the treatment of tuberculosis.

Ethambutol (2,2'-[ethylene diimino]-di-1-butanol dihydrochloride), a new drug for the treatment of tuberculosis, is often used in combination with INH. It is valuable in the treatment of disease caused by tubercle bacilli resistant to other tuberculostatic drugs and is effective against certain atypical mycobacteria.

Nitrofurans are a group of synthetic antimicrobial drugs with activity against gram-positive and gram-negative bacteria, fungi, and protozoa but little toxicity for tissues. Their mode of action is to interfere with basic enzyme systems within the microbial cell. An important one is *nitrofurazone* (Furacin), applied topically to prevent infections of wounds, burns, ulcers, and skin grafts. *Nitrofurantoin* (Furadantin) is used in bacterial infections of the urinary tract. *Furazolidone* (Furoxone) is effective against *Giardia lamblia* in intestinal infections; in a mixture with *nifuroxime* (Micofur) it is used in the treatment of vaginitis caused by bacteria, *Trichomonas vaginalis*, and *Candida albicans*.

Antibiotics

Experiments and observations carried on during the last 3 decades have proved that numerous organisms can produce substances with the power to inhibit the multiplication of other organisms or even to kill them. Such substances are known as *antibiotics*.* In 1941 the term "antibiotic" was defined by Waksman as follows: "An anti-

*"Antibiosis" was introduced in 1889 by Vuillemin, who said "No one considers the lion which leaps on its prey to be a parasite nor the snake which injects venom into the wounds of its victim before eating it. Here there is nothing equivocal; one creature destroys the life of another to preserve its own; the first is completely active, the second completely passive. The one is in absolute

TABLE 17-2. MICROBIAL SOURCES OF ANTIBIOTICS

Antibiotic	Group origin	Specific origin
Bacitracin	Bacteria	*Bacillus licheniformis*
Colistin		*Bacillus colistinus*
Polymyxin		*Bacillus polymyxa*
Tyrothricin		*Bacillus brevis*
Adriamycin (oncolytic antibiotic)	Streptomycetes	*Streptomyces peucetius*
Amphotericin B		*Streptomyces* species
Bleomycin (antineoplastic antibiotic)		*Streptomyces verticillatus*
Carbomycin		*Streptomyces halstedii*
Chloramphenicol		*Streptomyces venezuelae*
Chlortetracycline		*Streptomyces aureofaciens*
Cirolemycin (antineoplastic antibiotic)		*Streptomyces bellus* (var. *cirolerosis*)
Erythromycin		*Streptomyces erythraeus*
Fosfomycin (antibiotic)		*Streptomyces fradiae*
Kalafungin (antibiotic)		*Streptomyces tanashiensis*
Kanamycin		*Streptomyces kanamyceticus*
Lincomycin		*Streptomyces lincolnensis*
Mitocarcin (antineoplastic antibiotic)		*Streptomyces* species
Mithramycin (antineoplastic)		*Streptomyces plicatus*
Neomycin		*Streptomyces fradiae*
Nystatin		*Streptomyces noursei*
Oleandomycin		*Streptomyces antibioticus*
Oxytetracycline		*Streptomyces rimosus*
Paromomycin		*Streptomyces rimosus*
Ranimycin (antibiotic)		*Streptomyces lincolnensis*
Rifampin		*Streptomyces mediterranei*
Streptomycin		*Streptomyces griseus*
Streptozotocin		*Streptomyces achromogenes*
Tetracycline		Certain *Streptomyces* species under given conditions
Vancomycin		*Streptomyces orientalis*
Viomycin		*Streptomyces puniceus*
Zobramycin		*Streptomyces bikiniensis*
Ristocetin	Actinomycetes	*Nocardia lurida*
Cephalothin	Fungi	*Cephalosporium*
Fumagillin		*Aspergillus fumigatus*
Griseofulvin		*Penicillium griseofulvum* Dierckx, *Penicillium janczewskii* Zal., and *Penicillium patulum*
Nifungin (antifungal)		*Aspergillus giganteus*
Penicillin		*Penicillium notatum* and *chrysogenum*
Statolon (antiviral in animals)		*Penicillium stoloniferum*

opposition to the life of the other. The conception is so simple that no one has ever thought of giving it a name. This condition, instead of being examined in isolation, can appear as a factor in more complex phenomena. In order to simplify words we will call it antibiosis . . . ." (From Florey, H.: Antibiotics. Fifty-second Robert Boyle Lecture presented to the Oxford University Scientific Club, Springfield, Ill., 1951, Charles C Thomas, Publisher.)

biotic is a chemical substance produced by microorganisms which has the capacity to inhibit the growth of bacteria and even destroy bacteria and other microorganisms in dilute solution.*

Antibiotics may be produced by bacteria, fungi, actinomycetes, streptomycetes, or higher plants, including certain flowering ones. Some organisms produce more than one antibiotic, and certain antibiotics are produced by more than one organism. Table 17-2 gives sources of antibiotics. Although thousands of species of microorganisms elaborate antibiotics and hundreds of them are known, only a few are practically useful in the treatment of infectious disease.

Some antibiotics are protein; many are polypeptides. We consider them collectively, yet their physical, chemical, and pharmacologic properties are widely divergent.

■ **Spectrum of activity.** Some antibiotics affect mainly gram-positive bacteria, with little or no effect on gram-negative bacteria. Others seem to affect only certain species of bacteria, regardless of whether they are gram-negative or gram-positive. A few attack fungi. A few are active against the rickettsiae and the bedsoniae (chlamydiae).

The susceptibility (or resistance) of bacteria to different antibiotics may be determined by suitable laboratory procedures (p. 95) designated sensitivity tests (Fig. 7-13), and there are also satisfactory ways for determining the amount of certain antibiotics in the blood.

The *spectrum* of a compound is its range of antimicrobial activity. A *broad-spectrum* antibiotic indicates a wide variety of microorganisms affected, including usually both gram-positive and gram-negative bacteria. A *narrow-spectrum* antibiotic limits only a few. Table 17-3 lists some of the antibiotics falling into either category.

■ **Mode of action of antimicrobial drugs.** Several mechanisms explain in part the harmful or lethal effects that antimicrobial drugs have for bacteria and other microbes. Most of the time the mechanisms of injury are subtle ones, usually reflecting an inhibitory effect on some aspect of microbial physiology. The first mechanism studied was the role of the sulfonamides in competition for a place in the metabolic activities of the cell *(metabolic antagonism)*. A drug utilized by mistake because of a close chemical resemblance to a vital cytoplasmic substance could halt growth and function. The physiologic integrity of the microbial cell depends upon key macromolecules and the processes in which they are involved. Certain ones known to be vulnerable to clinically used antimicrobials include (1) bacterial cell wall synthesis, (2) nucleic acid synthesis, (3) protein synthesis, and (4) cell membrane function. Table 17-4 lists antimicrobial compounds as to category of action.

The rigid cell wall jackets the bacterial cytoplasm with its characteristically high osmotic pressure. If the cell wall is faulty in its construction and weakened because of the action of penicillin, the cytoplasmic membrane cannot withstand the internal pressure. In a medium of ordinary toxicity the bacterial cell explodes.

Antibiotics can interfere with the synthesis of the nucleic acids within the cell. Some complex with DNA and block the formation of the messenger RNA. One antiviral compound halts the synthesis of DNA. Other antibiotics inhibit the ribosomes (RNA) directly, thereby materially affecting the elaboration of proteins within the cell. As a result so-called nonsense proteins appear, indicating a chemical mixup.

*From Waksman, S. A.: Antibiotics today, Bull. N. Y. Acad. Med. **42**:623, 1966.

TABLE 17-3. ANTIBIOTICS – RANGE OF ACTIVITY

Narrow-spectrum antibiotics	Broad-spectrum antibiotics
Penicillin	Chloramphenicol
Streptomycin	Chlortetracycline
Dihydrostreptomycin	Demethylchlortetracyline
Erythromycin	Oxytetracycline
Lincomycin	Tetracycline
Polymyxin B	Kanamycin
Colistin	Ampicillin
Ristocetin	Cephalothin
Oleandomycin	Rifampin
Vancomycin	Gentamicin
Nystatin	
Spectinomycin	

TABLE 17-4. CELLULAR DISTURBANCES (MODE OF ACTION) RELATED TO ANTIMICROBIAL AGENTS

Inhibitory effect on					Metabolic antagonism
Cell wall synthesis	Nucleic acid synthesis	Protein synthesis		Cell membrane function	Competitive interference in cell metabolism
		Ribosome function	Other		
Bacitracin Cephalosporins Penicillins Ristocetin Vancomycin	Actinomycin D (antineoplastic antibiotic) Griseofulvin Idoxuridine Rifampin	Chloramphenicol Dihydrostreptomycin Gentamicin Kanamycin Neomycin Paromomycin Puromycin Streptomycin	Chlortetracycline Erythromycin Lincomycin Methacycline Oleandomycin Oxytetracycline Tetracycline	Amphotericin B Benzalkonium chloride Colistin Nystatin Polymyxins	Isoniazid Nitrofurans Para-aminosalicylic acid Sulfonamides

Since the biochemical activities relating to nucleic acid and protein synthesis are complicated, the possibilities for injury to the cell are many.

One of the antifungal antibiotics combines with sterols in the cell membrane of the susceptible fungus, destroying membrane function. Cell contents leak out and the cell dies. Some antibiotics alter the permeability of the plasma membrane. Benzalkonium chloride can disrupt a bacterial cell membrane by its detergent action.

An important factor in any consideration of the mode of action of antimicrobial compounds is that of *selective toxicity:* The agent, to be useful, must kill the microbial cells without damaging the cells of the animal host. The nonselective agents are the disinfectants and antiseptics that coagulate proteins in *all* living cells.

■ **Side effects.** There are four complications of antibiotic therapy: (1) hypersensitivity or allergy, (2) toxicity, (3) effects of the replacement flora, and (4) induction of bacterial resistance.

The turnover of drugs by the human body involves a series of complex, enzymatically mediated reactions. It is increasingly apparent that there can be vast individual differences. Signs of hypersensitivity to the antibiotics are variable in different persons but include skin rashes, joint pains, enlargement of lymph nodes, fall in the white blood cell count, and sometimes even hemorrhages. Examples of toxicity* to antibiotic therapy include the aplastic anemia induced by chloramphenicol and the deafness sometimes produced by streptomycin.

REPLACEMENT FLORA. The appearance of a replacement bacterial flora is quite troublesome at times. Not being affected by the drug themselves, these organisms tend to increase during the period of therapy. The newly prominent microorganisms are no longer held back by the organisms that succumbed to the antibiotic during the period of therapy. They can begin to multiply in an uncontrolled way, and if potentially pathogenic, they may start a disease process that is difficult to treat. The replacement flora may be quite resistant to standard antibiotics. Such secondary infections are often caused by *Proteus* or *Pseudomonas* species, and infections with *Candida albicans* are prone to follow prolonged and extensive antibiotic therapy, especially with broad-spectrum preparations. *Superinfection* as a complication is more likely with the broad-spectrum antibiotics than with the narrow-spectrum ones.

DRUG RESISTANCE. Bacteria vary in their ability to develop resistance to a given drug. Resistance to a drug means that it is no longer effective in suppressing the growth and multiplication of the microorganisms. Remember that resistance is always a definite possibility whenever a bacteriostatic drug is given, since such a drug merely limits the activities of the microbes so that the body mechanisms can more easily dispose of them. Many antibiotics are only bacteriostatic in their effects within the body. If only *bactericidal* (lethal) antibiotics were given, resistance would be less of a problem.

In food processing, antibiotics are used to control bacterial spoilage of foods. Farmers use antibiotics freely in treating mastitis in cows producing market milk. Some 16 antibiotics are used as additives in livestock feed. All these factors are believed to contribute significantly to the problem of bacterial resistance.

The best known mechanism of drug resistance is the production of the adaptive enzyme *penicillinase*, which inactivates penicillin G, by disease-producing staphylococci and coliform bacteria.

One of the most interesting mechanisms yet described concerns the *R* or *resistance* factor (also resistance transfer factor, RTF). It was first detected in Japan in 1959 and was later identified in the United States as a DNA-containing particle that could be passed in bacterial conjugation from one microorganism to another within a species or even from one member of a genus to another (Fig. 4-9). The particle replicates independently of the main genetic material, and the resistance is passed to successive generations. This bit of nuclear material, free floating in the cytoplasm, endows its possessor with an all-inclusive type of resistance to many antibacterial agents—sulfonamides, streptomycin, tetracyclines, chloramphenicol, penicillin, neomycin, even experimental antibiotics not in general use—with obviously different modes of antibacterial activity. About two dozen gram-negative bacteria transfer the R factor. From the order Enterobacteriaceae they include members of the genera *Enterobacter, Escherichia, Klebsiella, Salmonella, Shigella, Pseudomonas, Serratia, Proteus, Citrobacter,* and *Vibrio.*

---

*Some antibiotics in the doses ordinarily given are associated with psychiatric disturbances. Depression and hallucinations have been related to certain long-acting sulfonamides. Psychosis, especially in the elderly, can be precipitated with the antiviral amantadine. A buoyant state of well-being and elation has been noticed in isoniazid-treated patients with tuberculosis.

FIG. 17-1. *Penicillium chrysogenum,* giant colony. From this mutant form of the green mold almost all the world's supply of penicillin is obtained. (Courtesy Chas. Pfizer & Co., Inc., New York, N. Y.; from Arnett, R. H., Jr., and Braungart, D. C.: An introduction to plant biology, ed. 3, St. Louis, 1970, The C. V. Mosby Co.)

■ **The roster.** Short discussions of several well-known antibiotics follow.

Tyrothricin was one of the first antibiotics to be studied carefully. It was isolated from an aerobic motile nonpathogenic bacillus *(Bacillus brevis)* growing in the soil. From tyrothricin were obtained two products—*gramicidin* (named in honor of the bacteriologist Gram) and *tyrocidine,* both highly antagonistic to gram-positive cocci and very toxic. Tyrothricin is without effect when given by mouth and is very dangerous when given parenterally. It therefore must be used locally. It prevents bacterial growth and causes bacteria to lyse.

Penicillin, an organic acid and an important antibiotic accidentally discovered, is isolated from certain molds, the most notable of which are *Penicillium notatum* and *Penicillium chrysogenum* (Fig. 17-1). The first antibiotic to come into general use, it is still in many ways the best.* It may be administered intravenously, subcutaneously, into body cavities, by mouth, and locally. In contrast to many antibiotics that are bacteriostatic, it is also bactericidal.

It is one of the most important anti-infectives because it is bactericidal! Penicillin acts by interfering with synthesis of the bacterial cell wall and therefore affects actively growing bacteria. In the individual not allergic to penicillin its toxicity for human

*Penicillin is an inexpensive drug. The chances are that a course of penicillin will cost only a fraction of the cost of a course with another antibiotic.

tissue is almost nonexistent. It is an effective agent, with some exceptions, against staphylococci, streptococci (Group A), pneumococci, meningococci, and other organisms, and its use revolutionized the treatment of syphilis. It does not have an effect on most gram-negative bacilli with the usual doses given. Most of the microorganisms originally sensitive to penicillin still are, with the notable exceptions coming from the penicillinase-producing staphylococci and gonococci.

Penicillin has been isolated in six closely related forms called F, G, X, K, O, and V. Penicillin G is the most satisfactory type to manufacture and use. Therefore about 90% of commercial penicillin is of this type. The sodium, potassium, and procaine salts are the commonly used ones. Penicillin G is the agent of choice in the nonallergic individual for the treatment of infections caused by group A hemolytic streptococci.

The strength of penicillin is designated in *units*. A unit (International) corresponds to the activity of 0.6 microgram (0.000006 gm.) of pure sodium penicillin G.

The generic name "penicillin" includes all antibacterials isolated from the genus *Penicillium*. The natural penicillins are extracted from cultures.

Semisynthetic penicillins are the sequel to a basic observation on the chemical structure of penicillin made in Great Britain in 1959. When the chemist knows how the molecule can be modified, he can make changes to give the new product more desirable properties than those possessed by the natural substance. As a consequence of this, more than 500 new semisynthetic penicillins have been made within the last few years. Phenethicillin potassium (Syncillin), an oral preparation, was the first synthetic penicillin to be dispensed by prescription.

Although it is spoken of as a single substance, penicillinase is probably a class of comparable enzymes that account for penicillin resistance in both gram-positive and gram-negative microorganisms. Penicillin resistance is most important when the resistant microbe is a pathogenic staphylococcus. In a general way, the newer penicillins are indicated for pathogenic staphylococci resistant to penicillin G; otherwise the potency of these new drugs is less than that of the ordinary penicillin. Compounds designed to counteract staphylococcal penicillinase are methicillin sodium (Staphcillin), oxacillin sodium (Prostaphlin), nafcillin sodium (Unipen), cloxacillin (Tegopen), and sodium dicloxacillin monohydrate (Pathocil).

Ampicillin (Polycillin) is a new broad-spectrum semisynthetic penicillin with activity against a number of gram-positive and gram-negative bacteria, including *Salmonella, Shigella, Escherichia coli,* and *Proteus.* If disease of the biliary tract is not present, Ampicillin can be used to treat the typhoid carrier, and it has value in the treatment of intestinal amebiasis. It is ineffective in the presence of penicillinase.

Carbenicillin (Pyopen) is a British semisynthetic penicillin with reported activity against *Pseudomonas. Disodium carbenicillin* (Geopen) is active against the gram-negative *Pseudomonas* and *Proteus.*

Cross allergenicity exists among the penicillins; that is, a person sensitive to penicillin G will be found sensitive to one of the semisynthetic penicillins. It is estimated that 5% to 6% of the population is allergic to penicillin.

Cephalothin, cephaloridine, cephaloglycin, and *cephalexin* are cephalosporins, a group of antibiotics of an entirely new class, though closely related chemically to penicillin. They are semisynthetic derivatives from the fermentation products of the fungus *Cephalosporium* (Fig. 17-2), whose natural habitat is the seacoast of Sardinia in Italy. Sodium cephalothin (Keflin) is a broad-spectrum antibiotic bactericidal against

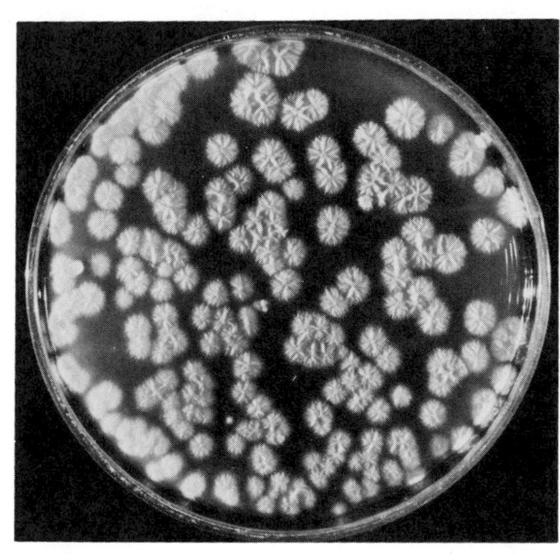

FIG. 17-2. *Cephalosporium*, colonies on agar plate. Important antibiotics are derived from this fungus. (From Profile of an antibiotic, Indianapolis, Ind., 1966, Eli Lilly & Co.)

gram-positive and gram-negative bacteria. It must be given parenterally; it is unaffected by penicillinase and is of low toxicity. Cephalordine (Loridine) is similar to cephalothin. Cephaloglycin (Kafocin) and cephalexin (Keforal) can be given orally. An advantage to the cephalosporins is the general lack of cross allergenicity with penicillin.

Streptomycin, an organic base, was isolated from *Streptomyces griseus*, an organism growing in the soil. Its action in a general way is bactericidal, like that of penicillin. It is active against gram-negative and acid-fast bacilli. Although seldom given intravenously, it may be administered by many routes. There are three disadvantages to streptomycin: (1) toxicity—its use may be followed by injury to the eighth cranial nerve, with deafness; (2) drug resistance—bacteria sensitive to it can quickly become resistant to it during the course of therapy; and (3) poor oral absorption—if the drug is given by mouth, it is not significantly absorbed. A unit of streptomycin corresponds to the activity of 1 μg. of pure crystalline streptomycin base.

Dihydrostreptomycin is a derivative of streptomycin and much like it. Its use also may be followed by neurotoxic symptoms, but, as a rule, it is less toxic than streptomycin.

Neomycin and *viomycin*, both polypeptides, are similar to streptomycin. Neomycin is derived from *Streptomyces fradiae*. Viomycin is derived from *Streptomyces puniceus*. Neomycin can damage the kidney. Viomycin is sometimes used in the treatment of tuberculosis. It is also nephrotoxic.

Kanamycin (Kantrex), a broad-spectrum antibiotic and an aminoglycoside obtained from *Streptomyces kanamyceticus*, must be given parenterally. It resembles streptomycin and neomycin. Its activity is directed against many aerobic gram-positive and gram-negative bacteria, including most strains of staphylococci and many strains of *Proteus*. The major toxic effect of this drug, working on the auditory portion of the eighth cranial nerve, can be partial or complete deafness. It is also nephrotoxic.

Gentamicin sulfate (Garamycin), an aminoglycoside, is a new broad-spectrum antibiotic given parenterally. It is chemically related to neomycin and kanamycin. Still an investigational drug, it is used topically and for systemic infections caused by gram-negative bacteria. It is nephrotoxic and ototoxic. Loss of hair including that of the eyebrows has been reported with its use.

Spectinomycin (Trobicin), a new member of the streptomycin-kanamycin family, is designed for the intramuscular one-dose treatment of gonorrhea. It is not believed to be nephrotoxic or ototoxic.

Erythromycin (Ilotycin) is derived from *Streptomyces erythraeus.* A valuable antibiotic, it is effective when given orally against gram-positive bacteria and some of the gram-negative bacteria. It is especially recommended for the treatment of infections caused by organisms that have become resistant to penicillin. Its use is seldom followed by toxic manifestations.

Lincomycin (Lincocin) is a new medium-spectrum antibiotic that can be administered orally or by injection. It inhibits the growth of gram-positive cocci, including that of staphylococci that produce penicillinase. Although its action is similar to that of erythromycin, erythromycin is antagonistic to it as are the cyclamate artificial sweeteners formerly used in various low calorie preparations.) Lincomycin is a parent drug to *clindamycin* (Cleocin), which may be the first antibiotic that is also an antimalarial drug.

Carbomycin is derived from several streptomycetes. One of these is *Streptomyces halstedii.* The action of carbomycin is much like that of erythromycin.

Polymyxin is a general name for a number of polypeptide antibiotics derived from different strains of *Bacillus polymyxa.* It is known as polymyxin A, B, C, and D. Polymyxin is toxic, especially to the kidney* and brain. It is used locally and as an intestinal antiseptic. Polymyxin B is usually effective against infections with *Pseudomonas aeruginosa.* However, *Proteus* resists it. It is important in topical preparations, but its systemic use is hazardous.

Colistin (Coly-Mycin), a polypeptide antibiotic isolated in Japan from *Bacillus colistinus,* is chemically similar to polymyxin. It is effective against gram-negative bacteria in general and, like polymyxin B, against *Pseudomonas aeruginosa* in particular. It is less toxic than polymyxin B, with fewer side reactions in the central nervous system and kidney. It is of special value in urinary tract infections and is given intramuscularly.

Bacitracin is a polypeptide antibiotic produced by *Bacillus licheniformis.* It inhibits the growth of many gram-positive organisms, including penicillin-resistant staphylococci, and the gram-negative gonococci and meningococci. It can only be used locally.

Vancomycin (Vancocin) is a new antibiotic that must be given intravenously. It is a bactericidal drug isolated from *Streptomyces orientalis* and advanced for the therapy of infections from antibiotic-resistant gram-positive bacteria, especially staphylococci.

Oleandomycin, isolated from *Streptomyces antibioticus,* is similar to erythromycin but less active. It can be given orally. It is hepatotoxic.

Ristocetin (Spontin) represents two antibiotics obtained from a species of actinomycetes, *Nocardia lurida.* Spontin, the mixture, is an intravenous drug with both

*The polypeptide antibiotics—polymyxin, neomycin, and bacitracin—are all nephrotoxic.

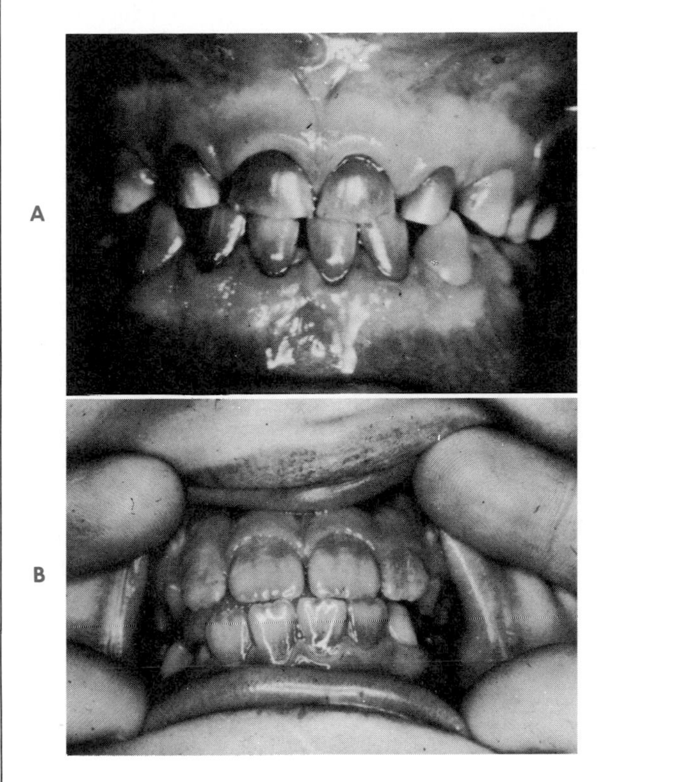

FIG. 17-3. Discoloration of teeth caused by tetracycline. **A,** Deciduous teeth (brownish gray). **B,** Permanent teeth (yellowish gray). (From Gorlin, R. J., and Goldman, H. M.: Thoma's oral pathology, ed. 6, St. Louis, 1970, The C. V. Mosby Co.)

bacteriostatic and bactericidal action. It is used for the treatment of infections with gram-positive bacteria, especially resistant staphylococci.

Rifampin (Rifadin, Rimactane) one of the rifamycins, is a semisynthetic broad-spectrum antibiotic taken by mouth, with bactericidal activity notable against the tubercle bacillus. It is effective in the treatment of carriers of the meningococcus and has been used in the treatment of leprosy. It has an effect against the causative bedsoniae (chlamydiae) of trachoma and inhibits replication of the poxvirus by blocking synthesis of viral protein. Adverse reactions are infrequent. The urine, feces, saliva, sputum, tears and sweat may take on an orange red color because of therapy with the drug.

Tetracycline is one of a group of broad-spectrum antibiotics and represents the chemical moiety common to the group. It is produced chemically from oxytetracycline and has been isolated from a streptomycete found in Texas. As a group tetracyclines sometimes irritate the alimentary tract. They are given orally to treat intestinal amebiasis. Tetracyclines rarely pigment the teeth. If tetracycline gains access to the circulation of an unborn baby because the drug is given to the mother, it is deposited in areas of developing bones and teeth. The damage results in staining of tooth enamel (Fig. 17-3).

Chlortetracycline (Aureomycin), a broad-spectrum antibiotic isolated from a streptomycete *(Streptomyces aureofaciens)*, is effective against a wider range of organisms than are such antibiotics as penicillin and streptomycin. It affects gram-positive bacteria, certain gram-negative bacteria, rickettsiae, and certain viruses. It may be given orally or intravenously.

Oxytetracycline (Terramycin) is derived from *Streptomyces rimosus*. Like chlortetracycline, it has a wide range of antibacterial activity. It is given orally and is comparatively nontoxic.

Demethylchlortetracycline (Declomycin) another member of the group of tetracyclines was first prepared chemically. It was later found in a mutant strain of *Streptomyces aureofaciens*. In general, its antibacterial activity is similar to that of the other tetracyclines, perhaps somewhat prolonged. Cases of photosensitization (sensitivity to light) have followed its use. *Methacycline* (Rondomycin) is similar to demethylchlortetracycline.

Doxycycline (Vibramycin), another of the tetracyclines, is similar in the range of clinical effectiveness and in the presence of untoward reactions to the other members of the group.

Chloramphenicol (Chloromycetin), the first of the broad-spectrum antibiotics, was isolated from *Streptomyces venezuelae* (found in the soil of Venezuela) but has now been crystallized. It is effective against certain bacteria, rickettsiae, and the bedsoniae (chlamydiae). It is the best antibiotic for the treatment of typhoid fever. It is usually given orally, but it may be given intramuscularly or intravenously. Two peculiar toxic disturbances occasionally complicate chloramphenicol therapy. One is hematologic; an aplastic anemia develops with cessation of blood cell formation. The other is cardiovascular collapse seen in babies 2 months of age and younger. It is known as *gray syndrome*. Since chloramphenicol is toxic, it should not be used in those cases where tetracyclines are effective.

Nystatin (Mycostatin), an antifungal, polyene antibiotic produced by *Streptomyces noursei*, is used with favorable results in infections caused by *Candida albicans*. It is often given with a broad-spectrum antibiotic to suppress the growth of the resistant intestinal fungi (replacement flora) that might produce lesions in the bowel after prolonged antibiotic therapy.

Amphotericin B (Fungizone), a broad-spectrum antifungal drug and another polyene obtained from *Streptomyces nodosus*, should be given intravenously. It is said to be effective in the treatment of the deep fungous infections, those involving the internal organs and widely spread within the body. It has no effect on bacteria. The initial infusion of this drug is regularly accompanied by a chill-fever reaction.

Fumagillin is produced during the growth of *Aspergillus fumigatus*. It is inactive against most bacteria, fungi, and viruses but is destructive to *Entamoeba histolytica* (the cause of amebic dysentery).

Griseofulvin (Fulvicin) is an oral antibiotic obtained from at least four kinds of *Penicillium*. A *fungistatic* drug, not a fungicidal one, it has a wide range of antifungal activity against practically all of the dermatophytic fungi *(Microsporum, Epidermophyton,* and *Trichophyton)*. Although systemically administered, it does not act on *Candida albicans*, on the deep systemic fungi, or on bacteria. The drug can be demonstrated to attack the advancing hyphal threads of invading fungi in skin, nails, and hair, causing these segments to shrivel and curl. Toxicity seems to be low; in anticipation of any effect on the blood-forming organs, periodic white blood cell counts are advised.

Paromomycin (Humatin), an amebicide, is a broad-spectrum antibiotic obtained from *Streptomyces rimosus.* It is not given by injection but orally, because of its nephrotoxicity. It is poorly absorbed from the gastrointestinal tract. It is effective against *Entamoeba histolytica* and holds back the secondary infection of the bowel that complicates amebiasis. In addition to paromomycin, the antibiotics with amebicidal activity are neomycin, kanamycin, and the tetracyclines.

Antivirals

Antiviral agents are still largely in the experimental stage. There is a great deal of interest, and many compounds are being studied.

Interferon, a naturally occurring nonimmune protein, was the first substance to be associated with an antiviral action. It is a diffusible substance produced by a living cell infected with virus that blocks further viral infection of that cell (see p. 409). It can also be induced by bacteria, bacterial products, bedsoniae (chlamydiae), and protozoa. Certain chemicals such as double-stranded RNA from a synthetic source can act as interferon inducers. Their use in enhancing resistance to viral infection is still experimental.

Idoxuridine (5-iodo-2′-deoxyuridine, or *IDU*), an antimetabolite and pyrimidine analog that inhibits normal cellular DNA synthesis, has been found effective against eye infection caused by the herpes simplex virus.

Methisazone (N-methylisatin-β-thiosemicarbazone) has been shown to prevent the disease smallpox in susceptible persons exposed in the epidemics studied in India.

Amantadine hydrochloride (Adamantanamine Hydrochloride) is a synthetic compound with a selective antiviral action. It has been used for the prevention of influenza due to A_2 strains of the virus only. It also has been used against rubella virus. Amantadine (Symmetrol) blocks penetration of the virus into the host cell. It stimulates the central nervous system and can cause convulsions in large doses.

QUESTIONS FOR REVIEW

1. Define disinfection, germicide, bactericide, antiseptic, asepsis, fumigation, bacteriostasis, preservative, antibiotic, iodophor, R transfer factor, penicillinase, interferon, detergent, cross allergenicity, enzyme detergent.
2. How do chemical disinfectants act?
3. Broadly classify antiseptics and disinfectants.
4. What is an antibiotic sensitivity test?
5. Name four sulfonamide compounds used in medicine.
6. What is meant by chemotherapy and chemotherapeutic agent?
7. Discuss ethylene oxide as a sterilizing agent.
8. What is meant by cold sterilization?
9. What are broad-spectrum antibiotics? Name five.
10. What agents are most effective in the disinfection of articles contaminated with *Mycobacterium tuberculosis?*
11. Name three chemotherapeutic drugs used in the treatment of pulmonary tuberculosis.
12. Briefly discuss harmful side effects of antibiotic therapy. What is meant by a replacement flora?
13. What is a narrow-spectrum antibiotic? Name three.
14. Briefly describe one antifungal agent and one amebicide.
15. Briefly indicate the mode of action of antibiotics.

REFERENCES. See at end of Chapter 18.

18 Practical technics

SURGICAL DISINFECTION AND STERILIZATION

Many different methods of sterilizing surgical instruments and supplies and of disinfecting wounds, the field of operation, and the hands of persons taking part in surgical operations are used in different hospitals and related medical or health field settings. An overview of the standard methods is presented in Table 18-1. In the following paragraphs some of the ones most commonly used are given:

The method of choice wherever applicable is sterilization by steam under pressure as usually carried out in a pressure steam sterilizer (preferably the modern high-

TABLE 18-1. METHODS FOR CONTROLLING MICROBIAL LIFE

Method	Temperature required	Minimum exposure time in minutes (0′ 2′ 10′ 12′ 15′ 20′ 90′ 120′ 150′ 180′)
*Sterilization**		
Saturated steam under pressure	285° F. (140° C.)	(Instantaneous)
	270° F. (132° C.)	2′
	250° F. (121° C.)	12′
Ethylene oxide gas (12% ETO 88% Freon-12)	130° F. (54° C.)	90′
	105° F. (40° C.)	150′
Hot air	320° F. (160° C.)	120′ or more
Chemical sporicide (solution)	Room temperature	180′ or more
Disinfection		
Steam—free flowing or boiling water	212° F. (100° C.)	2′
Chemical germicide (solution)	Room temperature	10′
Sanitization		
Water, detergents, or chemicals aided by physical means for soil removal	Maximum 200° F. (93° C.)	15′

*See pp. 239 and 240 for definitions.

Modified from chart copyrighted 1967, Research and Development Section, American Sterilizer Co., Erie, Pa.

TABLE 18-2. COMPARISON OF STEAM STERILIZATION WITH GAS STERILIZATION*

	Steam	Gas
Medical indications	Syringes, instruments, drapes, gowns, some plastics, rubber	Heat-sensitive plastics, rubber, bulky equipment (ventilators, telescopic instruments†)
Equipment	Automatic chamber (large and bulky)	Automatic chamber (large and bulky); also portable model
Penetration of material to be sterilized	Rapid	Very rapid
Time required	Minutes (15 minutes or less in some models)	Hours (2 to 12)
Efficiency	Excellent	100%
Technical snags	Removal of air	Prevacuum; correct humidification; temperature over 70° C. (158° F.)
Danger	None	Ethylene oxide undiluted explosive‡
Hazard	Sterilization inadequate if air not completely removed	Toxic residues if aeration inadequate
Contraindications	Many plastics; sharp instruments; rubber deteriorates	Previously irradiated plastics such as polyvinyl chloride

*From Rendell-Baker, L., and Roberts, R. B.: Gas versus steam sterilization: when to use which, Med. Surg. Rev. **5**:10 (Fourth Quarter), 1969.

†Modern day cements used on the optical lenses of such instruments have been sufficiently improved to permit sterilization with ethylene oxide gas.

‡The mixture of 12% ethylene oxide (ETO) and 88% Freon-12, which is generally used, is a safe one.

prevacuum steam autoclave) (Figs. 16-1, 16-3, and 18-1). There are exceptional circumstances in which chemical disinfection, including sterilization with ethylene oxide gas, has to be substituted for autoclave sterilization. Certain items are damaged by the environment of the pressure steam sterilizer, such as surgical instruments with optical lenses, instruments or materials made of heat-susceptible plastics, articles made of rubber, certain delicately constructed surgical instruments, and nonboilable gut sutures.

Generally, the pressure steam sterilizer (gravity air removal type) for most routine purposes of sterilization is maintained at 121° to 123° C. (250° to 254° F.), 15 to 17 pounds of pressure, for 30 minutes (Table 16-1). For certain items of rubber subjected to steam sterilization the time is shorter, 15 to 20 minutes. (Note the shorter time with the higher temperatures in the high-prevacuum sterilizer.)

■ **Gas sterilization.** Ethylene oxide is widely applied today to the sterilization of many items that cannot be steam sterilized. Gas sterilization is carried out in a specially devised chamber (Fig. 18-2). Table 18-2 compares this form with that of steam under pressure.

Surgical instruments and supplies

In Table 18-3 practical suggestions are given for the sterilization or chemical disinfection of specific instruments and supplies. (See also Table 18-4.)

TABLE 18-3. TECHNICS OF STERILIZATION (OR DISINFECTION) OF COMMONLY USED ITEMS IN THE CLINICAL SETTING

Item to be sterilized	Autoclave (method of choice)	Chemical disinfection (choice indicated)*	Other technics applicable
Surgical instruments and supplies, mechanically clean			
Noncutting instruments	Yes	1. Amphyl 2%, 20 min. 2. Bard-Parker Germicide, 10 min. to 2 hr. 3. Ethylene oxide gas, 2 to 12 hr.	Boil completely submerged 30 minutes
Sharp instruments	Yes, in certain instances	1. Amphyl 2%, 20 min. 2. Bard-Parker Germicide, 10 min. to 2 hr. 3. Ethylene oxide gas, 2 to 12 hr.	1. Bake in hot-air sterilizer, 150° C., 3 hr. 2. Boil 30 min. (the cutting part wapped with cotton and the instrument kept from tossing about; water should be boiling before instruments are placed in it†)
Surgical needles	Yes, in packages	1. Amphyl 2%, 20 min. 2. Isopropyl alcohol 70% to 90%, 20 min. 3. Ethylene oxide gas, 2 to 12 hr. Wash needles in sterile water	Hot air at 160° C., 2 hr.
Hypodermic syringes and needles	Yes	Cold sterilization inadequate to remove viruses of hepatitis	1. Dry heat, 160° C., 2 hr. 2. Boil 30 min.
Hinged instruments	Yes, in certain instances	1. Isopropyl alcohol 70% to 90%, 20 min. 2. Activated glutaraldehyde (2% aqueous), 20 min. to 10 hr. 3. Ethylene oxide gas, 2 to 12 hr.	–
Transfer forceps	Yes, for the container	Bard-Parker Germicide (change solution once a week)	–
Endoscopes and instruments with optical lenses (cystoscopes and others)	–	1. Ethylene oxide gas, 2 to 12 hr. (method of choice) 2. Aqueous formalin 20%, 12 hr. 3. Activated glutaraldehyde (2% aqueous), 10 hr.‡	–

*Sodium nitrite should be added to alcohols, formalin, formaldehyde-alcohol, quaternary ammonium, and iodophor solutions to prevent rusting. Sodium bicarbonate should be added to phenolic solutions to prevent corrosion. Antirust tables are available commercially.

†Boiling dulls knives and other sharp instruments because iron tends to enter solution in the ferrous state when heated. The site of corrosion—the part that is electropositive—is the sharp edge of the instrument. A "cathodic" sterilizer is so equipped electrically that such instruments are connected with the negative pole of a battery and corrosion prevented. This permits long-continued boiling without dulling.

‡Exposure to Nos. 2 or 3 for 15 min. eliminates tubercle bacilli and enteroviruses.

TABLE 18-3. TECHNICS OF STERILIZATION (OR DISINFECTION) OF COMMONLY USED ITEMS IN THE CLINICAL SETTING—cont'd

Item to be sterilized	Autoclave (method of choice)	Chemical disinfection (choice indicated)	Other technics applicable
Surgical instruments and supplies, clean—cont'd			
Polyethylene tubing	—	Immerse tubing carefully filled with disinfectant: 1. Bard-Parker Germicide, 12 hr. 2. Ethylene oxide gas, 2 to 12 hr. 3. Aqueous formalin 20%, 12 hr.	—
Rubber tubing, silk catheters, and bougies	—	1. Ethylene oxide gas, 2 to 12 hr. (method of choice) 2. Iodophor, 500 p.p.m. available iodine, 20 min.* 3. Amphyl 2%, 20 min.* 4. Activated glutaraldehyde (2% aqueous), 15 min.*	—
Inhalation equipment for anesthesia	—	1. Ethylene oxide gas, 2 to 12 hr. 2. Activated glutaraldehyde (2% aqueous), 10 hr.	—
Thermometers	—	1. Bard-Parker Germicide, 12 hr. 2. Ethylene oxide gas, 2 to 12 hr. (cold cycle only) 3. Isopropyl alcohol 70% to 90% plus iodine 0.2%, 15 min. (to eliminate tubercle bacilli and enteroviruses)	—
Surgical instruments and supplies, contaminated	Yes, wherever possible—as a rule, double exposure time, other conditions the same	Ethylene oxide gas, 2 to 12 hr.	—

*To eliminate tubercle bacilli and enteroviruses, tubing must be filled.

TABLE 18-4. RECOMMENDED METHODS OF PHYSICAL STERILIZATION TO ELIMINATE THE HEPATITIS VIRUSES*

Method	Time	Temperature
Autoclave†	30 min.	121° C. (15 lb. pressure)
Dry heat	2 hr.	170° C.
Boiling (active)	30 min.	100° C.

*Most human viruses are destroyed at 60° C. for 30 minutes, with the exception of the virus of serum hepatitis. At this temperature a 10-hour period is required for its inactivation.
†The autoclave is definitely the preferred method and strongly advised over the other two.

FIG. 18-1. Bank of recessed autoclaves in modern hospital. (Courtesy American Sterilizer Co., Erie, Pa.)

Syringes and needles contaminated by even minute traces of serum or plasma may be sterilized by one of the methods recommended in Table 18-4. Disposable needles, blood lancets, and syringes are available commercially, and their use is strongly recommended wherever feasible, because one important method of spreading the hepatitis viruses is thereby blocked. (Disposable items are designed to be used once only and should never be sterilized and reused.)

The proper disposition of disposables must be stressed! Disposable syringes and needles must be kept away from children and drug addicts and must not wound the hand of the garbage collector. Once used, they can be incinerated, deformed in boiling water, or crushed in a machine.

One method for safe discard of disposable syringes and needles is as follows: The needle is placed inside the syringe, needle point upward toward the piston. The piston is inserted into the syringe and jammed down, locking the needle tightly between the barrel of the syringe and the plunger. The unit may then be thrown into the trash.

It is hard to render instruments with optical lenses completely free of microbes. To kill completely all microbes present might well loosen the cement of the lens system in the process or corrode the metal in some of the instruments. A recently developed, heat-resistant cement makes it possible to carry out a type of pasteurization wherein

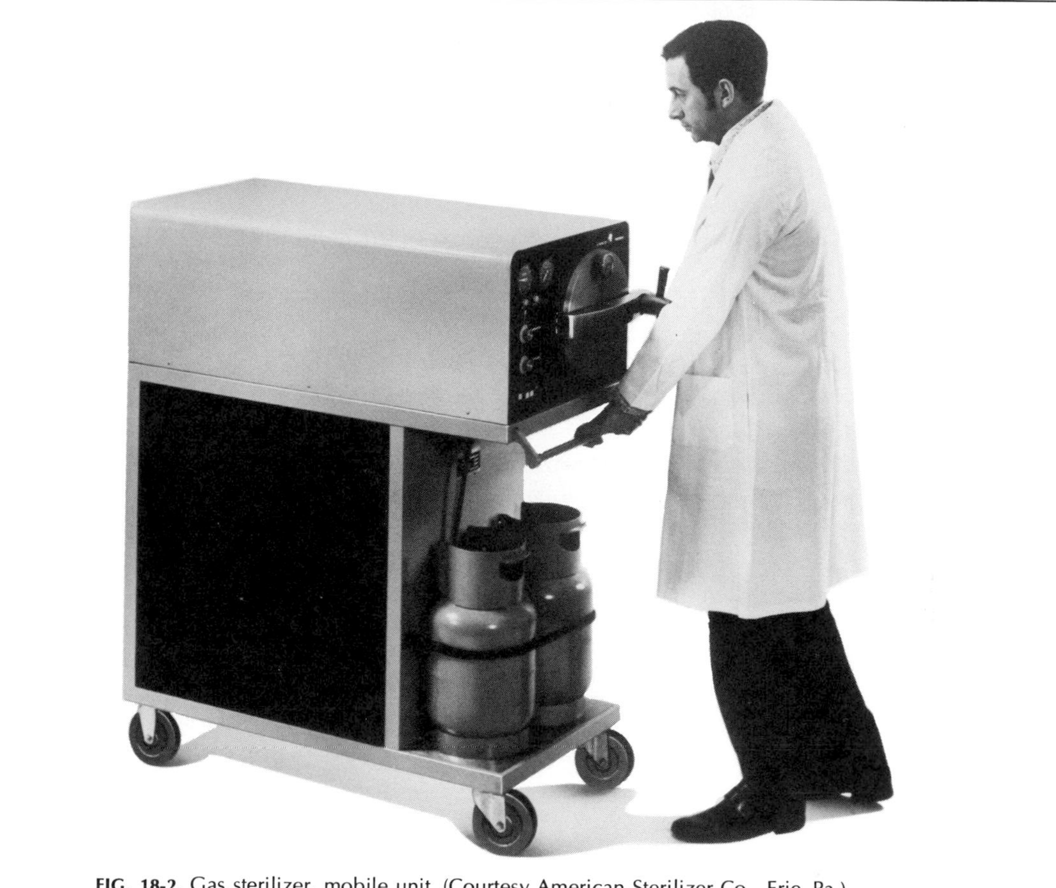

FIG. 18-2. Gas sterilizer, mobile unit. (Courtesy American Sterilizer Co., Erie, Pa.)

endoscopes are submerged in hot (not boiling) water at a temperature of 85° C. for 1 hour. This method kills most of the ordinary bacterial contaminants. Telescopic instruments with improved cements in the lens systems can also be sterilized with ethylene oxide.

Formerly the preparation and care of rubber tubing used for intravenous medication was a difficult and time-consuming task. The use of plastic disposable units *should completely eliminate* the need for rubber tubing for intravenous medication and blood transfusion.

■ **Dressings and linens.** Towels, gowns, and other articles of cloth may be sterilized by autoclaving at 15 pounds of pressure for 30 minutes. They should be arranged in packs 12 × 12 × 20 inches in size and wrapped in two layers of muslin or one layer of muslin and one or two layers of heavy paper.

■ **Reusable gloves.** When an operation is finished, the surgeon and his assistants should wash their gloves in cold water before removing them. After removal the gloves are washed outside and inside with tincture of green soap, thoroughly rinsed, and dried. They are then powdered, placed in glove envelopes, and wrapped. They are sterilized as follows: (1) Autoclave at 15 pounds of pressure for 20 minutes, (2) follow with a vacuum for 5 to 10 minutes, (3) open the door of the sterilizer slightly and allow gloves to dry in the sterilizer for about 5 minutes, and (4) remove and separate the packages to allow rapid cooling (this prevents sticking). *Caution:* Gloves are often ruined by being allowed to remain in the sterilizer too long. *Note:* In some sterilizer models all these steps will not be necessary.

By putting an iodine solution in the water used to wash off glove powder and by taking advantage of the well-known color change, one can detect holes in gloves. If there is even a very tiny hole, a brownish discoloration of the hand is seen after the glove is removed.

■ **Hand brushes.** To sterilize hand brushes, (1) clean thoroughly, rinse free of soap under running water, and submerge in 1:1000 aqueous benzalkonium chloride (Zephiran Chloride)* for 30 minutes (leave in the disinfectant until needed) or (2) autoclave and store in a sterile dry container. (Note the availability and practicality of prepackaged, sterile, disposable hand brushes.)

■ **Materials of fat or wax, oils, and petrolatum gauze.** To sterilize these heat-susceptible objects, place them in the sterilizing oven at 160° C. for 2 to 2½ hours.

■ **Sutures and ligatures.** To sterilize sutures and ligatures, (1) scrub the glass tubes with soap and water to remove greasy film, (2) rinse well with water to remove soap, (3) submerge in alcohol-formalin solution of the Bard-Parker type for 1 hour, (4) place in a sterile suture jar containing the foregoing antiseptic, and (5) remove with sterile forceps and rinse in sterile water before using.

Silkworm-gut, silk, horsehair, and metal ligatures may be sterilized by boiling. Surgical gut sutures in sealed tubes are available commercially. The tubing fluid is usually a mixture; one is isopropyl alcohol 90%, phenyl mercuric benzoate 0.025%, and water sufficient to make 100%. The tubes are handled and shipped in jars that contain a storage disinfectant. Such solutions are usually mixtures of isopropyl alcohol, formaldehyde, sodium nitrite, sodium bicarbonate, and distilled water and are of such a density that the tubes remain completely submerged and covered by the solution. (If hexachlorophene is added to this storage solution, the spore-killing time is reduced from 18 to 3 hours.)

■ **Water and saline solutions.** Water and saline solutions are placed in suitable containers and autoclaved at 15 pounds of pressure for 30 minutes. Great care should be exercised to prevent fibers of lint or other particles from getting into solutions that are to be given intravenously.

Operating room

The operative field is exposed to pathogenic bacteria floating in the air. These arise from the noses and throats of the occupants of the surgical suite and may be destroyed by ultraviolet radiation.

Terminal disinfection refers to the disinfection of the operating room at the end of an operative procedure and after the patient has been removed. One of the phenol derivatives is usually used for this.

Body sites

■ **Hands.** There are many methods of disinfecting the hands. The following method gives good results:

---

*Although only one disinfectant, such as benzalkonium chloride, may be recommended for various purposes in the pages of this chapter, it does not follow that the one cited is necessarily the best. There are certainly many other good ones.

1. Wash the hands and arms well with soap and water.
2. With soap and a sterile brush, scrub the nails and knuckles of both hands.
3. Scrub the hands, beginning with the thumbs and in succession scrubbing the inner and outer surfaces of the thumbs and fingers, giving several strokes to each area.
4. Scrub the palms and backs of the hands and then the forearms to 3 inches above the elbow for 3 minutes; rinse well with running water.
5. With a sterile orange stick, clean under each nail thoroughly.
6. With a second sterile brush, scrub the hands and arms for 7 minutes in the same manner as with the first brush.
7. Thoroughly rinse off all soap with water and then rinse the hands in 70% alcohol.
8. With one end of a sterile towel, dry one hand and, with a circular motion, dry the forearm to the elbow. With the other end of the towel, dry the other hand in the same manner.

When one is scrubbing the hands for a surgical operation, plenty of soap should be used, and the scrubbing should be done in running water. When one is using cake soap, the soap is held between the brush and the palm until the scrubbing is complete. During the entire procedure the hands should be held higher than the elbows in order to prevent contamination.

At the present time the methods of disinfection of the hands are shortened by the habitual use of detergent soaps containing germicidal substances. The following method, known as the Betadine (p. 247) surgical scrub regimen, is widely used today. Because some members of the surgical team as well as some patients are allergic to iodine, pHisoHex (p. 247) can still be found in the health field setting.

1. Clean under the nails. Keep the fingernails short and clean.
2. Wet the hands and forearms.
3. Apply to the hands two portions, about 5 ml., of Betadine Surgical Scrub from a dispenser. (Depress Betadine dispenser foot pedal twice.)
4. Without adding water, wash the hands and forearms thoroughly for 5 minutes. Use a brush if desired.
5. With a sterile orange stick, clean under the nails. Add a little water for copious suds. Rinse thoroughly with running water.
6. Apply 5 ml. from the dispenser to the hands. Repeat steps just given. Rinse with running water.

Actual sterilization of the hands is not possible. Microbes live not only on the surface of the skin but also in its deeper layers, in the ducts of sweat glands, and around the hair follicles. These resident microbes are chiefly nonpathogenic staphylococci and diphtheroid bacilli and are plentiful about the nails.

■ **Site of operation.** Although the deeper portions of the skin cannot be absolutely sterilized, enough bacteria can be removed to make infection unlikely. The selection of an antiseptic for preoperative skin preparation has had the attention of surgeons since the earliest times, and many different ones have been used.

The following outline represents a satisfactory method of skin preparation.

Preliminary preparation (to be carried out several hours before the operation).

1. Wash the part thoroughly with tincture of green soap; detergent soaps containing hexachlorophene may be used.
2. Shave the part.

There is no need to scrub the part with ether or alcohol during the preliminary preparation or to cover the area with sterile towels, as was once the custom, because this may increase the apprehension of the patient and it is not necessary from the surgical standpoint.

Some authorities object to the skin shave being done ahead of the time the patient is taken to the operating room. They feel the best time is right before the surgical operation, in the operating room, after the patient has been put to sleep and positioned. If lather from a dispenser can and a new razor blade are used, there is no danger of cross infection, and the lather traps the hair. When the skin shave is done many hours before surgery, there is a good chance bacteria will grow in the inflammatory reaction about small razor nicks. The incision the surgeon makes is then likely to go through minute abscesses.

Preparation at operation

1. Cleanse the umbilicus with an applicator stick containing soap or detergent if the skin of the abdomen is being scrubbed.
2. Scrub the skin with detergent or soap beginning with the imaginary line of incision and working laterally to the bed line.
3. Apply benzalkonium chloride 1:1000 to the umbilicus of abdomen with the applicator.
4. Paint the skin with benzalkonium chloride 1:1000 on a sponge. Begin at the imaginary line of incision and proceed as in step 2. (Tint a colorless solution such as this so that the area of skin prepared will be outlined for the surgeon.)

NOTE I. Tincture of iodine mixed with 70% to 90% isopropyl alcohol, tincture of Merthiolate 1:1000, tincture of Mercresin, tincture of iodine, and other disinfectants may be used instead of benzalkonium chloride. If tincture of iodine is used, the iodine must be removed from around the edges of the field with an alcohol sponge.

NOTE II. When any skin area is prepared for operation, the imaginary line of incision is never painted with the same sponge more than once.

The following method utilizes Betadine Surgical Scrub and can be modified as indicated:

1. Shave the field of operation. Wet with water.
2. Apply Betadine on a sponge to the umbilicus, if the skin of the abdomen is being scrubbed, and discard sponge.
3. Scrub the imaginary line of incision with Betadine on a sponge for 1 minute, using a rotating motion.
4. Carry the same sponge out from the line of incision and scrub the outer portions of the surgical area for 1 minute; again use a circular motion and discard the sponge.
5. Repeat this entire maneuver for five separate instances.
6. Use a little water to make suds. Use wet sterile gauze to rinse surgical area.
7. Pat dry with sterile towel.
8. Paint area with Betadine Solution. Allow to dry. The surgical area is now ready for the drapes.

The operating room personnel and those who prepare patients for operations should not let the fact that many surgical infections may be controlled by antibiotics lull them into satisfaction with careless antiseptic or aseptic technic.

■ **Other areas of skin.** For the rapid sterilization preceding routine hypodermic injections and venipunctures for intravenous medication, 70% to 90% ethyl or isopropyl alcohol or 2% iodine in 70% alcohol is effective when applied to a clean area of skin for

30 to 60 seconds. When lumbar puncture is to be done, the overlying skin should be cleansed with a hexachlorophene soap, rinsed, dried, and disinfected with a suitable antiseptic.

■ **Mucous membranes.** Absolute disinfection of mucous membranes is impossible. The mouth and throat may be cleansed with Dobell's solution, hydrogen peroxide, hexylresorcinol, Ceepryn Chloride 1:10,000 to 1:5000. The vagina may be irrigated with benzalkonium chloride 1:5000 to 1:2000.

■ **Wound disinfection.*** Antiseptics that may be used for disinfection of wounds are Ceepryn Chloride 1:1000, hexachlorophene, soap, and Furacin 1:500. Bacitracin in the form of an ointment and water-soluble chlorophyl derivatives such as Chloresium ointment (0.5%) are often applied to wounds. Chloresium acts as a deodorizer and, according to some observers, promotes the formation of granulation tissue, the tissue of repair. The chloramines (which slowly liberate chlorine) are sometimes used, as are such dyes as gentian violet, crystal violet, brilliant green, and malachite green because these dyes are effective against the gram-positive bacteria most often responsible for wound infections. The dyes also have a moderately destructive action on tissue cells. Favorable reports have been given on the efficacy of Medicinal Zinc Peroxide in the treatment of certain wounds (p. 244). *The use of antiseptic solutions should never be substituted for indicated surgical cleansing of infected wounds.*

DISINFECTION OF EXCRETA AND CONTAMINATED MATERIALS FROM INFECTIOUS DISEASES

A patient with a communicable disease is not properly supervised until every avenue by which the infectious agent may be spread from the patient to others has been closed. These avenues are not closed until all excreta and all objects that convey the infection have been properly disinfected. A patient who is being cared for in such a manner that all avenues by which the infection may spread to others are closed is said to be *isolated.* A ward patient with a conscientious and capable nurse is more strictly isolated from the microbiologic point of view than is the patient in a room with plastered walls and airtight doors who has a careless or incompetent nurse. Isolation refers to avenues of infection, not to walls and doors. *Remember that pathogenic microbes are most virulent when first thrown off from the body.*

Disinfection in infectious diseases is of two types: concurrent and terminal. *Concurrent disinfection* means the immediate disinfection of the patient's excreta or objects that have become contaminated by the patient or his excreta. *Terminal disinfection* means the final disinfection of the room, its contents, and environs after it has been vacated by the patient. The final chapter is written when it has been proved that patient and attendants are free of the agent that caused the disease, that is, they have not become carriers. In the following paragraphs are given methods of disinfecting excreta, materials, and objects by which infection is most often transmitted from the sick to the well.

■ **Hands of the nurse.** Unless properly disinfected, the hands of nurses, doctors, or other persons in contact with the patient are almost certain to transfer the infectious

*Remember that local treatment of a wound with sulfonamides or penicillin can result in the patient's developing a sensitivity to these antibacterial agents (p. 211).

agent to themselves or to others. Although disinfection of the hands to prevent the spread of communicable diseases is not so time consuming as for a surgical operation, it should be just as conscientiously done. The hands and lower arms are submerged in a basin of 1:1000 aqueous benzalkonium chloride and rubbed vigorously for 2 minutes. It is good to follow this by washing the hands with soap and running water. It is not necessary to use a brush for scrubbing the hands and arms, although in some institutions this is required. The cleansed hands should be dried thoroughly. Hand lotion keeps the skin in good condition. The nails should be kept clean at all times. The hands should be cleansed in this way every time they contact the patient for any purpose whatsoever. To allow one's hands to contact a person suffering from a communicable disease and then, without disinfection, to allow them to contact another person is little short of criminal neglect.

■ **Soiled linens and clothing.** All linens and clothing that have come in contact with the patient should be considered contaminated and should be kept in the patient's room until ready for final disposal. In the hospital they are wrapped in sheets or placed in pillow cases and put into a special bag, care being taken that the outside of the bag does not become contaminated. Before the bag is full, it is closed tightly, marked to indicate that it contains infectious material, and sent to the laundry. Upon reaching the laundry, the bag is emptied into the washer, after which the bag itself is put into the washer. The washer is then closed, and live steam is introduced for 15 to 20 minutes. After this the regular procedure of the laundry is carried out. In the home the linen or bedclothes are bundled and carried to the kitchen or home laundry to be placed in a boiler containing warm water and soapsuds and boiled from 10 to 15 minutes. After boiling, the process of laundering may be completed in the home or by a commercial laundry. On occasion, the boiler is carried to the patient's room to receive the soiled linen.

Mattresses, pillows, and blankets may be sterilized by heat in special sterilizing chambers, or they may be sterilized in a formaldehyde chamber. They also may be dried in the sunlight for a period of 24 hours. Mattresses may be encased in plastic covers that have been cleansed with antiseptic solutions.

■ **Shoes.** Disinfection of shoes is best accomplished by placing a pledget of cotton saturated with formalin in each shoe and allowing the shoes to stay in a closed box for 24 hours. Shoes may also be disinfected by being sprayed with formalin from an atomizer for three successive evenings.

■ **Feces and urine.** Where connections exist to modern sewage disposal systems, feces and urine are disposed of in the commode without disinfection. Otherwise, feces and urine, especially from patients with typhoid fever, infectious hepatitis, and similar disorders, should be mixed with either 5% phenol (carbolic acid) or preferably chlorinated lime. The volume of disinfectant is three times that of the material to be disinfected. Feces should be thoroughly broken up in order to ensure even penetration. After the urine and feces are thoroughly mixed with the disinfecting solution, the mixture is allowed to stand for a least 2 hours before disposal of it is made. Contaminated colon tubes and such may be disinfected in Amphyl 2% solution for 20 minutes.

■ **Discharges from the mouth and nose.** Discharges from the mouth and nose are received on squares of tissue and placed in a paper bag pinned to the side of the bed. After the bag is wrapped in several layers of clean newspaper, it is burned in the incinerator.

■ **Sputum.** Sputum should not be allowed to dry so as to permit particles to float off in the air. It should be received in a paper cup held in a special metal holder. The paper cup is discarded and the metal holder sterilized at least twice daily (more often if necessary). To discard a paper cup, remove the cup from the holder with forceps and set it on several thicknesses of newspaper. The cup is then filled with sawdust to decrease the amount of moisture and make the cup more easily burned. The paper is wrapped around the cup and tied securely; next, the packet is placed in the incinerator or in a special receptacle to be carried to the incinerator later. After this has been done, the metal holder is autoclaved or boiled for 30 minutes. Ambulatory patients often use collapsible cups.

Sputum can be received into a waxed paper cup with a tight-fitting lid that can be secured before the cup is discarded.

■ **Clinical thermometers.** A clinical thermometer must be sterile each time it is used. After a thermometer is used and the temperature recorded, the thermometer must be washed with soap or detergent and water (with friction applied). It may be disinfected as indicated in Table 18-3.

After sterilization a thermometer is wiped with sterile cotton and placed in a sterile container for later use. Preferably it is stored in alcohol, or in fresh daily solutions of tincture of iodine ½% or 150 ppm of an aqueous iodophor. Alcoholic solutions of the quats must be used, since the aqueous solutions do *not* kill tubercle bacilli. Thermometers are rinsed in cold water before use.

A patient with hepatitis should have his own thermometer, which is destroyed when he leaves the hospital.

■ **Eating utensils.** A special covered container is placed in the sickroom, and the dishes are placed therein, preferably by the patient himself. Care to prevent contamination of the outside of the container must be exercised. After the dishes, knives, and forks are placed in the container, it is carried to the kitchen and boiled in soapy water for 5 minutes. Sometimes the eating utensils are carried to the kitchen before they are placed in a container. Leftover bits of food may be placed in a paper sack, wrapped well with newspaper, and burned in the incinerator. If this is not practical, the food remains are placed in chlorinated lime or phenol solution and left for 1 hour.

■ **Terminal disinfection.** When a patient is removed from a room, the floors, walls, woodwork, and furniture should be scrubbed with soapsuds and thoroughly aired. The phenol derivatives are valuable and widely used in terminal disinfection. Other good housekeeping disinfectants are 1% sodium hypochlorite and Wescodyne or other iodophor (500 ppm available iodine).

DISINFECTION OF ARTICLES FOR PUBLIC USE

■ **Rest rooms and bathroom fixtures.** The woodwork, floors, sinks, basins, and toilets of public rest rooms should be washed frequently with hot soapsuds (or detergent) and rinsed thoroughly. Phenolic disinfectants or one of the iodophors such as Wescodyne are suitable for disinfection.

■ **Woodwork.** The woodwork of schoolhouses, churches, theaters, or other places of public assembly should be washed frequently with a detergent-germicide solution.

■ **Public conveyances.** All parts touched by the hands of passengers should be frequently washed with a detergent-germicide solution.

FUMIGATION OF ROOMS AND DISINFECTION OF AIR

■ **Fumigation.** In former years it was an almost universal procedure to put a sickroom "into fumigation" as soon as the room was vacated. Fumigation at that time meant terminal gaseous disinfection with formaldehyde, the gas of choice. This gaseous method of disinfection has been abandoned because it does not destroy insects, small animals, or pathogenic microbes. The scrubbing of walls and floors to cleanse them thoroughly has replaced fumigation in the control of communicable disease; natural drying of surfaces effects the destruction of bacteria.

The term "fumigation" now denotes the destruction of disease-carrying animals, insects, or vermin. The fumigant of choice is hydrocyanic acid, a deadly poison that must be handled with extreme care. Sulfur dioxide is a splendid insecticide, but because of its destructive action on metals and household goods, it is seldom used in homes, public buildings, or hospitals.

■ **Air disinfection.** Two methods of destroying bacteria and other disease-producing agents as they are transferred through the air are ultraviolet radiation and chemical disinfection by means of aerosols.

Aerosols are chemical vapors liberated into the air for disinfection. Two of the most potent ones are propylene glycol, effective in a dilution of one part in 100 million parts of air, and triethylene glycol, effective when diluted to one part in 400 million parts of air. The vapors may be introduced into rooms through ventilating systems, bombs, or special equipment. The importance of thorough cleansing and disinfection of air-conditioning units must be kept in mind.

STERILIZATION OF BIOLOGIC PRODUCTS

In the preparation of bacterial vaccines it is necessary to kill the bacteria at as low a temperature as possible because a temperature of more than 62° C. coagulates the bacterial protein and renders the vaccine worthless. It has been found that an exposure for 1 hour to a water bath temperature of 60°C. kills any of the ordinary bacteria used in making vaccines. Bacterial vaccines are sometimes sterilized by the addition of chemicals such as phenol or cresol. Mercury-vapor lamps as a source of ultraviolet radiation are used in the preparation of bacterial as well as viral vaccines, but ultraviolet radiation, ideal in theory, is practically ineffective at times. Ethylene oxide gas is also used for this purpose.

QUESTIONS FOR REVIEW

1. How may knives be sterilized?
2. What is the preferred method of sterilization for endoscopes? What is meant by pasteurization of endoscopes?
3. List three surgical items easily damaged by heat.
4. Give a method of preparing the skin for operation.
5. How may the air be kept free of pathogenic bacteria during the time of operation?
6. What is the difference between physical and microbiologic isolation?
7. How would you dispose of the feces and urine of a patient with typhoid fever?
8. Discuss the drugs used in wound disinfection.

9. Explain what is meant by concurrent disinfection. Terminal disinfection.
10. What precaution should be taken in handling sputum from a patient with tuberculosis? How is is sputum sterilized?
11. How are clinical thermometers sterilized?
12. Briefly discuss the use of hydrocyanic acid as a fumigant.
13. How may hepatitis viruses be eliminated?
14. How are biologic vaccines sterilized?
15. Compare steam sterilization with gas sterilization.

REFERENCES FOR UNIT FOUR

AMA Council on Drugs: Evaluation of a new antibacterial agent, J.A.M.A. **209:**549, 1969.

AMA Council on Drugs: Evaluation of a new antituberculous agent, Rifampin (Rifadin, Rimactane), J.A.M.A. **220:**414, 1972.

AMA Drug Evaluations, Chicago, 1971, American Medical Association.

Ballinger, W. F., II, Treybal, J., and Vose, A. B.: Alexander's Care of the patient in surgery, ed. 5, St. Louis, 1972, The C. V. Mosby Co.

Bell, A. N.: Personal communication, 1972.

Bishop, C., et al.: The use of ethylene oxide for sterilization of ventilators, Hosp. Topics **43:**112, (May) 1965.

Byrd, R. B., et al.: Isoniazid toxicity, J.A.M.A. **220:**1471, 1972.

Chloramphenicol induced bone marrow suppression, editorial, J.A.M.A. **213:**1183, 1970.

Cohen, S.: A decade of R factors, editorial, J. Infect. Dis. **119:**104, 1969.

Davies, J. E., and Rownd, R.: Transmissible multiple drug resistance in Enterobacteriaceae, Science **176:**758, 1972.

Dineen, P.: Duration of shelf life—an evaluation, A.O.R.N. J. **13:**63, (March) 1971.

Easley, J. R.: Personal communication, 1972.

Edgeworth, D.: Nursing and asepsis in the modern hospital, Nurs. Outlook **13:**54, (June) 1965.

Evans, M. J.: Some contributions to prevention of infections, Nurs. Clin. N. Amer. **3:**641, 1968.

Fason, M. F.: Controlling bacterial growth in tube feedings, Amer. J. Nurs. **67:**1246, (June) 1967.

Fox, F.: A review of the types and problems of sterilization, Hosp. Manage. **103:**135, (March) 1967.

Friedman, R. M.: Interferons and virus infections, Amer. J. Nurs. **68:**542, (March) 1968.

Garner, J. S., and Kaiser, A. B.: How often is isolation needed? Amer. J. Nurs. **72:**733, 1972.

Ginsberg, F.: Ethylene oxide gas is a useful and safe hospital sterilant, Mod. Hosp. **111:**114, (July) 1968.

Ginsberg, F.: Gas sterilizers: handle with care, Mod. Hosp. **111:**84, (Aug.) 1968.

Ginsberg, F.: Hair, long or short, must be covered in O.R., Mod. Hosp. **117:**106, (July) 1971.

Ginsberg, F.: Sterilizing anesthesia equipment is difficult but it is necessary, Mod. Hosp. **109:**108, (July) 1967.

Goodman, L. S., and Gilman, A., editors: Pharmacological basis of therapeutics, New York, 1970, The Macmillan Co.

Goth, A.: Medical pharmacology, ed. 6, St. Louis, 1972, The C. V. Mosby Co.

Grove, V. E., Jr.: Common drugs cause psychiatric illness, Texas Med. **67:**30, (March) 1971.

Heller, W. M.: The United States Pharmacopeia, its value to the professions, J.A.M.A. **213:**576, 1970.

Hitchings, G. H.: A quarter century of chemotherapy, J.A.M.A. **209:**1339, 1969.

Hobby, G. L., editor: Antimicrobial agents and chemotherapy, Ann Arbor, Mich. 1967, American Society for Microbiology.

Infection control in the hospital, Chicago, 1971, American Hospital Association.

Interferon, editorial, J.A.M.A. **213:**118, 1970.

Jackson, G. G.: Influenza; the present status of chemotherapy, Hosp. Pract. **6:**75, (Nov.) 1971.

Jeanes, C. W. L., et al.: Inactivation of isoniazid by Canadian Eskimos and Indians, Canad. Med. Ass. J. **106:**331, 1972.

Johnston, D. F.: Essentials of communicable disease: With nursing principles, St. Louis, 1968, The C. V. Mosby Co.

Kanof, N. M.: Who needs hexachlorophene? Hospitals and some people, editorial, J.A.M.A. **220:**409, 1972.

Keitel, H. G., and Soentgen, M. L.: Dental staining, tetracycline induced, Med. Sci. **17:**57, (Jan.) 1966.

Kimbrough, R. D.: Review of toxicity of hexachlorophene, Arch. Environ. Health **23:**119, 1971.

LaDu, B. N., Jr.: The genetics of drug reactions, Mod. Med. **40:**10, (Feb. 7) 1972.

Lal, S., et al.: Effects of rifampicin and isoniazid on liver function, Brit. Med. J. **1:**148, (Jan. 15) 1972.

Litsky, B. M. Y.: Are pathogenic spores a hospital threat? Hosp. Manage. **101:**74, (May) 1966.

Litsky, B. Y.: Environmental control: The operating room, A.O.R.N. J. **14:**39, (July) 1971.

Litsky, B. M. Y.: How do you know how well you are doing—in getting rid of the pathogens? Hosp. Manage. **103:**94, (March) 1967.

MacKenzie, A. R.: Effectiveness of antibacterial soaps in a healthy population, J.A.M.A. **211:**973, 1970.

Mathews, R.: TLC with the penicillin, Amer. J. Nurs. **71:**720, 1971.

Meeks, C. H., Pembleton, W. E., and Hench, M. E.: Sterilization of anesthesia apparatus, J.A.M.A. **199:**276, 1967.

Merigan, T. C., Jr.: Interferon and interferon inducers: The clinical outlook, Hosp. Pract. **4:**42, (March) 1969.

Modell, W., editor: Drugs of choice 1972-1973, St. Louis, 1972, The C. V. Mosby Co.

Moravec, D. F.: All drugs should be respected but not feared, Hosp. Manage. **106:**62, (Oct.) 1968.

Morse, R. A.: Environmental control in beehive, Sci. Amer. **226:**92, (April) 1972.

Nelson, J. P.: OR clean rooms, A.O.R.N. J. **15:**71, (May) 1972.

Nichols, G. A., and Dobek, A. S.: Sterility of items sealed in plastic, Hosp. Topics **49:**84, (Dec.) 1971.

Notes on air hygiene: Summary of a conference on air disinfection, Arch. Environ. Health **23:**473, 1971.

Notes on the package insert, editorial, J.A.M.A. **207:**1335, 1969.

Oviatt, V. R.: How to dispose of disposables, Med. Surg. Rev. **5:**57, (Second quarter) 1969.

Penicillin, editorial, J.A.M.A. **209:**1520, 1969.

Perkins, J. J.: Principles and methods of sterilization in health sciences, Springfield, Ill., 1969, Charles C Thomas, Publisher.

Physicians' desk reference to pharmaceutical specialties and biologicals, Oradell, N. J., 1973, Medical Economics, Inc., subsidiary of Litton Publications, Inc.

Poupard, J. A., et al.: A general reference chart for common bacterial agents, Lab. Med. **2:**26, (April) 1971.

Prioleau, W. H.: Fashion hides its head in the OR, Hospitals, J.A.H.A. **46:**107, (May 1) 1972.

Rees, R. B.: Cutaneous drug reactions, Texas, Med. **66:**92, (Feb.) 1970.

Resistance to antibiotics, editorial, J.A.M.A. **203:**1132, 1968.

Roberts, R. B., and Rendell-Baker, L.: Ethylene-oxide sterilization, Hosp. Topics **50:**60, (May) 1972.

Roberts, R. B., and Stark, D. C. C.: Sterile procedures in the doctor's office, Med. Surg. Rev. **6:**24, (June-July) 1970.

Schwabe, C. W.: Veterinary medicine and human health, Baltimore, 1969, The Williams & Wilkins Co.

Selwyn, S., and Ellis, H.: Skin bacteria and skin disinfection reconsidered, Brit. Med. J. **1:**136, 1972.

Shirkey, H. C.: The package insert dilemma, J. Pediat. **79:**691, 1971.

Spaulding, E. H.: Principles and application of chemical disinfection, A.O.R.N. J. **1:**36, (May-June) 1963.

Spectinomycin, new weapon against gonorrhea—with some old side effects, Nursing '72 **2:**17, (April) 1972.

Strominger, J. L.: The bactericidal mechanisms of antibiotic drugs, Hosp. Pract. **2:**54, (July) 1967.

Summary of the report of the *Ad Hoc* Advisory Committee on isoniazid and liver disease (current trends), Morbid. Mortal. Wk. Rep. **20**(26):231, 1971.

Swenson, O., and Grana, L.: A new, improved sterilization set-up for the operating room, A.O.R.N. J. **13:**80, (Jan.) 1971.

The cost of disposables, editorial, J.A.M.A. **217:**1859, 1971.

Weinstein, L., and Chang, T. W.: Interferon: Nonviral infections and nonviral inducers, editorial, Ann. Intern. Med. **69:**1315, (Dec.) 1968.

Yanis, B. M.: Part 2; What a disinfectant is supposed to do: the folly of buying by cost alone, Hosp. Manage. **99:**88, (March) 1965.

Yanis, B. M.: Part 3; Testing the killing power of disinfection detergents, Hosp. Manage. **99:**102, (April) 1965.

Yanis, B. M. Part 4; Testing the residual properties of disinfectant—detergents, Hosp. Manage. **99:**85, (May) 1965.

Yanis, B. M.: Part 5; Testing of surgical scrubs and sterilization of instruments, Hosp. Manage. **99:**73, (June) 1965.

LABORATORY SURVEY OF UNIT FOUR

PROJECT

Effects of physical agents on bacteria

Part A—Sterilization by filtration with demonstration by instructor and discussion

1. Method of filtration using either the Berkefeld or Seitz filter
2. Filtration of a test solution, which can be a broth culture of some non-pathogenic organism diluted 1:20 with sterile isotonic saline solution
3. Cultures taken before and after filtration of test material
4. Observations made on cultures after incubation period of 24 hours

Part B—Effects of heat on bacteria

1. Prepare five suspensions each of *Bacillus subtilis* and of *Escherichia coli* by placing in sterile cotton-plugged test tubes 1 or 2 ml. amounts of a bacterial suspension furnished by the instructor.
2. Treat a properly labeled tube of each suspension as follows:
   a. Autoclave at 15 pounds of pressure for 15 minutes.
   b. Stand in a pan of boiling water for 30 minutes.
   c. Expose to free-flowing steam in an Arnold sterilizer for 1 hour.
   d. Stand in a pan of water at 60° C. for 1 hour.
   e. Use the fifth pair as a control.
3. Transfer a loopful of the bacterial suspension from each tube to a culture tube of broth and incubate at 37° C. for 24 hours.
4. Observe growth in cultures made. Record growth obtained.

	Growth obtained	
Exposure to heat	*Bacillus subtilis*	*Escherichia coli*
Autoclave; 15 pounds of pressure, 15 min.		
Boiling water; 30 min.		
Free-flowing steam; 1 hour		
60° C.; 1 hour		
Control		

5. Which tubes of the heated suspensions of *Bacillus subtilis* show growth? What does this indicate? What is the difference between the growth of *Bacillus subtilis* and that of *Escherichia coli?* What does this indicate? What part does spore formation play in this experiment?

Part C—Effects of boiling

1. Obtain test specimens of bacteria.
   a. Dip three sterile cotton swabs in a 5-day broth culture of *Bacillus subtilis.*
   b. Dip three sterile cotton swabs into a 24-hour broth culture of *Escherichia coli.*
2. Expose test specimens to the effects of boiling water for 1 minute.
   a. Hold a swab from each suspension in a separate beaker of boiling water for 1 minute and then dip into a tube of nutrient broth.
   b. With the other swabs, repeat the procedure, using periods of 5 and 10 minutes.
   c. Incubate the six tubes of broth 24 to 48 hours.
   d. Indicate growth obtained in the following table:

	Period of boiling		
Organism	*1 min.*	*5 min.*	*10 min.*
Bacillus subtilis			
Escherichia coli			

   e. What do the foregoing findings indicate?

Part D—Operation of autoclave with demonstration by instructor and discussion

1. Discussion of the autoclave or pressure steam sterilizer given on pp. 232 to 235.
2. Demonstration of the parts of the instrument and its operation
3. Direct supervision of students as they operate the autoclave

Note: An autoclave or pressure steam sterilizer is filled with steam under pressure; therefore, if the pressure gets too high, there is always danger of an explosion.

PROJECT
Effects of chemical agents on bacteria

Part A—Effects of chemical disinfectants

1. Obtain from the instructor four sterile cotton-plugged tubes containing the following:
   a. 5 ml. of 70% isopropyl alcohol
   b. 5 ml. of 5% phenol
   c. 5 ml. of some widely advertised antiseptic (one or several may be used)
   d. Broth suspension of staphylococci
2. Proceed as follows:
   a. To each tube of disinfectant, add 0.5 ml. of the bacterial suspension. Do not let the broth run down side of tube. Mix by gently tapping the bottom of the tube.
   b. Place in water bath at 20° C.
   c. At the end of 5, 10, and 15 minutes, transfer a loopful of material from each tube of disinfectant to a tube of sterile broth.

d. Label the tube, noting the organism and the time held in the water bath, and incubate 48 hours.

e. Record growth on the accompanying chart, using the plus sign to indicate growth and the minus sign to indicate absence of growth.

3. Repeat the preceding experiment using *Bacillus subtilis* or another spore-forming organism instead of staphylococci. Record on chart.

	Staphylococcus			*Sporeformer*		
Disinfectant	*5 min.*	*10 min.*	*15 min.*	*5 min.*	*10 min.*	*15 min.*
70% isopropyl alcohol						
Phenol (5%)						
Commercial disinfectant						

Part B—Susceptibility of bacteria to penicillin

Note: Students may work in groups of three or four for this experiment.

1. Obtain from instructor the following:
   a. Petri dish of tryptose agar that has just been inoculated with an organism susceptible to the action of penicillin
   b. Petri dish of tryptose agar that has just been inoculated with an organism *not* susceptible to the action of penicillin
2. Use sterile forceps to drop filter paper disks* impregnated with different strengths of penicillin onto the inoculated surface of the agar plates. Place the disks about 2 cm. apart.
3. Invert plates and incubate for 24 to 48 hours.
4. Note the presence or absence of bacterial growth around the filter paper disks.

Part C—Determination of sensitivity of bacteria to different antibiotics

Note: Students may work in groups of three or four.

1. Obtain from the instructor the following:
   a. A Petri dish of tryptose agar that has just been inoculated with bacteria sensitive to some of the antibiotics to be used and not susceptible to other test antibiotics.
   b. Disks of filter paper* impregnated with different antibiotics to be used.
2. Use sterile forceps to place the antibiotic disks on the surface of the agar about 2 cm. apart.
3. Invert and incubate at 37° C. for 24 to 48 hours.
4. Examine and record which antibiotics inhibited bacterial growth and which did not.
5. Perform a susceptibility test on an agar plate inoculated with *Pseudomonas aeruginosa,* using a number of different antibiotics.
6. Note the resistance exhibited by this organism.

---

*Disks impregnated with graded amounts of penicillin or other antibiotics may be obtained from commercial sources.

PROJECT
Testing the efficiency of sterilization

Note: Students may work in groups of two or more to test the efficiency of some of the sterilizing processes outlined in this book.

Part A — Test of the effectiveness of methods for disinfecting the hands

1. Lightly touch the tips of the fingers of the recently disinfected hand to the surface of a Petri dish of nutrient agar.
2. Invert the dish and incubate. Examine the growth at the end of 24 hours.
3. Set up a control. Rub the palm of one hand before washing or disinfection with a moistened sterile cotton swab. Inoculate the swab onto sterile nutrient agar. Incubate for 24 hours and observe growth. Make comparisons.

Part B — Test of the sterility of recently sterilized solutions

1. Use a sterile pipette to remove a few drops of isotonic saline solution from a recently sterilized flask. (Sterile distilled water may be used. Commercially packaged solutions may also be used.)
2. Add the test solution to a tube of nutrient broth. Incubate for 24 hours and examine for growth.

Part C — Test of the sterility of a clinical thermometer

1. Dip the tip of a recently disinfected thermometer into a tube of nutrient broth.
2. Incubate the broth for 24 hours and examine for growth.

EVALUATION FOR UNIT FOUR

Part I

In the following exercise place in the blank space the letter from column B that represents the appropriate or preferred method of sterilizing or disinfecting each article named in column A.

COLUMN A—TASK

_____ 1. Silk sutures
_____ 2. Dressings and linens
_____ 3. Silk catheters and bougies
_____ 4. Cystoscopes
_____ 5. Noncutting instruments
_____ 6. Hypodermic syringes and needles
_____ 7. Hand brushes
_____ 8. Sharp cutting instruments
_____ 9. Rubber gloves
_____ 10. Water and saline solutions
_____ 11. Shoes
_____ 12. Sputum from patient with tuberculosis
_____ 13. Feces and urine
_____ 14. Air of nursery for newborn infants
_____ 15. Rectal thermometers
_____ 16. Hands of communicable disease nurse
_____ 17. Bacterial exotoxins
_____ 18. Glassware
_____ 19. Inoculating platinum wire loop
_____ 20. Bacterial and viral vaccines
_____ 21. Petrolatum gauze
_____ 22. Furniture
_____ 23. Floors, walls, and woodwork
_____ 24. Wounds
_____ 25. Mucous membranes
_____ 26. Plastic parts of heart-lung machine
_____ 27. Polyethylene tubing
_____ 28. Suture needles
_____ 29. Eating utensils
_____ 30. Skin site of an operation
_____ 31. Powders
_____ 32. Hinged instruments
_____ 33. Contaminated surgical instruments
_____ 34. Plastic Petri dishes

COLUMN B—PROCEDURE

(a) Pressure steam sterilization (in an autoclave)
(b) Chemical disinfection
(c) Dry heat sterilization
(d) Incineration
(e) Seitz filtration
(f) Gaseous chamber sterilization (as with ethylene oxide)
(g) Boiling
(h) Ultraviolet radiation

Part II

Place the letter from the column on the right in front of the corresponding definition on the left.

	COLUMN A	COLUMN B
________	1. Destroys or masks offensive odors (gas)	(a) Fumigant
________	2. Prevents growth of bacteria, but does not destroy them	(b) Disinfectant
________	3. Prevents deterioration of food	(c) Preservative
________	4. Destroys disease-producing agents and their products	(d) Deodorant
________	5. Destroys insects and small animals (gases or fumes)	(e) Germicide
________	6. Prevents multiplication of bacteria	(f) Antiseptic
________	7. Destroys bacteria (chemical)	(g) Bacteriostatic agent

Part III

Circle the one item that best completes the introductory statement.

1. In using a germicidal agent it is best to know
   (a) the volatility of the agent
   (b) the concentration needed and the time of contact
   (c) the exact number of bacterial spores present
   (d) the exact number of bacteria present
2. In the management of infection by an organism that has become highly resistant to tetracycline it would be best to
   (a) withdraw all antibiotic treatment
   (b) test the causative microorganism for sensitivity to other antibiotics
   (c) increase the dose
   (d) switch to penicillin
   (e) do nothing
3. The development of resistance to an antibiotic by a microorganism comes from an:
   (a) alteration of cell wall permeability
   (b) alteration of plasma membrane
   (c) alteration of capsule
   (d) alteration of mitochondria
   (e) none of these
4. The phenol coefficient of a germicide is determined by comparison of that germicide with phenol on the basis of
   (a) the concentration that kills the same number of bacteria
   (b) penetrating power of the agent
   (c) proportion of bacteria killed
   (d) zone of inhibition of bacteria
   (e) none of these
5. If 0.1 ml. of serum is added to 0.9 ml. of saline and mixed and if one half the mixture is transferred to a second tube containing ½ ml. of saline, the dilution in the second tube (after mixing) would be
   (a) 1:5
   (b) 1:10
   (c) 1:20
   (d) 1:40
   (e) 1:100
6. Which of the following antimicrobial agents interferes with cell wall synthesis?
   (a) Penicillin
   (b) Colistin
   (c) Isoniazid
   (d) Tetracycline
   (e) Actinomycin D

7. Which of the following antimicrobial agents does *not* inhibit protein synthesis?
   (a) Chloramphenicol
   (b) Tetracycline
   (c) Kanamycin
   (d) Polymyxin
   (e) Streptomycin
8. The antimicrobial action of the polymyxins and colistin is related primarily to:
   (a) Interference in cell membrane function
   (b) Inhibition of cell wall synthesis
   (c) Inhibition of nucleic acid synthesis
   (d) An unknown mechanism
   (e) Inhibition of protein synthesis

Part IV

When the paired statements are compared in a quantitative sense, one will be found to exert or imply a greater effect or value. Indicate which is the stronger statement in the following way:
"a" means that the first statement is the stronger one
"*b*" indicates that the second statement represents a greater effect

________ 1. Effectiveness of chemical disinfectant in saline (a)
Effectiveness of chemical disinfectant in serum (b)
________ 2. Lethal effect of temperatures 10 degrees above maximum growth temperature (a)
Lethal effect of temperatures 10 degrees below minimum growth temperature (b)
________ 3. Effect of chemical disinfectant at pH 7 (a)
Effect of chemical disinfectant at pH 4 (b)
________ 4. Time required for sterilization in autoclave (a)
Time required for sterilization in hot air sterilizer (b)
________ 5. Time required for sterilization in pressure steam sterilizer (a)
Time required for sterilization in high-prevacuum steam sterilizer (b)
________ 6. Time required for sterilization in pressure steam sterilizer (a)
Time required for sterilization in ethylene gas sterilizer (b)
________ 7. Sterilizing action of hot air at 170° C. for 15 minutes (a)
Sterilizing action of free flowing steam for 15 minutes (b)
________ 8. Disinfectant action of ethyl alcohol against tubercle bacilli (a)
Disinfectant action of mercurials against tubercle bacilli (b)
________ 9. Disinfectant action of iodophors against spores (a)
Disinfectant action of glutaraldehyde against spores (b)
________ 10. Likelihood of drug resistance in tuberculosis if streptomycin is used alone (a)
Likelihood of drug resistance if drug combinations are used in tuberculosis (b)
________ 11. Sensitivity of vegetative cells to chemical treatment (a)
Sensitivity of spores to chemical treatment (b)
________ 12. Duration of active immunity (a)
Duration of passive immunity (b)
________ 13. Temperature for sterilization by autoclaving (in pressure steam sterilizer) (a)
Temperature for sterilization by dry heat (in hot air sterilizer) (b)
________ 14. Variability in Gram-staining reaction of gram-positive organisms (a)
Variability in Gram-staining reaction of gram-negative organisms (b)
________ 15. Optimal growth temperature of a psychrophile (a)
Optimal growth temperature of mesophile (b)
________ 16. Spectrum of microbes sensitive to penicillin (a)
Spectrum of microbes sensitive to tetracycline (b)

UNIT FIVE

MICROBES: pathogens and parasites

19 The gram-positive cocci: pyogenic cocci

20 The gram-negative cocci: neisseriae

21 The aerobic gram-positive bacilli: diphtheria bacilli, anthrax bacilli

22 The anaerobic gram-positive bacilli: clostridia

23 The gram-negative enteric bacilli

24 The gram-negative small bacilli

25 Miscellaneous microbes

26 The spiral microorganisms

27 The viruses

28 Viral diseases

29 The rickettsiae and bedsoniae

30 The acid-fast bacteria

31 Fungi: medical mycology

32 Protozoa: medical parasitology

33 Metazoa: medical parasitology

Laboratory survey of unit five

Evaluation for unit five

19 The gram-positive cocci: pyogenic cocci

The pathogenic cocci are often called the *pyogenic* cocci because of their ability to cause pus formation. The important gram-positive cocci are the staphylococci, the streptococci, and the pneumococci (genera *Staphylococcus, Streptococcus,* and *Diplococcus,* respectively).

STAPHYLOCOCCUS SPECIES (THE STAPHYLOCOCCI)

■ **General characteristics.** Staphylococci occur typically in grapelike clusters (Fig. 19-1). Under special conditions they may occur singly (Fig. 19-2), in pairs, or in short chains. They are gram-positive, nonmotile, and nonspore-forming and grow luxuriantly on all culture media. Most of them grow best in the presence of oxygen, but they can

FIG. 19-1. Staphylococci, scanning electron micrograph to show typical grapelike cluster. (Courtesy Millipore Corporation, Bedford, Mass.)

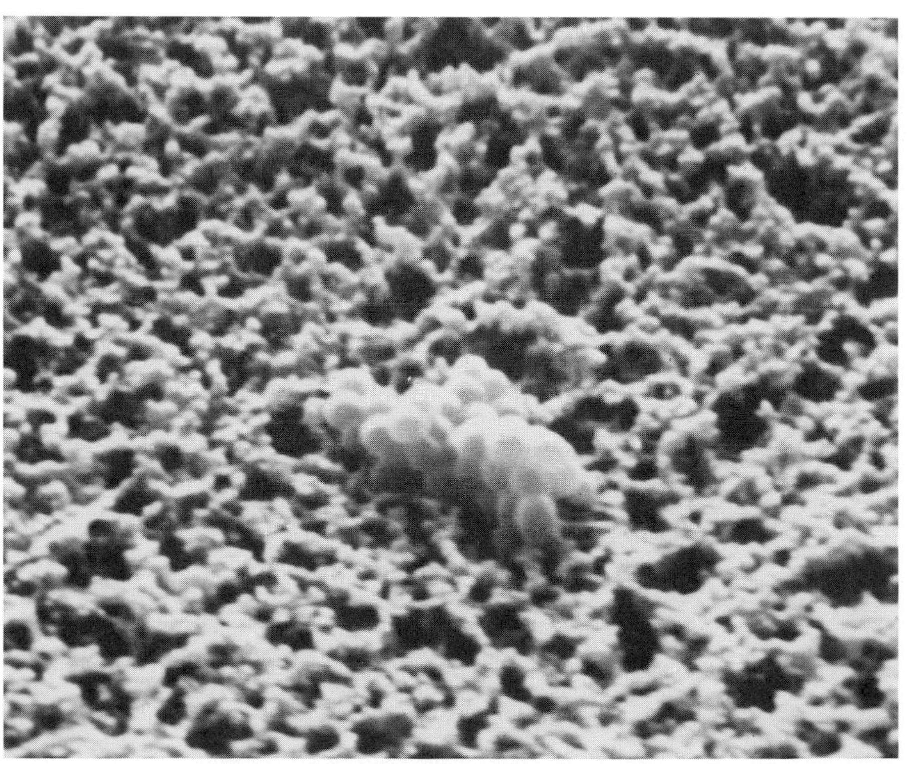

grow in its absence. A few are strictly anaerobic. They grow best between 25° and 35° C. but may grow at a temperature as low as 8° C. or as high as 48° C.

Staphylococci are the most resistant of all nonspore-forming organisms to the action of heat, drying, and chemicals. Although most vegetative bacteria are destroyed by a temperature of 60° C. for 30 minutes, staphylococci resist a temperature of 60° C. for 1 hour and some strains resist a temperature of 80° C. for 30 minutes. In dried pus they may live for weeks or months. Staphylococci also become resistant to the sulfonamides and antibiotics. They have adapted quickly and easily to such agents. Many (about 80%) are penicillin resistant.

When staphylococci are grown on blood agar, there is characteristic pigment production, the colors ranging from a deep golden to lemon yellow to white. The deep golden color observed with growth was originally responsible for the species name *aureus.* We now know that there are white variants of the golden staphylococci. The emphasis placed upon the pigmentation staphylococci produced in cultures was at one time responsible for a classification of staphylococci related to the color of the pigment so produced: *Staphylococcus aureus,* a deep golden pigment; *Staphylococcus citreus,* a lemon yellow color; and *Staphylococcus albus,* a white pigment. The ability of a staphylococcus to produce disease is of paramount importance. Since there is no consistent relation between the pathogenic properties of a staphylococcus and its color, this classification has been laid aside.

FIG. 19-2. Staphylococcus, electron micrograph. Note attached staphylophages. (Courtesy S. Tyrone, Dallas, Texas.)

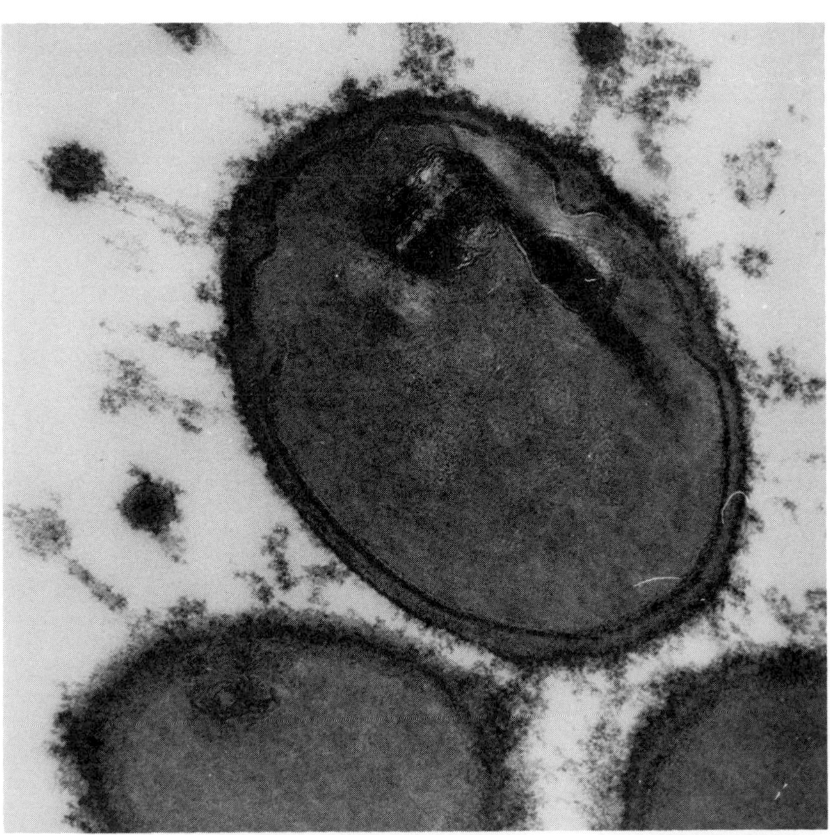

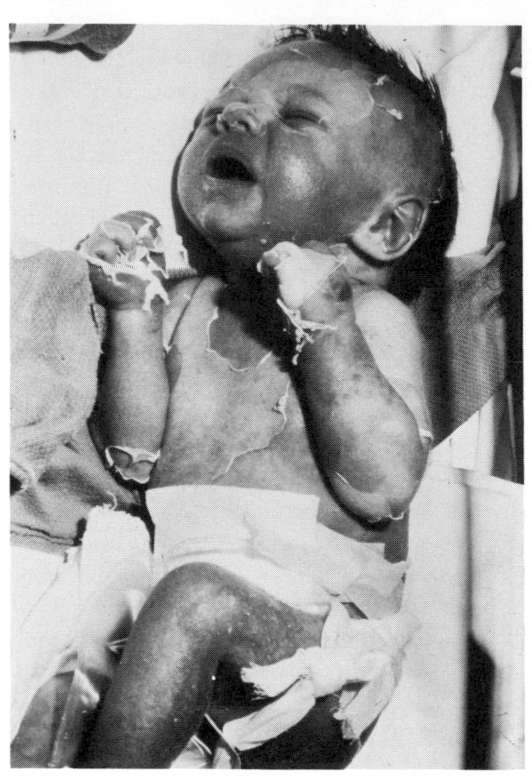

FIG. 19-3. Scalded skin syndrome in an infant. Note peeling of superficial skin layer. (From Melish, M. E., and Glasgow, L. A.: The staphylococcal scalded-skin syndrome, N. Eng. J. Med. **282:**1114, 1970.)

In *Bergey's Manual of Determinative Bacteriology,* two species are listed under the genus *Staphylococcus: Staphylococcus aureus* and *Staphylococcus epidermidis.* These two organisms are separated by a biochemical reaction and the presence of an enzyme. *Staphylococcus aureus* ferments mannitol, which the other staphylococcus does not, and *Staphylococcus aureus* is coagulase-positive. *Staphylococcus epidermidis* is coagulase-negative.

■ **Toxic products.** Several important metabolic products are elaborated by staphylococci. Among the ones with toxic properties are those that (1) destroy red blood cells (hemolysins, staphylolysins), (2) destroy leukocytes (leukocidin), (3) cause necrosis of tissue (necrotizing exotoxin), (4) produce death (lethal factor), and (5) cause gastroenteric symptoms (enterotoxin). Probably no single strain produces all of these poisons, and many produce none of them.

In addition, some staphylococci produce the enzyme *coagulase,* by means of which they cause the plasma of the blood to clot. The *coagulase test* indicates the presence of coagulase, which is generally considered the best single bit of evidence that a given staphylococcus is a pathogenic organism. Because of coagulase activity, a surface layer of fibrin accumulates on an individual staphylococcus and protects it from phagocytic attack. The production of coagulase correlates with the production of other toxic products. Staphylococci also produce staphylokinase, which dissolves fibrin, and hyaluronidase.

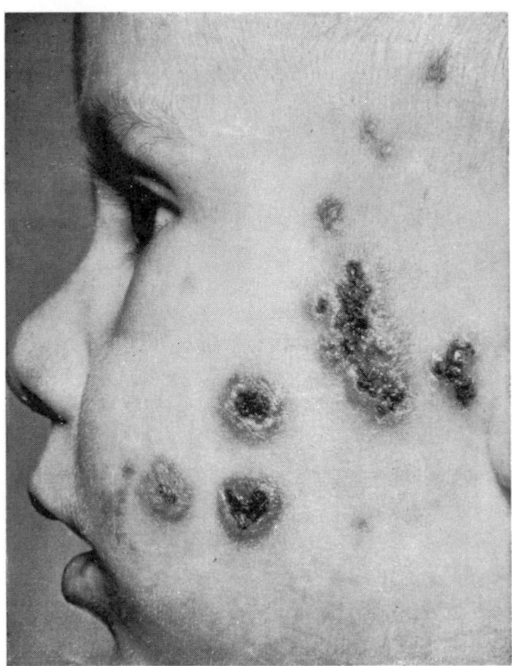

FIG. 19-4. Impetigo contagiosa, highly communicable skin disease. Note varying-sized lesions, some dark and encrusted. (From Top, F. H., Sr.: Communicable and infectious diseases, ed. 6, St. Louis, 1968, The C. V. Mosby Co.)

FIG. 19-5. Triangular area of face where staphylococcal infection is very dangerous.

■ **Pathogenicity.** Some staphylococci are nonpathogenic; others produce severe infections. Severe infections are caused by *Staphylococcus aureus. Staphylococcus epidermidis,* though sometimes responsible for very mild, limited infections, is generally considered nonpathogenic except under unusual medical circumstances.

The usual varieties of staphylococcal infections are those of the skin and superficial tissues of the body, the features of which are greatly influenced by the age of the patient. The scalded skin syndrome (Fig. 19-3) is an example of staphylococcal disease produced by certain strains with a predilection for the newborn and very young. Its manifestations are clearly depicted by the name. Staphylococci cause boils, pustules, pimples, furuncles, abscesses, carbuncles, paronychias, and infections of accidental or surgical wounds.

Staphylococci also produce systemic disease, and all organ systems in the body may be affected. They are one cause of pneumonia, empyema, endocarditis, meningitis, brain abscess, cystitis, and pyelonephritis. Staphylococci can infect the valve prosthesis in the heart. Staphylococcal pneumonia was a fatal complication to cases of influenza in recent epidemics. Staphylococci are the commonest cause of osteomyelitis and impetigo contagiosa (Fig. 19-4). Systemic staphylococcal disease is often acquired in the hospital, especially in patients already ill with a serious disease. Staphylococcal pneumonia as a superinfection is a threat after large doses of antibiotics have been given.

Staphylococcal septicemia may assume one of two forms. The first is a fulminating, profound toxemia with death a few days later. The second and more frequent is of longer duration and characterized by severe clinical disease, the formation of metastatic abscesses in different parts of the body, and a mortality not quite so high as for the first type. Staphylococcal septicemia may be a primary condition, but mostly it is a secondary invasion of the bloodstream by organisms from a localized site of infection in the skin (often a trivial one). It can come from an infected wound, a dental abscess, a pneumonia, or an infected intravenous catheter site. An indwelling intravenous catheter should not remain in place longer than 3 to 4 days because of the likelihood of serious infection. Boils about the nose and lip are easily so complicated, and for this reason they should not be traumatized.

There is an area of the face, triangular in shape, lying with its base along the opening of the mouth and its apex placed in the region above the upper part of the nose. It is called the "dangerous triangle" (Fig. 19-5) because of the threat to a person's life if infection originating there spreads backward into the cranial vault. It is a peculiar area in that anatomic factors operate within it to favor just such a disaster. (A comparable lesion in another part of the body would be inconsequential.) These factors include poor connective supports that provide no mechanical barriers, veins without valves connecting with veins that drain backward, and muscles of the face that are more or less constantly in motion.

If piercing of the ears (for cosmetic reasons) is done in an unsanitary setting, there is the danger of secondary infection with potentially deadly staphylococci. They can enter a wound that is kept open for several days, and in rare instances the spread of infection to the bloodstream has been tragic.

On occasion, implantation of staphylococci in the intestinal tract after antibiotic therapy results in enteritis or dysentery. Staphylococcal food poisoning is the most common type of food poisoning from ingestion of a bacterial toxin.

Staphylococci are an important cause of suppurative conditions in cattle and horses. Mastitis of staphylococcal origin in cows can be transmitted to other cows by the hands of the milker. Staphylococcal bacteremia may follow tick bites in lambs.

■ **Pathology.** The hallmark of staphylococcal infection is the *abscess,* the type lesion. It reflects the excellent pus-forming ability of staphylococci and their limited capacity for spread. Since the microbes reside on the skin, this is the area most frequently involved. A boil is a skin abscess. The type lesion is modified by anatomic location and degree of involvement. Staphylococcal pneumonia means multiple abscesses in the lungs. Pyelonephritis means multiple abscesses extending down the tubular system of the kidney. Staphylococcal septicemia is the development of multiple abscesses over the body; the word *pyemia,* literally saying "pus in the blood," is more appropriate. Staphylococci, as an important cause of wound infections, are responsible for pus formation therein.

■ **Hospital-acquired (nosocomial) infection.** Because of man's intimate contact with this microbe and because of its ubiquity, there are unique features to its infections at any time. Recently, staphylococcal infections have been especially troublesome in hospitals all over the world. In one regard, the problem is a complication to modern antibiotic therapy. To begin with there are several background factors. Antibiotics generally are bacteriostatic, *not* bactericidal. Important chemotherapeutic agents have been freely given. Staphylococci are well endowed for survival and consequently antibiotic-resistant strains developed. Moreover, with prepaid medical plans and medical advances favoring early detection of disease, more persons now are treated in hospitals. Modern surgery is expanded. Complicated surgical technics can be done with a greater exposure of tissue at operation for a longer period of time than ever before. In certain instances drugs that depress the patient's resistance to infection, such as the corticosteroids, are indicated and of value.

The hospital population is already large, complex, but growing with the increased demand for persons with specialized technical skills. Here, then, is a patient, sometimes seriously ill, confined within a large institution where the contacts with all kinds of persons are many and varied. Consider these factors in light of the fact that staphylococci are everywhere. Is it any wonder that in this kind of setting the staphylococcus can be quite mischievous!

Four major categories of disease caused by antibiotic-resistant staphylococci related to hospital-acquired infection are stressed:

1. Skin abscesses (impetigo and pyoderma) in newborn infants. Many of these infants develop a breast abscess. Fatal staphylococcal pneumonia or septicemia is prevalent. The nursing mother can pick up a virulent organism from her baby. Abscess formation in her breast may result.
2. Wound infections, especially of surgical wounds
3. Secondary staphylococcal infections in hospitalized elderly and debilitated persons
4. Gastroenteritis, as a result of a change in the bacterial flora of the intestinal tract

■ **Sources and modes of infection.** Staphylococci are normal inhabitants of the skin, mouth, throat, and nose of man. They live in these areas without effect, but once past the barrier of the skin and mucous membrane, they can cause extensive disease. They pass through the unbroken skin via the hair follicles and ducts of the sweat glands under certain conditions. The natural invasive characteristics of staphylococci and the resistance of the body are so well balanced that infection probably never occurs unless a highly virulent organism is encountered or body resistance is lowered. The first event of infection is a localized process such as an abscess or a boil. It usually heals without widespread dissemination of the bacteria. In some cases organisms do escape from the localized process, invade the bloodstream, and affect distant parts of the body.

Important to the spread of staphylococcal infections are the direct person-to-person contacts. For example, in the hospital the hands of the doctor or nurse or attendant may carry the infection from one patient to another. Nasal carriers are an important source. Virulent organisms may also be passed to man from livestock and household pets.

■ **Bacteriologic diagnosis.** A bacteriologic diagnosis of a staphylococcal infection is made easily by smears and cultures. When blood cultures for staphylococci are made, special pains must be taken to exclude those inhabiting the skin. The coagulase test (slide and test tube methods) is performed to differentiate nonpathogenic from pathogenic staphylococci. A freshly isolated staphylococcus is most likely to be a virulent pathogen if it produces a yellow pigment, hemolyzes blood, ferments mannitol, elaborates deoxyribonuclease, and is coagulase-positive.

The bacteriophage typing of staphylococci deserves special mention. Bacteriophages (p. 411) are viruses that attack bacteria and in certain instances dissolve the bacterial cell parasitized. The action of phages is specific; that is, only a certain phage or group of phages affects the given strain of bacteria. It has been found that specific bacteriophages (staphylophages, Fig. 19-2) react with about 60% of coagulase-positive staphylococci. Coagulase-negative strains of staphylococci are not so susceptible. Because of the specificity, phage typing of staphylococci can be done. For convenience, bacteriophages have been given identifying numbers, and the strains of staphylococci related to these particular phages are designated by the number of the bacteriophage or phages dissolving them. Twenty-nine are available commercially, but the most important ones are the 42B, 44A, 47C, 52, 80, and 81 phages.

■ **Immunity.** Man possesses considerable natural immunity to staphylococci. Patients with debilitating diseases, especially diabetes, are more susceptible to staphylococcal infections than are normal persons. Staphylococcal infections are not followed by any detectable immunity.

■ **Prevention and control of staphylococcal infection.** The most important measures of controlling staphylococcal infections are the maintenance of good housekeeping standards* and the adherence to strict aseptic technics. There is strong evidence that the currently pathogenic strains of staphylococci are as susceptible to the action of chemical germicides as are the ordinary nonpathogenic strains. The detection of nasal carriers, especially in the nurseries for newborn infants, is important.

The phage typing of coagulase-positive staphylococci has proved valuable in the epidemiologic study of hospital-acquired infections to track down the sources to carriers or other foci of staphylococcal disease. The most significant phage type has been the 42B/52/81 type, or the 80/81 type as it is designated by the International Typing Reference Center. This has been the epidemic strain largely instrumental in producing infections in hospitals over wide geographic areas. It is known to be resistant to penicillin, streptomycin, and the tetracyclines. In some cases it may be sensitive to other antibiotics. Vaccine therapy has not proved effective in dealing with this organism because of certain of its immunochemical peculiarities. A staphylococcal phage lysate designed to call forth cell-mediated immunity is more promising.

If one strain of coagulase-positive staphylococcus is biologically anchored to a given site in the human body, another coagulase-positive strain cannot implant itself there. The colonization of the second strain is blocked. There is *bacterial interference* between the two. Today in hospital nurseries practical application of this phenomenon is made to protect newborn infants against current epidemic strains of staphylococci.

GAFFKYA TETRAGENA

Gaffkya tetragena is a large gram-positive coccus that resembles the staphylococcus but occurs in groups of four. Sometimes found in the saliva of healthy persons, it is often associated with cavity formation in pulmonary tuberculosis.

STREPTOCOCCUS SPECIES (THE STREPTOCOCCI)

General considerations

The term "streptococcus" rests upon a morphologic basis and includes all cocci that occur in chains and are insoluble in bile. Within this group are found significant

*As Welton Taylor has said recently, "Perhaps the most important ally in the hospital's campaign against hospital-borne infection is the housekeeper, who merely has to employ the commonsense sanitation that any good housewife knows." (From Hosp. Tribune, 7:3 [Oct. 9] 1967.)

variations in cultural characteristics and disease-producing properties. Some produce deadly disease, whereas others do so only under special conditions, and still others are nonpathogenic. As a whole, streptococci are probably responsible for more illness and cause more different kinds of disease than any other group of organisms. They attack any part of the body and can as easily cause primary as well as secondary disease. They attack both man and lower animals. Some occur as saprophytes in milk and other dairy products.

A great deal of research is carried out on the biology of the streptococcus. The individual streptococci are being studied in great detail as are the reactions they induce upon contact with a given host.

■ **Characteristics.** Typical streptococci occur in varying sizes arranged in long or short chains. Long chains contain 50 or more organisms; short chains contain as few as 4 or 6, with the bacteria arranged in pairs within the chains. Chains form when bacteria divide in one plane and cling together after division. Streptococci are nonmotile, gram-positive organisms that do not form spores. Capsule formation is variable. Some species form distinct capsules; most do not.

The majority of streptococci grow best in the presence of oxygen, but may grow in its absence. A few species are strict anaerobes. They grow well on all fairly rich media, and visible growth usually appears within 24 to 48 hours. Growth is especially luxuriant on hormone media or media containing unheated serum, whole blood, or serous fluid. Streptococci multiply in milk. They grow best at body temperature but may grow through a temperature range of 15° to 45° C. They are not soluble in bile and do not ferment inulin, characteristics that differentiate them from pneumococci.

Streptococci may remain alive in sputum or other excreta for several weeks and in dried blood or pus for several months. They are destroyed at a temperature of 60° C. within 30 to 60 minutes. They are killed in 15 minutes by 1:200 phenol solution, tincture of iodine, or 70% isopropyl alcohol. Penicillin is the most effective antibiotic against most types, except the enterococci.* Streptococci readily acquire a resistance to the other antimicrobial drugs.

■ **Classification.** Streptococci may be classified on the basis of their action on blood agar, their biochemical properties, or their serologic behavior (agglutination and precipitation).

To determine their action on blood agar, plates of blood agar are prepared. Nutrient agar is melted and cooled to 45° C.; sterile defibrinated blood is added in the proportion 5 to 10 parts of blood to 100 parts of blood agar mixture. The mixture is poured into Petri dishes and allowed to cool. The plates are streaked with the material containing the streptococci and incubated at 37° C. for 24 hours or longer. Corresponding to their action on the blood agar plates, streptococci may be classified broadly as:

1. Alpha-hemolytic or viridans—colony surrounded by a green halo; hemolysis slight or incomplete
2. Beta-hemolytic—colony surrounded by a clear, wide, colorless zone of hemolysis (Fig. 19-6)
3. Gamma—colonies showing neither hemolysis nor color change (Fig. 19-7)

As a general rule, the beta-hemolytic streptococcus is the most virulent and is associated with acute fulminating infections; the viridans type is most often associated

*Against beta-hemolytic streptococci, penicillin is still the drug of choice.

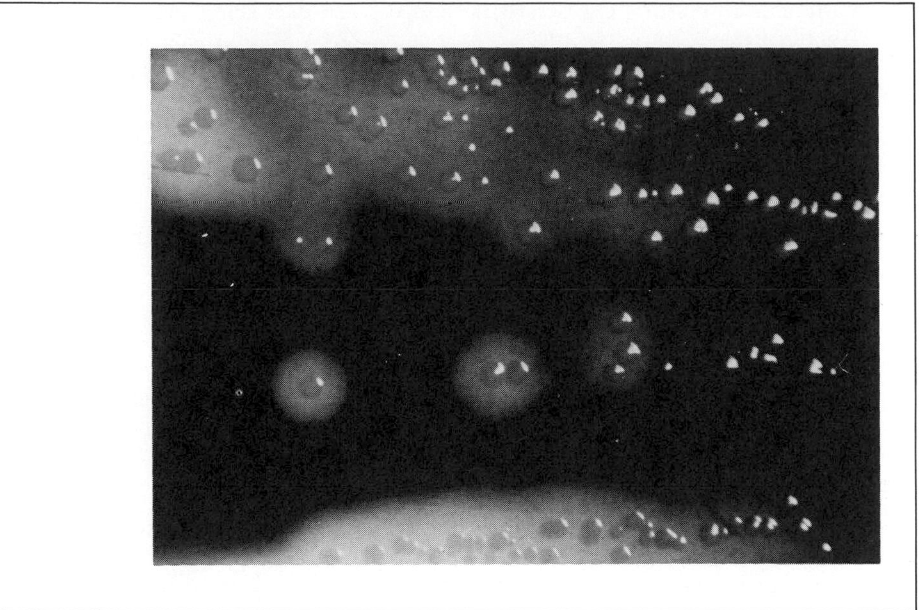

FIG. 19-6. Beta-hemolytic streptococci. Note wide clear zones about small gray colonies on routine blood agar plate.

with chronic low-grade infections in man. However, infections of the viridans type (example, subacute bacterial endocarditis) may be as serious as infections with beta-hemolytic streptococci. Alpha-hemolytic streptococci are also responsible for such nonlethal infections as tooth abscesses and sinus infections. Many, but not all, strains of streptococci of the gamma type are nonpathogenic.

Another classification divides the streptococci into (1) the pyogenic group, (2) the viridans group, (3) the lactic group, and (4) the enterococcus group (see Table 19-1). Enterococci, of which *Streptococcus faecalis* is an important species, are frequently responsible for infections of the genitourinary tract and occasionally for respiratory infections; they are rare causes of subacute bacterial endocarditis. The lactic group, found in sour milk includes *Streptococcus lactis* and *Streptococcus cremoris.* It is not pathogenic but is important in the dairy industry and in certain types of biologic assay. The important member of the viridans group is *Streptococcus mitis,* often referred to as *Streptococcus viridans.* It is responsible for 90% of the cases of subacute bacterial endocarditis. The type organism of the pyogenic group is *Streptococcus pyogenes,* one of the most important of pathogenic bacteria.

By serologic (precipitin) methods streptococci fall into 13 groups (the classification of Lancefield and others), which groups correspond in a general way to their pathologic action (see Table 19-2). These groups may be further subdivided into types given arabic numbers. For over 50 types in group A the antigenic substance determining type specificity is the *M protein.* Superficially attached to the cell wall, it protects the virulent streptococci of this group from the phagocytes of the host. Most human infections are due to group A streptococci (*Streptococcus pyogenes*). Certain members of groups B, C, D, H, K, and O and all members of group N are nonhemolytic.

■ **Toxic products.** Streptococci elaborate certain extracellular poisons, some of which

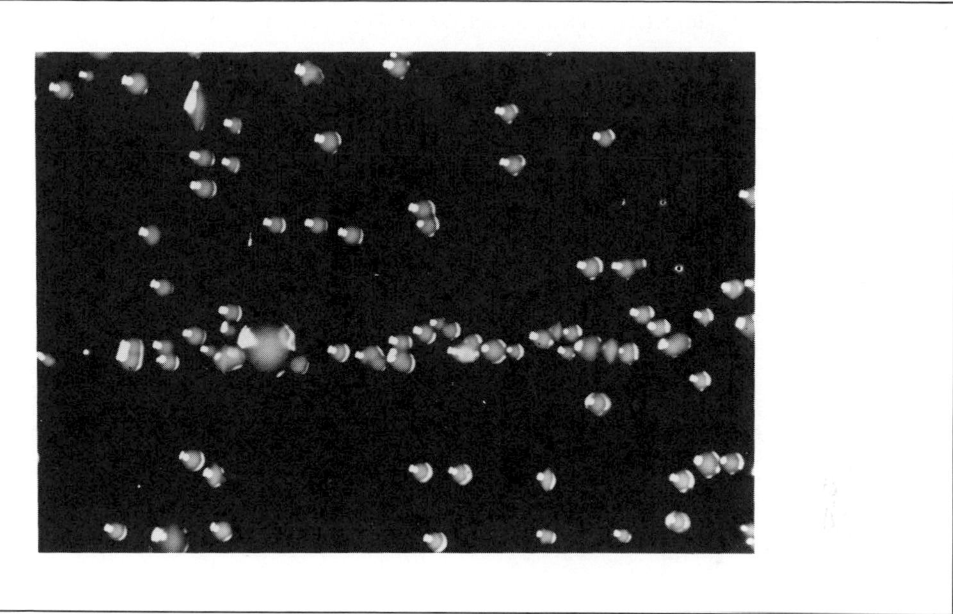

FIG. 19-7. Gamma-hemolytic streptococci. No hemolysis indicated about these colonies of streptococci.

can be called exotoxins. Among these are (1) hemolysins, (2) leukocidin, (3) streptokinase, (4) streptodornase, (5) hyaluronidase, and (6) erythrogenic toxin.

The hemolysins or *streptolysins* produced by streptococci are of two types, S and O. Streptolysin S is produced primarily by members of group A; streptolysin O is elaborated by most members of group A, by the "human" members of group C, and by certain members of group G. Most workers believe that the leukocidin of streptococci and streptolysin O are identical. *Streptokinase* or *fibrinolysin* is important in that it activates an enzyme that destroys fibrin (the component that forms the framework of blood clots). Blood clots play an important part in wound healing and in blocking the spread of local infections.

Streptodornase (streptococcal deoxyribonuclease) aids in the liquefaction of thick, tenacious exudates such as are seen in pneumonia. The enzymatic activity of the deoxyribonuclease is directed against the deoxyribonucleoprotein content of the exudate, the factor responsible for its viscosity. *Hyaluronidase* (the spreading factor, *invasin)* is a factor increasing the permeability of the tissues to bacteria and toxins by breaking down hyaluronic acid, one of the substances that cement tissue cells together. The *erythrogenic toxin* produces considerable erythema or redness when injected into the superficial layers of the skin, and with a large enough dose, a generalized rash. It is the scarlet fever toxin of Dick and occurs in two immunologic types, A and B.

■ **Pathogenicity.** Streptococci may be responsible for a localized inflammatory reaction, such as an abscess, or a generalized reaction, such as septicemia. The nature of the lesion depends upon the virulence of the streptococci, the number introduced into the body, the mode of introduction, the tissue invaded, and the resistance of the host.

TABLE 19-1. BIOLOGIC PROPERTIES OF STREPTOCOCCI

Division	Hemolytic streptococcus	Viridans streptococcus	Enterococcus	Lactic streptococcus
Serologic group	A, B, C, E, F, G, H, K, L, M, O	None	D	N
Hemolysis on blood agar	Usually beta	Usually alpha	Alpha, beta, or gamma	Alpha or gamma
Growth in 0.1% methylene blue milk	–	–	+	+
Growth in 6.5% salt broth	–	–	+	–
Growth on 40% bile blood sugar	–	±	+	+
Antibiotic sensitivity (bacitracin)	Usually sensitive	May be resistant	May be resistant	(nonpathogenic)

TABLE 19-2. SEROLOGIC GROUPS OF STREPTOCOCCI

Group	Species	Significance
A	*Streptococcus pyogenes*	Important human diseases (infection by beta hemolytic streptococci initiates acute rheumatic fever); group sensitive to penicillin
B	*Streptococcus agalactiae*	Bovine mastitis
C	*Streptococcus equi* *Streptococcus zooepidemicus* *Streptococcus equisimilis* *Streptococcus dysgalactiae*	Animal diseases; mild respiratory infections in man
D	*Streptococcus faecalis* *Streptococcus durans* *Streptococcus zymogenes* *Streptococcus liquefaciens*	(Enterococci) Genitourinary tract infections, endocarditis, wound infections in man; found in dairy products
E		Disease of swine; found in normal milk
F	*Streptococcus minutus*	Found in respiratory tract of man
G	*Streptococcus anginosus*	Mild respiratory infections in man; genital infections in dogs
H	*Streptococcus sanguis*	Found in respiratory tract of man
K	*Streptococcus salivarius*	Found in respiratory tract of man
L		Genital tract infections in dogs
M		Genital tract infections in dogs
N	*Streptococcus lactis* *Streptococcus cremoris*	(Lactic group) Found in dairy products
O		(Viridans group) Subacute bacterial endocarditis; found in upper respiratory tract in man
		(Microaerophilic streptococci)
		(Anaerobic streptococci—13 species)

Pathology. Pathologically, the type lesion of hemolytic streptococci is the diffuse, ill-defined, spreading lesion of *cellulitis.* The exudate is poorly cellular and consists largely of fluid with little fibrin. The toxic products of the microbes greatly aid their extension through both natural tissue and inflammatory barriers. They tend to infect the lymphatic vessels at the site of invasion. Many well-known forms of streptococcal infection can be designated as an expression of cellulitis. Erysipelas is cellulitis with an anatomic pattern; septic sore throat is cellulitis of the throat. Streptococcal bronchopneumonia with a comparable exudate is poorly limited. Streptococci are also important in wound infections.

Allergic manifestations may follow certain streptococcal infections. For instance, acute rheumatic fever and one form of kidney disease (glomerulonephritis) are considered as allergic reactions to streptococcal infection, usually in the upper respiratory tract.

As a rule, the more virulent an infection, the more virulent are the streptococci isolated from the lesions for animals of the same species. Usually their virulence is lowered when they are introduced into animals of another species. When transferred from one animal to another, streptococci tend to produce the same type of lesion in the new host as in the original one. Rabbits and white mice are more susceptible to streptococci of human origin than are other laboratory animals.

STREPTOCOCCI IN HUMAN DISEASES. In addition to being the cause of erysipelas and of two epidemic diseases, scarlet fever and epidemic sore throat, streptococci belonging to group A *(Streptococcus pyogenes)* are the most common cause of acute endocarditis, septicemia, and puerperal sepsis. They may cause pneumonia, boils, abscesses, cellulitis, peritonitis, tonsillitis, lymphangitis, infection of surgical wounds, osteomyelitis, and empyema. From an infection of the middle ear (otitis media), streptococci may spread to the mastoid cells and cause mastoiditis. From either the middle ear or the mastoid cells, spread of infection to the meninges means streptococcal meningitis.

Streptococcus pyogenes is responsible for the majority of bronchopneumonias complicating whooping cough, measles, and influenza. Such bronchopneumonias are often fatal and may reach epidemic proportions when outbreaks of whooping cough, measles, or influenza occur in communities containing numerous carriers of *Streptococcus pyogenes.* Streptococcal pneumonia is often the terminal phase of chronic diseases like tuberculosis and cancer. Streptococci as well as staphylococci (often acting together) are responsible for the highly communicable skin disease known as *impetigo contagiosa.*

STREPTOCOCCAL INFECTIONS IN LOWER ANIMALS. The majority of streptococcal infections in lower animals are caused by streptococci in Lancefield's groups B and C. Strangles, an acute communicable disease of the upper respiratory passages of horses, is caused by *Streptococcus equi.* Streptococcal mastitis, a serious disease of cows that renders milk unfit for use, is caused by *Streptococcus agalactiae* and is most likely spread from cow to cow by the hands of the milker. *Streptococcus agalactiae* does not affect man.

■ **Sources and modes of infection.** Streptococci are normal inhabitants of the mouth, nose, throat, and respiratory tract. They may be conveyed from person to person by direct contact or by contaminated objects, hands, and surgical instruments. The hands are important conveyers of infection in puerperal sepsis and wound infections. Milk can be an important source. Streptococcal diseases, especially scarlet fever and septic throat, may be spread by milk that has been contaminated with the mouth and nose secretions of a carrier or by milk from a cow with mastitis caused by *Streptococcus pyogenes.*

Streptococci usually enter the body by the respiratory tract or through wounds of the skin. Only a minute abrasion is necessary, and streptococcal infections in such abrasions have many times led to fatal septicemias in physicians and nurses. Streptococci leave the body by way of the mouth and nose and in the exudates from areas of infection. The nasal carrier is an important source of infection. The enterococci are normal inhabitants of the intestinal canal and are excreted in the feces.

■ **Laboratory diagnosis.** Streptococci are detected by smears and cultures from the site of disease. In septicemias from beta-hemolytic streptococci the organisms can usually be detected by blood cultures. In subacute bacterial endocarditis, since the viridans organisms escape into the blood intermittently, repeated cultures may have to be made.

A presumptive test of value in the identification of streptococci is the *bacitracin disk test.* Even in low concentrations this antibiotic appears to be specifically active against members of Lancefield group A streptococci; it is without effect on other groups. A paper disk containing a known unit of antibiotic is placed on a blood agar plate previously streaked with the streptococci in question. A zone of inhibition of growth found around the streptococci identifies them in group A. This method is practical and is as accurate as precipitin tests to catch group A.

When infection with streptococci that produce streptolysin O takes place, antibodies against the streptolysin O appear in the blood of the patient. The detection and quantitation of these antibodies form an important diagnostic procedure in streptococcal infections of a chronic and persistent nature. This evaluation is the *antistreptolysin O titer* (ASTO). It is useful in the diagnosis and management of acute rheumatic fever and acute hemorrhagic glomerulonephritis (allergic reactions to beta-hemolytic streptococci). The course of these may be related to the titer. The test helps to differentiate certain diseases similar to rheumatic fever where the streptococcus is not comparably implicated. The titer in these diseases is not significant.

■ **Immunity.** With the exception of scarlet fever, streptococcal infections are not followed by an immunity to subsequent attacks, and in scarlet fever the immunity is established only against scarlet fever toxin, not against the organisms.

■ **Prevention.** When caring for a patient with a streptococcal infection, the nurse should remember that she is dealing with an infection that may be most virulent and very easily spread. Physicians and nurses attending such infections, especially scarlet fever and erysipelas, should not attend an obstetric case or surgical operation until they are incapable of spreading the infection. Obstetric cases should be handled with strictest aseptic care because the recently emptied uterus is extremely vulnerable.

The buccal and nasal secretions from a patient with streptococcal bronchopneumonia should be handled in the same manner as those from a patient with diphtheria. A patient with streptococcal bronchopneumonia should be isolated from patients with measles or influenza. Conditions favoring contact infection should be avoided, and all wounds and abrasions on the body should be thoroughly disinfected.

The fact that more than 50 types exist among group A streptococci, all immunologically specific, explains the delay in the development of a clinically feasible vaccine. Vaccine production centers around the highly antigenic M protein in the streptococcal cell membrane that is significantly related to virulence.

Scarlet fever

Scarlet fever (scarlatina) is an acute infection of childhood characterized by sore throat, severe constitutional manifestations, and a distinct skin eruption with massive exfoliation. The rash results from an acute hyperemia of the skin with petechial hemorrhages. The nasopharynx and tonsils may be covered with a membrane. The lymph nodes, especially those of the neck, are swollen and inflamed. The white blood count ranges from 15,000 to 30,000 cells per cubic millimeter, of which 85% to 95% are neutrophils.* Scarlet fever is caused by streptococci that produce erythrogenic toxin. In almost all cases streptococci are of Lancefield's group A, and only on the rarest occasion is scarlet fever produced by streptococci belonging in groups C and D.

The present opinion is that scarlet fever and streptococcal sore throat are different manifestations of the same basic disease. If the streptococci causing the sore throat produce erythrogenic toxin and the recipient of the infection is not immune to the toxin, scarlet fever results. If the streptococci do not produce erythrogenic toxin or if the recipient is immune to the toxin, then only the sore throat is present. There is a close relation between the streptococci causing scarlet fever, erysipelas, and puerperal sepsis.

In the past, scarlet fever was a disease of great severity, and often fatal. Now it is relatively benign but may be complicated by suppurative otitis media and peritonsillar abscess or be followed by rheumatic fever and acute nephritis. It is much more common in European countries than in North America.

■ **Sources and modes of infection.** The sources of scarlet fever are the nose and throat secretions of patients or carriers and pus from infected lymph nodes, ears, and other lesions. The organisms are present throughout the course of the illness and may persist in the nose and throat or in the exudates for weeks or months thereafter. As long as a person harbors the organisms, he is a dangerous source of infection. The desquamated scales are not infectious.

The organisms usually enter the body through the mouth and nose, less often through wounds, burns, and the parturient uterus. Infection may be transmitted by direct contact or by contaminated objects, such as handkerchiefs, towels, pencils, toys and dishes.

■ **Erythrogenic toxin.** The erythrogenic toxin of scarlet fever streptococci is released by the organisms at the primary site, is absorbed into the body, and brings about the rash and other constitutional effects of the disease. It may be found in the blood in a concentration as high as 300 units per milliliter and also in the urine. It is prepared artificially. Scarlet fever streptococci are grown for 5 days in broth, after which the bacteria are removed by filtration. The filtrate contains the toxin. Toxin prepared in this manner causes scarlet fever when given in large doses. When injected in very small amounts into the skin of persons susceptible to scarlet fever, it gives rise to an inflammatory reaction, the basis of the *Dick test.* It can induce an active immunity and the formation of antitoxin when injected into a suitable animal. The unit of measurement of scarlet fever toxin is known as the skin test dose (STD), the smallest amount of toxin causing an inflammatory reaction in the skin of a susceptible person. It is necessary to use susceptible humans for the test because the majority of laboratory animals do not react.

*Normal white cell count is 5000 to 9000 cells per cubic millimeter with 55% to 65% neutrophils.

■ **Immunity.** Immunity to scarlet fever relates to scarlet fever antitoxin in the blood. Infants inherit an immunity from their mothers that is lost within a year, and susceptibility increases until the sixth year. After the sixth year susceptibility decreases until adult life, at which time the majority of people are immune. An attack of scarlet fever is usually followed by permanent immunity. Remember that although an immune person will not be harmed by scarlet fever toxin the streptococci themselves may invade his body and cause tonsillitis, abscesses, or otitis media. Immune persons may harbor the organisms for a long time and spread the infection widely.

DICK TEST. The Dick test is performed by injecting between the layers of the skin of the forearm 0.1 ml. of scarlet fever toxin so diluted that one STD is given. In immune persons the antitoxin in the blood neutralizes the injected toxin, and no reaction occurs. In susceptible persons without antitoxin the toxin injures the cells around the injection site, producing within 24 hours a zone of inflammation and redness at least 1 cm. in diameter. The test is positive at the beginning but becomes negative during the course of scarlet fever.

SCHULTZ-CHARLTON PHENOMENON. If a small amount of the serum of a person convalescent from scarlet fever or of an animal immunized against scarlet fever is injected intradermally into an area of scarlet fever rash, the skin blanches at the injection site because the toxin in the area is neutralized by the antitoxin. This test differentiates scarlet fever from measles, German measles, and other skin diseases.

■ **Prevention.** The patient with scarlet fever should be isolated, and the discharges from the mouth and nose, as well as all contaminated articles, should be disinfected. The disinfecting procedures are the same as those for diphtheria (p. 329). Persons attending a patient should exercise every precaution to prevent the spread of infection to others, especially to obstetric or surgical cases. The patient should remain isolated until the discharges from the mouth and nose are free of scarlet fever streptococci and all complications have healed. Remember that during epidemics of scarlet fever persons with rhinitis and sinusitis may just as effectively spread the disease as those with a skin eruption. Pasteurization prevents milk-borne epidemics.

Erysipelas

Erysipelas* (St. Anthony's fire) is an acute inflammation of the skin caused by hemolytic streptococci of Lancefield's group A or, infrequently, of group C or D. The organisms grow at the site of infection and elaborate toxic substances responsible for the constitutional state. They are present not in the central portion of the inflamed area but at the periphery.

■ **Mode of infection.** The portal of entry for the streptococci is a wound, fissure, or abrasion. The lesion begins at the site of infection and extends peripherally as a hard red thickening of the skin. The streptococci grow almost exclusively in the lymph channels of the inflamed area, and, as the disease progresses, they spread peripherally several centimeters beyond the line limiting the area of obvious inflammation. When streptococci enter the blood, the prognosis is bad. Erysipelas may be complicated by abscesses, pericarditis, arthritis, endocarditis, septicemia, and pneumonia. Patients with uncomplicated disease or without open wounds or super-

*Erysipelas should not be confused with erysipeloid, a localized infection of the skin caused by *Erysipelothrix insidiosa rhusiopathiae* (gram-positive rod with a tendency to form long filaments) and occurring in those who handle fish or meats.

ficial discharges will not transmit the infection to others. Erysipelas may be associated with other group A streptococcal infections.

■ **Immunity.** Instead of inducing immunity, an attack of erysipelas seems to render the patient more vulnerable to future attacks. Penicillin is the drug of choice in the treatment of erysipelas.

Streptococcal sore throat (septic sore throat)

Septic sore throat is an ulcerative inflammation of the throat with severe symptoms and a high mortality. This is the disease referred to as "strep" throat. It is caused by hemolytic streptococci belonging to Lancefield's group A or, in a small proportion of cases, group C. It may be transferred by direct contact, droplet, or airborne infection. However, direct transfer from person to person is not frequent.

Streptococcal sore throat may extend to the lungs to produce streptococcal pneumonia. Like scarlet fever, it may be followed by nephritis or rheumatic fever. Penicillin is the drug of choice in treatment.

Endocarditis

Endocarditis is inflammation of the lining of the heart (endocardium) including the reduplications or folds of endocardium making up the valves. Without qualification, the term "endocarditis" is used to mean inflammation of the valves of the heart. As a rule, the process selects one or more valves, mainly the valves of the left side of the heart and the mitral more often than the aortic, but it may incorporate the lining of the heart wall (mural endocarditis). It is an important disease both for its immediate effects and for its permanent injury.

Once in the bloodstream, certain streptococci may be key agents in etiology. Acute fulminant endocarditis, an overwhelming aspect of septicemia, is often caused by *Streptococcus pyogenes.* Subacute bacterial endocarditis with a slow clinical course is predominantly the consequence of *Streptococcus viridans.* (Both may be caused by various other microorganisms). Many patients with subacute bacterial endocarditis give a history of dental extraction just before onset of disease. Such a surgical event provides the occasion for a transient release of bacteria into the bloodstream, where they precipitate onto a heart valve. The deposit of bacteria on a valve injures the surface and causes it to swell. Fibrin, platelets, and cells from the blood collect at the site to form aggregates known as "vegetations" (Fig. 31-9). A vegetation in the flow of blood over it is prone to crumble, sending small infected fragments out into the stream. These travel to various areas of the body and lodge there, producing secondary foci of involvement. The severe injury of acute disease may result in destruction and even tearing of the valve. In the more slowly moving form of endocarditis, changes associated with repair of damage become prominent. As a consequence, scarring ensues. The valve becomes so thickened, shortened, and deformed that it cannot close properly, thereby materially affecting the pumping action of the heart.

Rheumatic fever

Acute rheumatic fever, the forerunner of rheumatic heart disease, has a clinical course similar to an acute infection. In its acute phase there is fever, an increased pulse rate (tachycardia), and a characteristic type of polyarthritis. Rheumatic fever is more common in the northern than in the southern climates and is infrequent in the tropics. It is most common between the sixth and twelfth years of life.

More than 50% of the attacks of acute rheumatic fever are preceded by tonsillitis or severe sore throat caused by hemolytic streptococci. At this time the events of rheumatic fever are interpreted as an allergic reaction to the previous streptococcal infection. An important diagnostic feature is a rise in the serum titer of antistreptolysin O, which occurs in more than 80% of the patients.

Characteristic of the disease is an inflammatory process in the heart (rheumatic carditis), in which small nodules, called *Aschoff bodies,* form in the heart valves and all through the layers of the heart. Rheumatic fever tends to be a disease of long standing, with repeated attacks of the acute process. This favors the development of complications that add up to serious heart disease.

To prevent the harmful effects of rheumatic heart disease is to stop the recurrent attacks of acute rheumatic fever. Patients are protected as much as possible from streptococcal infections and even are given antibiotics prophylactically at times of the year when the incidence of such infections is high.

Puerperal sepsis

Puerperal sepsis (puerperal septicemia) is usually caused by a hemolytic streptococcus from the nose and throat of the patient herself or those in close association with her. It reaches the uterus via contaminated hands or instruments. Most of such streptococci belong to Lancefield's group A of organisms exogenous to the generative tract, but from 20% to 25% of the cases come from anaerobic streptococci, which are normal inhabitants of the vagina.

DIPLOCOCCUS PNEUMONIAE (THE PNEUMOCOCCUS)

■ **General characteristics.** The organism *Diplococcus pneumoniae,* best known in relation to pneumonia, occurs typically as lancet-shaped diplococci with their broad ends in apposition. Within the animal body or in excretions each pair is enclosed within a capsule. They are gram-positive and nonmotile and do not form spores.

Pneumococci grow equally well in the presence or absence of oxygen. Their optimum temperature is 37° C., and they grow best in a slightly alkaline medium. Growth is most abundant on such enriched media as hormone agar and blood agar (Fig. 19-8). On the latter medium the green zone of slight hemolysis surrounding the colony is an important diagnostic feature. The power to form capsules is lost when pneumococci are cultivated for a long time on artificial media.

The pneumococcus is not a very hardy organism and has no natural existence outside the animal body. In the finely divided spray thrown off from the nose and mouth, pneumococci live about 1½ hours in the sunlight. In large masses of sputum they live for 1 month or more in the dark and about 2 weeks in the sunlight. They are more susceptible to ordinary germicides than are most other bacteria and are destroyed in 10 minutes by a temperature of 52° C.

■ **Types.** Although all pneumococci are very much alike microscopically and culturally, they show distinct immunological differences. This was discovered in 1910, when different cultures of pneumococci were used to immunize animals, and the serum of these animals was used to agglutinate pneumococci from various other sources. It was found that the majority of pneumococci fell into one of three rather distinct types (types I, II, and III) and that an antiserum prepared by immunizing

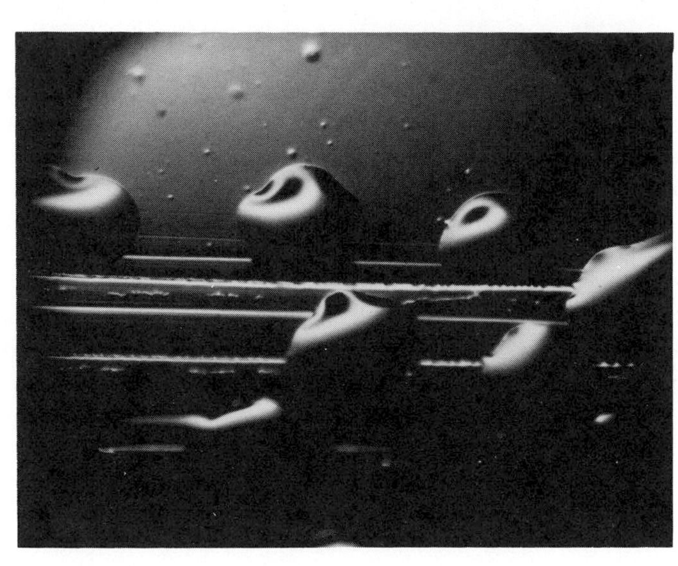

FIG. 19-8. Pneumococci. Greatly magnified colonies growing on routine blood agar plate.

an animal against pneumococci of one type agglutinated pneumococci of that type only. Pneumococci that did not fall into any of the three types were placed in group IV, which later was divided into 29 types. Other types were described in rapid succession until at the present time there are at least 79 types or subtypes. All can cause pneumonia, but 50% to 80% of adult cases are due to types I, II, and III. Type III pneumonias are more common in the aged. In the pneumonias of children the primary types of pneumococci are XIV, I, VI, V, VII, and XIX (in order of frequency).

Formerly, therapeutic serums made from rabbits were available for the treatment of pneumonia caused by at least 30 of the types known at that time, and the determination of the type of pneumococcus causing the infection was necessary before serum therapy could be instituted. Today serum therapy for pneumonia has been replaced by antimicrobial therapy.

Pneumococcus type III differs somewhat from other pneumococci, being characterized by its wide capsule and slimy growth on culture media. It is not lancet shaped. There seems to be a direct relation between capsule development and virulence, thus explaining why type III infections cause such a high mortality.

The substances that give pneumococci their type characteristics are soluble carbohydrates in the capsules. They can be detected by the precipitin reaction in broth cultures of pneumococci and in the blood and urine of patients with pneumonia. These polysaccharides are the *specific soluble substances* (SSS). In addition to these a somatic antigen is common to all pneumococci.

■ **Toxic products.** The clinical features of pneumococcal disease indicate that the disease is a toxemia, but a toxin similar to that elaborated by the diphtheria bacillus has never been found. Pneumococci do, however, produce hemolysins, leukocidins, and necrotizing substances. Many strains produce hyaluronidase.

■ **Pathogenicity.** *Pneumonia* is an inflammation of the alveoli (air sacs), bronchioles, and smaller bronchi of the lungs, in which these structures are filled with exudate. The pneumococcus is the most important cause of two kinds: lobar pneumonia and bronchopneumonia.

Lobar pneumonia is an acute disease, characterized by a severe toxemia and a massive inflammatory exudation that fills the air spaces of one or more lobes of the lungs (consolidation). Rapid shallow breathing, increased pulse rate, cyanosis, and nausea with vomiting are signs and symptoms related to the plugging of the air sacs with exudate. The blood count usually shows a leukocytosis (30,000 to 40,000 per cubic millimeter) with 90% to 95% polymorphonuclear neutrophils. With recovery and resolution, the exudate in the lungs liquefies and is removed partly by absorption and partly by expectoration. Air reenters the affected lobe or lobes, and the lung completely returns to its former efficiency. Pleurisy (inflammation of the pleura) is a part of the disease. Occasionally, delayed resolution leads to abscess formation or a chronic organizing pneumonia. The disease has yielded so dramatically to the use of the sulfonamides and antibiotics that the pathologic stages described years ago are rarely seen today.

Lobar pneumonia is a primary disease, and 95% of cases are pneumococcal. Of the rest, the majority are caused by the Friedländer bacillus, the influenza bacillus, or the streptococcus.

Bronchopneumonia, more often secondary than primary, is a most important complication to measles, influenza, whooping cough, and chronic diseases of the heart, blood vessels, lungs, and kidneys. It peaks in the early and late years of life and frequently is the terminal event in debilitating diseases of the very young or extremely old. (It has long been called the "old man's friend.") It may follow the administration of an anesthetic or the aspiration of infectious material into the lungs during an operation. In newborn infants it is often caused by the aspiration of infected amniotic fluid.

Although commonly pneumococcal, bronchopneumonia may be caused by any one of a number of microbes, among which are streptococci, staphylococci, and influenza bacilli, operating singly or in variable combinations. Unlike lobar pneumonia, bronchopneumonia consists of scattered small inflammatory foci, usually more numerous at the lung bases. The exudate consists of leukocytes, fluid, and bacteria, but little or no fibrin or red blood cells. The white blood cell count rises to 20,000 or even to 40,000 per cubic millimeter. Pleurisy and empyema are complications. This kind of pneumonia does not always resolve readily and chronic pneumonia often persists.

Hypostatic pneumonia is bronchopneumonia complicating the hypostatic congestion of heart failure.

Pneumococci may cause other diseases, such as empyema, endocarditis, meningitis, arthritis, otitis media, peritonitis, and corneal ulcers. Some of these complicate pneumonia; others occur as primary conditions. Pneumococcal peritonitis is primary in children. Pneumococcal otitis media tends to spread to the meninges, with resultant pneumococcal meningitis.

The most characteristic feature of pneumococcal infections is the presence of an abundance of fibrin in the areas of inflammation. In lobar pneumonia there is much in the lungs as there is also in the subarachnoid space of a patient with pneumococcal meningitis.

■ **Sources and modes of infection.** Lobar pneumonia is endemic in all centers of population. Epidemics seldom occur but may if conditions enhance exposure to infec-

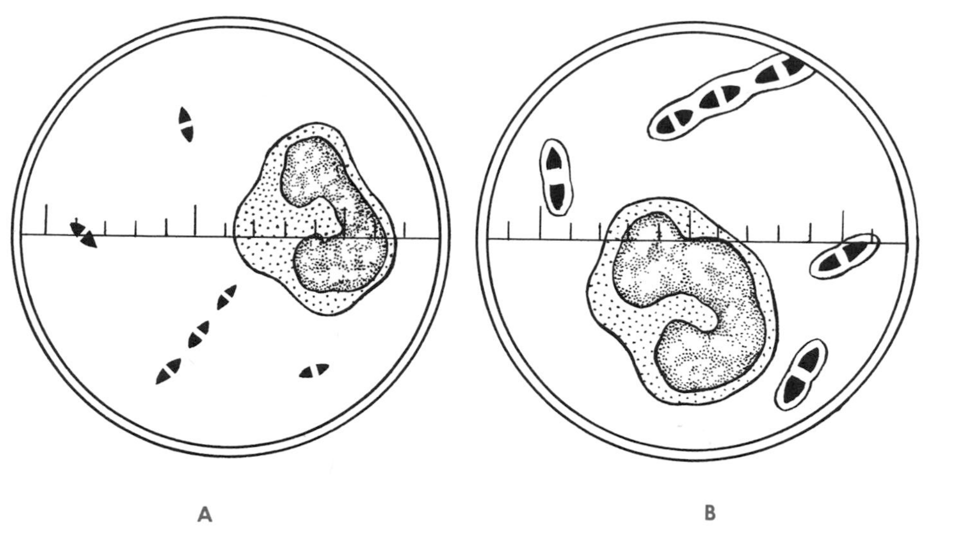

FIG. 19-9. Neufeld reaction. Capsular swelling in diagram. **A,** No detectable change in capsule of lancet-shaped diplococci. **B,** Swelling of pneumococcal capsule in presence of type-specific serum.

tion with concomitant lowering of host resistance. The sources of infection are active cases and carriers, but many times the connection between patient and source cannot be traced.

Pneumococci enter and leave the body by the same route, the mouth and nose. Infection is usually transferred directly, most often by droplets from the mouth and nose, but indirect transmission by contaminated objects may also occur.

Practically every person becomes a carrier of pneumococci for a short time during the year. Those who have contacted a patient often carry the organisms in their throats for a few days or weeks. Most carriers not in contact with pneumonia harbor comparatively avirulent pneumococci and are of little danger, although type III pneumococci may be found in such carriers. That carriers of type III pneumococci are common whereas type III infections are comparatively rare is difficult to explain.

■ **Laboratory diagnosis.** Pneumococci may be detected in sputum and other body fluids with some degree of certainty by direct microscopic examination of smears stained for capsules and by Gram's method. Confirmatory methods are cultures and the inoculation of white mice. (Rabbits and mice are very susceptible to pneumococci.) About 1 ml. of the emulsified sputum is injected into the peritoneal cavity of a white mouse. The pneumococci outstrip all other organisms, and the mouse becomes ill after 5 to 8 hours, at which time it is killed. The peritoneal cavity contains many pneumococci to be identified by microscopic, cultural, and typing methods. Animal inoculation demonstrates pneumococci when direct smears fail to do so, and it always demonstrates them more quickly than do cultures.

Since serum therapy of pneumonia has yielded to antimicrobial drugs, typing is done primarily to determine whether a highly virulent organism is present. There are several methods. All depend upon the action of agglutinating and precipitating serums (typing serums) prepared by immunizing animals against the different types of pneumococci. The method most often used is the one devised by Neufeld, in which flecks of sputum or other pneumococcus-containing material are mixed with the various type-specific serums. In the mixture in which the type of pneumococcus and that of the serum correspond, the capsules of the pneumococci swell (Fig. 19-9). If the sputum contains too few pneumococci or if the typing is otherwise unsatisfactory, some of the sputum may be injected into a white mouse, as previously described, and in the course of a few hours typing may be carried out on the yield from the peritoneal exudate.

■ **Differentiation of pneumococci from streptococci.** Pneumococci and streptococci bear such a close microscopic resemblance to each other that their differentiation is always an important laboratory problem. In fact, they are so closely related that some workers regard the pneumococcus as a species of streptococcus, calling it *Streptococcus pneumoniae.* The important differential points are the following: (1) on blood agar the colonies differ; (2) in animal tissues pneumococci have capsules, but streptococci seldom do; (3) when one part of bile is added to three parts of a liquid culture, pneumococci are dissolved, but streptococci are not; (4) pneumococci ferment inulin, but streptococci do not; (5) pneumococci are inhibited by optochin (ethylhydrocupreine hydrochloride), whereas streptococci are not, and (6) pneumococci are more pathogenic for mice than are streptococci. Further differential aids are agglutination and precipitation tests with specific antiserums.

■ **Immunity.** Recovery from a pneumococcal infection confers a 6- to 12-month immunity to the type of pneumococcus causing the infection, but not to other types. (Instances have been reported in which a person has had pneumonia more than a dozen times.) The natural resistance of man against the pneumococcus is comparatively high, and a person probably never contracts pneumonia unless his resistance is lowered. Negroes are more susceptible than are Caucasians, and men more so than women.

■ **Prevention.** The number of persons contacting a patient with pneumonia should be restricted. The discharges from the mouth and nose of the patient should be burned or disinfected. The hands and all objects, such as spoons, cups, and other utensils, that may be contaminated by the discharges of the patient should be disinfected. Measures should be taken to minimize droplet infection in the spray leaving the mouth of the patient while he talks or coughs.

Because of the antibiotic resistance encountered in pneumococci, vaccine therapy is coming back. The model is new. In progress is the design of a polyvalent vaccine given in a single injection. It incorporates the purified capsular polysaccharides from the 12 types of pneumococci most likely to cause disease.

QUESTIONS FOR REVIEW

1. Name important gram-positive cocci. Why are pyogenic cocci so called?
2. Characterize staphylococci.
3. List diseases caused by staphylococci.
4. Discuss briefly the localization and invasion of the body by staphylococci.
5. How is pathogenicity of a staphylococcus established?
6. Comment on the backbround factors in hospital-acquired infections. How are they studied epidemiologically?
7. Outline the streptococci according to the following categories: morphology, general features, classification, pathogenicity.
8. Name 10 diseases of man caused by streptococci. Name two diseases of animals. State the Lancefield groups responsible.
9. How are streptococci and pneumococci transmitted from person to person?
10. What is the Dick test? The Neufeld reaction? The bacitracin disk test? The coagulase test? The Schultz-Charlton phenomenon? The optochin test?
11. How is the antistreptolysin O titer used clinically?
12. Briefly discuss septic sore throat, puerperal sepsis, and endocarditis.
13. What is the relation between streptococcal disease and rheumatic fever?
14. Compare lobar pneumonia with bronchopneumonia.
15. Contrast the type lesion of staphylococcal and streptococcal infections. State the notable feature of pneumococcal injury.
16. What is the current status of vaccines against pyogenic cocci?

REFERENCES. See at end of Chapter 26.

20 The gram-negative cocci: neisseriae

The most important gram-negative cocci are the gonococcus *Neisseria gonorrhoeae* and the meningococcus *Neisseria meningitidis.* Certain other gram-negative cocci, particularly *Neisseria catarrhalis* and *Neisseria sicca,* are important, not because of their pathogenicity but because of their habitat in the mouth and upper respiratory passages. These members of the genus *Neisseria* are gram-negative diplococci with their adjacent sides flattened.

NEISSERIA GONORRHOEAE (THE GONOCOCCUS)

Neisseria gonorrhoeae is the cause of gonorrhea, one of the most prevalent diseases affecting man, the most common of the venereal diseases, and the number one communicable disease problem today in the United States. A disease known to the ancient Chinese and Hebrews, it was termed *gonorrhea* by Galen around A.D. 130. The gonococcus is sometimes called the diplococcus of Neisser after its discoverer, and its infections are referred to as neisserian infections.

■ **General characteristics.** The gonococcus is a gram-negative, nonmotile, nonspore-forming diplococcus (Fig. 20-1). In smears the opposing sides of the two cocci are flattened. The cocci are placed like two coffee beans lying with their flat sides together. Each pair is surrounded by a capsule.

Gonorrhea is a disease accompanied by a discharge from the genital tract that is at first thin and watery, then later purulent. The incubation period is 3 to 5 days. During the early stages the gonococci are found free in the serous exudate or attached to epithelial cells, but when the exudate becomes purulent, phagocytosis takes place, and the gonococci are found characteristically within the cytoplasm of the pus cells (polymorphonuclear neutrophilic leukocytes). A single white blood cell may contain from 20 to 100 microorganisms. These gonococci are not dead and are still infectious. In later stages of the disease they may be found outside the white blood cells, and when the disease becomes chronic, they often cannot be found at all.

The gonococcus will not grow on ordinary culture media and is somewhat difficult to cultivate even on media prepared especially for it. Gonococci grow best at or slightly below body temperature (35° to 37° C.) and in an atmosphere containing oxygen and carbon dioxide (3% to 10%). They produce the enzyme *oxidase* as do other members of the genus *Neisseria.* The *oxidase reaction* is used to identify colonies of neisserian species in cultures.

Although gonococci are difficult to destroy within the body, they possess little resistance outside it. They are killed in a very short time by sunlight and drying. Incorporated in pus or on clothing, in moist dark surroundings, they may live from 18 to 24 hours. They are very susceptible to disinfectants, especially silver salts, and are killed by a temperature of 60° C. within 10 minutes. Although gonococci are susceptible to the action of the modern antibiotics, drug resistance is an ever present problem, especially with the number of drug-resistant strains coming in from southeast Asia and the Philippine Islands. Gonococci are not exactly alike immunologically but are not typed in any standard way. There is only one defined strain.

■ **Pathogenicity.** The gonococcus is parasitic specifically for man. Spontaneous infection does not occur. Nothing comparable to any of the clinical forms of gonorrhea had been produced artificially in lower animals until recently when gonococcal urethritis was produced experimentally in the chimpanzee and the male-to-female animal transmission was demonstrated.

In typical cases of gonorrhea the sites of primary infection in the female are the urethra and cervix; in the male the site is the urethra. Gonococci injure columnar epithelium like that lining the cervix uteri and the rectum and the transitional (urothelial) epithelium lining the urinary tract. Vaginal infection is not seen because the epithelium lining the vagina of the adult woman is a cornified stratified squamous type resistant to infection with gonococci. Before the age of puberty, the vagina is lined with a softer, extremely susceptible epithelium. Gonorrheal vulvovaginitis in prepubertal girls may be epidemic and difficult to eradicate. The change in the epithelium with the onset of puberty usually eliminates a childhood infection completely. An important site of primary infection seen more and more today is the conjunctiva of the eye (gonorrheal conjunctivitis and keratitis), and the process is one that actively damages the tissues of the eye (Fig. 20-2). *Ophthalmia neonatorum,* gonorrheal conjunctivitis in newborn infants, results when the eyes are infected during the birth passage. A profuse purulent discharge in the eyes of a neonate can build up to a con-

FIG. 20-1. *Neisseria gonorrhoeae,* cross section, electron micrograph. (×100,000.) (Courtesy Technical Information Services, State and Community Services Division, Center for Disease Control, Health Services and Mental Health Administration, Department of Health, Education, and Welfare.)

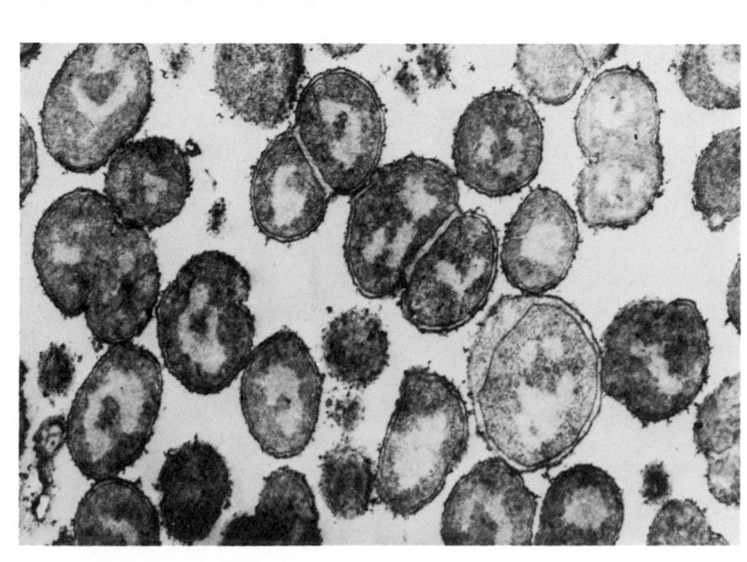

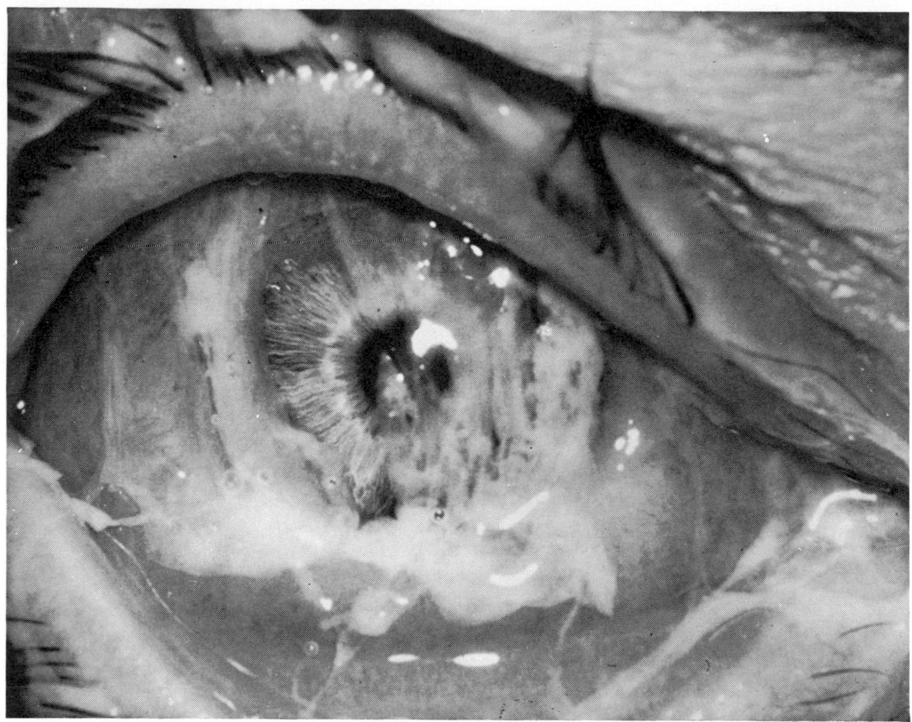

FIG. 20-2. Acute gonococcal conjunctivitis (gonorrheal ophthalmia of adults). Note inflammation, redness, and irritation of conjunctivae and copious amounts of pus. Smear of exudate showed pus cells with intracellular diplococci. Culture grew out gonococci. Patient also had gonococcal pelvic inflammatory disease. (From Donaldson, D. D.: Atlas of external diseases of the eye, vol. 1, St. Louis, 1966, The C. V. Mosby Co.)

siderable pressure behind the lids. If the lids are forced apart, pus may spurt out. The physician and attendants upon these babies must be careful to protect their own eyes. In babies or adults the infection may easily result in blindness or serious sight impairment because of the inflammatory distortion of the structures in the eye. Ophthalmia neonatorum can be prevented by the Credé method, as is required by law for all babies.

From the urethra of the male, gonococcal infection may spread by direct extension to the other parts of the male reproductive system. In the female it may likewise spread to other parts of the tract, especially to Bartholin's glands and the fallopian tubes. The lining of the uterus seems to resist the action of gonococci. Invasion of the fallopian tubes usually occurs with the first or second menstrual period after infection; however, in some cases it may not occur until later. Involvement of the fallopian tubes is associated with considerable distortion and scarring if the disease becomes chronic. Scarring of the urethra in the male may lead to stricture or closure of the urethral lumen at one or more focal points.

Gonococci sometimes pass from the genitourinary tract via the lymphatics or the bloodstream to set up distant sites of infection. Serious consequences may be an endocarditis or meningitis. Gonococcemia is associated with varied skin lesions from which organisms may be identified. An important manifestation of extragenital gonococcal infection is a purulent, destructive arthritis. As the overall incidence of gonorrhea increases, extragenital lesions become prominent.

■ **Sources and modes of infection.** Gonococci are never found outside the human body unless they are on objects very recently contaminated with gonorrheal discharges, and here they live only for a short time. Therefore, gonorrheal infections are practically always spread by direct contact, and the majority of times the mode of contact is sexual intercourse. It is not to be denied that gonorrhea can be transmitted indirectly by contaminated objects, but this is rare.

Gonorrheal ophthalmia of adults is usually accidental. Infection from the genitourinary tract is inadvertently transferred to the eyes by the hands of the same or a different person. Vulvovaginitis in children is spread by the use of common bed linen, bathtubs, toilets, and such. It has been known to result from the use of contaminated rectal thermometers. It usually occurs where children live in closely crowded quarters.

Untreated gonococcal infections tend to become chronic. Females, especially, who are untreated or inadequately treated become infectious carriers for years after manifestations of disease have disappeared.

■ **Laboratory diagnosis of gonorrhea.** The microbiologist has at his command several procedures applicable to the diagnosis of gonorrhea. Smears, cultures, and the oxidase reaction are the *presumptive* tests. To confirm the results of these and to establish the diagnosis of gonorrhea he uses fluorescent antibody technics and carbohydrate fermentation reactions.

Direct smears of genital discharges may be stained with the Gram stain. There are rare exceptions to the rule that for all practical purposes the finding of gram-negative diplococci within the pus cells of an exudate from a genital infection strongly *suggests* gonococci. This is especially true if the exudate is from the male urethra. In the male with typical acute purulent urethritis, the Gram-stained smear of the exudate containing the distinctive intracellular diplococci ordinarily makes the diagnosis. In the female, early in the disease, typical diplococci may be seen in smears of material from Skene's and Bartholin's glands, but even a working diagnosis cannot be made on this basis alone. The reasons for this are several—gram-negative diplococci other than gonococci occur outside cells. Gonococci occur outside cells, singly or in pairs, and gram-positive organisms having the morphology of gonococci occur within cells. All that can be said about gram-negative diplococci found outside cells is that they *may be* gonococci. Very infrequently are gram-negative diplococci other than gonococci found within the pus cells of a genital exudate, but it is possible. The smear prepared from gonorrheal exudates should be quite thin because gonococci react to the Gram stain in an erratic way if the smear is thick and uneven. The microbes usually are not seen in the exudate of chronic gonorrhea.

Cultural methods are of special value in the diagnosis of chronic disease and in the determination of a cure. Cultures are incubated under increased carbon dioxide tension (Fig. 20-3). Enriched media such as chocolate agar, Hirschberg's egg medium, and currently the Thayer-Martin medium* are used to cultivate the delicate gonococcus. Thayer-Martin (T-M) medium contains hemoglobin, certain chemicals to enhance gonococcal growth, and antimicrobial agents to inhibit selectively fungi, gram-positive organisms, and many other gram-negative ones. The plates of T-M medium to be inoculated must be streaked in a specified way so as to spread the organisms out of the associated mucus, which tends to lyse them (Fig. 20-4). A modification of Thayer-Martin

---

*Also called VCN medium for the three antibiotics it contains—vancomycin hydrochloride, colistimethate sodium, and nystatin.

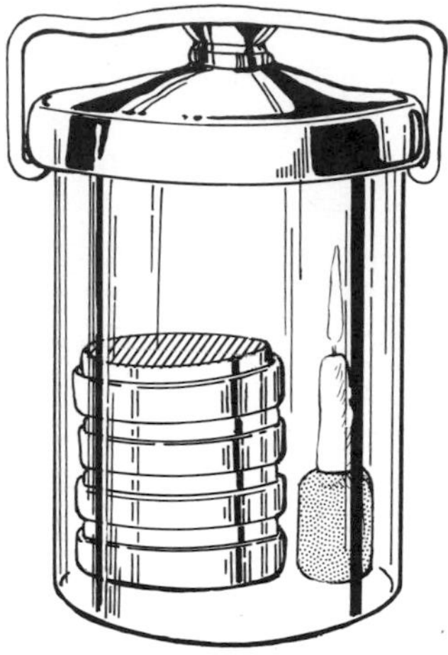

FIG. 20-3. Culture under increased carbon dioxide tension, a simple method of obtaining partially anaerobic conditions for growing cultures. The cultures and a lighted candle are placed in a container and the container is made airtight. The candle burns until oxygen is almost completely exhausted and then goes out.

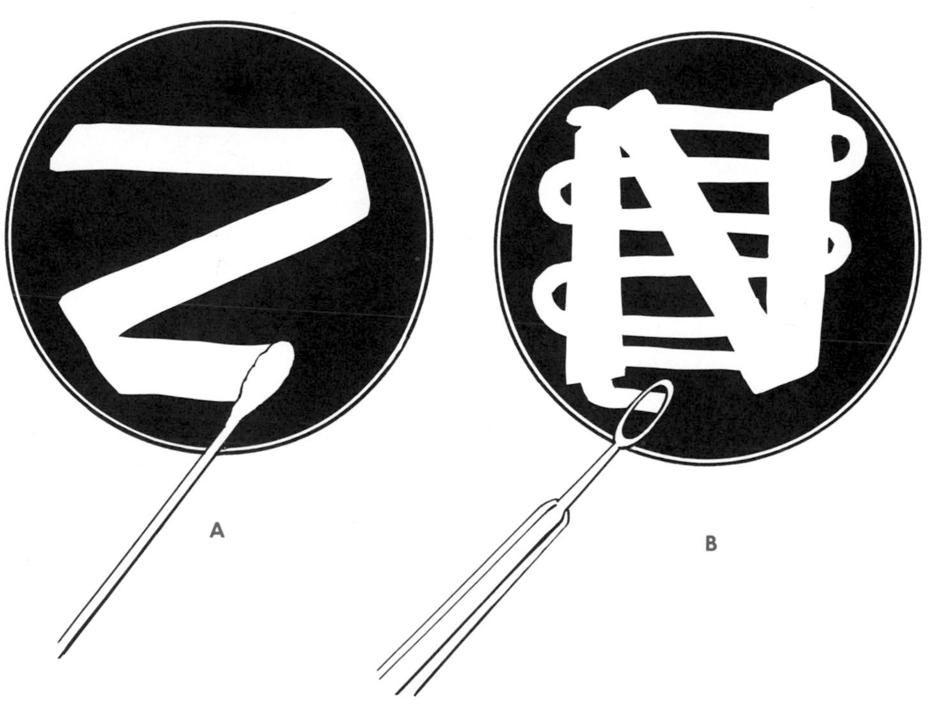

FIG. 20-4. Method of streaking plate for culture of gonococci. **A,** Cervical (or rectal) secretions on swab rolled gently onto plate in **Z** pattern. **B,** Cross streaking done with platinum loop as indicated in **A.**

medium, Transgrow, has been developed so that suspect cultures can be sent into central laboratories. Transgrow comes in a screw cap bottle containing a mixture of air and carbon dioxide. To inoculate Transgrow, one holds the bottle upright to prevent carbon dioxide from escaping and unscrews the cap. The swab with test material on it is rolled over the medium. The cap is returned and the bottle is ready for shipping. The unopened bottle can be incubated directly at the receiving laboratory. Gonococci survive and grow 48 to 96 hours in this medium.

Public health authorities today recommend that suspect material for culture be obtained from the anorectal area and pharynx as well as from the urogenital tract (anterior urethra, endocervical canal). Rectal gonorrhea is easily overlooked. Carefully taken swabs are important in male homosexuals, and in females both cervical and rectal swabs are needed since half the women infected harbor the gonococcus in the rectum. Infection may persist there after it has been eliminated in the cervix.

The next steps in the bacteriologic study of the gonococcus are the determination of the biochemical reactions and identification of the organisms with fluorescein-labeled antiserums. The biochemical reactions of *Neisseria* species discussed in this chapter are given in Table 20-1. The fact that the gonococcus possesses a specific K-type antigen is the basis for the fluorescent antibody test to detect gonococci in direct smears of exudate or in smears made from cultures.

Several serologic tests under study are still in pilot stages. A serologic test to catch the large number of women in the silent, communicable reservoir, although highly desirable, is associated with problems. One of these is that the gonococcus seems to share its antigens with other, harmless neisseriae.

It is never within the province of the laboratory alone to say that a person is cured of gonorrhea, and in many cases the efforts of both laboratory and clinician do not determine whether the disease is eradicated. In dealing with a case of gonorrhea, physicians and nurses should consider its medicolegal potential.

■ **Social importance of gonorrhea.** In the United States today gonorrhea, the single most prevalent bacterial infection, is increasing in frequency to epidemic and pandemic proportions. In a decade the number of reported cases has more than doubled. It is estimated that there are more than 2 million new cases annually, about one, it is said,

TABLE 20-1. CHARACTERISTICS OF THE NEISSERIAE

Organism	Growth on nutrient agar	Growth at 22° C.	Oxidase	Fermentation of sugars with acid only					Pigment production	Habitat
				Glucose	Maltose	Sucrose	Fructose	Mannitol		
Neisseria gonorrhoeae	−	−	+	+	−	−	−	−	−	Found only in genital infections of man
Neisseria meningitidis	−	−	+	+	+	−	−	−	−	Nasopharynx of man
Neisseria sicca	+	+	+	+	+	+	+	−	−	Nasopharynx of man
Neisseria catarrhalis	+	+	+	−	−	−	−	−	−	Nasopharynx of man
Neisseria flava	+	+	+	+	+	−	+	−	Greenish yellow	Nasopharynx of man

every 15 seconds. More than half of the cases occur in young people under 25 years of age. Generally three males are treated and reported for each female, since the disease in males nearly always produces sufficiently noticeable and painful symptoms to motivate them to seek medical attention. Since females are often asymptomatic or relatively so, many of them are reported and diagnosed only because of information from male consorts. The silent reservoir of asymptomatic females (an undetected 80% to 90% of the overall infected) constitutes a primary obstacle to control of disease, and if widespread screening of the female population is not done, it may remain so. It is recommended by some health authorities that a culture for *Neisseria gonorrhoeae* be considered an essential part of prenatal care and that more frequent cultures be taken at the time a pelvic examination is done in any woman.

The most destructive piece of misinformation that has been handed down from generation to generation is that gonorrhea is no worse than a cold. Such a fallacy both underestimates the danger of gonorrhea and creates the impression that colds are of little importance. The medical, social, psychologic, and even medicolegal implications of gonorrhea are sizable.

Gonorrhea is said to be the most common cause of sterility in both sexes. In women, sterility results from occlusion of the fallopian tubes by scar tissue formed during the healing of a gonorrheal salpingitis. In men, it results from occlusion of the vasa deferentia by a similar process of gonorrheal inflammation and healing, with scar tissue formation.

■ **Immunity.** An attack of gonorrhea confers little, if any, immunity to subsequent attacks.

■ **Prevention.** The general public should be warned of the dangers of gonorrhea and the difficulty of its cure. It is unfortunate that use of the wonder drugs has engendered an attitude that the disease is unimportant. The dangers of quack doctors and folk remedies must be stressed. The patient must not allow his discharges to soil toilets or articles used by others. He must be warned of the danger of transferring the infectious material to his eyes by means of his hands.

Prophylactic treatment in newborn babies is as follows: Immediately after birth, the eyelids of the baby are cleansed with sterile water. A different piece of cotton is used for each eye, and the lids are stroked (or irrigated) from the nose outward. Next, the lids are opened, and one or two drops of 1% silver nitrate solution are instilled into each eye, care being taken that the conjunctival sac is completely covered with the solution. After 2 minutes the eyes are irrigated with isotonic saline solution. (A mild irritation of the lining membranes of the eye may be produced but is short lived.) This is known as *Credé's* method, a procedure so important that it is required by law in most of the 50 states. It appears that penicillin and other antibiotics are as effective as silver nitrate in the prevention of ophthalmia neonatorum, but where the law requires silver nitrate, such antibiotics are not used.* Some states have recently passed laws allowing a suitable antibiotic to be used instead of silver nitrate.

Vulvovaginitis in children may be prevented by proper care of bed linen, bathtubs, night clothes, and wash water. All children should be examined for gonorrhea before admission to children's institutions or hospital wards with other children. It

---

*The National Society for the Prevention of Blindness (New York) is currently recommending that the silver nitrate method be continued as standard procedure.

has recently been found that vulvovaginitis in children spontaneously regresses within a few weeks in more than 85% of the cases.

The United States Public Health Service is presently concerned with the development of a vaccine for gonorrhea. Because of the nature of the diseases and the mode of transfer, every person who has gonorrhea should be serologically tested for syphilis.

■ **The VD pandemic.** Every 2 minutes somewhere in the United States a teen-ager contracts a venereal disease. Because of changing factors in our society forms of disease contracted primarily through intimate sexual contact and some that may be contracted this way are on the rise. This is a matter of great concern to public health officials and agencies. Table 20-2 gives a list of diseases so transmitted with causative agents.*

TABLE 20-2. THE VENEREAL DISEASES

Disease		Agent
The "classic five"		
Gonorrhea	Bacterium	*Neisseria gonorrhoeae*
Syphilis	Spirochete	*Treponema pallidum*
Chancroid (soft chancre)	Bacterium	*Haemophilus ducreyi*
Granuloma inguinale	Bacterium (Donovan body)	*Calymmatobacterium granulomatis*
Lymphopathia venereum	Bedsonia	
Newcomers		
Herpes progenitalis	Virus	Herpesvirus hominis type II
Molluscum contagiosum	Virus (molluscum body)	Poxvirus group
Mycoplasmosis	Mycoplasma—T strains	
Candidiasis	Fungus	*Candida albicans*
Trichomoniasis	Protozoan	*Trichomonas vaginalis*
Condyloma acuminatum	Virus	

NEISSERIA MENINGITIDIS (THE MENINGOCOCCUS)

The meningococcus *Neisseria meningitidis*† is the cause of meningococcal septicemia with or without localization in the leptomeninges to produce epidemic cerebrospinal meningitis. Epidemic cerebrospinal meningitis is known also as cerebrospinal fever, spotted fever, and meningococcal meningitis and is one form of acute bacterial meningitis. During World War II more soldiers in the Army of the United States died as a result of meningococcal infection than of any other infectious disease.

■ **General characteristics.** Meningococci are gram-negative diplococci with a striking resemblance to gonococci but with more irregularity in size and shape. They are nonmotile and do not form spores. Capsules are not seen in ordinary smears but may be demonstrated by special methods. Meningococci produce endotoxins liberated when the cocci disintegrate; these toxins are partly responsible for the manifestations of meningitis. In the cerebrospinal fluid meningococci appear both within and without

*With the exception of gonorrhea, these diseases are discussed in subsequent chapters.

†In 1887, Weichselbaum isolated and described this organism from the cerebrospinal fluid of a patient with meningitis.

the polymorphonuclear neutrophilic leukocytes. (It is necessary to know the source of the specimen to determine whether gram-negative intracellular diplococci found in a smear are meningococci or gonococci.)

Meningococci grow best at body temperature in an atmosphere containing 10% carbon dioxide. They do not grow at room temperature. Growth requires special media containing enriching substances such as whole blood, serum, or ascitic fluid. Agar containing laked rabbit blood and dextrose supports the growth of meningococci especially well, as does enriched hormone agar. Different strains of meningococci vary considerably in the ease with which they grow on artificial culture media. Salt has a toxic effect on meningococci and should be left out of media on which they are to be cultivated. Like gonococci, they are oxidase-positive.

Meningococci are such frail organisms that they survive only a short time outside the body. Sunlight and drying kill them within 24 hours. Away from the body they are easily killed by ordinary disinfectants, but in the nasopharynx they are very resistant. They are very susceptible to heat and cold. Meningococci quickly lyse in cerebrospinal fluid removed from the body. Therefore, specimens of suspect fluid should be examined as quickly as possible.

■ **Meningococci and gonococci compared.** Meningococci and gonococci are alike in that (1) both are strict parasites and cause disease only in man, (2) they show little difference in resistance to injurious agents, (3) their distribution in the inflammatory exudate is the same, and (4) they grow on artificial media with difficulty. Their disease-producing properties are comparable. Skin lesions of gonococcemia are similar to those of meningococcemia, and gonococci have caused the Waterhouse-Friderichsen syndrome. It may be said that they are morphologically, physiologically, and immunologically much alike. The immunologic resemblance results from the presence of a common antigenic substance in both organisms, known as substance C. In addition to substance C, substance P is common to meningococci, and they also contain substances that give them group specificity.

■ **Groups of meningococci.** By serologic reactions, including a capsular swelling test similar to that used in typing pneumococci, meningococci are classified into four main groups—A, B, C, and D.* (The United States Public Health Service Center for Disease Control in Atlanta, Georgia, recognizes seven different types of meningococci with specific antigenic and epidemiologic differences—A, B, C, D, X, Y [E], and Z.)

Groups A and C are encapsulated and possess a specific capsular polysaccharide. There is a polysaccharide-polypeptide component in group B but usually no capsule. Group A meningococci are responsible for 95% of cases of epidemic meningitis; group A was especially important during World War II. Groups B and C are the endemic organisms figuring in sporadic outbreaks between major epidemics. Until the mid sixties, group B was the prevalent one. Since then group C is causing an increasing number of epidemics at military bases. There are few meningococci of group D in this country.

■ **Pathogenicity.** The meningococcus is not very pathogenic for the lower animals, and typical epidemic cerebrospinal meningitis occurs only in man. Invasion of the body by meningococci occurs in three steps: (1) implantation in the nasopharynx, (2) entrance into the bloodstream with the production of a septicemia, and (3) localization in the meninges (cerebrospinal meningitis).

---

*Groups A, B, C, and D correspond respectively to groups I, II, II alpha, and IV in the formerly used classification.

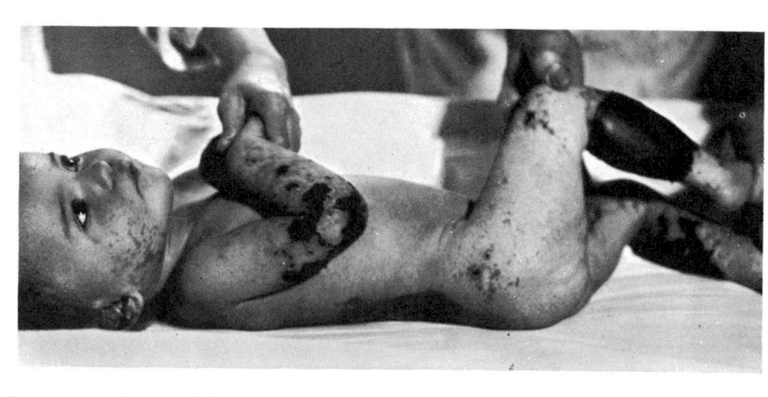

FIG. 20-5. Acute fatal meningococcal infection in a 14-month-old child. Note the vacant stare on the face of the child and the large areas of hemorrhage over the child's body. One day after the onset of sore throat small pinpoint hemorrhages (petechiae) appeared in the skin. Three days later larger areas of skin became hemorrhagic, necrotic, and then gangrenous. Meningitis developed on the twelfth day of illness. (From Top, F. H., Sr., and Wehrle, P. F.: Communicable and infectious diseases, ed. 6, St. Louis, 1972, The C. V. Mosby Co.

In the majority of patients the invasion ends with implantation in the nasopharynx (the carrier state).

About one third of meningococcal infections are septicemias of such severity that without treatment the patient dies before the meningeal infection can occur. Meningococci can proliferate as massively in blood and in tissues as though growing in laboratory broth culture. The events of meningococcal septicemia come together under the designation *Waterhouse-Friderichsen syndrome* (see Fig. 20-5). This is an acute fulminating condition characterized by many small areas of hemorrhage in the skin and within a period of hours death in peripheral circulatory failure.* At autopsy large areas of hemorrhage are seen in the adrenal glands. The blood in meningococcal infections shows a distinct leukocytosis (15,000 to 60,000 per cubic millimeter; neutrophils 85% to 90%).

In only a few patients do bacteria in the bloodstream infect the meninges or other body tissues. The result in the meninges is a purulent inflammation. An exudate composed of leukocytes, fibrin, and meningococci forms in the subarachnoid space, the area between the two meningeal coverings, the arachnoid and the pia mater. Most prominent along the base of the brain, the exudate extends into the cerebral ventricles and down the spinal subarachnoid space. The cerebrospinal fluid may be so purulent that it scarcely runs through the lumen of the spinal puncture needle.

Among the complications of epidemic cerebrospinal meningitis are arthritis, hydrocephalus, otitis media, retinitis, deafness from involvement of the eighth nerve, pericarditis, endocarditis, conjunctivitis, pneumonia, and blindness.

■ **Sources and mode of infection.** Since meningococci are such frail organisms and seldom found outside the body, the source of infection must be a patient or a carrier. The organisms reside in the nasopharynx of both patients and carriers and leave the body in the nasal and buccal secretions. The mode of transfer is by close

*The rapid course of events possible with meningococcal disease—collapse and death within an hour—is terrifying to the social group. The disease upsets people to the point of mass hysteria.

contact, and the organisms enter the body by the nose and mouth. Venereal transmission is reported. When meningococci reach a new host, they localize in the nasopharynx and multiply. Infection traced to articles recently contaminated by infected nasal and buccal secretions seldom occurs.

Meningococcal infections are not highly communicable, and there seems to be a high degree of natural immunity to them. Many persons exposed become carriers but few develop disease. It is estimated that about one carrier in 1000 develops meningococcemia or meningitis. In the absence of immunity, the great number of carriers in the general population would maintain an epidemic at all times. Since meningococcal infection is endemic in all densely populated centers, it may become epidemic under overcrowded conditions, as in army camps and barracks. Overcrowding concentrates the carriers and promotes factors that lower individual resistance. When troops are mobilized, it is one serious disease to which the men seem to be exceedingly vulnerable.

Under ordinary conditions probably from 2% to 5% of the general population are carriers. When epidemics occur, the number of carriers, especially of group A meningococci, increases. In military establishments, 50% of the personnel may become carriers. About two thirds of those who contract meningitis harbor the organisms for a variable time after convalescence. The carrier state lasts about 6 months. However, healthy persons who have never had meningitis but nevertheless harbor meningococci in their nasopharynx play a more important part in the spread of meningitis than do those with the organisms after recovery from the disease.

■ **Laboratory diagnosis.** In order to give the patient the advantage of early treatment, which means so much in meningococcal infections, septicemia should be diagnosed clinically. The diagnosis may be confirmed by blood cultures. A valuable diagnostic adjunct is the demonstration by smears or cultures of meningococci in the hemorrhagic skin lesions associated with septicemia. When cultures for meningococci are made, the specimen must not cool before inoculation onto media warmed to room temperature. The inoculated cultures are promptly incubated.

In epidemic meningitis the cerebrospinal fluid is first turbid and then purulent. In smears of cerebrospinal fluid, the typical arrangement of gram-negative diplococci within the pus cells is sufficient for practical purposes to make the tentative diagnosis of epidemic meningitis. However, cultures must confirm the diagnosis. Because of the tendency of the organisms to undergo autolysis, the specimen must be examined as quickly as possible after it is withdrawn.

Fluorescent antibody technics detect the organisms, and immunoelectrophoresis can be used to test for the presence of meningococcal antigen in cerebrospinal fluid and serum. Group typing is done by immunoelectrophoresis.

■ **Immunity.** That epidemics of meningeal infection are not more common stems from the fact that whereas nasopharyngeal infection is prevalent (low immunity to localization), resistance to invasion of the bloodstream and body tissues is high. Some observers believe that a moderate degree of immunity results from an attack of meningitis. Others believe that no protection is acquired. Immunity seems to increase with age because the disease is more prevalent in children than in adults.

Before the advent of the antimicrobial compounds the mainstay in the treatment of epidemic meningitis was antimeningococcal serum of two types, bactericidal and antitoxic. Although serum therapy reduced the mortality in epidemic meningitis, its results did not equal those now obtained with the antimicrobial drugs. The death rate was 40% to 50%. It is now 5% to 10%.

TABLE 20-3. MICROBES RELATED TO MENINGITIS

	Organism	Comment
A. Primary meningitis:		
1. Common offenders	*Neisseria meningitidis*	Cause of epidemic cerebrospinal fever
	Haemophilus influenzae	Most frequent cause of meningitis in infants and children when meningococci not implicated
	Diplococcus pneumoniae	Common cause of endemic meningitis in adults; less so in children
2. Less common	*Mycobacterium tuberculosis*	Cause of nonpurulent meningitis; infection may be mixed
	Other *Neisseria* species	
	Mima species	Organisms may be confused with meningococcus
	Escherichia coli	Cause of meningitis in newborn infants
	Other coliforms	
	Listeria monocytogenes	
	Streptococcus pyogenes	
	Alkaligenes species	
	Klebsiella-Enterobacter-Serratia organisms	
	Salmonella species	
	Bacterioides species	
	Pseudomonas species	
	Proteus species	
	Nocardia asteroides	
B. Secondary meningitis		
1. Common offenders	*Staphylococcus aureus* *Pseudomonas* species *Escherichia coli* and other coliforms *Proteus* species	Organisms are introduced directly into meninges because of recent trauma, surgical procedure, lumbar puncture, or congenital malformation (for example, spina bifida)
2. Less common	Any pathogenic microbe with capacity to invade human tissue (examples, *Streptococcus pyogenes* and *Pasteurella multocida*)	
C. Meningitis part of generalized infection:		
1. Common offenders	*Neisseria meningitidis*	Part of fulminating meningococcal septicemia
	Diplococcus pneumoniae	
	Mycobacterium tuberculosis	Part of miliary tuberculosis
2. Less common	*Neisseria gonorrhoeae* *Salmonella typhi* *Brucella* species *Staphylococcus aureus* *Bacillus anthracis* *Nocardia asteroides* *Cryptococcus neoformans* (fungus)	Almost any generalized infection with bacteremia may be associated with meningitis

■ **Prevention.** The patient with a meningococcal infection should be isolated until cultures fail to show meningococci in his nasopharynx. All discharges from the mouth and nose and articles soiled therewith should be disinfected. The urine occasionally contains the organisms and therefore should be disinfected. The physician and nurse should use every precaution to avoid infection or the carrier state. The nurse should exercise care lest her hands convey the infection. Dishes used by the patient should be properly sterilized. Persons who have been in close contact with a patient should not mingle with other persons until bacteriologic examination has proved them to be free of meningococci.

General preventive measures include the proper supervision of carriers. The wholesale isolation of carriers does not eliminate infection, and the trend presently is to isolate only those in immediate contact with a patient. Nose sprays are probably of little value. The antibiotic rifampin is used to eliminate the carrier state.

Field testing continues with the expectation of a polyvalent vaccine against groups A, B, and C. Polysaccharide antigens are used. Results are good.

The control of epidemic meningitis is yet to be accomplished. For reasons that we do not understand, it suddenly becomes virulent in a community, attains epidemic form, persists for a time, and then disappears.

■ **Microbial infection of the meninges.** Meningococci are responsible for the *epidemic* form of meningitis. *Nonepidemic meningitis* is caused by other organisms as outlined in Table 20-3. Bacteria reach the meninges (1) by penetrating wounds; (2) by passage from the nasopharyngeal mucosa through lymph spaces; (3) by extension of regional infections in the middle ear, mastoid process, bony sinuses, or cranial bones; and (4) by the bloodstream (bacteremia or septicemia). The diagnosis is made by the demonstration of the causal organisms in smear and culture of cerebrospinal fluid.

OTHER GRAM-NEGATIVE COCCI

■ **Neisseria catarrhalis.** The organism *Neisseria catarrhalis* is a normal inhabitant of the mucous membranes, especially of the respiratory tract. It is important to the microbiologist because it may be confused with the meningococcus or the gonococcus. Like these organisms it is a gram-negative, biscuit-shaped diplococcus and on rare occasions may assume the intracellular position. Differentiation depends upon agglutination tests and the ability of *Neisseria catarrhalis* to grow on ordinary culture media at room temperature.

■ **Other neisseriae.** Other gram-negative diplococci that may lead to errors in microbiologic diagnosis are *Neisseria flava* and *Neisseria sicca,* both normal inhabitants of the pharynx and nasopharynx.

QUESTIONS FOR REVIEW

1. Name and describe the microbe causing gonorrhea.
2. What is the social importance of gonorrhea?
3. Briefly discuss the venereal disease pandemic.
4. List 10 diseases venereally transmitted. Give the causative organism for each.
5. Outline the laboratory diagnosis of gonorrhea.
6. What are the sources and modes of infection in gonorrhea?
7. How is gonorrheal ophthalmia contracted in adults? In infants?
8. What is Credé's method?
9. Compare gonococci with meningococci.
10. Name and describe the microbe causing epidemic meningitis.
11. What are the sources and mode of infection in epidemic cerebrospinal meningitis?
12. State the importance of the carrier in meningococcal infections.
13. Give the nursing precautions in epidemic meningitis.
14. How is the diagnosis of meningococcal infection made in the laboratory?
15. List 10 microorganisms causing meningitis.
16. Explain what is meant by the Waterhouse-Friderichsen syndrome.

REFERENCES. See at end of Chapter 26.

21 The aerobic gram-positive bacilli: diphtheria bacilli, anthrax bacilli

CORYNEBACTERIUM DIPHTHERIAE (THE BACILLUS OF DIPHTHERIA)

Six or seven decades ago diphtheria, or "membranous croup," as it was called, was a major cause of death. Since the causative organism, *Corynebacterium diphtheriae,* was first discovered by Edwin Klebs (1834-1913) in 1883, the epidemiology has become known, methods of producing a permanent immunity have been devised, and the mortality has been reduced from nearly 50% to such a level that death seldom occurs in patients who are adequately treated during the early days of the disease. Few diseases have been so well studied and are today so well understood as diphtheria.

The causative organism belongs to the genus *Corynebacterium* of gram-positive, unevenly staining bacteria with clubbed or pointed ends. (*Coryne* signifies "clubbed.") *Corynebacterium diphtheriae* is often called Klebs-Löffler's bacillus because it was identified by Klebs and first grown in 1884 in pure culture by F. A. J. Löffler (1852-1915). The word *diphtheria* is derived from a Greek word meaning "leather." The disease was so named because of the leathery consistency of the diphtheritic membrane.

■ **General characteristics.** The organism *Corynebacterium diphtheriae* is distinct for its variation in size, shape, and appearance (pleomorphism). The bacteria may be straight or curved and may be swollen in the middle or clubbed at one or both ends. When division occurs, the bacteria remain attached at the point of separation in a V-shaped pattern (snapping). Stained organisms have a granular, solid, or barred appearance, and deeply staining granules (metachromatic granules) are characteristic. Granules at the ends of the bacilli are known as *polar bodies.*

Corynebacterium diphtheriae is a gram-positive, nonmotile, aerobic organism that does not form spores. It grows best at 35° C. on almost all ordinary media, but growth is most luxuriant on Löffler's blood serum and media containing potassium tellurite. Potassium tellurite inhibits the growth of many organisms found in the throat, and colonies of diphtheria bacilli assume a typical appearance on media containing it. The morphologic features of the organisms are best preserved on Löffler's blood serum.

Based upon cultural characteristics, fermentation tests, and immunologic reactions, *Corynebacterium diphtheriae* may be divided into three groups: *gravis* (meaning "severe"), *intermedius* (meaning "of intermediate severity"), and *mitis* (meaning "mild"). Epidemics are most often the *gravis* type, and the manifestations are the most severe. Next in importance is the *intermedius* type. There is thought to be a relation between the type of *Corynebacterium diphtheriae* and its virulence.

For men and some laboratory animals *Corynebacterium diphtheriae* is highly pathogenic. The incubation period of diphtheria is 2 to 5 days.

Diphtheria bacilli are fairly resistant to drying but are easily destroyed by heat and chemical disinfectants. Boiling destroys them in 1 minute. They may remain alive in bits of diphtheritic membrane for several weeks.

■ **Extracellular toxin.** Diphtheria bacilli owe their pathogenicity to their extracellular toxin. They do not invade the tissues but grow superficially in a restricted area, usually on a mucous membrane. The exotoxin liberated is absorbed into the bloodstream and circulated. Because it blocks the cellular synthesis of cytochrome b, an enzyme important in tissue respiration, diphtheria toxin produces cellular injury, systemic disturbances, and degenerative changes in the body organs. Its action is directed toward certain nerves, the heart muscle, the kidneys, and the cortex of the adrenal gland. The features of the disease stem directly from the exotoxin. Diphtheria is a molecular disease.

When diphtheria bacilli are grown in a suitable liquid, the soluble exotoxin is released into the medium. After they are filtered out, the toxin is concentrated to yield the diphtheria toxin of commerce. Such preparations contain 200 to 1000 MLD (p. 597) per milliliter. The toxin can be separated from the medium and purified. It deteriorates with age and is destroyed by a temperature of 60° C.

Not all diphtheria bacilli produce toxin. Toxigenic organisms cannot be distinguished from nontoxigenic ones by microscopic appearance or cultural characteristics. Differentiation is by animal inoculation. A small amount of a liquid culture of the bacilli is injected either beneath or between the superficial layers of the skin of two guinea pigs, *only one* of which has been protected with a dose of diphtheria antitoxin. If the bacilli are toxigenic, a zone of inflammation appears at the site of inoculation in the *un*protected pig. There is no reaction at the site in the protected pig. If the bacilli do not produce toxin, neither pig shows a reaction. This is the *guinea pig virulence test.* Virulence tests may be carried out on rabbits. The technic differs somewhat from that in guinea pigs, but the underlying principle is the same. Eight or ten tests may be carried out at one time on the same rabbit. There is also a cultural method whereby the virulence of diphtheria bacilli may be determined.

■ **Pathogenicity.** Diphtheria is an acute infectious disease induced by the extracellular toxin of *Corynebacterium diphtheriae.* There is a characteristic type of inflammatory change at the site of infection, and systemic disturbances accompany it. Pathologically the type lesion of diphtheria is the *pseudomembrane,* a superficial lesion occurring on mucous membranes (Fig. 21-1). The first stage in its formation is degeneration of the epithelial cells of the affected area. This is followed by an abundant fibrinous exudation onto the surface. As the fibrin precipitates, it entraps leukocytes, red blood cells, bacteria, and dead epithelial cells to form a thick, tough membranelike structure anchored to the underlying tissues. If the pseudomembrane is pulled off, a raw bleeding surface is left, but a new pseudomembrane soon forms. In the absence of antitoxin it persists for 1 week or 10 days and then disappears. It may obstruct breathing, and in some patients the tracheotomy or intubation is required to prevent suffocation.

The most frequent sites for pseudomembrane formation are the tonsils, pharynx, larynx, and nasal passages. The diphtheritic membrane usually begins on one or both tonsils and spreads to the uvula and soft palate. Less often, diphtheria attacks the vulva, conjunctiva, middle ear, and skin, and infects wounds. Skin infections were notable among soldiers of World War II. Organisms other than *Corynebacterium diphtheriae* may

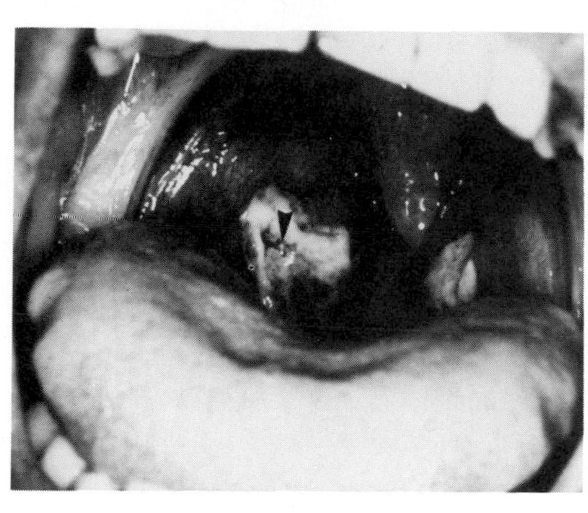

FIG. 21-1. Diphtheria in a 43-year-old man; fourth day of disease. Arrow points to membrane over tonsillar area of throat. (From McCloskey, R. V., Eller, J. J., Green, M., Mauney, C. U., and Richards, S. E. M.: The 1970 epidemic of diphtheria in San Antonio, Ann. Intern. Med. **75**:495, 1971.)

produce pseudomembranes, and in a few patients with diphtheria pseudomembranes do not form.

Diphtheria occurs in three clinical forms: (1) *faucial,* in which the membrane appears on the tonsils and spreads to other parts of the pharynx, (2) *laryngeal,* in which the membrane may easily cause suffocation, and (3) *nasal,* in which the membrane is often not associated with severe disease because the toxin is poorly absorbed by the lining of the nose. Bronchopneumonia is an important complication.

The really serious effects of diphtheria stem from the action of the toxin. The damage it does to the heart may precipitate heart failure, even after other manifestations of the disease have subsided, and sometimes in comparatively mild cases. Sudden death has been reported several weeks after apparent recovery. The degeneration of peripheral nerves induced by toxin leads to late paralysis, particularly of the soft palate.

■ **Sources and modes of infection.** The sources of infection in diphtheria are persons with typical cases, persons with mild undetected cases, and carriers. Ordinarily the bacteria enter and leave the body by the same route, the mouth and nose. They may be transferred directly from person to person by droplets expelled from the mouth and nose or indirectly by cups, toys, pencils, dishes, and eating utensils contaminated

TABLE 21-1. BACTERIOLOGIC DIAGNOSIS OF DIPHTHERIA

Organism	Morphology	Growth on blood tellurite	Hydrolysis of urea
Corynebacterium diphtheriae	Pleomorphic	Gray to black colonies	−
Corynebacterium pseudo-diphtheriticum (example of a diphtheroid)	More uniform	Opaque grayish	+

by the buccal or nasal secretions. The direct method of transfer is the more important.

Nasal diphtheria is an important source of infection because it is often overlooked. Diphtheria bacilli can be spread from skin infections appearing to be less serious ailments. In recent epidemics such skin sources figured significantly. A few milkborne epidemics have been reported. Such are usually caused by the contamination of the milk by a person working in the dairy who is either a carrier or who may have a mild case of disease. He transfers his buccal or nasal discharges by his hands. The rather prevalent idea that cats convey diphtheria is not true.

Over the last decade or so in this country diphtheria has continued to occur in the economically depressed areas of the South where living conditions are crowded, medical care lacking, hygiene poor, and immunization inadequate.

DIPHTHERIA CARRIERS. In about one half of patients with diphtheria, the bacilli leave the body within 3 days after the membrane disappears, and in four fifths of the patients they have disappeared within 1 week, but sometimes they persist and the patient becomes a carrier. Also, certain contacts of a diphtheria patient or a carrier become carriers themselves without contracting the disease. It has been estimated that from 0.1% to 0.5% of the population are carriers of virulent diphtheria bacilli. The percentage increases during epidemics and in crowded communities during cold weather.

Most carriers harbor the bacilli a short time only (from a few days to a few weeks), but a few harbor them permanently and remain carriers in spite of intensive treatment. Because a high percentage of the organisms with the morphology and cultural characteristics of diphtheria bacilli in normal throats are nontoxigenic and, therefore, not dangerous, virulence tests should always be done on diphtheria bacilli from suspected carriers. If a suitable antibiotic, such as erythromycin or penicillin, is given in conjunction with antitoxin during the acute stage of the disease and continued during the period of convalescence, the carrier rate is reduced.

■ **Bacteriologic diagnosis.** The laboratory diagnosis of diphtheria is made by culture of the organism (see Table 21-1). Some of the material from the pseudomembrane is removed with a sterile swab; a slant of Löffler's serum or tellurite medium is inoculated and incubated from 12 to 14 hours. Smears made from the culture are stained with Löffler's methylene blue or one of the special stains for diphtheria bacilli. A diagnosis may be made sometimes by finding the bacilli in smears made directly from the pseudomembrane, but failure to find them in no manner indicates that the patient does not have the disease. Antiseptics, gargles, or mouthwashes must not be used before cultures are taken because they may prevent the proper growth of the bacteria. Antibiotics given 5 to 7 days previously may inhibit bacterial growth

Reduction of nitrate	Fermentation of carbohydrates				Toxigenicity
	Trehalose	Glucose	Maltose	Sucrose	
+	−	+	+	−	+
+	−	−	−	−	−

in culture. If possible, cultures should be taken before antibiotics are given. Care should be taken to bring the swab in contact only with the site of disease and cultures should be made from both throat and nose.

In a patient with an ulcerative or membranous inflammation of the throat both cultures for *Corynebacterium diphtheriae* and smears for the organisms of Vincent's angina should be made. The two conditions may be easily confused. If only a culture is made, an infection with the organisms of Vincent's angina could be missed because these organisms do not grow in cultures, and diphtheria could be missed if smears alone are examined. Moreover, the two diseases can exist together.

The finding of diphtheria bacilli in the throat does not necessarily mean the patient has diphtheria; he may be a carrier. Remember that membranous infections of the throat may be caused by organisms other than *Corynebacterium diphtheriae* and that nonmembranous infections with *Corynebacterium diphtheriae* occasionally occur. If streptococcal infection is suspected, a blood agar plate is inoculated.

DIPHTHEROID BACILLI. The diphtheroid bacilli bear a close microscopic resemblance to *Corynebacterium diphtheriae* and are sometimes confused with it on throat smears. They are quite numerous and have been isolated from many different sources, such as the skin, nose, throat, urethra, bladder, vagina, and prostate gland. They reside in soil and water as saprophytes, do not produce toxins, and are nonpathogenic for man except under exceptional circumstances. They have recently been reported as rare causes of wound infections, meningitis, osteomyelitis, and hepatitis.

■ **Immunity.** Immunity to diphtheria results from the presence of diphtheria antitoxin in the blood. Newborn babies of immune mothers receive a passive immunity because of the transfer of antitoxin from the maternal to the fetal circulation via the placenta. In breast-fed infants this immunity is augmented by antibodies in the milk of the mother. This immunity is usually lost by the end of the first year, and from this time to the sixth year most children are susceptible. Minor infections with *Corynebacterium diphtheriae* reestablish an immunity during late childhood.

About 60% of adults are immune, a figure once considerably higher. The reduction in the incidence of diphtheria, the decreased likelihood of minor infections, and the institution of vaccination during childhood (which does not give the permanent immunity that repeated minor infections do), has meant a decrease in adult immunity. The presence of from $1/500$ to $1/250$ unit of diphtheria antitoxin per milliliter of blood renders a person immune.

The *Schick test* purports to determine whether a person has sufficient diphtheria antitoxin in his blood for immunity. One fiftieth MLD of diphtheria toxin (0.1 ml. of diluted toxin) is injected into the skin of one arm of a subject and a control dose of toxoid into the other. The skin sites on both arms are read 4 days later. In principle, if the subject's blood contains sufficient antitoxin for protection against diphtheria, no reaction occurs (negative test). An insufficient amount of antitoxin (positive test) is indicated by the appearance within 24 to 36 hours of a firm red area, 1 to 2 cm. in diameter, persisting 4 or 5 days. In the past the Schick test has been used to detect persons lacking immunity and to determine the efficacy of active immunization once given. Since the highly purified adult type of toxoids have become available, the need for routine Schick testing is largely eliminated. Active immunization is desirable, and difficulties encountered with the Schick test are bypassed.

An attack of diphtheria is usually followed by a fair degree of immunity, which may be temporary or permanent. In a few patients immunity does not seem to develop. Most carriers of virulent diphtheria bacilli are immune to the toxin.

■ **Prevention and control of diphtheria.*** All persons ill of diphtheria should be isolated, and neither they nor their close contacts should be released until it is proved they harbor no virulent diphtheria bacilli in their noses and throats. The mouth secretions and all objects so contaminated should be disinfected. The patient's eating utensils should be boiled. The nurse must be careful that she does not contaminate her hands with the mouth and nose secretions from the patient so as to infect herself. Persons known or presumed to be susceptible who contact a diphtheria patient should receive diphtheria toxoid (for active immunization) and antibiotics. If they cannot be seen daily by the physician, they may be temporarily immunized with 10,000 units of diphtheria antitoxin intramuscularly. Remember that the administration of antitoxin can set the stage for an anaphylactic reaction should another agent containing horse serum be given that person.

The general measures that cause the greatest reduction in the incidence of diphtheria are (1) the detection and treatment of carriers and (2) the production of an active immunity in all susceptibles—children under 6 years of age and all older children and adults, especially physicians and nurses not previously immunized.

BACILLUS ANTHRACIS (THE ANTHRAX BACILLUS)

Known since antiquity, anthrax is an acute infectious disease caused by *Bacillus anthracis.* It is primarily a disease of lower animals, especially cattle and sheep, but it is easily communicated to man. Horses and hogs may become infected. Anthrax is worldwide in distribution but, with the exception of certain restricted areas, is unusual in the United States.

In man anthrax presents in two forms: *external* (malignant pustule or carbuncle and anthrax edema) and *internal* (pulmonary anthrax or woolsorters' disease and intestinal anthrax). External anthrax is the more common. Because of the fever and enlargement of the spleen that accompany the disease, anthrax is often called splenic fever. The incubation period for external anthrax is 1 to 5 days.

In the body of an animal dead of acute disease, there is little change other than the dark blood and a swollen spleen.

■ **General characteristics.** The organism *Bacillus anthracis* is quite large, in fact one of the largest pathogens. It forms spores and is gram-positive. The swollen and concave ends of the bacilli give the chains of bacilli the appearance of bamboo rods. The organism grows in the presence of oxygen and also in its absence. It is the only sporeforming aerobic pathogenic bacterium and, unlike most sporeforming aerobic bacteria, is nonmotile. Since spores are not formed in the absence of oxygen, they are not formed in the animal body. In the blood and tissues of infected animals, the anthrax bacillus is surrounded by a capsule. Growth is luxuriant on all ordinary culture media. Spores retain their vitality for years and are extremely resistant to heat and chemical disinfectants in the ordinarily used concentrations. They are fairly susceptible to sunlight.

*For immunization procedures in diphtheria, see pp. 598, 599, 607, 617-619, and 627.

Bacillus anthracis is of historic interest because it was the first pathogenic organism to be seen under the microscope, the first one proved to be the cause of a specific disease, the first one to be grown in pure culture, and the organism that Pasteur used in his classic experiments on artificial immunization.

■ **Modes of infection.** Animals usually become infected via the intestinal route while grazing in infected pastures. Buzzards carry the infection long distances and contaminate soil and water with organisms on their feet and beaks. Dogs discharge spores in their feces after eating the carcasses of infected animals. This disease is sometimes transmitted from animal to animal and from animal to man by greenhead flies and houseflies.

In man, anthrax is primarily an occupational disease confined to those who handle animals and animal products, such as hair and hides. Imported animal products are of special danger. The most common route of infection is through wounds or abrasions of the skin (cutaneous route of infection). In numerous cases of anthrax, infection of the face by the bristles of cheap shaving brushes has been reported. The pulmonary form of anthrax is caused by inhalation of dust containing the spores (respiratory route of infection) and endangers those who handle dry hides, wool, or hair. The intestinal form of anthrax is acquired by ingestion of infected milk or insufficiently cooked food (intestinal route of infection). The bacilli leave the body in the exudate of the local lesion (malignant pustule) and in the sputum, feces, and urine.

■ **Prevention.** Patients with anthrax should be isolated. The dressings of external lesions should be burned, and the person doing the dressings should wear gloves. The feces, urine, sputum, and other excreta should be disinfected at once to prevent the formation of spores. The disinfectant should be strong and applied for a long time. Five percent phenol is probably the best one. The local lesions of anthrax should not be traumatized as by squeezing because this may cause the bacilli to invade the bloodstream.

The possibility of confusing a malignant pustule with an ordinary boil is so important that a description of the former is not out of place. A malignant pustule begins as a small, hard, red area; a small vesicle soon develops in the center. The area increases in size, vesicles develop at the periphery, and the surrounding tissues become swollen. The center of the lesion softens, and a dark eschar forms. Pain is not present. The draining lymph nodes are swollen.

Infected animals should be separated from the herd. The bodies of animals dead of anthrax should be burned if possible. Their blood should not be allowed to escape, and their bodies should not be opened for autopsy except by an experienced veterinarian because the bacilli form spores when exposed to the air. If the body is buried, it should be packed with lime and buried at least 3 feet in the ground. When infection occurs in a herd, the pasture must be changed, and the herd quarantined. A sharp watch should be kept for new cases, and the quarantine should not be lifted until 3 weeks after the last case. Milk from infected herds should not be used. Hides, hair, and shaving brushes should be sterilized. If the soil becomes contaminated, it remains infectious for years. The best soil disinfectant is lye. Anthrax spores from buried animals have been brought to the surface by earthworms.

Prophylactic vaccination of animals against anthrax must be carried out with vigor. Cattle and sheep may be actively immunized with a vaccine that contains living but attenuated bacteria. Dead bacteria are without effect. A therapeutic serum has been prepared by immunizing horses against anthrax bacilli. An anthrax vaccine is now available for human beings.

QUESTIONS FOR REVIEW

1. What does the word *diphtheria* mean?
2. Name and describe the bacterium causing diphtheria.
3. What are the three clinical forms of diphtheria?
4. Explain the pathogenicity of diphtheria bacilli. What is the nature of the disease?
5. Briefly discuss virulence tests used in diagnosis of diphtheria.
6. Characterize the pseudomembrane found in diphtheria.
7. State the role of the nurse in the prevention and control of diphtheria.
8. What are diphtheroid bacilli? What is their importance?
9. How is diphtheria transmitted?
10. Outline the laboratory diagnosis of diphtheria.
11. Name the clinical types of anthrax. How is each contracted?
12. Why is the anthrax bacillus of historic interest?
13. Describe the organism causing anthrax.
14. Discuss the proper disposal of the bodies of animals dead of anthrax.

REFERENCES. See at end of Chapter 26.

22 The anaerobic gram-positive bacilli: clostridia

CLOSTRIDIUM TETANI (THE BACILLUS OF TETANUS)

Tetanus or lockjaw is a disease, worldwide in distribution, with manifestations in the central nervous system produced by the potent exotoxin of *Clostridium tetani.* There are classical muscular spasms in tetanus, involving the face, neck, and other parts of the body, that are provoked by the slightest sort of stimulation (noise, movement, touch). These are very painful. Severe ones may compromise respiration. The name *lockjaw* comes from the muscular spasms that rigidly close or lock the jaws together. The corners of the mouth are turned up and the eyebrows are peaked so that a characteristic facial expression is produced and referred to as *risus sardonicus* (sardonic grin).*

*The spasms are characterized by a violent rigidity, usually sudden in onset but sometimes working up to a crescendo, with every single voluntary muscle in the body thrown into intense, painful tonic contraction. The eyes start, the jaw clenches, the tongue is bitten, the neck is retracted, the back arched, and opisthotonus is extreme. Often there is a muffled inspiratory cry, as the diaphragm contracts and draws air through the apposed vocal cords. Finally laryngeal spasm becomes complete, the chest fixed, and respiration ceases from muscle spasm. At the same time there is a gross outpouring of secretion, with profuse perspiration and foaming at the mouth. (From Ablett, J. J. L.: Tetanus and the anaesthetist; review of symptomatology and recent advances in treatment, Brit. J. Anaesth. **28**:288, 1956.)

TABLE 22-1. BACTERIOLOGIC PATTERNS OF CLOSTRIDIA

Organism	Growth in litmus milk	Growth in cooked meat	Gelatin liquefaction
Clostridium tetani	Soft clot	Gas, slow blackening	+ (blackening)
Clostridium perfringens	Stormy fermentation	Gas	+ (blackening)
Clostridium novyi	Acid, no clot	Gas	+
Clostridium septicum	Slow clot, gas	Gas	+ (gas)
Clostridium botulinum	Acid	Gas, blackening, digestion	+

The tetanus bacilli do not invade the body but remain at the site of infection, where they elaborate their powerful poison. The disease comes from infection of a wound and is not transmitted from person to person.

■ **General characteristics.** The tetanus bacillus is an anaerobic sporeformer. Biologically a saprophyte, it probably never infects a wound that does not contain dead tissue. *Clostridium tetani* stains with ordinary stains and is gram-positive. The vegetative forms are slightly motile; the sporulating forms are nonmotile. The spore is situated in one end, giving the bacillus an appearance of a roundheaded pin or a drumstick. Growth is fairly luxuriant on all ordinary media if anaerobic conditions are maintained (Table 22-1). The optimum temperature for growth is 37° C. In both cultures and infected wounds the presence of certain aerobic bacteria accelerates the growth of tetanus bacilli. Young cultures contain numerous vegetative bacilli. Old cultures are composed chiefly of sporulating organisms. The bacilli are seldom demonstrated in cultures from infected wounds because so few are present.

The vegetative bacilli are no more resistant to destructive influences than are other vegetative organisms. The spores, however, are very resistant. They withstand boiling for several minutes, pass through the intestinal canal unaffected, and when protected from the sunlight remain infective for years.

■ **Distribution.** Tetanus bacilli are common residents of the superficial layers of the soil. Since they are normal inhabitants of the intestines of horses, cattle, and other herbivora, they are always found where manure is freely used as fertilizer, and barnyard soils are heavily contaminated. They are found in the intestinal canal of about 25% of human beings. Tetanus spores may be spread over a wide area by flies and high winds.

■ **Extracellular toxin.** If it were not for their toxin, infection with tetanus bacilli would be without effect. Tetanus toxin is one of the most powerful poisons known, second only to that of botulism in potency. This explains why a comparatively few bacilli at the site of infection can induce such profound changes. The toxin of tetanus

Nitrate reduction	Fermentation			Hemolytic	Spores	Motility
	Glucose	Lactose	Sucrose			
−	−	−	−	+	Round, terminal	+
+	+	+	+	+	Rare	−
−	+	−	−	+	Ovoid, eccentric	+
+	+	+	−	+	Ovoid, eccentric	+
−	+	−	−	±	Ovoid, eccentric	+

is a simple protein of a single antigenic type with a molecular weight of 67,000. It is a neurotoxin, that is, one with a specific affinity for the tissues of the central nervous system. From the site of infection the toxin travels to the central nervous system along the axis cylinders of the motor nerves, where it is specifically and avidly bound to the gray matter. Its action is central and on the peripheral nerve endings.

Besides the important neurotoxin, the tetanus bacilli elaborate small amounts of a toxic substance that destroys red blood cells and injures the heart. Taken orally, tetanus toxin is harmless.

■ **Pathogenicity.** Tetanus in man is highly fatal. Horses, cattle, sheep, and hogs may become infected.

Tetanus occurs in two clinical forms: one form associated with a short incubation period (from 3 days to 2 weeks), an abrupt onset, severe manifestations, and a high mortality; the other characterized by a longer incubation period (from 4 to 5 weeks), less severe manifestations, and a lower mortality. Rapidly fatal tetanus is more likely to follow wounds about the head and face, that is, near the brain.

■ **Pathology.** The action of the toxin of *Clostridium tetani* is directly on the central nervous system. The characteristic muscle spasms, from which the disease got its name, and the convulsive seizures originate in the central nervous system, and the impulses are transferred to the muscles by the motor nerves.

In this disease process, the pathologic changes are functional, not structural. There is no typical lesion. Even after death, no organic lesions are seen and the cerebrospinal fluid is normal.

■ **Sources and modes of infection.** Tetanus is practically always caused by spores introduced into a wound. Whether a given wound is complicated by tetanus depends upon the type of wound, the chance of contamination with tetanus spores, the presence of dead tissue, and secondary infection. Deep puncture wounds are dangerous because they provide anaerobic conditions for the growth of the bacilli. In lacerations, gunshot wounds, compound fractures, wounds containing foreign bodies, and infected wounds, the presence of dead tissue and certain other bacteria favors the growth of tetanus bacilli. If such wounds are contaminated by soil, especially heavily manured soil, the chances of tetanus infection are greatly increased. It is not surprising that war wounds are often followed by tetanus. Rusty nail wounds may be so complicated, not because the nail is rusty but because rusty nails are usually dirty nails and tetanus spores are likely to be in the dirt.

In narcotic addicts injection-related tetanus is a public health problem. The rapidly progressive and highly fatal disease in the drug addict is another, even more terrifying expression of tetanus. It is seen in the large metropolitan centers of the United States, especially in Negro women. The narcotic, usually heroin, has been given by the subcutaneous route of injection. Quinine, the adulterant, favors the growth of the bacilli by promoting anaerobic conditions in the tissues and enhances the disease.

Puerperal tetanus exists in tropical countries. Infection from bacilli inhabiting the intestinal canal rarely occurs after intestinal operations. Tetanus of the newborn, *tetanus neonatorum,* is caused by infection through the navel and is still seen in the United States in certain Negro and Spanish-speaking groups where neonatal medical care remains primitive. At the extremes of life the mortality is very high. Elderly persons are exposed to tetanus-prone injuries in their gardening activities. Although tetanus is more likely to develop under the conditions just enumerated, it may follow trivial wounds inflicted by apparently clean objects.

■ **Prevention of tetanus.*** Any wound suggesting the least danger from tetanus should be treated surgically. Puncture wounds should be widely opened and thoroughly cleansed for two reasons: Tetanus bacilli and other organisms are removed, and oxygen, antagonistic to the growth of tetanus bacilli, is allowed access. Mangled wounds should be thoroughly cleansed and all dead tissue removed. Proper wound care cannot be overstressed, since tetanus organisms do not multiply in a surgically clean, aerobic wound with a good blood supply. Antibiotics help to control associated pyogenic infections that damage tissues. *Clostridium tetani* is sensitive to penicillin, but no amount of antibiotic has any effect at all upon toxin released from the wound site.

Active immunization for every member of the community cannot be too strongly recommended. In special need of protection are those workers in industry or agriculture whose occupation predisposes them to wounds easily contaminated with tetanus bacilli, all allergy-prone individuals, and even recipients of a driver's license. All athletes must be fully immunized. (Football players are suffering more abrasions and minor injuries from sliding and falling on artificial turf than on sod fields, although artificial turf is a less likely source of tetanus spores.)

Active immunization is practiced on a large scale in the armed forces of both the United States and England. That this has been effective is proved by the fact that only 12 cases of tetanus appeared among 2,785,819 hospital admissions of military personnel for war wounds and injuries during World War II. Of the 12 patients, six had not been adequately immunized and two failed to get the booster dose of toxoid at time of injury.

CLOSTRIDIUM PERFRINGENS, NOVYI, AND SEPTICUM (THE CLOSTRIDIA OF GAS GANGRENE)

■ **The organisms.** Gas gangrene is a highly fatal disease caused by the contamination of wounds with one alone or any combination of certain anaerobic, toxin-producing, sporeforming bacteria (Table 22-1). (Less than half the time there is only one organism.) All exist as saprophytes in the soil and as normal inhabitants of the intestinal canal of man and animals. Infection happens when soil contaminated by feces gets into a wound. Most important are *Clostridium perfringens (Clostridium welchii,* Welch's bacillus†), *Clostridium novyi (Clostridium oedematiens),* and *Clostridium septicum* (vibrion septique). *Clostridium sporogenes,* considered a nonpathogenic organism, is often present also. The single most significant of the clostridia is *Clostridium perfringens* related to 75% of all cases of gas gangrene and responsible for practically all instances in civilian life. In a third of cases it is the sole pathogen.

Aerobic pus-producing microbes and certain proteolytic organisms without effect on clean wounds often produce considerable destruction of tissue in wounds infected with gas bacilli.

---

*See also pp. 599, 600, 604, 605, 607, 617-618, and 627.

†William Henry Welch (1850-1934), first dean of the medical school of Johns Hopkins University, was well known for his original research in microbiology and pathology, especially for his studies (in 1892) of *Clostridium perfringens* and its relation to gas gangrene.

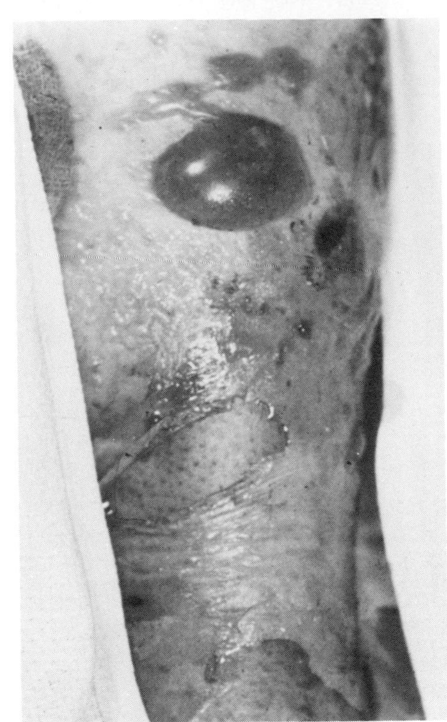

FIG. 22-1. Gas gangrene in a 16-year-old boy complicating compound fractures of leg sustained in motorcycle accident. The organism *Clostridium perfringens*. Note blisters and discoloration. (From Altemeier, W. A., and Fullen, W. D.: Prevention and treatment of gas gangrene, J.A.M.A. **217:**806, 1972.)

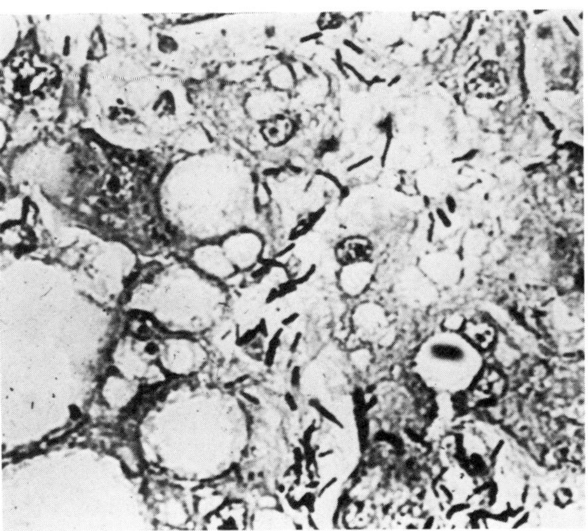

FIG. 22-2. Gas gangrene is Gram-stained tissue section of liver. Large air-filled spaces disrupt structure. Large gram-positive rods are present. Patient had leukemia. (Courtesy Dr. R. C. Reynolds, Dallas, Texas.)

■ **The disease.** The organisms responsible for gas gangrene grow in the tissues of the wound, especially in muscle, releasing exotoxins and fermenting the muscle sugars with such vigor that the accumulated gas pressure tears the tissues apart. The air-filled (emphysematous) spaces of the wound give the name gas gangrene (Figs. 22-1 and 22-2). The exotoxins cause swelling and death of tissues locally, breakdown of red blood cells in the bloodstream, and damage to various organs over the body. The bacteria enter the blood just before death. Clinically there is a profound toxemia.

Clostridium perfringens elaborates several hemolysins and an extracellular proteolytic enzyme, collagenase, that aids the spread of gas gangrene organisms through tissue spaces, since it is active against the fibrous protein supports of the body.

Lacerated wounds, compound fractures, wounds with extensive dead tissue, and war wounds are especially likely to be sites of gas gangrene. Injuries acquired in the vicinity of railroad tracks tend to be so complicated. Clostridial infection rarely complicates gangrenous appendicitis, strangulated hernia, and intestinal obstruction in the abdominal cavity.

Skin samples taken from the thighs, groins, and buttocks of hospitalized patients sometimes yield a heavy growth of *Clostridium perfringens.* To eliminate this kind of contamination, compresses of povidone-iodine or 70% alcohol may be applied. The presence of the organisms of gas gangrene in a wound does not invariably mean that gas gangrene will develop.

■ **Clostridial infection and malignancy.** *Clostridium septicum* is important in wartime gangrene but rarely complicates infection in otherwise healthy civilians. If the person's defenses are weakened, the situation may be reversed. This organism is acquiring some notoriety currently by complicating cases of leukemia and intestinal cancer with clostridial septicemia.

■ **Prevention.*** Prevention of gas gangrene depends upon the proper surgical care of wounds. With established disease, there must be free incision to open the wound as widely as possible, all devitalized tissue must be excised, foreign bodies must be removed, and adequate drainage of the wound instituted. Gas gangrene antitoxin has been prepared against the organisms important in causing gas gangrene, but its efficacy today is challenged. Toxoids for *Clostridium perfringens* and *Clostridium novyi* are experimental.

CLOSTRIDIUM BOTULINUM (THE BACILLUS OF BOTULISM)

■ **Nature of botulism.** Botulism is a specific intoxication caused by the ingestion of foods in which *Clostridium botulinum* has grown and excreted its toxin. The toxin, not the bacilli, is responsible for the disease. Like tetanus, it is a poisoning instead of an infection. The foods most often responsible for botulism are sausage (the disease derives its name from the Latin word, *botulus*, for "sausage"), pork, and canned vegetables such as beans, peas, and asparagus. Cases have been traced to ripe olives and tuna. Most of the foods incriminated have shared one feature: They were processed (improperly) in the home by canning or pickling weeks or months before.

In the last decade there have been 78 outbreaks of botulism in the United States, plus nearly 200 individual cases. The most serious ones of our times, those in 1963, resulted from the consumption of commercially processed foods. From eating such foods as liver paste, tuna fish, salmon eggs, and smoked whitefish products, 46 persons were poisoned, and 14 died of a disease ordinarily considered quite rare.

The disease develops as follows: Since the organism is widely distributed in nature, food is easily contaminated. If it is not canned or otherwise preserved properly, the spores are not destroyed. In the interval between preservation and use, the spores revert to the growing stage. The bacilli multiply and excrete their toxin into the food. (Toxin remains potent in canned foods for 6 months or more.) If the contaminated food is not heated sufficiently for consumption, toxin has not been inactivated. When such food is ingested, the toxin is absorbed through the intestinal wall to exert its effects. Toxic signs usually appear within 24 hours afterwards, consisting of generalized weakness, disturbance of vision (often double vision), thickness of speech, nausea and vomiting, and difficulty in swallowing. There is no fever. Death from asphyxia usually occurs between the third and seventh day. The mortality ranges from 50% to 100%.

Natural food poisoning of this type exists among certain wild animals (if toxin is swallowed with their food). Examples are limberneck of chickens, fodder disease of horses, and duck sickness.

■ **The organism.** The microbe *Clostridium botulinum* is an anaerobic, gram-positive,

*See also p. 599.

sporeforming bacillus (Table 22-1). A common inhabitant of the soil, the usual source from which foods are contaminated, *Clostridium botulinum* is primarily a saprophyte. (Infections of laboratory animals have been caused experimentally.) Generally, the organism is unable to grow inside the body of a warm-blooded animal. *Clostridium botulinum* and its toxin are destroyed in 10 minutes at the temperature of boiling water, but the "hard shell' spores must be held at a temperature of 120° C. (249° F.) for 15 minutes to be killed. Spores can withstand more than 2000 times the radiation lethal to a human being.

The toxin of *Clostridium botulinum* is the most deadly of poisons, the mere tasting of food having been known to cause death. One ounce could exterminate all the people in the United States, and a mere half pound of botulinal toxin could wipe out the population of the world. (Botulinal toxin has been classified as an agent for chemical warfare.) It is 10,000 to 100,000 times more potent than diphtheria toxin or animal venoms. It differs from diphtheria and tetanus toxins in that it causes disease when swallowed and is more resistant to heat. A fast-acting neurotoxin, it affects the central nervous system, paralyzing the muscles of vision, swallowing, and respiration. It dilates blood vessels, and hemorrhages occur in different parts of the body. The toxin's effect is a specific one at the neuromuscular junction. Mental faculties are not impaired, and there are no sensory disturbances.

There are six types of *Clostridium botulinum* from the standpoint of toxigenicity: A, B, C, D, E, and F. All types resist the intestinal juices, and the action of type E toxin can be increased 50 times by trypsin. In man, botulism is almost always the result of the ingestion of types A, B, E, and F toxins. C and D toxins cause disease in animals. Most A and B toxins are found in home-preserved vegetables and fruit (food of plant origin). Recent outbreaks have pointed to the presence of *Clostridium botulinum*, type E, in water and in marine wildlife. Therefore type E is found in fish and marine products. Type E spores can germinate at refrigerator temperatures and form toxin. Recently type F has been reported in home-processed venison jerky.

The laboratory diagnosis of botulism is made by (1) finding the toxin in the patient's serum (toxemia may persist for prolonged periods), (2) isolating the microbes, and (3) identifying the toxin in the food ingested. The mouse is a suitable test animal for inoculation. Being highly susceptible to the effects of the toxin, it succumbs to even the small amounts of circulating toxin in the blood sample received from the patient with the disease. Identification of the toxin as to type in suspected foods may be made by mouse tests.

■ **Prevention.*** Prevention of botulism depends primarily upon (1) heating foods to a temperature of 120° C. for at least 15 minutes in the canning process (or steam cooking under adequate pressure, especially for low acid foods)† to destroy any spores present, (2) boiling canned foods for 15 minutes immediately before they are eaten to destroy the heat-labile toxins, and (3) proper refrigeration of foods after cooking. In the canning of high acid foods, such as tomatoes and fruits, botulinus spores can be killed by a temperature of 120° C. for 15 minutes. For the processing of low acid foods (beets, beans, corn, and meats) steam-pressure methods are preferable.

Do not so much as taste canned food until it has been fully heated. Remember the poison is odorless and tasteless!

---

*See p. 598 for passive immunization.

†If food is boiled, at least 3 hours are required.

If fowls that have been eating discarded food develop limberneck and if the responsible food can be determined, persons known to have eaten the same food should receive botulinus antitoxin. If a case develops among a group of people who have eaten the same food, the members in the group should be given the antitoxin.

Round the clock help—available antisera, consultation, laboratory assistance to establish the diagnosis—can be obtained from the United States Public Health Center for Disease Control in Atlanta, Georgia (telephone: (404)-633-3311).

QUESTIONS FOR REVIEW

1. Name and describe the causative organisms for tetanus, gas gangrene, and botulism.
2. What is the derivation of the word *botulism?*
3. What is the effect of the tetanus toxin? Of the diphtheria toxin? Of the botulinum toxin? Compare the three as to potency.
4. List the types of wounds likely to become infected with *Clostridium tetani.*
5. List the types of wounds likely infected with the clostridia of gas gangrene.
6. What measures, other than immunization, help to prevent tetanus and gas gangrene?
7. Briefly characterize the disease tetanus.
8. What are the circumstances producing botulism? What are the main preventive measures?

REFERENCES. See at end of Chapter 26.

23 The gram-negative enteric bacilli

The *enteric bacilli,* a large group making up the family Enterobacteriaceae, include several closely related genera of short, aerobic, nonsporeforming, gram-negative rods that inhabit or produce disease in the alimentary tract. Here are the nonpathogenic bacteria that normally inhabit the intestinal canal and the highly pathogenic bacteria that invade and injure it. Heading any list of enteric pathogens are the consistent trouble-makers, the *Salmonella* and *Shigella* species. Normal residents include many species in a number of genera. If given the right opportunity, these supposedly benign bacteria can be awesome in their change of character.

Today, members of this family of bacteria are notorious as causes of urinary tract infection and are recovered from a variety of clinical specimens taken from diseased foci other than in the gastrointestinal tract.

The enteric microorganisms of this chapter are defined broadly in the classification most widely used in microbiologic laboratories. The starting point is Bergey's Family IV, Enterobacteriaceae. The breakdown as proposed by W. H. Ewing and approved by the subcommittee of the American Society for Microbiology is as follows:*

	Genera
Tribe I: *Escherichieae*	*Escherichia (Escherichia coli,* including *Alkalescens-Dispar)* *Shigella*
Tribe II: *Edwardsielleae*	*Edwardsiella*
Tribe III: *Salmonelleae*	*Salmonella* *Arizona*† *Citrobacter* (including Bethesda-Ballerup)
Tribe IV: *Klebsielleae*	*Klebsiella* *Enterobacter* (formerly *Aerobacter,* including *Hafnia*) *Pectobacterium*‡ *Serratia*
Tribe V: *Proteeae*	*Proteus* *Providencia*

---

*Note departure from Bergey's classification on p. 6.

†The term *paracolon bacilli* was formerly used for enteric bacilli resembling *Escherichia coli* but fermenting lactose much more slowly. They were placed in the genus *Paracolobactrum* and separated into the Bethesda-Ballerup group, the Arizona group, the Providence group, and the Hafnia group. In the new scheme, paracolon bacilli are regrouped and the term *paracolon* is discarded.

‡A genus of plant pathogens (soft rot coliforms), it is not implicated in human infection. As might be expected, it is the only genus in Family IV liquefying sodium pectate.

TABLE 23-1. SIMPLIFIED SCHEME FOR THE SEPARATION OF GRAM-NEGATIVE ENTERIC BACILLI

Step 1. Fermentation of lactose

	IMViC reaction*			
	Indole	Methyl red	Voges-Proskauer	Citrate
Rapid—coliforms				
Escherichia coli	+	+	—	—
Klebsiella	—	—	+	+

Slow—coliforms

Acid only	Elaboration of DNase†
Serratia	+
Enterobacter	—
Acid with gas	Growth in KCN medium
Citrobacter	+
Arizona	—

None—*Salmonella*
Shigella
Proteus
↓

Step 2. Fermentation of glucose

Acid with gas—*Proteus*, *Salmonella* → **Step 3. Splitting of urea (urease test)**

Proteus	+
Salmonella	—

Acid only—*Salmonella typhi*, *Shigella* → **Step 4. Production of hydrogen sulfide**

Salmonella typhi (motile)	+
Shigella (nonmotile)	—

None—*Alcaligenes* (nonpigmented saprophyte in feces; rarely a feeble pathogen; not classified in Enterobacteriaceae)

*The pattern of the IMViC reaction is important to the differentiation of the coliform bacilli (normal residents of the enteron). It refers to four biochemical color reactions. The first, "I," is the production of indole from tryptophane (pink or red color). The next two, the methyl red reaction, "M," and the Voges-Proskauer reaction, "Vi," indicate a difference in the fermentation of glucose in a special test medium after 2 to 4 days' incubation. If the test organism ferments glucose with the accumulation of acid end products, the methyl red indicator gives a red color (pH below 4.5), a positive test. If glucose has been fermented with the accumulation of neutral end products, the Voges-Proskauer test is positive, a red-orange color. This reaction shows the presence of acetylmethylcarbinol. The fourth reaction, "C," refers to the utilization of Simmons citrate agar and indicates that the test organism can use the citrate as its sole source of carbon.

†Deoxyribonuclease.

Enteric bacilli may be studied on the basis of disease production, physiologic behavior, and immunologic reactions. The practical identification of the enteric gram-negative bacilli, however, involves the use of an elaborate array of biochemical reactions. To demonstrate biologic activities of test organisms is usually a complex ma-

TABLE 23-2. BIOCHEMICAL PATTERNS FOR SALMONELLA SPECIES

Organism	Production of hydrogen sulfide	Reduction of nitrate	Production of indole	Liquefaction of gelatin
Salmonella typhi	+	+	−	−
Salmonella paratyphi	−	+	−	−
Salmonella schottmuelleri	+	+	−	−
Salmonella hirschfeldii	+	+	−	−
Salmonella typhimurium	+	+	−	−
Salmonella choleraesuis	Variable	+	−	−
Salmonella enteritidis	+	+	−	−
Salmonella gallinarum	Variable	+	−	−
Salmonella pullorum	+	+	−	−

neuver in the clinical laboratory. In a greatly simplified scheme, Table 23-1 presents key reactions for the bacteriologic separation of some of the enteric organisms to be discussed.

SALMONELLA SPECIES*

■ **General characteristics.** The genus *Salmonella*† comprises the causative organisms of the salmonelloses. It includes some 250 species of motile enteric bacilli that are found everywhere that there are animals and man. Of the more than 1200 strains known, all can produce disease in their natural hosts.

Although the members of the different species are similar in morphology, staining reactions, and cultural characteristics, they can be separated by fermentation reactions‡ (Table 23-2) and agglutination tests. Serologic typing is the important last step in their bacteriologic identification. More than 800 serotypes have been identified. The immunologic classification of *Salmonella* is complex but important in tracing epidemiologic patterns of disease. Salmonellae are susceptible to the action of heat, disinfectants, and radiation. At the temperature of the refrigerator, they do not multiply but remain viable.

■ **Sources and modes of infection.** Since only a few species, such as *Salmonella typhi*, are indigenous to man, most infections are the result of species from lower animals reaching man. In nature, salmonellae occur in the intestinal tract of man and animals. Therefore they must be spread by feces either directly or indirectly from one host to another. Excreta contain bacilli because the infected host has obvious disease or in-

*The *Salmonella* genus was named for Daniel Elmer Salmon (1850-1914), American veterinary pathologist who first isolated the organisms in 1885.

†Dr. Ewing recognizes only three species in genus *Salmonella*, namely *Salmonella choleraesuis* as type species, *Salmonella typhi* as unique pathogen of man, and *Salmonella enteritidis* to comprise all other serotypes and serobiotypes. By this schema, *Salmonella enteritidis* would now be *Salmonella enteritidis* ser Enteritidis, and *Salmonella typhimurium* would be *Salmonella enteritidis* ser Typhimurium. Since this system is not yet widely accepted, we shall continue with traditional species names.

‡In keeping with their role as intestinal pathogens, salmonellae do not ferment lactose.

Fermentation of carbohydrates

Glucose	Lactose	Maltose	Sucrose	Mannitol	Dulcitol
Acid	–	Acid	–	Acid	–
Acid, gas	–	Acid, gas	–	Acid, gas	Variable
Acid, gas	–	Acid, gas	–	Acid, gas	Variable
Acid, gas	–	Acid, gas	–	Acid, gas	Acid, gas
Acid, gas	–	Acid, gas	–	Acid, gas	Variable
Acid, gas	–	Acid, gas	–	Acid, gas	Variable
Acid, gas	–	Acid, gas	–	Acid, gas	Variable
Acid	–	Acid	–	Acid	Acid
Acid, gas	–	Acid, gas	–	Acid, gas	–

apparent infection or is a carrier. The principal vehicle is water, milk, or food contaminated by feces. The most dangerous link in a chain of infection is the food handler who is a carrier.

If the food ingested contains only a small number of salmonellae, the acid in the stomach of the host destroys them readily. If the food is heavily contaminated, some of the bacilli escape the effects of the gastric acid to enter and injure the small bowel.

■ **Pathogenicity.** Infection with *Salmonella* is salmonellosis, the major site of which is the lining of the intestinal tract. Because of their toxic properties, every known strain of *Salmonella* can cause any one of three types of salmonellosis: (1) acute gastroenteritis of the food infection type, (2) septicemia or acute sepsis similar to pyogenic infections, and (3) enteric fever such as typhoid or paratyphoid fevers.

Most of the time, however, *Salmonella* infection is acute gastroenteritis. This condition is characterized by fever, nausea, vomiting, diarrhea, and abdominal cramps. Sometimes gastroenteritis is the forerunner of septicemia. Once the bacilli have spread, they can cause a wide range of lesions in different parts of the body. Examples are abscesses of various organs, arthritis, endocarditis, meningitis, pneumonia, and pyelonephritis. Salmonelloses may be severe and death dealing, or they may be mild, even inapparent.

With regard to their natural host, salmonellae are divided into the following categories: (1) those that primarily affect man, (2) those that primarily affect the lower animals but may cause disease in man, and (3) those that affect lower animals only.

Salmonella typhi (the typhoid bacillus)

The organism *Salmonella typhi** causes typhoid fever, which as the natural infection occurs only in man.

■ **General characteristics.** A short, motile, nonencapsulated bacillus, *Salmonella typhi* grows luxuriantly on all ordinary media. It grows best under aerobic conditions but may grow anaerobically. The temperature range for growth is from 4° to 40° C., and the optimum, 37° C. It ferments glucose, with the production of acid but not gas. Typhoid bacilli can survive outside the body. They live about 1 week in sewage-

*Also known as *Salmonella typhosa, Eberthella typhosa, Bacterium typhosum,* and *Bacillus typhosus.*

contaminated water and not only live but multiply in milk. They may be viable in fecal matter for 1 or 2 months. Their pathogenic action is related to endotoxins.

■ **Portal of entry.** Any person who contracts typhoid fever has swallowed typhoid bacilli. If the bacilli enter the body by a route other than the alimentary tract, infection does not occur. The incubation period is from 7 to 21 days.

■ **Pathogenicity.** Typhoid fever is an acute infectious disease with clinically continuous fever, skin eruptions, bowel disturbances, and profound toxemia. Except in the first few days, leukopenia is always present in uncomplicated cases of typhoid fever, probably because typhoid bacilli depress the bone marrow. Leukocytosis in the course of the disease suggests a complication.

Typhoid bacilli first penetrate the intestinal lining to attack the lymphoid tissue in the wall. Peyer's patches in the small bowel bear the brunt of the attack. From the intestinal wall the bacilli pass to the mesenteric lymph nodes draining the area, which they colonize. Passing on to the thoracic duct, they enter the bloodstream. Here many bacilli are lysed. Liberation of endotoxins brings about the manifestations of the disease. The bacilli that escape bacteriolysis tend to localize in the gallbladder, bone marrow, and spleen and have disappeared from the bloodstream beyond the first week of illness. The rose spots seen on the skin of the abdomen early in the illness contain many bacilli. As the disease progresses, the bacilli confine their activities to the intestinal wall.

Pathologically, typhoid fever is a granulomatous inflammation that involves the lymphoid tissue of the body. The hyperplastic lymphoid masses swell and plateau above the level of the intestinal mucosa. In most cases necrosis and sloughing take place, leaving an oval ulcer of varying depth. If a blood vessel undergoes necrosis, hemorrhage may occur, and if the ulcer extends deep enough, perforation of the intestinal wall occurs. The mesenteric lymph nodes are swollen and may suppurate. The hyperplastic spleen is soft and enlarged two or three times its normal size. The liver and kidneys show degenerative changes, and the gallbladder may be inflamed. Toxic injury to the myocardium may precipitate acute heart failure and death.

Fully three fourths of the deaths in typhoid fever come from some complication. *Hemorrhage* occurs most often during the third week; the passing of a tarry stool or clotted blood may be the first sign. *Perforation* of the bowel is usually single but may be multiple and occurs most often in the lower 18 inches of the small intestine. The mortality is very high. *Cholecystitis* may follow the disease, and typhoid bacilli may be found in the gallbladder years later.

■ **Portal of exit.** Typhoid bacilli leave the body via feces and urine and are found therein usually after the second week of illness.

■ **Spread.** The fundamental basis of every typhoid infection is the same: *Typhoid bacilli from the feces or urine of a carrier or a person ill of typhoid fever have reached the mouth of the victim.*

Typhoid carriers are of two types: fecal and urinary. In the more common fecal carriers the bacilli multiply in the gallbladder and are excreted in the feces. (Infection of the gallbladder with stagnation of bile predisposes to the formation of gallstones.) In urinary carriers the organisms multiply in the kidney and are excreted in the urine. From 40% to 45% of patients become convalescent carriers for 3 to 10 weeks. About 5% carry the bacilli for 1 year, and about 2% become permanent carriers. Women are carriers more often than men are. The average carrier is female and more than 40 years of age.

The most dangerous factor in the spread of typhoid fever is the carrier food-handler who prepares foods that are served raw. Carriers contaminate their fingers with their discharges and then contaminate food with their fingers. It is hard to prevent the carrier state, and treatment for it can be difficult. Removal of the gallbladder is curative in selected cases.

With improvements in sanitation, widespread epidemics of typhoid fever from contamination of water supplies with sewage have become almost unknown. Strict laws regulate the cultivation of oysters and other shellfish eaten raw to eliminate contamination of these foods with sewage-laden water.

■ **Laboratory diagnosis.** The laboratory offers four procedures for the diagnosis of typhoid fever: (1) blood culture, (2) the Widal test, (3) stool culture, and (4) urine culture. Blood cultures are positive during the first week in 75% to 80% of the patients. The percentage falls until not more than 10% of the patients have positive blood cultures by the fourth week. The diagnosis of infection with *Salmonella typhi* is made with the isolation of the organism.

After 7 to 10 days of infection, agglutinins against typhoid bacilli appear in the patient's blood, increasing during the second and third weeks of the disease. They may persist for weeks, months, or years after the patient recovers. Agglutinins also appear in the blood after typhoid vaccination and are usually found in the blood of carriers.

When an animal is immunized with typhoid bacilli and certain other actively motile organisms, two types of agglutinins are formed. One, acting on the flagella of the bacterium, is the *H (flagellar)* agglutinin. The other, acting on the body of the bacterium, is the *O (somatic)* agglutinin. The *O* agglutinins are thought to appear earlier in the disease and to indicate actual infection with typhoid bacilli or *closely related organisms.* If they are present in the serum of persons vaccinated it is in small quantities and only for a short time. In addition to H and O antigens typhoid bacilli responsible for active disease or the carrier state contain a third antigen, the *Vi (virulence)* antigen. Vi antibodies are not thought to occur significantly after vaccination. *Salmonella* species other than *Salmonella typhi* possess O and H antigens; some possess Vi antigens. The species may be divided into groups on the basis of their O antigens and further subdivided on the basis of their H antigens.

The Widal test* is used to detect the presence of these agglutinins. It is said to be positive in 15% of typhoid patients during the first week of illness, with the incidence of positive tests rising to 90% or more during the third week. However, there are considerable difficulties in interpretation of the Widal test. Many authorities are thoroughly disillusioned with it.

Stool and urine cultures are of value in the detection of carriers and indicate when a given patient ceases to be a source of infection. Typhoid bacilli may be cultivated from the stool in 50% of the cases by the third week. Positive urine cultures are obtained at some time during the course of the disease in 25% to 50% of the patients. Convalescent carriers excrete the bacilli in their feces or urine during convalescence only, but permanent carriers continue to excrete them.

■ **Immunity.** In about 98% of the cases an attack of typhoid fever renders the patient immune for the remainder of his life. The production of an artificial immunity by the

---

*The agglutination test, devised in 1896 by Fernand Widal (1862-1929), French clinician and bacteriologist.

administration of typhoid vaccine has been one of the most important factors in the control of typhoid fever.*

■ **Prevention.** The prevention of typhoid fever is twofold—community and personal. Community prevention means those measures taken by the community as a whole to prevent the spread of disease among its members; personal prevention means those measures taken to prevent the spread of infection from a person ill of the disease. The most important factors in the community program are (1) a supply of clean pasteurized milk, (2) a pure water supply, (3) the efficient disposal of sewage, (4) the proper sanitary control of food and eating places, (5) the detection and isolation of carriers, especially food handlers, (6) the destruction of flies, and (7) vaccination.†

Personal prevention depends upon isolation of the patient. Isolation does not mean merely putting the patient in a room and shutting the doors but must include closure of all the routes by which bacteria may be transmitted to others.

Preferably the patient is hospitalized and kept there until he is no longer infectious. A nurse attending a patient with typhoid fever should consider every secretion and excretion of the patient to be a living culture of typhoid bacilli and should use every means to protect herself, members of the patient's family, and people in the community. She should have nothing to do with the preparation of food, and her hands should be disinfected after each contact with the patient or anything that either the patient or his secretions have touched. Feces and urine can be disinfected with chlorinated lime, 5% phenol, or 2% Amphyl (see also p. 274). Sputum should be received on disposable paper tissues and burned. Linen and bedclothes should be sterilized. Food remains should be burned and dishes boiled. The patient's bath water can be sterilized with chlorinated lime. The sickroom should be screened, and flies that accidentally gain access to it should be killed. All rugs, curtains, and similar fabric materials should be removed from the room. Pets should not be allowed to enter. Disinfection should continue through convalescence, and no patient should be discharged as cured until repeated cultures of feces and urine have failed to show typhoid bacilli.

That the procedures just outlined, together with vaccination, have materially reduced the incidence of typhoid fever is proved by the following facts: In 1900 the death rate in the United States from typhoid fever was 35.9 per 100,000 persons; today around 100 cases are reported annually. That universal vaccination alone will materially reduce the incidence of typhoid fever is proved by the experience of the United States Navy during the years 1911, 1912, and 1913, a period in which no revolutionary developments in sanitation took place. In 1911 there were 361 cases of typhoid fever per 100,000 men. In 1912 universal vaccination of naval personnel was put into effect. In 1913 the rate had fallen to 34 per 100,000, a reduction of more than 90%.

The paratyphoid bacilli

The paratyphoid bacilli are so called because they have been isolated from paratyphoid fever in man, a disease resembling typhoid fever but milder in its mani-

---

*For immunization procedures against typhoid fever, see pp. 607, 625, 627, and 629.

†Currently the United States Public Health Service is *not* recommending typhoid immunization within the United States. Selective immunization may be carried out if there is close contact with a known typhoid carrier, if there is an outbreak of typhoid in the community, or if travel is planned to an endemic area. The Public Health Service Advisory Committee does not indicate typhoid vaccination for persons going to summer camp or for those surviving a flood disaster.

festations and shorter in duration. Although associated with enteric fever, each of the three causes other forms of salmonellosis.

The three are *Salmonella paratyphi, Salmonella schottmuelleri,* and *Salmonella hirschfeldii,* formerly known respectively as the paratyphoid bacilli A, B, and C. Infections with *Salmonella paratyphi* occur almost exclusively in man; *Salmonella schottmuelleri* occasionally causes infections in lower animals. *Salmonella hirschfeldii* is rarely found in the United States; infections are frequent in Eastern Europe and mice are said to be its natural host. *Salmonella paratyphi* resembles *Salmonella typhi* more closely than does *Salmonella schottmuelleri,* but infections with the latter are more prominent.

The mode of infection, the sources, laboratory diagnosis, nursing precautions, and prevention are the same as those in typhoid fever. The immunity produced by a paratyphoid infection or vaccination is uncertain.

Other salmonellae

Among the many salmonellae that primarily affect the lower animals but may cause disease in man are *Salmonella typhimurium* (cause of a typhoidlike disease of mice; may cause gastroenteritis in man); *Salmonella choleraesuis* (once thought to be the cause of hog cholera; can cause septicemia in man); and *Salmonella enteritidis,* known also as Gärtner's bacillus (found in hogs, horses, mice, rats, and fowls; causes infections in man).

Organisms primarily affecting animals are *Salmonella gallinarum* (cause of fowl typhoid; differs from other *Salmonella* species in that it is not motile) and *Salmonella pullorum* (variant of *Salmonella gallinarum,* causing white diarrhea in chicks). Attacks of enteric infection are on occasion traced to salmonellae from unusual sources. Recently salmonellae were demonstrated in the small turtles given as pets to children who subsequently developed salmonellosis. Contaminated baby chicks have been a similar source for small children.

SHIGELLA SPECIES (THE DYSENTERY BACILLI)

Dysentery (Greek *dys,* 'painful'; *enteron,* 'intestine') is a painful diarrhea accompanied by the passage of blood and mucus and associated with abdominal pain and constitutional symptoms. It may occur as a primary or a secondary disease. When primary, it may be of protozoal, bacterial, or viral origin. Protozoal dysentery caused by an ameba is amebic dysentery (p. 511). Dysentery caused by members of the *Shigella* genus is known as bacillary dysentery. Bacillary dysentery may be epidemic or endemic, is usually acute but may be chronic, and is an important form of summer diarrhea of infants.

■ **General characteristics.** The *Shigella* are classified into four species, each of which is divided into several types: *Shigella dysenteriae* (seven types), isolated by Shiga in the Japanese epidemic in 1898; *Shigella flexneri* (six types), isolated by Flexner in the Philippine Islands in 1900; *Shigella boydii* (eleven types); and *Shigella sonnei* (six types). *Shigella flexneri* and *Shigella sonnei* are worldwide in distribution. In recent years infection with *Shigella dysenteriae* has been limited to the Orient.

Dysentery caused by the Shiga bacillus *(Shigella dysenteriae)* is much more severe than is that from the other organisms because this bacillus produces a powerful exotoxin-like substance in addition to an endotoxin. There is no evidence, however, that

TABLE 23-3. BIOCHEMICAL REACTIONS OF SHIGELLA SPECIES

Organism	Production of hydrogen sulfide	Liquefaction of gelatin	Reduction of nitrate	Production of indole	Fermentation of carbohydrates				
					Glucose	Lactose	Sucrose	Mannitol	Dulcitol
Shigella dysenteriae	−	−	+	−	Acid	−	−	−	−
Shigella flexneri	−	−	+	+	Acid	−	−	Acid	−
Shigella boydii	−	−	+	Variable	Acid	−	−	Acid	Variable
Shigella sonnei	−	−	+	−	Acid	−	Acid	Acid	−

this substance is actively excreted by the bacterial cell. Rather, it seems to be liberated by the disintegration of the cell. As a neurotoxin the exotoxin-like substance acts upon the nervous system, causing paralysis in the host, and the endotoxin irritates the intestinal canal.

The dysentery bacilli are gram-negative, nonspore-bearing rods that grow on all ordinary media at temperatures from 10° to 42° C. but best at 37° C. They are aerobic and facultatively anaerobic. Unlike most other members of the enteric group, they are nonmotile. Table 23-3 gives biochemical reactions for *Shigella* species.

■ **Pathogenicity.** Dysentery is a human disease, and natural infections of the lower animals do not exist. The incubation period is 1 to 7 days. Epidemic dysentery is primarily an intestinal infection. Unlike typhoid bacilli, the organisms do not invade the bloodstream and are seldom if ever found in the internal organs or excreted in the urine. They are excreted in the feces.

Pathologically, bacillary dysentery is recognized as diffuse inflammation with ulceration of the large intestine and sometimes the lower portion of the small intestine. Early in the disease shallow ulcers or extensive raw surfaces that may be covered by a pseudomembrane form in the mucous membrane. In mild cases healing is complete, but with severe cases there is extensive scarring.

■ **Mode of infection.** The mode of infection is practically the same as for typhoid fever; the bacilli enter the body of the victim by way of the mouth, having been transferred there from the feces of carriers or patients. Contaminated food, water, fingers, or other objects are the vehicles of spread. Contact with persons who have symptomless infections is especially important in the spread of bacillary dysentery. Flies are an important factor.

■ **Laboratory diagnosis.** The only practical method of laboratory diagnosis is the cultivation of the organisms from the stool. Unfortunately this can be done only during the first 4 or 5 days of the disease.

■ **Immunity.** An attack probably confers some degree of immunity, but the same person has been known to have two episodes in a single season.

■ **Prevention.** Bacillary dysentery may be checked by the sanitary measures that control typhoid fever, but it remains ready to rise in epidemic form when people are crowded together in unsanitary conditions. The feces and everything contaminated by the patients should be handled exactly as for typhoid fever. Food or milk should not be carried from the premises, and persons attending the patient should not handle food for others. The patient should not be dismissed as cured until repeated feces cultures have failed to show the causative organism. After the patient has recovered, the sickroom should be thoroughly cleaned.

Preventive vaccination is not yet available. An oral attenuated vaccine against *Shigella flexneri* is being tested. One such is a "mutant hybrid," produced by "mating" *Escherichia coli* with a nonvirulent strain of *Shigella flexneri.* There is no serum therapy.

ESCHERICHIA COLI AND RELATED ORGANISMS (THE COLIFORM BACILLI)

■ **General characteristics.** Besides *Escherichia coli,* coliform bacteria comprise members of the genera in the following groups: (1) *Klebsiella-Enterobacter-Serratia,* (2) *Arizona-Edwardsiella-Citrobacter,* and (3) the "Providence" group.

As normal inhabitants of the intestinal tract,* these microbes share certain traits. They are gram-negative, short rods that do not form spores. They grow at temperatures from 20° to 40° C. but best at 37° C. A few species are surrounded by capsules. Most are motile. They are not so susceptible to the bacteriostatic action of dyes as are many other bacteria. Therefore, media for the isolation of coliform bacilli can contain inhibitory dyes. Table 23-4 presents bacteriologic features of coliforms.

In a sanitary water analysis the presence of coliform bacilli indicates fecal pollution of the water supply being tested.

■ **Pathogenicity.** Coliform bacilli usually do not penetrate the intestinal wall to produce disease unless one of the following supervenes: (1) the intestinal wall becomes diseased, (2) the resistance of the host is lowered, or (3) the virulence of the organisms increases greatly. Under one of these conditions, coliforms may pass to the abdominal cavity or enter the bloodstream. Once outside the intestinal canal and in the tissues of the body, their virulence is remarkably enhanced (Fig. 23-1).

Among the diseases that these bacilli may cause are pyelonephritis, cystitis, cholecystitis, abscesses, peritonitis, and meningitis. They play a part in the formation of gallstones; they are found in the cores of such stones. In the peritonitis occurring after intestinal perforation, the coliform group is joined by other organisms, such as streptococci and staphylococci. From any focus of inflammation coliform organisms may enter the bloodstream and produce a septicemia.

■ **Escherichia coli.** An important member of the coliform group and long known as the colon bacillus because of its natural habitat in the large bowel, *Escherichia coli* is unabashed as an opportunist when out of its natural setting. It is one of the most frequent causes of pyelonephritis and urinary tract infections and is an important cause of epidemic diarrhea in nurseries for newborn infants. The immunologic pattern for *Escherichia* is as complex as that for *Salmonella.* Immunologic subdivision is made on the basis of O (somatic) antigens, K (capsular) antigens, and H (flagellar) antigens. At least 137 serologic types are known, of which 11 have been correlated with infantile diarrhea.

INFANTILE DIARRHEA. Infectious diarrhea is an extremely important cause of infant death. In underdeveloped areas of the world over 5 million infants succumb to

*In the adult, 30% of the dry weight of the feces is made up of bacteria.

TABLE 23-4. BIOCHEMICAL REACTIONS OF THE COLIFORMS AND PROTEUS BACILLI

Organism	Production of indole	Methyl red test	Voges-Proskauer reaction	Production of hydrogen sulfide	Liquefaction of gelatin (22° C.)
Coliforms:					
Escherichia coli	+	+	−	−	−
Klebsiella species	−	−	+	−	−
Enterobacter species	−	Variable	Variable	−	Variable
Serratia marcescens	−	−	+	−	+
Arizona species	−	+	−	+	Slow
Edwardsiella tarda	+	+	−	+	−
Citrobacter species	−	+	−	+	−
Providencia species	+	+	−	−	−
Proteus bacilli:					
Proteus vulgaris	+	+	−	+	+
Proteus morganii	+	+	−	−	−

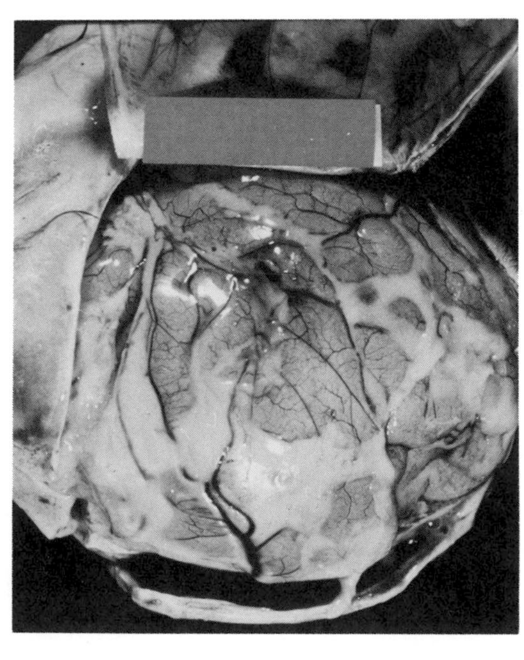

FIG. 23-1. *Escherichia coli* as cause of meningitis in 24-day-old premature infant. Note immature brain with coating of pus over outer surface and prominent vascular markings. Skull flaps are laid back; brain is seen in cranial cavity.

Use of Simmons' citrate	Splitting of urea	Fermentation of carbohydrates				
		Glucose with gas	Lactose	Sucrose	Mannitol	Dulcitol
−	−	+	+	+	+	Variable
+	+	+	+	+	+	−
+	−	+	Variable	+	+	−
	(Variable)			(Variable)		
+	Variable	+	Very slow	+	+	−
+	−	+	Slow	−	+	−
−	−	+	−	−	−	−
+	Variable	+	+	Variable	+	+
+	−	Variable	−	Variable	Variable	−
Variable	+	+	−	+	−	−
−	+	+	−	−	−	−

it each year. Infantile diarrhea including summer diarrhea is especially prevalent during the hot weather among bottle-fed babies who are reared in unhygienic surroundings, but it is also a problem in modern hospital nurseries. In institutional outbreaks (hospitals and orphanages) half the cases are caused by *Shigella* (especially *Shigella sonnei*), *Salmonella,* and enteropathogenic *Escherichia coli.* Strains of enteropathogenic *Escherichia coli* have recently been shown to release an exotoxin that acts on the wall of the small intestine to cause movement of fluid and electrolytes across the mucous membrane into the lumen. The presence of an "enterotoxin" correlates with severe watery diarrhea. Other causative agents include *Proteus, Pseudomonas, Staphylococcus,* cholera-like vibrios, and enteroviruses. A significant number of cases is related to the enteroviruses.

In the hospital nursery, care in the preparation of infant formulas and the sterilization of bottles and nipples are essential to prevent this disorder, which is uncommon in the breast-fed baby. Fluorescent antibody technics aid in the detection of the serotypes of *Escherichia coli* in stools or rectal swabs from the babies and in tracing carriers among the personnel.

■ **Klebsiella-Enterobacter-Serratia.** The biologic characteristics of *Klebsiella* indicate a relationship to the coliforms; the disease-producing capacities indicate a kinship to organisms of respiratory infections. *Klebsiella* are nonmotile, gram-negative, aerobic organisms surrounded by a broad, well-developed polysaccharide capsule. They are differentiated from *Escherichia coli* biochemically (Table 23-4). The type organism is *Klebsiella pneumoniae* (also known as *Bacillus mucosus capsulatus,* Friedländer's bacillus, and the pneumobacillus). It was discovered by Carl Friedländer (1847-1887) prior to the discovery of the pneumococcus and was thought by him to be the sole cause of pneumonia. Actually, it is responsible for less than 10% of the cases.

A likely hazard to the chronic alcoholic, pneumonia caused by Friedländer's bacillus may be either lobar or lobular and is very severe and frequently fatal. Middle ear infections and meningitis may complicate it. Friedländer's bacillus is responsible for diseases other than pneumonia. One of these is septicemia.

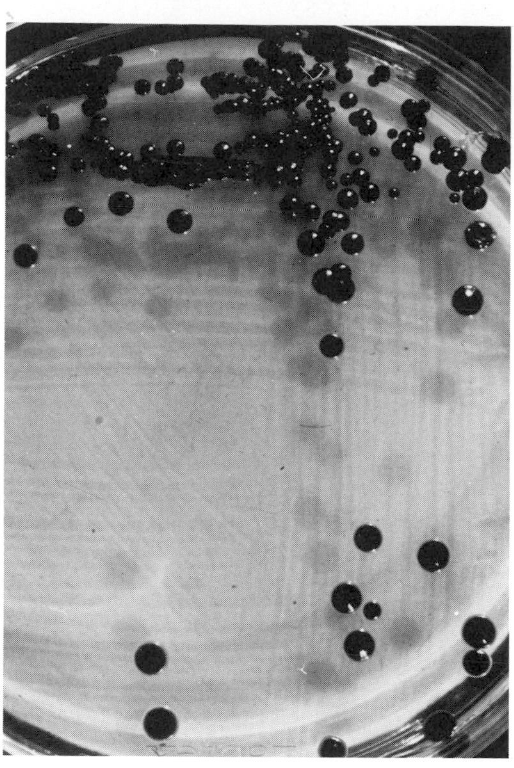

FIG. 23-2. *Serratia marcescens*. Colonies of pigmented strain on nutrient agar plate. Characteristic red suggested by dark appearance.

The genus *Enterobacter* includes the organisms formerly known as *Aerobacter aerogenes*. These bacilli are found in the soil, in water, in dairy products, and in the intestines of animals, including man. Opportunists, they figure in urinary tract infections and in sepsis. The type species is *Enterobacter cloacae*.

Serratia marcescens is a small, free-living, ubiquitous rod celebrated for the red pigment produced in cultures (Fig. 23-2). Since it has emerged as a sometime formidable pathogen, we have found that only a small number of the organisms is chromogenic. The pigment-producing strains do so at room temperature, seldom at incubator temperatures. As is true for other coliforms, infections with this organism (serratiosis) complicate surgical procedures and antibiotic therapy, especially in elderly, debilitated patients in the hospital. The results may be urinary tract infections, pneumonia, empyema, meningitis, or wound infections, to mention but a few. Bacteriologically the organisms in the genera *Klebsiella*, *Enterobacter*, and *Serratia* are similar. When they are grown on a modified deoxyribonuclease (DNase) agar containing toluidine blue as indicator, colonies of *Serratia* may be distinguished from those of the other organisms. A bright pink zone around *Serratia* indicates that this organism elaborates deoxyribonuclease. The clear blue color of the other colonies indicates that these organisms do not.

■ **Arizona-Edwardsiella-Citrobacter organisms.** Formerly known as paracolon bacilli, these organisms ferment lactose very slowly, if at all.

Arizonae are short rods similar to the salmonellae. From 7 to 10 days may be required for them to ferment lactose. The type species is *Arizona hinshawii* (old name: *Arizona arizonae.*) They are found in human infections and cause disease in dogs, cats, and chickens.

Because of their pattern of biochemical reactivity, certain motile rods were separated into a new genus *Edwardsiella*. The type species *Edwardsiella tarda* was given a name implying a kind of biochemical inactivity. *Edwardsiella tarda* has been isolated from man and animals, especially snakes. In fact, snake meat is postulated to be a source of infection. In man the infections are similar to the salmonelloses: (1) bacteremia with coexisting localization away from the gastrointestinal tract, (2) gastroenteritis, (3) typhoid-like disease, (4) wound infections, and (5) the establishment of a carrier state.

Citrobacter are motile rods fermenting lactose. The type species is *Citrobacter freundii* (formerly *Escherichia freundii*). They may be confused with *Salmonella* and *Arizona* organisms. Certain strains possess the Vi antigen found in *Salmonella typhi*. *Citrobacter* is differentiated from *Arizona* and *Salmonella* by its ability to grow in the presence of potassium cyanide (positive KCN test).

■ **"Providence" organisms.** Also known as paracolons, these organisms are free-living, lactose-negative, and biochemically related to *Proteus* bacilli. The type species is *Providencia alcalifaciens*. Ordinarily nonpathogenic, they have been recovered from cases of human diarrhea and urinary tract infections.

PROTEUS SPECIES (THE PROTEUS BACILLI)

The *Proteus* bacilli comprise a genus of motile organisms that resemble and yet differ from the other enteric bacilli in many ways (Table 23-4). They are gram-negative, do not form spores, and are normal inhabitants of feces, water, soil, and sewage. They grow rapidly on ordinary media and, being very actively motile (Fig. 23-3), tend to "swarm" over the surface of cultures on solid media. As a consequence colonies do not remain discrete. Growth from the colonly edge in time extends all over the available surface area as a thin, translucent, bluish film. The bacilli can be seen microscopically to break away, migrating over the agar surface. This property poses problems to the isolation of *Proteus* bacilli from mixed cultures, although generally with care it can be done.

Proteus is pathogenic for rabbits and guinea pigs. The primary pathogenicity of these organisms is slight, but as secondary invaders they are vigorous. The type species is *Proteus vulgaris*, which is remarkably resistant to antimicrobial drugs. *Proteus bacilli* in the stool often increase in number after diarrhea caused by other organisms, especially when antibiotics have been given for the diarrhea.

Proteus bacilli rank next to *Escherichia coli* in importance as a cause of urinary tract infections and pyelonephritis. They are important in wound infections and are a rare cause of peritonitis. *Proteus morganii* is thought to cause infectious diarrhea in infants.

It is a peculiar fact that although *Proteus* bacilli do not cause typhus fever nor act as secondary invaders the serum of a patient with typhus fever will agglutinate certain strains of *Proteus* bacilli. This is the basis of the Weil-Felix reaction for typhus fever.

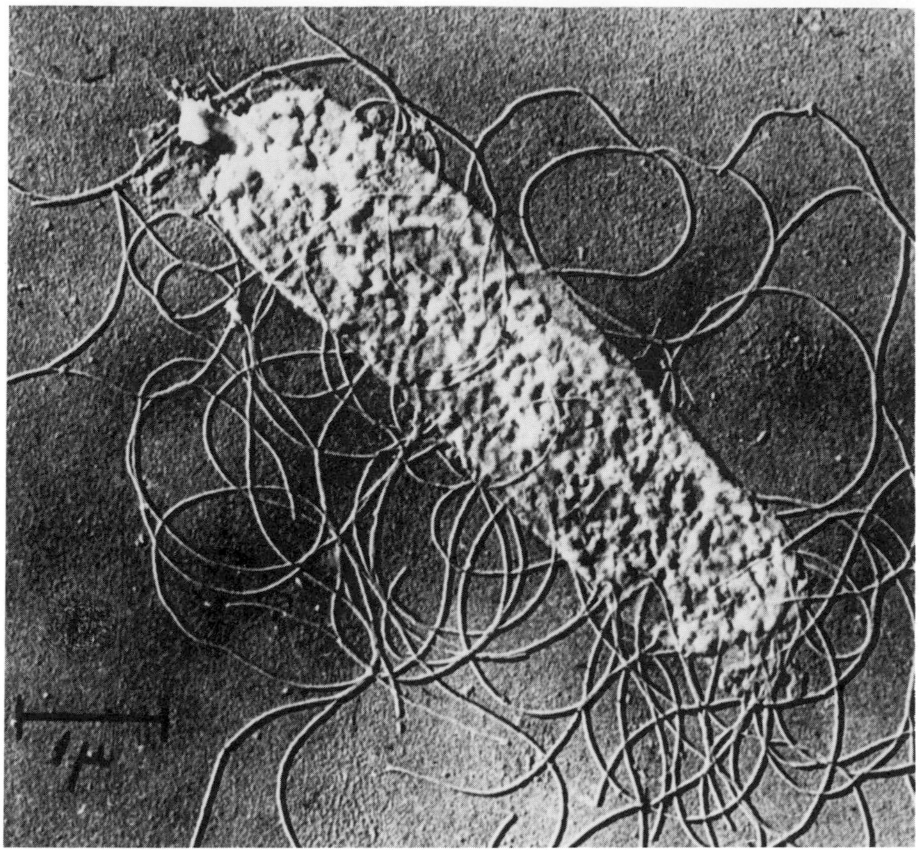

FIG. 23-3. *Proteus vulgaris.* Electron micrograph of actively motile bacterium. Note flagella over the bacterial cell (Courtesy J. B. Roerig Division, Chas. Pfizer & Co., Inc., New York, N. Y.)

ENDOTOXIN SHOCK

Endotoxin shock (bacteremic shock, gram-negative shock, gram-negative sepsis), is a very serious complication of infection with gram-negative enteric bacilli, notably *Pseudomonas aeruginosa** (85% mortality), *Proteus* bacilli, and coliforms of the genera *Klebsiella, Enterobacter,* and *Serratia.* The mortality is over 50%. The dramatic clinical picture is drawn by the precipitous development of chills, fever, nausea, vomiting, diarrhea, and prostration. There is a sharp drop in blood pressure, and the state of shock is profound. The combination of events appears to result directly from the action of the lipopolysaccharide endotoxin of a gram-negative enteric organism that has gained access to the bloodstream. Endotoxemia has been proved to exist.

This kind of bacteremia is especially prone to complicate infections where enteric bacilli are already entrenched, such as those of the genitourinary tract, the lungs, the intestinal tract, and the biliary tract. Instrumentation or an operative procedure on the

*See p. 371.

infected area favors the release of organisms into the bloodstream. In diseases such as leukemia where host resistance is low the organisms flood the bloodstream from no apparent focus.

ENTERIC BACILLI IN HOSPITAL-ACQUIRED (NOSOCOMIAL) INFECTIONS

In a recent study it was estimated that around 6% of all patients admitted to the modern hospital of today would acquire an infection there. Since around 30 million persons are admitted to hospitals annually, this means that more than a million and a half hospital patients develop an infection they did not themselves bring into the hospital.

The predominating organism in hospital-acquired infections has been *Staphylococcus aureus.* (It is still important, accounting for about one-fifth.) Today most observers unanimously agree that the prime offender is gram-negative enteric bacilli as a group (*Escherichia coli,* other coliforms, *Proteus, Pseudomonas*). Such enteric bacilli

FIG. 23-4. Sources of infection in the hospital setting. Note the hazards in the operating room. How many areas in this picture can you identify as potential foci of infection? What precautions would you advise?

easily account for two-thirds of nosocomial infections. Other important agents are enterococci, pneumococci, and streptococci.

Most hospital-acquired infections with enteric bacilli seem to occur in the very young or the very old patient. They tend to complicate chronic debilitating diseases, diseases that alter the patient's resistance to infection, and diseases treated with antibiotics to which the gram-negative bacilli have or easily acquire resistance. Most infections are endogenous (that is, caused by organisms available from a plentiful supply within the gastrointestinal tract, the respiratory tract, or the urinary tract of a given patient).

Around 90% of the time there is a surgical wound infection (especially where the operation was done on the alimentary tract), a urinary tract infection, or respiratory tract infection such as bronchopneumonia.

It is to be reemphasized that varied pathogens are everywhere in the hospital environment, in foci one would least suspect—faucets, soaps, and lotions, for examples. (See Fig. 23-4.) The role of the more sophisticated items indigenous to a hospital is often overlooked—urethral catheters, intravenous catheters, respirators, reservoir nebulizers, and so on. Much current thought is going to measures designed to reduce the hazards.

QUESTIONS FOR REVIEW

1. Classify the enteric bacilli.
2. Comment on the various ways to identify enteric bacilli.
3. Briefly characterize organisms of the genus *Salmonella.*
4. Name the three major types of salmonellosis.
5. Name and describe the organism causing typhoid fever.
6. Draw a diagram tracing the spread of typhoid infection from person to person.
7. Discuss the nursing precautions in typhoid fever and other infections of the alimentary tract.
8. What is the laboratory diagnosis of typhoid fever?
9. Name and describe the organisms causing bacillary dysentery.
10. Which of the *Shigella* causes the most severe disease and why?
11. Briefly comment on the pathogenicity of *Shigella* species.
12. Name the members of the coliform group of microorganisms.
13. Under what conditions can coliform bacilli invade the tissues of the body?
14. Discuss infantile diarrhea. Give causes.
15. What is the basis of the Weil-Felix reaction for typhus fever?
16. Compare *Proteus* bacilli with coliforms.
17. How does enteropathogenic *Escherichia coli* produce disease?
18. Briefly discuss nosocomial infections.
19. What is meant by gram-negative sepsis?
20. Make a list of opportunist pathogens you have studied.

REFERENCES. See at end of Chapter 26.

24 The gram-negative small bacilli

HEMOPHILIC BACTERIA

The hemophilic bacteria (Table 24-1) include several species varying considerably but with unifying features such as (1) their growth depends upon, or is aided by, hemoglobin, serum, ascitic fluid, or certain growth-accessory substances; (2) they are small; (3) they are gram-negative and nonmotile; and (4) they are strict parasites. Two factors in blood aiding in their growth are X and V. Factor X is associated with hemoglobin and withstands the temperature of the autoclave. Since factor V is heat labile, it must be supplied by potatoes or yeasts in culture media. Some of the hemophilic bacteria depend upon both factors for continued multiplication, others depend upon only one, and still others depend on neither. The genus names of these organisms are *Haemophilus* (blood loving) and *Bordetella.*

Haemophilus influenzae (the influenza bacillus)

In 1892, R. F. J. Pfeiffer (1858-1945) described a bacillus, now known as Pfeiffer's bacillus or *Haemophilus influenzae,* which he found in the sputum of patients with influenza. Until the 1918-1919 pandemic this bacillus was regarded as the sole cause of influenza. It is now known that the cause of influenza is a virus. *Haemophilus influenzae* maintains an important position as a secondary invader in influenza and many other respiratory diseases. It may be the primary cause of a purulent meningitis in young children and of subacute bacterial endocarditis. It may be one of the primary

TABLE 24-1. BIOLOGIC FEATURES OF SOME HEMOPHILIC BACTERIA

Organism	Hemolysis	Capsule	Production of indole	Factor required for growth	
				X	V
Haemophilus influenzae	−	+	+ or −	+	+
Haemophilus ducreyi	Slight	−	−	+	−
Haemophilus aegyptius	−	−	−	+	+
Haemophilus parainfluenzae	−	+	+ or −	−	+
Bordetella pertussis	+	+	−	−	−

pathogens in mixed infections of the upper respiratory tract and bronchopneumonia.

■ **General characteristics.** The smallest known pathogenic bacillus, *Haemophilus influenzae* grows best in the presence of oxygen but may grow without it. Nonmotile and nonspore-forming, it does not occur outside the body and is very susceptible to destructive influences. There are two forms, the encapsulated and the nonencapsulated. The encapsulated form has been divided into types a, b, c, d, e, and f. Most infections are caused by encapsulated type b organisms.

Influenza bacilli grow only on special media such as chocolate agar or hemoglobin oleate agar. Both factors V and X are required for growth, which is often more luxuriant about an organism such as *Staphylococcus aureus.* Some staphylococci and certain other bacteria produce factor V. The vigorous growth of one organism in proximity to colonies of another is the *satellite phenomenon. Haemophilus influenzae* forms a toxic substance that resembles an exotoxin though not a true one. It is moderately pathogenic for the lower animals, especially the rabbit.

By technics of immunofluorescent microscopy, the organisms may be identified in cerebrospinal fluid taken from children with meningitis. Diagnostic serologic tests include complement fixation and agglutination.

Haemophilus influenzae is found in the throats of about 30% of normal persons. Adult human beings show bactericidal substances for *Haemophilus influenzae* in their blood; such substances are not found in young children. Therefore infection is much more common and much more serious in children. *Haemophilus influenzae* meningitis, the most frequent form of nonepidemic meningitis, occurs in children between 2 months and 3 years of age in 80% of the cases. Of these cases, 90% are due to type b bacilli, against which an anti–*Haemophilus influenzae* serum has been made in rabbits. Modern methods of treatment have considerably reduced the death rate in influenzal meningitis, which until recent years was uniformly fatal. *Haemophilus influenzae* is also responsible for an obstructive inflammation of the larynx, trachea, and bronchi that begins suddenly with great severity. Breathing is blocked. If the obstruction is not relieved, death may occur in 24 hours.

A vaccine to *Haemophilus influenzae* type b is being field tested.

Bordetella pertussis (the Bordet-Gengou bacillus)

Whooping cough (pertussis), a communicable disease of 1 or 2 months' duration, affects children chiefly. It is a catarrhal inflammation of the respiratory tract with a paroxysmal cough that ends in a whoop. (The whoop may be absent in adults and in very young children.) The incubation period is 5 to 21 days. A widespread and dangerous disease, it is one of the major causes of death in young children; 90% occur in children under 5 years of age. The danger lies in the frequency with which bronchopneumonia, malnutrition, or chronic diseases such as tuberculosis follow in its wake. In the 25% of children who have whooping cough before they are 1 year old death is usually from bronchopneumonia.

The causative organism was first observed in the sputum of patients with whooping cough by Jules Jean Baptiste Bordet and Octave Gengou in 1900. Often spoken of as the Bordet-Gengou bacillus, it is properly known as *Bordetella pertussis.*

In possibly 5% of the patients a disease clinically indistinguishable from whooping cough has been related to an organism other than *Bordetella pertussis* known as *Bordetella parapertussis.*

■ **General characteristics.** A small gram-negative bacillus, *Bordetella pertussis*

(Haemophilus pertussis) closely resembles *Haemophilus influenzae.* It shows polar staining, often forms capsules, is nonmotile, and does not form spores. It grows best in the presence of oxygen but may grow anaerobically. When freshly isolated from the body, all pertussis bacilli are alike, but when cultivated, they resolve into four phases. Phase I represents the freshly isolated and pathogenic organism, phase IV is the completely nonpathogenic form, and phases II and III are intermediate.

Freshly isolated from the body, *Bordetella pertussis* grows only on special media containing blood. One especially suitable is glycerin-potato-blood agar, on which it grows slowly, with 2 or 3 days being required for visible growth. If repeatedly cultured on media in which the amount of blood is decreased stepwise, *Bordetella pertussis* becomes able to grow on blood-free media. With prolonged cultivation, *Bordetella pertussis* loses its dependence upon both the X and V factors, and when this occurs, it loses its pathogenicity.

■ **Pathogenicity.** In the respiratory system of man the bacilli aggregate in large masses among the cells lining the wall of the trachea and bronchi. They do not invade the bloodstream. The action of *Bordetella pertussis* is, at least in part, the effect of a toxic substance it produces. This substance appears to be both endotoxin and exotoxin. One fraction is thermolabile, being destroyed by a temperature of 56° C. for 30 minutes. The other is thermostable. *Bordetella pertussis* is distinctly pathogenic for lower animals, especially rabbits and guinea pigs.

■ **Modes of infection.** Whooping cough is one of the most highly communicable of all diseases. It is usually transmitted by droplet infection and by direct contact but may be transmitted by recently contaminated objects. The disease is communicable during any stage and often during convalescence. The child is most likely to spread the infection just before the whoop appears and for about 3 weeks thereafter. The bacilli enter the body by the mouth and nose and are thrown off in the buccal and nasal secretions and in the sticky sputum that is coughed up. With the exception of those persons closely associated with the disease or patients convalescent from the disease, carriers do not exist. Convalescent carriers become noninfectious soon after the disease subsides.

■ **Laboratory diagnosis.** Two culture methods for the bacteriologic diagnosis of whooping cough are the cough plate method and the nasal swab method. In the first, an open Petri dish containing glycerin-potato-blood agar is held in front of the child's mouth during a paroxysm of coughing. The organisms are sprayed on the medium in droplets from the mouth and nose. In the second, a nasal swab is passed through the nose until it touches the posterior pharyngeal wall. After the swab is withdrawn, it is passed several times through a drop of penicillin solution on the surface of a plate of Bordet-Gengou medium. The penicillin solution destroys many species of contaminating bacteria. Then the material is spread over the surface of the medium with a platinum loop. The cultures made by either method are incubated for about 3 days for visible growth. Identification of the bacilli may be made also in nasopharyngeal smears stained by fluorescent antibody technics.

The white blood count found in pertussis (leukocytes 15,000 to 30,000 per cubic millimeter, of which 80% are lymphocytes) is peculiarly characteristic and in many cases suggests the correct diagnosis.

■ **Immunity.** Man possesses no natural immunity to whooping cough. However, an attack is usually followed by permanent immunity. Second attacks, except in the aged, are usually mild. A mother, even though immune herself, does not transmit immune bodies to her offspring. Therefore the newborn child is not protected against whooping cough. Some degree of immunity may be conferred upon the newborn child if the mother is immunized during the latter months of pregnancy.

■ **Prevention.*** Whooping cough itself cannot be prevented completely by our present public health methods, but its death rate can be reduced to a remarkable degree. Mothers should be taught how deadly a disease whooping cough is among young children and the value of immunization. The unvaccinated child should be protected from exposure. The infected child should be isolated for several weeks after the whoop has disappeared. However, the child need not be kept in bed but may be allowed to play in the sunshine and fresh air.

Haemophilus ducreyi (Ducrey's bacillus)

Haemophilus ducreyi, or Ducrey's bacillus, is the cause of *chancroid,* a local, highly contagious, venereal ulcer with no relation to the chancre, the initial lesion of syphilis. It begins as a pustule that ruptures, exposing an ulcer with undermined edges and a gray base. There are usually multiple ulcers that spread rapidly. Unlike chancres, they do not have indurated edges; hence chancroids are spoken of as soft chancres. The infection spreads to the inguinal lymph nodes to form abscesses known as buboes. *Haemophilus ducreyi* may be detected in smears made directly from the edges of the lesions or cultivated from the lesions if some of the exudate is inoculated into sterile rabbit blood. But smear and culture sometimes fail to demonstrate the organisms.

The chancroid must be differentiated from the chancre of syphilis, and the buboes from those of lymphopathia venereum. It is not uncommon for a chancroid and a syphilitic chancre to occupy the same site. The chancroid appears first and heals, after which the chancre presents, or the chancre may appear before the chancroid heals. About one half of venereal ulcers are syphilitic chancre mixed with chancroid. Chancroid is usually transmitted by sexual intercourse but may be transmitted by surgical instruments or dressings.

An intradermal test for the diagnosis of chancroid has been devised and used extensively in European countries. It consists of the intradermal injection of a saline suspension of *Haemophilus ducreyi.* A positive result is an area of redness and induration at the site of injection. The reaction reaches it maximal intensity at the end of 48 hours.

Haemophilus aegyptius

The Koch-Weeks bacillus, as *Haemophilus aegyptius* is commonly known, is the cause of pinkeye, a highly contagious conjunctivitis prone to occur in epidemics. It is well named because of the intense inflammation of the conjunctival linings, which imparts a brilliant pink color to the white of the eyes. There is intense itching, but rubbing the eyes aggravates the situation. A yellow discharge forms, dries, and crusts on the eyelids. The disease is transferred by hands, towels, handkerchiefs, and other objects that contact the face and eyes. Other names for the bacillus are *Haemophilus Koch-Weeks* and *Haemophilus conjunctivitidis.* Since 1950, it has been renamed *Haemophilus aegyptius.*

Haemophilus suis

Haemophilus suis in combination with a virus causes swine influenza (p. 433).

*For immunization, see pp. 604-606, 617, and 618.

Haemophilus parainfluenzae

Very similar to *Haemophilus influenzae, Haemophilus parainfluenzae* may cause subacute bacterial endocarditis.

Moraxella lacunata (The Morax-Axenfeld bacillus)

A subacute or chronic inflammation of the conjunctiva, eyelid, and cornea may be caused by *Moraxella lacunata.*

BRUCELLA

The genus *Brucella* includes three intracellular parasites that attack lower animals primarily, from which they are transmitted to man to cause *brucellosis.* One, known as *Brucella melitensis,* produces Malta fever in goats and sheep, a disease with prolonged fever, arthritis, and a tendency to abort. Another, known as *Brucella abortus,* causes contagious abortion, or Bang's disease, in cattle. Contagious abortion of cattle is characterized by a tendency to abort, retention of placentas, sterility, and death of newborn offspring. The third, known as *Brucella suis,* causes contagious abortion in hogs. Male hogs are affected by brucellosis even more often than are female hogs, and in the females the tendency to abort is less than in cows. Brucellosis sometimes attacks animals other than the ones just mentioned. The disease known as poll evil in horses is caused by brucellae. A newly recognized strain, *Brucella canis,* infects dogs.

Brucella melitensis was discovered in Malta in 1887 by Sir David Bruce (1855-1931), a surgeon in the British army, and *Brucella abortus* was discovered in Denmark by B. L. F. Bang (1848-1932), in 1897. In 1918 and 1925, Alice C. Evans proved that these organisms are closely related. *Brucella suis,* discovered in infected hogs in the United States, was found to be closely related to the other two, and all three were placed in the genus *Brucella,* honoring Bruce, the discoverer of *Brucella melitensis. Brucella melitensis* and *Brucella suis* probably represent variations of *Brucella abortus,* the result of its adaptation to the goat and hog, respectively.

■ **General characteristics.** Brucellae are small, nonmotile, ovoid, gram-negative, nonspore-forming coccobacilli. Some form capsules. Brucellae elaborate an endotoxin. As is true of other gram-negative organisms, their cell wall is made up of a lipoprotein-carbohydrate complex. Being so closely related to each other, they cannot be differentiated by ordinary cultural methods. Special methods using to advantage the action of dyes such as basic fuchsin and thionine on the different species of *Brucella* effect separation. The production of hydrogen sulfide is another differential point. Brucellae grow best on enriched media, but growth is slow. *Brucella melitensis* and *Brucella suis* grow in atmospheric oxygen. For the growth of some strains of *Brucella abortus* the presence of 5% to 10% carbon dioxide in the atmosphere is necessary. The agglutination test separates *Brucella melitensis* from *Brucella abortus* and *Brucella suis* but not *Brucella abortus* from *Brucella suis.* Table 24-2 summarizes important differential characteristics.

Brucellae are destroyed within 10 minutes by a temperature of 60° C. Fortunately they are quickly destroyed by pasteurization. They can remain alive and virulent for as long as 4 months in dark, damp surroundings.

■ **Pathogenicity.** *Brucella melitensis* is more pathogenic for man than is *Brucella suis,* and the latter is more pathogenic than *Brucella abortus.* Animals may be infected

TABLE 24-2. DIFFERENTIAL PATTERNS OF BRUCELLA SPECIES

	Organism		
Characteristics	*Brucella abortus*	*Brucella melitensis*	*Brucella suis*
Culture in medium containing dyes			
1. Thionine	Growth inhibited	No effect	No effect
2. Basic fuchsin	No effect	No effect	Growth inhibited
Exposure to atmosphere of 10% carbon dioxide	Required for best growth	Not required	Not required
Production of hydrogen sulfide	For 2 to 4 days	None	For 6 to 10 days
Hydrolysis of urea (urease test)	None (or very slowly)	None (or very slowly)	Rapid
Lysis by phage	Yes	None	None at routine test dilution (RTD) Yes at 10,000×RTD
Agglutination reaction	Differentiates this organism from *Brucella melitensis* but not from *Brucella suis*	Differentiates this one from other two	Differentiates this one from *Brucella melitensis* but not from *Brucella abortus*
Biotypes	Nine	Three	Four
Host reservoir	Cattle	Goats (also sheep)	Pigs (also hares, reindeer)

with any one of the three. Pregnant domestic animals are more likely to be infected than are young animals or nonpregnant adults, and abortion is probable when the infection is acquired during the pregnancy.

Brucellae can attack every organ and tissue of the human body. For this reason a person with brucellosis may seek as his initial consultant almost any medical specialist—an internist, a surgeon, or other.

Infections in man follow five patterns. The most common form of brucellosis is characterized by a long-continued fever or cycles of fever alternating with afebrile periods, considerable weakness, and profuse sweating. The other four types are variants. The incubation period varies from 5 to 30 days (sometimes longer). *Melitensis* infections are common in the Mediterranean basin. Imported goats brought the infection from this area to the southwestern United States. *Brucella abortus* and *Brucella suis* infections are widespread in the United States.

Various names are applied to the disease in man—Malta fever (because of its prevalence on the island of Malta), undulant fever (because of its clinical fever curve—Fig. 24-1), and brucellosis. In southwest Texas it is often spoken of as goat fever and Rio Grande fever.

■ **Sources and modes of infection.** Infection is widespread among goats, cattle, and hogs. In cattle, brucellosis ranks with tuberculosis as a source of economic loss. Infection is transmitted from animal to animal by food contaminated with the urine, feces, or lochiae of infected animals and by contact with infected placentas or fetuses. Suckling animals may be infected by the milk of infected mothers. Cows become carriers and excrete the organisms in their milk for as long as 7 years.

Human infections are common but often overlooked, and in man, as in animals, infection enters via the gastrointestinal tract or through the skin. Most times the infectious material is derived from the excreta of living animals or the blood or tissues

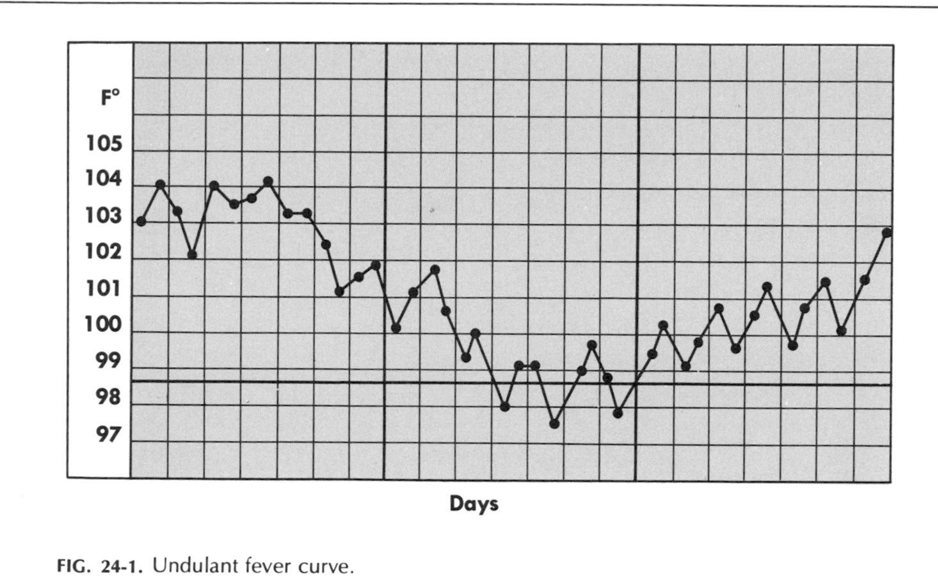

FIG. 24-1. Undulant fever curve.

of dead animals, and the organism penetrates cuts or abrasions in the skin. Packinghouse workers, butchers, farmers, stockmen, and veterinarians can get the infection by handling infected cows or hogs and the meat of these animals. Man can become infected by eating dairy products from infected cows. Infection with *Brucella melitensis* is contracted by the consumption of unpasteurized milk from infected goats. Dust containing the organisms may be infectious. Transfer of the infection directly from person to person is very rare.

■ **Laboratory diagnosis.** Laboratory aids in the diagnosis of undulant fever are an agglutination test, blood cultures (about one-third positive), animal inoculation, feces cultures, urine cultures, cultures of bone marrow (taken by sternal puncture), and cultures of material aspirated from lymph nodes. The last two are important because, after the bacilli have spread through the body, they localize in the bone marrow, liver, spleen, and lymph nodes. Positive cultures confirm the diagnosis, but even in the presence of known disease *Brucella* often cannot be so recovered. Therefore one must usually rely on the agglutination test done with a carefully standardized antigen. (It also may be negative, even with definite disease.) Persons who have received cholera vaccine may develop agglutinins against *Brucella,* a point to be noted in the investigation of suspected cases of undulant fever.

Patients recovering from one attack of brucellosis may continue to show agglutinins for months or years. If symptoms reappear, and the disease seems to be recurring, the determination of the nature of that person's immunoglobulins is significant. If immunoglobulins are 7S, even in low titers, active disease is a good bet. If they are 19S in type, it is not. Also used in the laboratory diagnosis of brucellosis are a modified Coombs' test (helpful in chronic cases), a complement fixation test, and fluorescent antibody technics. Veterinarians use the milk ring test, an agglutination test using whole milk instead of serum as the source of antibodies.

Guinea pigs may be inoculated with blood or cream from a suspect cow. The skin test detecting hypersensitivity to *Brucella* is formed by injecting brucellergen, a crystalline polypeptide extract of killed organisms, between the layers of the skin. If positive, the skin test indicates infection at some time past or present. A negative skin test tends to eliminate the possibility of infection.

■ **Immunity.** Children under 10 years of age seldom contract brucellosis. Men are more often attacked than are women (4:1), probably because of greater exposure. One attack may confer permanent immunity, although the disease is not self-limited. After an acute attack a person may be well, but there is reason to think that the organisms remain alive but quiescent in lymph nodes and the spleen.

■ **Prevention.** Vaccines being developed may provide temporary protection for human beings but are still experimental. An avirulent live strain of *Brucella* can be used to vaccinate cattle.

All milk (including goat's milk) and milk products should be pasteurized. All dairy cattle should be tested for brucellosis and infected animals removed from the herd. Those who work with animals must know how the disease is spread and what precautions to take to avoid contact with the flesh or excreta of infected animals. The strict application of sanitary and hygienic measures helps to minimize the occupational hazards.

The excreta of a patient with undulant fever should be disposed of as though from a patient with typhoid fever.

PASTEURELLA

The genus *Pasteurella* (named for Louis Pasteur) includes *Pasteurella pestis* (the cause of plague), *Pasteurella (Francisella) tularensis* (the cause of tularemia), *Actinobacillus (Pseudomonas) mallei* (the cause of glanders), and a group of pasteurellas causing hemorrhagic septicemia in lower animals but seldom attacking man. Two members of this genus, *Pasteurella pestis* and *Pasteurella (Francisella) tularensis,* are transmitted by infected insects.

Pasteurella pestis (the plague bacillus)

Plague is an infectious disease of rodents, especially rats, and is transferred from them to man. First described in Babylon, it has been a devastating pestilence for more than 3000 years. Pandemics in the past have swept over great areas of the world, with a terrifying mortality (50% or more). Hundreds of years ago the Chinese related the disease to rats by using the term "rat pestilence." That this disease still exists over the world must be remembered in these days of extensive travel and commerce. In our own country, plague has been encountered in Texas, Louisiana, and many states in the far West. The disease is endemic (enzootic) among the wild rodents of the southwestern United States, especially ground squirrels, and this focus of infection is an ever present and increasing threat. Among wild rodents it is known as sylvatic plague (Latin *silva,* 'forest').

■ **General characteristics.** The organism *Pasteurella pestis* (Table 24-3). (discovered in 1894) is small, aerobic, gram-negative and does not form spores. It grows on all ordinary media. Growth on agar containing from 3% to 5% salt is a specific feature in its identification.

TABLE 24-3. BIOCHEMICAL IDENTITY OF PASTEURELLA

Organism	Optimal temperature for growth	Fermentation of sugars				Production of indole	Production of hydrogen sulfide	Oxidase reaction
		Glucose	Maltose	Sucrose	Lactose			
Pasteurella pestis	28° C.	Acid	Acid	–	–	–	–	–
*Pasteurella tularensis**	37° C.	Acid	Acid	–	–	–	+	–
Pasteurella multocida	37° C.	Acid	–	Acid	–	+	+	+

*Note again the hazard of handling cultures of this organism. In routine identification, biochemical differentiation is not necessary.

Plague bacilli may live in the carcasses of dead rats, in the soil, and in sputum for some time. They retain their vitality for months in the presence of moisture and in the absence of light. Five percent phenol or 2% Amphyl solution destroys them in 20 minutes.

Plague bacilli are highly pathogenic for many animals—monkeys, rats, mice, guinea pigs, and rabbits. They owe their action to toxic substances that partake of the nature of both endotoxins and exotoxins.

■ **Clinical types of plague.** There are three clinical types: bubonic, septicemic, and pneumonic. The first is the most frequent. In bubonic plague the organisms enter the skin and are carried by the lymphatics to the lymph nodes draining the site of infection, commonly in the inguinal region. Here the bacilli multiply and form abscesses of the nodes (buboes). The bubo is extremely painful; a severe cellulitis surrounds it. Secondary buboes are formed in the nodes draining the primary buboes, and the bacilli finally move into the bloodstream, causing septicemia. The hemorrhages in bubonic plague that result in black splotches in the skin gave the name "black death" to the plague during the Middle Ages. (From A.D. 1347 to 1349 the black death killed 25 to 40 million persons.)

Pneumonic plague is manifestly a bronchopneumonia, and the bacilli are abundant in the sputum. Only a very small percentage of the cases in the average epidemic take the pneumonic form (or that of a meningitis), but epidemics strictly of pneumonic plague may occur. Septicemic plague is a highly virulent form in which the patient dies before buboes have time to develop.

■ **Modes of infection.** Although many rodent species may be infected by *Pasteurella pestis,* the most important source of primary infection is the rat, and, as a rule, epidemics of human plague closely follow epizootics of rat plague. Plague is transmitted from rat to rat by the bite of the rat flea. Man contracts the bubonic form of plague by the bite of a flea that has previously fed on an infected rat or, rarely, on an infected person. The reason for epidemics of human plague occurring after epizootics of rat plague is that an epizootic destroys the rat flea's food of first choice—the rat. The flea must then seek his food of second choice—man. Unless infection is mechanically transferred by blood adherent to its mouthparts, the flea spreads the infection 4 to 18 days after biting an infected animal or person. Plague bacilli are found in the intestine of the flea, where they live for a long time and multiply rapidly, and the bacilli are introduced into the human body by material regurgitated from the flea while it is biting. The flea may die of the infection or become a carrier. The incubation period in man is 3 to 8 days.

Pneumonic plague may result from involvement of the lungs during the course of bubonic plague (primary pneumonic plague) or from the inhalation of particles of sputum thrown off by a person with pneumonic plague (secondary pneumonic plague). Epidemic pneumonic plague is of the secondary type.

Plague may occasionally be acquired by the handling of infected rodents. The disease is transmitted from locality to locality by infected rats. Dozens of wild and domestic rodents other than rats may contract plague naturally—ground squirrels, voles, prairie dogs, chipmunks, marmots, and guinea pigs. The black house rat is more dangerous to man than are other types because this one lives in dwellings and, therefore, its fleas are more likely to bite a human being. The common gray sewer rat and the Egyptian rat can be infected.* Bedbugs and the human flea may transmit plague from person to person.

Recently carriers (without symptoms) have been found to harbor *Pasteurella pestis* in their throats.

■ **Laboratory diagnosis.** The important methods of laboratory diagnosis are the agglutination test on the patient's serum and the demonstration of the bacilli in the lesions by smears, cultures, and animal inoculation. There is a fluorescent antibody test to identify the organisms in sputum.

■ **Immunity.** An attack of plague usually brings about a state of permanent immunity.

■ **Prevention.**† The prevention of plague depends upon the eradication of rats and rat fleas. Patients with plague should be isolated in well-screened, vermin-free rooms. (The technic of quarantine was first used with plague.) Although seldom do patients with bubonic or septicemic plague transmit the infection to others, no chance should be taken. They should be nursed with the same precautions as are patients with typhoid fever. The sputum from patients with pneumonic plague should receive special care. The sickroom should be dusted with the insecticide DDT to eliminate the vector flea. Attendants should be protected with plague vaccine.

Measures designed to block the transport of rats from infected to noninfected localities by trains, aircraft, and ships are imperative. Rat-infested ships should be fumigated with a potent rodenticide. Maritime rat control measures are in force in most parts of the world.

Pasteurella (Francisella) tularensis

Pasteurella (Francisella) tularensis is the cause of tularemia (rabbit fever), an acute infectious disease of wild animals, especially rodents, transferred from animal to animal by the bite of an insect. It may be transmitted from animal to man by insect bite, but human infections come from contamination of the hands or conjunctivae by the tissues or body fluids of an infected animal or insect.

■ **General characteristics.** The organism *Pasteurella tularensis* derives its name from Tulare County, California, where the disease was first observed. *Pasteurella tularensis* is a small, gram-negative, nonspore-forming, nonmotile organism varying consider-

---

*Today in Southeast Asia conditions are favorable for outbreaks of plague. The large infected rat population with the right number of rat fleas exists alongside a dense nonimmunized human population in a humid, wet (not hot) climate. During the wet season, the monsoons cancel out the effect of insecticides and force the rats into human shelters. The natives kill the rats with sticks and the fleas move onto their new, human hosts.

†For immunization procedures, see pp. 606, 627, and 629.

ably in size. Enormous numbers fill the phagocytic cells of the body. It grows on cystine agar (not on ordinary agar) and is easily demonstrated in the lesions by animal inoculation.

■ **Clinical types of tularemia.** Tularemia is a febrile disease accompanied by severe constitutional manifestations, pain, and prostration. The incubation period is usually 1 to 10 days. The clinical types are (1) the ulceroglandular (there is an ulcer at the site of infection with involvement of regional lymph nodes), (2) the oculoglandular (the pattern is similar to the ulceroglandular type with the conjunctiva the site of primary infection), (3) the glandular (the lymph nodes are involved but ulceration is absent), and (4) the typhoid (neither ulceration nor lymph node involvement is present). Death occurs in about 5% of the patients. Those who recover are incapacitated for weeks or months. (Clinically tularemia may be confused with cat-scratch disease.)

■ **Modes of infection.** Tularemia is most prevalent in cottontail rabbits, jack rabbits, snowshoe rabbits, and ground squirrels. Certain birds have tularemia, and the disease is not unknown in cats and sheep. The water of streams inhabited by infected animals (beavers and muskrats), once contaminated, may remain a source of infection for a long time. Epidemics caused by pollution of streams have been reported. In Russia human infections are often contracted from a species of fur-bearing water rat. More than 45 species of wild animals are known to be infected. Tame rabbits are susceptible, but ordinarily they do not contract the disease since they do not harbor the small parasites that transmit the infection among animals.

The insects that transmit tularemia in animals are wood ticks, rabbit ticks, lice, horseflies, and squirrel fleas. Wood ticks, dog ticks, horseflies, and squirrel fleas may sometimes transmit the infection to man. Rabbit ticks, rabbit lice, and mouse lice, important agents in transferring the disease from rodent to rodent, do not bite man.

In our country, man usually contracts tularemia by handling infected rabbits; of these, about 70% are the cottontails. Cold-storage rabbits remain infectious for 2 to 3 weeks, and if kept frozen, they may be a source of infection for 3½ years. The disease may be contracted if insufficiently cooked rabbit meat is eaten.

Inhalation of organisms can result in disease. The ocular type of tularemia usually results from a person rubbing his eye with contaminated fingers. Tularemia is not transmitted directly from man to man.

Pasteurella tularensis is the most easily communicable of all organisms, and many laboratory workers who have investigated tularemia have contracted the disease. Some have died.

■ **Laboratory diagnosis.** The laboratory diagnosis of tularemia depends upon the agglutination test, immunofluorescence (Fig. 24-2), Foshay's antiserum test, the tularemia skin test, and the inoculation of guinea pigs or rabbits with material from the lesions. Cultures are too dangerous to be handled routinely.

■ **Immunity.** An attack of tularemia is followed by permanent immunity.

■ **Prevention.** The nature of the infection and its modes of transfer render the preventive measures obvious.

Pasteurellas of hemorrhagic septicemia

Certain members of the genus *Pasteurella* cause hemorrhagic septicemia (pasteurellosis) a disease of cattle, horses, swine, bison, and poultry, characterized by septicemia, petechial hemorrhages of mucous membranes and internal organs, edema, and changes in the lungs. The mortality is high. The most important members of the hemor-

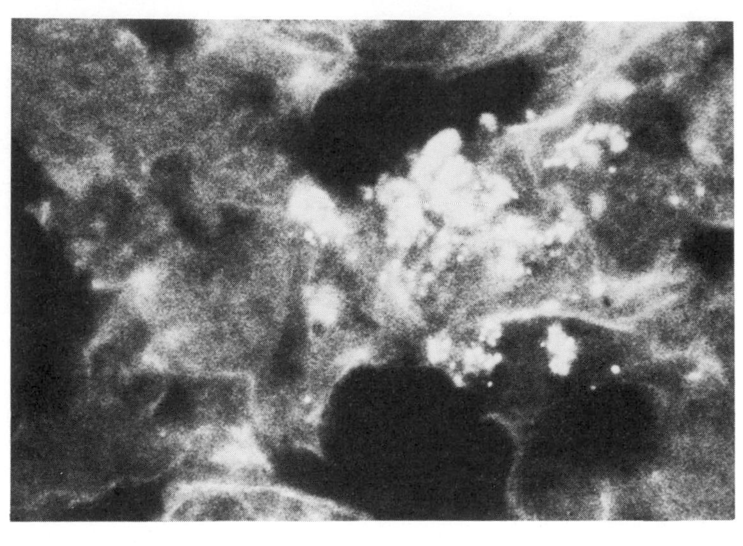

FIG. 24-2. *Pasteurella tularensis* identified by fluorescent antibody staining. Glow of bacterial immunofluorescence apparent even in black and white photomicrograph. (From White, J. D., and McGavran, M. H.: Identification of *Pasteurella tularensis* by immunofluorescence, J.A.M.A. **194:**294, 1965.)

rhagic septicemia group are *Pasteurella multocida* (cause of fowl cholera and shipping fever of cattle), *Pasteurella haemolytica* (cause of pneumonia in sheep and cattle), and *Pasteurella pseudotuberculosis* (cause of hemorrhagic septicemia in guinea pigs, rabbits, and mice). *Pasteurella multocida* and *Pasteurella pseudotuberculosis* can infect man.

■ **Microbiology of dog and cat bites.** Over half a million persons, principally children, are bitten by animals each year in the United States. Most of the bites come from cats and dogs; cats head the list. Generally such wounds heal without complications, but they can be infected by microorganisms passed by the animal in biting.

Since it is normally found in the nose and throat of cats and dogs, *Pasteurella multocida* is an offender. Other organisms encountered, usually in mixed culture, include *Staphylococcus aureus* and alpha- and beta-hemolytic streptococci. The lesion produced may be a cellulitis, an abscess, or, about the extremities, a tenosynovitis. Osteomyelitis can complicate a deep-seated bite, and there is always the danger of bacteremic spread. Resolution of the tissue injury from the bite can be slow. Scarring results from the damage to tissue, and the distortion from scar about the head and neck is disfiguring.

ACTINOBACILLUS MALLEI

Glanders (farcy) is primarily a disease of horses, mules, and donkeys, but it may sometimes be transmitted to man. Cattle are immune. It is characterized by the formation of ulcerating tubercle-like nodules that involve the lungs, superficial lymph nodes, and mucous membranes. If the lymph nodes of animals are affected, the disease is known as *glanders;* if the mucous membranes are affected, it is *farcy*. With the replacement of the horse by the automobile, this once prevalent disease has become an uncommon one.

The causative organism, *Actinobacillus (Pseudomonas) mallei,* is a narrow, sometimes slightly curved bacillus that is gram-negative, nonmotile, nonencapsulated, and nonspore-forming and with little resistance to physical and chemical agents. It is a most dangerous organism, and material from the lesions and cultures must be handled with extreme care. Man is easily infected if he contacts tissues or excreta of diseased animals since these contain the bacilli. The organisms enter the body by a wound, scratch, or abrasion of the skin.

■ **Laboratory diagnosis.** Glanders is diagnosed in man and animals by the isolation of the bacilli from the lesions or in blood cultures. There is a complement-fixation test. The skin test utilizes *mallein,* a product of *Actinobacillus mallei.* In animals, mallein is injected subcutaneously, or mallein tablets are placed in the conjunctival sac. If infectious material is injected intraperitoneally into a male guinea pig, testicular swelling and generalized manifestations develop within 3 or 4 days. This is the *Straus reaction*.

■ **Prevention.** There is no immunity to glanders. Control of the disease depends upon the destruction of animals that have it in its clinical or occult form and upon the disinfection of stables, blankets, harnesses, and drinking troughs used by sick animals.

FIG. 24-3. Donovan bodies in cytoplasm of phagocytic cell. Note safety-pin appearance of dark bipolar bodies. Clear surrounding areas correspond to capsules about microorganisms. (From Davis, C. M.: Granuloma inguinale, a clinical, histological, and ultrastructural study, J.A.M.A. **211:**632, 1970.)

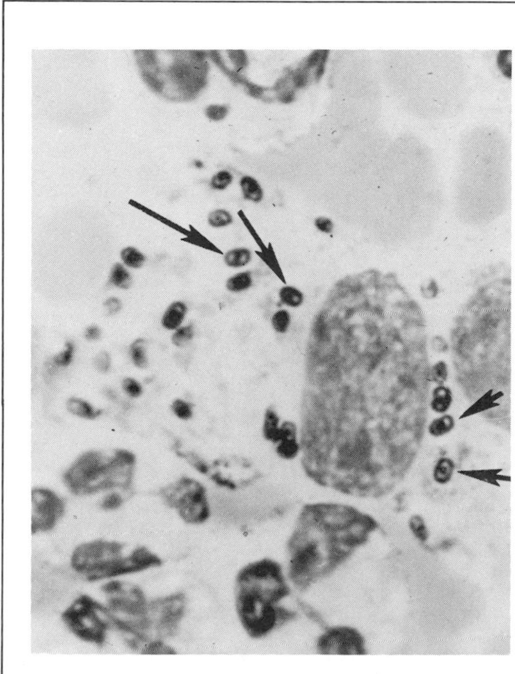

CALYMMATOBACTERIUM GRANULOMATIS

Calymmatobacterium granulomatis, an organism of the only species in its genus, is the cause of granuloma inguinale, one of the venereal diseases. This is an ulcerative infection in the skin and subcutaneous tissues of the groin and genitalia. In the lesions, encapsulated, oval, rodlike bodies with a distinctive "safety pin" appearance are found within the cytoplasm of large mononuclear phagocytic cells (Fig. 24-3). These are the causative agents, also known as the Donovan bodies.

The laboratory diagnosis of the disease rests on the microscopic demonstration of the Donovan bodies in Wright-stained smears made of material from the lesions.

QUESTIONS FOR REVIEW

1. List the features shared by the hemophilic bacteria.
2. What is the importance of factors V and X?
3. Characterize *Haemophilus influenzae* and *Bordetella pertussis*.
4. Briefly discuss the pathogenicity of the influenza bacillus.
5. How is whooping cough transmitted? What are the dangers of the disease?
6. Give the laboratory diagnosis of pertussis.
7. What is a chancroid? What causes it?
8. Describe the condition of pinkeye. What organism usually causes it?
9. Name the species of *Brucella*. How are they alike and how do they differ?
10. Comment on the laboratory diagnosis of brucellosis making mention of difficulties encountered.
11. Relate microbes in the genus *Pasteurella* to disease produced.
12. List five animals (other than man) that contract tularemia. Name five vectors associated.
13. Give the clinical types of plague and of tularemia.
14. What is the reservoir of plague? How is it conveyed to man?
15. What are the Donovan bodies? Where found?
16. Briefly discuss the microbiology of dog and cat bites.
17. State the method of spread of glanders. How may this disease be controlled?

REFERENCES. See at end of Chapter 26.

25 Miscellaneous microbes

PSEUDOMONAS SPECIES

Pseudomonas aeruginosa

The organism of blue pus, *Pseudomonas aeruginosa* is an actively motile, gram-negative, nonspore-forming bacillus (Fig. 25-1). When stained with simple stains, it bears a resemblance to *Corynebacterium diphtheriae*. It is characterized by the diffusion of a blue pigment (pyocyanin and fluorescein) through the medium on which it is grown, and the same color is seen in purulent discharges it produces in disease. Green pigment is a characteristic of several species of bacteria, but the blue color seems to be specific for *Pseudomonas aeruginosa.**

Pseudomonas aeruginosa is widely found in water and soil and often as a normal inhabitant of the skin and intestinal tract in man and animals. Ordinarily it is but mildly pathogenic and causes primary infections only when the resistance of the host is lowered. It is a problem in young infants (Fig. 25-2) and the aged. It is often a secondary invader and frequently delays the repair of wounds. It possesses an endotoxin, a collagenase, and a hemolysin. Among the diseases that may be caused by *Pseudomonas aeruginosa* are otitis media, abscesses, wound infections, sinus infections and bronchopneumonia (Fig. 25-3). It is an important cause of bacteremic shock, urinary tract infections, and the sepsis complicating severe burns.

This organism is more resistant to the chemical disinfectants usually employed than are other gram-negative bacilli, but it can be killed by exposure to a temperature of 55° C. for 1 hour. In the treatment of *Pseudomonas* infections, polymyxin and colistin, rather toxic drugs, are often the only antibiotics to which the organisms are susceptible. Since *Pseudomonas aeruginosa* is resistant to the action of most antibiotics, it tends to become the dominant organism in a diseased area after extended antimicrobial therapy has eliminated the primary ones. Then it takes on a vigorous disease-producing capacity and is responsible for significant destruction of tissue. This is especially true with lowered host resistance. Such infections with *Pseudomonas aeruginosa* are quite difficult to deal with. Pathologically, a massive overgrowth of the organisms is seen microscopically in dead tissue.

---

*Pathogenic pseudomonads produce a great deal of fluorescein. With the Wood's ultraviolet light, this pigment is readily detected, even under the black, silver-stained eschar of a burn wound that has been treated with silver nitrate.

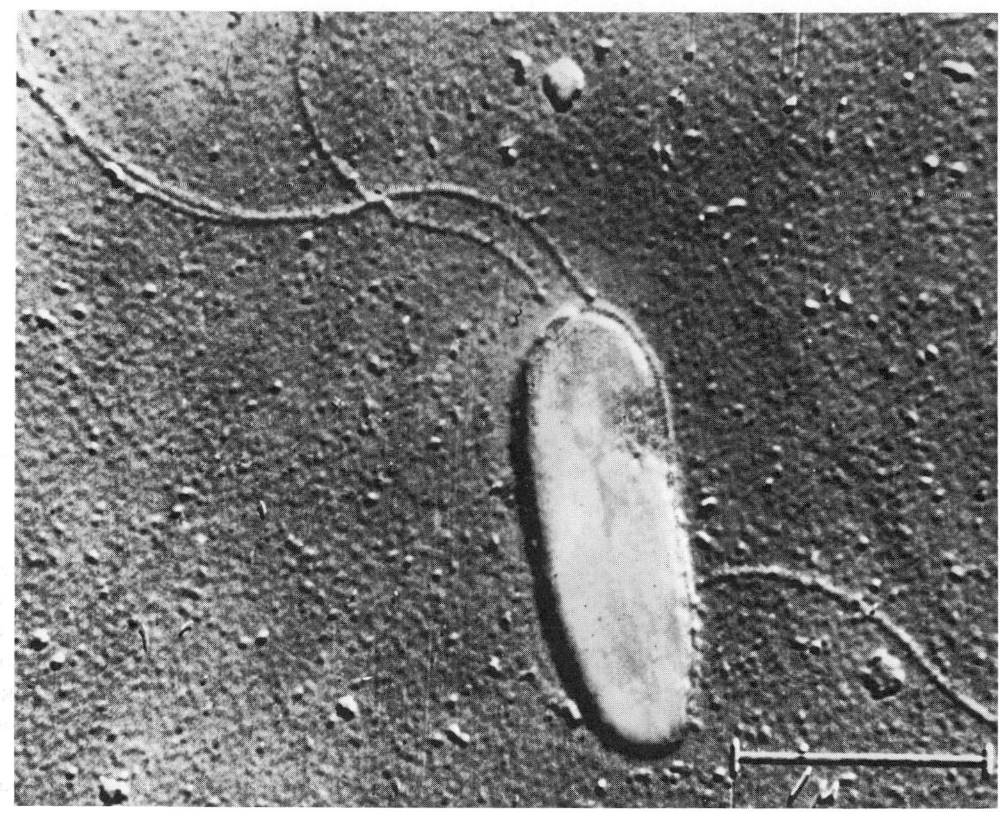

FIG. 25-1. *Pseudomonas aeruginosa,* electron micrograph. Note flagella. (Courtesy J. B. Roerig Division, Chas. Pfizer & Co., Inc. New York, N. Y.)

The sources of *Pseudomonas aeruginosa* in hospital-acquired infections are many, but especially the pieces of equipment that are hard to clean and sterilize, such as face masks, certain kinds of humid rubber tubing, water containers, nebulizers, and parts of the intermittent positive-pressure breathing machines. A source easily overlooked is the hospital supply of distilled water. This is a serious hazard because of the widespread need for distilled water in preparation of various hospital solutions—detergents, disinfectants, and even parenteral medications. Pseudomonads have been demonstrated not only to survive but to multiply rapidly in *distilled* water and seem to have a greater resistance to antimicrobial agents than do the ones recovered from laboratory media.

Patients with 40% or more burned surface area are susceptible to *Pseudomonas* sepsis. Administration of a septivalent *Pseudomonas* vaccine to these seriously burned persons is a great advance in the management of their injury. It dramatically reduces the mortality.

Pseudomonas pseudomallei

Pseudomonas pseudomallei is a gram-negative, motile, nonpigmented bacillus that produces *melioidosis*, an uncommon tropical disease of man and animals. Its pulmonary manifestations are easily confused with tuberculosis. There may be a septicemia, with widespread lesions over the body. Nicknamed the "Vietnamese time bomb," the disease can be dormant for a number of years only to appear suddenly, become active, and produce death within days or weeks. The disease is endemic in Southeast Asia, where the organism is found in surface waters, soil (notably the

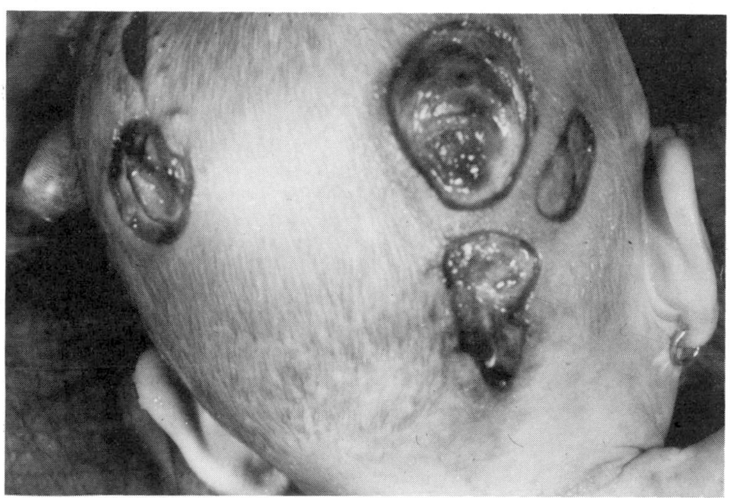

FIG. 25-2. Skin ulcers on head of young child with systemic *Pseudomonas aeruginosa* infection.

FIG. 25-3. Bronchopneumonia caused by *Pseudomonas aeruginosa*. A well-defined, solidified, darkened portion is sharply distinct in thin slice of air-containing lung. The focus of disease is hemorrhagic and partly necrotic.

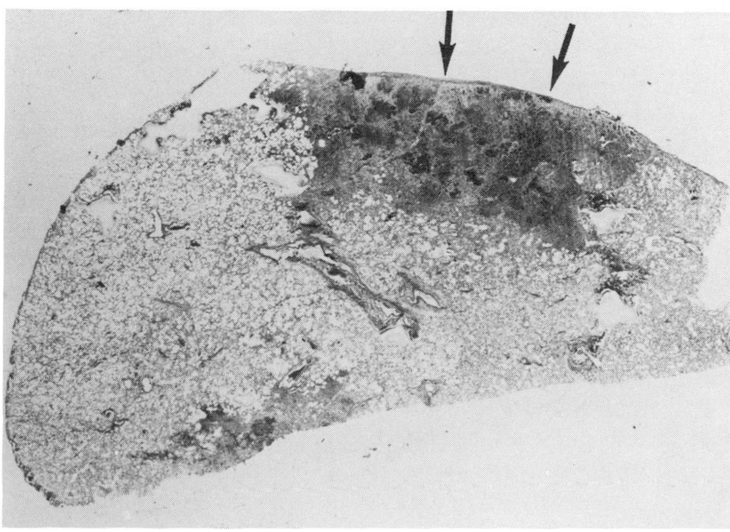

rice fields), and organic matter. The epidemiology and transmission of the disease are not known. The organisms are thought to enter the body by way of the mouth and nose or through open wounds in the skin.

Growth of the bacilli is typical in cultures. A hemagglutination test is available for diagnostic study.

BACTEROIDES SPECIES

Bacteria of the genus *Bacteroides,* on the whole, are nonspore-forming, strictly anaerobic, very pleomorphic rods, usually gram-negative, that are normal inhabitants of the colon, genital tract, and upper respiratory tract. *Bacteroides* species are said to make up a large percentage of the bacterial content of the stool. Outside their normal habitat, they are astounding pathogens. They are found in ulcers of mucous membranes and may form foul-smelling abscesses in the lungs and other organs, often together with other microorganisms. They cause sepsis. Their infections are especially prone to complicate abdominal surgery.

The bacteroides are fastidious as to their growth needs; anaerobic conditions must be strict.

Fusobacterium fusiforme (Bacteroides fusiformis) is a member of a closely related genus in family VI—Bacteroidaceae. It is associated with a spirochete in Vincent's angina. (See p. 389.)

LACTOBACILLUS SPECIES (THE LACTOBACILLI)

The lactobacilli are gram-positive, microaerophilic rods that produce lactic acid from rather simple carbohydrates and are able to live in a more highly acid environment than most other bacteria can. They are widely distributed in nature. Many are normal inhabitants of the alimentary tract. Most are nonmotile. Since they require many essential growth factors for their metabolism, they can be used commercially in the bioassay of the B complex vitamins and certain amino acids. They are a more sensitive index than are current chemical methods. They also are used extensively in the fermentation and dairy industries.

The significant members of this genus are *Lactobacillus acidophilus, Lactobacillus bifidus, Lactobacillus bulgaricus,* and *Lactobacillus casei. Lactobacillus acidophilus,* a normal inhabitant of the intestinal tract, is increased by a diet rich in milk or carbohydrates. The *Boas-Oppler bacillus,* found in the stomach in gastric cancer, is most likely the same organism. *Lactobacillus bifidus* is found in the intestines of breast-fed infants. *Lactobacillus bulgaricus* was isolated from Bulgarian fermented milk and is the organism that Metchnikoff thought would prevent intestinal putrefaction, thereby increasing length of life. *Lactobacillus casei* is used in the bioassay of vitamins and amino acids. *Lactobacillus* species make up the normal flora of the vagina during a woman's active reproductive years, where they are referred to collectively as *Döderlein's bacilli.*

Some of the lactobacilli and certain acid-producing streptococci are found in the mouth in relation to dental caries (tooth decay). Most observers believe that they play a definite etiologic role. The first attack on tooth structure is postulated to come from these acid-producing organisms beneath food particles. They produce an enzyme,

FIG. 25-4. Early dental decay (plaque), microscopic section. Microbial growth is associated with debris, mucus, and desquamated cells. The action of lactobacilli on sugars in the food results in an accumulation of lactic acid, believed to trigger the process of decay by decalcifying the enamel of the tooth. (From Bhaskar, S. N.: Synopsis of oral pathology, ed. 3, St. Louis, 1969, The C. V. Mosby Co.)

dextran sucrase, that creates a sticky film on tooth surfaces, providing a haven for bacteria to grow and from which to attack the enamel of the tooth. Subsequently, tooth substance is broken down, and the characteristic cavity forms. Fragments of tooth and food debris found in the cavity furnish a medium for more bacterial growth, thus perpetuating the process of decay (Fig. 25-4).

LISTERIA MONOCYTOGENES

Listeria monocytogenes is a small (less than 2μ long), gram-positive, nonsporeforming, motile bacillus that affects man and a great variety of animals. It causes disease *(listeriosis)* characterized by an increase in large mononuclear leukocytes (monocytes) in the bloodstream. The bacillus produces an encephalitis or encephalomyelitis, especially in ruminant animals. In man it may bring about purulent meningitis, conjunctivitis, endocarditis, urethritis, and a type of infectious mononucleosis. Involving the fetus and newborn infant, *granulomatosis infantiseptica* is an intrauterine listerellal infection associated with widespread necrosis of the internal organs and hemorrhagic areas in the skin. The newborn infant usually dies within 2 or 3 days. Although human listeriosis is rare, the number of cases in man and in poultry and livestock is increasing.

From specimens of blood, spinal fluid, and pus the organism can be identified by cultural methods in the laboratory. It produces beta-hemolysis and is novel in that it can grow at the temperature of the refrigerator. A highly specific test for identification of the organism utilizes the rabbit; the everted eyelid of the animal is swabbed with the culture. Typically, the reaction to the presence of *Listeria* is the development of purulent keratoconjunctivitis. The transmission of listeriosis in man and animals is not known.

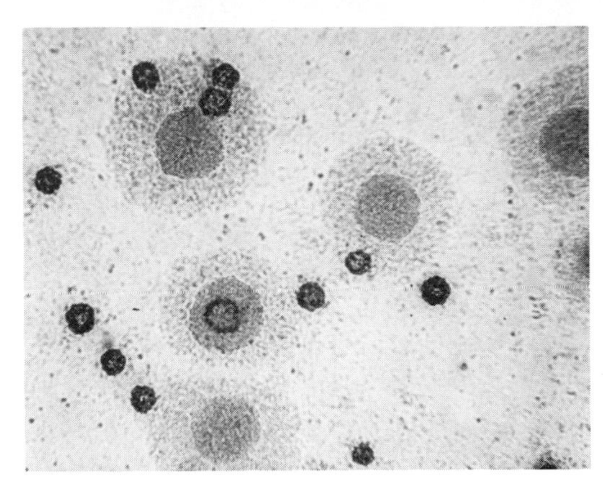

FIG. 25-5. Mycoplasmas. Mixture of classical mycoplasma *(Mycoplasma hominis)* colonies and smaller, very dark colonies of T (T-strain) mycoplasmas. The direct test for urease (that of Shepard and Howard) has been applied to this preparation. Colonies of T-strain mycoplasmas are urease-positive showing the dark bronze color of the reaction product. Note fried-egg appearance of classical colonies. (×300.) (From Shepard, M. C., and Howard, D. R.: Identification of "T" mycoplasmas in primary agar cultures by means of a direct test for urease, Ann. N. Y. Acad. Sci. **174:**809, 1970.)

MIMA (HERRELEA) SPECIES

Bacteria of this category are a poorly defined group of tiny, gram-negative, nonmotile, nonspore-forming, aerobic coccobacilli, about which there is taxonomic confusion. They are widespread in nature and have been recovered from various body sites and fluids. They have recently been implicated in hospital-induced infections, particularly meningitis, and in sepsis. The sources of infection are incompletely understood. They are readily cultured but are biochemically somewhat inactive. When smeared out from growth on solid media, they appear as diplococci, simulating either meningococci or gonococci.

MYCOPLASMA SPECIES

■ **General characteristics.** Mycoplasmas (pleuropneumonia-like organisms, PPLO's) are smaller than ordinary bacteria and about the size of the larger viruses; in fact, they are the smallest microorganisms that can be cultivated on cell-free media (Fig. 25-5). They possess soft, fragile cell bodies and lack a rigid cell wall.* Hence they are highly pleomorphic. They contain both DNA and RNA and some metabolic enzymes. Like viruses, they are filtrable and, in the animal body, intracellular parasites. Also they are sensitive to ether and insensitive to many antibiotics. Their ability to induce cytopathogenicity in tissue cultures is a nuisance. The type species is *Mycoplasma mycoides,* the cause of pleuropneumonia in cattle.

■ **Pathogenicity.** In addition to bovine pleuropneumonia, mycoplasmas cause contagious mastitis in sheep and goats and respiratory disease in poultry. Dogs, rats, and

*Since they have no cell wall, they would have to be gram-negative.

mice may become infected with these organisms. In man they are linked to infections of the genitourinary tract and of the upper and lower respiratory passages, and rarely to other systemic disorders.

PRIMARY ATYPICAL PNEUMONIA. *Mycoplasma pneumoniae* (the Eaton agent*) is the smallest known pathogen that can live outside of cells. It causes primary atypical pneumonia, an acute febrile, self-limited disease of man that begins as an upper respiratory infection and spreads to the lungs. It was among the most common respiratory infections of World War II, and today the Eaton agent can be recovered from a large number of military recruits. The manifestations of the disease, including headache and malaise, are fairly severe. The cough is paroxysmal, but little sputum is raised. The lungs may be diffusely involved, although in most patients only one lobe is affected. The attack lasts 2 to 3 weeks. Recovery is gradual. The disease, though prevalent, does not occur in widespread epidemics. Epidemics occur among persons living in crowded conditions and have been reported among newborn infants. Milder respiratory disease or even inapparent infection ("walking pneumonia") may be associated with *Mycoplasma pneumoniae.*

Primary atypical pneumonia is not very contagious, although apparently it can spread directly from one person to another. Very likely the disease is transmitted by oral or nasal secretions, and the portal of entry is the upper respiratory tract. *Mycoplasma pneumoniae* is easily recovered from the throats of convalescent patients. The disease is not always followed by immunity.

The classic case of primary atypical pneumonia is associated with an increased number of cold agglutinins† for red blood cells in the patient's serum. In recent studies over half of mycoplasmal pneumonias had demonstrable cold agglutinins. In the sera of some convalescent patients agglutinins develop to *Streptococcus MG* (from the name McGuinness, the patient from whom it was originally recovered). There is also a complement fixation test.

An alum vaccine prepared from formalin-killed organisms has been tested experimentally in animals and man. A live mycoplasma vaccine is also in process.

■ **T-strain mycoplasmas.** "T" strains of mycoplasmas, "tiny forms," or "T form" PPLO colonies represent human mycoplasmas with a distinct growth pattern. T strains produce very small colonies on their agar media (only a few microns in diameter). Colonies of the "classic" organisms, first studied, are much larger, many microns in diameter, even up to 0.75 mm. (Both "tiny form" and "large form" colonies must be viewed with a microscope.) Moreover, T strains are unique among mycoplasmas in being sensitive to therapeutic agents not affecting the classic ones and in possessing an active urease system. Postulated to be agents of venereal infection, T strains cause nongonococcal urethritis (NGU) in the male and in the female contribute to infertility and premature births. Classic mycoplasmas are also implicated in so-called reproductive failure.

---

*The Eaton agent was discovered by Monroe Davis Eaton of Harvard University in 1944. Originally grown on chick embryo lung culture, it was thought to be a virus until it was cultured on an artificial agar medium. Dr. Eaton suspected its relation to primary atypical pneumonia, but it was only after fluorescent antibody technics were applied years later that the Eaton agent was proved a cause of this disease.

†These are agglutinins that clump human type O, Rh-negative erythrocytes at low temperatures (5° to 20° C.) but not at body temperature. They are found in several unrelated diseases.

■ **Laboratory diagnosis.** The laboratory diagnosis of the presence of mycoplasmas is made by cultural methods adapted to their size and peculiar growth requirements. The morphology of the colonies is distinct, mycoplasmas proliferate also in tissue culture and in the developing chick embryo. Serologic identification is practical since mycoplasmas are vulnerable to neutralizing antibodies and induce hemagglutination and hemadsorption reactions. Since many normal persons demonstrate circulating antibodies to the organisms, it is the *rising* titer that is of diagnostic significance.

BARTONELLA SPECIES

The intracellular parasites of the genus *Bartonella* are classified by Bergey in the same order as rickettsiae and bedsoniae. They are small bacteria-like rods, sometimes called coccobacilli, flagellated, and very pleomorphic. All infect red blood cells as well as certain tissue cells. They are transmitted by the night bite of *Phlebotomus* (the sand fly). One member of this group, *Bartonella bacilliformis,* is the cause of a South American disease known as Carrión's disease (Oroya fever, Peruvian verruca). Carrión's disease is described by an acute anemia with fever that is followed by a verrucous or wart-like skin eruption. *Salmonella* species are prone to complicate it. *Bartonella* species are easily identified in stained films of peripheral blood because of their unique appearance. Infection with *Bartonella* is endemoepidemic in the Andean valleys of the Peruvian Sierra.

QUESTIONS FOR REVIEW

1. What is the significance of *Pseudomonas aeruginosa* in clinical medicine? Why is it so hard to deal with?
2. What is melioidosis? Name and describe the causative organism.
3. Briefly characterize *Bacteroides* species.
4. List diseases caused by *Pseudomonas.*
5. Comment on the relation of the lactobacilli to dental caries.
6. State the practical importance of the lactobacilli. List the important members of the genus *Lactobacillus.*
7. Define listeriosis. What forms of human listeriosis are important?
8. Briefly discuss *Mima* species.
9. Characterize mycoplasmas. How is the laboratory diagnosis of mycoplasmas made?
10. Compare mycoplasmas with viruses.
11. Give the role of T strain mycoplasmas in human disease.
12. Outline the salient features of primary atypical pneumonia.

REFERENCES. See at end of Chapter 26.

26 The spiral microorganisms

Spirochetes are actively motile, flexible, spiral microorganisms found in contaminated water, sewage, soil, decaying organic matter, and in the bodies of animals and man (see Fig. 26-1). They vary in length from only a few microns to 0.5 mm. Movement is in a rotating sinuous manner. There are no flagella. Although the organisms have features of both bacteria and protozoa, most microbiologists consider them more closely related to bacteria. They differ from bacteria in their susceptibility within the body to arsenic, antimony, mercury, and bismuth compounds. Like many bacteria, some of the spirochetes are susceptible to the action of penicillin and certain other antibiotics.

FIG. 26-1. Spiral microorganisms, comparative morphology. **A,** *Borrelia.* **B,** *Treponema.* **C,** *Leptospira.* **D,** *Spirillum.*

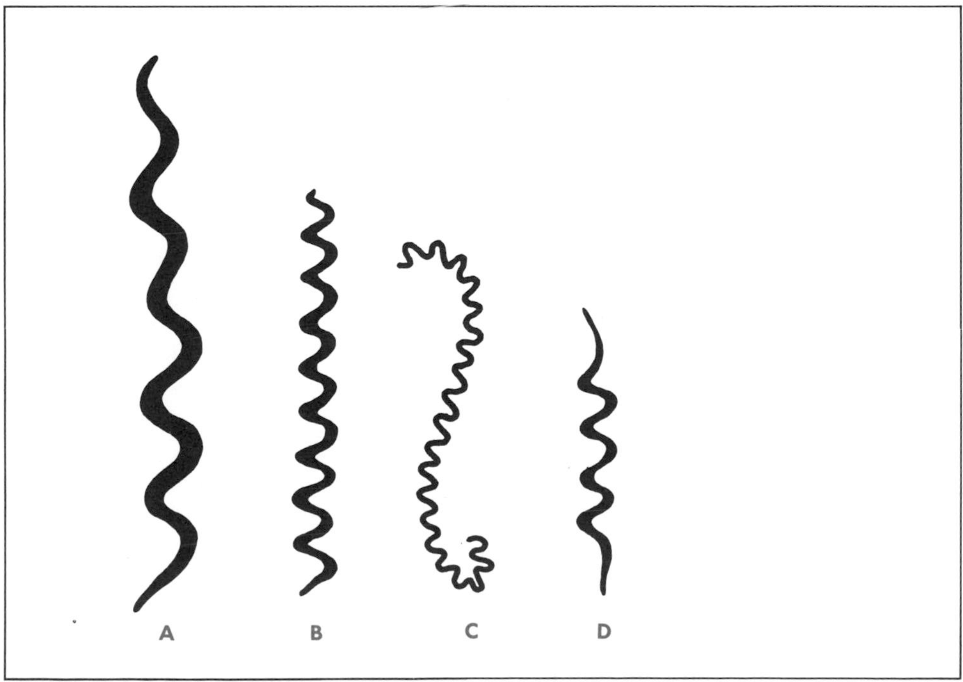

Within Bergey's Order IX, Spirochaetales, of slender, spiral bacteria, there are six genera of note in Family III, Treponemataceae. Within the three genera of nonpathogenic spirochetes are those often found on the mucous membranes of the mouth, about the teeth, and on the genitals. Within the three genera of pathogenic spirochetes are *Treponema*, which includes the organisms responsible for syphilis and yaws; *Borrelia*, which includes the organisms responsible for relapsing fever; and *Leptospira*, which includes the organisms responsible for infectious jaundice (Weil's disease) and other forms of leptospirosis.

Other spiral-shaped microorganisms with slightly different properties fall into another order and family of Bergey's classification. In Order I, Pseudomonadales, of soil and water bacteria, Family VII, Spirillaceae, are the vibrios, comma-shaped rods, and the spirilla, spirally twisted rods. Both vibrios and spirilla are rigid, possess one flagellum or a tuft of flagella, and are actively motile. They may be saprophytes or pathogens. Two genera are discussed: *Vibrio*, containing the bacteria causing cholera, and *Spirillum*, containing the organisms causing one form of rat-bite fever.

TREPONEMA PALLIDUM (THE SPIROCHETE OF SYPHILIS)

Syphilis (lues venerea) is an infectious disease caused by *Treponema pallidum*. Clinically it may be either *acquired* or *congenital*, that is, incurred after or before birth; the former is the more frequent. The historic origin of syphilis is a debated question. Many observers believe that Columbus' sailors introduced it into Europe on their return from the New World. About that time the disease did spread over certain parts of Europe in virulent epidemic form.

Treponema pallidum (the pale spirochete) is an actively motile, slender, corkscrew-like organism that, when properly searched for, can be found in practically every syphilitic lesion (Fig. 26-2). It is especially abundant in chancres and mucous patches (lesions of the skin and mucous membranes found early in the disease). It has 6 to 14 spirals and rotates on its long axis. Usually rigid, it may bend on itself. In addition to its rotary motion, it has a slow progressive one.

Treponema pallidum is very difficult to stain with ordinarily used bacteriologic dyes* and is best demonstrated by dark-field microscopy in the syphilitic chancres and mucous patches. Since the organisms are examined in the living state by this method, their two most important diagnostic features, motility and shape, are seen.

Treponema pallidum can live outside the body under suitable conditions for 10 to 12 hours but is killed within 1 hour by drying. It does not occur outside the body except very briefly on objects contaminated with syphilitic secretions. In whole blood or plasma at refrigerator temperature it remains viable for 24 hours but dies within 48 hours.

Under natural conditions *Treponema pallidum* infects man only, but anthropoid apes, monkeys, rabbits, and guinea pigs may be infected artificially. The disease in rabbits and monkeys resembles the human disease in many but not all respects.

*Fritz Schaudinn (1871-1906), parasitologist, and Eric Hoffmann (1868-1959), clinician, described the causative agent of syphilis, *Treponema pallidum*, in 1905. Their discovery of an almost invisible parasite was the result of incomparable skill in technic and staining methods.

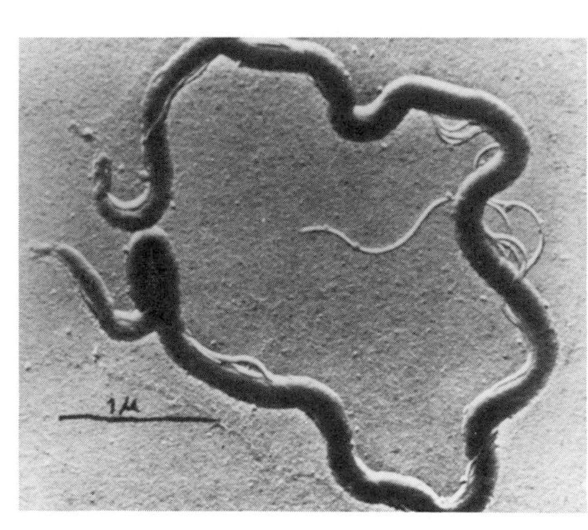

FIG. 26-2. *Treponema pallidum,* electron micrograph. (Courtesy Technical Information Services, State and Community Services Division, Center for Disease Control, Health Services and Mental Health Administration, Department of Health, Education, and Welfare.)

Acquired syphilis

■ **Modes of infection.** Syphilis is caused by a microorganism that is not borne by food, air, water, or insect. Man is its only reservoir. To contract the disease, a human being must be in close and intimate contact with an infectious person. Acquired syphilis, therefore, is contracted in the great majority of instances by sexual intercourse. In a few cases it is contracted by other types of direct contact of skin or mucous membranes, such as kissing, if the patient has lesions in his mouth. It is very rarely spread by contaminated objects, such as drinking cups or towels. Physicians, dentists, and nurses may become infected in the examination of a syphilitic patient.

With modern blood banks and blood-banking methods, the danger of transferring syphilis by blood transfusion is minimal. Blood that comes into a modern blood bank is routinely tested by a standard serologic test for syphilis. If the result is positive, that blood is not released for transfusion.

■ **Evolution of a typical case.** The course of syphilis is outlined in Table 26-1.

INCUBATION STAGE. *Treponema pallidum* is a highly invasive organism. When introduced into the human body, it begins to multiply. Many of the treponemas migrate via the lymphatics to the regional lymph nodes. From the thoracic duct they enter the bloodstream and rather quickly spread over the entire body. Some continue to multiply at the original site. From 2 to 6 weeks later an inflammatory reaction at the site of inoculation develops, and the primary lesion, the *chancre* (Fig. 26-3), forms.

PRIMARY STAGE. The chancre is the first clinical sign of infection. It usually appears on the genitals, where in the male it is readily detected. In the female the chancre, if it is on the cervix, is so located as to escape notice. Some 10% or more of chancres are extragenital, being found on the face, lips, tongue, tonsils, breasts, or fingers.

The chancre presents beneath the mucous membrane or skin as a small nodule having the feel of a shot. It breaks down, forming a shallow ulcer with indurated edges and a hard, clean base. There is little pain or discharge unless secondary infection

TABLE 26-1. EVOLUTION OF A TYPICAL CASE OF SYPHILIS

Stage	Duration	Clinical disease	Activity of Treponema pallidum	Diagnosis	Tissue change
Incubation	2 to 6 weeks (most often 3 to 4 weeks)	None	Spirochetes actively proliferate at entry site, spread over body	Identification of *Treponema pallidum:* a. Dark-field microscopy b. Fluorescent antibody technic	Chancre appears at inoculation site
Primary	8 to 12 weeks	1. Chancre present at inoculation site 2. Regional lymphadenopathy	Chancre teeming with them	1. Dark-field microscopy of chancre 2. STS* become positive	Chancre present
Primary latent	4 to 8 weeks	None	Inconspicuous	STS positive	None demonstrable; chancre has healed with little scarring
Secondary	Variable over period of 5 years (latent periods with recurrences)	1. Skin and mucosal lesions ("mucous patches") 2. Generalized lymphadenopathy	Skin and mucosal lesions rich in spirochetes (highly infectious)	1. Dark-field microscopy of lesions 2. STS positive	1. Infection active: a. Vascular changes b. Cuffs of inflammatory round cells about small blood vessels 2. Resolution spontaneous—little scarring
Latent	Few months to a lifetime (average 6 to 7 years)	None	Inconspicuous	STS positive (can be negative)	
Tertiary	Variable—rest of patient's life	Related to organ system diseased and the incapacity thereof	Paucity of spirochetes in classic lesions	1. STS positive or negative 2. Special silver stains of tissue lesions may show spirochetes	1. *Gumma* 2. Definite predilection to heal in lesions 3. Scarring 4. Tissue distortion and abnormal function

*Serologic tests for syphilis.

occurs. Chancres are usually single but may be multiple. They vary in size but are seldom more than one-half inch in diameter. After a few days the lymph nodes draining the site enlarge. Pain is absent, and the nodes show no tendency to suppurate.

After the chancre has persisted for 4 to 6 weeks, it heals. For the next 4 to 8 weeks (prior to the secondary stage) the patient shows no signs of disease. This is the *primary latent period.*

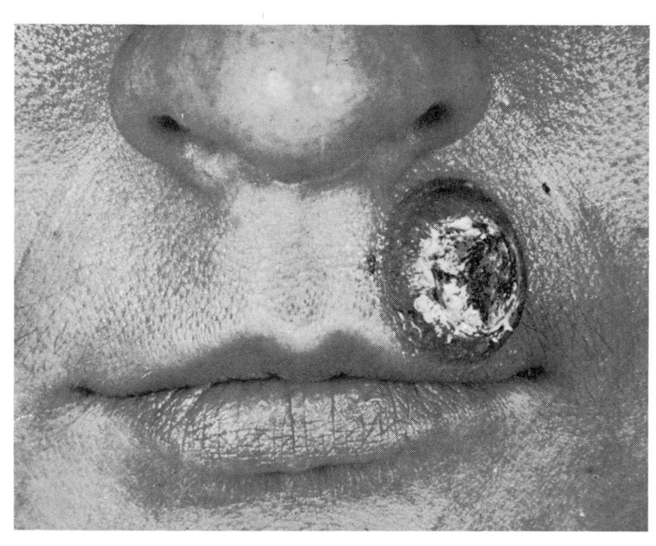

FIG. 26-3. Chancre, primary lesion of syphilis. Note extragenital location on lip. (From Top, F. H., Sr. and Wehrle, P. F.: Communicable and infectious diseases, ed. 7, St. Louis, 1972, The C. V. Mosby Co.)

SECONDARY STAGE OR STAGE OF SYSTEMIC INVOLVEMENT. The manifestations of secondary syphilis may recur over a period of 5 years. They are (1) skin lesions, (2) lesions of the mucous membranes, (3) generalized rubbery lymphadenopathy, and (4) an influenza-like syndrome. The skin eruption, a variable one, is usually symmetrically arranged, macular, and copper colored, seldom itching or burning. A patchy loss of hair, even that in the eyebrows, is associated. Known as mucous patches and most often found in the mouth, the mucosal lesions are painful superficial ulcers with a white raised surface. They are swarming with treponemas.

LATENT STAGE. After the secondary stage is a period during which the patient shows no signs, with the disease being recognized only by serologic tests. This stage may last a few months, but in 25% to 50% of the patients it lasts a lifetime. It may at anytime become active.

In some patients there is no latent period, because the tertiary stage appears right after systemic manifestations have disappeared. In those patients in whom the chancre and secondary manifestations are not present or not detected, the whole course of the disease may be of the latent type. Lesions of primary and secondary syphilis may be atypical and inconspicuous. Many patients are unaware or ignore the signs of the disease in these stages. Latent syphilis probably represents a biologic balance between the disease-producing powers of *Treponema pallidum* and the defensive forces of the body.

TERTIARY STAGE. The most characteristic lesion of the tertiary syphilis is the destructive *gumma*, a firm, yellowish white central portion surrounded by fibrous tissue. The tertiary lesions involve the deeper structures of the body and interfere materially with the functions of the internal organs. Syphilis seems to have a predilection for the cardiovascular system and the central nervous system, but other organ systems are vulnerable. Treponemas are sparse in tertiary lesions, and the tissue reaction is usually attributed to some form of allergic response.

■ **Cardiovascular syphilis.** Syphilitic involvement of the heart and blood vessels results in inflammation of the aorta (syphilitic aortitis), aneurysmal dilatation of the thoracic aorta, and distortion of the aortic valve to produce aortic insufficiency (one form of valvular disease of the heart). Syphilis also predisposes to arteriosclerosis in the aorta.

■ **Neurosyphilis.** Neurosyphilis is syphilitic involvement of the central nervous system. When syphilis becomes generalized during its early stages, the central nervous system seldom escapes invasion. In many patients the treponemas die without causing any damage there but in some 30% to 40% of patients they remain alive and initiate tertiary stage changes that are apparent weeks, months, or years later. Depending upon the anatomic site, neurosyphilis takes three forms: (1) syphilitic meningitis or meningovascular syphilis, (2) tabes dorsalis, and (3) general paresis. In addition, gummas occur as isolated lesions in various parts of the central nervous system.

A very common manifestation of neurosyphilis, *syphilitic meningitis* (more properly called *meningovascular syphilis*) is inflammation of the meninges and blood vessels of the central nervous system with or without involvement of the brain itself. The meninges are variably thickened, and typical changes occur in the walls of blood vessels. Although not all patients with syphilis develop meningovascular disease, a good percentage does. It usually occurs within the first 5 or 6 years after infection but may appear as early as 2 months or as late as 40 years.

Tabes dorsalis or locomotor ataxia is a degeneration of the posterior columns of the spinal cord and the posterior nerve roots and ganglia caused by *Treponema pallidum*. Since the sensory pathways of the spinal cord are involved, the disease is chiefly one of muscular incoordination and sensory disturbances.

General paresis (general paralysis of the insane; paralytic dementia) is a diffuse meningoencephalitis characterized by progressive mental deterioration, insanity, and generalized paralysis, terminating in death. About 3% of patients with syphilis are affected. It prefers the highly civilized to the primitive races, better educated to less well-educated persons, and is 5 times more frequent in males than in females.

Neurosyphilis may simulate almost any disease of the central nervous system, but a careful history with a correct interpretation of laboratory tests on the blood and spinal fluid usually makes the differentiation.

■ **Immunity.** Although antibody formation comparable to that in typhoid fever does not occur in syphilis, a person develops some kind of resistance. A person with syphilis is not susceptible to superimposed syphilitic infection. If he is completely cured, however, susceptibility becomes as great as ever. The serum of a patient with syphilis (cerebrospinal fluid in neurosyphilis) does contain an antibody substance not found in normal blood (or spinal fluid) that combines with an antigen prepared as a lipid extract of normal tissues, for example, the heart muscle of an ox (cardiolipin). Upon this phenomenon are based the serologic tests for syphilis, since this kind of antigen-antibody combination can be demonstrated to destroy complement or to cause visible aggregation in a colloidal suspension of antigen.

There has been much speculation about this antibody substance. Because of its uncertain nature, it is called *reagin*. According to some observers a syphilitic infection causes a breakdown of the patient's tissue cells, releasing fatty substances. These combine with the protein of the treponemas to form an antigen that stimulates the formation of antibodies to both lipids and treponemas. This explains why lipid extracts of heart muscle may be used as the antigen in the different serologic tests for syphilis and why it is possible to use treponemas as the antigen in the recently developed *Treponema pallidum immobilization test* (TPI). In this test, syphilitic serum blocks or interferes with the distinctive movements of organisms obtained from the testicles of a rabbit.

There is no definitive method of producing artificially an active or passive immunity against syphilis. Inadequate early treatment may be of greater harm than good because it fails to effect a cure and at the same time retards the establishment of any degree of immunity.

■ **Laboratory diagnosis.** The laboratory has at its disposal two important approaches to the diagnosis of syphilis: (1) the demonstration of the organisms in the lesions by dark-field microscopy or by immunofluorescence, and (2) serologic testing for syphilis (STS), notably the use of complement fixation and flocculation tests. The serologic tests include nontreponemal ones to look for reagin (performed on serum and cerebrospinal fluid) and treponemal tests to detect the antigens of *Treponema pallidum* (performed on serum as the direct source of the specific antibodies).

The dark-field microscope for demonstration of *Treponema pallidum* is most practical in the investigation of suspected chancres and mucous patches. Fluorescent antibody technics can also be used on exudates from these.

During the first few days of the chancre, the organisms are found in almost all patients. As it gets older, *Treponema pallidum* gradually disappears. Serologic tests are seldom positive during the first few days of the chancre but usually become positive before the appearance of the secondary stage. In neurosyphilis, serologic tests on both blood and cerebrospinal fluid are positive in the majority of patients.

The complement-fixation test, regardless of the technic, is commonly referred to as the Wassermann test because August von Wassermann (1866-1925) first applied in 1906 the principle of complement fixation to the diagnosis of syphilis. The test has been so much improved that, of the original, only the principle and the name remain. Two modifications of the Wassermann test in common use are the Kolmer test and the Eagle test.

Precipitation (or flocculation) tests are the Kline, Kahn, Eagle, Hinton, Mazzini, RPR (rapid plasma reagin), and VDRL (Venereal Disease Research Laboratory) tests. The VDRL is probably the most widely used and exclusively so for testing cerebrospinal fluid. The Kahn presumptive test and the Kline exclusion test are highly sensitive and can detect amounts of syphilitic reagin much less than the smallest amount that the so-called standard or diagnostic tests can. These sensitive tests have the disadvantage that positive results may often be obtained in perfectly healthy persons or persons who at least do not have syphilis. Therefore they are more significant when negative than when positive and are used for *screening*. A positive result is followed by other serologic tests, and for the laboratory diagnosis of syphilis positive results are required from two or more tests less sensitive than the screening test.

Flocculation and complement-fixation tests for syphilis may give negative results in the presence of syphilis and sometimes give positive results in its absence.* Such results are spoken of as false negative and false positive reactions. False negatives

*A serologic test for syphilis cannot be considered diagnostic of the disease. It only gives immunologic information, and none devised so far is absolutely specific.

may be caused by technical errors or by undetectably small amounts of syphilitic reagin in the blood. False positives come from technical error, or they may be biologic false positives (BFP). BFP results are occasionally or uniformly found in certain diseases, such as yaws, infectious mononucleosis, malaria, leprosy, and rat-bite fever. Partly to eliminate the biologic false positive, developments in serologic research have largely been concerned with the production of tests utilizing either the organisms themselves, dead or alive, or chemical extracts of the organisms as antigens.

The TPI test, in which *Treponema pallidum* itself is immobilized in the presence of guinea pig complement and serum from a person with syphilis, is negative with serums giving BFP reactions. Its clinical usefulness depends upon this fact. As specific as and more sensitive than the TPI test is the FTA-ABS (fluorescent treponemal antibody absorption) test, considered to be the best of the treponemal tests today. As the name indicates, test serum is absorbed of nonspecific (confusing) antibodies. It is then brought into contact with *Treponema pallidum* and fluorescein-tagged antihuman globulin in a special way so that the combination may be viewed microscopically. Antibodies to the spirochetes, if present, attach to the organism, with the antihuman globulin in turn uniting with them. When seen through the ultraviolet microscope the result glows beautifully.

Other treponemal tests are the *Treponema pallidum* complement fixation test and the Reiter protein complement fixation test (RPCF).

Remember that the diagnosis of syphilis must be made only after careful evaluation of both clinical features and laboratory findings.

■ **Prevention.** The patient is most likely to convey syphilis during the primary and secondary stages because chancres and mucous patches are practically living cultures of syphilitic organisms. Attendants of patients with these lesions should be careful not to contact the patient's secretions. Certain manual examinations should always be made with gloves because a patient may show no evidence of syphilis but at the same time be capable of transmitting the infection.

People should still be educated as to the universal prevalence of syphilis and the danger of syphilis not only to the person who has it but also to his marriage partner and his children. Today a growing public health problem, the disease is significant among young adults and especially teen-agers, in whom there has been better than a 200% increase in the incidence of syphilis over the last decade.

Considerable thought today is going to the development of a syphilitic vaccine. One under investigation is the gamma-irradiated Nichols strain of *Treponema pallidum,* nonvirulent but antigenic.

Congenital (prenatal) syphilis

Congenital syphilis refers to the disease acquired before birth. As a result of the upsurge in the incidence of venereal disease including syphilis, over 2000 babies will be born this year in the United States with congenital syphilis. For this form of syphilis to occur, the mother must be infected. The treponemas are blood borne to the maternal side of the placenta and deposited there. Syphilitic foci develop, and the organisms spread through to the fetal circulation. A syphilitic father can transmit the infection to his child only indirectly, that is, by infecting the mother.

The shorter the time elapsing between infection of the mother and conception, the more likely will the unborn baby be infected. If adequate treatment of the mother is instituted before the fifth month of pregnancy, the child should be born free of

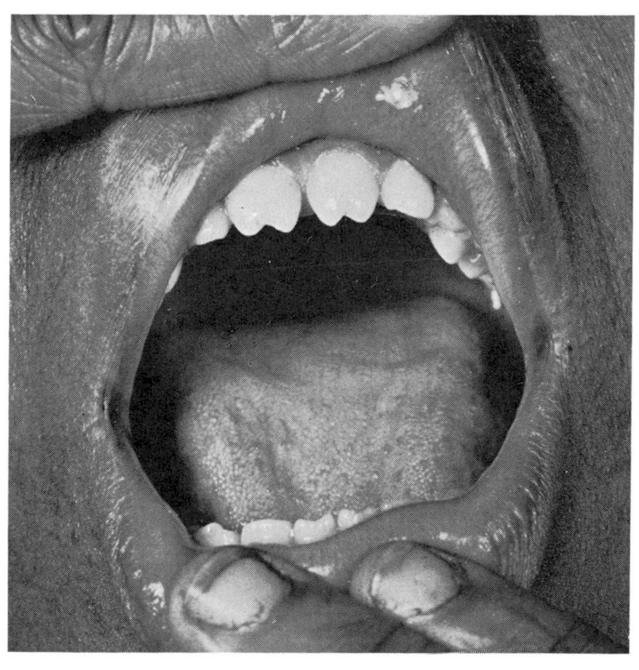

FIG. 26-4. Congenital syphilis, Hutchinson's teeth. Note notching. (From Top, F. H., Sr. and Wehrle, P. F.: Communicable and infectious diseases, ed. 7, St. Louis, 1972, The C. V. Mosby Co.)

syphilis. The requirements by law in many states for premarital examination for syphilis and for prenatal serologic tests for syphilis have been important public health measures in the prevention of this form of the disease.

Congenital syphilis usually appears at birth or within a few weeks thereafter, but it may appear years later. The child of a syphilitic mother may be (1) born dead, (2) born alive with syphilis, (3) born in apparently good health but show evidence of syphilis several weeks or months later, or (4) entirely free of the disease.

The placenta of a syphilitic child is large for the weight of the child. In children stillborn, the lungs fill the entire thoracic cavity, are grayish white, and are incompletely developed. This condition, known as white pneumonia or *pneumonia alba,* is pathognomonic for congenital syphilis. Congenital neurosyphilis simulates the acquired form and may appear in early life or be delayed to adolescence.

Infants born with active syphilis are undersized and have an appearance strikingly like that of an old man. A vesicular skin eruption and a persistent nasal discharge *(snuffles)* are often present. The child may have linear scars at the angles of the mouth *(syphilitic rhagades).* Among the late manifestations of congenital syphilis are poorly developed, small, peg-shaped permanent teeth. The upper central incisors are wedge shaped and show a central notch *(Hutchinson's teeth)* (Fig. 26-4). Other late manifestations are interstitial keratitis, anterior bowing of the tibia (saber shin), dactylitis, and neurosyphilis.

To detect congenital syphilis in a newborn baby with no outward signs, a modified FTA-ABS test is applied. Specific antihuman immunoglobulin tagged with fluorescein indicates the presence of 19S immunoglobulins. Unlike the smaller 7S immunoglobulins, which are formed by the mother, these macroglobulins cannot cross the intact placenta and must be formed by the baby. Their presence indicates the baby's reaction to his disease.

Syphilis and yaws

Yaws or frambesia, a tropical disease closely akin to syphilis, is caused by *Treponema pertenue*, an organism that cannot be differentiated from *Treponema pallidum* and that, like *Treponema pallidum*, is susceptible to arsenic and penicillin. It is positive to the same serologic tests. Some believe that yaws is but a special form of syphilis; however, most observers believe that they are distinct.

Yaws is neither venereal nor congenital. Skin and bone lesions are prominent. When the yellowish crusts covering the large pustules are removed, the slightly bleeding surface looks exactly like a raspberry stuck on the skin—hence the name frambesia or "raspberry" disease.

BORRELIA RECURRENTIS

■ **The disease.** Relapsing fever is an acute infectious disease caused by several varieties of spirochetes of the genus *Borrelia* and characterized clinically by alternating periods of febrile illness with apparent recovery. The incubation period is 3 to 10 days. Most important is *Borrelia recurrentis*. The spirochetes are found in the peripheral blood during the fever and are transferred from man to man by body lice and from rodent to man by ticks. This gives rise to two types of relapsing fever—the louse borne and the tick borne. The former occurs in epidemics with the general pattern of louse-borne diseases. The latter does not occur in epidemics and is the only type found in North America. Infected ticks may transmit the infection to their offspring for generation after generation. Body lice do not transmit the infection. Rodents, such as the ground squirrel and prairie dog, are the main reservoir of the tick-borne disease. The disease has been contracted from ticks found in caves in Texas.

■ **Laboratory diagnosis.** Actively motile spiral organisms, *Borrelia* differ from *Treponema pallidum* in taking the usual laboratory stains. They may be detected in the Wright-stained smears of peripheral blood or by dark-field illumination. Blood from a patient may be inoculated into a white mouse or young rat. Wright-stained films made from the tail blood 1 to 4 days later will show the organisms. *Borrelia* species can be grown in the chick embryo.

LEPTOSPIRA SPECIES

■ **The disease.** Leptospirosis is an acute disease caused by certain spirochetes of the genus *Leptospira*, the best known of which is *Leptospira icterohaemorrhagiae*. There are various names for this condition, including swamp fever, swineherd's disease, infectious jaundice, spirochetal jaundice, and Weil's disease. Weil's disease represents the most severe form and is usually caused by *Leptospira icterohaemorrhagiae*, although other types of leptospiras as well can cause serious infection.

Clinically, leptospirosis is characterized by high fever, muscular pains, redness of the conjunctivae, jaundice (not invariable), and aseptic meningitis. An attack is followed by a lasting immunity. Inapparent infections also occur.

■ **The organisms.** Leptospiras are tightly coiled spiral organisms with one end bent typically to form a hook. Eight known serologic types cause disease in man, and several others cause disease in animals.

Leptospiras are found in wild and domestic animals all over the world. Disease is mainly in animals; man is only accidentally infected. In animals these spirochetes localize in the kidney and are excreted profusely in the urine. They can survive if urine is discharged to neutral or slightly alkaline water, sewage, or mud. Man probably acquires the infection through the skin from soil contaminated with the urine of infected rats or through the mouth by food or water that has been contaminated in the same way. The organisms enter the animal or human body through the abraded skin and the mucous membranes of the eye, nose, and mouth. Especially vulnerable to the disease are workers in rat-infested mines, rice fields, and sewage disposal plants.

■ **Laboratory diagnosis.** The spirochetes are found in the blood early in the disease and in the urine after the seventh day. Leptospiras of distinctive shape and motility are seen by dark-field examination of blood and urine. They may be recovered from blood cultures and from the peritoneal cavity of a guinea pig inoculated with blood or urine. An agglutination test is also available.

OTHER SPIROCHETES AND RELATED ORGANISMS

Borrelia buccalis and *Treponema microdentium* are nonpathogenic organisms found in the mouth. Their presence may cause confusion in the examination of material from the mouth thought to contain the spirochete of syphilis.

Borrelia refringens is found in the mouth and associated with *Treponema pallidum* in various syphilitic lesions.

Fusospirochetal disease

In fusospirochetal disease there is a grayish white pseudomembrane that forms in the throat or mouth beneath which ulceration occurs. When the gums and mouth are primarily involved, the disease is known as *trench mouth.* When the throat and tonsils are ulcerated with extensive tissue involvement, it is called *Vincent's angina.* The disease is important within itself and because the membrane may be mistaken for a diphtheritic one. The extensive ulceration may cause the disease to be confused with syphilis.

Associated together in the lesions are two organisms: one a gram-negative, cigar-shaped anaërobic bacterium, *Fusobacterium fusiforme,* and the other a spirochete, *Borrelia vincentii.* Early, the bacilli are more numerous; late, the spirochetes are. These organisms do not cause this condition, and most workers believe that they are only secondary invaders. The fusospirochetal organisms growing symbiotically in this condition are opportunists. Since everyone already has the organisms in his mouth, merely their presence is not enough to cause disease. This must be triggered by some unusual circumstance such as an injury to the mouth or decreased oral resistance, as with malnutrition, viral infection, or poor oral hygiene. Fusospirochetal organisms are identified directly on a stained smear of the membranous exudate.

VIBRIO SPECIES

Vibrio cholerae (the comma bacillus)*

■ **The organism.** Cholera is caused by *Vibrio cholerae (Vibrio comma),* the comma bacillus, a small, comma-shaped, motile, gram-negative rod. It multiplies rapidly in the lumen of the small bowel and produces a powerful exo-enterotoxin that acts on the lining of the bowel to induce copious loss of water and essential salts. This explains the fecal discharges being described as "rice water"—clear, not malodorous, with flecks of mucus. Organisms abound in the stools. There are two immunologic types.

The cholera vibrio grows aerobically on routine laboratory media. If a few drops of sulfuric acid are added to a growth of cholera vibrios in nitrate-peptone broth, a red color develops. This is the *cholera red reaction.*

■ **The disease.** Asiatic cholera is a specific infectious disease that affects the lower portion of the intestine and is described by violent purging, vomiting, burning thirst, muscular cramps, suppression of urine, and rapid collapse. Untreated, it can be a terrifying disease with a 70% or more mortality. Only plague causes as much panic. With massive diarrhea the patient's fluid losses are enormous—10 to 20 quarts a day. With severe rapid dehydration death comes within hours. Four great pandemics spread over the world during the eighteenth century, and on two occasions in the nineteenth century the disease invaded the United States: in 1832 in New York City and in 1848 in New Orleans, whence it spread up the Mississippi Valley. The disease is endemic in India and China. The "scourge of antiquity," it originated in India in the vicinity of Calcutta and on the delta of the Ganges River, where it was known at the time of Alexander the Great.

■ **Mode of infection.** The disease is contracted by the ingestion of water or food contaminated by the excreta of persons harboring the bacilli.† Man is the only host. The bacilli leave the body in the feces, urine, and secretions of the mouth. As a rule, the feces become free of bacilli during the last days of the disease, but some patients become convalescent carriers. Permanent carriers do not occur.

Cholera today is rampant in the areas of the world occupied by better than half the world's population where, because of low standards of sanitation, human excrement easily contaminates the waterways and surface wells that are sources of drinking water. Vibrios may live in water for as long as 2 weeks.

As a disease cholera should be only a historical note. Modern sanitation can eliminate waste-contaminated water supplies, the primary source of infection, and modern medical treatment can effectively deal with it.

Vibrio Eltor

A biotype to *Vibrio cholerae,* more resistant to physical and chemical agents and in itself quite virulent, is *Vibrio Eltor.* Unlike the cholera vibrio, the Eltor vibrio produces a hemolysin that lyses the red cells of sheep and goats. (See Table 26-2 for a comparison of the two.) It causes cholera, and in recent times has been more significant

**Vibrio comma* of Bergey's classification.

†John Snow (1813-1858), English epidemiologist, was the first to recognize the transmission of cholera by contaminated water, and by the middle of the nineteenth century he had formulated his views. He correctly observed in 1854 that the Broad Street Pump in London was a source of infection in an epidemic of cholera killing some 11,000 persons.

TABLE 26-2. "CLASSIC" CHOLERA VIBRIO COMPARED WITH "ELTOR" BIOTYPE

	Voges-Proskauer test	Production of indole	Liquefaction of gelatin	Production of hydrogen sulfide	Cholera red reaction	Fermentation of glucose	Fermentation of lactose	Hemolysis of sheep (or goat) cells	Hemagglutination of chicken cells	Phage lysis
Vibrio cholerae	−	+	+	−	+	+	Slow	−	−	Susceptible
Vibrio cholerae, biotype Eltor	+	+	+	−	+	+	Slow	+	+	Resistant

than the classic vibrio as the cause of pandemics in South and Southeast Asia. The current one is the seventh resulting from this biotype.

■ **Immunity.** The immunity occurring after an attack of cholera is short lived. The risk of reinfection is only slightly less than that of the initial infection. Although vaccination is done against cholera,* there is to date no truly effective vaccine universally available. An experimental "toxoid" vaccine has been put to field trial.

Other vibrios and vibriosis

Vibrio parahemolyticus, a vibrio like the classic one, is a major cause of a gastroenteritis with manifestations similar to the classic disease. It is widespread in its distribution and in countries such as Japan is responsible for 50% to 60% of cases of "summer diarrhea." It is a small halophilic vibrio giving a negative cholera red reaction. Present vaccines are of no effect against it.

Vibrio fetus, a microaerophilic vibrio, is an important cause of infectious abortion in cattle and sheep and is isolated rarely from man.

SPIRILLUM MINUS

There are two similar clinical entities known as rat-bite fever that are conveyed by the bite of a rat or other rodent. One, known in Japan as sodoku, is caused by *Spirillum minus (Spirochaeta muris).* This small and rigid spiral organism is found primarily in wild rats and is spread from rat to rat and from rat to man by the bite of the rat. The features of the disease are ulceration of the bite, fever, and a skin eruption. The spirochetes may be seen in material from the ulcer examined in stained smears or by darkfield illumination, and a guinea pig may be inoculated with blood or tissue from lymph nodes to isolate them. Rat-bite fever of this type is uncommon but is worldwide in distribution. The other type of rat-bite fever is not caused by a spirochete but by an actinomycete-like organism (p. 505).

*For cholera immunization, see pp. 606, 627, and 629.

QUESTIONS FOR REVIEW

1. Name and describe the causative agent of syphilis.
2. Outline the evolution of a typical case of syphilis.
3. Characterize congenital syphilis. What are its hazards?
4. How is syphilis transmitted?
5. Outline the laboratory diagnosis of syphilis. Briefly discuss the serologic tests. State the criteria for the laboratory diagnosis.
6. Give a general classification of spiral microorganisms. Present their salient features.
7. Comment on the pathogenicity of *Treponema pallidum.*
8. Name the causative agent of yaws. How does it compare with the spirochete of syphilis?
9. Name two vectors of relapsing fever and possible rodent reservoirs.
10. Characterize leptospirosis. What spiral microorganism causes Weil's disease?
11. Give the names of the two microbes causing rat-bite fever.
12. How is Vincent's angina diagnosed bacteriologically?
13. Compare the classic cholera vibrio with its biotype Eltor.
14. How is Asiatic cholera transmitted? How can it be prevented?
15. Explain the pathogenic mechanism for the diarrhea in cholera. State the serious consequence of this.

REFERENCES FOR CHAPTERS 19 TO 26

Allison, F., Jr., and Sanders, C. V.: The extragenital manifestations of gonococcal infection, Southern Med. Bull. **59:**38, (April) 1971.

Altemeier, W. A.: The significance of infection in trauma, A.O.R.N. J. **15:**92, (March) 1972.

Altemeier, W. A., and Fullen, W. D.: Prevention and treatment of gas gangrene, J.A.M.A. **217:**806, 1971.

Anderson, D. R.: The organisms called *Mycoplasma,* Bull. Path. **9:**2, 1968.

Anderson, P., et al.: Immunization of humans with polyribophosphate, capsular antigen of *Haemophilus influenzae,* type b, J. Clin. Invest. **51:**39, (Jan.) 1972.

Anthony, B. F., et al.: Nursery outbreak of staphylococcal scalded skin syndrome, Amer. J. Dis. Child. **124:**41, (July) 1972.

Armstrong, D., and Kaplan, M. H.: Sepsis: A pictorial guide to some important clinical signs and initial laboratory studies, Hosp. Med. **8:**22, (July) 1972.

Artenstein, M. S., and Ellis, R. E.: The risk of exposure to a patient with meningococcal meningitis, Milit. Med. **133:**474, (June) 1968.

Barnett, J. A., and Sanford, J. P.: Bacterial shock, J.A.M.A. **209:**1514, 1969.

Barson, A. J.: Fatal *Pseudomonas aeruginosa* bronchopneumonia in a children's hospital, Arch. Dis. Child. **46:**55, 1971.

Bartlett, R. C.: Infection by opportunistic bacteria, Bull. Path. **9:**107, 1968.

Barton, F. W.: Venereal disease, editorial, J.A.M.A. **216:**1472, 1971.

Battle, C. U., and Glasgow, L. A.: Reliability of bacteriologic identification of β-hemolytic streptococci in private offices, Amer. J. Dis. Child. **122:**134, 1971.

Belliveau, R. R., Grayson, J. W., Jr., and Butler, T. J.: A rapid, simple method of identifying Enterobacteriaceae, Amer. J. Clin. Path. **50:**126, 1968.

Benenson, A. S.: Cholera today, editorial, Arch. Environ. Health **18:**305, 1969.

Benenson, A. S., editor: Control of communicable diseases in man, Washington, D. C., 1970, American Public Health Association, Inc.

Blount, J. H.: A new approach for gonorrhea epidemiology, Amer. J. Public Health **62:**710, 1972.

Boris, M., Shinefield, H. R., Romano, P., et al.: Bacterial interference, Amer. J. Dis. Child. **115:**521, 1968.

Bowmer, E. J.: Salmonellae: Ubiquitous pathogens of animals and man, editorial, Canad. J. Public Health **59:**408, (Oct.) 1968.

Braude, A. I.: Anaerobic infection: diagnosis and therapy, Hosp. Pract. **3:**42, (Feb.) 1968.

Brown, W. J.: Acquired syphilis, drugs and blood tests, Amer. J. Nurs. **71:**713, 1971.

Brown, W. J.: Trends and status of gonorrhea in the United States, J. Infect. Dis. **123:**682, 1971.

Bugg, R.: Fighting the masked crippler: Rheumatic fever, Today's Health **4:**37, (March) 1968.

Busch, L. A.: Human listeriosis in the United States, 1967-1969, J. Infect. Dis. **123:**328, 1971.

Caldwell, J. G.: Congenital syphilis: A nonvenereal disease, Amer. J. Nurs. **71:**1768, 1971.

Carpenter, C. C. J., Jr., and Hirschhorn, N.: Pediatric cholera: Current concepts of therapy, J. Pediat. **80:**874, 1972.

Caten, J. L., and Kartman, L.: Human plague in the United States, J.A.M.A. **205:**333, 1968.

Charles, A. G., et al.: Asymptomatic gonorrhea in prenatal patients, Amer. J. Obstet. Gynec. **108:**595, 1970.

Cherubin, C. E., Millian, S. J., Palusci, E., et al.: Investigations in tetanus in narcotic addicts in New York City, Amer. J. Epidem. **88:**215, (Sept.) 1968.

Christy, J. H.: Pathophysiology of gram-negative shock, Amer. Heart J. **81:**694, 1971.

Crawford, R. P., et al.: Human infections associated with waterborne leptospires, and survival studies on serotype *pomona,* J. Amer. Vet. Med. Ass. **159:**1477, 1971.

Crowder, J. G., et al.: Type-specific immunity in Pseudomonas diseases, J. Lab. Clin. Med. **79:**47, (Jan.) 1972.

Dajani, A. S.: The scalded-skin syndrome: Relation to phage-group II staphylococci, J. Infect. Dis. **125:**548, 1972.

Davies, J. W., et al.: Typhoid at sea: Epidemic aboard an ocean liner, Canad. Med. Ass. J. **106:**877, 1972.

Dillon, H. C.: Impetigo contagiosa: Suppurative and non-suppurative complications, Amer. J. Dis. Child. **115:**530, 1968.

Domingue, G. J., Dean, F., and Miller, J. R.: A diagnostic schema for identifying Enterobacteriaceae and miscellaneous gram-negative bacilli, Amer. J. Clin. Path. **51:**62, 1969.

Drewett, S. E., et al.: Eradication of *Pseudomonas aeruginosa* infection from special-care nursery, Lancet **1:**946, 1972.

Dryden, G. E.: Sources of infection in the surgical patient, A.O.R.N. J. **13:**64, (May) 1971.

Duma, R. J.: Management of hospital infections, GP **39:**106, (Jan.) 1969.

Duma, R. J.: *Pseudomonas* and water, with a new twist, editorial, Ann. Intern. Med. **76:**506, 1972.

Dunton, E. A., and Raymond, O. C.: The ramifications of salmonellosis, Canad. J. Public Health **59:**246, 1968.

Edwards, P. R., and Ewing, W. H.: Identification of Enterobacteriaceae, Minneapolis, 1972, Burgess Publishing Co.

Edwin Klebs (1834-1913), peripatetic bacteriologist, editorial, J.A.M.A. **204:**729, 1968.

Eickhoff, T. C., Brachman, P. S., Bennett, J. V., et al.: Surveillance of nosocomial infection in community hospitals: I. Surveillance methods, effectiveness, and initial results, J. Infect. Dis. **120:**305, 1969.

Epidemic plague, editorial, J.A.M.A. **205:**871, 1968.

Experimental induction of glomerulonephritis with group A streptococci, editorial, J.A.M.A. **204:**922, 1968.

Favero, M. S., Carson, L. A., Bond, W. W., et al.: *Pseudomonas aeruginosa:* Growth in distilled water from hospitals, Science **173:**836, 1971.

Feinstein, A. R.: A new look at rheumatic fever, Hosp. Pract. **3:**71, (March) 1968.

Fekety, F. R., Jr., et al.: Bacteria, viruses and mycoplasmas in acute pneumonia in adults, Amer. Rev. Resp. Dis. **104:**499, 1971.

Feldman, H. A.: Some recollections of the meningococcal diseases. J.A.M.A. **220:**1107, 1972.

Finkelstein, R. A., and LoSpalluto, J. J.: Crystalline cholera toxin and toxoid, Science **175:**529, 1972.

Fleming, W. L., et al.: Clinical gonorrhea in the female: laboratory findings, Southern Med. J. **65:**890, 1972.

Gangarosa, E. J., and Faich, G. A.: Cholera: The risk to American travelers, Ann. Intern. Med. **74:**412, 1971.

Gilardi, G. L.: Practical schema for the identification of nonfermentative gram negative bacteria encountered in medical bacteriology, Amer. J. Med. Tech. **38:**65, (March) 1972.

Goodheart, B.: Winter's unvanquished foe—pneumonia, Today's Health **48:**50, (Nov.) 1970.

Greenberg, J. H., and Madorsky, D. D.: Young physicians' knowledge of venereal disease, editorial, J.A.M.A. **220:**1736, 1972.

Greenwood, B. M., Whittle, H. C., and Dominic—Rajkovic, O.: Counter-current immunoelectrophoresis in the diagnosis of meningococcal infections, Lancet **2:**519, 1971.

Guilt of *Streptococcus* proved, editorial, J.A.M.A. **209:**261, 1969.

Hardman, J. M.: Fatal meningococcal infections; the changing pathologic picture in the 60's, Milit. Med. **133:**951, (Dec.) 1968.

Harner, R. E., Smith, J. L., and Israel, C. W.: The FTA-ABS test in late syphilis, J.A.M.A. **203:**545, 1968.

Hofer, J. W., and Davis, J.: Survival and dormancy of Clostridia spores, Texas Med. **68:**80, (Feb.) 1972.

Isenberg, H. D.: The ecology of nosocomial disease, ASM News **38:**375, 1972.

Janeff, J., et al.: A screening test for streptococcal antibodies, Lab. Med. **2:**38, (July) 1971.

Jewett, J. F., Reid, D. E., Safon, L. E., et al.: Childbed fever—a continuing entity, J.A.M.A. **206:**344, 1968.

Johnston, D.: Essentials of communicable disease with nursing principles, St. Louis, 1968, The C. V. Mosby Co.

Jordan, G. W.: Bacteriologic and clinical features of *Edwardsiella tarda,* Bull. Path. **10:**249, 1969.

Kadis, S., Montie, T. C., and Ajl, S. J.: Plague toxin, Sci. Amer. **220:**92, (March) 1969.

Kahn, S. P., et al.: Clostridia hepatic abscess: Unusual manifestation of metastatic colon carcinoma, Arch. Surg. **104:**209, 1972.

Kampmeier, R. H.: The matter of venereal disease in 1971, editorial, Ann. Intern. Med. **75:**793, 1971.

Kaplan, K., and Weinstein, L.: Diphtheroid infections of man, Ann. Intern. Med. **70:**919, 1969.

Kellogg, D. S., Jr., and Mothershed, S. M.: Immunofluorescent detection of *Treponema pallidum,* J.A.M.A. **207:**938, 1969.

Klein, J. O.: Mycoplasmas, GU tract infections and reproductive failure, Hosp. Pract. **6:**127, (Jan.) 1971.

Krolls, S. O., et al.: Oral manifestations of syphilis, Hosp. Med. **8:**14, (July) 1972.

Krugman, S., and Ward, R.: Infectious diseases of children, ed. 4, St. Louis, 1968, The C. V. Mosby Co.

Kushner, I.: Gonococcal arthritis: An important complication of gonorrhea, Res. Staff Phys. **18:**37 (March) 1972.

La Force, F. M., Young, L. S., and Bennett, J. V.: Tetanus in the United States (1965-1966): Epidemiologic and clinical features, New Eng. J. Med. **280:**569, 1969.

Lancefield, R. C.: Current problems in studies of streptococci, J. Gen. Microbiol. **55:**161, 1969.

Lenz, P. E.: Women, the unwitting carriers of gonorrhea, Amer. J. Nurs. **71:**716, 1971.

Lerro, S. J.: Human brucellosis in Texas, epidemiology 1960-1969, Texas Med. **67:**60, (Oct.) 1971.

Levin, J., et al.: Detection of endotoxin in the blood of patients with sepsis due to gram-negative bacteria, New Eng. J. Med. **283:**1313, 1970.

Mackey, D. M., Price, E. V., Knox, J. M., et al.: Specificity of the FTA-ABS test for syphilis, J.A.M.A. **207:**1683, 1969.

Maisel, J. C., Babbitt, L. H., and John, T. J.: Fatal *Mycoplasma pneumoniae* infection with isolation of organisms from lung, J.A.M.A. **202:**287, 1967.

Maniar, A. C. and Fox, J. G.: Techniques of an "in vitro" method for determining toxigenicity of *Corynebacterium diphtheriae* strains, Canad. J. Public Health **59:**297, 1968.

McCue, C. M.: Acute rheumatic fever today, Res. Staff Phys. **16:**51, (Nov.) 1970.

Melish, M. E., and Glasgow, L. A.: Staphylococcal scalded skin syndrome: Expanded clinical syndrome, J. Pediat. **78:**958, 1971.

Meyers, B. R., et al.: Infections caused by microorganisms of the genus *Erwinia,* Ann. Intern. Med. **76:**9, 1972.

Michaelson, M.: Wyoming's war on deadly strep, Today's Health **49:**40, (March) 1971.

Mudd, S.: Resistance against *Staphylococcus aureus,* J.A.M.A. **218:**1671, 1971.

Nosocomial bacteremias associated with intravenous fluid therapy—USA (epidemiologic notes and reports), Morbid Mortal Wk. Rep. **20**(9) (Special Suppl.):1, 1971.

Office Veterinary Public Health Services, Center for Disease Control, report of subcommittee on public health: Brucellosis in the United States, 1970, Arch. Environ. Health **25:**66, (July) 1972.

Owen, R. L., and Hill, J. L.: Rectal and pharyngeal gonorrhea in homosexual men, J.A.M.A. **220:**1315, 1972.

Palmer, D. L., et al.: Clinical features of plague in the United States, 1969-1970 epidemic, J. Infect. Dis. **124:**367, 1971.

Pariser, H., and Farmer, A. D.: Diagnosis of gonorrhea in the asymptomatic female: Comparison of slide and culture technics, Southern Med. J. **61:**505, 1968.

Piggott, J. A., and Hochholzer, L.: Human melioidosis, Arch. Path. **90:**101, 1970.

Problems in leptospirosis, WHO Chron. **22:**159, (April) 1968.

Puerperal fever today, editorial, J.A.M.A. **206:**367, 1968.

Quinn, R. W.: Epidemiology of gonorrhea, Southern Med. Bull. **59:**7, (April) 1971.

Rammelkamp, C. H., Jr.: Rheumatic fever: Advances and problems, editorial, Hosp. Pract. **3:**9, (March) 1968.

Rammelkamp, C. H., Jr., and Monson, T. P.: Strep infections: Time is not on your side. Res. Staff Phys. **17:**44, (July) 1971.

Read, S. E., and Reed, R. W.: Electron microscopy of the replicative events of A_{25} bacteriophages in group A streptococci, Canad. J. Microbiol. **18:**93, 1972.

Rensberger, B., and Roueché, B.: When Americans are a swallow away from death, Today's Health **49:**40, (Sept.) 1971.

Rhamy, R. K.: Gonococcal complications in genitourinary system of the male, Southern Med. Bull. **59:**29, (April) 1971.

Rhodes, L. M.: Why cholera is the disease nations try to hide, Today's Health **49:**51, (June) 1971.

Rodman, M. J.: Drugs for treating tetanus, RN **34:**43, (Dec.) 1971.

Rosebury, T.: Microbes and morals: The strange story of venereal disease, New York, 1971, Viking Press.

Rowley, D.: Endotoxins and bacterial virulence, J. Infect. Dis. **123:**317, 1971.

Rudolph, A. H.: Control of gonorrhea, guidelines for antibiotic treatment, J.A.M.A. **220:**1587, 1972.

Sanders, D. Y.: Complications of ear piercing, Woman Physician **26:**459, 1971.

Sanders, W. E.: Diphtheria: "From miasmas to molecules" revisited, editorial, Ann. Intern. Med. **75:**639, 1971.

Sayeed, Z. A., et al.: Gonococcal meningitis, J.A.M.A. **219:**1730, 1972.

Schmale, J. D.: Serologic detection of gonorrhea, editorial, Ann. Intern. Med. **72:**593, 1970.

Schroeter, A. L., and Pazin, G. J.: Gonorrhea, Ann. Intern. Med. **72:**553, 1970.

Shepard, M. C.: Nongonococcal urethritis associated with human strains of "T" mycoplasmas, J.A.M.A. **211:**1335, 1970.

Small, N. N.: Evaluation of PathoTec™ strips in diagnostic bacteriology, Amer. J. Med. Tech. **34:**65, (Feb.) 1968.

Smith, B. B.: After the nail it's too late, Today's Health **48:**30, (Aug.) 1970.

Smith, D. T., et al.: Zinsser microbiology, New York, 1968, Appleton-Century-Crofts.

Smith, J. W.: Tetanus, Texas, Med. **66:**52, (Sept.) 1970.

Sommers, H. M.: Are we recognizing most cases of clinical botulism? Lab. Med. 1:41, (Aug.) 1970.

South, M. A.: Enteropathogenic *Escherichia coli* disease: New developments and perspectives, J. Pediat. **79:**1, (July) 1971.

Spitz, B.: In the hospital a bug is no joke, Hosp. Topics **50:**42, (Jan.) 1972.

Stein, F., and Cohen, J. R.: Acute *Listeria* meningitis, GP **38:**88, (July) 1968.

Strage, M.: VD: The clock is ticking, Today's Health **49:**16, (April) 1971.

Sullivan, R. J., Jr., et al.: Adult pneumonia in general hospital, Arch. Intern. Med. **129:**935, 1972.

Syphilis: The same old disease, editorial, Southern Med. J. **65:**250, 1972.

Thatcher, R. W., and Pettit, T. H.: Gonorrheal conjunctivitis, J.A.M.A. **215:**1494, 1971.

The two faces of pathogenic *Escherichia coli,* editorial, J.A.M.A. **218:**248, 1971.

Tillotson, J. R., and Lerner, A. M.: *Bacteroides* pneumonias: Characteristics of cases with empyema, Ann. Intern. Med. **68:**308, 1968.

Tindall, J. P., and Harrison, C. M.: *Pasteurella multocida* infections following animal injuries, especially cat bites, Arch. Derm. **105:**412, 1972.

Today's VD control problem 1972, New York, 1972, American Social Health Association.

Tong, M. J.: Septic complications of war wounds, J.A.M.A. **219:**1044, 1972.

Tooth decay thought to be infectious disease, Today's Health **46:**76, (April) 1968.

Top, F. H., Sr., and Wehrle, P. F.: Communicable and infectious diseases, ed. 7, St. Louis, 1972, The C. V. Mosby Co.

Torphy, D. E.: Bacteriology of dog and cat bites, Bull. Path. **10:**276, 1969.

Vandermeer, D. C.: Meet the VD epidemiologist, Amer. J. Nurs. **71:**722, 1971.

VanHeyningen, W. E.: Tetanus, Sci. Amer. **218:**69, (April) 1968.

Van Ness, G. B.: Ecology of anthrax, Science **172:**1303, 1971.

Venereal disease rampant, editorial, J.A.M.A. **218:**731, 1971.

Von der Muehll, E., et al.: A new test for pathogenicity of staphylococci, Lab. Med. **3:**26, (Jan.) 1972.

Washington, J. A., II: Useful laboratory procedures: gram-negative bacilli other than Enterobacteriaceae, Bull. Path. **10:**316, 1969.

Weed, L. A.: Useful laboratory procedures: Brucellosis, Bull. Path. **10:**120, 1969.

Whitby, J. L., and Rampling, A.: *Pseudomonas aeruginosa* contamination in domestic and hospital environments, Lancet **1:**15, (Jan. 1) 1972.

Wilkowske, C. J., Washington, J. A., 2d., Martin, W. J., et al.: *Serratia marcescens:* Biochemical characteristics, antibiotic susceptibility patterns and clinical significance, J.A.M.A. **214:**2157, 1970.

Williams, C. P. S., and Oliver, T. K., Jr.: Nursery routines and staphylococcal colonization of newborn, Pediatrics **44:**640, 1969.

Wood, W. H., Jr., and Wolinsky, E.: Treatment of staphylococcal disease, Hosp. Med. **7:**87, (March) 1971.

World Health Organization Technical Report Series No. 464, Joint FAO/WHO Expert Committee on Brucellosis, 5th report, Geneva, 1971, World Health Organization.

Ziegler, P.: The Black Death, London, 1969, William Collins Sons & Co., Ltd.

27 The viruses

GENERAL CONSIDERATIONS

■ **Definition.** Viruses are agents that cause infectious disease and can only be propagated in the presence of living tissues. If we say that the least requirement for life is the ability of a living thing to duplicate or reproduce itself, then viruses are the smallest known living bodies. Most of them are so small that they cannot be seen with an ordinary light microscope, and some are so small that they approximate the size of the large protein molecules. Viruses seem to lie in a partially explored twilight zone between the cells the biologist studies, on the one hand, and the molecules with which the chemist deals, on the other, being smaller than the smallest known bacterial cells

TABLE 27-1. RELATIVE SIZES OF VIRUSES

Biologic unit	Approximate diameter (or diameter × length) in mμ*
RED BLOOD CELL	7,500
BEDSONIAE (CHLAMYDIAE)†	300-800
Poxvirus	230×300
Rhabdovirus	60×225
Coronavirus	80-160
Paramyxovirus	100-300
Myxovirus	80-120
Herpesvirus	110-200
Bacterial virus	25-100
Adenovirus	70-80
Leukovirus	100
Reovirus	70-75
Rubella virus (togavirus)	50-60
Papovavirus	40-55
Arbovirus	40
Picornavirus	18-30
Parvovirus (picodnavirus)	18-24
Certain plant viruses	17-30
SERUM ALBUMIN MOLECULE	5

*1 mμ = 1/1000 μ = 1/1,000,000 mm.
†See p. 463.

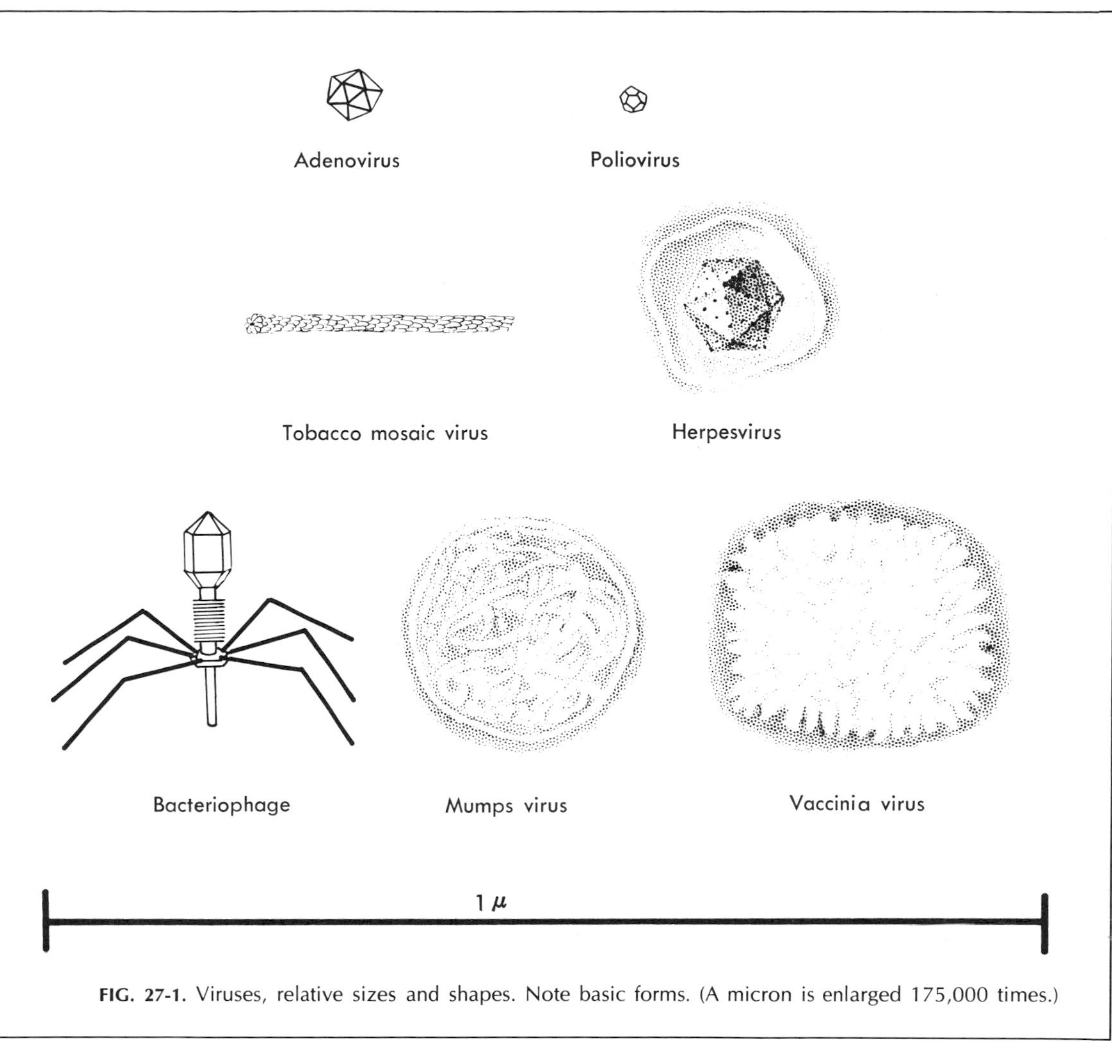

FIG. 27-1. Viruses, relative sizes and shapes. Note basic forms. (A micron is enlarged 175,000 times.)

and just larger than the largest protein molecules.* (It would take 2500 poliovirus particles to span the point of a pin.) Most viruses can pass through filters that retain all ordinary bacteria and for many years have been called filtrable viruses.

Our knowledge of the nature and structure of viruses has increased greatly in the last several decades, especially since the invention of the electron microscope, the introduction of the ultracentrifuge, the development of modern tissue culture technics, and the applications of immunofluorescence and cytochemistry. As a result, many viruses previously unknown have been isolated and identified—in the last 10 or more years, over a 100. Also, more has been learned about the natural history or *life cycle* of a virus.

■ **Structure.** Viruses are particulate and vary considerably in size (Table 27-1) and in shape (Fig. 27-1). Generally, the viruses of man and animals are spherical, those

*Dr. Wendell Meredith Stanley (1904-1971), biochemist and virologist, defined viruses as follows: "The virus is one of the great riddles of biology. We do not know whether it is alive or dead, because it seems to occupy a place midway between the inert chemical molecule and the living organism." (From Alvarez, W.: Mod. Med. **35**:78, Jan. 30, 1967.)

TABLE 27-2. COMPARISON OF VIRUSES WITH OTHER INFECTIOUS MICROBES*

Agent	Presence of nucleic acids	Reproduction	Possession of metabolic enzymes	Obligate intracellular parasite	Rigid cell wall	Antibiotic suscepti-bility
Viruses	Either DNA or RNA (not both)	Replication (synthesis, then assembly of subunits)	None	Yes	None	None
Bedsoniae (chlamydiae)	DNA and RNA	Binary fission	Yes (parts of enzyme systems present)	Yes	May be present	Yes
Rickettsiae	DNA and RNA	Binary fission	Yes	Yes	Yes	Yes
Mycoplasmas	DNA and RNA	Binary fission	Yes	No	None	Yes
Bacteria	DNA and RNA	Binary fission	Yes	No	Yes	Yes

*(After Moulder, J. W.: The life and death of the psittacosis virus, Hosp. Practice **3:**35 June 1968.)

of plants are rod shaped or many sided, and those of bacteria (bacterial viruses or *bacteriophages*) are tadpole shaped. Once thought to be fairly simple, viruses are known to be highly complex structures in which viral components are fitted together into rigid geometric patterns with mathematical precision. In fact, a highly purified preparation of virus is referred to as a virus crystal (Fig. 28-2). There are three basic forms in viral structure—the icosahedral, the helical, and the complex. The *icosahedral* is a solid many-sided geometric form—with 20 triangular faces and 12 apexes. The *helical* form indicates a spiral tubular structure bound up to make a compact mass. The *complex* is variable in size and shape.

In the simplest viruses there is a long coil of nucleic acid, the *chromosome* of the virus, tightly folded and packed within a protein coat called the *capsid.* The capsid is made up of subunits or *capsomeres,* arranged in precise fashion around the nucleic acid core. In the larger viruses a more complicated chemical structure is seen wherein fatty substances and complex sugars (but no enzymes) are linked to the protein coat. The viral particle as a unit is referred to as a *viron.**

Cells of bacteria, plants, and animals contain both deoxyribonucleic acid (DNA) and ribonucleic acid (RNA). True viruses, however, contain only one. DNA serves as the genetic material in complex viruses, such as vaccinia and bacteriophage, as it does for all other living things. RNA is the genetic material not only in intricate viruses such as the influenza viruses but also in some of the simplest and smallest such as the polioviruses.

Table 27-2 highlights some of the differences between viruses and other microbes.

*A *viroid* is a new kind of agent recently described. Said to be like a virus, it is 1/80th the size of the smallest known virus. It may be the responsible agent in some forms of cancer and for such diseases as hepatitis (p. 449).

■ **Life cycle.** The life cycle of a virus may be briefly sketched as follows: A virus contacts and parasitizes a susceptible cell in a given host. In some way the protein coat of the virion seems to be able to find just the right cell. Besides protecting the viral particle during its extracellular existence, the protein coat can attach to the cell under attack. Next the protein coat is stripped off, and viral nucleic acid only (the chromosome of the virus) penetrates the parasitized cell, probably aided by enzymes from that cell. This kind of phagocytosis is called *viropexis.* The dissembled virus appears to vanish into the cytoplasmic maze of the cell.

Invasion of the cell by the virus is an act of piracy. Because a virus contains such a small number of proteins, it cannot do much, if any, metabolic work. It is not capable of independent reproductive activity. It must seek out an environment in which chemical building blocks and energy for work are provided, conditions existing only inside a cell. A virus must take nourishment from the parasitized cell, and it can multiply only within the confines of that cell. Since its genetic material is carried into the cell, the information needed to make new viruses inside that cell is available.* In fact, viral RNA can initiate the life cycle of the virus from which it was isolated even when chemically separated from the protein coat. The entry of viral nucleic acid into the host cell results in an infected cell that is then immune; it cannot be reinfected by the same or related viruses.

A variable period of time (the eclipse) elapses after the resources of the cell have been commandeered before new viral particles are released. At the end of this phase, swarms of full-grown viruses appear and escape from the cell. (In the case of poliovirus [Fig. 27-2] in only a few hours a single parasitized cell has produced 100 thousand poliovirus particles.) When released, the new generation of viruses is able to survive outside the cell until it can reach the susceptible cells of another host, but this new generation of viruses will have to find new living quarters.

■ **Pathogenicity.** In many instances, little or no harm is occasioned in the host by the life cycle of a virus, and many viral infections are silent, inapparent ones. For example, unrecognized infection with polioviruses is hundreds of times more common than is the clinical disease. Some infections are latent ones, undetectable until activated by the proper stimulus; for example, infections with herpes simplex. However, if the cells are damaged by the viral attack, then disease exists, and the signs of viral infection naturally reflect the anatomic location of the cells.

The pathology of viral diseases then relates primarily to the visible effects of intracellular parasitism. When cells that have been specifically attacked by a virus show changes that directly reflect cell injury, the sum total of these is designated the *cytopathogenic* or *cytopathic effect* (CPE), an especially useful concept in tissue culture work.

In some cells small, round, or oval bodies known as *inclusion bodies* will be formed as a manifestation of the cytopathic effect. Inclusion bodies are typical in size, shape, and intracellular position for a given virus. They are emphasized by the pathologist since they are usually easily visualized in the compound light microscope. In stained preparations inclusion bodies are commonly about the size of a red blood cell and rounded. Some viruses produce inclusion bodies in the cytoplasm of the parasitized cell and some in the nucleus; others produce them in both cytoplasm and nucleus. At times within the inclusion bodies still smaller units known as *elementary bodies* are seen. The elementary bodies are the virus particles, and an inclusion body with its

*Viruses act as independently existing genes.

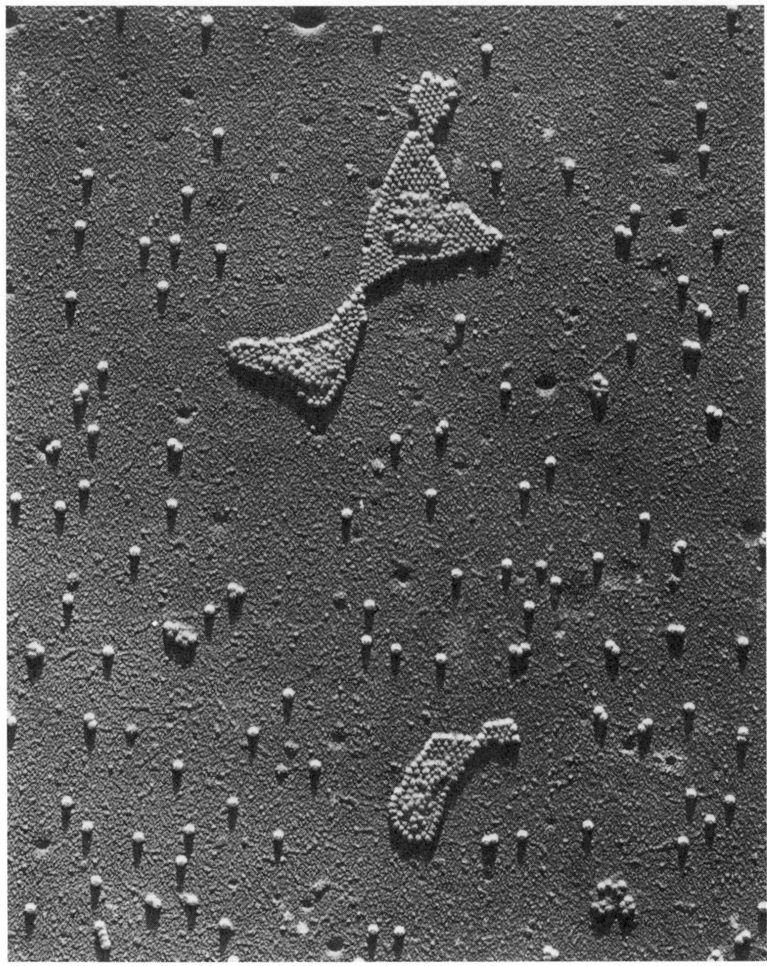

FIG. 27-2. Poliomyelitis virus crystal (type 1 strain purified and crystallized), electron micrograph. Individual viral particles are 28 mμ in diameter. In the absence of any immunity one crystal of poliovirus contains enough particles to infect the population of the world. (×64,000.) (Courtesy Parke, Davis & Co., Detroit, Mich.)

elementary bodies may be likened to a colony of bacteria. Table 27-3 lists some infections where an intracellular parasite is related to an inclusion body.

Another reflection of cytopathogenic effect of virus is the formation of a giant cell or syncytium. This is prominent in measles and cytomegalic inclusion disease (Fig. 27-3).

Mechanisms of viral injury to the infected cell are not understood. Certain viruses, such as the poliovirus, activate and release enzymes contained in the lysosomes of the cell. Progressive injury to a parasitized cell may mean that the cell is either partly or completely destroyed.

There is an opposite effect that viruses at times produce on cells parasitized, an effect not seen in relation to any other known living agent. Viruses can stimulate

TABLE 27-3. INCLUSION BODIES IN INFECTIONS WITH INTRACELLULAR PARASITES

Disease	Etiologic agent	Location of inclusion*	Name of inclusion body or comment
Rabies	Virus	Cytoplasm	Negri bodies
Yellow fever	Virus	Cytoplasm	Councilman bodies—areas of necrosis in cell
Cytomegalic inclusion disease	Virus	Nucleus, also cytoplasm	Cytomegalic inclusions; prominent in enlarged cells
Molluscum contagiosum	Virus	Cytoplasm	Molluscum bodies; elementary bodies are Lipschütz bodies
Varicella-zoster	Virus	Nucleus	Prominent inclusions
Herpes simplex	Virus	Nucleus	Prominent inclusions
Vaccinia-variola	Virus	Cytoplasm	Guarnieri bodies; elementary bodies are Paschen bodies
Adenovirus pneumonia	Virus	Nucleus	Inclusions of rosette type
Measles giant cell pneumonia	Virus	Nucleus, also cytoplasm	Prominent within syncytial giant cells
Trachoma	Bedsonia	Cytoplasm	Prominent
Psittacosis	Bedsonia	Cytoplasm	Psittacosis bodies
Lymphopathia venereum	Bedsonia	Cytoplasm	Gamna-Favre bodies
Granuloma inguinale	Bacterium	Cytoplasm	Donovan bodies

*Note that what is called an inclusion body is a body visible with the *compound light microscope.* The electron microscope visualizes viral particles throughout the cell even in instances where there is a characteristic inclusion body in only one area.

FIG. 27-3. Inclusion bodies produced by measles virus within nucleus of multinucleated giant cell. Such are seen in tissue culture and in lymphoid tissues of patients with measles. (This is a microscopic section of lung from patient with rare measles giant cell pneumonia.) Arrows spot giant cell. Note many nuclei piled upon one another in murky cytoplasm. In each, chromatin is displaced to nuclear membrane by inclusion body. Viral particles in cytoplasm are not shown in routine hematoxylin-eosin stain. (×800.)

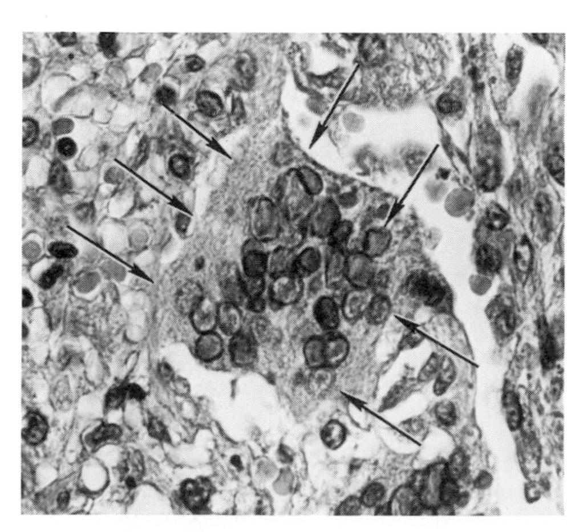

the parasitized cells to increase their numbers in a way the cells would not do if they did not contain virus. Such an increase in the number of cells is referred to as *hyperplasia* (virus induced). It is this tissue response of cells parasitized by certain viruses that implicates viruses as a possible cause of cancer in human beings. Viruses may not only destroy cells or stimulate them to proliferate but may also induce a reaction in parasitized tissue cells that combines features of both destruction and hyperplasia.

Two factors influencing the response of cells to viral injury are (1) the rapidity with which a virus produces its effects—the more rapid its action, the more likely for cells to be damaged severely, and (2) the power of the cells injured to multiply. Nerve cells regulating sensory and motor function in the body cannot duplicate themselves *(regenerate)*; if parasitized, they tend to be killed in the process. Cells lining the skin possess great powers of regeneration; therefore, viral lesions in the lining cells of the skin may result in a piling up of increased numbers of epithelial cells at the site of injury.

If not followed by any kind of cellular deterioration, virus-induced hyperplasia is not accompanied by any detectable inflammation.

There is usually little in the way of measurable inflammatory response in viral infection. Where there is cell destruction, the tissues may show a minimal to moderate infiltration of inflammatory cells (lymphocytes and mononuclear cells, a few neutrophils) and vascular changes (edema and hyperemia) of limited extent.

Viruses are responsible for more than 80 diseases of plants and many diseases of the lower animals. Practically all plants of commercial importance are attacked by them.* Viruses also attack insects. Of more than 550 viruses known, better than 200 produce over 50 viral diseases in man, some of which are the most highly communicable and dangerous diseases known. Today the prevalent infections in man are viral.

■ **Classification.** At present, most virologists seem to favor a classification system based upon major biologic properties, such as nucleic acid present, size, structure (geometric pattern of capsid and number of capsomeres), sensitivity to physical and chemical agents (ether and chloroform), immunologic aspects, epidemiologic features, and pathologic changes. Consistent with this, most animal viruses fall into 13 groups, eight with a DNA core (deoxyriboviruses) and five with an RNA one (riboviruses). The important groups (corresponding to genera) are outlined as follows:

1. Poxviruses
2. Herpesviruses
3. Adenoviruses
4. Papovaviruses
5. Parvoviruses (picodnaviruses)
6. Myxoviruses
7. Paramyxoviruses
8. Rhabdoviruses
9. Arboviruses
10. Reoviruses
11. Picornaviruses
12. Leukoviruses
13. Coronaviruses
14. Unclassified viruses

POXVIRUSES. These fairly large brick-shaped DNA viruses of complex structure form characteristic inclusions in the cytoplasm of parasitized cells. There are 22 members whose action may either destroy cells or stimulate them to proliferate. The skin is a prime target and viruses included are those of smallpox (variola), vaccinia, molluscum contagiosum, and cowpox.

*Viruses have been known to attack the fungus from which penicillin is derived, thereby interfering with its manufacture.

HERPESVIRUSES. These medium-sized ether-sensitive DNA viruses, like poxviruses, often parasitize lining cells of skin. The characteristic viral inclusions are found in the nucleus. The protein shells of these viruses show cubic symmetry, with 162 capsomeres. Latent infections with these agents may endure a lifetime. There are twenty members including varicella-zoster virus, herpes simplex virus types I and II, Epstein-Barr (EB) virus, and cytomegaloviruses. (Some oncogens are found here.)

ADENOVIRUSES. These medium-sized ether-resistant DNA viruses persist for years in human lymphoid tissue. Adenoviruses have cubic symmetry, with 252 capsomeres. There are 31 human, 12 simian, and 2 bovine types. Adenoviruses produce characteristic cytopathogenic changes in tissue culture, and their inclusion bodies are within the nucleus. These agents produce a range of diseases; some cause tumors in animals.

PAPOVAVIRUSES. These small ether-resistant DNA (circular) viruses are notable for producing neoplasms in animals (oncogenic). Their growth cycles are relatively slow, and they replicate in the nucleus of the parasitized cell. Capsid symmetry is cubic. This category includes the *pa*pilloma viruses of rabbits and man, the *po*lyoma virus of mice, and the *va*cuolating virus of monkeys (SV/40), all known oncogenic (tumor-producing) agents. There are at least 11 of them.

PARVOVIRUSES (PICODNAVIRUSES). These are small ether-resistant DNA viruses. Capsid symmetry is cubic. Certain adenoassociated or adenosatellite viruses (defective viruses not able to replicate without adenovirus) and certain viruses of hamsters, rats, and mice are found here.

MYXOVIRUSES. These medium-sized, spherical ether-sensitive RNA viruses have helical symmetry. Viral particles are pleomorphic, sometimes filamentous. The members of the group agglutinate red blood cells of many mammals and birds. This hemagglutination is associated with the viral particle itself and inhibited by antibody acting against it. Some contain an enzyme capable of splitting neuraminic acid from mucoproteins. Myxoviruses include the influenza viruses A, B, and C; the virus of swine influenza; and that of fowl plague.

PARAMYXOVIRUSES. These medium-sized, ether-sensitive RNA viruses have helical symmetry. They are similar in appearance too but somewhat larger than myxoviruses. These viruses appear to be antigenically stable, and certain ones hemagglutinate red blood cells, as do myxoviruses, with or without hemolysis. Replication in cell cultures occurs within cytoplasm. Some cause multinucleated giant cells to form in tissue cultures and sometimes in human tissues (for example, measles virus). Distinctive inclusion bodies are seen with certain ones. Within this category are the parainfluenza viruses (four types) respiratory syncytial virus, and the viruses of measles and mumps in man, and Newcastle disease and distemper in animals.

RHABDOVIRUSES. These ether-sensitive RNA viruses have helical symmetry. Members of this group have an unusual appearance. Mature virions are shaped like bells or bullets. Intracytoplasmic inclusions, the Negri bodies, are seen with the rabies virus. Here are classified the viruses of rabies and vesicular stomatitis of cattle, some insect viruses, and three important plant viruses.

ARBOVIRUSES. These small ether-sensitive RNA viruses (*ar*thropod*bo*rne) have a complex life cycle involving biting insects, especially mosquitoes and ticks. Arboviruses can multiply in many species—man, horses, birds, bats, snakes, and insects. With the exception of dengue fever and urban yellow fever, man is only an accidental host. They are most prevalent in the tropics, notably in the rain forests. Three disease patterns are prominent; one is fairly mild, the other two are severe and often fatal. They include

(1) fever (dengue-like), with or without skin rash, (2) encephalitis, and (3) hemorrhagic fever, as its name suggests, with skin hemorrhages and visceral bleeding.

Presently, on the basis of antigenicity, arboviruses are laid out in groups A, B, C, the Bunyamwera group, and certain other minor ones (total: 27). Groups A and B are best known. Included in group A are the viruses of western equine encephalitis, eastern equine encephalitis, and Venezuelan equine encephalitis, and, in group B, Japanese B encephalitis, St. Louis encephalitis, Murray Valley encephalitis, West Nile fever, dengue fever, and yellow fever. As arboviruses are more carefully studied, a rearrangement seems indicated. Rubella virus is similar in size and appearance to arboviruses. It has been advocated that groups A and B viruses plus rubella virus be placed in a separate category designated *togavirus.*

The ecologic hodgepodge of this group with over 220 members is reflected in the names of many exotic diseases, such as Bwamba fever, Singapore splenic fever, Kyasanur Forest disease of India, and O'nyong-nyong infection in Uganda.

REOVIRUSES. These are medium-sized ether-resistant RNA (double-stranded) viruses. Capsid symmetry is cubic, with 92 capsomeres. These viruses replicate within the cytoplasm. The three members in this group were originally so named because of their presence in the respiratory tract and in the enteric canal and because of their orphan status (*re*spiratory *e*nteric *o*rphans). (Orphans are not known to produce disease.) The term *diplornavirus* is suggested to take in the reoviruses of mammals, wound tumor virus of plants, Colorado tick fever virus, and certain arboviruses with double-stranded RNA. A number of reovirus strains have been recovered from patients in Africa with Burkitt's lymphoma.

PICORNAVIRUSES. These include the smallest known, simplest, and most readily crystallizable RNA viruses. The term *picorna* was coined to designate enteric and related viruses. *Pico* is for very small viruses, and *rna* indicates their nucleic acid. In addition, *p* is for polioviruses, the first known members of the group; *i* is for insensitivity to ether, a distinguishing feature of the group; *c* is for coxsackieviruses; *o* is for orphan or echoviruses; and *r* is for rhinoviruses, further indicating the membership of the group. Capsid symmetry is cubic, with 32 capsomeres.

The nearly 200 members are subdivided into (1) enteroviruses, including polioviruses (three types), coxsackieviruses (30 types), and echoviruses (29 types); and (2) rhinoviruses (100 types), the major causes of the common cold. A rhinovirus in cattle causes foot-and-mouth disease. Picornaviruses produce a wide range of diseases in many areas of the human body, the best known of which is poliomyelitis.

LEUKOVIRUSES. These small ether-sensitive RNA viruses of known structure are member viruses of the avian leukosis complex and the murine and feline leukemias. No such viruses have yet been identified in man.

CORONAVIRUSES. Ether-sensitive RNA viruses, the members of this group are similar to myxoviruses. Symmetry is helical, and the surface projections are petal shaped. Replication is cytoplasmic. In this newly defined group are included the viral agents of avian infectious bronchitis, mouse hepatitis, and certain human respiratory viruses.

UNCLASSIFIED VIRUSES. These include viruses on which pertinent information is lacking. This category includes the agents of well-known diseases such as rubella, infectious hepatitis, serum hepatitis, and lymphocytic choriomeningitis. Rubella virus, a small, ether-sensitive RNA virus, does not seem to fit readily into existing schemes.

When viruses are emphasized as to their source, the following three categories are mentioned: (1) enteroviruses, or those isolated from the alimentary tract, (2) respiratory viruses, or those isolated from the respiratory tract, and (3) arboviruses (arthropod borne), or those isolated from insects.

Sometimes, in the classification of viruses, emphasis is given to the anatomic area in the body where the virus produces its dramatic effects. Viruses so distinguished include (1) those whose lesions appear on the skin and mucous membranes—*dermotropic viruses* of smallpox, measles, chickenpox, and herpes simplex, (2) those related to acute infection of the respiratory tract—*pneumotropic viruses* of the common cold, influenza, and viral pneumonia, (3) those that primarily affect the central nervous system—*neurotropic viruses* of rabies, poliomyelitis, and encephalitis, and (4) those in a miscellaneous group with no common organ system (each virus having its own special affinity for a given organ)—*viscerotropic viruses* of infectious hepatitis (liver), mumps (salivary glands), and other diseases. This restricted scheme of classification, although of some interest to the pathologist, is far from rigid since certain viruses or groups of viruses induce various disease processes. In some instances, a virus may invade the body without necessarily attacking the part that it ordinarily does. Further knowledge of viruses is revealing the fact that certain ones thought limited in their preference for certain cells can infect a number of different cells in man and animals. The viruses of poliomyelitis, thought to be highly neurotropic for many years, are now known to be more flexible; with all three types growing well in tissue cultures of many different cells.

Viral diseases may be either generalized or localized. In generalized infection the virus is disseminated by the bloodstream throughout the body but without significant localization, even though a skin rash may be present. Examples of generalized infections include smallpox, vaccinia, chickenpox, measles, yellow fever, rubella, dengue fever, and Colorado tick fever. In some viral diseases there is restriction of viral effect to a particular organ, to which the virus travels by the bloodstream, peripheral nerves, or other body routes. Examples of localized virus infections include poliomyelitis, the encephalitides, rabies, herpes simplex, warts, influenza, common cold, mumps, and hepatitis.

■ **Cultivation.** To repeat, viruses multiply only inside living cells. Viruses in virus-containing material may be artificially brought into contact with living cells by (1) inoculation of a susceptible animal, such as a mouse (Fig. 27-4), (2) inoculation of tissue cultures, and (3) inoculation of the membranes or cavities of the developing chick embryo (Fig. 27-5).

The chick embryo method is carried out in the following manner: A fertile egg is incubated until it is 7 to 15 days old. With a syringe and needle (Fig. 27-6), the virus-containing material is injected into the membranes of the embryo or into one of the cavities connected with its development (yolk sac, amniotic cavity, or allantoic sac). This may be done through a window made in the shell of the egg over the chick embryo or through a hole drilled in the shell. Of course, *aseptic technic is mandatory.* After inoculation, the egg is incubated 48 to 72 hours, with time being allowed for the virus to multiply. The material in which the virus has multiplied is then removed, used for the study of the virus, or sometimes purified, processed, and used as a vaccine.

One of the most widely used laboratory methods of growing viruses today is the tissue culture method, which has come into its own within the last 3 decades. It has been made possible because of discoveries in basic technics, such as the preparation of improved culture media for living cells. The use of antibiotics has eliminated bacterial contamination, and the recognition of the cytopathogenic effect has furnished

Fig. 27-4 Mouse inoculation with virus-containing material, intranasal route.

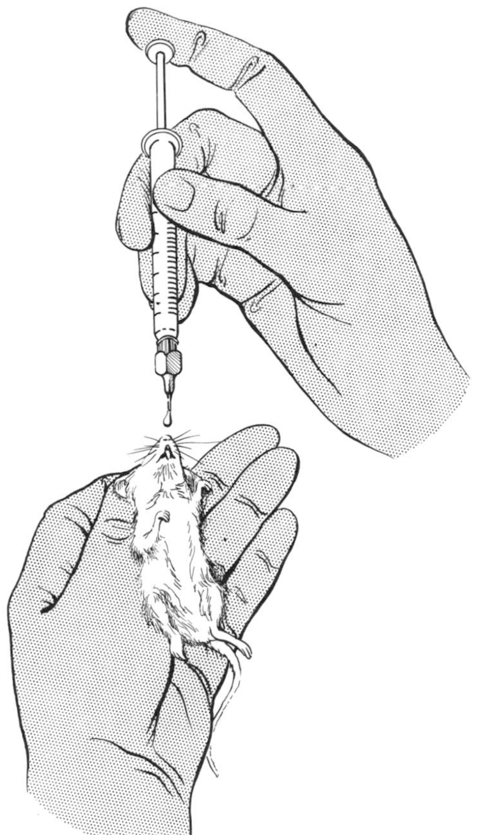

FIG. 27-5. Injection of yolk sac in hen's egg containing embryo, schematic drawing.

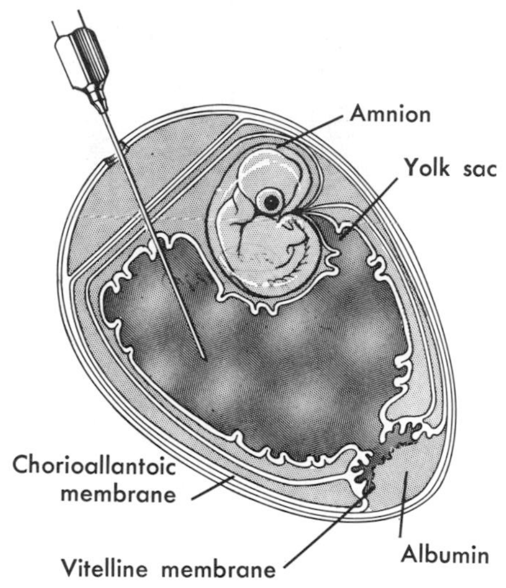

FIG. 27-6. Cultivation of influenza virus in fertile hen's egg. Injection site on top of each embryo is disinfected. A needle and syringe deliver live virus. (Courtesy J. E. Baily, Parke, Davis & Co., Detroit, Mich.)

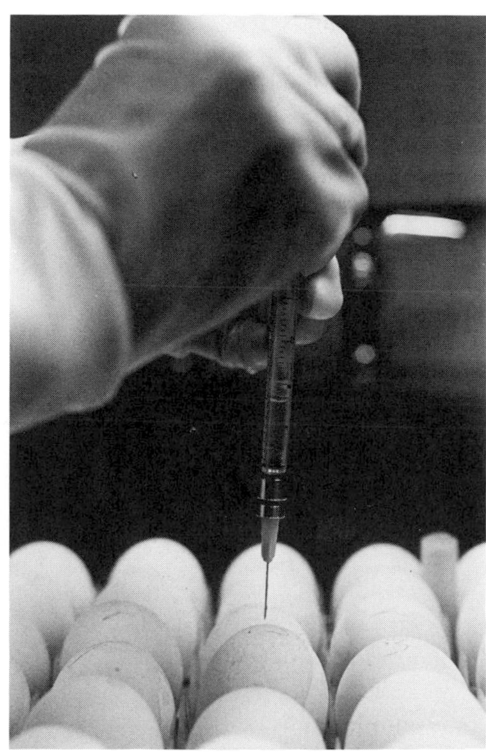

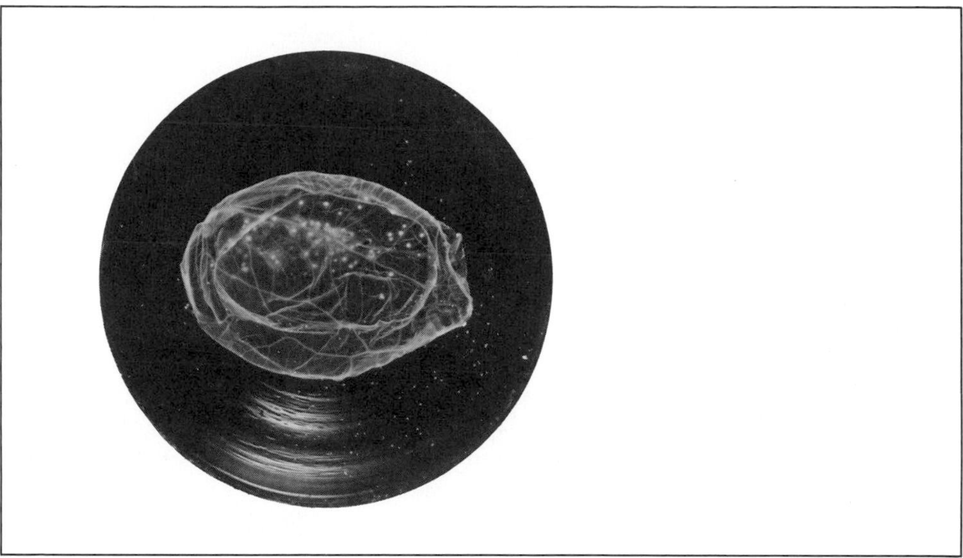

FIG. 27-7. Chorioallantoic membrane, showing whitish plaques (pocks) produced by smallpox virus. (From Hahon, N., and Ratner, M.: J. Bact. **74:**696, 1957.)

a measure of viral injury in growing cells. Cytopathic changes indicate that virus is living and growing in the culture. Both the chick embryo method and the tissue culture method enable the laboratory worker to obtain large quantities of virus for study and for vaccine production.

■ **Laboratory diagnosis of viral infections.** Many procedures are available for the laboratory diagnosis of viral infections, although some are most efficiently carried out in specially equipped laboratories. These include (1) technics for isolation and identification of the virus, (2) serologic tests done serially during the course of infection, (3) fluorescent antibody technics to detect viral antigen directly in lesions, (4) microscopic examination of the lesions by smears, imprints, scrapings, or diseased tissues, and (5) skin tests.

Test material may be inoculated into the embryonated hen's egg, tissue culture, or a susceptible laboratory animal (mouse, guinea pig, cotton rat, rabbit, monkey), from which the virus is recovered and identified (Fig. 27-7). The appearance of typical lesions or inclusion bodies may be diagnostic.

Important in the identification of viruses are several serologic procedures, of which the virus neutralization, hemagglutination, complement-fixation, and precipitin reaction tests are especially valuable. For the serologic evaluation of a patient's serum it is important that *paired samples* of blood be taken. The first is collected as early as possible in the acute stages of disease, and the second is drawn 2 to 4 weeks later during convalescence. The significant finding is at least a fourfold increase in the content of antibodies between the specimens.

Viruses contain good antigens in their makeup (that is, in the capsid), and, as a result, stimulate antibody formation. One of the most important antibodies formed in man and animals with viral infection is the virus-neutralizing antibody, an antibody that neutralizes or obliterates the destructive capacity of the virus. This humoral antibody is an important constituent of immune serum given for viral infection. To measure its activity, suitable mixtures of serum and virus preparation are inoculated into white mice susceptible to the pathogenic effect of the virus, and the protective quality of the test serum is titrated. To gauge the protective effects of the combination, either tissue cultures or chick embryos may also be used. This is the virus neutralization test.

The hemagglutination test, important in the identification of a given virus, is made valid because most disease-producing viruses exert a direct or indirect action on red blood cells of man and other animals to clump them. Viral hemagglutination may be blocked by the action of specific antibodies found in immune or convalescent serum. This is the basis of the hemagglutination-inhibition (HI) test.

Many viruses induce the formation of complement-fixing antibodies.

Precipitin reactions associated with several viruses are analogous to those observed with soluble products and toxins of bacteria. Precipitation is measured by immunodiffusion, commonly carried out by the Ouchterlony technic (double diffusion in agar). Microslides are covered with a suitable gel diffusion medium. The center well is filled with serum (human or animal) containing known antibodies. Test serums are placed in wells opposite, and the microslide is incubated in a moist chamber for 24 to 48 hours. Combination of specific antibody with antigen from a given serum is read in the appearance of a precipitation line. The precipitin in agar technic may be used at times as a simple screening test for more rapid identification and differentiation of closely related strains.

The fluorescent antibody technic can be used to demonstrate viral antigens in specimens from patients or diseased animals. A diagnosis may sometimes be made within a few hours. In rabies, for example, the fluorescent antibody is applied to thin sections or smears of brain tissue or salivary gland, and fluorescence is observed within hours if viral antigen is present.

Direct microscopic examination may be made of suitable preparations such as bodies (viral aggregates) and cellular reactions are easily recognized in thin sections of diseased tissue with the compound light microscope (see Table 27-3). The electron microscope is being used for an increasing number of viral diseases to show viral particles both in body fluids and secretions and in the *ultra*thin sections of tissue. Negative staining with an electron-dense material such as phosphotungstic acid is a technic used in the preparation of electron micrographs of viruses (see Fig. 5-8).

A skin test is available to aid in the diagnosis of some viral (and bedsonial) diseases as indicated in Table 27-4.

Table 27-5 indicates how the laboratory is used in the diagnosis of viral (and bedsonial) disease.

■ **Spread.** The most significant of the viral diseases are caused by organisms that have as their natural host, man himself. Spread from man to man is either by direct or indirect contact, especially by means of nose and throat secretions, fecal material, and articles so contaminated. Droplet infection is very common. Some viral diseases are spread to man by insect vectors (flies, cockroaches, mosquitoes, or ticks) from a reservoir of infection in either a lower animal or in the insect. One viral disease (rabies) is transmitted directly by the bite of the infected animal; another (cat scratch disease) is transmitted by the bite or scratch of an infected cat. Viral diseases are transmitted by water (infectious hepatitis, enterovirus infections), and instances of milk- or food-borne epidemics are occasional (infectious hepatitis, poliomyelitis, and enterovirus infections). Human carriers, except in enteric infections, are not important in the spread of viral diseases.

TABLE 27-4. SKIN TESTING IN VIRAL AND BEDSONIAL DISEASES—DEMONSTRATION OF ALLERGY TO AGENT

Disease	Test material	Interval between injection and evaluation	Interpretation
Mumps	Infected allantoic fluid	24 and 48 hours	Previous or present infection
Herpes simplex	Infected amniotic fluid	48 hours	Previous or present infection
Smallpox	Vaccinia virus vaccine	48 hours	Immunity
Cat scratch disease*	Lymph node extract	48 hours	Previous or present infection
Lymphopathia venereum*	Lymph node extract or chick embryo vaccine	48 hours	Previous or present infection

*A bedsonial disease.

■ **Immunity.** In many viral diseases (mumps, measles), one attack of the disease confers lifelong immunity. In others (common cold), no immunity of any appreciable duration is produced. Where immunity is short lived, the incubation period has been short, viruses have not circulated in the bloodstream, and the antibody-forming tissues have failed to receive adequate stimulation. In most persons in whom immunity lasts, antibodies in the serum of that person may be demonstrated for many years, and it has been postulated that virus, inactivated and non-disease-producing, has remained within his body.

The mechanisms of natural resistance are poorly understood. Newborn mice are extremely susceptible to coxsackievirus and easily infected, whereas adult mice are quite resistant under ordinary conditions. The virus of chickenpox is pathogenic only for man; other animals are completely resistant.

An *interference* phenomenon unlike immunologic mechanisms for bacteria and other microbes is observed with viruses. A plant or animal cell exposed to a given virus subsequently develops a resistance to infection by a closely related strain of the same virus or another closely related virus. There is this kind of interference between the viruses of yellow fever and dengue fever in the body of their insect vector, the mosquito *Aedes aegypti.* Such a mosquito cannot spread more than one of these diseases at the same time.

Interferon is a broad-spectrum antiviral agent, a soluble, nontoxic, nonantigenic protein, smaller in size than antibodies, that is elaborated in small amounts by a normal body cell under attack from an invading virus. Interferon is cell specific (including species specificity), *not* virus specific. Present within the specific cell, it is able to block the effect of the virus by stopping the synthesis of nucleic acid for the virus and thereby breaking into its life cycle. The action of interferon is a manifestation of viral interference, and it is a most important part of the body's defense against viral infection. A patient with agammaglobulinemia who is plagued with repeated infections caused by bacteria seems to recover uneventfully from those coming from viruses, perhaps because of interferon. Practically every class of animal virus has been associated with its formation.

TABLE 27-5. LABORATORY DIAGNOSIS OF SELECTED VIRAL (AND BEDSONIAL) DISEASES IN MAN

		Practical diagnostic tests								
Virus	**Test specimen**	**Animal inoculation**	**Chick embryo culture**	**Tissue culture—cytopathic effect**	**Complement-fixation**	**Hemagglutination-inhibition**	**Virus neutralization**	**Fluorescent antibody**	**Inclusion bodies**	**Other**
Viruses in respiratory disease										
Influenza viruses	Nose and throat secretions		x		x	x		x		
Adenoviruses	Throat secretions, fecal material, eye fluid, cerebrospinal fluid			x	x	x	x			
Viruses in central nervous system disease										
Polioviruses	Throat secretions, rectal swabs, blood, urine, cerebrospinal fluid	x		x	x		x	x		
Encephalitis viruses	Blood, throat swabs, cerebrospinal fluid, urine, brain, other tissues if illness fatal	x		x	x	x	x			
Rabies virus	Saliva, throat swabs, eye fluid, cerebrospinal fluid	x			x		x	x	x	
Viruses in skin disease										
Variola virus	Material from vesicle, pustule, or scab		x	x	x	x	x	x	x	
Rubeola virus	Blood, urine, throat swabs, eye fluid			x	x	x	x			
Rubella virus	Blood, urine, throat swabs	x	x		x		x	x		
Viruses in other diseases										
Colorado tick fever virus	Blood, throat swabs, fecal material	x	x	x	x	x	x			
Lymphopathia venereum bedsonia	Lymph nodes, buboes				x		x			Frei test

There is a great deal of interest in interferon inducers, substances stimulating the endogenous production of interferon and thereby active against viral infection. A well-known one in the investigational field is poly I·poly C or PI:C (polyinosinio-polycytidylic acid), a synthetic, double-stranded RNA.

■ **Prevention of viral disease.** Immunization procedures prevent viral diseases (Chapters 37 and 38) as do proper technics of sterilization (Chapter 18). Effective disinfectants to inactivate or destroy viruses are formalin, dilute hydrochloric acid, organic iodine, and phenol 1%. Roentgen rays and ultraviolet light destroy viruses, but the effective dose varies with different viruses. Most viruses, except those of hepatitis, are destroyed in pasteurized milk. Influenza viruses are readily destroyed by soap and water.

Except to treat bacterial complications that are prone to follow viral diseases, the antimicrobial drugs are generally of no value in the management of diseases caused by true viruses.

Bacteriophages (bacterial viruses)

■ **General characteristics.** In 1917 d'Herelle* discovered that a bacteria-free filtrate obtained from the stools of patients with bacillary dysentery contained an agent that when added to a liquid culture of dysentery bacilli caused the bacteria to dissolve. If but a minute portion of the dissolved culture were added to another culture of dysentery bacilli, the bacteria in the second culture were likewise dissolved. This transfer from culture to culture could be kept up until the bacteria in hundreds of cultures were dissolved, which proved that the agent was not consumed in the process but apparently increased in amount. This agent d'Herelle called *bacteriophage* meaning "bacteria eater" (Fig. 19-2). We know bacteriophages today as viruses that attack bacteria—or *bacterial viruses!*

In recent years viruses infecting many strains of bacteria have been isolated, and it has been demonstrated that a given phage acts only on its own particular species or group of species of bacteria. In fact, their highly specific nature has made them useful to the epidemiologist in typing or classifying bacteria; for example, the phage typing of pathologic staphylococci is important in the epidemiologic study of hospital-acquired staphylococcal infections.

Tending to occur in nature with their specific hosts, bacteriophages are found most plentifully in the intestinal discharges of man and the higher animals or in water and other materials contaminated with these discharges. They are also found in pus and even in the soil. When phages are named, reference is made to the specific hosts, as with coliphages, staphylophages, cholera phages, and typhoid phages. A very few bacteria such as the pneumococci do not possess phages. Of all the viruses known, bacterial viruses are the most easily studied in the research laboratory, and the ones most thoroughly investigated have been those related to the enteric group of microorganisms. As obligate intracellular parasites, they resemble closely the other viruses in their biologic properties.

■ **Life cycle.** With the aid of the electron microscope, phages are seen to be tiny tadpole units possessing a head, which is either rounded or many sided, and a tail, which is a specialized structure for attachment (Fig. 27-8). Like other viruses, a phage particle is composed of nuclear material (DNA) encased in a protein coat (Fig. 27-9).

---

*Two investigators independently discovered bacteriophage—F. W. Twort (1877-1950) in 1915 and F. H. d'Herelle (1873-1949) in 1917.

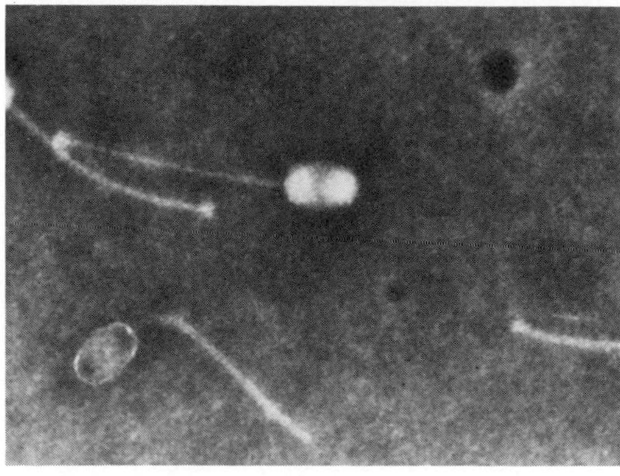

FIG. 27-8. Staphylophages, demonstrated by negative staining and magnified more than 50,000 times in the electron microscope. (Courtesy S. Tyrone, Dallas, Texas.)

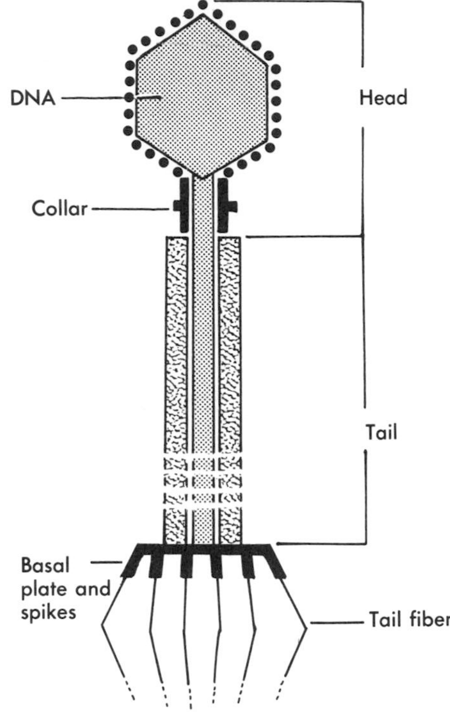

FIG. 27-9. Phage (mycobacteriophage), diagram. Large dots about head are protein coat. Measurements of this phage: head about 950 Å, tail about 3000 Å, and tail fibers 1500 Å. (From Arnett, R. H., Jr., and Braungart, D. C.: An introduction to plant biology, ed. 3, St. Louis, 1970, The C. V. Mosby Co.)

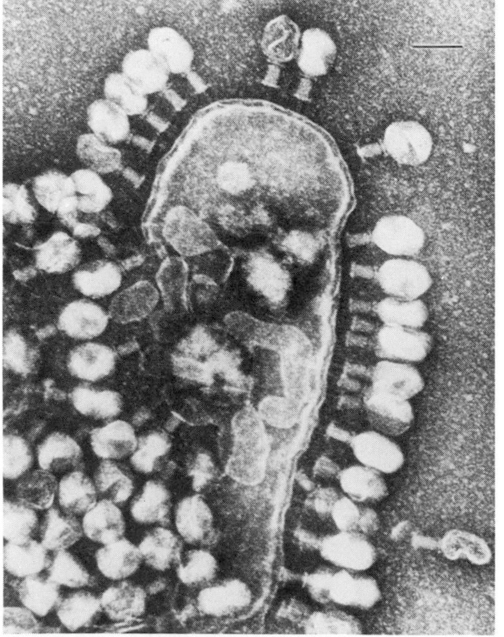

FIG. 27-10. Bacteriophage infection of bacteria. Numerous specific phages are adsorbed on *Escherichia coli*. The distance between baseplates of phage particles and the cell wall is 300 to 400 Å. (Negative staining.) (From Simon, L. D., and Anderson, T. F.: Virology **32:**279, 1967.)

Although appearing to be harmless, a bacteriophage in its virulent form can literally blow its bacterium to bits in a matter of minutes. The attack on the specific bacterium occurs in a fantastic series of steps. Seeking out the susceptible bacterial cell, the virus fixes itself tail first to the cell (Fig. 27-10). By chemical action it drills out a tiny hole in the cell wall and the tail penetrates the plasma membrane into the interior of the cell. The head changes shape and soon the DNA of the virus flows through the tail into the cell. The bacteriophage seems to work like the world's smallest syringe and needle when it injects its nuclear DNA into the bacterial cell. Once the fatal injection is made, drastic changes occur. The DNA takes command of the vital forces of the microbe and in a matter of minutes imposes the synthesis of hundreds of bacterial viruses exactly like the one originally invading the injured bacterial cell. The deranged cell swells and shatters, setting free a multitude of new viruses able to seek out a new host and repeat the cycle. This rapid destruction of a bacterium is *lysis.*

Lysis is not the invariable result of bacteriophage action. When certain phages infect, they do not destroy but seem to be able to establish a relatively stable symbiotic relation with the host they have parasitized. The host bacterium continues to grow and multiply, carrying the virus in its interior in a noninfective condition more or less indefinitely, and the virus meantime multiplies at the same rate. This kind of phage is referred to as a *prophage* and this condition of mutual tolerance is termed *lysogeny* or *lysogenesis.* Lysogeny is widespread in nature. That the process is not without effect is seen in changes in physiologic characteristics in the infected bacteria. Phages may be responsible for the conversion of an avirulent strain to a virulent one.

■ **Laboratory diagnosis.** Bacteriophages may be isolated and studied in the laboratory. If they are taken from a source in nature, they must be separated from bacteria by filtration. When they are added to a growing culture of specific bacteria, their action to block bacterial growth may be nicely shown. In a liquid culture medium, clearing indicates bacterial lysis. On a solid culture medium, lysis by virulent phages is seen in zones, usually circular, where bacterial growth has disappeared (clearing of bacterial growth). These zones are called *plaques* (Fig. 27-11). Each plaque contains many particles, which in turn can form plaques. Pure preparations of phage may be obtained by picking material from well-isolated plaques.

■ **Importance.** Bacteriophages may be of practical importance. In certain of the fermentation industries in which the commercial product is dependent upon bacterial action (streptomycin, acetone, and butyl alcohol, for examples), an "epidemic" of viral infection in the large vats used to grow the microorganisms can be of grave economic concern.

Viruses and teratogenesis

Currently the role of viruses in teratogenesis (the production of physical defects in the offspring in utero) is being carefully studied. When the pregnant woman contracts a viral infection accompanied by a viremia, the infection is often passed across the placenta to the susceptible embryo or fetus. Notable examples are measles, smallpox, vaccinia, western equine encephalitis, chickenpox, poliomyelitis, hepatitis, and coxsackievirus infection. Viral infections involving the offspring within the uterus may be more common than suspected, especially in the lower socioeconomic brackets. Although the possibilities for fetal infection are many, only three viruses have been fully qualified as teratogens. For rubella virus, cytomegalovirus, and herpesvirus, the evidence is clear cut.

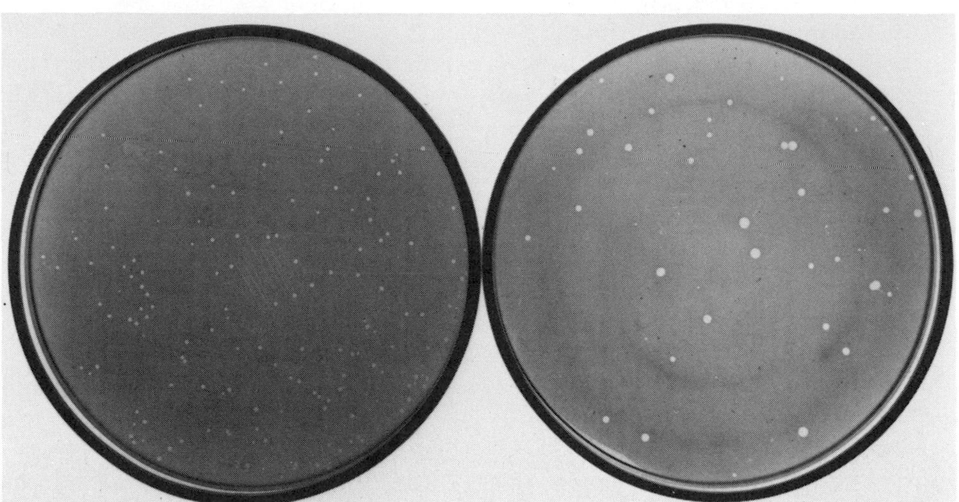

FIG. 27-11. Phage plaques. The small plaques (zones of clearing) in the assay plate on the left are *Staphylococcus aureus* phages; those on the right are *S. epidermidis* phages. In preparation of an assay plate, a mixture of test phages and microorganisms is poured over a nutrient base. After the plate has been incubated, the plaques are evaluated and counted. (Courtesy Drs. E. D. Rosenblum and B. Minshew, Dallas, Texas.)

Birth defects with rubella virus (see also p. 421) have been more thoroughly examined than those with the other viruses, and most of what is known about viral teratogenesis comes from such investigations. In rubella, the teratogenic mechanism seems to stem from a direct interaction between virus and parasitized cell that continues throughout the length of gestation. That interaction plus viral damage to blood vessels disrupts normal organ development.

The cytomegalovirus (salivary gland virus) produces a mild infection in the mother but can cause extensive and widespread damage in the neonate. (See also p. 430.) In a few instances herpesvirus has crossed the placenta to produce generalized infection with documented malformations in the central nervous system and in the eye. (See also p. 428.)

Viruses and cancer

Experimental evidence indicates that filtrable viruses do cause several kinds of cancerous growths in lower animals, that is, in rabbits (Fig. 27-12), mice, chickens, hamsters, rats, dogs, frogs, monkeys, horses, squirrels, and deer (Table 27-6). There are more than 60 known viral oncogens. Cancer is induced in all major groups of animals (including subhuman primates). Table 27-7 emphasizes certain features of viruses known to cause tumors.

Leukemia is an important form of virus-induced neoplasm in animals. Within the last decade the polyoma virus has been shown to cause many different solid cancers in the skin, breast, lungs, gastrointestinal tract, and salivary glands of several species of laboratory animals. Oncogenic viruses are known to lie dormant for a long

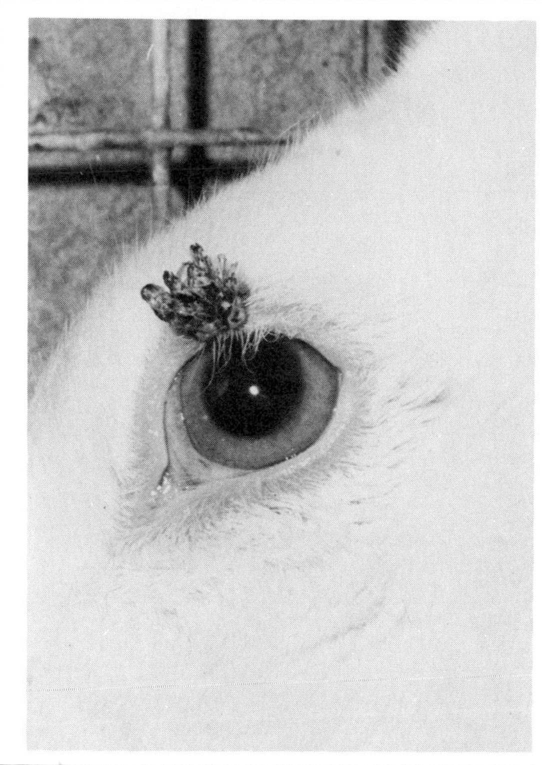

FIG. 27-12. Example of virus-induced neoplasm. Wartlike growth on eyelid of a wild cottontail rabbit. Eyelids and ears are common sites for such tumors. An enzootic disease of wild rabbits, rabbit papillomatosis is also seen in domestic ones. Virus transmitted in nature by rabbit tick and in laboratory by mosquito. (From Hagen, K. W.: Bull. Path. **8:**308, 1967.)

TABLE 27-6. SOME EXAMPLES OF VIRAL NEOPLASMS IN ANIMALS

Animal species	Date reported	Cancerous growth (neoplasm)
Chicken	1908	Chicken leukemias
Chicken	1910	Rous sarcoma
Rabbit	1933	Shope papilloma-carcinoma
Mouse	1936	Breast carcinoma*
Frog	1938	Kidney carcinoma
Mouse	1951	Spontaneous leukemia
Mouse	1957	Salivary gland tumors of polyoma virus
Hamster, rat, and rabbit	1958	Great variety of solid tumors produced by polyoma virus
Hamster	1962	Chest and liver tumors (adenovirus responsible isolated from human cancer)

*This most common form of cancer in the most commonly used laboratory animal was shown by John Bittner to be related to a virus. The agent has long been referred to as Bittner's milk factor. Since Bittner's discovery, the role of viruses in the induction of cancerous growths has been increasingly emphasized.

time in animals before they are activated. It is postulated that either external factors such as X rays and other forms of radiant energy or internal factors of metabolic or hormonal nature trigger the mechanism of neoplasia (new growth). John Bittner demonstrated that breast cancer in mice is caused by an interplay of the virus (milk factor), hormones, and hereditary background.

TABLE 27-7. PATTERNS OF TUMOR VIRUSES

Category	Nucleic acid	Approximate size	Member viruses
Papovaviruses	DNA	45 mμ	Polyoma
			SV 40
			Papilloma Human Rabbit Bovine Dog
Poxviruses	DNA	250 × 200 mμ	Yaba Fibroma
			Molluscum
Adenoviruses	DNA	80 mμ	Types 3, 7, 12, 18, 31
Herpesviruses	DNA	100 mμ	Lucké
			Marek's disease
			Herpesvirus saimiri
			Herpesvirus of rabbit EB (Epstein-Barr)
			Herpesvirus hominis, type II
Myxovirus-like particles	RNA	70-110 mμ	Rous sarcoma
			Avian leukosis complex Murine leukemia complex: 14 strains, including Gross, Friend, Graffi, Rauscher, Moloney Milk factor or Bittner virus (mouse mammary tumor)

The role of viruses in human neoplasms is unknown, but recent experimental reports strongly suggest that viruses are indeed involved. In man, the common wart, verruca vulgaris, a benign or harmless tumor, has been clearly demonstrated to be caused by a virus. Viruslike particles have been isolated from human cancers, most recently from human breast cancer and from the milk of breast cancer patients. The

Natural host	Tumors produced	Experimental hosts
Mouse	Solid tumors in many sites of the animal body	Mouse, hamster, guinea pig, rat, rabbit, ferret
Rhesus monkey	Sarcomas (malignant tumors of connective tissue)	Hamster
Man	Common warts (papillomas)	Man
Rabbit	Papillomas	Rabbit
Cow	Papillomas	Cow, hamster, mouse
Dog	Papillomas	Dog
Monkey	Benign histiocytomas	Monkey
Rabbit, squirrel, deer	Fibromas, myxomas	Rabbit, squirrel, deer
Man	Molluscum contagiosum (wartlike skin lesion)	Man
Man	Sarcomas, malignant lymphomas	Hamster, mouse, rat
Leopard frog	Renal adenocarcinoma	Leopard frog
Chicken	Neurolymphoma (Marek's disease)	Chicken
Squirrel monkey	Melendez' lymphoma of owl monkey	Owl monkey, marmoset
Rabbit	Hinze's lymphoma	Cottontail rabbit
Man	Burkitt's lymphoma(?), nasopharyngeal carcinoma (?)	
Man	Carcinoma of cervix of uterus (?)	
Chicken	Sarcomas, leukemias, adenocarcinoma of kidney	Chicken, turkey, rat, monkey, hamster, guinea pig
Chicken	Leukemias, sarcomas	Chicken
Mouse	Leukemias, malignant lymphoma	Mouse, rat, hamster
Mouse	Mammary carcinoma	Mouse

TABLE 27-8. VIRUSES INDICTED FOR ONCOGENICITY IN MAN

Virus or virus-like particle	Tumor
Type C and related particle*	Leukemia Lymphoma Sarcomas (rhabdomyosarcomas) Bowen's disease (precancerous skin cancer)
Type B particle (RNA particle) (mouse mammary tumor virus)	Breast cancer
EB virus (Epstein-Barr virus, herpes type of virus)	Burkitt's lymphoma Hodgkin's disease Leukemia Lymphoma
Herpesvirus hominis type 2 (genital strain of herpes)	Carcinoma of cervix uteri
Reovirus type 3	Burkitt's lymphoma

*Viruses with an RNA core and a double membrane constituting a complex of oncogenic viruses, the first viruses isolated from animal neoplasms.

particles closely resemble those producing breast tumors in mice. Table 27-8 presents viruses indicted as oncogens in man.

■ **Burkitt's lymphoma.** Burkitt's African lymphoma* is a malignant disease of the jaw and abdomen affecting children between the ages of 2 and 14 years. The striking feature of the disease is its sharp geographic distribution. It is found in Central Africa limited to a malarious belt where the conditions of rainfall, temperature, altitude, vegetation, and humidity are the same. It also appears to be a geographic equivalent of lymphomas of children in other parts of the world. But, unlike lymphomas elsewhere, it has a pattern for a specific infectious disease, strongly suggesting that it is caused by an agent such as a virus, and the case for its viral etiology is getting stronger. Several viruses have been found in tumor tissue and in tissue culture made of tumor. The Epstein-Barr virus was first isolated from such a tissue culture.

At first it was postulated that the viral agent causing the disease was spread by a vector mosquito. The lymphoma is prevalent in areas where malaria is endemic and is rare outside tropical Africa. Now it is thought that the mosquito is no more implicated than in the transmission of malaria, for it is felt that the damage to the lymphoid system from chronic malaria is the factor that determines in some way the oncogenic potential of a virus. The EB virus implicated here is also found in non-neoplastic disease elsewhere in the world.

*Lymphoma is a malignant neoplasm (cancer) of the lymphoid tissue.

QUESTIONS FOR REVIEW

1. State briefly the salient features of viruses. Compare them with other microbes.
2. Sketch the life cycle of viruses (including bacterial viruses.)
3. What are inclusion bodies? Their importance? Their specificity?
4. What is meant by the cytopathic effect of viruses? How is this used in virology?
5. Give the two major pathologic effects viruses produce on cells they parasitize.
6. How may viruses be cultivated in the laboratory?
7. Discuss briefly the spread of diseases caused by viruses.
8. Outline the laboratory diagnosis of viral infections.
9. Comment on the nature and importance of interferon and viral interference.
10. Classify broadly the viruses. Indicate the most widely used system.
11. Discuss the role of viruses in teratogenesis.
12. What is the importance of the nucleic acids in viruses?
13. State the case for viral oncogenesis.
14. Define or briefly explain: bacteriophage, arbovirus, enterovirus, virion, viropexis, capsid, picorna, virology, lysogeny, papovaviruses, viral hemagglutination, plaque, icosahedron, hyperplasia, dermatropic, prophage, capsomere, life cycle, helix, regeneration, virus neutralization.

REFERENCES. See at end of Chapter 29.

28 Viral diseases

SKIN DISEASES

Measles (rubeola)*

Measles is an acute communicable disease associated with a catarrhal inflammation of the respiratory passages, fever, constitutional symptoms, a skin rash, and a distinct predilection for grave complications. Among these are streptococcal or pneumococcal bronchopneumonia, encephalitis, otitis media, and mastoiditis. The incubation period is 9 to 14 days. One of the most common diseases, measles is said to be responsible for about 1% of the deaths occurring in the temperate zones.

Only one immunologic type of measles virus is known. The virus of measles is thrown off in the lacrimal, nasal, and buccal secretions and enters the body by the mouth and nose. It is found in the blood, the urine, the secretions of the eyes, and the discharges of the respiratory tract. The infection is transmitted directly from person to person. Healthy carriers are unknown. The time that an object contaminated with the secretions of a patient with measles remains infectious is short. Measles is not transferred by the scales from the skin. Measles virus may be spread a considerable distance through the air. Measles is most highly communicable during the 3 or 4 days preceding the skin eruption. It is not transmitted after the fever has subsided. Epidemics tend to recur every 2 or 3 years and are likely to break out when young adults from rural communities come together in large groups, such as when armies are mobilized. The disease is especially virulent in populations native to tropics or in primitive races anywhere with no ethnic history of previous exposure. Children of mothers who have had measles are immune to the disease until they are about 6 months old. An attack of measles almost invariably produces a permanent immunity.

Measles depresses certain allergic conditions and immune processes. It renders the tuberculin test and agglutination tests less positive or even negative. The Dick and Schick tests may become more strongly positive, and eczema and asthma often disappear during or after the attack.

The measles patient should be isolated and protected against streptococcal infections, staphylococcal infections, and common colds. Discharges from the nose, mouth, and eyes should be disinfected. When measles appears in army camps, daily inspections of personnel should be made, and persons having conjunctivitis, colds, or fever should be isolated. The vulnerability to pneumonia and other serious complications of patients with measles should always be kept in mind.

*For a discussion of measles vaccines and immunization procedures, see pp. 602, 603, 610, 618, and 626.

Rubella (German measles)*

German measles (3 day measles) is a mild but highly prevalent disease of viral origin described by fever, enlargement of lymph nodes, mild catarrhal inflammation of the respiratory tract, and a skin rash similar to that of measles (rubeola) or scarlet fever (Fig. 28-1). The incubation period is 14 to 21 days. Transmission is airborne from person to person. A common source is someone with an inapparent infection. The cause is a medium-sized, spherical unit, an extremely pleomorphic virus that, isolated from throat washings and the blood of patients, has been grown in tissue cultures. This peculiar virus with an RNA core is about the size of a myxovirus but it behaves more like an arbovirus. There is no insect vector. Only one antigenic type is known. The virus is as yet unclassified.

Rubella is important because approximately one out of four children born to mothers contracting German measles during the first 4 months of pregnancy has congenital defects or is malformed. If rubella is contracted during the first 4 weeks of pregnancy, approximately half the children born will be deformed.

From the mother's blood, the rubella virus crosses the placenta into the developing tissues of the new individual, where it can persist throughout gestation and into the neonatal period. The unborn child is infected at the same time as his mother, and since the viral injury leads to irregularities in the development of one or more organs, malformations result. The defects are common in the eyes, ears, heart, and brain. Examples are microcephaly (extremely small head), deaf-mutism, cardiac defects, and cataracts† (Fig. 28-2).

Viral injury in rubella is unique. In keeping with the mild nature of the infection, the virus neither destroys nor invigorates. It merely slows things down. The cells it parasitizes continue to grow and multiply but at a decreased rate. The embryo and fetus of the first 3 to 4 months of pregnancy are in a period of development during which the anatomic units define their shapes, take their places, and lay out their interconnections. Timing here is as critical as it is with a trapeze artist.

When infection with this virus persists after birth, the baby is born with *congenital rubella,* the manifestations of which may be mild or severe. There may be a single serious defect or multiple ones in a small undersized baby. The *rubella syndrome* indicates active infection in the newborn, which is easily demonstrated by recovery of the rubella virus from nasopharyngeal washings, conjunctivae, urine, and spinal fluid.

■ **Laboratory diagnosis.** As a routine, virus isolation is impractical. The following serologic procedures are available for the laboratory diagnosis of rubella infection: virus neutralization, complement fixation, hemagglutination inhibition (HI), and immunofluorescence.

The hemagglutination inhibition test is the most sensitive and most widely done. With a single specimen of serum this test indicates immunity in the presence of HI antibody. With paired serums spaced a week or two apart and a fourfold increase in the HI titer, it can make the diagnosis. If a woman is exposed to a possible case of rubella during her pregnancy, the HI test may be used to determine both in the pregnant woman and in the contact whether the exposure is to rubella and what the woman's

---

*For immunization, see pp. 602, 603, 611, 618, and 621.

†As a result of the epidemic of 1964 in the United States, there were 20,000 stillbirths and 30,000 babies born with congenital anomalies to mothers whose pregnancies were complicated early by rubella.

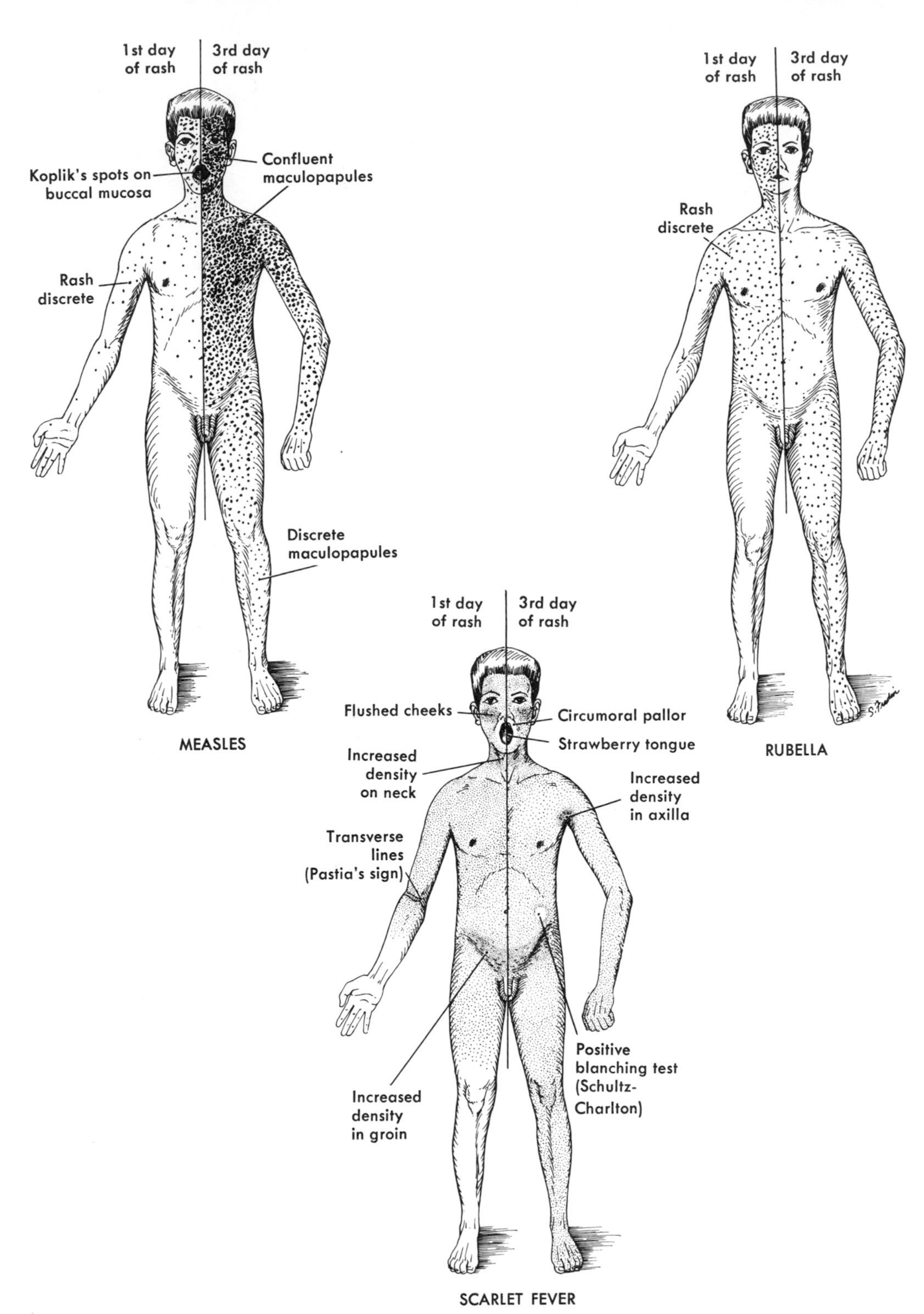

FIG. 28-1. Skin rash patterns in measles, rubella, and scarlet fever, schematic drawings. (From Krugman, S., and Ward, R.: Infectious diseases of children, ed. 5, St. Louis, 1973, The C. V. Mosby Co.)

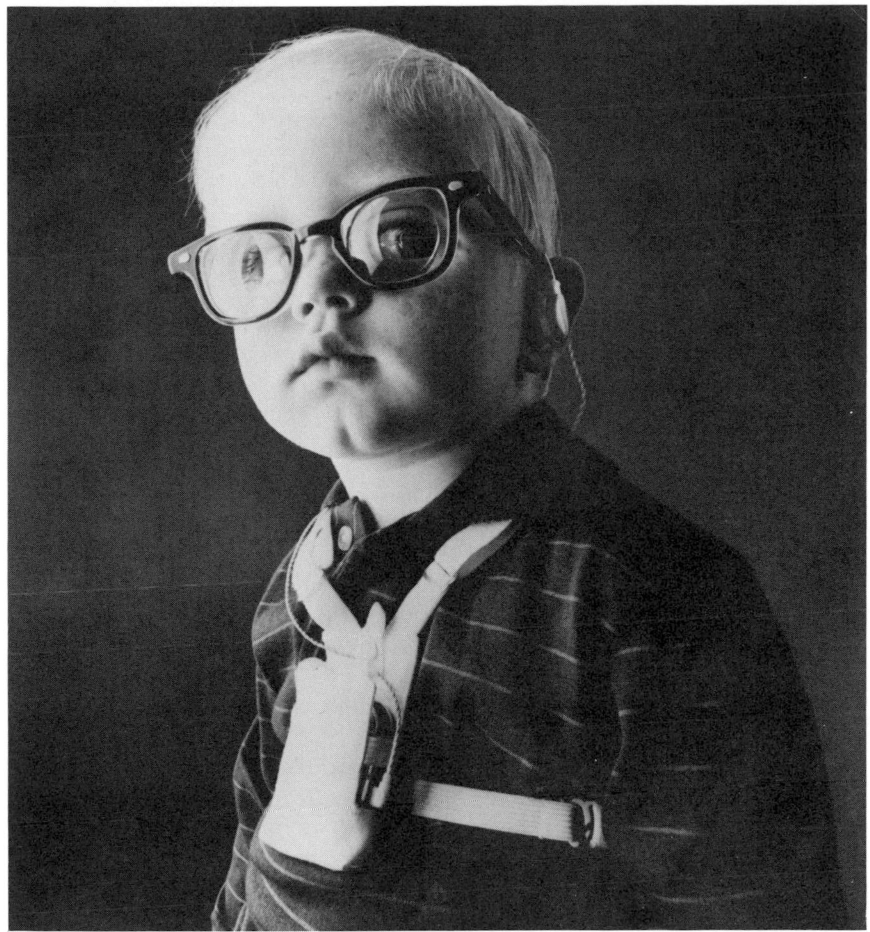

FIG. 28-2. Rubella, two important congenital defects: partial blindness caused by cataracts, and deafness. (Note glasses with thick lenses and hearing aid.) (Courtesy Metropolitan Life Insurance Co., New York, N. Y.)

immune status is to rubella. For the diagnosis of congenital rubella there is the determination of rubella-specific immunoglobulin M in the baby. (This is an antibody that does not cross the placenta and that therefore could not be derived from the baby's mother.)

■ **Immunity.** An attack of German measles is followed by permanent immunity. Girls should have German measles if possible before childbearing years, and a pregnant mother should avoid exposure to the disease. This is true even in women supposed to have had an attack because an erroneous diagnosis is often made. Although the newborn baby possesses virus-neutralizing antibodies in high titer, he may continue to harbor virus and to shed it, even for several years, thus constituting an important reservoir and source of infection to susceptible individuals in his environment.

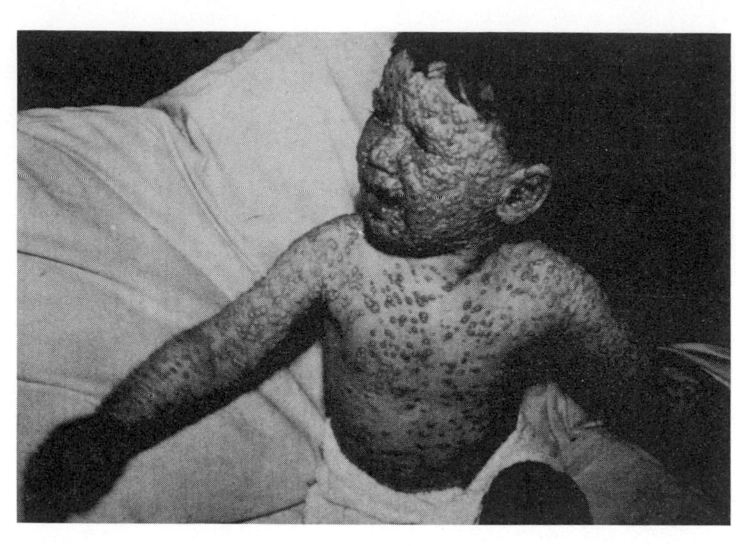

FIG. 28-3. Smallpox in an unvaccinated 2½-year-old boy. The picture was taken on the eighth day of illness. The attack was severe, but the boy recovered. (Courtesy Dr. Derrick Baxby, University of Liverpool, England.)

Smallpox (variola)*

Smallpox, one of the most highly communicable diseases, is characterized by severe constitutional symptoms and a rash (Fig. 28-3) that goes through a typical evolution to become hemorrhagic in the severest cases. The incubation period is 12 days. Before the days of vaccination, smallpox spread over the world in great epidemics. It has been estimated that in some of these 95% of the population were attacked and 25% died.

Smallpox is the most infectious of diseases, and the death rate in unprotected individuals is high. Although today in the United States death from smallpox is almost unheard of, thousands of persons succumb each year to the disease in countries such as India and Pakistan. Recently an unsuspected case in a Pakistani girl initiated an epidemic of smallpox in Great Britain, with at least 13 deaths resulting. With the modern jet plane, an area in which smallpox is prevalent is less than a day away from this country.

There are two types of smallpox virus. One causes a severe form of the disease, *variola major*. Infection caused by the other type is mild and is spoken of as *variola minor*, or *alastrim*.

■ **Transmission.** Smallpox is transmitted directly from person to person by droplet infection. Man alone carries the infection. It may sometimes be conveyed by objects, such as handkerchiefs, and pencils, that have been contaminated with the nasal and buccal secretions of a person ill of the disease. It may also be transmitted from the pustules by the hands. Persons immune by virtue of vaccination or an attack of the dis-

*For a discussion of immunization, see pp. 604, 605, 612, 618, 619, and 627.

ease may become contact carriers and disseminate virus for a short period. The infectious agent is generally believed to enter the body by the respiratory tract and leave by the buccal and nasal secretions. The virus may be found in the blood, skin lesions, and secretions of the mouth and nose. It may remain active for a long time in the dried crusts of the skin lesions. Even the dead body may be a source of infection.* A mother with smallpox may infect her child in utero and the child be born with a typical smallpox skin eruption.

An attack of smallpox usually renders the patient immune for the remainder of his life.

■ **Laboratory diagnosis.** The elementary bodies of the smallpox virus may be seen microscopically within cells of a stained preparation of scrapings taken from a lesion. In material suitably prepared the virus may also be identified in an electron micrograph, usually within a few hours. A smallpox gel precipitin test has been developed.

■ **Prevention.** Patients with smallpox should be isolated. The nurse should be isolated and of course vaccinated or revaccinated at once. All objects in contact with the patient should be sterilized, preferably by heat (burning, high-pressure steam, or boiling). If this cannot be done, they should be soaked in a 2.5% saponated cresol solution. The feces and urine should be disinfected with chloride of lime. Sputum and discharges from the mouth and nose should be received on tissues and burned. The patient should not be released until desquamation is complete.

■ **Vaccinia.** Smallpox and vaccinia or cowpox are caused by viruses with similar biologic properties. Both can be grown on the chorioallantoic membrane of the chick embryo. The virus of vaccinia produces a mild disease either in man or in cattle (its natural host), and its importance is that it can be used to produce immunity against the more severe disease, smallpox. (The importance of vaccination against smallpox cannot be overstressed.)

Molluscum contagiosum

Molluscum contagiosum is a skin disease characterized by the presence of small pink wartlike lesions on the face, extremities, and buttocks. It is spread from person to person by direct and indirect contacts. The cause is a large poxvirus that produces very dramatic intracytoplasmic inclusions in the squamous cells lining the affected skin site.

Chickenpox and shingles (varicella—herpes zoster)

The two diseases chickenpox (varicella) and shingles (herpes zoster) represent two phases of activity of a single virus, referred to as the varicella-zoster, or VZ virus. The first invasion of the body by the virus produces chickenpox, the generalized infection. Shingles, a localized infection, represents the reinvasion of an immune host.

Varicella, usually a mild disease of childhood, presents a typical (teardrop) vesicular rash, which, though generalized in the skin and mucous membranes, is concentrated on the trunk (Fig. 28-4). The incubation period of chickenpox is 14 to 16 days. Herpes zoster, on the other hand, is a disease of adults. It is characterized by the appear-

---

*In the recent smallpox epidemic in the United Kingdom, the body of the victim was thoroughly sprayed with 5% Lysol, placed in a plastic bag, wrapped in a rubber sheet, placed on a Lysol-soaked bed of sawdust, and sealed in an airtight and watertight casket. Cremation was urged. The health officers caring for the body wore full protective clothing.

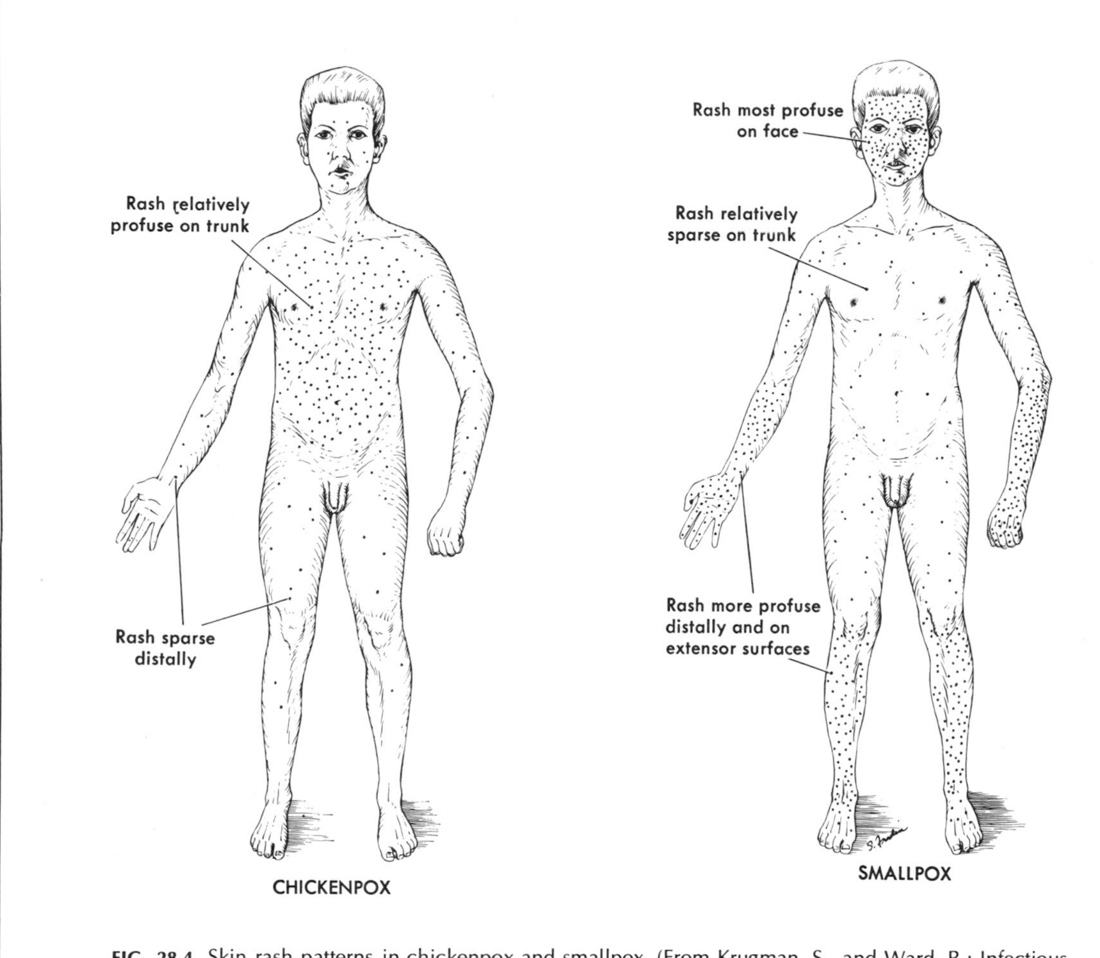

FIG. 28-4. Skin rash patterns in chickenpox and smallpox. (From Krugman, S., and Ward, R.: Infectious diseases of children, ed. 5, St. Louis, 1973, The C. V. Mosby Co.)

ance of a vesicular eruption, the individual blister-like lesions of which are similar to those of varicella, but their distribution is quite different. In shingles, vesicles occur on one side of the chest, following the course of the peripheral nerves supplying that part of the chest. They may occur in other parts of the body, but since the primary involvement is in the ganglion of the posterior nerve root, they always follow the course of the nerve or nerves supplying the affected part. Shingles may be the cause of prolonged suffering because the skin lesions are associated with intense burning pain.

In the blisterlike lesions of the skin in either disease there are typical reddish inclusion bodies in the nuclei of injured epithelial cells. The virus is present in the fluid in the vesicles of either disease (Fig. 28-5). The moist crusts of chickenpox are infectious, whereas the dry ones are not. Chickenpox is transmitted by direct contact with a patient. Frequently it has been observed that herpes zoster in adults has served as a source of varicella in children.

No immunity for chickenpox is conferred by the mother on her newborn infant, and convalescent serum is of little value in preventing or modifying the disease. A vaccine for chickenpox is in the experimental stage, but any licensed vaccine goes far into the future. There is no test for immunity in chickenpox, no vaccine for prevention and no medicine for cure. A gel precipitin test has recently been developed for the diagnosis of the disease.

Herpes simplex

Infection with herpes simplex virus *(Herpesvirus hominis)* is related to either of two recognized serotypes, each with unique biologic features. The more common type I is responsible for the familiar fever blisters and cold sores found about the mouth and lips *(herpes labialis)*. Type I lesions include recurrent labialis, gingivostomatitis, corneal lesions in the eye, and eczema herpeticum. This type is probably spread via the respiratory route and occurs in older children and adults.

Type II is found about the genital organs. Spread as a venereal disease through sexual contact, it produces skin and mucosal lesions below the waist *(herpes progenitalis)*. (Type I tends to produce lesions above it.) Because patients with cancer of the mouth of the womb (cervix uteri) show strong evidence for having had type II infection previously, this herpesvirus has recently been implicated in the causation of this form of cancer.

In most instances the first infection with herpesvirus is an inapparent one; clinical signs of disease occur in only about 15% of persons infected. Saliva and genital secretions are probably sources of infection. The incubation period is 2 to 12 days. It is believed that thereafter the virus exists in the body as a latent infection and that when

FIG. 28-5. Virus of chickenpox, electron micrograph. Numerous viral particles are seen in a segment of squamous cell from skin. (**A,** ×15,000. **B,** ×30,000.)

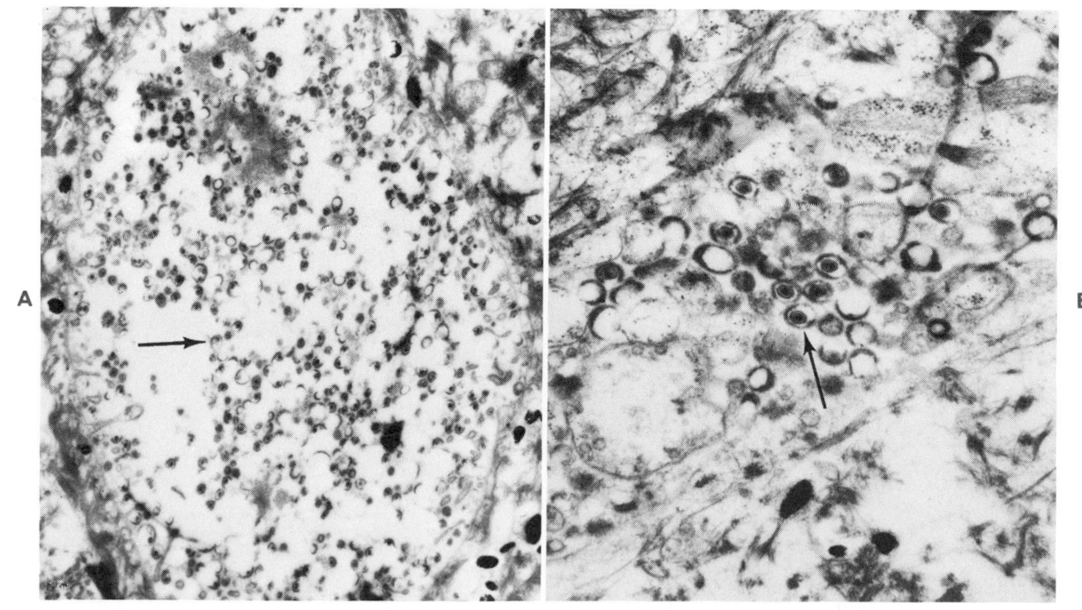

the body resistance is weakened from any cause the virus is reactivated. Recurrent herpes simplex often accompanies febrile illness, exposure to cold or sunlight, fatigue, mental strain, or menstruation (see Table 28-1). Unlike other viral infections, the hosts's humoral antibodies apparently give no protection against subsequent lesions.

One very serious complication of herpes progenitalis is infection of the newborn. By passing against a herpetic lesion in his mother's birth canal, the baby acquires the infection. In 1 to 3 weeks he becomes gravely ill with generalized herpesvirus infection and usually succumbs. Herpetic lesions can be found in all the organs including the brain.

A

B

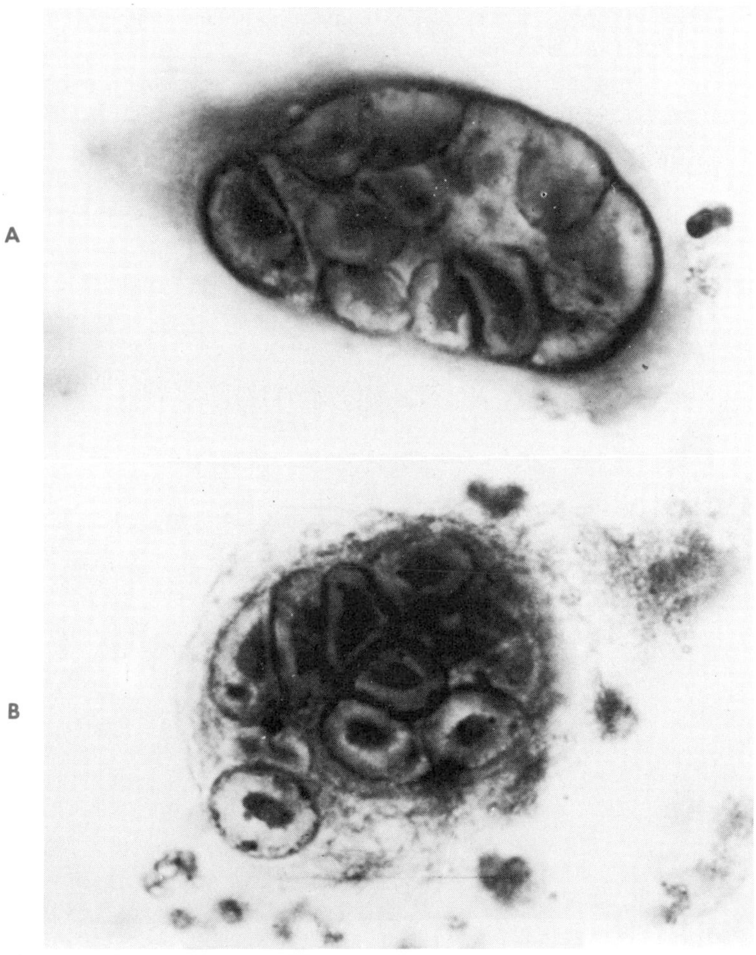

FIG. 28-6. Herpesvirus seen with light microscope, high-power photomicrograph. Inclusion bodies within the nuclei of a multinucleated squamous cell from skin. **A,** Early stage in formation of inclusion bodies. **B,** Inclusion bodies well defined.

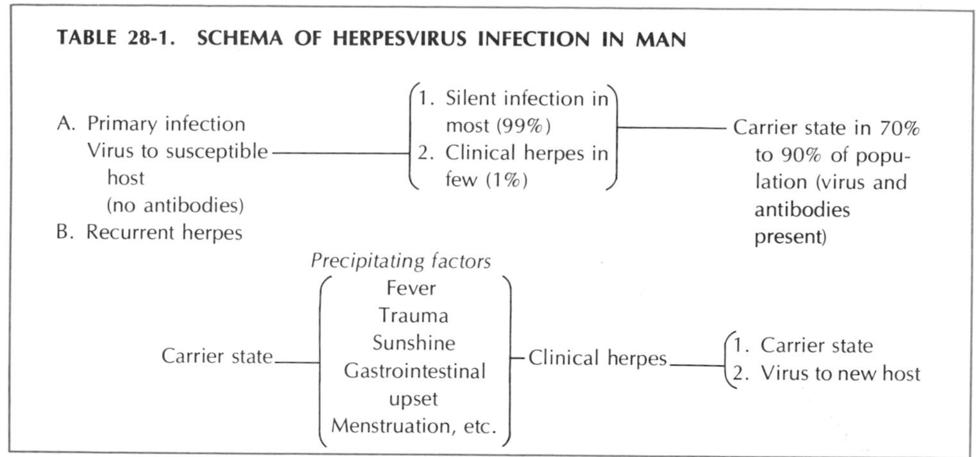

TABLE 28-1. SCHEMA OF HERPESVIRUS INFECTION IN MAN

A. Primary infection
Virus to susceptible host (no antibodies) — 1. Silent infection in most (99%); 2. Clinical herpes in few (1%) — Carrier state in 70% to 90% of population (virus and antibodies present)

B. Recurrent herpes
Carrier state — *Precipitating factors*: Fever, Trauma, Sunshine, Gastrointestinal upset, Menstruation, etc. — Clinical herpes — 1. Carrier state; 2. Virus to new host

The virus can be cultivated in tissue cultures and forms classic intranuclear inclusion bodies that may be identified in suitable preparations in the fluid from the characteristic superficial vesicles either by the light or the electron microscope (Figs. 28-6 and 28-7). Fluorescent antibody technics are used in the diagnosis of this infection, and there is a complement fixation test.

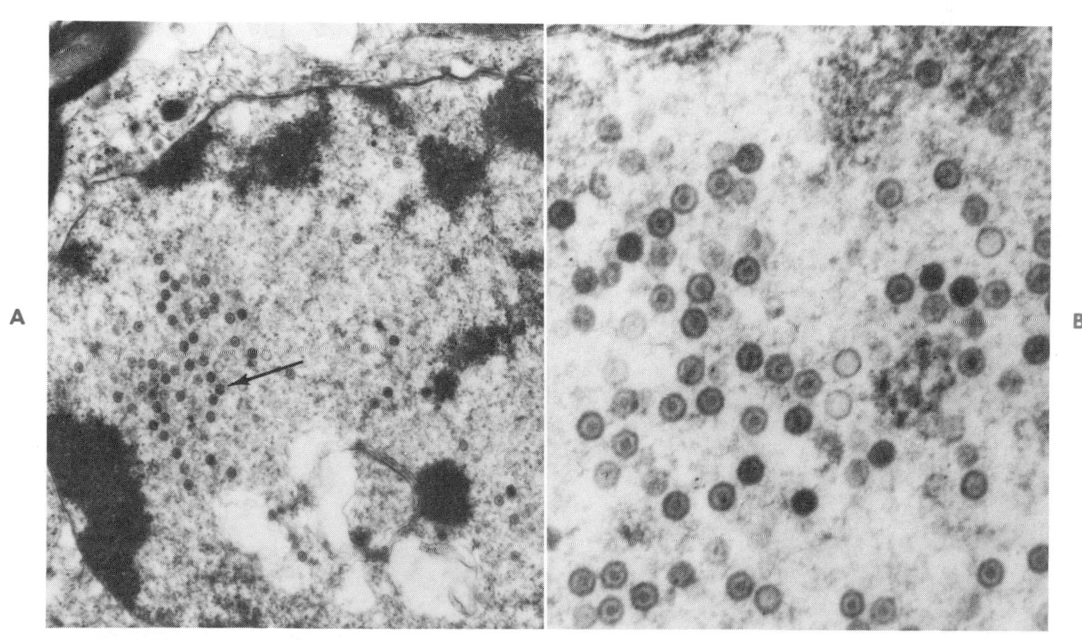

FIG. 28-7. Herpesvirus, electron micrograph. Numerous viral particles in one nucleus of squamous cell from skin. (**A,** ×20,000. **B,** ×56,000.)

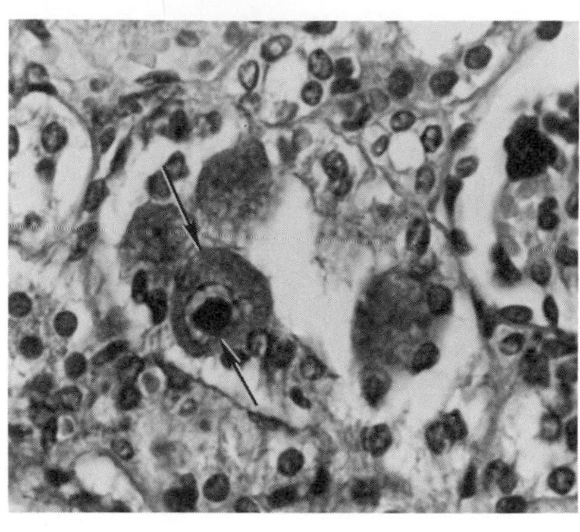

FIG. 28-8. Cytomegalic inclusion disease of kidney, microscopic section. Arrows spot enlarged renal tubular cell. Note large intranuclear inclusion body of cytomegalovirus. (×800.)

Cytomegalic inclusion disease

Cytomegalic inclusion disease (salivary gland disease) is infection caused by the ubiquitous cytomegaloviruses (salivary gland viruses). The pathologic lesions are striking by their nature. Large well-defined viral inclusions are seen in the nucleus, and smaller ones are found in the cytoplasm of the injured cells, which are enlarged (*cytomegaly* means 'cell enlargement') (Fig. 28-8). In fatal cases the cell changes are seen in the gastrointestinal tract, lung, liver, spleen, and other organs. In nonfatal cases, inclusions may be found in epithelial cells from the kidney shed into the urinary sediment.

In its overt form, cytomegalic inclusion disease is seen in the newborn as a congenital infection acquired from a mother who was probably asymptomatic and in an older individual as a complication of a preexisting disease state. In the infected newborn cytomegalovirus is the cause of birth defects, especially in the central nervous system (for example, microcephaly, hydrocephaly, blindness). Malformed babies exhibit the viral inclusions. In modern clinical medicine the patient on immunosuppressive therapy after organ transplant and the leukemic patient receiving cancer chemotherapy are vulnerable for cytomegalovirus pneumonia. Cytomegalic inclusion disease is associated with the postperfusion syndrome. This is an infectious mononucleosis–like syndrome occurring after perfusion of fresh blood in patients undergoing open-heart surgery with cardiopulmonary bypass.

The viral inclusions are readily visualized and diagnostic. Serologic technics used are those of complement fixation and immunofluorescence.

RESPIRATORY DISEASES

Influenza

Influenza* is a highly communicable disease occurring in epidemics that are characterized by explosive onset, rapid spread, involvement of a high percentage of the population, and frequency of serious secondary bronchopneumonia. An in-

*For immunization, see pp. 608-610, 623, and 627.

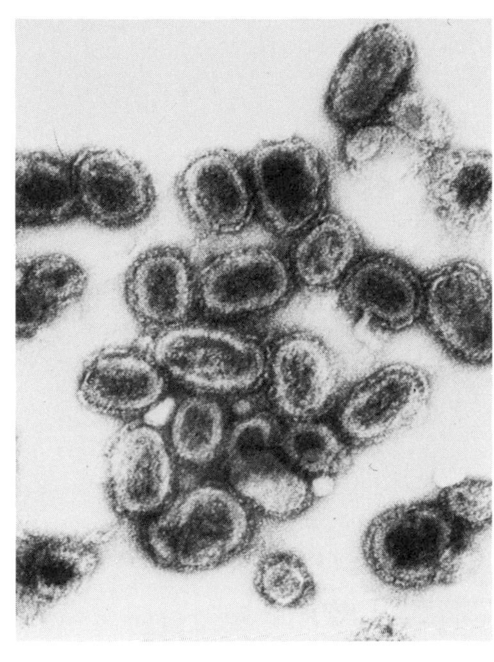

FIG. 28-9. Influenza virus negatively stained in electron micrograph. This is the A_2/Aichi/68 or "Hong Kong" virus. (×303,100.) (Courtesy C. A. Baechler, Virus Research, Parke, Davis & Co., Detroit, Mich.)

fluenza pandemic occurs about every 10 years. The name *influenza* comes from Italian astrologers of long ago who believed that the periodic appearance of the disease was in some way related to the "influence" of the heavenly bodies.

One of the world's greatest catastrophes was the pandemic of 1918-1919. Although other epidemics have had higher death rates, on the basis of total number of deaths this was the worst pestilence civilization had ever experienced. It is estimated that there were 200 million cases with 20 million deaths over the world. In the United States alone there were 850,000 deaths, a figure greater than that for the combined battlefield losses of World War I, World War II, and the Korean War. And this occurred within the twentieth century!*

Influenza is caused by a virus† (so tiny that 29 to 30 million of them could rest comfortably on the head of a pin), and, like many viral diseases, it seems to potentiate serious bacterial infection, mostly bronchopneumonia. The mortality of the 1918 pandemic was largely the result of the severe, complicating bronchopneumonia produced by virulent streptococci.

■ **The agent.** Three types of influenza virus are known today: A, B, and C. Of these, A and B have been best studied. They are alike in many ways but differ serologically. Each includes numerous distinct strains. Influenza A strains (Fig. 28-9) have figured

*Only two places in the world allegedly escaped the pandemic—St. Helena, an island in the South Atlantic, and Mauritius, one in the Indian Ocean.

†It was originally believed that *Haemophilus influenzae* was the cause of influenza. This organism is an important *secondary* invader in this and other respiratory diseases.

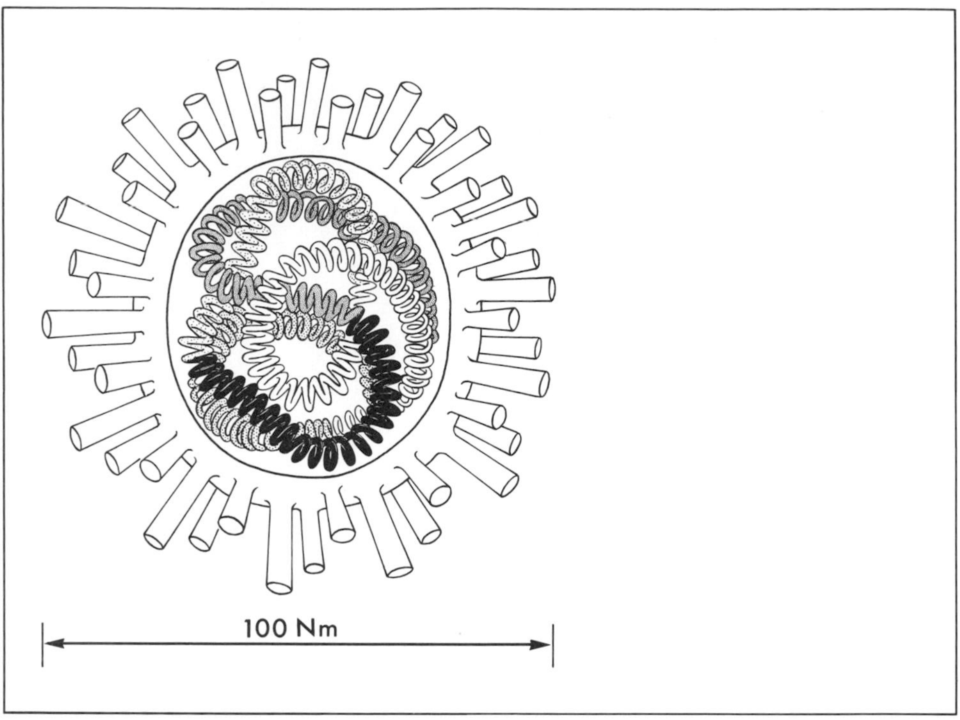

FIG. 28-10. Influenza virus, schematic drawing to show unusual nature of viral nucleic acid. Note that five discrete segments compose viral ribonucleic acid. This may account for variants of virus types emerging periodically to cause disease.

much more frequently in epidemics than those of type B, the strains of which are considered to be less virulent than those of type A. The pandemic of 1918-1919 is thought to have been caused by type A. The epidemic of 1947 was from a strain of virus A designated as A prime, Asian strains of virus A were responsible for the pandemic of 1957, and a variant also in the A group—A_2 Hong Kong/68—caused the Hong Kong pandemic of 1958.

■ **Epidemiology.** The infectious agent of influenza enters the body by the mouth and nose and leaves by the same route. Recently influenza virus has been recovered from anal swabbings. Influenza has been produced artificially in chimpanzees, ferrets, and mice by injection of filtrates of nose and throat washings from known human cases.

The disease is probably spread by direct and indirect contact, including droplet infection. An epidemic is of short duration and quickly subsides, to be followed after several weeks by a secondary wave, a free period, and a tertiary wave. Never does the infection spread faster than people travel. What happens to the virus between epidemics still remains a mystery. It is believed that in the interval the virus resides in animals, multiplies, and changes genetically (Fig. 28-10), thus becoming infective for man again. An epidemic seems to depend upon the emergence of the new strain; since this organism has a dramatic tendency to change its chemical and genetic nature, that is, to mutate, new variants of the virus types arise constantly. If the majority of persons do not possess the immune mechanisms to meet the new variation, then far-reaching epidemics and pandemics are a threat.

The explosive outbreak of an epidemic may be explained by the high communicability of the disease, the great number of susceptible people, and the fact that during the early days of the attack the patient is not confined to bed but mingles freely with other people. The average duration of an epidemic in a community is from 6 to 8 weeks. In secondary and tertiary outbreaks the number of people attacked is less than in a primary outbreak, but the disease tends to be severe, complications are more common, and the mortality is higher. Interepidemic cases are usually mild.

■ **Recent epidemics.** In early 1957 influenza appeared out of north China and within the next several months spread all over the world. There were millions of cases over a wide area of the Far East. Referred to as Asian flu, it quickly spanned the extent of the United States.

The virus was found to comprise new strains of influenza virus, type A, designated the Asian strains, and cultures were established in embryonated eggs. On the whole, the mortality in this pandemic was low; in areas where notable, it was associated with severe staphylococcal pneumonia.

In the summer of 1968 another pandemic of influenza came out of the Orient, starting in Hong Kong with close to half a million cases. Hong Kong influenza was caused by a virus not completely different from the Asian flu virus of 1957, and the clinical disease was similar to that of the prior decade. During the last 3 months of 1968 it was estimated that 30 million persons were stricken. There were more deaths with Hong Kong influenza than with Asian influenza, especially among chronic invalids. *Staphylococcus aureus* and *Pseudomonas aeruginosa* were the complicating bacterial infections.

Asian strains could dominate the influenza picture for a period of time, with epidemics every few years, until the level of immunity in the general population has risen significantly.

■ **Immunity.** Part of the population apparently possesses natural immunity to influenza because during an epidemic some persons do escape infection. There is reason to think that natural immunity acquired from an attack of the disease protects against a repeat attack by the same or very closely related strain. However, there is no protection against the new strains that are appearing. Little success has been obtained in producing passive immunity by means of convalescent or immune serum.

■ **Laboratory diagnosis.** The diagnosis of influenza can be confirmed only by isolation of the virus or serologic studies, usually available at a state health laboratory or medical center.

■ **Prevention.** When an epidemic of influenza strikes, all the methods known to preventive medicine fail to check it. Wholesale isolation seems to be of little value and is almost impossible to achieve. Nurses attending patients with influenza should disinfect the mouth and nasal secretions of the patient and should avoid exposing themselves to droplet infection.

■ **Swine influenza.** Swine influenza is a severe respiratory infection of hogs caused by the combined action of a virus and a bacterium *Haemophilus suis.* The bacterium is closely related to *Haemophilus influenzae,* and the virus is closely related to the virus of human influenza. The diseases differ in that the human virus can produce the complete disease, whereas the virus of swine influenza alone produces only a mild form. Swine may become infected with the strains of type A influenza virus that happen to be prevalent in the human population at a given time. The Asian strains of the 1957 epidemic were recovered from hogs in Japan.

Acute respiratory syndromes

As reflected in a national health survey, acute viral infection of the respiratory tract is one of the most common causes of illness in man. It is responsible for approximately one third of all days lost from work and two thirds of days missed from school. Acute respiratory infection of viral type is not a single disease but a spectrum of such entities as rhinitis, pharyngitis, tonsillitis, laryngitis, and bronchitis (see Table 28-2). These inflammatory processes can occur singly, but combinations are frequent.

■ **Clinical features.** Viral infection of the lining of the nose results in a reddened mucous membrane and a hypersecretion of mucus—hence the well-known "runny nose." A sensation of stuffiness comes from the nasal congestion and blockage. When the throat is involved, it is reddened, tonsils and other lymphoid masses are enlarged, and there is soreness. Hoarseness may result from inflammatory swelling of the larynx. Other clinical features associated in whole or in part with acute respiratory disease include low-grade fever, cough, pain and discomfort in the chest, headache, sensations of chilliness, and sometimes enlargement of cervical lymph nodes. Droplet infection accounts for spread, respiratory diseases are prevalent in cool months, and the epidemic pattern is a familiar one.

The disease influenza was originally considered to be part of the general complex

TABLE 28-2. RESPIRATORY ILLNESSES RELATED TO KNOWN RESPIRATORY VIRUSES

Illness	Virus
Influenza	Influenza viruses A, B, C
Common cold of adults	Rhinovirus (Salisbury)—100 or more types Coronavirus—3 types
Acute respiratory syndromes* Common cold syndrome in children, some adults Rhinitis Pharyngitis and nasopharyngitis Tonsillitis Laryngitis Tracheitis Bronchitis Bronchiolitis Bronchopneumonia Pneumonitis Croup (acute laryngotracheobronchitis)	Influenza viruses A, B, C Parainfluenza viruses 1, 2, 3, 4 Respiratory syncytial virus Adenovirus—10 types Echovirus—6 types Coxsackievirus A—24 types Coxsackievirus B—6 types Poliovirus, types 1, 2, 3 Reovirus, types 1, 2, 3
Viral pneumonia (primary atypical pneumonia)	Influenza viruses A, B, C Parainfluenza viruses 1, 2, 3, 4 Adenovirus—2 types Respiratory syncytial virus (in children) Measles virus Varicella virus Vaccinia virus Rubella virus
Pharyngitis (part of poliomyelitis)	Poliovirus, types 1, 2, 3
Pharyngitis (part of infectious mononucleosis)	Agent of infectious mononucleosis (Epstein-Barr virus?)

*No specific constant relation between given disorder and viral agent. See text.

of respiratory disorders. With a typical pattern for recurrent epidemics, influenza could be easily separated when the causative virus was discovered. Primary atypical pneumonia emerged in part from the respiratory assortment in World War II because of the characteristic findings in the lungs of the soldiers. A pleuropneumonia-like organism, *Mycoplasma pneumoniae,* not a virus, is responsible for a significant number of these cases (p. 377). In Great Britain certain viruses have been recovered from adults with the common cold (acute coryza) and the infection transferred to the chimpanzee, the only animal susceptible to the particular agents. The experimental studies suggest that the common cold may well have unique features.

■ **Relation of viruses.** By the early 1950's, tissue culture technology was developed to a high degree of efficiency. Viruses were recognized and recovered in such rapid-fire succession as to incur a kind of "viral explosion." So many were identified that proper assimilation and definition of medical importance are not yet complete. Regarding respiratory infections, it soon became clear that an array of viral agents was associated with a variety of clinical infections with no fixed or specific relationship between the ailment and the virus. Viruses were recovered singly or together from the respiratory tract, both in health and in disease. Many observers prefer the term "syndrome," the sum total of the clinical features of an illness, to emphasize the lack of specific correlation between any precise agent and the clinical manifestations. In the light of present knowledge of viruses associated with respiratory disease, the best we can do is to present certain viruses as being found in respiratory secretions and tissues and to point out the variety of associated syndromes. Table 28-2 lists the viral agents recovered from acute respiratory syndromes.

■ **Respiratory viruses.** From the known acute respiratory infections, viral agents have been recovered in a majority of instances. Viruses that cause respiratory disease are spoken of collectively as *respiratory viruses,* and as an entity they encompass several major groups of viruses, including the myxoviruses (influenza viruses*), the paramyxoviruses (parainfluenza viruses and the respiratory syncytial virus), the picornaviruses (enteroviruses and rhinoviruses), the adenoviruses, and the reoviruses.

Parainfluenza viruses, classified as paramyxoviruses, were first recognized in 1957. The four antigenic types are more stable than the influenza viruses; they do not mutate so often. Parainfluenza viruses are present in the community most of the year. The typically mild first contact infection usually comes early in life. These viruses enter by way of the respiratory tract, and except in infants and young children, inflammation is usually confined to the upper part. They cause a number of illnesses, primarily in infants and young children, ranging in severity from mild upper respiratory infection to croup and pneumonia. The clinical features of their infections are not distinct; the majority are clinically inapparent.

Respiratory syncytial (RS) virus, also a paramyxovirus, infects a child before the age of 4 years. It is probably the most important cause of acute respiratory disease in infants and very young children. A formalin-killed vaccine containing the respiratory syncytial and types 1, 2, and 3 parainfluenza viruses from kidney cell cultures is being field tested.

Among the *picornaviruses,* certain of the *enteroviruses* are respiratory pathogens. All of the coxsackieviruses and some of the echoviruses (see p. 447) are believed to cause respiratory disease. Coxsackieviruses have been isolated from nasal and pharyngeal

*See p. 431.

secretions, and coxsackievirus A, type 20, has been studied in its relation to an acute upper respiratory cold–like illness. The respiratory agents among the echoviruses probably infect through the respiratory tract. They are known to multiply in the pharynx, producing a pharyngitis, and most likely are transmitted in respiratory secretions. Upper tract infections of echovirus are generally mild; the syndrome is many times that of the common cold. Echovirus types 11 and 20 have been specifically observed in such infections.

*Adenoviruses,** also known as adenoidal-pharyngeal-conjunctival (A.P.C.) viruses, were first discovered in adenoids removed surgically from persons without any detectable clinical disease. To date, 31 types have been isolated from human beings. Although these viruses may be recovered from persons without disease, some do cause acute infection of the respiratory and conjunctival mucous membranes. Adenoviruses are related to a striking variety of clinical conditions, such as acute respiratory disease, febrile pharyngitis or pharyngoconjunctival fever (especially in children), acute follicular conjunctivitis, epidemic keratoconjunctivitis (shipyard eye), tracheo-

*For immunization, see p. 608.

FIG. 28-11. Adenovirus crystal in tissue culture, electron micrograph. (×68,000.) (Courtesy C. A. Baechler, Virus Research, Parke, Davis & Co., Detroit, Mich.)

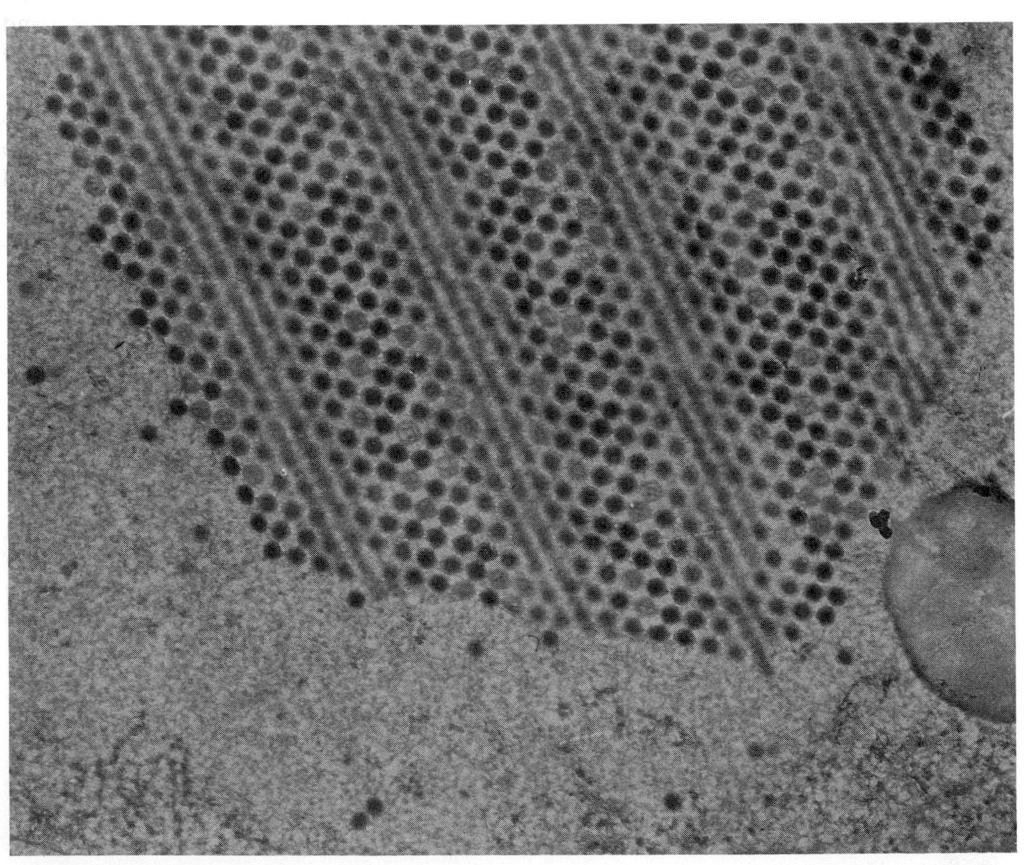

bronchitis, bronchiolitis, and pneumonitis. Types 3, 4, and 7 are related to epidemics of respiratory disease. Type 8 virus is the major cause of epidemic keratoconjunctivitis. Type 3, sometimes found in swimming pools, is the cause of pharyngoconjunctival fever.

In civilian groups, only a small percent of acute viral respiratory diseases is caused by adenoviruses, but this group, particularly types 3, 4, 7, 14, and 21, is of special importance to the Armed Forces. It incapacitates many of the recruits in the fall and winter about 8 weeks after induction. Immunization of military inductees therefore is highly desirable.

Man is the only known reservoir of adenoviruses, and the virus is transmitted in respiratory and ocular secretions. The fact that the conjunctival inflammation often precedes the respiratory infection indicates the eye to be a significant portal of entry. Epidemics of pharyngoconjunctival fever and conjunctivitis have been traced to dissemination of the virus in swimming pools.

The respiratory and ocular involvement, often combined in adenovirus infection, is associated with enlargement of the lymphoid tissue in the respiratory passages and of the lymph nodes in the neck.

Adenoviruses cannot be studied by their effects in animals since they do not produce disease in the routinely used laboratory animals. Nor do they grow on the membranes of the fertile egg. However, they can be demonstrated nicely in tissue cultures in which certain human and monkey cells are used (Fig. 28-11). In virus-infected cells examined with the electron microscope, viral particles can be seen. This group shares a common antigen demonstrable in the serum of an infected person by means of a complement-fixation test. The breakdown of the group into the different types is done by means of the virus-neutralization test.

Reoviruses, viruses of undetermined pathogenicity and ubiquitous in nature, have been recovered from the respiratory and alimentary tract in man in health and disease. Respiratory disease from all three types is usually low grade.

In the early stages of certain viral diseases, for example, poliomyelitis and infectious mononucleosis, involvement of the upper respiratory tract figures prominently.

■ **Influence of age.** The patterns of respiratory disease in adults and children are basically similar in many respects. Influenza, for example, is much the same at any age. Yet there are points of contrast. These are presented in Table 28-3.

TABLE 28-3. COMPARISON OF VIRAL INFECTION OF THE RESPIRATORY TRACT IN CHILDREN AND ADULTS

	Children	Adults
Type of infection	Primary contact often	Inactivation of latent virus often; rarely, primary contact
Degree of involvement	Severe; infection may be fatal; mild cases also seen	Mild (except in aged and debilitated)
Clinical picture	Spectrum of respiratory disease	Common cold and related syndromes prominent
Causative viruses	Parainfluenza viruses, adenoviruses, and respiratory syncytial virus ("viruses of childhood")	Cold viruses (coryza and rhinoviruses) important; parainfluenza viruses, adenoviruses, and enteroviruses implicated

Common cold (acute coryza)

The common cold is said to be the most commonly occurring ailment of mankind and to disable temporarily more people than any other infectious disease. The characteristic nasal discharge of only a few days' duration is spread by direct contact, and a cold is most communicable in the early stages. The incubation period is 2 to 3 days. The agent enters the body by way of the upper respiratory tract. In sneezing and coughing, the affected person spreads not only his disease but also the bacteria he may be carrying in his throat. As with other viral diseases, the cold is often complicated by more serious bacterial diseases such as pneumonia.

Some persons appear comparatively resistant to colds, whereas others are relatively susceptible. Factors said to predispose to colds are exposure to chilling and dampness, sudden changes in temperature, dusty atmospheres, drafts, loss of sleep, overwork, and lowered bodily resistance. Most colds occur between October and May and preschool children have the greatest number. Immunity after an attack is brief. An individual may average two to four colds a year. The development of a vaccine is still experimental.

In patients with the syndrome of the common cold, adenoviruses, certain enteroviruses, respiratory syncytial virus, influenza viruses, and parainfluenza viruses have at times been incriminated specifically. A true cold may be caused by one or more than one virus.

■ **Cold viruses.** Echovirus 28, reclassified as rhinovirus, type 1 (p. 435), was the first to be implicated as a "cold virus." Later, pathogenic strains referred to as *rhinoviruses* or *Salisbury viruses* (100 rhinoviruses recognized, 75 characterized) were found in persons with colds and were grown in tissue cultures containing cells from lungs of human embryos. Rhinoviruses are responsible for more colds than are any other known agent. Another group of cold viruses, the *coronaviruses,* is made up of at least 10 serotypes. It is so named because of a segmented ring about the virion. The cold caused by coronaviruses has a longer incubation period and a shorter sequence of illness than is seen with rhinoviruses.

Viral pneumonia*

Viral pneumonia is a clinical syndrome—*not* a single, specific disease—acute, infectious, and self-limited. It is similar in its manifestations to primary atypical pneumonia, also a syndrome. Viruses implicated include the influenza viruses, the parainfluenza viruses, the adenoviruses, and, in infants, the respiratory syncytial virus.

DISEASES OF THE CENTRAL NERVOUS SYSTEM

Rabies (hydrophobia)†

■ **The disease.** Rabies is an acute, paralytic, ordinarily fatal, infectious disease of warm-blooded animals, including man. Rabies has been recognized for 2000 years and has changed little in that period of time. It is primarily a disease of the lower animals, and dogs are chiefly responsible for its propagation in civilized communities. Other domestic animals that contract rabies are cats, horses, cows, sheep, goats, and hogs. When the disease occurs in wildlife, it is referred to as sylvatic rabies or wildlife rabies.

---

*See the discussion on primary atypical pneumonia, p. 377.
†For immunization, see pp. 601, 610-611, and 621-623.

Wild animals that often contract it are wolves, foxes, skunks, coyotes, raccoons, and hyenas. The rabid skunk is most dangerous to man or other animals, since the saliva of the skunk carries 100 to 1000 times the amount of virus carried by the dog. Skunks are more susceptible than are dogs and spread the disease among themselves. The skunk is said to be the most common rabies carrier in the United States today. In certain parts of the world (South America, West Indies, Central America, and Mexico) the vampire bat transmits the disease to horses and other livestock. This is a serious veterinary problem because the vampire bat differs from other animals in that it does not succumb to the disease but becomes a symptomless carrier, remaining infective a long time. The disease has recently been recognized in certain insect-eating bats in parts of the United States (Florida, Texas, Pennsylvania, California, and Montana), in the Caribbean area, and in Europe. No instance of the transmission of the disease from man to man has been recorded. Rabies in dogs is declining in incidence but the disease in wild animals is definitely on the increase. It is rare in rodents (rats, mice, squirrels) and almost unheard of in laboratory animals (hamsters, gerbils, guinea pigs).

Rabies in animals occurs in two forms, the *furious* and the *dumb*. In the former, a stage of increasing excitability is followed by a stage of paralysis ending in death. The animal fearlessly* attacks anything that it encounters; it drools saliva because its throat is paralyzed. In dumb rabies, paralysis and death supervene without a preceding stage of excitement. The animal neither bites nor attacks. Except that its lower jaw drops down, it may show little sign of illness. It often hides and may be found dead. In either form of rabies there is no fear of water, and the animals do attempt to drink. The human patient, however, avoids fluids because of painful spasms in the throat muscles induced with the act of swallowing. This explains the name *hydrophobia* for the disease. It has been said that the agony of the spasms of the victim of rabies possibly exceeds all other forms of human suffering.

■ **The virus.** The bullet-shaped rabies virus (Fig. 28-12) is found in the nervous system and saliva of infected animals. It is transmitted in the saliva introduced into the body through a wound that is usually the bite of a rabid animal. If a wound, such as a cut or abrasion, becomes accidentally contaminated with the saliva of a rabid animal, infection is as likely to occur as if the animal had inflicted the wound. Rabid animals can transmit the infection to others several days before they show signs of disease. When a person or animal is infected, the infectious agent passes from the site of inoculation along nerve trunks to the central nervous system. When the virus reaches the brain, the manifestations of rabies appear. About 30% to 40% of persons and 50% of dogs bitten by rabid animals develop the disease. *But once the disease develops, it is almost invariably fatal.*†

Rabies virus in the brains of animals naturally infected is known as street virus. If some of the emulsified brain containing street virus is inoculated into the brain of a rabbit, that rabbit will develop rabies within 15 to 30 days. If some of the emulsified brain of this rabbit is injected into a second rabbit, the second rabbit will contract rabies in a slightly shorter time. When the virus has been passed through a series of some 90 rabbits, it becomes highly virulent for the rabbit, producing rabies in 6 days. At the same time it has to some extent lost its virulence for all other species of

---

*Absence of fear in a wild animal is abnormal.

†The first time anyone has survived the disease has been in a recent case given extensive coverage by the news media.

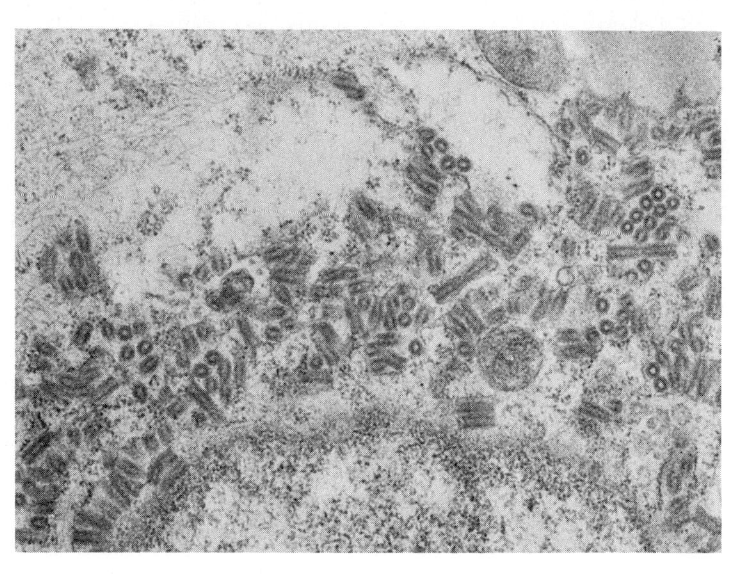

FIG. 28-12. Bullet-shaped rabies virus particles from hamster kidney cells shown in clusters in this electron micrograph. (×32,000.) (Courtesy Dr. Klaus Hummeler, Philadelphia, Pa.)

animals. It is now *fixed* virus because of this loss of virulence. Further passage through rabbits does not reduce the time required for rabies to occur. This virus, treated with phenol, is the Semple Vaccine used in the Semple method of antirabies vaccination. It has been replaced by improved preparations today.

■ **Incubation period.** The period of incubation of rabies is remarkable for its length. In man it can vary from 10 days to 2 years, with an average of 2 to 6 weeks. In dogs it varies from 8 days to 1 year, with an average of 2 to 8 weeks. The nearer the site of inoculation to the brain, the shorter the period of incubation. It is also shorter in children than in adults.

■ **Laboratory diagnosis.** Adelchi Negri (1876-1912) discovered certain bodies in the brain cells of animals with rabies. The *Negri bodies* store ribonucleoprotein and virus antigen and are found in almost 100% of the patients when disease is fully developed. By finding them the laboratory diagnosis of rabies is most quickly made. The fluorescent antibody examination of brain tissue is the preferred diagnostic method.

■ **Prevention.** *If a person is bitten by an animal suspected of having rabies, the animal should not be killed but should be placed in the hands of a competent veterinarian for observation.* If the animal has rabies, the disease will be sufficiently developed within a few days for a definite diagnosis to be made. If the dog is well 10 days after biting a person, the person is in no danger. If, on the other hand, the animal is destroyed at the time of the bite, examination of the brain may fail to show Negri bodies because they are often absent in the early stages of the disease.

If examination of the brain of the animal fails to show Negri bodies, six or eight mice should be inoculated intracerebrally with an emulsion of the brain. If the animal had rabies, the mice may show Negri bodies in their brain tissue as early as 6 days and will exhibit signs of the disease on the seventh or eighth day. If the mice survive the twenty-first day, the animal may be considered not to have had rabies.

Local treatment of wounds inflicted by rabid animals is very important in prevention of the disease. Adequate first aid treatment is crucial. Immediate and thorough cleansing of superficial wounds with a tincture of green soap or benzalkonium chloride solution may inactivate the virus. (Even tap water has merit.) An antiseptic may be applied and the wound dressed. If a wound is deeply placed and washing by soap and water is not feasible, it is then considered advisable to cauterize the wound with fuming (concentrated) nitric acid. Antiserum may be injected into the base of the wound. Dogs and cats bitten by rabid animals should be destroyed at once or kept under observation for 6 months.

Persons who work with this disease are often confronted with the question of what course to take when a person drinks the milk or eats the flesh of a rabid animal. Eating or handling infected flesh or drinking contaminated milk can produce rabies if there is an open lesion on the skin or in the alimentary tract.

The nursing precautions in rabies are rather simple. All that is necessary is to sterilize the secretions from the mouth and nose of the patient and articles so contaminated. Rabies can be eradicated from civilized communities by measures designed for intensive control of stray dogs, immunization of resident dogs, and elimination of any reservoir of infection in the wild animals of the area.

Viral encephalitis and encephalomyelitis

The term "encephalitis" (*plural,* encephalitides) means inflammation of the brain, but by common usage it is applied to inflammatory conditions of the brain accompanied by degenerative changes instead of suppuration (pus formation). When the brain is involved, the spinal cord usually is also, hence the term "encephalomyelitis." Encephalitis and encephalomyelitis indicate primary disease of the central nervous system.

Animal viruses carried in the body of an insect vector are *arboviruses,* a simplification of the term "*ar*thropod-*bo*rne viruses." Arboviruses are the most numerous of the viruses infecting man. Important forms of encephalitis caused by neurotropic arboviruses are the *arboviral encephalitides.* Two other noteworthy forms of encephalitis are *epidemic encephalitis,* or *encephalitis lethargica,* and *postinfection encephalomyelitis.* Neither is due to a neurotropic virus nor transmitted by an insect.

Arboviral encephalitides

Arbovirus infections occur principally in mammals or birds, and for most, man is only an accidental host. Arbovirus infections are the largest group of zoonoses. The mosquito is an important insect vector although arbovirus is found in ticks and mites. Although the arboviral encephalitides are similar in many respects, the causative agents are immunologically distinct, and the geographic distribution of disease is typical.

Equine encephalomyelitis, a disease of horses and mules, secondarily of man, occurs in three types, each with its own virus: The eastern type (EEE) is seen in the southern and eastern United States, the western type (WEE) in the western United States and Canada, and the Venezuelan type (VEE) in South America and Panama. St. Louis encephalitis (SLE), so named because it was first recognized in an epidemic in the vicinity of St. Louis in 1933, is a disease widespread in the United States. It occurs only in man. Japanese B encephalitis is found in the Far East, and Murray Valley encephalitis is found in Australia.

The eastern type of equine encephalitis is a severe form of disease with a mortality up to 70%. The death rate is much lower in the western form and in St. Louis encephalitis. Young persons, particularly infants, have an increased susceptibility to western equine encephalomyelitis, and St. Louis encephalitis has its greatest incidence in persons of middle age or older. The incubation period for the different encephalitides ranges from 5 to 15 days.

The epidemiologic pattern for the different forms of arboviral encephalitis is reasonably consistent, with but slight variations. Arbovirus resides and multiplies in wild birds, its natural and primary hosts, and occasionally in domestic fowl. Venezuelan equine encephalomyelitis multiplies best in a reservoir of mammals; the Eastern equine encephalomyelitis virus multiplies in both mammals and birds. From its natural reservoir arbovirus is transmitted from fowl to fowl (or mammal to mammal) and from bird to horses and man (terminal hosts) mainly by the mosquitoes of the genus *Culex*. There is no known instance of direct person-to-person transmission of encephalitis.

Encephalitis is a disease of warm weather and the summertime. In nontropical areas snakes and other cold-blooded animals harbor the virus for the cold months of the year. When snakes come out of hibernation, they seek out areas that also shelter large numbers of wild birds. Virus circulates in the bloodstream of snakes, as it does in birds, so that in the early spring mosquitoes can pick up the virus from such an overwintering host and carry it straight to the wild bird reservoir. Once infected, a mosquito remains so for life.

For mosquito-borne encephalitis to reach epidemic proportions, a combination of factors is required, including a large wild bird population, favorable breeding sites for mosquitoes in stagnant pools and puddles, a high temperature, and susceptible terminal hosts.

The laboratory diagnosis of encephalitis is made upon recovery of the virus from suitable specimens and its proper identification. Clotted whole blood or serum, throat washings, cerebrospinal fluid, urine, and, in fatal cases, brain tissue are specimens for virus isolation. Young or newborn mice are inoculated and observed. Serologic tests for identification of virus or its presence are the neutralization test, the complement-fixation test, and a hemagglutination-inhibition test.

Effective vaccines made by growing the equine viruses in a chick embryo confer immunity in horses lasting from 6 months to 1 year or longer. Vaccines suitable for use in human beings are still in the experimental stage. The practical approach to the control of the disease is eradication of the mosquito vector.

Epidemic encephalitis (von Economo's disease)

Epidemic encephalitis is probably an old disease, but it did not become prominent in medicine until a great outbreak followed the influenza pandemic of 1918-1919. Sometimes called sleeping sickness, it is described by lethargy and drowsiness that gradually pass into coma. The varied nervous system manifestations are severe. Although the early epidemics seemed to follow in the wake of influenza, the relation of the two diseases is not known. It is assumed that the disease is caused by a virus, but one has never been identified. It is believed that epidemic encephalitis is spread by droplets of buccal and nasal secretions and that the virus gains access to the central nervous system via the nasopharynx. The disease may be epidemic or sporadic, and institutional outbreaks occur. Peculiarly, the patient's family is not especially vulnerable although the incidence among doctors and nurses is rather high. There is no known treatment for established disease and no known method of preventing or controlling an outbreak. Fortunately no epidemics have occurred since 1926.

Postinfection (demyelinating) encephalomyelitis

Postinfection encephalomyelitis (allergic encephalomyelitis) is an acute disease of the central nervous system that occasionally arises during convalescence from infectious diseases or occurs after vaccination against such. The diseases are most often viral, important among which are measles, German measles, smallpox, and influenza. Vaccination for smallpox and rabies is complicated by this type of encephalomyelitis. Postinfection encephalomyelitides resemble the viral encephalitides, not from the pathologic but from the clinical standpoint.

Most encephalomyelitides occurring after smallpox vaccination (postvaccinal encephalomyelitis) occur in children and young adults who have never been vaccinated. Infants seem to be resistant. The incidence of postvaccinal encephalomyelitis is less than 1 time in 33,000 vaccinations.

The following theories are offered to explain postinfection encephalomyelitis: (1) It is caused by the virus that produced the primary disease, (2) vaccination activates some latent virus in the body, and (3) the disease reflects an allergic reaction to the virus causing the preceding infection or to the patient's own nervous tissue now altered by a virus.

Currently this last, the autoimmune approach, is favored. The cause has not been demonstrated to be a virus and still remains undetermined.

Slow virus infections

In some chronic diseases of animals there is experimental evidence that the end-stage changes in the tissues, particularly in the central nervous system, are the result of a slowly progressive and damaging proliferation of a virus. In such diseases there is a long incubation period (years), a slow start, no demonstrable formation of antibodies, no fever, and no inflammatory changes. *Scrapie*, a slowly developing, fatal, neurologic disease of sheep, is an example of a slow virus infection. In one such chronic disorder, subacute sclerosing panencephalitis, an agent resembling measles virus has recently been cultured from affected brain tissue. Slow viruses may also be implicated in arthritic and rheumatic diseases. Slow viruses are so designated because of their very long developmental cycles during which they are masked or inapparent. The current interest in slow infections stems from the suggestions they give as to the cause of many poorly understood central nervous system diseases in man. An example in man is *kuru*, a neurologic disease found in certain cannibals in New Guinea. This disease is transmitted by the ingestion of undercooked brain tissue containing the virus.

Poliomyelitis (infantile paralysis)*

■ **Poliovirus infection.** Poliomyelitis is an acute infectious disease that in its severest form affects the brain, spinal cord, and certain nerves (Fig. 28-13).

*For immunization, see pp. 602, 603, 610, 617, 618, 624-625, and 627.

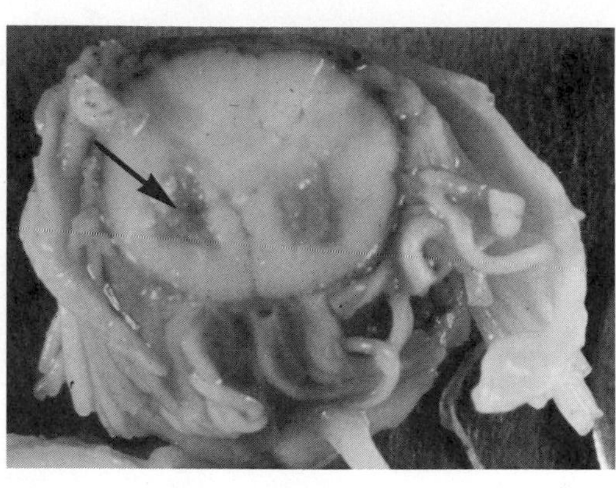

FIG. 28-13. Poliomyelitis, cross section of spinal cord from patient with paralytic poliomyelitis. (Paralysis results when motor neurons of the anterior horns of the spinal cord are destroyed by poliovirus.) Arrow points to anterior horn on left. It is a softened, depressed area of dead tissue, focally hemorrhagic. The anterior horn on the right is similarly involved. (Courtesy Dr. B. D. Fallis, Dallas, Texas.)

Poliomyelitis occurs in four forms:

1. Silent or asymptomatic infection

   Disease is not apparent clinically. Infection of the human alimentary tract by the poliovirus is exceedingly common all over the world. Any symptoms present are so mild as to be largely overlooked by the infected person. Such a person is a healthy carrier. From the alimentary tract, the virus at times enters the blood or lymph stream. Silent infection exists among the members of a family in which a clinical case of poliomyelitis is present.

2. Abortive infection

   Findings referable to the nervous system are absent, although the affected person has a brief febrile illness, a mild respiratory infection with headache or a simple gastrointestinal upset with nausea and vomiting. Most cases are abortive.

3. Nonparalytic infection

   Findings point to the nervous system, but no residual paralysis develops.

4. Paralytic infection

   Paralysis persists.

■ **Polioviruses.** There are three serologic types of poliomyelitis virus, designated type I (Brunhilde), type II (Lansing), and type III (Leon). These viruses, among the smallest in size, may be grown in monkeys and chimpanzees.

Tissue cultures of monkey kidney sustain a generous growth, as do certain human cell cultures. Polioviruses are pathogenic for man, monkeys, chimpanzees, and apes. As far as we know, man is the only animal subject to the disease; no reservoir of infection has been demonstrated in animals.

Polioviruses can be inactivated by ultraviolet radiation, by drying, and, if in a *watery* suspension, by a temperature of 50° to 55° C. for 30 minutes. Inactivation is slow with disinfectant alcohol and unsatisfactory with many bacterial germicides in common use.

■ **Transmission.** The incubation period ranges from 3 to 35 days. The infected person is most likely to pass the virus to a noninfected person during the latter part of the incubation period and the first week of the clinical illness, the time at which virus is present in his throat. Before the onset of symptoms in infected persons, poliovirus is found in the secretions from the mouth and throat and in the feces. Today the poliovirus is believed to enter the body at the upper part of the alimentary tract and to leave through either the upper or the lower end. Poliovirus has been recovered in large amounts from sewage, and milk-borne epidemics have been recorded. But the epidemiologic pattern for poliomyelitis is not that for the enteric infections. The most important mode of spread seems to be the direct one from person to person. It has been said that poliomyelitis "travels with a crowd." Houseflies, filth flies, and cockroaches can be contaminated with virus, especially during an epidemic, but the role of these agents in transmission is undefined.

■ **Epidemiology.** Poliomyelitis occurs sporadically but tends to be epidemic. Infantile paralysis is not a good name for the disease because it may occur in adults and paralysis may be absent. In the early epidemics the disease chiefly attacked children less than 5 years of age, but then, in time, the children affected were older, and the number of adult cases increased.

The disease is thought to be increasing in severity and is worldwide. All races and classes of people are affected. In temperate climates, poliomyelitis appears in early summer. The number of patients and the severity of the disease increase; a peak is reached in late summer and early fall; and the disease subsides after the first frost. Occurrence in more than one member of a family is frequent. There is an increased incidence in pregnant women since they are more susceptible to the disease, and there have been cases of congenital poliomyelitis reported in which the mother had the disease late in pregnancy.

Radical changes have taken place in the epidemiology of poliomyelitis with the use of the polio vaccines. In 1957, for the first time, all states and territories of the United States were free of epidemics. For the year 1957, just less than 6000 cases of poliomyelitis were reported to the United States Public Health Service, of which about 2500 were paralytic. A decade later in 1967, there were 44 cases (29 were paralytic), mostly in unimmunized or inadequately immunized children. Compare these figures with over 57,000 cases (21,000 paralytic) reported for the prevaccine year of 1952!

■ **Immunity.** Infection with one type of virus does not confer immunity for the other types. Poliomyelitis confers lasting immunity only for the virus type responsible. A high percentage of adults have virus-neutralizing substances in their blood. Infants can inherit an immunity from their mothers by placental transfer.

■ **Prevention.** Preferably, the patient with poliomyelitis is isolated although, with the widespread distribution of the virus, real insulation from contacts is impossible. Cross infections between patients are probably inconsequential in hospitals where no attempt to isolate is made. However, the disease is apparently sometimes spread from patient to attendant, nurse, or physician. On the whole, contacts, especially between children, should be minimized during epidemics, and in an epidemic, such public health measures as the closing of public swimming pools are not inadvisable. Tonsillectomies should not be done during epidemics or in the season of the year when the incidence of the disease is highest. There is a greater risk for the severe form of the disease to develop in persons who have recently had their tonsils removed. Healthy children may carry the virus in their throats, and virus has been demonstrated in surgically removed tonsils.

OTHER ENTEROVIRUS DISEASES

Infections with coxsackieviruses

The coxsackieviruses, a subdivision of enteroviruses, are worldwide in distribution and have been frequently associated with epidemics of poliomyelitis. They were named after the town in New York where the first virus was identified as research work was done on poliomyelitis. They resemble the viruses of poliomyelitis in many respects, including their epidemiology. They are found in the nasopharynx and feces and at times in other parts of the body. They have never been found in the cerebrospinal fluid.

They are divided into groups A and B. Within each group is a number of types, of which at least 30 are known. These viruses possess an unusual pathogenicity for infant mice and hamsters but none for the adult animals. Group B is the more important in man. Coxsackieviruses produce such conditions as aseptic meningitis, epidemic pleurodynia (group B), herpangina (group A), vesicular pharyngitis, encephalitis, hepatitis, orchitis, an influenza-like disease, "three-day fever," pericarditis, and a severe form of myocarditis in infants. Group A viruses usually are associated with infections of the mouth. Most illnesses produced by coxsackieviruses present in children. Neutralizing antibodies to the virus are found in the convalescent serum of patients recovering from the diseases listed.

Modern classification of viruses assigns the coxsackieviruses to the category of picornaviruses, along with the polioviruses and the echoviruses. Most strains of coxsackieviruses and echoviruses can cause a disease closely resembling either paralytic or nonparalytic poliomyelitis.

FIG. 28-14. *Aedes aegypti* mosquito, larval forms in Petri dish. (From Med. World News **6:**168, Nov. 12, 1965; courtesy Eli Lilly & Co., Indianapolis, Ind.)

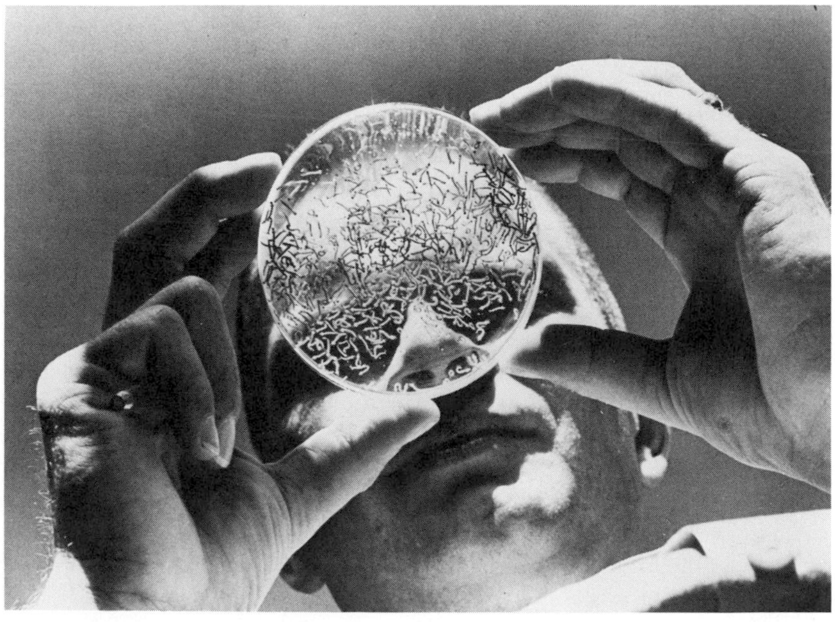

Infections with echoviruses

Echoviruses were found by accident during epidemiologic studies of poliomyelitis. They were recovered from human fecal material in tissue cultures. Their destructive effect on tissue culture cells and the fact that any disease to which they were related was unknown at first led to the designation of *echo*—*e*nteric, *c*ytopathogenic, *h*uman *o*rphan viruses. Some 29 members of the subdivision have been so far classified. Many of these viruses may be harmless parasites, but some cause epidemics of aseptic meningitis, summer diarrhea in infants and young children, and febrile illnesses, with or without rash. To date, the echoviruses have been found as etiologic agents only in clinical syndromes (not specific diseases) that may be produced also by a number of other viruses (and bacteria as well). For example, aseptic meningitis, their chief central nervous system manifestation, is a syndrome, not a distinct disease.

OTHER ARTHROPOD-BORNE VIRAL DISEASES

Yellow fever

Yellow fever,* a mosquito-borne hemorrhagic fever, is an acute infectious disease defined by an abrupt onset, a rapid course, and a high mortality. The most characteristic feature is the rapid and extensive destruction of the liver. Prominent features are jaundice (yellow color to skin and mucous membranes), albumin in the urine, hemorrhage, and vomiting.

The virus of yellow fever is transmitted by the mosquito *Aedes aegypti* (Fig. 28-14), which also transmits dengue fever. The mosquito bites a person during the first few days of illness and becomes infected. The virus multiplies in the body of the mosquito and reaches the salivary glands at the end of about 12 days; the mosquito remains infective the remainder of *her* life. (Only the female transmits the disease.) When a nonimmune person is bitten by an infected mosquito, manifestations of yellow fever develop in 3 to 5 days. The virus is found in the blood during the first 3 days of the disease.

Yellow fever still remains endemic in many parts of the world. We must always guard against it because an epidemic requires only a person ill of the disease, mosquitoes to be infected, and susceptible individuals. The transport of infected mosquitoes by airplanes and ships must be prevented.

A source of infection in tropical Africa and Latin America is the jungle yellow fever found in monkeys. It is transmitted from animal to animal and from animal to man by forest mosquitoes (genus *Haemagogus*) living high in treetops. Epidemics appear in monkeys. The causative virus seems to differ in no respect from that producing ordinary yellow fever. If man becomes infected, the infection is transmitted from him by *Aedes aegypti.*

---

*For immunization, see pp. 612, 627, and 629. Whenever yellow fever (once known as yellow jack) is discussed, eight names are brought to mind: Carlos Juan Finlay (1833-1915), who first accused the mosquito of spreading yellow fever; Walter Reed (1851-1902); James Carroll, Jesse W. Lazear, and Aristides Agramonte, who, in 1900, formed the Commission of the United States Army to study yellow fever in Cuba; Privates John R. Kissinger and John J. Moran, who permitted themselves to be inoculated with yellow fever; and William C. Gorgas, who applied the knowledge obtained to make the tropics more habitable for man. The members of the commission went to Cuba, lived in the tents of those who had had yellow fever, wore their clothes, and were bitten by infected mosquitoes. Carroll and Lazear contracted the disease; Lazear died. To these men, all who live in tropical and temperate zones owe a debt of gratitude.

An active immunity to yellow fever can be established. The value of convalescent serum is undetermined.

Dengue fever

Dengue, or "breakbone" fever, is an acute disease lasting 10 days, and depicted by sudden onset of paroxysmal fever, intense joint pain, skin rash, and mental depression. It is caused by a virus that is found in the bloodstream during the early days of the attack and transmitted by the *Aedes aegypti* mosquito, which, once infected, remains so the rest of *her* life. When dengue fever is introduced into a community, a high percentage of the population contracts the disease. An attack is followed by an immunity that may persist for 1 to 2 years or the remainder of the patient's life.

There are four distinct immunologic types of the virus, of which types 1 and 2 are most often responsible for the disease in the Western Hemisphere. The dengue viruses can be grown in tissue cultures. Type 1, live virus, weakened by repeated passages through mouse brain is an effective vaccine in the clinical trials to date. Neutralizing and complement-fixing antibodies may be demonstrated in the serum of a patient after infection.

Colorado tick fever

An arboviral infection transmitted by an insect vector other than the mosquito is Colorado tick fever, caused by a single small virus transmitted by the wood tick, *Dermacentor andersoni*. A mild febrile, diphasic illness, it is found largely in the Western states, the natural habitat of the tick vector. Patients usually have been in a tick-infected area 4 to 6 days before becoming ill, and ticks are even found still attached to their bodies. The onset of disease is sudden, with chills, fever, and bodily aches that continue for 2 days or so. Then follows a period wherein the patient is essentially free of clinical manifestations. The fever returns to last for several days more before the disease has run its course. There are no complications. The disease has not been fatal, and the pathology is not known.

■ **Laboratory diagnosis.** Colorado tick fever virus may be recovered from the patient in specimens of blood, throat washings, and stool. It grows in tissue culture and in the fertile hen's egg. When inoculated into young mice, it produces paralysis. There are complement-fixation, virus-neutralization, and hemagglutination-inhibition tests for serologic diagnosis.

■ **Immunity and prevention.** Infection is thought to produce lasting immunity. Prevention depends upon the avoidance of tick-infested areas or the wearing of adequate protective clothing. The body should be regularly inspected for ticks, and any present should be detached. Effective vaccines have been prepared by growing the virus in chick embryos.

DISEASES OF THE LIVER (VIRAL HEPATITIS)*

There are two viral diseases of the liver of special interest. They closely resemble each other, yet because of certain supposed differences in their epidemiology and mode of onset they have been defined as distinct clinical entities. One has been

*Viruses that may cause hepatitis include arbovirus, myxovirus, adenovirus, herpesvirus, coxsackievirus, reovirus, cytomegalovirus, and the agent of infectious mononucleosis.

known as infectious hepatitis (also catarrhal jaundice, epidemic jaundice, short incubation hepatitis), and the other as serum hepatitis (also homologous serum jaundice, transfusion jaundice, posttransfusion jaundice, long incubation hepatitis). The virus postulated to cause infectious hepatitis has been referred to as virus A or IH virus, and the virus for serum hepatitis as virus B or the SH virus. As a frame of reference a traditional consideration of these two is presented herein (Table 28-4). With the impact of new discoveries related to the diseases and their causative agents, our concepts in the very near future are going to be drastically changed.

Conventionally both are acute or subacute infections of the liver found only in man.* There are clinical similarities, and the pathologic changes in the injured liver are the same. The true overall incidence of viral hepatitis is unknown since subclinical infection occurs without jaundice. It has been estimated that 80% to 90% of cases go unrecognized. Immunity from an attack of the one form of disease has been believed to result in immunity to that form only.

One uncertainty continues, that is, whether these clinical forms of hepatitis are caused by one virus, two distinct viruses, perhaps a whole family of viruses, or even by a virus agent at all.

■ **Australia antigen.** As he was working in Australia with the blood of an Australian aborigine, a population geneticist in 1964 found a new antigen, which he called the *Australia antigen.* As it turned out, his was a most significant discovery.†

Although its true nature is as yet undefined, the Australia antigen (Au) was shortly visualized with the electron microscope as a very small, spherical, viruslike particle about 200 Å in diameter. With knoblike subunits on its surface, it looks like picornavirus. It has been isolated and can be agglutinated. Its identity as a virus is challenged because it has neither an RNA or a DNA backbone and does not grow in tissue culture. There is speculation that here may be a new kind of infectious agent replicating without nucleic acid.

More important still is the fact that Australia antigen is tightly linked to acute and chronic hepatitis (although it is also found in diseases such as Down's syndrome and leukemia, wherein immune mechanisms have gone awry). Its role in hepatitis rests on observations such as the following: More than half the patients with serum hepatitis have the antigen in their blood. It can be recovered from liver cell nuclei, urine, and feces. If it is given experimentally to a group of human beings, the majority will come down with hepatitis (posttransfusion type). Antibodies develop in some persons who remain asymptomatic.

Some observers feel the Australia antigen is exclusively associated with serum hepatitis, but the widely used term *hepatitis-associated antigen (HAA)* provides a loophole. (It is also termed hepatitis B antigen or HBAg.) Clearly not all viral hepatitis is associated with this antigen. All agree that there is an Au-positive and an Au-negative hepatitis. Interestingly, another antigen has been found, this time in the blood of a patient with Au-negative hepatitis. It is termed the *Milan antigen.*

The immediate application of all of this is in the blood bank to screen blood donors. The incidence of viral hepatitis among patients receiving blood containing Australia antigen is five times greater than it is among those patients given Au-negative

*Recently it has been reported that serum hepatitis has been transmitted to nonhuman primates.
†The research from which this discovery "fell out" was a basic type, not goal-oriented at all. The importance of the finding and the way in which it was made reemphasize the continuing need for the "basic" approach to scientific matters.

TABLE 28-4. COMPARISON OF THE TRADITIONAL TWO TYPES OF VIRAL HEPATITIS

	Infectious hepatitis	Serum hepatitis
Agent		
Virus	A (IH) (?)	B (SH) (?) (see discussion)
Epidemiology		
Transmission	Fecal-oral route; also parenteral	From blood and blood products—primarily parenteral; can be oral and contact
Experimental inoculation	Oral, parenteral	Parenteral
Incubation period	Around 30 days	50 to 180 days
Seasonal peak	Fall, winter	None
Age preference	Children, young adults	All ages
Duration of infectivity	Virus in feces and blood 1 to 2 weeks before disease; remains 3 to 4 weeks longer	Virus in blood 3 months before disease; asymptomatic carrier for as long as 5 years
Virus present	Feces, blood	Blood (thought to be in feces)
Clinical features*		
Onset	Acute	Slow, usually insidious
Fever	Common before jaundice	Less common
Jaundice	Rare in children, more frequent in adults	Rare in children, more frequent in adults
Severity of disease	Less severe	More severe
Prognosis	Good	Less favorable
Pathologic changes in liver	Same for both forms	
Laboratory evaluation		
Thymol turbidity†	Increased	Normal
Abnormal SGOT‡	Transient, 1 to 3 weeks	Prolonged; 1 to 8 or more months
IgM levels	Increased	Normal
HAA (Australia antigen) in blood (see text)	Not present	Present during incubation period and acute phase—occasionally persists
Prevention and control		
Prophylactic effect of gamma globulin	Good	Present but poor (questionable)
Nursing precautions and control	1. Isolation of patient 2. Sterilization of contaminated items	1. No isolation of patient needed 2. Sterilization of permanent medical apparatus in contact with blood 3. Use of disposables (needles, syringes, tubing) for all patients

*Many clinical features are the same.
†Test of liver function.
‡The enzyme serum glutamic oxaloacetic transaminase is elevated with liver disease.

blood. Already a number of tests have been designed to detect the Australia antigen in blood for transfusion. In rapid succession many are coming onto center stage. Technics include those of complement fixation, immunoelectrophoresis, and hemagglutination inhibition. At last something can be done to minimize a very serious risk.

Infectious hepatitis

Infectious hepatitis is usually spread directly from person to person. The principal pathway is the fecal oral route. Poor sanitation favors its spread. Water, milk, and food can be sources of infection. Recent outbreaks have occurred in persons who have eaten raw shellfish contaminated by sewage. The cockroach and fly may be vectors.

The virus of infectious hepatitis is postulated to be the smallest, smaller even than the poliovirus. It is present in the feces and blood of infected persons.

The mortality of infectious hepatitis is not high. Widespread immunity does exist, probably gained from inapparent childhood infection. Gamma globulin given as late as 6 days before the onset of the disease may protect a person for as long as 6 or 8 weeks.

The patient with recognized disease should be isolated. Diligent handwashing, wearing of protective gowns, and autoclaving of articles contaminated by the patient are necessary.

Serum hepatitis

Serum hepatitis (homologous serum jaundice) is carried by human serum (or plasma) and may complicate blood transfusion, or the administration of convalescent serum, vaccines, and other biologic products containing human serum. Needles, syringes, and tubing sets for transfusions and stylets for finger puncture are important conveyors of the infection when soiled by blood or blood products.* *As little as 0.000025 ml. of blood contaminated with SH virus (0.01 ml. with IH virus) has been shown to cause disease.* For this reason disposable needles, disposable stylets for finger puncture, and disposable syringes, all now commercially available, are strongly recommended. Disposable units of plastic tubing suitable for blood transfusion are now in wide use.

Serum hepatitis represents an occupational hazard among medical personnel working with blood or serum, including laboratory technologists, blood bank workers, physicians, nurses, and dentists. Persons so exposed should observe rigid asepsis. The attack rate is high in the renal dialysis unit, and, paradoxically, the hepatitis is much severer in the nurses and other personnel than it is in the patients. The disease is also prevalent among drug addicts who share their unsterilized and contaminated hypodermic needles. Heroin addicts are especially notorious for "passing the needle."

MISCELLANEOUS VIRAL INFECTIONS

Mumps (epidemic parotitis)†

Mumps or epidemic parotitis is an acute contagious disease prevalent in winter and early spring, accompanied by a painful inflammatory swelling of one or both parotid glands. It is easily recognized because of the typical appearance of the patient.

*For sterilization technics to prevent the spread of the hepatitis viruses, see p. 267.
†For immunization, see pp. 602, 603, 610, 618, and 621.

(A child may look like a chipmunk with nuts in its cheeks.) Caused by a paramyxovirus, it occurs most often between the fifth and fifteenth years. Adult epidemics may erupt in military organizations and are extremely difficult to control. The period of contagion begins before the glandular swelling and persists after it has subsided. The incubation period is usually 14 to 28 days. Mumps is mostly transferred directly from person to person by droplets of saliva, but indirect transfer by contaminated hands or inanimate objects occurs. An estimated 30% to 40% of persons infected with the mumps virus have a silent infection followed by permanent immunity. During the course of the inapparent infection, these individuals may pass the virus to others.

The virus is believed to enter the body by way of the mouth and throat and probably reaches the salivary glands via the bloodstream. Viremia is responsible for such complications as orchitis, oophoritis, encephalitis, and pancreatitis. Inflammation of the testicle (orchitis), usually one sided, occurs in 20% of adult males with this disease. If, as is rarely the case, both testicles are involved, sterility may result. The mumps virus is an important cause of the aseptic meningitis syndrome. Inflammation in the central nervous system can be present with and without parotitis.

■ **Laboratory diagnosis.** Mumps virus may be recovered from the urine, saliva, or blood; in cases of central nervous system disease, it is found in the cerebrospinal fluid. The chick embryo and tissue cultures are used for its identification. The virus agglutinates red blood cells of the chicken and of the human blood group O. A skin test, not widely used, is available. The antigen for the test is killed virus from fertile eggs. Inflammation at the site of injection 18 to 36 hours later indicates immunity in the absence of sensitivity to egg protein.

■ **Immunity.** A passive immunity is given by an immune mother to her baby and persists for about 6 months. Convalescent serum given from 7 to 10 days after exposure protects a high proportion of children from infection. An attack of mumps is usually followed by permanent immunity, and, contrary to lay belief, the same immunity follows unilateral as follows bilateral involvement.

Infectious mononucleosis

Infectious mononucleosis, or glandular fever, is an acute infectious disease characterized by enlargement of the lymph nodes and spleen, sore throat, and mild fever. It is transmitted as an airborne infection. The total number of leukocytes is increased, and characteristic white blood cells (lymphocytes of unusual or atypical appearance) appear in the peripheral circulation. Rarely fatal, it may occur in epidemic form or sporadically. It usually attacks children and young adults (to 30 years of age). The incubation period is unknown but is believed to be 4 to 14 days or longer. The disease may last 1 to 6 weeks, sometimes longer.

Infectious mononucleosis is important because it may be mistaken for diphtheria (because of sore throat) and such a serious disease as lymphatic leukemia. Nonsyphilitic patients may give positive serologic tests for syphilis during the disease and for several weeks after recovery. The heterophil antibody or Paul Bunnell test is important in the diagnosis.

■ **Epstein-Barr virus (EB virus).** Infectious mononucleosis has long been considered a viral disease, since the course of the disease suggests that it should be. But a specific virus was never pinpointed.

In the mid-1960s a brand new member of the herpesvirus group was found under circumstances linking it first to Burkitt's lymphoma (see p. 418) and then even more

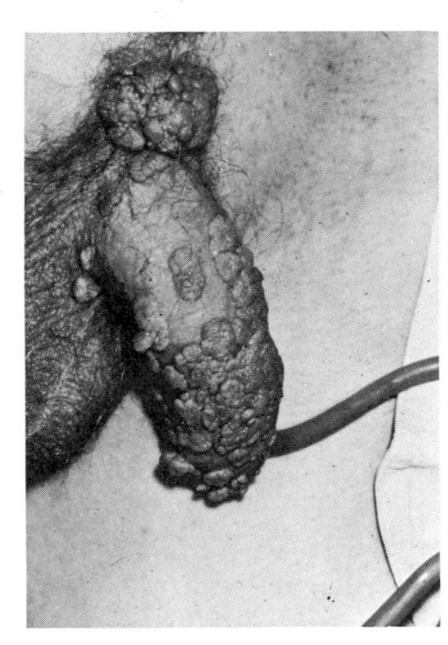

FIG. 28-15. Condylomata acuminata of skin of penis. Note confluence of the wartlike growths. A giant condyloma acuminatum is present at penile-abdominal junction. (From Winter, C. C.: Practical urology, St. Louis, 1969, The C. V. Mosby Co.)

significantly to infectious mononucleosis. The virus was named Epstein-Barr virus after the tissue culture cell line from which it was first isolated. (It is freely called the EB virus.)

There are good reasons for thinking that the EB virus causes infectious mononucleosis and a few reservations. For one thing, serologic studies comparing antibodies to EB virus with the diagnostic heterophil antibodies in the course of the disease correlate well. One can demonstrate that antibodies to EB virus and heterophil antibodies come in and peak at about the same time in a given case. Three months later heterophil antibodies drop out; EB virus antibodies remain. One important reservation is that Koch's postulates have not been satisfied. Infectious mononucleosis has not as yet been transmitted to healthy human volunteers.

Generally infection with EB virus must be widespread. The virus is everywhere and very common. Most infections are postulated to occur early in life, presenting themselves as either silent inapparent ones or at most as mild upper respiratory ailments. There are immunofluorescent and complement fixation tests for EB virus, but these are not offered in routine clinical laboratories.

Condyloma acuminatum

Condyloma acuminatum is a cauliflower-like mass of coalescent warty excrescences formed on the external genitalia and about the anal region in either sex (Fig. 28-15). The growths are often multiple and vary much in size. The lesion is called a venereal wart and is transmitted as a venereal disease. The incubation period is 1 to 6 months. Large and bulky warts cause considerable discomfort and are associated with a disagreeable odor. The cause of this condition is a virus, either the same or an allied strain of the one causing the human wart so familiar in its extragenital setting.

Foot-and-mouth disease (rhinovirus of cattle)

Foot-and-mouth disease, worldwide in distribution, is an acute infectious process in animals characterized by the formation of vesicles in the mouth, on the udder, and on the skin about the hoofs. One of the most contagious of all diseases, it primarily affects cloven-footed animals (cattle, sheep, pigs, and goats). Horses are immune, and man seldom contracts the disease. Within the last few years the disease has been present in Mexico.

The disease is readily transmitted by direct contact or indirectly by contamination of fodder, with infectious discharges. The disease has been transferred to man by contact with infected animals, by the milk of infected animals, and by contaminated material.

Control of the disease depends upon the slaughter of all exposed animals, with burial or cremation of their bodies, disinfection of pens, and proper quarantine. There is no specific treatment. A diagnosis may be established by injecting suspicious material into the footpads of a guinea pig. Typical lesions will be produced if the rhinovirus is present.

Distemper of dogs and cats

Canine distemper and feline enteritis (cat distemper) are important diseases of young dogs and cats. They are caused by two entirely different viruses. The distemper of dogs affects primarily the respiratory and nervous systems. In cat distemper, the intestinal tract is primarily involved. Canine distemper is spread by means of food and water contaminated with the virus and to a lesser degree via the respiratory tract. Feline distemper is spread by direct contact and probably by fleas. Canine distemper affects dogs and related species. Feline distemper affects cats and racoons. Neither affects man. Highly successful vaccines have been prepared against both diseases.

Additional diseases caused by viruses

The following are lists of viral diseases that affect respectively man, animals, plants, and insects:

Man

- Hemorrhagic fevers (Omsk, Crimean, Bolivian, Indian, and Korean types, hemorrhagic nephrosonephritis, Kyasanur Forest disease)
- Warts (verrucae)
- Lymphocytic choriomeningitis
- Phlebotomus (sand fly) fever
- Inclusion conjunctivitis
- Bwamba fever
- Rift Valley fever
- West Nile fever
- Semliki Forest disease
- Louping ill
- Russian spring-summer encephalitis

Animals

- Pox of horses, sheep, goats, swine, buffalo, and camels
- Rinderpest
- Ectromelia
- Vesicular stomatitis
- Orf

Infectious canine hepatitis
Aleutian disease of mink
Hog cholera
Rift Valley fever of sheep
Louping ill of sheep
Fowl plague
Fowlpox
Newcastle disease of chickens
Mengo fever
Visna in sheep

Plants

Tobacco mosaic
Tomato bushy stunt
Tobacco necrosis
Peach yellows
Curly top of sugar beet
Swollen shoot
Wound tumor
Tobacco rattle
Barley stripe mosaic
Turnip yellow mosaic
Tomato spotted wilt
Lettuce necrotic yellows
Potato X
Striate mosaic of wheat
Alfalfa mosaic

Insects (mostly the larval stage actively diseased)

Chronic bee paralysis
Silkworm jaundice
Tipula iridescence
Polyhedrosis of spruce sawfly
Other polyhedroses
Granuloses

QUESTIONS FOR REVIEW

1. Make a chart showing the causative agent, clinical features, laboratory diagnosis, transmission, and prevention of smallpox, measles, German measles, mumps, influenza, poliomyelitis, rabies, yellow fever, dengue fever, Colorado tick fever, infectious mononucleosis.
2. Discuss briefly the relation of chickenpox to shingles.
3. Outline the logical procedure to follow when a person is bitten by an animal suspected of having rabies.
4. Briefly compare infectious hepatitis with serum hepatitis.
5. Explain the occurrence of fever blisters.
6. Briefly characterize adenoviruses, coxsackieviruses, echoviruses, arboviruses, rhinoviruses, reoviruses, enteroviruses, coronaviruses, cytomegaloviruses, herpesviruses, cold viruses.
7. Give the significance of the Australia antigen.
8. What is meant by a syndrome? Briefly discuss acute respiratory syndromes.
9. Define or briefly explain: alastrim, variola major, variola minor, microcephaly, pustule, coryza, rhinitis, encephalomyelitis, yellow jack, silent infection, and SGOT.
10. What is the current position of the Epstein-Barr virus?
11. What serious effect may rubella have in a pregnant woman? Explain congenital rubella.
12. It is superstition among laymen that if the eruption of shingles encircles the body death occurs. Explain why the eruption does not do this.
13. Discuss sylvatic or wildlife rabies and its implications for the spread of the disease.
14. What is meant by slow virus infection? What is the implication?

29 The rickettsiae and bedsoniae

RICKETTSIA SPECIES (THE RICKETTSIAE)

■ **Special properties.** Rickettsiae are small, pleomorphic bacillary or coccobacillary bodies occupying an intermediate position between the bacteria and the viruses. They are named in honor of Howard Taylor Ricketts,* of Chicago. They occur singly, in pairs, in chains, or in irregular clusters. The electron microscope indicates that they have an internal structure much like that of bacteria and that they divide by binary fission. Most rickettsiae will not pass through bacteria-retaining filters. They are about the size of the bedsoniae (0.3μ in their smallest dimension). They are nonmotile and gram-negative, and they stain with difficulty. They are intracellular parasites. With rare exception, they do not multiply in the absence of living cells; the pathogenic forms grow only in the cells of infected animals. Some grow only in the cytoplasm; others grow in both the cytoplasm and the nucleus of the infected cell. The fragments of enzyme systems that rickettsiae possess allow them a range of metabolic activity. Their resistance to deleterious influences such as heat, drying, and chemicals is about the same as that of bacteria.

The rickettsial diseases are transmitted to man by insects, the natural and primary hosts of the rickettsiae.

■ **Pathogenicity.** When rickettsiae invade man, they attack the reticuloendothelial system, colonizing the endothelial lining cells of the walls of small blood vessels, small arteries, arterioles, and capillaries. Within these cells they induce a vasculitis (inflammation of the blood vessel) that is distributed all over the body, just as the small blood vessels are. The consequence is disease in many anatomic areas in the body. A characteristic skin rash and a whole host of pathologic changes occur. Rickettsial diseases can be quite serious and many times life threatening in spite of the best available therapy.

An attack of rickettsial disease is usually followed by a lasting immunity.

*Dr. Ricketts (1871-1910) first observed rickettsia bodies in a case of Rocky Mountain spotted fever in 1909, and he demonstrated that Rocky Mountain spotted fever was transmitted by the wood tick. Several years later he showed that tabardillo (Mexican typhus) was transmitted by the body louse, but in the investigation he contracted the disease and died, thus becoming a martyr to the disease process that he was studying.

■ **Laboratory diagnosis.** Rickettsiae can be cultivated in the yolk sac of the chick embryo (provided that the chicken did not receive antibiotics) or in tissue cultures. Recovery of the microorganisms may be made from the typical sites when they (or blood and suitable specimens from a patient) are inoculated into laboratory animals, such as guinea pigs, mice, and rabbits.

The most important rickettsial diseases are associated with a positive *Weil-Felix reaction.* This is an agglutination test similar to the Widal test for typhoid fever except that the test serum is mixed with different types of *Proteus* bacilli. The Weil-Felix reaction is a heterophil antibody reaction since *Proteus* bacilli *do not* cause any of the rickettsial diseases or even act as secondary invaders. Complement-fixation tests are of value in differentiating the rickettsial diseases.

■ **Classification.** Rickettsial diseases may be divided into the following groups:

1. Typhus fevers
2. Spotted fevers
3. Scrub typhus
4. Q fever
5. Trench fever

Table 29-1 indicates epidemiologic features of these groups. Rocky Mountain spotted fever, rickettsialpox, and Q fever are the three prevalent in North America.

RICKETTSIAL DISEASES

Typhus fever group

■ **General considerations.** The typhus fever group includes epidemic typhus, murine typhus, and Brill-Zinsser disease. Clinically the different types of typhus fever closely resemble each other but vary in severity. They usually begin with severe headache, chills, and fever. A rash develops about the fourth day and persists throughout the course of the disease. Mentally the patient is dull and stuporous. This feature of the disease gave origin to the name "typhus," which is derived from the Greek word *typhos,* meaning 'vapor' or 'smoke.' The course of the disease varies from 3 to 5 weeks. The mortality varies from 5% to 70%.

EPIDEMIC (CLASSIC, EUROPEAN, OLD WORLD, LOUSE-BORNE) TYPHUS. Epidemic typhus is caused by *Rickettsia prowazekii** and is spread from person to person by the body louse *(Pediculus humanus corporis).* Head lice *(Pediculus humanus capitis)* can transmit the disease but seldom do so. No animal reservoir has been found. Epidemic typhus is an acute and severe disease with a high mortality (10% to 40%) that has been known to spread over the world in devastating epidemics. It is a disease of overcrowding, famine, filth, and war. (It is also known as jail fever, war fever, and famine fever.)

When a louse bites a person with epidemic typhus, the rickettsiae are taken into the stomach of the louse and invade the cells lining the intestinal tract. They multiply to such an extent that the cells become greatly swollen, burst, and liberate the rickettsiae into the feces of the louse. When a louse bites a person, it defecates at the same time. The site of the bite itches, and that person introduces the infectious material into his skin by scratching. Infected lice die within 8 to 10 days after infection from intestinal obstruction caused by parasitism of the lining cells.

---

*The species name *Rickettsia prowazekii* is derived from Stanislas von Prowazek (1876-1915), an early investigator who lost his life in the study of typhus.

TABLE 29-1. RICKETTSIAL INFECTIONS IN MAN*

Group	Agent	Weil-Felix reaction			Complement fixation*
		OX-19	OX-2	OX-K	
Typhus fevers					+
Epidemic typhus	*Rickettsia prowazekii*	++			(Specific)
Murine typhus	*Rickettsia typhi (mooseri)*	++			+ (Specific)
Brill-Zinsser disease	*Rickettsia prowazekii*				
Spotted fevers					
Rocky Mountain spotted fever	*Rickettsia rickettsii*	+	+	−	+
Rickettsialpox	*Rickettsia akari*	−	−	−	+
Fièvre boutonneuse	*Rickettsia conorii*	+	+	+	
Scrub typhus	*Rickettsia tsutsugamushi*	−	−	++	
Q fever	*Coxiella burnetii*	−	−	−	+
Trench fever	*Rickettsia quintana*				

*Cross reactions between Rocky Mountain spotted fever and rickettsialpox.

MURINE (ENDEMIC, NEW WORLD, FLEA-BORNE) TYPHUS. Murine typhus fever, caused by *Rickettsia typhi,* is relatively mild. A natural infection of rats, less often of mice, it is transmitted from rat to man by the rat flea and from rat to rat by the rat louse and the rat flea. The rats do not die of the disease, nor do the fleas or lice that infest them. Murine typhus may be transmitted from man to man by the body louse and become epidemic in louse-infested communities. In man the mortality is about 2%. The endemic typhus fever of Mexico is known as *tabardillo.*

BRILL-ZINSSER DISEASE (RECRUDESCENT TYPHUS). Brill-Zinsser disease is a mild form of typhus existing along the Atlantic Coast and occurring in persons who have had classical typhus fever many years before.*

■ **Laboratory diagnosis.** The typhus fevers all are identified by a positive Weil-Felix reaction (agglutination of the bacteria known as *Proteus OX-19* in serum). Although the Weil-Felix reaction may be detected as early as the fourth day, it is strongly positive about the eighth day. The Weil-Felix reaction does not differentiate the typhus fevers

*The disease was first observed by Nathan E. Brill (1860-1925) in 1898 in immigrants to the United States from various countries of Eastern Europe. The exact nature of the disease was not known until 1934, when Hans Zinsser (1878-1940) advanced the hypothesis that the disease was a mild second attack of classic typhus, an idea now amply confirmed.

Reservoir in nature	Vector	Geography
Man	Body lice	Worldwide
Rats	Rat fleas	Worldwide
Man	—	Worldwide
Wild rodents	Ticks	Western hemisphere
House mice	Mites	United States, Russia, Korea
Small wild mammals	Ticks	Mediterranean coast, Middle East
Wild rodents	Mites	Asia, Australia, Pacific Islands
Ruminants, small mammals, domestic livestock	Ticks—animals Airborne animal products—man	Worldwide
Man	Body lice	Europe, Mexico

from Rocky Mountain spotted fever, but it does differentiate them from certain other rickettsial infections. Fluorescent antibody technics can be used in the diagnosis of the typhus fevers. The epidemic and endemic forms of the disease may be distinguished and the rickettsiae demonstrated if blood from a patient is inoculated into a male guinea pig.

■ **Immunity.** If a person recovers from either epidemic or endemic typhus, he is immune to both types, but a vaccine prepared against one type of the disease protects against only that type. Vaccines seem to be rather effective in preventing the typhus fevers, and when disease does occur in spite of vaccination, the severity is lessened and the mortality reduced.* In only a few cases is recovery from epidemic or endemic typhus not followed by permanent immunity.

■ **Prevention.** Prevention of epidemic typhus fever depends upon (1) isolation of the patient in a vermin-free room, (2) use of insecticides on clothing and bedding, with destruction of insect eggs attached to hair (nits), (3) quarantine of all susceptible persons if many lice are in the vicinity of the patient, (4) systematic delousing and vaccination of susceptible persons, and (5) general improvement of living and sanitary conditions.

*For immunization, see p. 612.

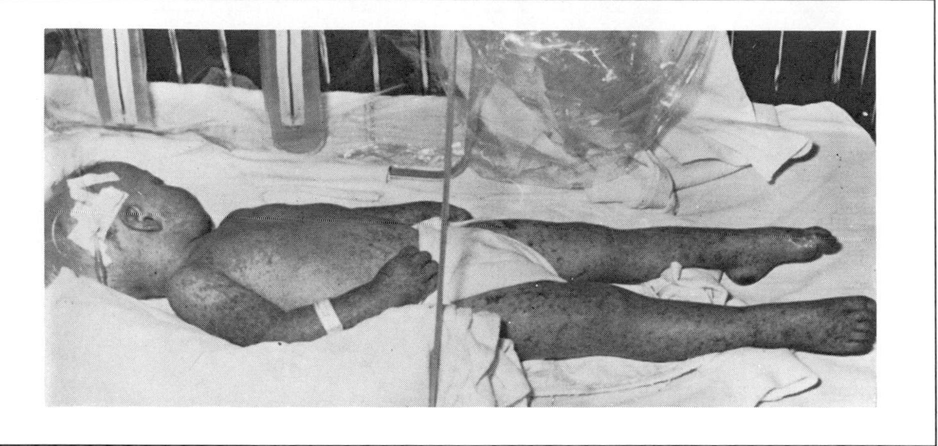

FIG. 29-1. Rocky Mountain spotted fever in a child. Note skin rash. (From Slotkin, R. I. In Shirkey, H.: Pediatric therapy, ed. 4, St. Louis, 1972, The C. V. Mosby Co.)

The control of the endemic form depends upon ratproofing buildings and destroying rodents with their louse and flea populations.

Spotted fever group

■ **Rocky Mountain spotted fever.** Recognized as one of the most severe of all infectious diseases, Rocky Mountain spotted fever (Fig. 29-1) is like typhus fever and is caused by a similar organism, *Rickettsia rickettsii.* In areas where it is well known it is often referred to as tick fever. (This should not cause it to be confused with relapsing fever, which also is known as tick fever.) It occurs chiefly in the Rocky Mountain states, the great majority of cases being found in Idaho and Montana. It occurs also on the Atlantic seaboard, chiefly east of the Appalachian Mountains. For this reason two types of the disease are designated: the western type (the original Rocky Mountain spotted fever) and the eastern type. The difference rests mainly on geographic distribution and mode of spread. The disease does show, however, great variation in severity. In Idaho the mortality varies from 3% to 10%, whereas in the Bitter Root Valley of Montana it may reach 90%. In the eastern states the disease is comparatively mild, and the mortality is about 20%.

The western variety is transmitted by the wood tick *(Dermacentor andersoni),* and the eastern variety is transmitted by the dog tick *(Dermacentor variabilis).* In the southwestern United States the disease is transmitted by the Lone Star tick *(Amblyomma americanum).* The male tick can infect the female, and the female is able to transmit the infection to her offspring. With fluorescent antibody technics, it can be shown that infected female ticks may pass rickettsia to all of their progeny. This is an example of transovarian infection. The infection may be retained in the ticks through many generations. The tick is therefore able to maintain the disease in nature without either rodent or human help. A reservoir of infection appears to exist among a variety of rodents, particularly jack rabbits and cottontail rabbits. The disease is transmitted from rabbit to rabbit by the rabbit tick, a tick that does not bite man.

The rickettsiae of Rocky Mountain spotted fever differ from those of typhus fever in that they invade both the cytoplasm and nucleus of the infected cell. An attack renders a person immune for life. The serum from persons with Rocky Mountain spotted fever agglutinates *Proteus OX-19* (positive Weil-Felix reaction), but the agglutination is not so strong as in typhus fever. *Proteus OX-2* also may be agglutinated. A complement-fixation test is highly specific. The patient's blood may be inoculated into a male guinea pig and organisms recovered from the animal.

Suspect ticks may be examined microscopically for the presence of rickettsiae. A tick engorged from feeding is dissected from its shell, fixed in a suitable solution, and stained with Giemsa stain. If infectious, the tick will contain rickettsiae within the nuclei of cells of the salivary glands.

Health officials advise persons going into tick-infested areas to wear protective clothing such as high boots and to examine clothing and one's body carefully afterward for the presence of ticks. Dogs and other pets should be inspected daily.

■ **Rickettsialpox.** Rickettsialpox is an acute mild febrile disease characterized by the appearance of a red papule at the site of inoculation followed later by a skin rash. The disease is caused by *Rickettsia akari.* The house mouse is the reservoir of infection, and the bite of the house mouse mite conveys the infection from mouse to mouse and from mouse to man. The serum of those ill with or convalescing from the disease does *not* give a positive Weil-Felix reaction. Clinically, rickettsialpox may be mistaken for chickenpox. The primary lesion at the site of inoculation often resembles the vesicle resulting from a smallpox vaccination. The destruction of mice and their mites eliminates the disease. Rickettsialpox has been found in apartment dwellers of the northern United States.

■ **Other fevers.** Included with the spotted fevers are tick-borne typhus fevers—fièvre boutonneuse, Queensland tick typhus, South African tick-bite fever, and tick-bite rickettsioses of India.

These illnesses are nonfatal and only moderately severe. They are caused by rickettsiae related to the agent of spotted fever. Each can be traced to the bite of an ixodid tick. Fièvre boutonneuse is the most widespread and the prototype; it is caused by *Rickettsia conorii.* In this disease there is an initial lesion, the *tache noire* (an ulcer covered over by a black crust is the "black spot"), which is followed by a generalized rash. The Weil-Felix reaction is positive but not strongly so.

Scrub typhus

Scrub typhus, known also as tsutsugamushi (Japanese, "dangerous bug") disease and mite typhus, is so called because the mites that transmit the disease to man are found in scrubland, land covered by a stunted growth of vegetation. The disease attacked thousands of soldiers in the Southwest Pacific area during World War II. It is an acute febrile disease with a rash and a mortality of 0.5% to 60%. The disease resembles Rocky Mountain spotted fever clinically, except that in scrub typhus there is a primary sore at the site of the mite bite. The name of the causative rickettsia is *Rickettsia tsutsugamushi.*

Scrub typhus is a disease primarily of mites and wild rodents. It is transmitted from rodent to rodent and from rodent to man by mites. Two types of mites are known to transmit the disease to man. The rodents and mites that take part in the propagation of the disease probably differ in various parts of the world. The mites grow on short blades of grass and in the top layers of the soil. Persons who contact the soil, therefore, are likely to contract the infection. There is no evidence that the disease spreads from man to man. The serum of persons who have the disease agglutinates *Proteus OX-K* but not *OX-19*.

Prevention depends upon the destruction of the mites in the soil by insecticides and the application of insecticidal chemicals to clothing. There is as yet no effective preventive vaccine.

Q fever

Q fever is a febrile disease of short duration caused by *Coxiella burnetii*. The usually mild respiratory symptoms of the disease make it similar to primary atypical pneumonia or influenza. It differs from most of the other rickettsial diseases in that it is not transmitted to man by an insect vector, there is no rash, and the serum of an infected person does not agglutinate *Proteus* bacilli. The disease was first observed in Queensland, Australia, but the place of discovery did not give origin to the term "Q fever" (Q indicates 'query')

Q fever is worldwide in distribution and is prevalent in the United States, especially in certain areas along the West Coast. It is a disease of animals (a zoonosis). Man becomes infected by residence or occupation that brings him in contact with infected livestock—cattle, sheep, and goats. Most infections are acquired by breathing contaminated air, such as the dust from dairy barns and lambing sheds. Infected cattle and sheep may shed infected placentas and fluids at the time of parturition. About one half the cows harboring this rickettsia discharge it in their milk. Ingestion of the milk of infected cows or goats is a source of infection. Although the rickettsiae of Q fever are fairly resistant to heat, more so than other rickettsiae, they are effectively destroyed in milk by the high-temperature, short-time method of pasteurization. Infection is especially common among packing house employees and laboratory workers. In some instances mass aerial infection of large groups has occurred. Such an instance happened to more than 1600 soldiers who were returning from Italy to the United States and who apparently received their infection from winds blowing at an airport.

A capillary tube agglutination test is used to detect the presence of antibodies in serum from human beings and animals. Blood from a patient may be inoculated into a male guinea pig.

Trench fever

Trench fever is a remittent or relapsing fever that affects soldiers on trench duty, but it may occur in civilian life when people live under conditions comparable to those of the trenches. It was prevalent during World War I but has since almost passed out of existence. It was infrequent during World War II. *Rickettsia quintana*, the causative agent, is transmitted by the body louse. Although trench fever is never fatal and recovery is usually complete, it may recur, and in some cases it is followed by a state of chronic ill health with pain in the limbs, mental depression, and circulatory disorders. An attack does not confer immunity.

Rickettsial diseases in animals

The following is a list of rickettsial diseases in animals:

Heartwater disease of cattle
Rickettsiosis of cattle
Febrile disease of dogs and sheep
Salmon disease of dogs and foxes

BEDSONIAE (CHLAMYDIAE)

■ **Descriptive qualities.** The name *bedsoniae,* or *chlamydiae,** is applied to a large group of microorganisms (psittacosis group) of comparable design and makeup and causing several important diseases. (See Tables 27-1 to 27-5.) Bedsoniae multiply within the cytoplasm of the cell attacked and form characteristic cytoplasmic inclusion bodies. The infectious particle is known as an *elementary body,* and the mature inclusion body contains a host of them. Their life cycle is a complex one.

For a long time bedsoniae were thought of as large viruses (of the psittacosis–lymphopathia venereum group), so classified and referred to as basophilic viruses and mantle viruses. Like viruses they are obligate intracellular parasites, but there are important points of difference (see Table 27-2). Unlike viruses they possess enzymes and both DNA and RNA. They have no capsid symmetry, and they are susceptible to some of the antimicrobial drugs. The patterns of disease they produce differ in many ways from those of viruses.

Bedsoniae are comparable to rickettsiae but are still distinct as a group. Recent studies indicate that these parasites are better separated from true viruses and considered as small bacteria (Fig. 29-2). The name *Bedsonia* was given in honor of Sir Samuel Bedson (1886-1969), who studied them extensively. (Some microbiologists feel the genus name should be *Chlamydia,* and these organisms are often called chlamydiae.) They are probably derived from gram-negative bacteria. They are nonmotile and reproduce by binary fission. Their outer cell wall is chemically similar to that of gram-negative bacteria. They are very sensitive to tetracyclines.

■ **Laboratory diagnosis.** Many of the procedures used in the laboratory diagnosis of viral infection are applicable to these organisms (p. 407). They grow in the fertile hen's egg and express in tissue culture their distinct cytopathogenicity. There are diagnostic serologic tests. Because of their unique staining qualities, they can be readily visualized in suitable specimens by technics of light or fluorescent microscopy. They are readily stained with basophilic dyes. The Frei test for lymphopathia venereum is a well-known diagnostic skin test.

■ **Importance.** In this category are bedsonial (or chlamydial) agents of psittacosis, ornithosis, trachoma, and lymphopathia venereum. The agent of cat scratch disease may also fit in here.

INFECTIONS WITH THE BEDSONIAE (CHLAMYDIAE)

Psittacosis (parrot fever) and ornithosis

Psittacosis (parrot fever) is a bedsonian infection so named because it affects primarily the psittacine birds, most often parrots and parakeets. Ornithosis is the corresponding disease in domestic fowl and birds other than those of the parrot family —chickens, turkeys, pigeons, and sea birds. Psittacosis is important because of its worldwide distribution. The number of human beings attacked, however, is small.

■ **Transmission.** Psittacosis is widespread among the birds just named, the sources

*From the Greek *chlamys,* 'cloak.' Each organism is surrounded by a dense cell wall in part of its life cycle.

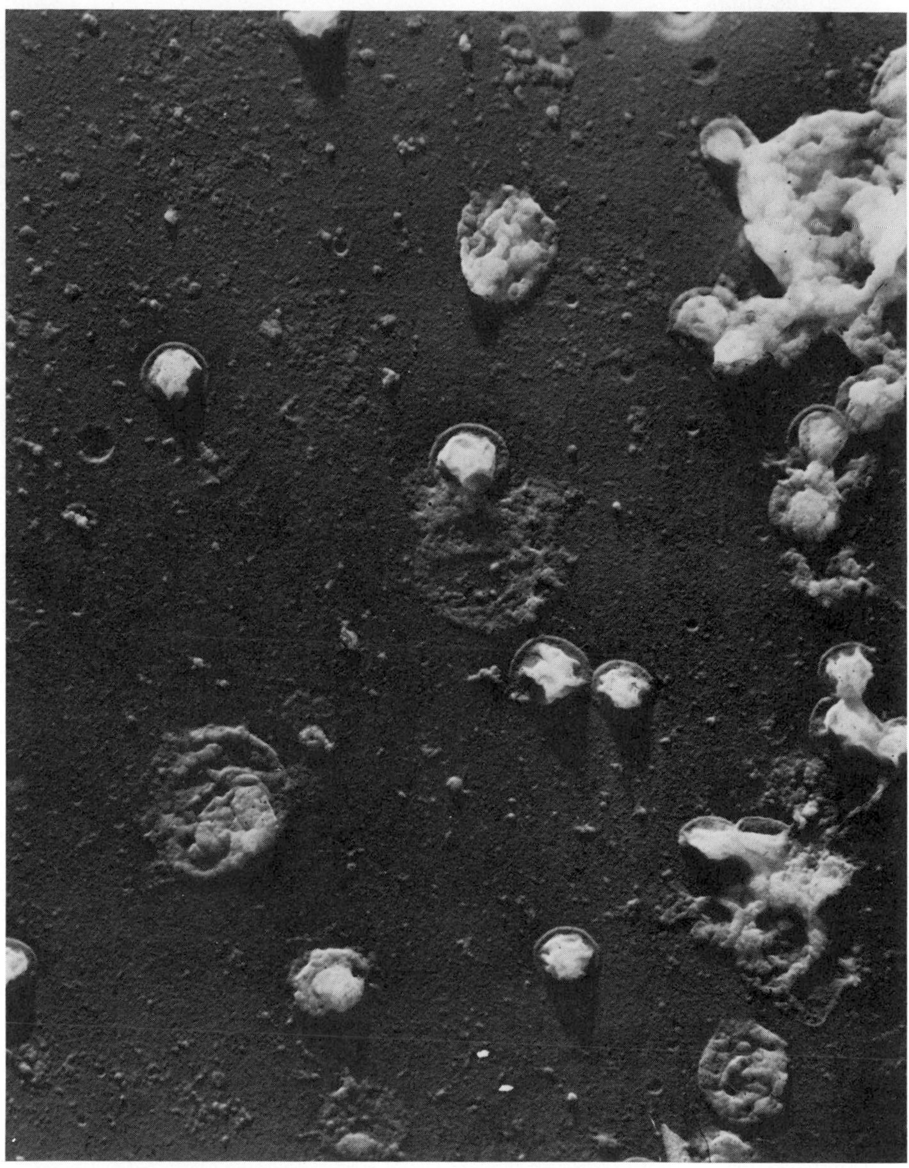

FIG. 29-2. Bedsoniae (meningopneumonitis agent), electron micrograph. Extensive studies on this agent were important to the conclusion that these basophilic organisms, now termed bedsoniae (chlamydiae), were not viruses. (×28,000.)

of most human infections. Man may contract the disease from birds through: (1) transmission of the agent through the air, (2) handling of sick birds or their feathers, (3) contact with materials contaminated by infected birds, and (4) bites or wounds inflicted by sick birds. Sick birds show the agent in their nasal discharges and feces. Birds may become carriers. In birds the gastrointestinal tract is primarily affected. In man the infection enters by way of the respiratory tract, and infection is localized there. In a few

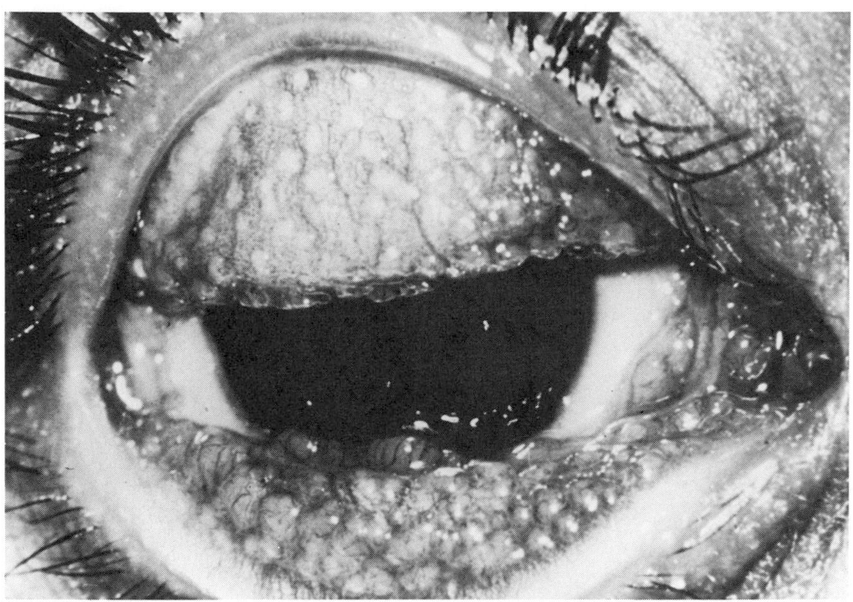

FIG. 29-3. Trachoma. Note the inflammatory nodules spread over the thickened conjunctival membrane of the eye of a Pima Indian (from a tribe of southern Arizona and northern Mexico). Trachoma is prevalent among the Indians of the Southwest. (Courtesy Dr. Phillips Thygeson, Los Altos, Calif.)

cases the infection is transmitted directly from man to man. The onset of the disease in man may be sudden, and the mortality high. Explosive outbreaks sometimes occur in poultry-processing plants.

■ **Laboratory diagnosis.** The disease may be diagnosed by injecting some of the patient's sputum into a mouse and observing the development of the disease in the mouse. A complement-fixation test is available.

■ **Prevention.** The sick patient should be isolated and all discharges disinfected. A vaccine has been prepared against psittacosis, but how effective it is in the control of the disease is undetermined.

Trachoma

Although rare in the United States, trachoma, one of the oldest diseases known to man, afflicts more than 500 million persons in this world (one seventh the world's population). The causative organism, one of the bedsoniae, specifically attacks the lining cells of the cornea and the conjunctival membrane of the eye. Growth in these cells produces very distinctive inclusion bodies. As is often the case with viral or bedsonian diseases, bacterial infection is superimposed to intensify the injury already done. The inflammatory charges are pronounced, and the scar tissue formed over the cornea, ordinarily transparent, results in impaired vision. The name "trachoma" is derived from a Greek word meaning 'rough' to set forth the pebble-like appearance of the infected conjunctival membrane (Fig. 29-3). This disease is the world's leading cause of blindness (in some 20 million with the disease).

■ **Spread.** Trachoma is endemic in the large underprivileged areas of the tropics

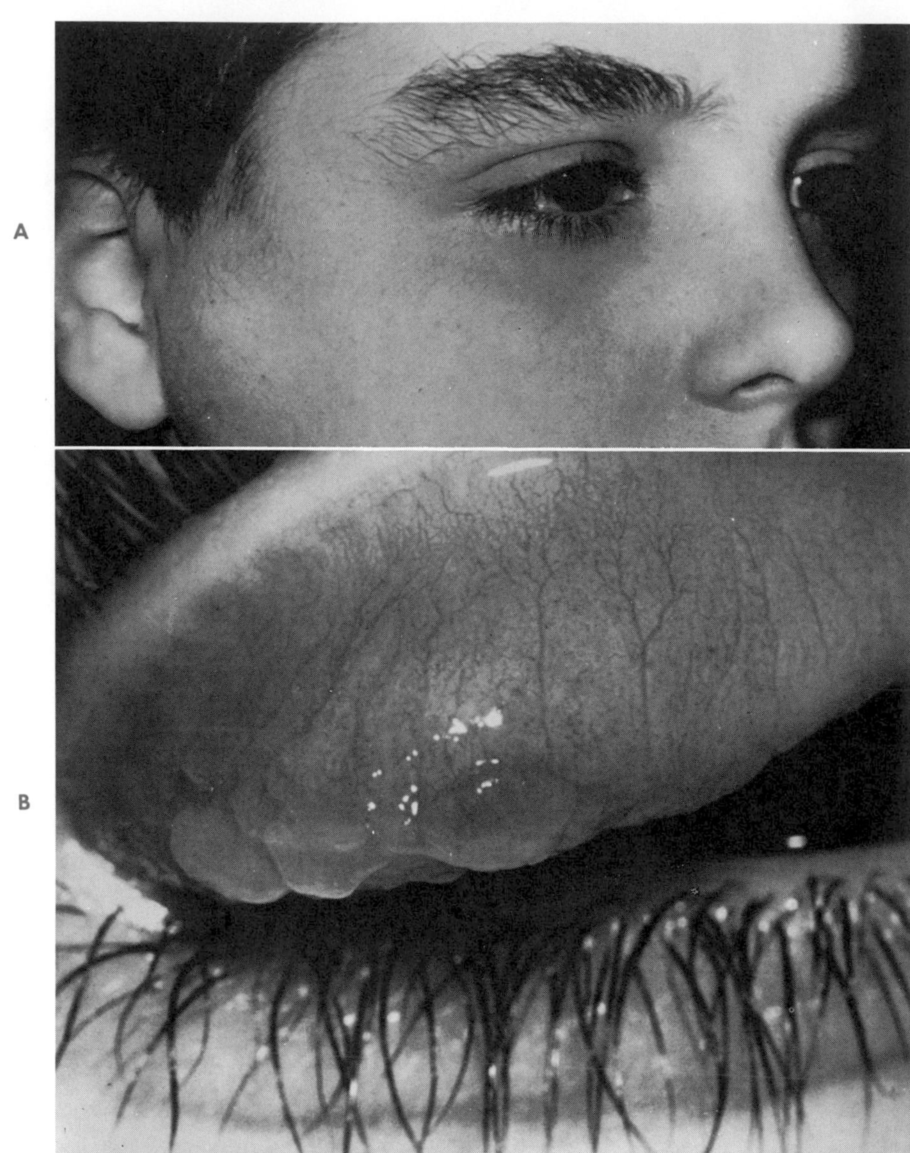

FIG. 29-4. Cat scratch disease. **A,** Lymph node regional to inflamed right eye (preauricular node) is quite swollen and tender. **B,** Upper eyelid everted to show inflammatory swelling and congestion of blood vessels. A purulent exudate is present. Not only did this patient have a cat, but his cat-scratch antigen test was positive. (From Donaldson, D. D.: Atlas of external diseases of the eye, vol. 1, St. Louis, 1966, The C. V. Mosby Co.)

and subtropics; in some countries 75% of the population may suffer. Transmission requires close personal contact, and for this reason the disease is often passed from mother to child. Flies carry the infection mechanically. Bedsoniae may be spread by contaminated fingers and articles of clothing.

■ **Laboratory diagnosis.** The diagnosis is made by identification of the characteristic inclusion bodies in scrapings made from the conjunctival membranes and by means of a complement-fixation test.

■ **Control.** The agent has been grown in fertile eggs, and an effective vaccine has been prepared. The prognosis in this disease has been greatly improved with the use of antibiotics alone. The incidence of the disease has been shown to drop simply with improvements in living conditions.

Lymphopathia venereum

Lymphopathia venereum (venereal lymphogranuloma, lymphogranuloma venereum, lymphogranuloma inguinale), formerly known as climatic bubo, is a venereal disease of bedsonian etiology. It seems to be increasing in this country and all over the world.

■ **Pathology.** In the male, it starts with a primary ulceration on the external genital organs. The infection extends to the inguinal lymph nodes where buboes are formed. In the female, infection extends from the primary lesion to the lymph nodes within the pelvis, where the chronic inflammation set up may lead to stricture of the rectum. The disease involves mainly the lymphatics in each sex. The scar tissue associated with long-standing lymphangitis results in striking deformity of the external genitalia in both sexes.

Lymphopathia venereum should not be confused with granuloma inguinale, also a venereal disease, but one caused by a different etiologic agent (see p. 370).

■ **Laboratory diagnosis.** Lymphopathia venereum is detected by the Frei test, which consists of the intradermal injection of material prepared by growing the agent in the yolk sac of a developing chick embryo. The development of a bright red papule at the site of inoculation constitutes a positive test. There is a diagnostic complement-fixation test. Inclusion bodies may be seen in suitably stained smears of pus from the buboes.

Cat scratch disease

Cat scratch disease (benign lymphoreticulosis) is a febrile illness that may come after scratching, biting, or licking by a cat. It is characterized by a purulent lesion at the site of infection, and 2 or 3 weeks later the lymph nodes draining the area enlarge (Fig. 29-4). Incubation time, from injury to illness, is usually 3 to 10 days. As a rule, the disease is mild. Cats are mechanical vectors and do not show related illness.

It is possible that animals other than cats carry the infection, including birds and insects, but the monkey is the only animal to be artificially infected.

■ **Laboratory diagnosis.** Antigen for the skin test is prepared from the purulent material aspirated from the lesions. The causative bedsonian agent has not been isolated, but elementary bodies similar to those of psittacosis may be seen in stained sections of affected lymph nodes.

■ **Prevention.** The only preventive measure known is the avoidance of cats.

Bedsonial diseases in animals

The following list gives examples of bedsonial diseases in animals:

- Feline pneumonitis
- Conjunctivitis and keratitis of sheep, cattle, goats, and chickens
- Pneumonitis in mice
- Polyarthritis (one form of stiff-lamb disease)
- Sporadic bovine encephalomyelitis

QUESTIONS FOR REVIEW

1. What are rickettsiae? How did they get their name? Give their characteristics.
2. What is the Weil-Felix reaction? Its importance?
3. Into what five groups may rickettsial diseases be classified?
4. Give the laboratory diagnosis of rickettsial infections.
5. Briefly discuss Q fever.
6. Explain the pathologic changes in rickettsial diseases.
7. State special properties of bedsoniae. Compare them with other microbes.
8. Give examples of bedsonial diseases with diagnostic inclusion bodies in parasitized cells.
9. What disease is responsible for most of the world's blindness? Give its salient features.

REFERENCES FOR CHAPTERS 27 TO 29

Adams, J. M.: Persistence of measles virus and demyelinating disease, Hosp. Prac. **5:**87, (May) 1970.

Bauer, H.: Guide for laboratory diagnosis and postexposure treatment of rabies, Arch. Environ. Health **19:**868, 1969.

Bedson, S., et al.: Virus and rickettsial diseases of man, Baltimore, 1967, The Williams & Wilkins Co.

Behbehani, A. M.: Human viral, bedsonial, and rickettsial diseases, Springfield, Ill., 1972, Charles C Thomas, Publisher.

Behbehani, A. M.: Laboratory diagnosis of viral, bedsonial and rickettsial disease, Springfield, Ill., 1972, Charles C Thomas, Publisher.

Behbehani, A. M., and Marymont, J. H., Jr.: The role of the hospital laboratory in the diagnosis of viral diseases. I. Basic concepts, Amer. J. Clin. Path. **53:**43, 1970.

Bell, J. F., et al. Nonfatal rabies in an enzootic area: Result of a survey and evaluation of techniques, Amer. J. Epidem. **95:**190, 1972.

Bugg, R.: The flu that few will forget, Today's Health **47:**24, (April) 1969.

Bugg, R.: On the way: Vaccines to snuff out sniffles, Today's Health **47:**36, (Dec.) 1969.

Burkitt, D. P., and Wright, D. H.: Burkitt's lymphoma, Baltimore, 1970, The Williams & Wilkins Co.

Carithers, H. A., Carithers, C. M., and Edwards, R. O., Jr.: Cat-scratch disease: Its natural history, J.A.M.A. **207:**312, 1969.

Carper, J.: The case of the yellow killer, Today's Health **48:**53, (Sept.) 1970.

Culliton, B. J.: Cancer virus theories: Focus of research debate, Science **177:**44, 1972.

Dietzman, D. E., et al.: The occurrence of epidemic hepatitis in chronic carriers of Australia antigen, J. Pediat. **80:**577, 1972.

Dmochowski, L.: Viruses and breast cancer, Hosp. Pract. **7:**73, (Jan.) 1972.

Dueñas, A., et al.: Herpesvirus type 2 in a prostitute population, Amer. J. Epidem. **95:**483, 1972.

Fairchild, G. A., and Roan, J.: Atmospheric pollutants and pathogenesis of viral respiratory infection. 1. Evaluation of murine influenza as an infectious disease model, Arch. Environ. Health **25:**51, (July) 1972.

Fisher, M. M., and Steiner, J. W., editors: Proceedings of the Canadian Hepatic Foundation 1971 International Symposium on viral hepatitis, Canad. Med. Ass. J. **106:**417, 1972.

Garibaldi, R. A., et al.: Hospital-acquired serum hepatitis, J.A.M.A. **219:**1577, 1972.

Gillette, R.: VEE vaccine: Fortuitous spin-off from BW research, Science **173:**405, 1971.

Gocke, D. J.: A prospective study of posttransfusion hepatitis, the role of Australia antigen, J.A.M.A. **219:**1165, 1972.

Gross, L.: Oncogenic viruses, New York, 1970, Pergamon Press.

Grossman, R.A.: Influenza, Milit. Med. **134:**8, (Jan.) 1969.

Hakosalo, J., and Saxén, L.: Influenza epidemic and congenital defects, Lancet **2:**1346, 1971.

Hazard, G. W., Ganz, R. N., Nevin, R. W., et al.: Rocky Mountain spotted fever in the eastern United States, New Eng. J. Med. **280:**57, 1969.

Heath, R. B., and Waterson, A. P., editors: Modern trends in medical virology, New York, 1970, Appleton-Century-Crofts.

Henle, G., and Henle, W.: EB virus in the etiology of infectious mononucleosis, Hosp. Pract. **5:**33, (July) 1970.

Henry, J. B., editor, and Widmann, F. K.: The Australia antigen: Where do we stand? Part I, Postgrad. Med. **50:**167, (Dec.) 1971; Part 2, **51:**257, (Jan.) 1972.

Hepatitis virus at last? editorial: J.A.M.A. **208:**1694, 1969.

Hill, D. A., et al.: Evaluation of an interferon inducer in viral respiratory disease, J.A.M.A. **219:**1179, 1972.
Holmes, I. H., Wark, M. C., and Warburton, M. F.: Is rubella an arbovirus? Virology **37:**15, 1969.
Horstmann, D. M.: Rubella: The challenge of its control, J. Infect. Dis. **123:**640, 1971.
Horta-Barbosa, L., et al.: Chronic viral infections of the central nervous system, J.A.M.A. **218:**1185, 1971.
Hummeler, K., and Koprowski, H.: Investigating the rabies virus, Nature **221:**418, (Feb. 1) 1969.
Hunter, J., et al.: Australia (hepatitis-associated) antigen among heroin addicts attending London addiction clinic, J. Hyg. (Camb.) **69:**565, 1971.
Jacobs, J. W., Peacock, D. B., Corner, B. D., et al.: Respiratory syncytial and other viruses associated with respiratory disease in infants, Lancet **1:**871, 1971.
Janeway, C. A.: Prevention of posttransfusion hepatitis, editorial, J.A.M.A. **214:**749, 1970.
Kaplan, M. M.: Epidemiology of rabies, Nature **221:**421, (Feb. 1) 1969.
Katz, S. L., and Griffith, J. F.: Slow virus infections, Hosp. Pract. **6:**64, (March) 1971.
Krech, U., et al.: Cytomegalovirus infections of man, White Plains, N. J., 1971, Albert J. Phiebig, Inc.
Krugman, S.: Viral hepatitis and Australia antigen, J. Pediat. **78:**887, 1971.
Langmuir, A. D.: Influenza: its epidemiology, Hosp. Pract. **6:**103, (Sept.) 1971.
Lennette, E. J., and Schmidt, N. J.: Diagnostic procedures for viral and rickettsial diseases, Washington, D. C., 1969, American Public Health Association, Inc.
Lindsay, M. I., Jr., and Morrow, G. W., Jr.: Primary influenzal pneumonia, Postgrad. Med. **49:**173, (May) 1971.
Lindsay, M. I., Jr., et al.: Hong Kong influenza, J.A.M.A. **214:**1825, 1970.
Lucas, C. J., et al.: Measles antibodies in sera from patients with autoimmune diseases, Lancet **1:**115, (Jan. 15) 1972.
Luria, S. E., and Darnell, J. E., Jr.: General Virology, New York, 1967, John Wiley & Sons, Inc.
Macintyre, E. H.: Oncogenic viruses and human neoplasia, J. Amer. Med. Wom. Ass. **23:**520, 1968.
Marymont, J. H., Jr.: Diagnostic virology in the community hospital, Bull. Path. **10:**13, 1969.
Marymont, J. H., Jr., and Behbehani, A. M.: The role of the hospital laboratory in the diagnosis of viral diseases. II. Technics, Amer. Clin. Path. **53:**51, 1970.
Maugh, T. H., II.: Hepatitis: A new understanding emerges, Science **176:**1225, 1972.
Merigan, T. C., Jr.: Interferon and interferon inducers: The clinical outlook, Hosp. Pract. **4:**42, (March) 1969.
Monif, G. R. G., et al.: The correlation of maternal cytomegalovirus infection during varying stages in gestation with neonatal involvement, J. Pediat. **80:**17, (Jan.) 1972.
Morley, D.: Severe measles in the tropics. I, Brit. Med. J. **1:**297, (Feb. 1) 1969; II, **1:**363, (Feb. 8) 1969.
Musher, D. M.: Q fever, J.A.M.A. **204:**863, 1968.
Phillips, D. F.: Hepatitis—2: The scientific advances, Hospitals **45:**48, (Jan. 1) 1971.
Probert, M., and Epstein, M. A.: Morphological transformation in vitro of human fibroblasts by Epstein-Barr virus; preliminary observations, Science **175:**202, 1972.
The public health implications of the presence of hepatitis B antigen in human serum (current trends), Morbid. Mortal. Wk. Rep. **21**(no. 16): 133, 1972.
Purcell, R. H., et al.: Seroepidemiological studies of transfusion associated hepatitis, J. Infect. Dis. **123:**406, 1971.
Rabies 1968 model, editorial, J.A.M.A. **203:**420, 1968.
Rhodes, A. J., and Van Rooyen, C. E.: Textbook of virology, Baltimore, 1968, The Williams & Wilkins Co.
Rogers, R. S., III, and Tindall, J. P.: Herpes zoster in the elderly, Postgrad. Med. **50:**153, (Dec.) 1971.
Rose, H. M.: Influenza; the agent, Hosp. Pract. **6:**49, (Aug.) 1971.
Rosen, P., and Hajdu, S.: Cytomegalovirus inclusion disease at autopsy of patients with cancer, Amer. J. Clin. Path. **55:**749, 1971.
Rubella: epidemic in retrospect, Hosp. Pract. **2:**27, (March) 1967.
Sauer, G. C.: Skin diseases due to viruses, Hosp. Med. **7:**82, (Aug.) 1971.
Sever, J. L.: Viral teratogens: a status report, Hosp. Pract. **5:**75, (April) 1970.
A shedding of light, editorial, J.A.M.A. **212:**1057, 1970.
Sutnick, A. I., et al.: Australia antigen: a genetic basis for chronic liver disease and hepatoma? Ann. Intern. Med. **74:**422, 1971.
Vianna, N. J., and Hinman, A. R.: Rocky Mountain spotted fever on Long Island: epidemiologic and clinical aspects, Amer. J. Med. **51:**725, 1971.
Weller, T. H.: Cytomegaloviruses: the difficult years, J. Infect. Dis. **122:**532, 1970.
Wenzel, R. P., et al.: Acute respiratory disease; clinical and epidemiologic observations of military trainees, Milit. Med. **136:**873, 1971.

30 The acid-fast bacteria

Nonacid-fast bacteria are easily stained with basic dyes, but the color is quickly removed when the microorganisms are treated with acid alcohol. Acid-fast bacteria, on the other hand, are so resistant to the penetration of basic dyes that in order to color them the stain has to be either gently heated or applied for a length of time. Once stained, however, they are not easily decolorized with acid alcohol.

These bacteria are more closely related to the fungi than to any other bacteria and are classified in the genus *Mycobacterium,* the members of which are straight or curved acid-fast rods that may show branching or irregular forms. The prefix *myco* suggests their relation to the fungi.

The acid-fast bacteria of greatest importance and those pathogenic to man are *Mycobacterium tuberculosis,* the cause of tuberculosis, *Mycobacterium leprae,* the cause of leprosy, and certain atypical "unclassified" mycobacteria, the cause of mycobacteriosis and various other infections. In addition to these acid-fast organisms, 40 or more species exist. Most of them are saprophytes and nonpathogenic, but they may gain access to milk, butter, or other dairy products, and be mistaken for *Mycobacterium tuberculosis.* Some are pathogenic for lower animals; for example, one causes a granulomatous enteritis in cattle known as Johne's disease.

MYCOBACTERIUM TUBERCULOSIS (THE TUBERCLE BACILLUS)

■ **Importance.** The acid-fast *Mycobacterium tuberculosis,* or tubercle bacillus of common parlance, causes tuberculosis, which attacks all races of men* and several species of domestic animals, especially cattle, swine, and chickens. Wild animals living in their natural surroundings do not have tuberculosis but may contract it when placed in captivity.

In 1900, tuberculosis was the leading cause of death in the United States as it was throughout the civilized world. (In the tropics it ranked second to malaria.) Since that time it has dropped from first place in this country but still remains an important cause of chronic disability and ill health produced by a communicable disease. In the United States each year slightly less than 40,000 new cases and some 5000 deaths are reported.

Tuberculosis is a health hazard of lower socioeconomic groups, densely populated areas, persons over 50 and under 5 years of age, and the chronically ill. Over the world it

*Tuberculosis has claimed the lives of many great writers, painters, and musicians—Keats, Chopin, Goethe, Poe, Gauguin, Paganini, Molière, and many others. In fact, no other disease has had such a significant relation to literature and the arts.

is still a principal cause of death, ranking with malaria and malnutrition; there are at least 15 million persons with active tuberculosis, and 3 million die annually of the disease. Throughout the world as well as in this country, it is the most frequent *infectious* cause of death. The mortality is highest in the Orient, Asia, Africa, and Latin America.

The infectious nature of tuberculosis was suspected 5 centuries before the tubercle bacillus was discovered by Koch in 1882, and it had been produced by artificial inoculation 40 years before that time.

■ **General characteristics.** The organism *Mycobacterium tuberculosis* is a slender, rod-shaped, nonmotile, nonspore-forming bacillus, often beaded or granular in appearance on acid-fast smears. Tubercle bacilli are more resistant than are other nonspore-forming organisms to the deleterious effects of drying and germicides. They remain alive in dried sputum or dust in a dark place for weeks or months and in moist sputum for 6 weeks or more. Direct sunlight kills them in 1 or 2 hours. Sufficiently susceptible to heat, however, they are destroyed by the temperature of pasteurization. Five percent phenol kills them in sputum in 5 or 6 hours. Tubercle bacilli are not affected by routinely used antibiotics but are susceptible to streptomycin, dihydrostreptomycin, para-aminosalicylic acid (PAS), rifampin, and isoniazid (INH). *Mycobacterium tuberculosis* tends to develop a resistance to these agents, and toxic effects are sometimes observed after their use. However, their value in treatment and in modifying the course of the disease has been remarkable.

Mycobacterium tuberculosis grows only on special media. Even then growth is slow; 2 to 4 weeks often elapse before any growth is visible (ordinary bacteria display colonies on routinely used media within 24 to 48 hours). Body temperature is best (37° C.), but the tubercle bacilli may grow at a temperature as low as 29° or as high as 42° C. Although tubercle bacilli are aerobic, 5% to 10% CO_2 enhances growth, and the presence of glycerin in the medium accelerates it for the human bacillus. A bacteriostatic dye, such as malachite green, added to the culture medium suppresses the more rapid growth of contaminants. The human tubercle bacillus is more readily cultivated than is the bovine bacillus.

■ **Species.** Three species or variants of *Mycobacterium tuberculosis* resemble each other closely but may be differentiated by animal inoculation and cultural procedures.* They are *Mycobacterium tuberculosis* (primary host man), *Mycobacterium bovis* (primary host cattle), and *Mycobacterium avium* (primary host birds) — the human, bovine, and avian variants.

Tuberculosis in human beings is caused by either the human or the bovine microbe. Human infections from bovine bacilli were once very common in children and young people and affected the cervical lymph nodes (scrofula), intestines, mesenteric lymph nodes, and bones. The infection was acquired from the milk of tuberculous cows. Routine tuberculin testing of cattle by the U. S. Department of Agriculture, with elimination of infected cattle from the herd, plus the pasteurization of milk, has practically eliminated bovine infection in the United States. In countries where no such safeguards exist, the bovine bacillus is responsible for 15% to 30% of the cases of tuberculosis in the young ages. Bovine bacilli cause pulmonary tuberculosis in cattle

*In 1898 Theobald Smith (1850-1934) of Albany, New York differentiated the human and bovine (cattle) forms of *Mycobacterium tuberculosis*. America's foremost bacteriologist, he made several important contributions to microbiology.

and can cause it in man. However, essentially all pulmonary infections in children and adults come from the human bacillus.

The bovine bacillus is more pathogenic for ordinary laboratory animals than is the human bacillus. When inoculated into a rabbit, the bovine bacillus kills the animal in 2 to 5 weeks, whereas the human bacillus kills it in about 6 months, and in some cases death does not occur at all. Guinea pigs are very susceptible to both the human and the bovine organisms but are unaffected by avian bacilli.

■ **Toxic products.** Tubercle bacilli do not produce exotoxins, hemolysins, or comparable substances, but poisonous products partly responsible for the clinical features of tuberculosis are liberated when the bacilli disintegrate.

When *Mycobacterium tuberculosis* is grown artificially, the culture medium contains a product known as *tuberculin,* without effect on a nontuberculous animal (no history of contact with the tubercle bacillus) but with powerful effects in the body of a tuberculous animal. These effects come from surprisingly small doses, and if the dose is a large one, the results are disastrous.

There are more than 50 methods of preparing tuberculin, and the nature of each tuberculin depends to some extent upon the method. The best known and, in the past, the most extensively used tuberculin is Koch's original (or old) tuberculin, often spoken of simply as O.T. and prepared from a culture of tubercle bacilli in 5% glycerin broth that is concentrated and filtered. The bacteria-free filtrate of tuberculin contains bacterial disintegration products, substances formed by the action of the bacilli on the culture medium, and concentrated culture medium.

The tuberculin used today in the tuberculin test is known as P.P.D. (purified protein derivative). It consists chiefly of the active principle of tuberculin without extraneous matter. It is more stable and less variable in results than O.T.

■ **Sources and modes of infection.** The sources of infection in tuberculosis are the sputum of patients with pulmonary disease, discharges from other tuberculous foci, and the milk of tuberculous cows. The patient discharging tubercle bacilli in the sputum is by far the most important reservoir of infection. The bacilli have no natural existence outside the body and are transmitted from source to destination by some form of direct or indirect contact. Droplet infection is the most common mode of spread. Droplets from the mouth of the tuberculous patient or dust containing the partly dried but still living bacilli are inhaled. The next most common mode is the transfer of the bacilli to the mouth by contaminated hands, handkerchiefs, and objects. The infant crawling on the floor may contract tuberculosis by contaminating his hands with tuberculous material and then placing them in his mouth. Unless properly washed and sterilized, the eating utensils used by a tuberculous patient may be a source of infection. The milk of a tuberculous mother may rarely convey the infection to her nursing child.

In the cow, tuberculosis usually attacks the lungs, but since the cow swallows her sputum, the bacilli are excreted in the feces. The udder and flanks of the cow become contaminated, and the bacilli gain access to the milk. Consumed unpasteurized, it becomes a source of infection. Sometimes the bacilli are excreted in the milk as it comes from the udder. This may occur without demonstrable tuberculous lesions in the udder. A small percent of hogs slaughtered for food show evidence of tuberculosis. Hogs are susceptible to the bovine, avian, and human bacillus.

The avenues of exit for the tubercle bacilli depend upon the part of the body infected. In pulmonary tuberculosis the bacilli are cast off in the sputum, although tubercle bacilli may be found in the feces from swallowed sputum. In intestinal tuberculosis they are discharged into the feces, and in tuberculosis of the genitourinary system they appear in the urine. Tubercle bacilli may be found in exudates from abscesses and lesions of the lymph nodes, bones, and skin.

■ **Pathogenicity.** Tuberculosis, a lifelong disease, is a chronic granulomatous infection caused by *Mycobacterium tuberculosis.* Although almost any tissue or organ of the body may be affected, the parts most often involved are the lungs, intestine, and kidneys in adults and the lungs, lymph nodes, bones, joints, and meninges in children.

■ **Pathology.** The reticuloendothelial system of the body provides the main line of cellular defense, and the term *tuberculosis* is derived from the small nodules of reticuloendothelial cells (tubercles), the unit lesions produced by *Mycobacterium tuberculosis.*

PRIMARY TUBERCULOSIS (CHILDHOOD TYPE). When tubercle bacilli first enter the body, the ensuing infection lasts several weeks, during which time the individual develops considerable immunity and a state of specific hypersensitivity to the tubercle bacillus. As a result his tuberculin test becomes positive and remains so for his lifetime. The sequence of events in the first infection is termed the *primary complex,* and it may occur in the lungs (most commonly), in the intestinal tract, in the posterior pharynx, or in the skin (rarely). In the average person the primary complex is benign on the whole. The lesions in the lung heal or become latent without treatment. Most of these cases show residual calcification.

A feature of the primary complex is the extension of tubercle bacilli to the regional lymph nodes. With a primary complex in the lung the chest nodes are infected as are the mesenteric nodes from the mucosal focus of the small bowel and the cervical lymph nodes from the mucosal focus of the tonsils, throat, or nasopharynx. Tuberculosis of the lymph nodes of the neck, or *scrofula,* is typically produced by the bovine tubercle bacillus as part of the primary complex it sets up in this area.

CHRONIC PULMONARY TUBERCULOSIS (ADULT TYPE). Chronic tuberculosis, adult type, from exogenous reinfection is considered to be quite rare. It may come from the progression of the primary infection. In a few persons, after a period of latency following the first infection, there is reactivation of dormant foci to produce the pattern of chronic disease. According to recent studies 5% of children with inactive untreated primary infection will develop the chronic pulmonary or extrapulmonary infection after the age of puberty, especially during adolescence.

Chronic tuberculosis is common in the lungs but may involve other sites in the body.

THE LESIONS. The reactions obtained in a positive tuberculin test and the responses in sensitized tissues depend upon the hypersensitive (allergic) state developed. As a consequence, lesions that are typical of tuberculosis and not exactly duplicated in any other disease are produced.

The reticuloendothelial system responds to the presence of tubercle bacilli by sending out macrophages to engulf them and to form restricting barriers around them. In their attack on the bacilli the macrophages form firm, round or oval, white, gray, or yellow nodules from 1 to 3 mm. in diameter, which are the *tubercles.* Microscopically a tubercle contains tubercle bacilli, epithelioid cells, and giant cells surrounded by a narrow band of lymphoid cells and encapsulated by fibrous tissue. Epitheloid cells are macrophages altered on contact with the fatty substances contained in the tubercle bacillus. The merging together of many macrophages results in a giant cell.

A tubercle may enlarge singly, or a number of small tubercles may coalesce to form a *conglomerate* tubercle. Because of the cellular reaction at the periphery of the

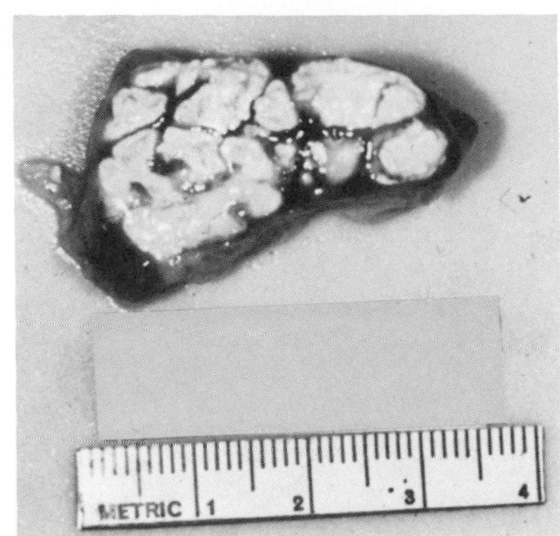

FIG. 30-1. Caseation in a lymph node. Note cheese-like quality.

tubercle, blood vessels are compressed and the nutrition of the area is disturbed. This, plus the lethal action of the tubercle bacilli themselves, leads to a peculiar type of change in the center of the tubercle, known as *caseation* (Fig. 30-1), from the dry, granular, cheesy quality of the dead tissue.

Such a caseous focus may remain unchanged for a long time, or it may be altered pathologically. It may calcify. Calcification, the deposit of lime salts in dead tissue, is a reparative process. The caseous nodule may be organized, in which event the dead tissue is at least partially replaced by connective tissue. The end stage of organization is a scar. On the other hand, the caseous nodule may soften, and the dead (necrotic) material may dissect through the tissues to form a so-called abscess. A caseous focus in the lung may release its contents into the lumen of a bronchus, forming a *cavity*. If a caseous area is situated near a body surface or mucous membrane, the discharge of caseous material onto that surface leaves a *tuberculous ulcer.*

SPREAD. Tubercle bacilli may be spread in the lymph or bloodstream to different parts of the body. Infection may permeate adjacent tissues, move along natural passages (from the kidneys to the bladder via the ureters), and expand over a surface. Spread by blood and lymph is the usual way. Occasionally, material heavily laden with tubercle bacilli (example, the liquefied center of a tubercle) is discharged into a blood vessel and infection is seeded widely over the body. Many small tubercles resembling millet seeds result in the lungs, spleen, liver, and various organs throughout the body. This is *miliary tuberculosis.* Discharge of tuberculous material into the pulmonary artery seeds one or both lungs; extrusion into the portal vein seeds the liver. In these areas miliary tuberculosis is *localized.*

TUBERCULOSIS OF THE LUNGS. *Pulmonary tuberculosis* (Fig. 30-2) is the common form of the disease caused by the human bacillus. In hypersensitive adults, tuberculosis involves the upper and posterior portions of the upper lobes, especially on the right side. This is *apical tuberculosis.*

Cavity formation is an important development in the course of pulmonary tuberculosis in adults. A tuberculous cavity may be only 1 or 2 cm. in diameter, or it may excavate a whole lobe of the lung. A cavity is formed when caseous material from a tuberculous focus in the lung is released into the lumen of a bronchus to be either

FIG. 30-2. Pulmonary tuberculosis. There is a large cavity at the apex as well as numerous tuberculous foci throughout the rest of the lung. (From Anderson, W. A. D., and Scotti, T. M.: Synopsis of pathology, ed. 8, St. Louis, 1972, The C. V. Mosby Co.)

coughed up or dispersed in the tracheobronchial tree. With a communication established between the excavated tuberculous focus and a bronchus, the newly formed cavity enlarges and becomes secondarily infected. Secondary infection is most often caused by streptococci or staphylococci and is frequently responsible for the hectic fever occurring in tuberculosis. Hemorrhage comes from the erosion of a blood vessel in the wall of the cavity.

The body's attempts to check and heal even a progressively destructive tuberculous focus such as a cavity are reflected in a buildup of scar tissue in the adjacent pulmonary tissues. Scar tissue replaces functioning pulmonary tissue, and contraction of the scar effects great shrinkage of the affected lung.

An acute type of pulmonary tuberculosis, *tuberculous pneumonia,* results from the sudden spilling of tuberculous exudate into the sensitized air sacs of a large lung area. It is a manifestation of allergy.

Tuberculous pleurisy or *pleuritis* is secondary to tuberculosis of the lungs. Pleural effusion (collection of fluid in the pleural cavity) often accompanies it. *All unexplained pleural effusions should be considered tuberculous until proved otherwise.*

TUBERCULOSIS OF THE INTESTINES. Tuberculosis of the intestines and regional mesenteric lymph nodes in adults is secondary to pulmonary tuberculosis. It develops because human tubercle bacilli are swallowed in the sputum raised from the tuberculous focus in the lung. (In children it is usually a primary infection from consumption of bovine bacilli in unpasteurized milk.) Intestinal tuberculosis affects the lower end of the ileum and the cecum. Tubercles and caseous foci form in the solitary lymphoid follicles and in Peyer's patches. The overlying epithelium breaks down, forming ulcers that tend to encircle the intestine. Perforation is not common, but the peritoneal surfaces of the ulcers become adherent to adjacent structures. Contraction of the scars formed in the healing of ulcers may lead to obstruction of the intestine.

TUBERCULOSIS MENINGITIS. In the meninges, tuberculosis is a well-marked, acute allergic inflammation associated with the formation of tubercles. It is usually a disease of childhood but may occur in adults. It may appear after generalized miliary tuberculosis or result from bacilli brought to the meninges by the bloodstream from distant foci. Tubercle bacilli may extend directly to the meninges from adjacent tuberculous foci in the brain or the bones of the skull and spinal column.

As a rule, the tubercles are more numerous on the meninges covering the base of the brain, but the meninges of the entire brain may be involved, and tubercles may be found on the lining of the ventricles and within the substance of the brain. An accumulation of excess fluid in the ventricles in tuberculous meningitis is so common that the disease has been called *acute hydrocephalus.*

■ **Laboratory diagnosis.** Direct microscopic examination of suspect material after it has been stained with an acid-fast stain is the first method to be used in the laboratory diagnosis of tuberculosis. *Mycobacterium tuberculosis* is the only acid-fast organism commonly found in sputum. The failure to find tubercle bacilli in sputum does not rule out pulmonary disease, since the bacilli may not appear in the sputum until the disease is advanced. Moreover, they may occur plentifully in one specimen and be scanty or absent in the next. Tubercle bacilli are especially hard to find in smears of urine, cerebrospinal fluid, pleural fluid, joint fluid, and pus because of the relatively small number of organisms present, sometimes even with advanced disease. Various methods of concentration are applied to the test material, especially sputum, if direct smears fail to reveal the bacilli.

The second step in the laboratory diagnosis of tuberculosis is the culture of suspicious material on special media devised for the growth of the tubercle bacillus, regardless of whether acid-fast organisms were demonstrated in the stained smear. The best media are the egg-containing ones, which are opaque, and those made of oleic acid agar, which are translucent. Earlier detection of colonies is possible on translucent media. Cultural methods require a number of days or weeks for the dry, crumbly, colorless colonies to form, sometimes even 10 to 12 weeks. When bacilli are not demonstrated directly in suitably prepared smears, they may frequently be found in cultures or by animal inoculation, the third step in laboratory diagnosis.

Some of the test material is injected subcutaneously into the groin of the extremely susceptible guinea pig. If tubercle bacilli are present—the guinea pig becomes infected, the tissues about the site of inoculation thicken and may ulcerate, the inguinal lymph nodes enlarge, a generalized tuberculosis develops, and the animal dies about 6 weeks after inoculation. As a rule, the guinea pig is not allowed to die but is killed at a stated time and an autopsy performed when the disease is known to be far advanced.

Infection of a guinea pig with a given acid-fast organism may be used to indicate the virulence of that organism; this is the *guinea pig virulence test.* If it produces disease in the animal, it is the pathogenic *Mycobacterium tuberculosis.* If it fails, it is a nonpathogenic acid-fast organism, with certain important exceptions (p. 480).

There is presently a fluorescent antibody test in human tuberculosis that is applied to the serum (indirect fluorescent antibody).

■ **Tuberculin tests.** The *tuberculin* test depends upon the fact that persons infected with tubercle bacilli are allergic to the tubercle bacillus and its products (tuberculin). If a scratch is made on the arm of a person at some time infected with tubercle bacilli and tuberculin rubbed into the scratch (the *von Pirquet test*) or if some of the diluted tuberculin is injected *between the layers* of his skin (the *Mantoux test*), an area of redness and swelling appears at the site of inoculation. If the person has never been infected, no reaction occurs. In the patch test (the *Vollmer test*) a drop of ointment containing tuberculin or a small square of filter paper saturated with tuberculin is placed on the properly cleansed skin and is held in place with adhesive tape for 48 hours. A positive test is indicated by redness and papule formation at the site of application.

The multiple puncture technic (tuberculin tine test) applies tuberculin *transcutaneously*. Tuberculin four times the standard strength of old tuberculin is dried onto the tines of a small, specially constructed metal disk, backed by a plastic holder. It is packaged commercially as a sterile disposable unit. After the skin has been cleansed, the disk is applied briefly to the test area of skin with firm downward pressure, allowing the prongs to pierce the skin (Fig. 30-3). Reactions to this technic are seen in Fig. 30-4.

The Mantoux test performed with serial dilutions of tuberculin, beginning with a high dilution (less tuberculin) and descending to lower dilutions (more tuberculin), is considered the most dependable of the tuberculin tests. The dose of tuberculin used in the Mantoux may be given by jet gun. The intradermal wheal produced should be 6 to 10 mm. in diameter. In case-finding, multiple puncture tests are practical and satisfactory for screening large groups, but a positive tine test, unless strongly reactive, should be confirmed with a Mantoux test.

The appearance of a positive tuberculin reaction indicates that tuberculous infection has occurred. Since many reach adulthood today in this country with a negative reaction, the circumstances under which there is conversion to a positive reaction may be known and the time that the infection was incurred approximated. The timing of the infection is important because it is in the early stages that tubercle bacilli proliferate very actively. Since infection *is* a recent event, a positive tuberculin reaction is a serious finding in the infant or young child. For all persons of any age with a positive tuberculin reaction, the best thinking is that a course of antituberculous therapy is advisable for at least 1 year. After a preliminary chest film, the individual is given isoniazid (see also p. 252) either singly or in combination with pyridoxine

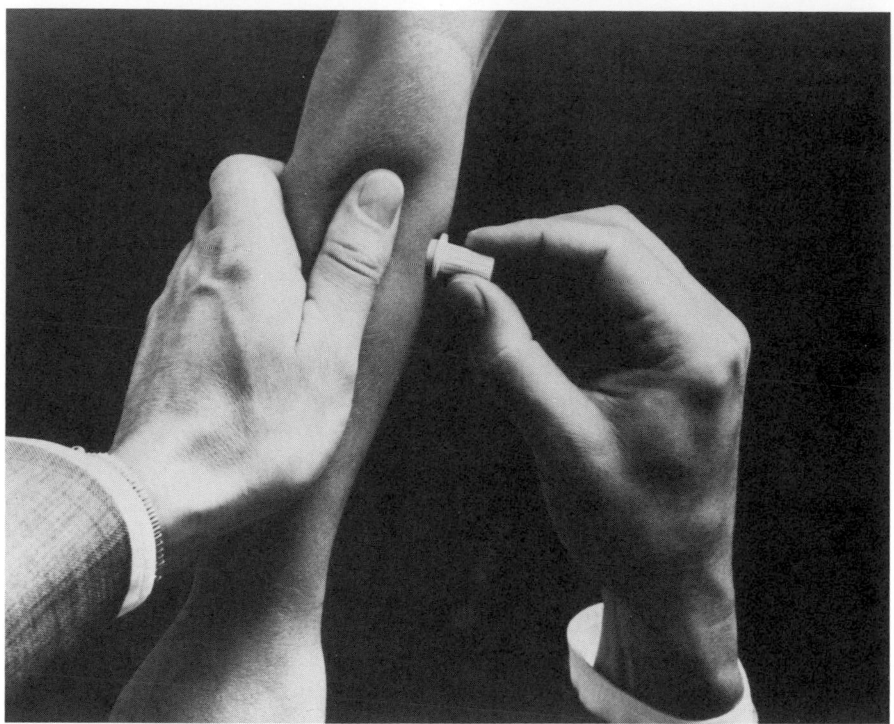

FIG. 30-3. Tine test (multiple puncture test). (Courtesy Lederle Laboratories, Pearl River, N. Y.)

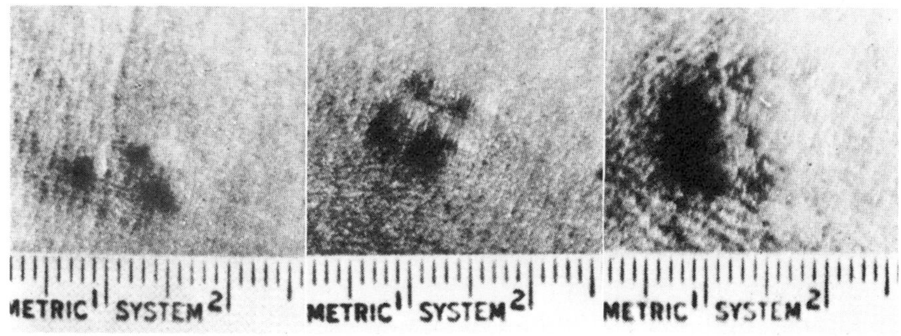

FIG. 30-4. Typical reactions—tuberculin, old, tine test. At 48 to 72 hours, reading should be made in a good light with forearm slightly flexed. By inspection and gentle palpation, induration is determined and the diameter of the largest single area taken in millimeters. Results: 5 mm. or more of induration—positive reaction, 2 to 4 mm.—doubtful reaction, and 2 mm. or no induration—negative. (Courtesy Lederle Laboratories, Pearl River, N. Y.)

hydrochloride. In an adult where the presence of calcified lymph nodes in the chest indicates a process of long standing, such therapy is recommended because of the 5% to 10% hazard that this person will have active tuberculosis during his lifetime.

The tuberculin reaction is of inestimable value in detecting tuberculous cows, and such cows are properly eliminated from the herd. Positive reactions in the tests just described are known as *local* reactions. If tuberculin is given *beneath the skin,* two other reactions may occur, the *focal* and the *constitutional.* By focal reaction one means acute inflammation around a tuberculous focus in the body. This may have serious consequences; for instance, if the tuberculous focus is in the lung, a hemorrhage may result. By constitutional reaction one means systemic reaction, with a sharp rise in temperature and a feeling of malaise lasting for several hours.

■ **Immunity.** Caucasians possess considerable inherent resistance to tuberculosis but never a complete immunity. The incidence in Negroes is high, averaging in proportion to population eight cases to one in the Caucasian race. Not only is this true, but when the Negro contracts tuberculosis, the average time that he will live, if untreated, is about one sixth that of the Caucasian. The American Indian and Mexican are also very susceptible. The incidence of tuberculosis tends to be increased among doctors, nurses, and persons working directly with the disease.

Persons who live in isolated communities seem to be more vulnerable when exposed for the first time during adult life than do those who have been reared in closely crowded and highly infected communities.

Not very long ago it was thought that tuberculosis was hereditary and ran in families, but tuberculosis runs in families because closely associated members of the family pass it to each other. Children of tuberculous parents may inherit certain predisposing factors, but a child born of tuberculous parents and at once removed to infection-free surroundings has a better chance of escaping tuberculosis than one born of healthy parents but reared in contact with tubercle bacilli.

The defensive factors that operate in a given infection depend on whether infection has occurred before. Most observers think that a degree of protection against subsequent infections is afforded the individual by the first infection. Whether a given infection means active disease depends upon (1) the number of bacilli, (2) their virulence, (3) the state of allergy, and (4) the resistance of the subject. Conditions that lower body resistance are malnutrition, crowded housing, and predisposing diseases such as measles, whooping cough, and diabetes mellitus.

■ **Prevention.*** Since tuberculosis sputum is the chief source of infection with human bacilli, it should be disposed of carefully. It should not be allowed to dry, for then it may be blown from place to place spreading the germs over a wide area. Sputum should be received in suitable, covered containers and burned. Promiscuous spitting should be forbidden. The patient must cover his mouth when he coughs. A tuberculous mother should not nurse her child. The woodwork of a room occupied by a tuberculous patient can be washed with soap and water, and a suitable disinfectant, usually one of the phenol derivatives, applied.

Eradication of tuberculous cows and the universal use of pasteurized milk controls infection with the bovine bacillus.

In communities with the best type of health supervision there has been a decided decrease in the incidence of tuberculosis because of several factors, among which are better living conditions and medical prophylaxis. Mass surveys of the population by chest x-ray examination have been of great value in detecting pulmonary tuberculosis and other lung lesions as well. Radiographs of the chest at regular intervals are recommended for individuals whose work brings them into contact with active cases of tuberculosis.

---

*For BCG vaccination, see pp. 606 and 623.

ANONYMOUS (UNCLASSIFIED) MYCOBACTERIA*

■ **The organisms.** In tuberculosis hospitals over the United States unusual acid-fast bacilli are found in the sputum of patients with a typical clinical setting for tuberculosis and cavitary disease by x-ray examination. Although they cause pulmonary disease mimicking tuberculosis *(mycobacteriosis)*, these acid-fast organisms differ sharply from the tubercle bacillus. They also cause infection in lymph nodes and skin. They are referred to as anonymous or unclassified because many of them have not been adequately identified to permit assignment to a definite species.

■ **Classification.** Unclassified mycobacteria are differentiated from *Mycobacterium tuberculosis* by their colonial characteristics, biochemical reactions,† susceptibility to antituberculous drugs, and animal pathogenicity. A separation into four groups has been made (Table 30-1). In group I the photochromogens produce yellow-orange colonies in the light. *Mycobacterium kansasii* ("yellow bacillus"), the type member of this group, is a respiratory pathogen. Group I members are most closely associated with human pulmonary disease.

TABLE 30-1. PATHOGENICITY OF ANONYMOUS MYCOBACTERIA

Group	Common name	Pathogenicity for		
		Man	Guinea pig	Mouse
I	Photochromogens	+	−	+
II	Scotochromogens	+	−	−
III	Nonphotochromogens	+,−	−	+,−
IV	Rapid growers	+,−	−	+,−

The scotochromogens of group II produce pigment in the dark as well as in the light. *Mycobacterium scrofulaceum* ("tap water bacillus") is the type member.

Not affected by light, the nonphotochromogens of group III form buff-colored colonies, soft in consistency by comparison with the rough ones of the tubercle bacillus. These include the Battey bacilli *(Mycobacterium intracellulare)* also respiratory pathogens.

The rapid growers of group IV form colonies on simple media within a short time even at room temperature. Their nonpigmented colonies most closely resemble those of the tubercle bacillus. The pathogen of note here, among many saprophytes, is *Mycobacterium fortuitum*.

Mycobacterium smegmatis (the smegma bacillus) is an acid-fast bacillus widespread in dust, soil, and water. It is inconsequential as a pathogen, but because it is a normal resident of the prepuce and vulva, it may gain access to urine and be mistaken for *Mycobacterium tuberculosis* unless the specimen is properly collected. Smegma bacilli are sometimes found in feces. They are differentiated from tubercle bacilli by

---

*Atypical mycobacteria, "MOTT" bacilli (*m*ycobacteria *o*ther *t*han *t*ubercle bacilli), "other mycobacteria."

†In the clinical setting, production of niacin (p. 110) by an acid-fast bacillus identifies it as the human tubercle bacillus.

their growth characteristics and by the fact that they do not produce disease in guinea pigs. They are now placed in group IV.

The pathogenicity of members of groups I, III, and IV is established; that of group II is equivocal. All the mycobacteria except those in group II cause disease in the mouse; none is pathogenic for the guinea pig.

The tuberculin test in mycobacteriosis is negative or weakly positive. A tuberculin reaction less than 10 mm. in diameter may represent the cross reaction to "unclassified" mycobacterial sensitivity.

■ **Transmission.** Much remains to be learned of the mode of spread and the source of these mycobacteria. Their tolerance for adverse environments and their survival in sparse media suggests they are free living. A wide host range of wild and domestic animals may exist. Some of the group I mycobacteria are found in raw milk and can survive the temperature of pasteurization. The Battey bacilli may spread to man from the soil. Mycobacteriosis is not contagious.

MYCOBACTERIUM LEPRAE (THE BACILLUS OF LEPROSY)

Leprosy is a chronic communicable disease caused by *Mycobacterium leprae** (Hansen's bacillus) that involves skin, mucous membranes, and nerves. Leprosy has affected man from the beginning of history, and descriptions of it occupy a prominent place in the Old Testament and other ancient writings. It is estimated that presently there are about 15 million lepers in the world with approximately 2000 in the United States. No animal shares man's natural susceptibility.

■ **The organism.** *Mycobacterium leprae* (Fig. 30-5) is an acid-fast organism closely resembling *Mycobacterium tuberculosis.* It occurs abundantly in the lesions of leprosy. Recently the bacillus of leprosy has been cultivated in the ears and rear footpads of white

*Discovered in Norway by Gerhard A. Hansen (1841-1912) in 1874, *Mycobacterium leprae* was actually the first bacterium identified as a cause of human disease. For nearly a century it remained the only one known to infect man that could not be cultured in the laboratory and that did not produce progressive disease if inoculated into a test animal.

FIG. 30-5. *Mycobacterium leprae,* fluorescent staining. Note morphology. (×1000.) (Courtesy Dr. R. E. Mansfield, Canton, Ill.)

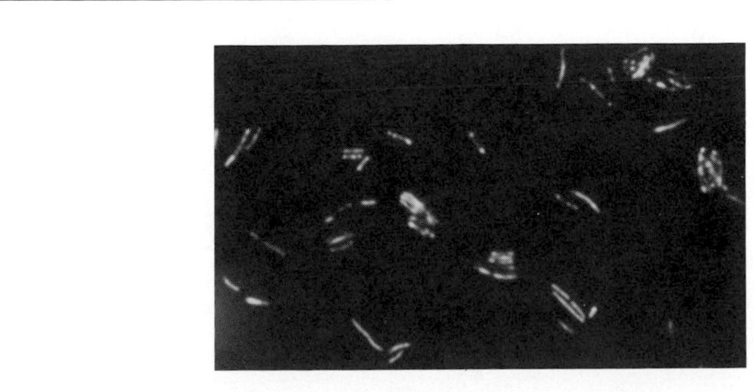

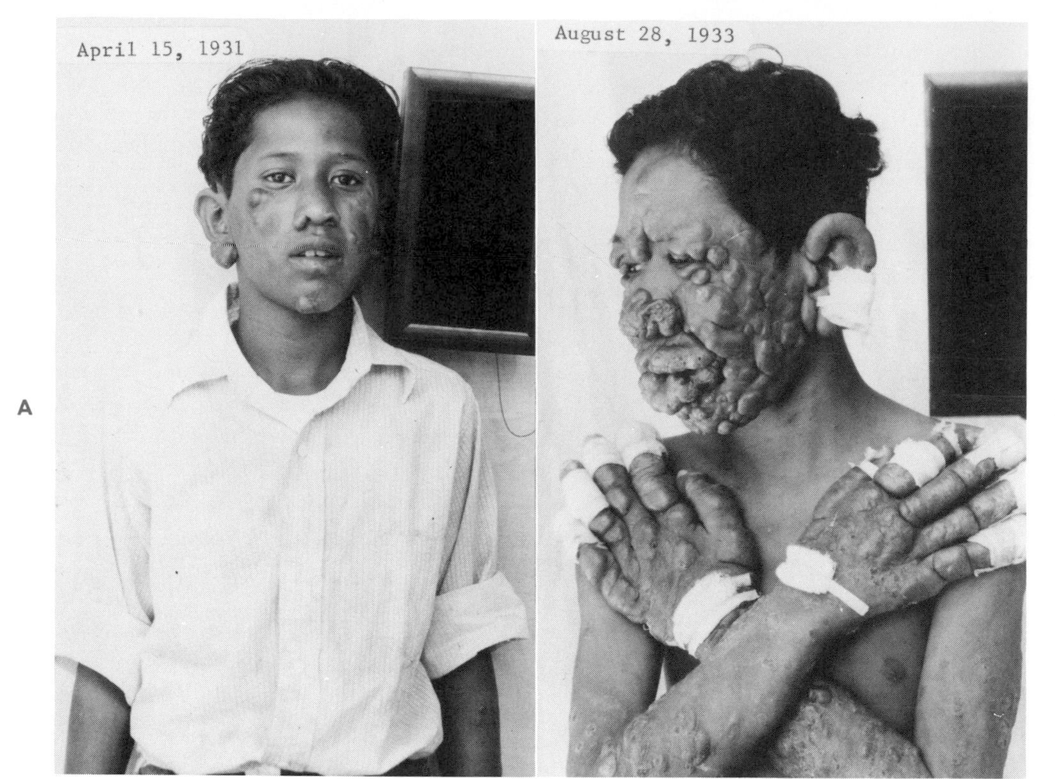

FIG. 30-6. Leprosy. Progression of disease in a teen-aged Filipino. Note dates; this occurred before sulfone drugs were available. (Courtesy Dr. C. Binford, Washington, D. C.)

mice and hamsters, sites chosen as the largest and coolest parts of the experimental animal. After many months mild changes detectable with the microscope occur at the sites of inoculation, but there is no gross advanced disease. Currently the armadillo is used as a test animal in leprosy research. The bacillus has also been grown in tissue cultures.

■ **Mode of infection.** The imagination of man has clothed leprosy with many attributes that it does not possess. One is that it is a highly communicable disease. This is not true. Man contracts leprosy only after prolonged and intimate contact, and even then he often escapes infection. The bacilli leave the body in great numbers from degenerating lesions of the skin and mucous membranes, and spread is directly from person to person. Nose and mouth discharges are especially dangerous because lesions are very common in these locations. The portals of entry are probably skin and mucous membranes. Small breaks in the skin may admit organisms discharged from a patient with the disease. Although the bacteria prefer certain cool sites, they are spread widely over the body. They may be found in feces and urine. The disease is not hereditary, but since children are unusually susceptible, the percentage of infection in children associated with leprous parents is very high (30% to 40%). The highly variable incubation period is estimated at from 5 to 15 years.

■ **Pathogenicity.** Leprosy belongs with the infectious granulomas (diseases identified by a defensive multiplication of reticuloendothelial cells at the sites of infection).

Although it is generalized with a variety of changes, there are two conspicuous patterns. One, the *lepromatous* or *nodular* form, is marked by tumorlike overgrowths of the skin and mucous membranes. The other, *tuberculoid* or *anesthetic* leprosy, is manifest by involvement of peripheral nerves with localized areas of skin anesthesia. The end stages of the disease are associated with extensive deformity and destruction of tissue (Fig. 30-6).

The earliest lesion in leprosy may be diagnostic in that it contains acid-fast bacilli. Since it may not indicate the subsequent course, it is referred to as an *indeterminate* lesion. *Dimorphous* leprosy refers to the coexistence of the two forms—tuberculoid and anesthetic.

■ **Laboratory diagnosis.** The organism *Mycobacterium leprae* occurs abundantly in the lesions of leprosy. A diagnosis of leprosy is most often made with scrapings from the nasal septum or biopsies of the skin of the ear, which have been stained for acid-fast bacilli. Fluorochrome staining (Fig. 30-5) can be done on leprae bacilli. The bacilli in large numbers are found within phagocytic cells packed together like packets of cigars. Bacilli may also be demonstrated in acid-fast stained material from lesions elsewhere in the skin and lymph nodes. Leprosy and tubercle bacilli are separated by guinea pig inoculation because leprosy bacilli have no effect on the animal.

■ **Prevention.** Leprosy is best prevented by good living conditions. Prolonged and intimate contact with lepers is hazardous. The strict isolation and segregation of lepers has not proved completely successful as a method of preventing the disease, but all lepers in the United States found to be discharging *Mycobacterium leprae* are sent to the National Leprosarium at Carville, Louisiana.

A patient who has improved and has failed to discharge the bacilli for a period of 6 months may be paroled. Paroled patients should be examined twice yearly. Back home the patient should be semisegregated with his own room, linens, and dishes. His discharges, cooking utensils, and linens should be carefully disinfected. No children or young people should live in the house. Children of lepers should be separated from their parents at birth because the chances of infection are much greater in infants and young children. All persons who contact a patient should be examined every 6 months.

Results of clinical trials in eastern Uganda in Africa indicate that BCG vaccine protects children against leprosy. The possiblity of such a surrogate vaccine strengthens the case for a strong cross-immunity among mycobacteria.

QUESTIONS FOR REVIEW

1. What are the salient features of an acid-fast microorganism?
2. Name and describe the acid-fast bacillus causing tuberculosis.
3. Classify acid-fast bacteria and indicate their pathogenicity.
4. Discuss the sources of infection in tuberculosis.
5. Outline the laboratory diagnosis of tuberculosis.
6. Give the technics and significance of the tuberculin test.
7. How should tuberculin tine tests be evaluated?
8. Compare the tubercle bacillus with anonymous mycobacteria.
9. Compare the tubercle bacillus with the lepra bacillus.
10. Briefly characterize leprosy.
11. What provision does the United States Public Health Service make for the care of patients with leprosy?
12. Define briefly: tuberculin, tubercle, scotochromogen, cavity, infectious granuloma, scrofula, miliary, "MOTT" bacilli, smegma bacilli, primary complex, Battey bacilli, "yellow bacillus," tuberculoid leprosy, and lepromatous leprosy.

31 Fungi: medical mycology

FUNGI IN PROFILE

Molds, yeasts, and certain related forms constitute the organisms in the plant kingdom known as *fungi.* From this group come those whose presence is a common sight on stale bread, rotten fruit, or damp leather. Be it fuzzy or sooty, green, black, or white, the growth on moldy food and clothing is familiar to everyone. Fungi (100,000 species or more*) are among the most plentiful forms of life—they powder the earth and dust the atmosphere. Their science is *mycology.*

Fungi do not contain chlorophyl and are probably degenerate descendants of chlorophyl-bearing ancestors, most likely the algae. Being unable to make their own food by photosynthesis as the higher plants do, they must either exist on other living organisms as parasites or avail themselves of the dead remains as saprophytes. Within the protoplasm of saprophytic fungi are elaborated chemical substances and enzymes that diffuse into the environment, changing what complex substances are there (wood, leather, clothing, bread, and dead organic plant or animal matter) to simpler substances that can be used for their food. The chemical processes of digestion are completed outside the organism, and the end products are then absorbed by the fungus.

■ **Structure.** Fungi vary in size. Some, the mushrooms and toadstools, are large and easily visible to the naked eye, but most of the medically important ones are microscopic or at least so small that the microscope is necessary for their complete investigation. A given fungus may be a single cell, or it may be composed of many cells laid out in a definite pattern. This distinction is not always a sharp one. Certain important pathogenic fungi produce disease in the body tissues as single cells but when cultured in the laboratory present a complex multicellular arrangement. In this chapter to describe fungi, we designate the unicellular forms as *yeasts* and *yeastlike* organisms and the multicellular ones as *molds* and *moldlike* forms.

The unicellular fungi or yeasts are nonmotile, round or oval organisms most of which reproduce by a characteristic process of budding. They vary considerably in size, depending upon age and species, but all are microscopic. Nuclei may be demonstrated by a suitable stain.

The multicellular fungi, the molds and related forms, present typical structures associated with nutrition and reproduction. Most molds are made up of a *mycelium*

*Less than 100 can invade man or animals, and less than a dozen can infect and kill.

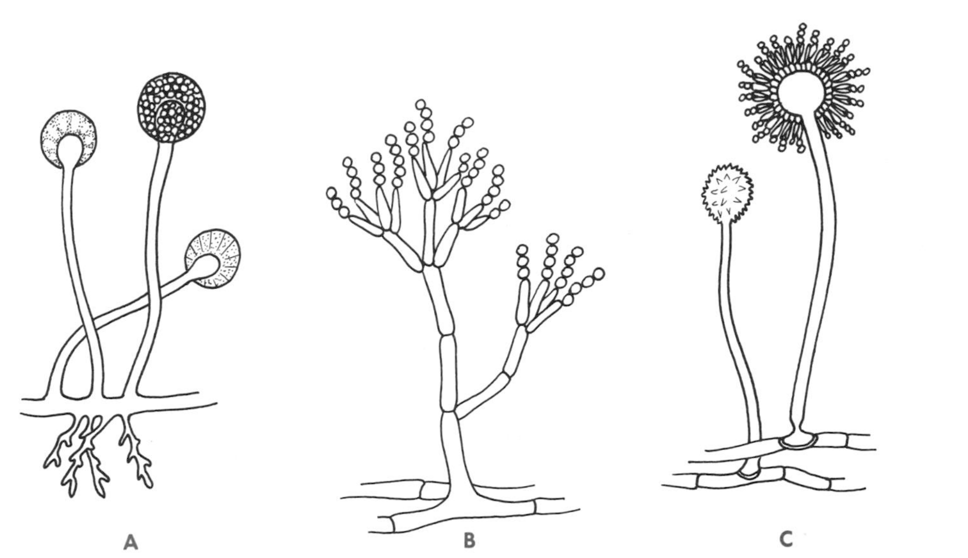

FIG. 31-1. Well-known saprophytic fungi, in sketch. Note fruiting bodies in which asexual reproductive spores form. **A,** Genus *Rhizopus,* black mold. **B,** Genus *Penicillium,* green mold. **C,** Genus *Aspergillus,* gray-green mold.

(Fig. 31-1), a network or matlike growth of branched threads bearing fruiting bodies. A rudimentary plant known as a *thallus* (see also p. 4) (no root, stem, or leaf) is formed. The individual threads of the mycelium are known as *hyphae.* In some fungi nonseptate hyphae consist of single threads containing many nuclei spaced along the thread. In most, septate hyphae are divided by cross walls, or *septa,* into distinct cells, each containing a nucleus. The hyphae have thin walls, thus allowing ready absorption of food and water. This fact helps to account for the rapid growth of fungi. The portion of the mycelium concerned with nutrition is referred to as the *vegetative* mycelium. The part that usually projects into the air is the *aerial* or *reproductive* mycelium.

■ **Reproduction.** Multicellular fungi reproduce by the conversion of a spore into a vegetative fungus. Spores are formed in a great variety of ways from the reproductive mycelium, depending upon the species. In some molds the spores are simply attached to the hyphae; in some the hyphae bear little pods or sacs in which the spores rest (Fig. 31-1*A*); in others the hyphae are branched to form brushlike processes, each of which bears a spore (Fig. 31-1, *B*); in still others the hyphae produce heads from which radiate fine chains of spores (Fig. 31-1, C). In most cases the spores separate from the hyphae before reproduction.

If a spore is formed after the nuclei of two hyphae have contacted and fused or after an association of a specialized structure on the mycelium with the nucleus of another specialized structure developed close by, that spore is designated a *sexual spore.* When there is no fusion of nuclei and the spore is simply formed as a swollen body at the end of the hypha, it is called an *asexual* spore. Asexual spores are formed in great numbers, sexual spores only occasionally. The fungi that produce only asexual spores are referred to as *Fungi Imperfecti,* or imperfect fungi. (Fungi producing sexual spores are perfect fungi.) The Fungi Imperfecti are important because the pathogenic fungi are concentrated in this division.

FIG. 31-2. Black bread mold, *Rhizopus nigricans*. Each tiny dot is a sporangium (fruiting body) containing hundreds of spores. (Growth on nutrient agar.) (From Beaver, W. C., and Noland, G. B.: General biology, ed. 8, St. Louis, 1970, The C. V. Mosby Co.)

Spores* are resistant to drying, cold, and moderate heat and may maintain their vitality for a long time. They are constantly present in the air. When they contact food or other material supplying the necessary elements for growth, they develop into new fungi.

Unicellular fungi thought of as yeasts may reproduce by spore formation. The cell enlarges slightly, and the nucleus becomes converted into a definite number of spores. However, *budding*, considered by some as a simple type of spore formation, is the usual way. In the process of budding the nucleus moves toward the edge of the cell and divides into two daughter nuclei. A knoblike protrusion of the cytoplasm forms at this point, and one of the nuclei passes into it. The protrusion increases in size and becomes constricted at its base until there is only a narrow connection between protrusion and the parent cell. Finally, the two separate, and the protrusion or bud, now a small yeast cell, continues to grow until it reaches full size; then the process is repeated.

Multiplication by simple fission, characteristic of some yeasts, resembles the process in bacteria. The yeast develops to its full size, and a membrane forms across the middle of the cell. A dividing wall forms, and the two parts separate.

■ **Conditions affecting growth.** Fungi grow best under much the same conditions as do bacteria, that is, in warm, moist surroundings. They grow in the presence of much acid and large amounts of sugar, which bacteria cannot do. Most do not grow in the absence of free oxygen, and large amounts of carbon dioxide are harmful to them. Many grow best at temperatures somewhat lower than body temperature. At

*The spores of fungi should not be confused with bacterial spores, the resistant bodies formed by bacteria for survival, not for reproduction. Only one spore forms from one bacterial cell.

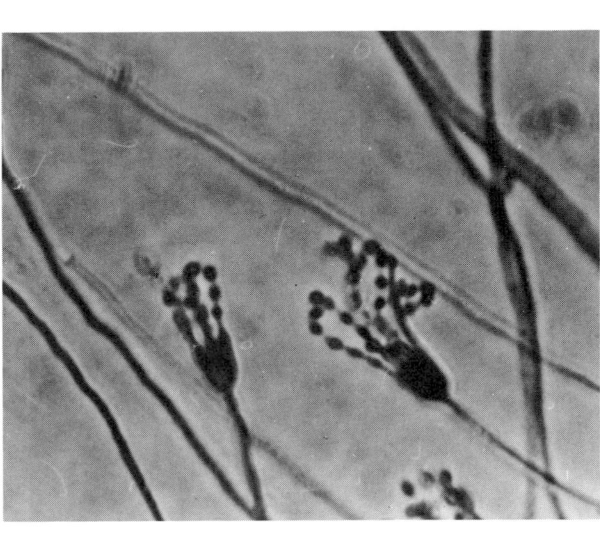

FIG. 31-3. Blue-green mold *Penicillium* (wet mount examined microscopically). Note conidia (specialized hyphae) and conidiospores (asexual spores). (From Beaver, W. C. and Noland, G. B.: General biology, ed. 8, St. Louis, 1970, The C. V. Mosby Co.)

low temperatures metabolic activities may be slowed, but the organisms do not die. In fact, they are rather resistant to cold; many of the commonly encountered ones survive freezing temperatures for long periods of time (months or years). Some can even grow at temperatures below freezing. To prevent the growth of mold, meats and certain food products must be refrigerated at temperatures less than 20° F. (−6.67° C.). On the other hand, fungi are quite susceptible to heat, being easily killed at high temperatures. Most species are little affected by light, being able to grow either in the light or in the dark.

■ **Classification.** The *Eumycetes* (p. 4) are the true fungi, the organisms without chlorophyl ordinarily thought of as yeasts, molds, and related forms. On the basis of morphology—the appearance of the colony, spores, and mycelium—the true fungi are placed in four classes:

1. *Phycomycetes* or *water molds.* These grow on water plants and fish. Included are the common bread molds of genera *Rhizopus* (Fig. 31-2) and *Mucor* and widespread contaminants. (The hyphae of the water molds are nonseptate; in the other three classes the hyphae of the mycelium, where present, are septate.)
2. *Ascomycetes* or *sac fungi.* These constitute the largest order. Sexual spores are formed within a specially developed sac or *ascus.* Included here are single-celled yeasts of the genus *Saccharomyces,* crucial to the brewing, baking, and wine industries, as well as the multicellular molds *Aspergillus* and *Penicillium* (Fig. 31-3). The common contaminant *Aspergillus* is a sometime pathogen, and *Penicillium* species are the source of the antibiotic penicillin. Certain pathogenic ascomycetes can cause maduromycosis. (See p. 505.)
3. *Basidiomycetes* or *club fungi.* These encompass the large fleshy toadstools,

puffballs, and mushrooms as well as the small plant smuts and rusts. Certain members produce poisons toxic to man. None is infectious.

4. *Fungi Imperfecti* or *Deuteromycetes*. These lack sexual spores. Asexual spores are formed on or from filaments in a special way. The important pathogens of this class will be discussed in their role as disease producers.

■ **Laboratory study.** Fungi may be studied and identified in a number of ways.

DIRECT VISUALIZATION. Stained or unstained material may be observed directly with the microscope. Usually a wet mount is prepared. The specimen, a bit of fungous growth, sputum, pus, skin scrapings, or infected hairs, is placed in a drop of mounting fluid on a glass slide, covered with a cover glass, and examined under the microscope. It is important that the illumination of the specimen be reduced by lowering the intensity of light from the source.

Yeasts may simply be suspended in water. Material containing molds is commonly placed in a 10% to 20% solution of potassium hydroxide. The specimen is cleared when left in contact with the alkali for 10 to 15 minutes or longer; that is, it becomes more transparent and is more easily defined microscopically.

Microscopic identification of a given fungus rests on the study of its structure, especially the kind of spores and their relation to the hyphae.

The phenomenon of fluorescence is important in the study of superficial fungous infections. For example, infected hairs often fluoresce under a filtered source of ultraviolet light.

CULTURE. Both pathogenic and saprophytic fungi are highly resistant to acid environments, and both prefer large amounts of sugar in their food supply. Consistent with these requirements, the French mycologist Sabouraud, around the turn of the century, devised a culture medium of maltose, peptone, and agar still widely used today. A fungous culture on Sabouraud's agar is incubated at room temperature (20° C.). Blood agar may also be inoculated but is incubated at body or incubator temperature (37° C.). Littman's oxgall agar is frequently used. Fungi do particularly well when portions of raw or cooked vegetables are added to nutrient media. Potato and carrot combinations with cornmeal agar are valuable. Antibiotics added to fungous culture media suppress bacterial growth. Fungi grow slowly and cultures must be kept a week or two.

In a liquid culture medium bacterial growth is dispersed; few bacteria form the characteristic sheet or pellicle on the surface of the liquid that is seen with the growth of fungi.

If set up as a slide culture, fungous growth may be observed daily under the microscope. If the very center of the agar medium in a Petri dish is inoculated, a giant colony grows out to the edge, covering the surface of the dish. The gross appearance of the colony is usually typical for a specific organism.

As with bacteria, fermentation reactions are important in identification of fungi, and animal inoculations are performed.

IMMUNOLOGIC REACTIONS. Serologic studies in the laboratory evaluation of fungous diseases include agglutination, precipitation, complement-fixation tests, and immunofluorescent and immunoelectrophoretic technics. These are generally positive with active disease. Skin testing is of great value.

■ **Pathology of fungous disease.** Fungi are important causes of disease in man and animals and one of the chief causes in plants. In man, *mycoses* (fungal infections) are of two types: superficial and systemic. Superficial fungi (in the skin, hair, and nails) causing the *dermatomycoses* spread from animal to man, or man to man, even cause epidemics, but do not invade. Systemic fungi contact man from his environment—from the soil, the vegetation, bird droppings, and so on. Ordinarily they are very insidious in their approach, gain a foothold in the body, and progress rather leisurely. (Regression seems slow also in mycoses.)

The body's response to the intrusion is granulomatous inflammation, that is, a reaction in which the reticuloendothelial system is the main participant. Tissue damage comes after the state of allergy has been set up in the host to the proteins of the fungus. In many respects the pathology of the mycoses is similar to that of tuberculosis. Usually the causative agent can be demonstrated in sections of infected tissues or in fluids therefrom; *its presence makes the diagnosis.* Systemic mycoses fall into three categories: (1) *primary* infections, usually with a geographic pattern; (2) *secondary* infections or "superinfections," and (3) *complicating* infections. Most of the fungous diseases discussed in this chapter are systemic, including the important primary mycoses. The four major ones are blastomycosis, coccidioidomycosis, cryptococcosis, and histoplasmosis.

Secondary mycoses develop during the course of a bacterial or viral disease for which antibiotics are being given. (Bacteria help to control fungi in nature.) The single most important example is candidiasis. Superinfection of this kind, often hospital acquired, is produced by both fungi and a number of bacteria—many gram-negative bacilli, including the enterics, and staphylococci.

Complicating infections appear after special therapeutic procedures such as peritoneal dialysis or the prolonged use of a catheter indwelling a blood vessel. They are a distinct hazard in the management of the "compromised host," that patient with severe, chronic, debilitating disease for which antibiotics, steroid hormones, or immunosuppressive agents are required. They also follow close upon disturbances of the immune mechanisms such as are seen in cancer of the lymphoid system and bone marrow failure.

Certain fungi (for example, *Cryptococcus*) are readily pathogenic either as primary or secondary invaders. Others, for example *Candida*, *Aspergillus* (Fig. 31-4), *Mucor*, and *Rhizopus*, benign in nature, seize upon the "opportunity" with an inordinate vigor.

■ **Importance.** Fungi are important in the processes of nature, agriculture, manufacturing, and medicine. Commonly encountered molds can injure woodwork and fabrics and spoil food. They destroy food during its growth, in the process of manufacture, and after it has reached the consumer. Foods most vulnerable are bread, vegetables, fruits, and preserves. Molds are an important factor in the decay of dead animal and vegetable matter, complex organic compounds broken down into simple ones are returned to the soil to be used as food by the green plants. Some fungi (for example, certain mushrooms) are food for man. Molds are used commercially in the manufacture of beverages and to give flavor to cheeses (Roquefort, Camembert). Penicillin, an effective antibacterial substance, is derived from a common mold, *Penicillium notatum.*

Yeasts are economically important because they ferment sugars. The fact that they convert sugars by enzymatic action into alcohol and carbon dioxide is practically applied in the manufacture of alcoholic beverages and in baking. In the manufacture of alcoholic beverages carbon dioxide is a by-product, whereas in baking it is the essential factor (see p. 573). In the preparation of commercial yeast the cells are grown in a suitable liquid medium, separated out from the liquid portion by centrifugation, mixed with starch or vegetable oil, and then molded and cut into cakes. Yeast is a source of vitamin B and of ergosterol, from which vitamin D is obtained.

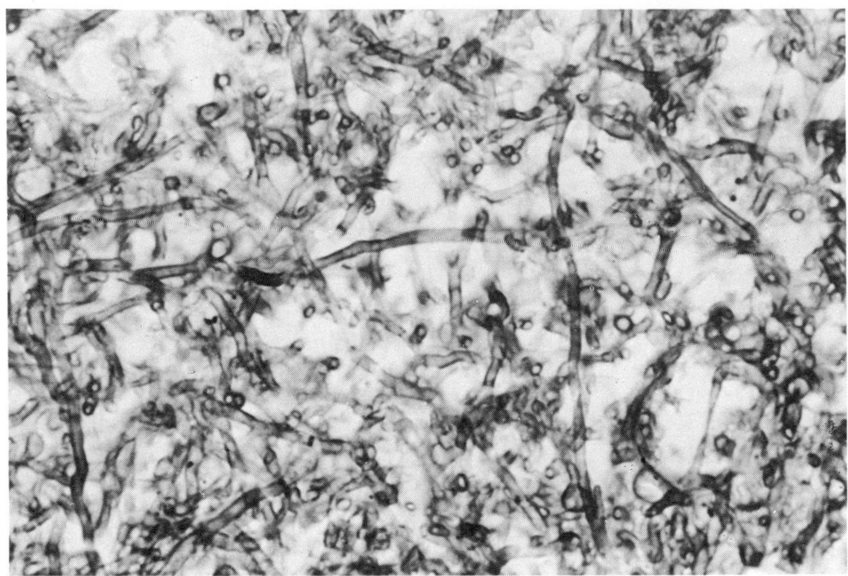

FIG. 31-4. *Aspergillus* demonstrated in microsection of lung with special stain. Note abundant growth of this opportunist and typical right-angle branching of hyphae. Patient was a diabetic. (From Zugibe, F. T.: Diagnostic histochemistry, St. Louis, 1970, The C. V. Mosby Co.)

THE ACTINOMYCETES

Although usually considered to occupy an intermediate position between the true bacteria and the true fungi, the actinomycetes are often included in medical mycology. They are classified in the same class—Schizomycetes (Bergey's classification)—as bacteria. The order Actinomycetales, composed of organisms resembling both bacteria and fungi, includes the three families: Mycobacteriaceae, Actinomycetaceae, and Streptomycetaceae. The first family includes the ones of which *Mycobacterium tuberculosis* is the type species. In the other two families are the actinomycetes and the streptomycetes. They resemble the fungi with their mycelium composed of masses of branched hyphae, but their hyphae are much slenderer than those of true fungi. The hyphae of actinomycetes fragment into spherical or rod-shaped segments that function as spores and in turn develop into new hyphae. Spore formation comparable to that in bacteria is seen in the hyphae of the genus *Streptomyces.*

Often called the higher bacteria, actinomycetes are nonmotile and gram-positive. They are killed by a temperature of 60° to 65° C. for 1 hour. There are two genera of note in the family Actinomycetaceae, *Actinomyces* and *Nocardia.* Species of the genus *Actinomyces* are anaerobic or microaerophilic, whereas species of the genus *Nocardia* are aerobic. The genus *Actinomyces* is nonacid-fast. The genus *Nocardia* sometimes partially acid-fast may be confused with *Mycobacterium tuberculosis.*

The actinomycetes are medically important because of the two or more genera causing disease. The streptomycetes are medically important because they are the source of many important antibiotics (p. 253). Both are widely distributed in nature and play a most vital part in changes in organic material of the soil. This activity of the actinomycetes is more crucial to man than is their disease-producing capacity.

DISEASES CAUSED BY FUNGI

Medical mycology treats of the fungi that bring about disease.

Superficial mycosis

Dermatomycoses

Superficial fungous infections of the skin, hair, and nails, generally called *ringworm* or *tinea,* are *dermatomycoses* or *dermatophytoses.* Fungi causing dermatomycoses and showing no tendency to invade the deeper structures of the body are called *dermatophytes.* There are three important genera: (1) *Microsporum,* (2) *Trichophyton,* and (3) *Epidermophyton.* The dermatophytes are closely related botanically.

The genus *Microsporum* is the most frequent cause of ringworm of the scalp and may give rise to ringworm in other parts of the body. Hairs removed from the affected regions are surrounded by a coat of spores, and scales of skin show many branched mycelial threads. *Trichophyton* causes ringworm of the scalp, beard, skin, or nails. The organisms are found as chains of spores, inside or on the surface of affected hairs, or as hyphae and characteristic spores in skin scrapings. *Trichophyton schoenleini* is the cause of almost all cases of favus.* The spores and mycelial threads are found in the favus crusts. Hairs in the affected areas are filled with vesicles and channels from which the mycelia have disappeared. *Epidermophyton* is largely responsible for ringworm of the body, hands, and feet. Epidermophyta appear as interlacing threads in the skin and do not invade the hairs.

Ringworm of the scalp *(tinea capitis)* seen most often in children is a common and highly communicable disease. It may be spread directly from person to person or by articles of wearing apparel. It occurs in domestic animals, from which it may be transmitted to man. Ten to thirty percent of ringworm infections occurring in cities and 80% of those in rural areas are thus transmitted, either by direct or indirect contact with the animal. Pets (dogs and cats) readily pass the ringworm fungi onto their human masters.

Favus usually affects the scalp with the formation of yellowish cup-shaped crusts or *scutula* about the mouths of the hair follicles. These crusts consist of masses of spores and mycelial threads mixed with leukocytes and epithelial cells. Favus may be transmitted directly or indirectly from person to person and tends to run in families.

Ringworm of the beard *(tinea barbae)* is known as *barber's itch.* Ringworm of the groin is known as *tinea cruris* or *dhobie itch.* Ringworm of the feet is known as *tinea pedis* or *athlete's foot.* It has been believed for years that athlete's foot is contracted from footwear, lockers, and floors, but experiments indicate that exposure to the pathogenic dermatophytes in public swimming pools or shower stalls plays a minor role. These fungi are everywhere, even on the feet of noninfected individuals. The lesions of athlete's foot most probably appear because of decreased resistance of the skin of the feet on contact with the causative fungi.

---

*In 1839, *Johann Lukas Schönlein* (1793-1864) isolated the fungus causing favus. A leading figure in German clinical medicine of the early nineteenth century, Schönlein is better known for his clinical teaching and his general influence on German medicine than for his individual contributions to medical progress.

Laboratory diagnosis of the dermatomycoses depends upon demonstration of fungi in the hair and skin scrapings by direct microscopic examination or by cultural methods.

Control of the dermatomycoses is very difficult. It consists of the proper sterilization of clothing, bathing suits, and objects subjected to frequent handling. Hygiene of the feet is extremely important.

Systemic mycosis

Aspergillosis

Aspergillosis is an infection most often produced by *Aspergillus fumigatus*, a gray-green mold growing in the soil. This fungus may cause various types of infection in chickens, ducks, pigeons, cattle, sheep, and horses. Animals usually contract the disease from moldy feed.

In man the disease usually is an infection of the external ear (otomycosis). The infection may be superficial and mild, or it may cause ulceration of the membrane lining the ear and perforate into the middle ear. Infection most likely comes from fungi living saprophytically on the earwax. Other types of infection in man are pulmonary infections, sinus infections, and infections of the subcutaneous tissues. Aspergillosis as a superinfection complicating antibiotic therapy affects the lungs (Fig. 31-4). (Aspergilli can no longer be passed off as weeds in the laboratory likely to contaminate any culture.)

Aspergillosis may be caused by aspergilli other than *Aspergillus fumigatus*. Among those are *Aspergillus nidulans*, *Aspergillus niger*, and *Aspergillus flavus*. Certain strains of *Aspergillus flavus* elaborate *aflatoxins*, toxic substances that can cause cancer in the liver of animals ingesting them. Animals contact these mycotoxins (fungal toxins) in the mold produced by *Aspergillus* on peanuts and other animal foods. Aflatoxins are exceedingly potent in this respect, cancer formation requiring an amount no more than 0.05 ppm.

Serologic aids to the diagnosis of aspergillosis include a complement fixation test, a double diffusion in agar gel technic, and the indirect fluorescent antibody determination.

Blastomycosis

There are two types of blastomycosis—the *North American* and the *South American*. North American blastomycosis, known as *Gilchrist's disease* after its discoverer, is a granulomatous and suppurative inflammation. Multiple abscesses form in the skin and subcutaneous tissues (blastomycetic dermatitis) or in the internal organs of the body (systemic blastomycosis). The cutaneous form is more often seen than the systemic (Fig. 31-5). The lesions of blastomycetic dermatitis may be mistaken for cancer or tuberculosis. With systemic disease pulmonary blastomycosis is most likely, closely resembling pulmonary tuberculosis.

North American blastomycosis, confined almost exclusively to the United States and Canada, is caused by *Blastomyces dermatitidis*. The organisms are demonstrated by direct microscopic examination of pus from the lesions or by cultural methods (Fig. 31-6). In pus they are round or oval granular yeast forms, varying from 8μ to 15μ in diameter. They are surrounded by a thick refractile wall, which makes them doubly contoured, and single budding forms are present (Fig. 31-7). The organisms grow characteristically in the mycelial phase on all media, but isolation is often complicated by the overgrowth of bacterial contaminants from the lesion.

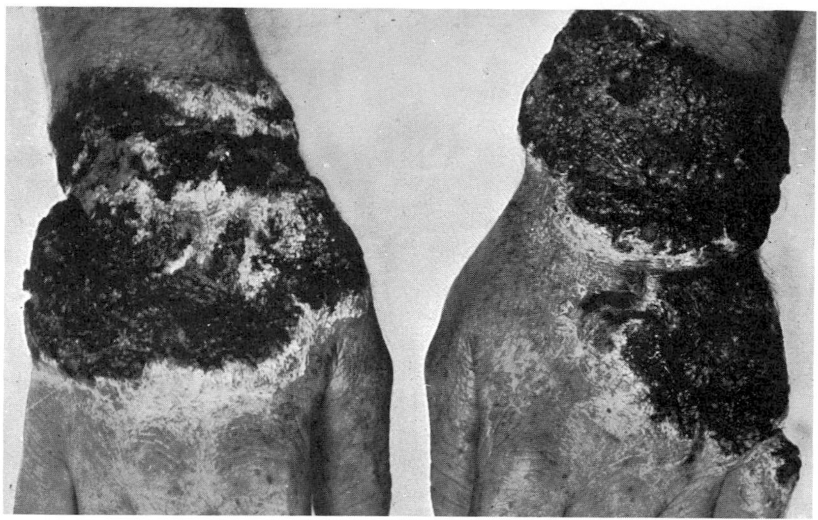

FIG. 31-5. Blastomycosis in a farmer from Minnesota. Note warty character of the skin disease with its sharp margins. (Courtesy Dr. John Butler; from Sutton, R. L., Jr.: Diseases of the skin, St. Louis, 1956, The C. V. Mosby Co.)

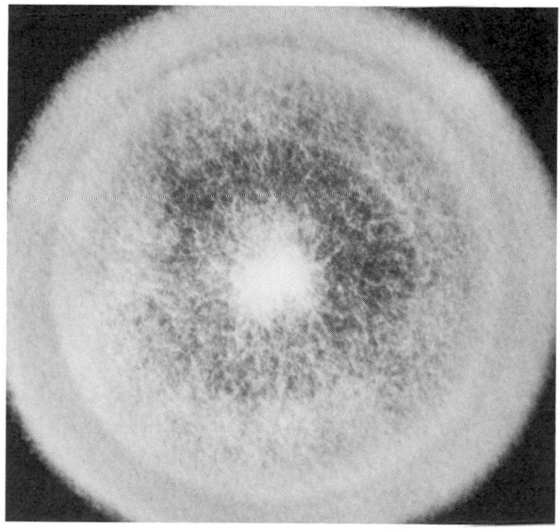

FIG. 31-6. *Blastomyces dermatitidis,* giant colony. Note abundant fluffy mycelial growth. (From Musgnug, R. H.: Med. Trib. **3:**16, May 28, 1962.)

Transmission of human infection is unsolved. Accumulated evidence indicates that cutaneous blastomycosis results from infection through wounds. The finding of a primary focus in the lungs in systemic blastomycosis indicates that systemic infections are probably acquired via the respiratory tract. In some cases of systemic blastomycosis, infection may come from the skin.

Although fungi are weak antigens, infections with these organisms are nevertheless associated with a degree of allergy useful in the laboratory diagnosis. For North

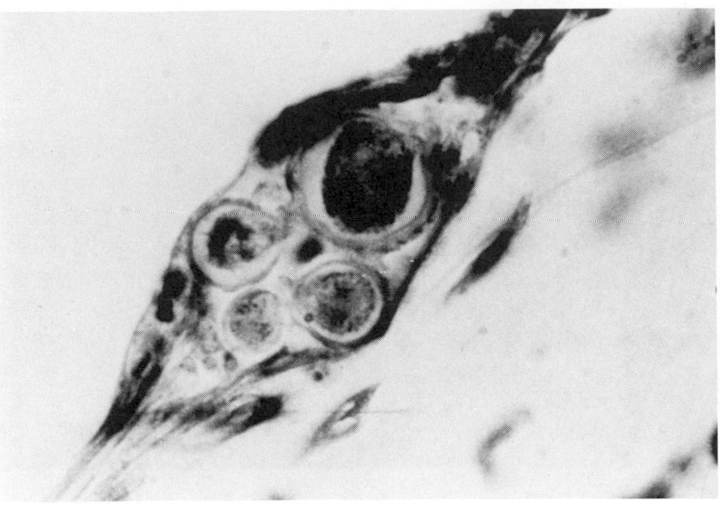

FIG. 31-7. *Blastomyces dermatitidis,* microscopic appearance of budding yeast forms in stained smear of sputum.

American blastomycosis there is both a complement-fixation test and a skin test (the blastomycin test). The first test is specific. Cross reactions may occur with the skin test for blastomycosis and that for histoplasmosis. (An individual with histoplasmosis may appear to give a positive test for blastomycosis.) Skin testing may be done with *Blastomyces dermatitidis* vaccine, considered a more effective antigen than blastomycin.

South American blastomycosis is similar to the North American type. Caused by *Blastomyces brasiliensis,* it is known also as *paracoccidioidal granuloma.* The fungus enters the body by way of the mouth, where it localizes and causes ulcers and granulomas. From these lesions the fungi spread to the lungs and other parts of the body. The disease may terminate fatally. Most cases have been reported in Brazil.

Candidiasis (moniliasis)

Candida (Monilia) albicans is a budding yeastlike organism, worldwide in distribution. Its relationship to disease is often hard to determine. It is found on the mucous membranes of the mouth, intestinal tract, and vagina and on the skin of normal persons with no disease.* It is found also in association with known pathogens in persons with known illness but in whom there is no reason to suspect its pathogenicity. In making the diagnosis of candidiasis, *Candida albicans,* the chief pathogen of the genus, must be repeatedly isolated from the lesions to the exclusion of better defined agents.

Two infections of *Candida albicans* have been with us for a long time. One, on the mucous membranes of the mouth, is known as *thrush.* The other, involving the mucous membranes of the female genitalia, is *vulvovaginitis* or *vaginal thrush.*

Thrush is characterized by many small milk-like flecks that may coalesce and cover the mucous membrane of the entire mouth. Beneath these patches on the inside

*Only a few fungi, such as *Candida albicans* and *Actinomyces israelii,* are normal inhabitants of the human body; most belong to the environment.

of the lips, on the hard palate, and on the tips and edges of the tongue are areas of catarrhal inflammation. Thrush is an especially troublesome infection in newborn infants in hospital nurseries. It is believed that the baby acquires the organism from the vagina of the mother during the birth process, and it is believed that the organisms can be spread from person to person by contaminated fingers, utensils, and nipples. Thrush tends to be rare after the newborn period in healthy subjects of any age but is comparatively common in poorly nourished children and in chronically ill and aged adults.

Vulvovaginitis is a thrushlike infection associated with a typical vaginal discharge. The sugar content of the urine in pregnancy and uncontrolled diabetes may be a contributing factor. It is associated with oral contraceptives. Today vaginal thrush is assuming the proportions of a venereal disease.

Candidal infections are found in those individuals, such as fruit canners, who by occupation must keep their hands constantly soaked in water. *Candida albicans* is an important cause of chest disease. The manifestations of bronchopulmonary or pulmonary candidiasis vary from those of a mild inflammation to those of a severe infection like tuberculosis.

Systemic candidiasis is increasingly encountered today. It is prone to follow persistent skin or mucosal lesions in a person with lowered resistance or altered immunologic mechanisms (Fig. 31-8). *Candida* easily gains the ascendancy when dosage with a broad-spectrum antibiotic is prolonged, since the normal bacterial flora is thereby depressed.

Overgrowth of *Candida* for any reason is an ominous event. If the organisms circulate in the bloodstream (a fungemia), they set up a serious toxic reaction and are widely disseminated in the body (Fig. 31-9). The patient at greatest risk is the one with leukemia, some kind of bone marrow failure, or an organ transplant. Other pathologic conditions over which the threat of candidiasis hangs are diabetes, chronic alcoholism, endocrine disorders, malnutrition, and certain kinds of cancers. Under the right circumstances *Candida* strikes as a formidable pathogen (Fig. 31-10).

The laboratory demonstration of *Candida (Monilia) albicans* is easily made either by direct microscopic examination of unstained or stained material or by culture. In the exudate from a lesion or in sputum the organisms appear as oval, budding yeasts with scattered hyphal segments.

Serologically there are the hemagglutination test, the precipitin reaction, and an indirect immunofluorescent technic. Inoculation of the chorioallantoic membrane of the chick embryo, with visible lesions appearing 72 to 96 hours later, is a good test for the pathogenicity of *Candida*. With oidiomycin, an extract of the organism, a skin test may be done to indicate past or present infection.

Coccidioidomycosis (coccidioidal granuloma)

Coccidioidomycosis, one of the most infectious of the fungous diseases, exists in two forms—the primary (usually self-limited) and the progressive. In the primary form the lesions are confined to the lungs, giving pulmonary symptoms of varying severity. Sometimes there is cavitation. As a rule, the infection ends in recovery, but in a small percentage of cases the process spreads from the lungs to produce the progressive form. This happens more often in Negroes and the more darkly pigmented races than in Caucasians.

In the progressive form the disease spreads to the skin, subcutaneous tissues, bones, meninges, and internal organs. The lesions in the skin resemble those of

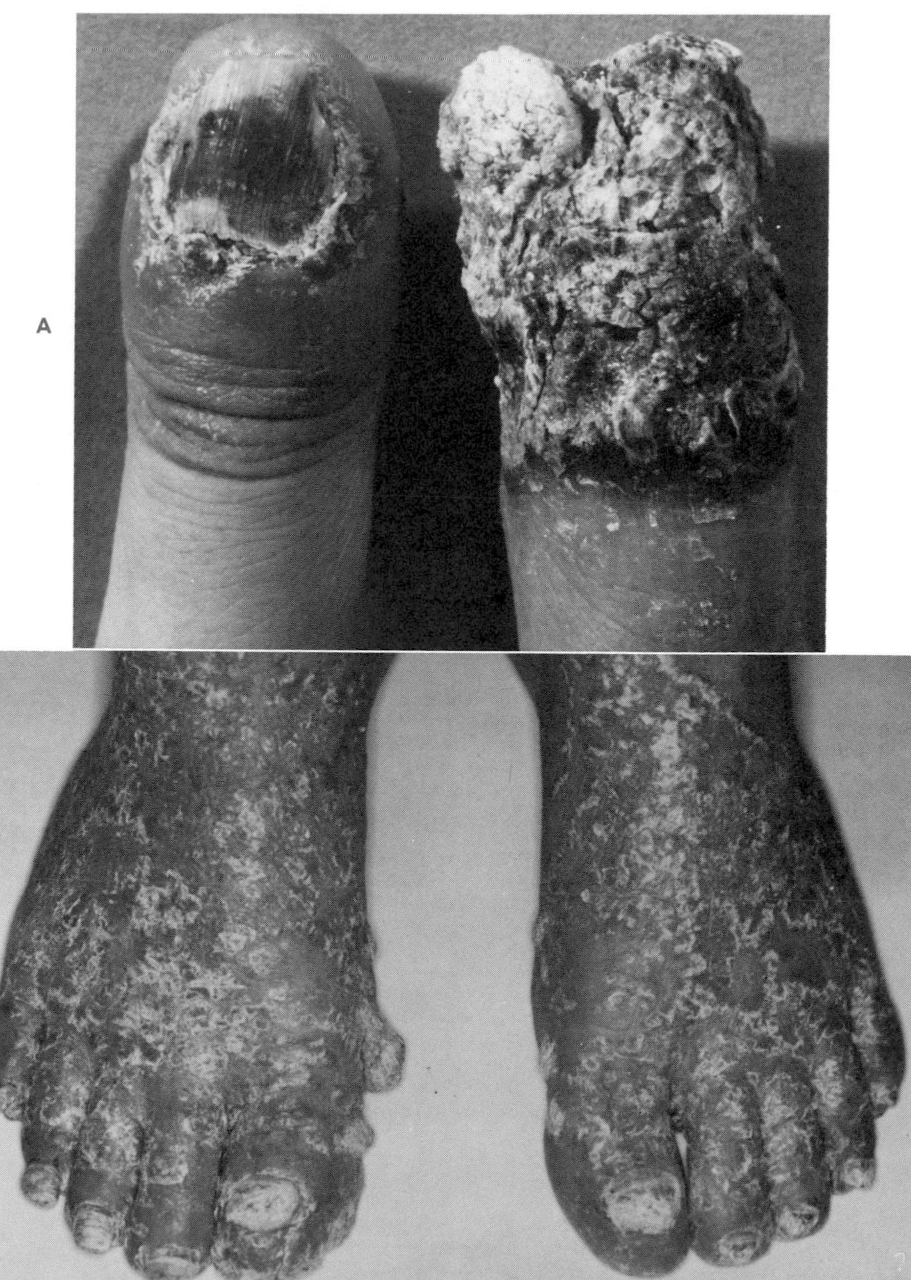

FIG. 31-8. Candidiasis of long standing in patient with immunologic disorder. **A,** Thumbs. Note destruction of nails with accumulation of horny material. **B,** Feet. Note changes in skin and nails. (From Kirkpatrick, C. H., Rich, R. R., and Bennett, J. E.: Ann. Intern. Med. **74:**955, 1971.)

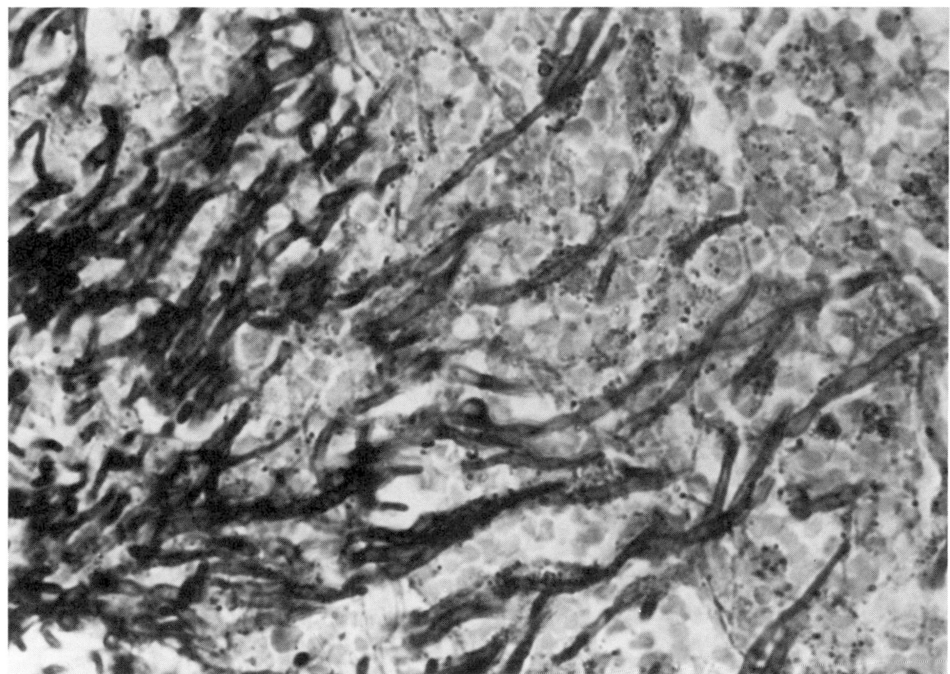

FIG. 31-9. Candidiasis in stained microsection of spleen in patient with disseminated disease. Note abundant mycelial filaments of *Candida albicans*. (×800.)

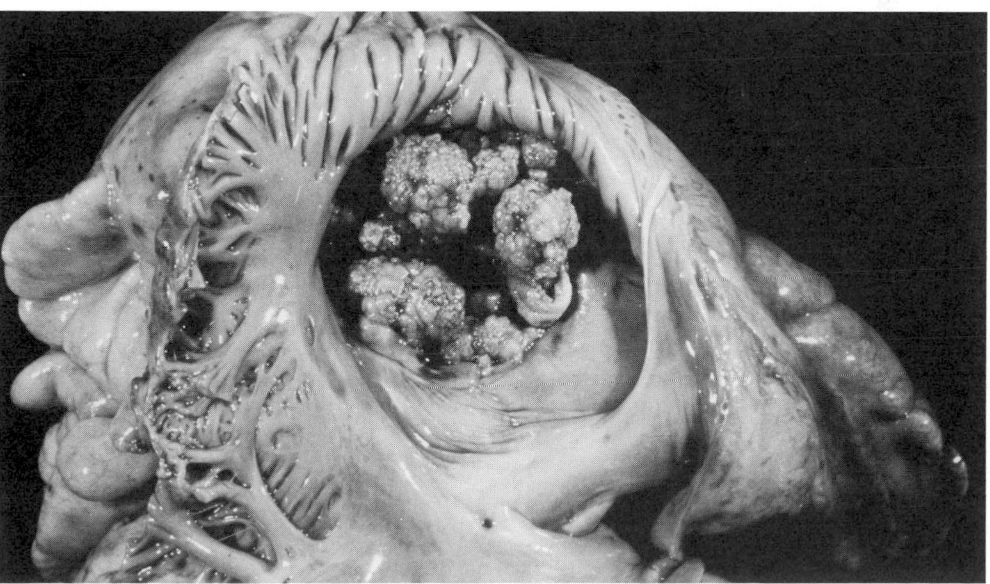

FIG. 31-10. Candidiasis of heart valve (candidal endocarditis). The atrium of the heart has been opened to expose the valvular opening. Note large cauliflower excrescences (vegetations) on leaflets of tricuspid valve. This was a complication in a drug addict.

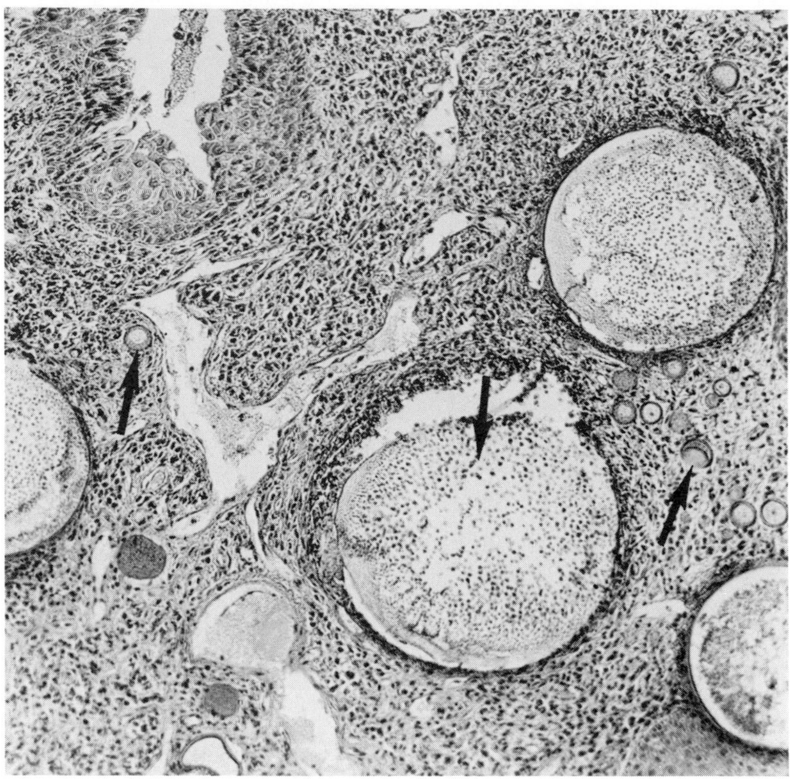

FIG. 31-11. *Coccidioides immitis,* spherules in tissue microsection; reproduction by endosporulation. The large balloonlike structures contain myriad endospores that they release into tissue spaces. A doubly refractile capsule is seen about varying-sized spherules.

blastomycosis. In other parts of the body they resemble the lesions of tuberculosis. The mortality is high. This is the form of the disease designated *coccidioidal granuloma.*

Coccidioidomycosis is endemic in the desert valleys of California and the dry dusty areas of Southwestern United States; especially in Arizona, New Mexico, Texas, and parts of Mexico. It is most likely that man and animals are infected by the inhalation of spore-bearing dust. The infection is found in cattle, sheep, dogs, and certain wild rodents. There may be a reservoir of infection in small wild rodents, and these animals could pass the spores in their feces to contaminate the soil, after which the spores would spread by the wind. The rather frequent dust storms of the Southwest might carry these organisms long distances.

The cause is *Coccidioides immitis.* Diagnosis is made by finding the characteristic fungi in the disease. The appearance of the fungi in the lesions differs from its appearance on culture media. In the body tissues and exudates one sees yeastlike forms, large thick-walled, nonbudding spherules, 20μ to 60μ in diameter, filled with endospores, 2μ to 5μ in diameter (reproduction by endosporulation) (Fig. 31-11). As many as 1000 endospores may be released from a single spherule. In laboratory cultures growth is that of a mold, and one sees a fluffy, cottony white colony. The *coccidioidin test,* a test of skin sensitivity to an extract of the organism, is of value. (Cross reactions may occur with the skin tests for histoplasmosis and blastomycosis, however.) Immunofluorescence can be used to identify the infection as can precipitation, latex agglutination, complement-fixation, and quantitative immunodiffusion. These serologic tests are used to follow the course of the infection in a given patient. Currently a coccidioidal vaccine made from formalin-inactivated spherules has been successful in trial runs on human beings.

Cryptococcosis (torulosis)

Cryptococcus neoformans (Torula histolytica) is a yeastlike organism that usually infects the lungs and central nervous system but may attack other parts of the body. *Cryptococcus* is the only encapsulated yeast to invade the central nervous system. In the disease multiple small nodules form, with the gross and microscopic appearance of tubercles. In infection of the central nervous system, the meninges are thickened and matted together, and the brain is invaded. Man becomes infected through the skin, mouth, nose, and throat. Transmission from man to man or from animal to man has not been recorded. This fungus, saprophytic in nature, has been found in cattle, horses, dogs, and cats. Birds are not its hosts, but the organism is a saprophyte in pigeon droppings, and cases of cryptococcal meningitis have been traced to the vast pigeon populations found in many large cities. Pigeons are mechanical vectors; they carry the organisms on their feet and beaks. Pigeons are not affected probably because of their high body temperature. *Cryptococcus neoformans* has a definite predilection for pigeon droppings, which are rich in creatinine. Creatinine is assimilated by this organism only and not by other species of cryptococci or other fungi.

Since *Cryptococcus neoformans* is known also as *Torula histolytica,* infections with it may be referred to as either *cryptococcosis* or *torulosis.* Torulosis or cryptococcosis can be diagnosed with certainty only by finding the budding organisms in the affected tissues, pus, sputum, or cerebrospinal fluid (Fig. 31-12). In wet mounts the cryptococci are ovoid to spherical, budding yeast forms 5μ to 15μ in diameter. With fluorescent antibody technics, the diagnosis may be made within hours. Agglutination tests are sensitive and specific.

Histoplasmosis

Histoplasmosis, sometimes called *Darling's disease,* is an infection caused by *Histoplasma capsulatum,* a diphasic organism—a single budding yeast at body temperature and a mold at room temperature and in nature. The fungus attacks primarily the reticuloendothelial system, by parasitizing the component cells. Like coccidioidomycosis, the disease exists in the primary and the progressive forms. The primary form involves the lungs but usually heals, leaving many small calcified foci in the lungs and lymph nodes of the chest. In the progressive disseminated form, ulcerating lesions are found in the nose and mouth, and there is enlargement of the spleen, liver, and lymph nodes. The progressive form is generally fatal.

Man probably contracts the disease by inhalation of spores from fungi growing in the soil. Infection of the soil comes from the excreta of a variety of bats in which the microbes have been found. No intermediate host is identified. Spores may be carried by prevailing winds and even by tornadoes. Outbreaks of the disease have been traced to the inhalation of dust from caves. Histoplasmosis is referred to as cave sickness or *speleonosis.* Victims of histoplasmosis and blastomycosis as well tend to be outdoor

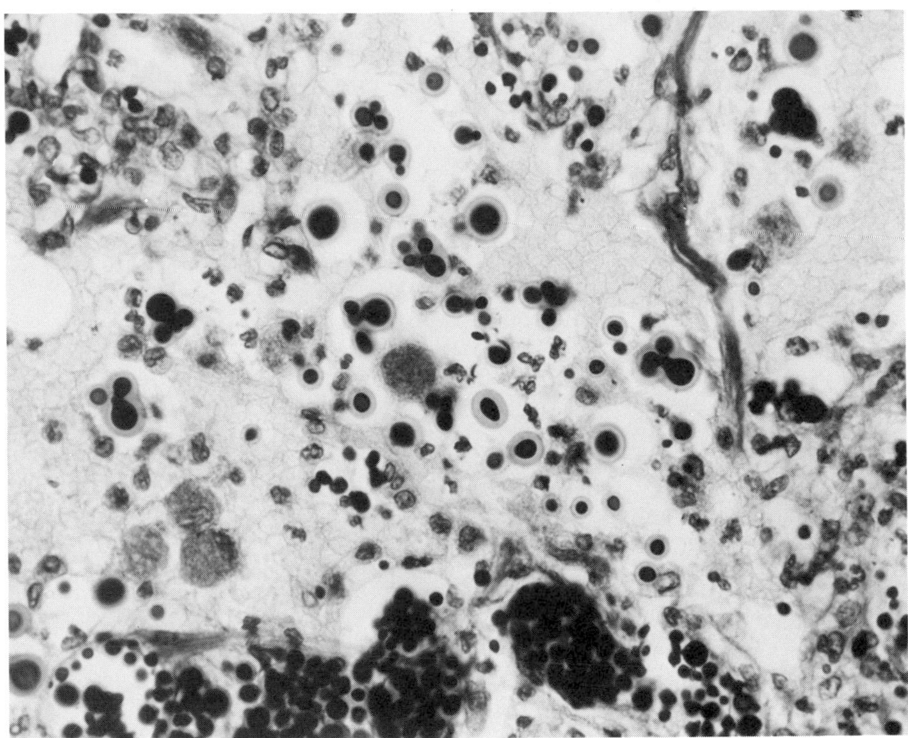

FIG. 31-12. *Cryptococcus neoformans (Torula histolytica),* microscopic appearance of budding yeast forms among partially hemolyzed red cells in air sacs of lung. Wide gray capsules encompass the microbes in this specially stained preparation. (From Kent, T. H., and Layton, J. M.: Amer. J. Clin. Path. **38:**596, 1962.)

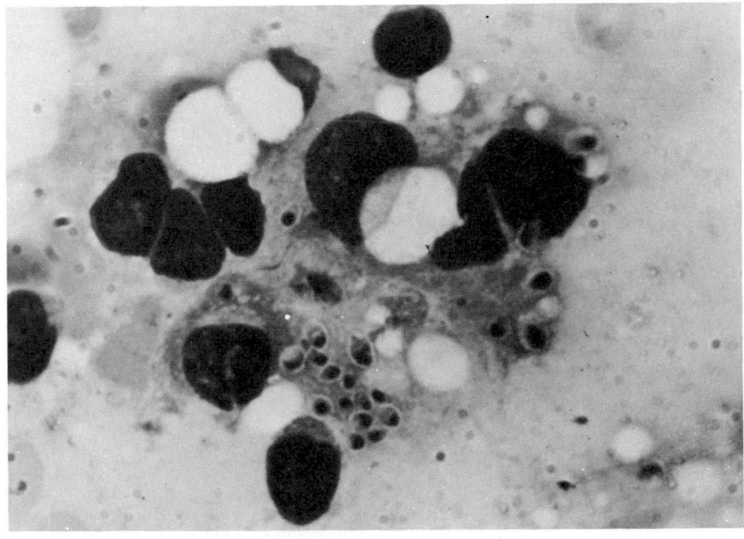

FIG. 31-13. Histoplasmosis of bone marrow, stained smear examined microscopically. (Wright's stain; ×950.) (From Anderson, W. A. D.: Pathology, ed. 6, St. Louis, 1971, The C. V. Mosby Co.)

types—construction workers, farmers, spelunkers, and so on. The disease is encountered in the Central Mississippi Valley and the Ohio Valley in the United States, and in widespread areas of the world.

To identify the organisms, stained smears and imprints (as well as cultures) are made of peripheral blood, bone marrow, aspirated material from lymph nodes, and sputum (Fig. 31-13). To aid in the diagnosis and follow-up of histoplasmosis, the laboratory offers a skin test (the histoplasmin test or the histoplasmin tine test) and five serologic tests—precipitation, agglutination, complement-fixation, fluorescent antibody detection, and immunodiffusion.

Phycomycosis

Phycomycosis or *mucormycosis* can be an overwhelmingly acute and fatal infection. It is caused by species of *Mucor* and *Rhizopus* and other normally harmless phycomycetes of the soil and decaying organic matter.

Diabetes, untreated and out of control, is the most important forerunner; the ketoacidosis rather than the hyperglycemia is thought to trigger the process. Along with disease in the lungs and central nervous system, an intraorbital cellulitis is a prominent feature. In the tissues, the hyphae abound, very broad and branching, especially within walls and lumens of blood vessels. No spores are seen, and there is little if any inflammation. In tissue section, hyphae are easily identified as belonging to the phycomycetes because they are nonseptate and coenocytic, that is, contain many nuclei within a continuous mass of cytoplasm. The term *mucormycosis* is often used simply to indicate infections in which such hyphae are seen.

Sporotrichosis

Sporotrichosis is a fungous disease caused by *Sporotrichum schenckii.* It may affect man, lower animals, or plants. *Sporotrichum* is widely distributed in nature as a saprophyte on vegetation. Man usually acquires the infection from plants, especially barberry shrubs and certain mosses that seem to harbor the fungi. The fungi are introduced into wounds (inoculation infection) by infected plants or vegetable matter. The agent is thought to be a normal inhabitant of the alimentary and respiratory tracts in man, and in a few instances the disease has been transferred from man to man. Transmission of the infection from lower animals to man by bites or indirect routes has been noted. Animals most often affected are horses, mules, dogs, rats, and mice. The majority of cases occur in the United States, especially in the Missouri and Mississippi River valleys.

The disease presents as a chronic infection usually limited to the skin and underlying tissues and is accompanied by the formation of nodular masses that slowly undergo softening and ulceration. In typical cases the first evidence of the disease is seen at the site of some trivial injury, usually on the fingers. The wound does not heal, and an ulcer appears, followed by nodular swellings in chain formation up the forearm. The disease seldom extends farther than to the regional lymph nodes, but secondary foci sometimes crop up in other parts of the body such as the lungs, spleen, liver, and other organs.

The oval or cigar-shaped organism of sporotrichosis, resident within mononuclear cells, is very rarely found in smears from the pus of a skin lesion or in sections of tissue taken from the lesions. As a rule, it is demonstrated only in culture (Fig. 31-14). Fluorescent antibody technics detect the microorganisms in exudates from the lesions.

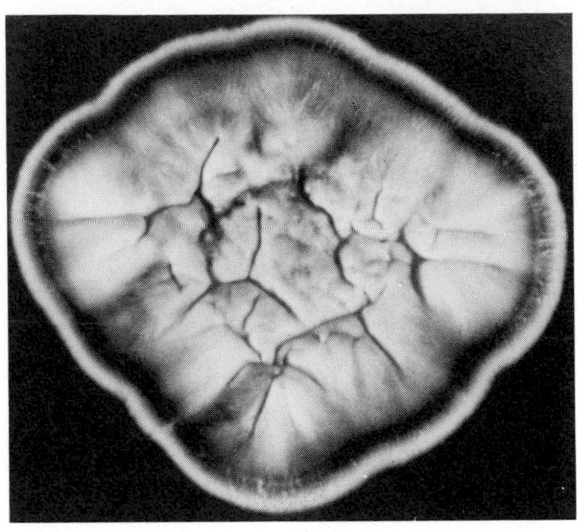

FIG. 31-14. *Sporotrichum schenckii,* giant colony. (Courtesy Dr. R. H. Musgnug, Haddonfield, N. J.)

Other fungous diseases of man

Other fungous infections, infrequent in man, are the following:

1. Rhinosporidiosis (caused by *Rhinosporidium seeberi*)
2. Geotrichosis (caused by one or more species of *Geotrichum*)
3. Chromoblastomycosis (caused by three different fungi)
4. Penicilliosis (caused by certain species of *Penicillium*)
5. Piedra (caused by two different species of fungi)
6. Otomycosis (caused by species of *Penicillium* and other fungi)
7. Erythrasma (caused by *Nocardia minutissima*)

Fungous diseases of lower animals

Fungous infections are very important to the veterinary microbiologist. The following are some of the common ones:

1. Ringworm of the horse (caused by *Trichophyton equinum*); may occur in man
2. Ringworm of horses, cattle, dogs, and possibly sheep and hogs (caused by *Trichophyton mentagrophytes*); infection possible in man
3. Ringworm of cats and dogs (caused by *Microsporum canis*)
4. Epizootic lymphangitis of horses (caused by *Blastomyces farciminosus*)
5. Favus of chickens (caused by *Trichophyton gallinae*)
6. Coccidioidomycosis of cattle, dogs, horse, and sheep (caused by *Coccidioides immitis*); same organism causes the disease in man
7. Aspergillosis of wild and domestic fowls (caused by *Aspergillus fumigatus*)
8. Candidiasis of poultry (caused by *Candida albicans*)
9. Histoplasmosis of dogs, cats, cattle, sheep, swine, poultry, and horses (caused by *Histoplasma capsulatum*)
10. Cryptococcosis (mastitis) of cattle (caused by *Cryptococcus neoformans*)

Fungous diseases of plants

Molds and yeasts are economically important for they cause so many diseases of plants. Plants and vegetables imported from other countries are rigidly inspected upon arrival in our seaports and points of entry. If a fungous infection is found, the plant or plant product is not allowed entry. Among important plant diseases caused by molds are:

1. Brown rot of peaches and plums
2. Chestnut blight
3. Mildew of grapes
4. White pine blister rust
5. Rust of oats, wheat, and barley
6. Smuts of various grains
7. Potato rot
8. Corn leaf blight

Ergotism

Ergot, a drug whose derivatives are widely used to check hemorrhage after childbirth, is composed of several alkaloidal poisons (mycotoxins) produced by the growth of a mold *Claviceps purpurea* in the grains of rye, wheat, and barley. The fungus (also referred to as ergot) grows as a purple-black, slightly curved mass that replaces the infected grain and converts it to a black sclerotium, from which the drug is extracted. An enzyme secreted by the hyphae is contained in a thick honeydew that attracts insects and helps to spread the fungus. When bread made from infected grain is eaten, the condition known as *ergotism* develops. Ergotism is characterized by gangrene of the extremities, abortion, and convulsions. It was at one time most prevalent in Central Europe.

A specific component of the pharmacologically potent alkaloids produced by the fungus ergot is lysergic acid. A well-known derivative is the hallucinogen lysergic acid diethylamide, or LSD.

DISEASES CAUSED BY ACTINOMYCETES

Actinomycosis

Actinomycosis is an infectious disease of lower animals (especially cattle) and man caused by several species of fungi belonging to the genus *Actinomyces*. Of these, *Actinomyces bovis* and *Actinomyces israelii* are the most important. Actinomycosis is typified clinically by the formation of nodular swellings that soften and form abscesses, discharging a thin pus through multiple sinuses. The disease in man is in three forms: (1) the cervicofacial, (2) the thoracic, and (3) the abdominal. The cervicofacial type is described by swelling and suppuration of the soft tissues of the face, jaw, and neck. It is the usual (about half the cases) and the least dangerous type. It appears to have a special association with dental defects. The thoracic type is characterized by multiple small cavities and abscesses in the lungs. The abdominal type usually begins about the appendix or cecum. In advanced thoracic and abdominal disease, sinus tracts extend to the surface. The disease in cattle is known as lumpy jaw.

The causative fungus is found in the pus or in the walls of the abscesses as small yellow granules about the size of a pinhead, known as *sulfur granules* (Fig. 31-15). When a sulfur granule is placed on a slide and a cover glass is pressed down upon it, a characteristic microscopic picture is seen—namely, a central threadlike mass of fungus from which radiate many clublike structures. For this reason *Actinomyces* are often spoken of as *ray fungi*. The clubbed appearance is not very pronounced in cul-

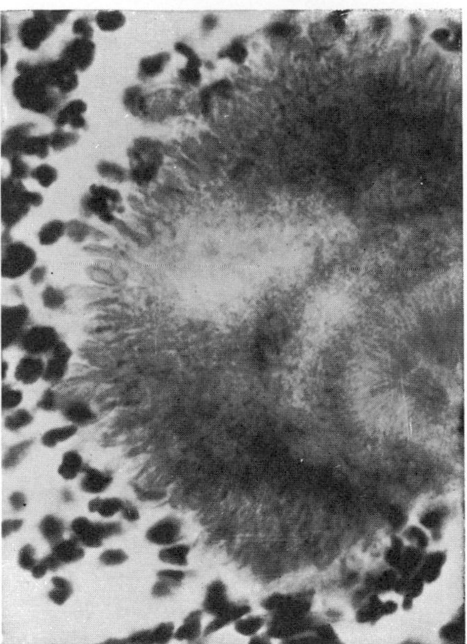

FIG. 31-15. *Actinomyces*. Sulfur granule (colony) in a microscopic field of pus. Club-shaped processes are seen at periphery. (From Bauer, J. D., Ackermann, P. G., and Toro, G.; Bray's clinical laboratory methods, ed. 7, St. Louis, 1968, The C. V. Mosby Co.)

tures. The laboratory diagnosis of actinomycosis is made by finding objects in the discharges that are both grossly and microscopically sulfur granules or by demonstrating the organisms in sections of tissue taken from the lesions. Cultural methods help, but serologic and skin tests contribute little.

Pathogenic actinomycetes are normal inhabitants of the mouth, probably existing in an attenuated state. Infection occurs in the event of injury to the mouth, with tooth decay, or with some other abnormal state. Such conditions favor invasion of the tissues. There is no evidence of direct transmission from animal to animal or from animal to man. There is also no evidence that the organism leads a saprophytic existence outside the body. People who pursue nonagricultural occupations are as likely to contract the disease as are farmers and stockman.

Nocardiosis

Nocardia, an aerobic actinomycete found free in nature, is responsible for a variety of infections in man. It is one cause of *mycetoma,* or "Madura foot," a localized involvement usually of the foot (Fig. 31-16). After initial injury in this condition, the organisms invade to produce in time a network of interlocking abscesses and granulomas in the soft tissues and bones of the foot. The organisms of actinomycotic mycetoma (or maduromycotic mycetoma, if certain other fungi are operative) can be identified in the purulent discharge from the draining sinus tracts.

Nocardia may be responsible for infections of the lungs and other internal organs. Brain abscess commonly complicates systemic infection. Nocardiosis (like cryptococosis), with an affinity for the lungs, accompanies malignant diseases of the reticuloendothelial system—Hodgkin's disease, lymphosarcoma, and leukemia. Steroid hormones enhance the infection.

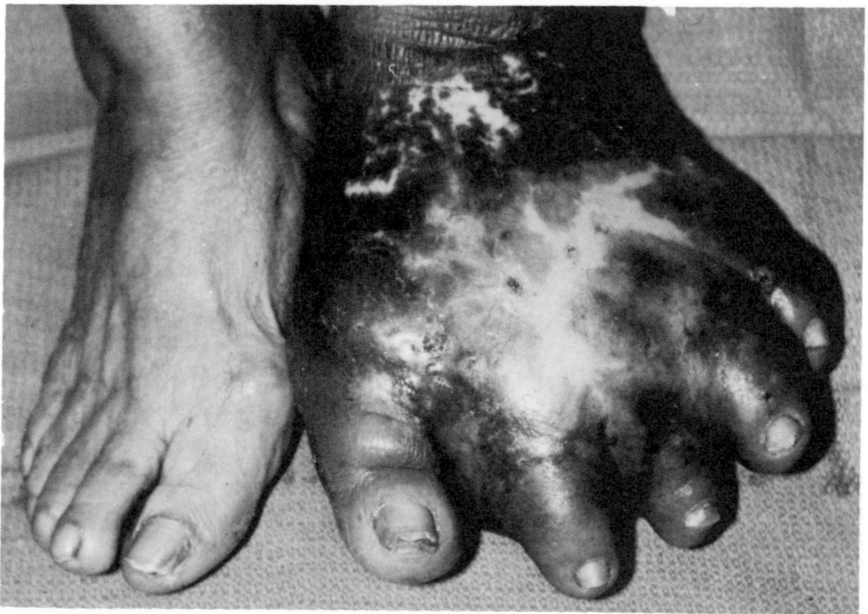

FIG. 31-16. Madura foot.

Rat-bite fever

There are two diseases known as rat-bite fever. One is caused by *Spirillum minus* (p. 391) and the other by the actinomycete-like organism *Streptobacillus moniliformis.* This is an inhabitant of the nasopharynx and mouth of normal rats, among which it causes widespread epidemics. Man contracts the infection from the bite of a rat. The disease is a febrile one with manifestations similar to those of rat-bite fever caused by *Spirillum minus.* Both resemble tularemia clinically. A milk-borne epidemic of this disease occurred in Haverhill, Massachusetts, in 1926. This gave origin to the name Haverhill fever.

QUESTIONS FOR REVIEW

1. Give the general characteristics of fungi.
2. Name the science that treats of fungi.
3. What is a fungous infection called?
4. Make the distinction between molds and yeasts. Is it a clear one?
5. How do fungi perpetuate themselves?
6. What are the dermatomycoses? Name three major dermatophytes.
7. Outline the laboratory diagnosis of fungous disease.
8. What are actinomycetes? Their importance?
9. Name and briefly describe the chief diseases caused by fungi.
10. What is ergotism? What is the pharmacologic nature of LSD?
11. Classify fungi.
12. Make pertinent comments regarding the pathology of fungous disease. What threat does fungous infection pose to the "compromised host"? Cite the common offenders.
13. Briefly define what is meant by: speleonosis, torulosis, sulfur granules, mycotoxin, Madura foot, tinea, superinfection, complicating infection (in fungous disease), mycelium, aflatoxins, favus, otomycosis, thallus, cave sickness.

REFERENCES FOR CHAPTERS 30 AND 31

Ashley, M. J., and Wigle, W. D.: Epidemiology of active tuberculosis in hospital employees in Ontario, 1966-1969, Amer. Rev. Resp. Dis. **104:**851, 1971.
Baker, R. D. (editor), et al.: Human infection with fungi, actinomycetes, and algae, New York, 1971, Springer Publishing Co.
B. C. G. and the tuberculin test, Lancet **1:**192, (Jan. 25) 1969.
Beware, the Sporothrix, editorial, J.A.M.A. **215:**1976, 1971.
Branson, D.: Timely topics in microbiology, mycobacteria, 1968-1971, Amer. J. Med. Tech. **38:**13, 1972.
Candidiasis: colonization vs. infection, editorial, J.A.M.A. **215:**285, 1971.
Caplan, R. M.: Medical uses of the Wood's lamp, J.A.M.A. **202:**1035, 1967.
Carper, J.: Exposing "histo"—disease in disguise, Today's Health **47:**30, (May) 1969.
Chapman, J. S.: The atypical mycobacteria, Amer. J. Nurs. **67:**1031, 1967.
Chapman, J. S.: Ecology of atypical mycobacteria, Arch. Environ. Health **22:**41, 1971.
Conant, N. F., et al.: Manual of clinical mycology, Philadelphia, 1971, W. B. Saunders Co.
Diagnostic standards and classification of tuberculosis, New York, 1969, National Tuberculosis and Respiratory Disease Association.
Dubos, R. J., and Hirsch, J. G., editors: Bacterial and mycotic infections of man, Philadelphia, 1965, J. B. Lippincott Co.
Dull, H. B., Herring, L. L., Calafiore, D., et al.: Jet injector tuberculin skin testing: a comparative evaluation, Amer. Rev. Resp. Dis. **97:**38, 1968.
Emmons, C. W., Binford, C. H., and Utz, J. P.: Medical mycology, Philadelphia, 1970, Lea & Febiger.
Field, M. H.: Opportunistic fungal infections, J.A.M.A. **23:**529, 1968.
Freedman, S. O.: Tuberculin testing and screening: a critical evaluation, Hosp. Pract. **7:**63, (May) 1972.
Goldblatt, L. A.: Aflatoxin, scientific background, control, and implications, New York, 1969, Academic Press, Inc.
Hand, W. L., and Sanford, J. P.: *Mycobacterium fortuitum*—a human pathogen, Ann. Intern. Med. **73:**971, 1970.
Hatcher, C. R., Jr., Sehdeva, J., Waters, W. C., 3d, et al.: Primary pulmonary cryptococcosis, J. Thorac. Cardiovasc. Surg. **61:**39, (Jan.) 1971.
Hughes, A. E., Bertonneau, D., and Enna, C. D.: Nurses at Carville, Amer. J. Nurs. **68:**2564, (Dec.) 1968.
Hull, F. E.: Tuberculin tine test, J.A.M.A. **203:**562, 1968.
Johnston, W. W.: The cytopathology of mycotic infections, Lab. Med. **2:**34, (Sept.) 1971.
Kelser, R. A., and Schoening, H. W.: Manual of veterinary bacteriology, Baltimore, 1969, The Williams & Wilkins Co.
Kepron, M. W., et al.: North American blastomycosis in central Canada: Review of 36 cases, Canad. Med. Ass. J. **106:**243, 1972.
Kinnear-Brown, J. A., et al.: BCG vaccination of children against leprosy in Uganda: Results at end of second follow-up, Brit. Med. J. **1:**24, (Jan. 6) 1968.
Luby, J. P., et al.: Jet injector tuberculin skin testing: A comparative evaluation, Amer. Rev. Resp. Dis. **97:**46, 1968.
Maddy, K. T.: Epidemiology and ecology of deep mycoses of man and animals, Arch. Derm. **96:**409, 1967.
Mansfield, R. E., Storkan, M. A., and Cliff, I. S.: Evaluation of the earlobe in leprosy, Arch. Derm. **100:**407, 1969.
Mayer, J.: Nutrition and tuberculosis, Postgrad. Med. **50:**53, (Dec.) 1971.
McClement, J. H.: The tuberculin test, editorial, Hosp. Pract. **7:**11, (May) 1972.
Meyers, W. M.: Leprosy—some immuno-pathologic considerations, Int. Path. **9:**57, (July) 1968.
Moore-Landecker, E.: Fundamentals of the fungi, Englewood Cliffs, N.J., 1972, Prentice-Hall, Inc.
Moss, E. S., and McQuown, A. L.: Atlas of medical mycology, Baltimore, 1969, The Williams & Wilkins Co.
Orr, E. R., and Riley, H. D., Jr.: Sporotrichosis in childhood: Report of 10 cases, J. Pediat. **78:**951, 1971.
Parkhurst, G. F., and Vlahides, G. D.: Fatal opportunistic fungus disease, J.A.M.A. **202:**279, 1967.
Perkins, J. E.: Airborne infection and tuberculosis, Arch. Environ. Health **16:**738, 1968.
Richards, R. N., and Talpash, O. S.: Sporotrichosis, Canad. Med. Ass. J. **106:**1097, 1972.
Richter, H. S.: Coccidioidomycosis, a report of 300 new cases, GP **39:**89, (Feb.) 1969.

Rosenthal, S. R., Nikurs, L., Yorky, E., et al.: Tuberculin tine test in infection and disease, Southern Med. J. **60:**1336, 1967.

Runyon, E. H.: Whence mycobacteria and mycobacterioses? editorial, Ann. Intern. Med. **75:**467, 1971.

Schaefer, W. B.: Incidence of the serotypes of *M. avium* and atypical mycobacteria in human and animal diseases, Amer. Rev. Resp. Dis. **97:**18, 1968.

Smith, D. T.: Isoniazid prophylaxis and BCG vaccination in control of tuberculosis, Arch. Environ. Health **23:**235, 1971.

Smith, J. W.: Coccidioidomycosis, a review, Texas Med. **67:**117, (Nov.) 1971.

Smith, J. W.: Leprosy, Texas Med. **67:**58, (July) 1971.

Smith, J. W., and Utz, J. P.: Progressive disseminated histoplasmosis, Ann. Intern. Med. **76:**557, 1972.

Snell, W. H., and Dick, E. A.: A glossary of mycology, Cambridge, Mass., 1971, Harvard University Press.

Sutaria, M. K., et al.: Focalized pulmonary histoplasmosis (coin lesion), Chest **61:**361, 1972.

Taschdjian, C. L., et al.: Serodiagnosis of candidal infections, Amer. J. Clin. Path. **57:**195, 1972.

Tempel, C. W.: Tuberculosis prevention; the child-centered program, GP **37:**99 (May) 1968.

Walter, J. E., and Coffee, E. G.: Control of *Cryptococcus neoformans* in pigeon coops by alkalinization, Amer. J. Epidem. **87:**173, (Jan.) 1968.

Weg, J. G.: Tuberculosis and the generation gap, Amer. J. Nurs. **71:**495, 1971.

Werner, S. B., et al.: Epidemic of coccidioidomycosis among archeology students in northern California, New Eng. J. Med. **286:**507, 1972.

Wiegand, S.: Ink blue agar for recognition of dermatophytes, Bull. Path. **10:**68, 1968.

32 Protozoa: medical parasitology

Parasites are generally defined as organisms that require living matter for their nourishment; that is, they must live within or on the bodies of other living organisms. According to this definition, a parasite may be a bacterium, virus, rickettsia, protozoon, a larger plant (example, mistletoe) or an animal. However, by common usage, *medical parasitology* refers to animal parasites of medical interest and their diseases.

The animal within or on which a parasite lives is the *host.* All stages of the parasite's development may take place in the same animal host. On the other hand, a parasite may have one or more hosts. It undergoes its larval stage in the *intermediate* host and its adult stage in the *definitive* host. A parasite that lives within the body of the host is known as an *endoparasite;* one that lives on the outside of the body is an *ectoparasite.* A tapeworm is an example of an endoparasite; a louse is an example of an ectoparasite.

PROTOZOA

General characteristics

The animal kingdom is divided into two great divisions, the *Protozoa,* unicellular organisms and the lowest form of animal life, and the *Metazoa,* multicellular organisms (see Chapter 33). Protozoa occupy the same relative position in the animal kingdom that bacteria do in the plant kingdom. They are more complex in their functional activities than bacteria or the average cell of a multicellular organism. Each is a complete unit within itself with special structures known as *organelles* to carry out such functions as nutrition, locomotion, respiration, excretion, and attachment to objects. The majority are of microscopic size. As a rule, the pathogenic ones are smaller than the nonpathogenic ones. They may be spherical, spindle, spiral, or cup shaped. In medical parasitology, identification of a given animal parasite is of paramount importance. Practically speaking, this is largely the recognition of its own specific structural (morphologic) features. There are many species of protozoa, but only about 30 affect man.

■ **Structure.** Protozoa are units of protoplasm differentiated into cytoplasm circumscribed by the cell or plasma membrane and a nucleus encased by the nuclear membrane. Some have more than one nucleus. The cytoplasm is separated into a homogeneous *ectoplasm* and a granular *endoplasm.* The ectoplasm helps form the various organs of locomotion, contraction, and prehension, such as pseudopods, flagella, cilia, and suctorial tubes. In certain species of protozoa the ectoplasm contains a definite opening or portal for intake of food. The endoplasm digests food materials and surrounds the nucleus.

Many protozoa, especially the pathogenic ones, absorb fluid directly through the plasma membrane. The majority take in solid particles, such as small animal or

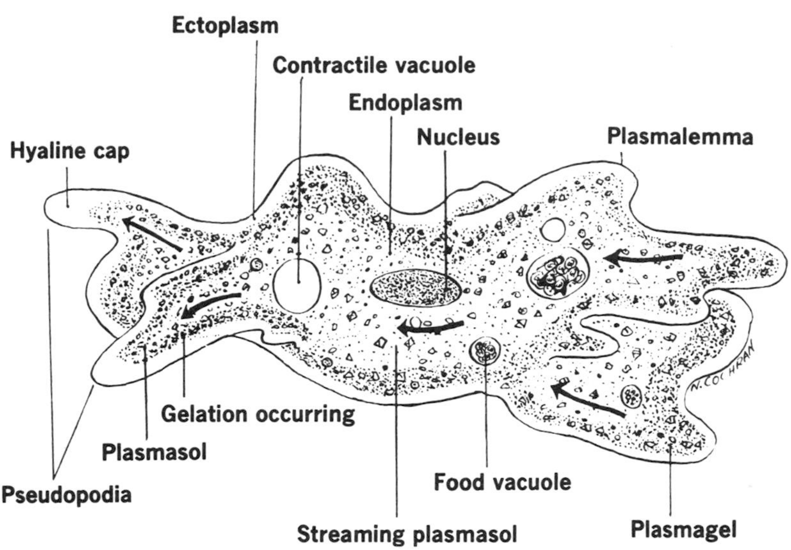

FIG. 32-1. Active locomotion sketched in an ameba. Note direction of arrows. (From Hickman, C. P.: Integrated principles of zoology, ed. 4, St. Louis, 1970, The C. V. Mosby Co.)

vegetable organisms, and digest them enzymatically. Because their food consists chiefly of bacteria, protozoa may be important in limiting the bacterial population of the universe. Waste material is excreted through the cell membrane or, in some cases, through an ejection pore.

■ **Locomotion.** All protozoa possess some type of motility. It may be by pseudopod formation (see Fig. 32-1) or by the action of flagella or cilia. For locomotion by *pseudopod* (false foot) formation, a sharp or blunt ectoplasmic process flows forward, pulling the rest of the organism after it. *Flagella* are whiplike prolongations of protoplasm that propel the organism by their lashing motions. Some protozoa have only one flagellum; others have several. Some of the flagellate protozoa also have an *undulating membrane* to help in locomotion. This is a fluted membranous process attached to one side of the organism. *Cilia* are similar to flagella except that they are shorter, more delicate, more plentiful, and are attached to the entire outer surface of the microbe. Individually they are less powerful than flagella, but the synchronous action of the many cilia accomplishes the most rapid motion of which unicellular organisms are capable.

■ **Cyst formation.** When protozoa are subjected to adverse conditions, they become inactive, assume a more or less rounded form, and surround themselves with a resistant membrane (cell wall) within which they may live for a long time and resist various destructive agents in their environment. This is *cyst* formation. When conditions suitable for growth are reestablished, the cyst imbibes water, and the protozoan returns to the vegetative state. Sometimes cyst formation precedes reproduction. Since vegetative protozoa are very susceptible to deleterious influences and cysts are very resistant, it is the cysts that are usually responsible for the spread of protozoan infections.

■ **Reproduction.** In protozoa, reproduction may be either sexual or asexual. In some (example, *Plasmodium* of malaria) the sexual cycle occurs in one species of animal and the asexual cycle in another. The sexual cycle occurs in the *definitive* host; the asexual

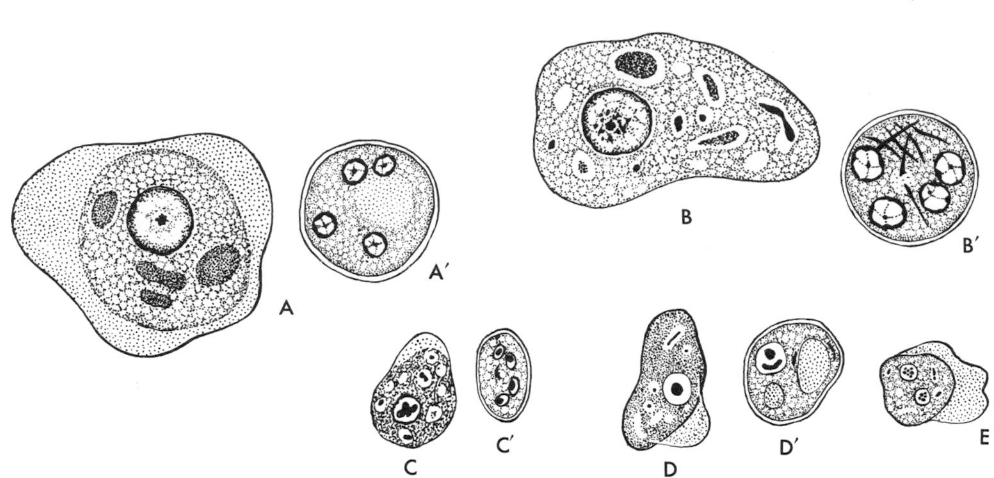

FIG. 32-2. Protozoa—amebas, pathogenic and nonpathogenic, in sketch. Note rounded cyst to right of trophozoite. **A-A′**, *Entamoeba histolytica* (the pathogen). **B-B′**, *Entamoeba coli.* **C-C′** *Endolimax nana.* **D-D′**, *Iodamoeba bütschlii (williamsi).* **E,** *Dientamoeba fragilis.*

cycle occurs in the *intermediate* host. Protozoan cells capable of sexual reproduction are known as *gametes.* The cell formed by the union of two gametes is a *zygote.* Asexual reproduction occurs in amebas and flagellates. Lengthwise or crosswise division of the protozoon yields two new members of the species.

■ **Classification.*** In the phylum Protozoa there are six classes of organisms of medical interest in man. The method of locomotion varies in each.

1. *Rhizopodea.* Locomotion is characterized by pseudopod formation. The cytoplasm is divided into ectoplasm and endoplasm. This class includes the pathogenic and nonpathogenic amebas (Fig. 32-2).
2. *Zoomastigophorea* (commonly called flagellates). Movement is by means of flagella and an undulating membrane. Flagellates have two nuclei, and the cytoplasm is not differentiated into endoplasm and ectoplasm. Cell bodies are often pear shaped and fixed in outline. The most important flagellates medically are in the genera *Trypanosoma, Leishmania, Trichomonas, Giardia,* and *Chilomastix* (Fig. 32-3, *A* to *C*).
3. *Telosporea.* There are no external organs of locomotion. These organisms live within the cells, tissues, cavities, and fluids of the body and are represented by the *Plasmodium* of malaria.
4. *Ciliatea.* Cilia are present for locomotion. The only pathogenic member of this group is *Balantidium coli* (Fig. 32-3, *D*).
5. *Toxoplasmea.* There are no external organs of locomotion. The protozoa move by bending and gliding movements of their bodies. The representative pathogen here is *Toxoplasma gondii.*
6. *Haplosporea.* There are no flagella but pseudopodia may form. The parasite here is *Pneumocystis carinii.*

---

*The classification is taken from Faust, E. C., Russell, P. F., and Jung, R. C.: Craig and Faust's clinical parasitology, ed. 7, Philadelphia, 1970, Lea & Febiger.

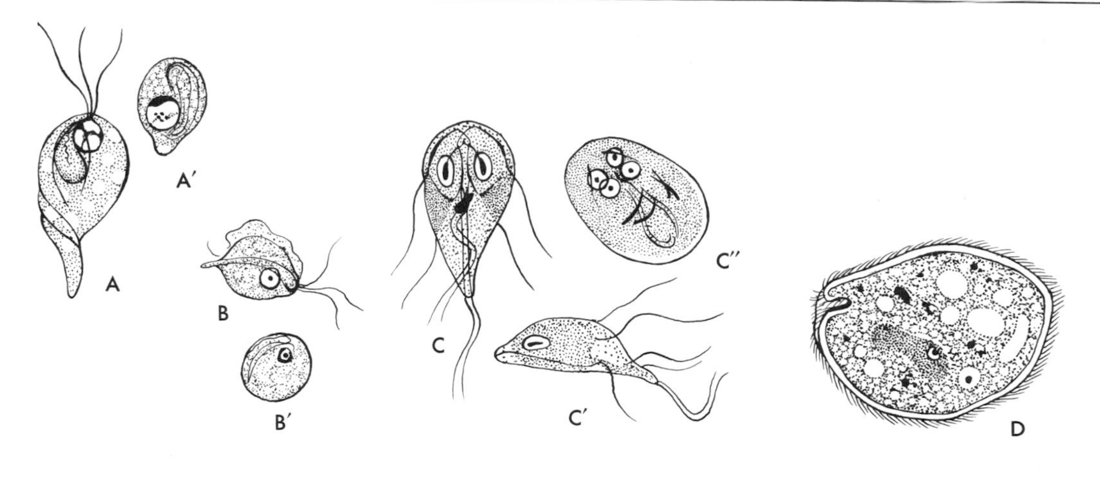

FIG. 32-3. Protozoa—flagellates and ciliate, in sketch. Note rounded or ovoid cyst with trophozoite. **A-A′,** *Chilomastix mesnili.* **B-B′,** *Trichomonas hominis (intestinalis).* **C** to **C″,** *Giardia lamblia.* **D,** *Balantidium coli.*

Protozoan diseases

Amebiasis

The term "amebiasis" indicates an infection with *Entamoeba histolytica.* The disease occurs in two forms: acute amebiasis (amebic dysentery), characterized by an intense dysentery with bloody, mucus-filled stools; and chronic or latent amebiasis, described by vague intestinal disturbances, muscular aching, loss of weight, even constipation. In some cases of chronic amebiasis, manifestations are absent. The chronic form is more common than the acute. An estimated 5% to 10% of persons in the United States is affected.

■ **The organism.** The organism *Entamoeba histolytica* exists as a vegetative ameba, or *trophozoite,* and as a *cyst.* Vegetative trophozoites possess an active type of ameboid motion on a warm microscopic stage. Microscopically one sees the pseudopods, a distinctive nucleus, and red blood cells within the cytoplasm of the trophozoite. Vegetative amebas are very susceptible to injurious agents. In an unfavorable environment they quickly succumb, therefore they do little to transmit the disease. The cysts are smaller than vegetative amebas, nonmotile, and surrounded by a resistant wall.

■ **Life history.** The life cycle of *Entamoeba histolytica* begins with the cysts by which the disease is transmitted from person to person. After the cysts are passed in the feces, they remain infectious for several days, if not destroyed by heat and drying. When the cysts are swallowed by a new host, they pass through the stomach unchanged. The shells are dissolved by juices of the small intestine, and the vegetative forms are liberated. The trophozoites pass to the large intestine to attack the mucous membrane and produce ulceration (Fig. 32-4). The vegetative amebas multiply in the ulcers; some escape into the lumen of the intestine. If diarrhea is present, they are swept out of the intestinal tract. If diarrhea is not present, they multiply one or more times and then encyst. Encystment does not occur outside the body. Cysts are excreted in the feces.

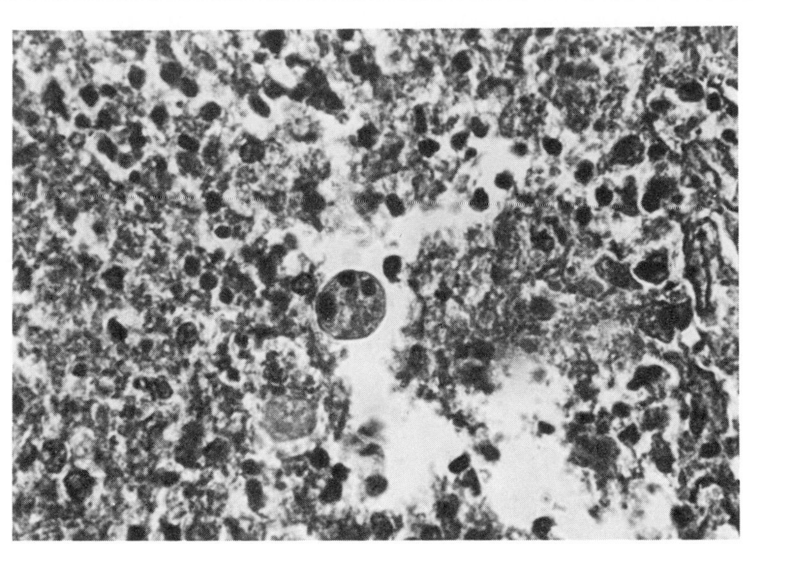

FIG. 32-4. *Entamoeba histolytica* in intestinal ulcer, photomicrograph of tissue section. Note zone of lysis about circular trophozoite in the center of the exudate.

■ **Sources and modes of infection.** The life cycle of *Entamoeba histolytica* reveals three facts: (1) Infection can be acquired only by swallowing cysts, (2) infection comes from the feces of a person excreting cysts, and (3) acute cases are of little danger. The feces of patients with acute amebiasis contain largely vegetative parasites that die quickly; they could not survive the acid gastric juice should they accidentally be ingested. Infection is usually acquired by eating uncooked food contaminated with feces containing cysts. The most important single source of infection is the food handler with chronic amebiasis, especially the one preparing uncooked foods. Other sources of infection are vegetables fertilized with human excreta and drinking water contaminated with sewage. Apparently the latter was the cause of the Chicago epidemic of 1933 in which there were 1409 cases with 98 deaths. The water in two hotels had been contaminated by sewage. Flies and other insects may spread the cysts mechanically.

■ **Lesions.** In the majority of cases there seems to be a state of balance between the amebas and the host. The patient experiences mild disturbances or none at all and is able to repair the ulcers almost as fast as they are formed. This is chronic or latent amebiasis.

If the resistance of the host is lowered or massive infection occurs, the host is unable to repair the ulcers as fast as they are formed, and the increasing ulceration causes a violent dysentery in which the stools may consist entirely of blood and mucus. This is acute amebiasis or amebic dysentery. Occasionally, intestinal perforation occurs.

Sometimes amebas penetrate deeper into the intestinal wall and enter tributaries of the portal vein to be carried to the liver, where they produce amebic hepatitis or liver abscess. Amebic abscesses can occur in the lungs or brain (Figs. 32-5 and 32-6).

■ **Laboratory diagnosis.** The laboratory diagnosis of amebiasis necessitates the examination of *fresh warm* stools for vegetative amebas and the examination of ordinary specimens for cysts. The examination of iron hematoxylin–stained smears of specimens

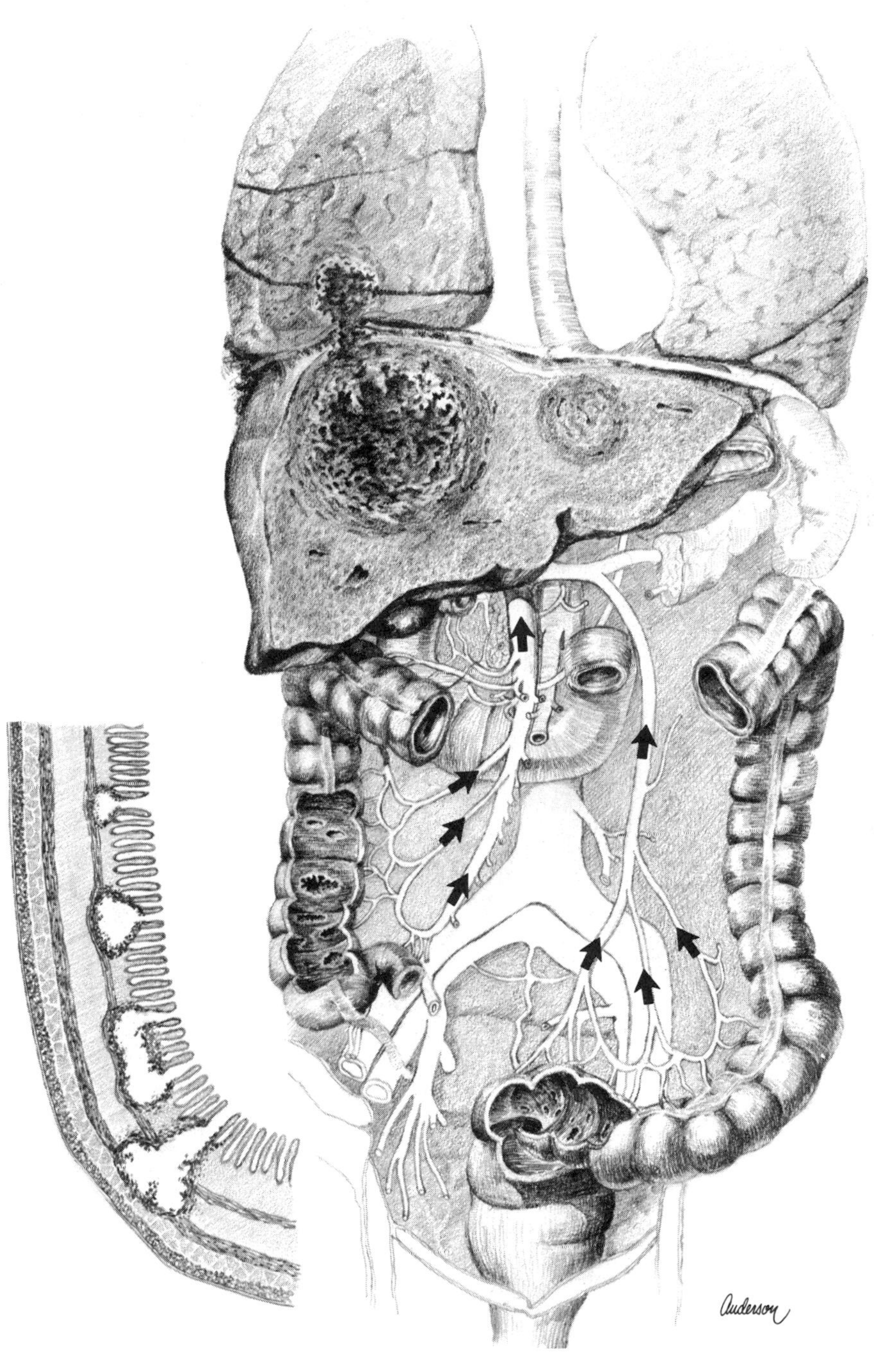

FIG. 32-5. Major pathology of amebiasis: Invasion of intestinal mucosa occurs most commonly in the cecum and next most commonly in rectosigmoid area. Passage of trophozoites via portal circulation may result in liver abscess formation. Metastasis through diaphragm may result in secondary abscess formation in lungs. Trophozoites carried in bloodstream may cause foci of infection anywhere in body. (From Beck, J. W., and Barrett-Connor, E.: Medical parasitology, St. Louis, 1971, The C. V. Mosby Co.)

FIG. 32-6. Amebic abscess of liver, cross section of the organ.

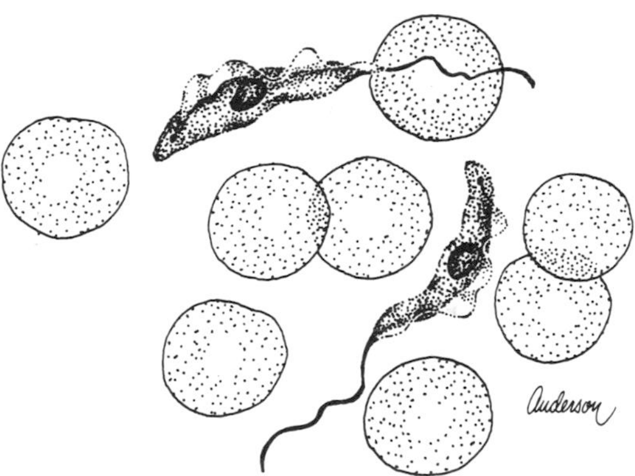

FIG. 32-7. *Trypanosoma gambiense* sketched in blood smear. (Beck, J. W., and Barrett-Connor, E.: Medical parasitology, St. Louis, 1971, The C. V. Mosby Co.)

is helpful. Although morphologic recognition of *Entamoeba histolytica* under the microscope is the prime concern of the laboratory, serologic tests are used to advantage to identify this parasite. These include a complement fixation test, hemagglutination tests, an indirect fluorescent antibody test, and an agar gel double diffusion technic.

■ **Prevention.** The prevention of amebiasis depends upon the proper control of carriers, proper sanitary supervision of foods, and general cleanliness.

In addition to *Entamoeba histolytica,* several other amebas may be found in the intestinal canal, but *Entamoeba histolytica* is the only one that causes disease. *Entamoeba coli* is notable because it must be distinguished from *Entamoeba histolytica.*

Trypanosomiasis

Trypanosomes (Fig. 32-7), of which there are many species, are spindle-shaped protozoa that enter the bloodstream of many different species of animals. They are found in the plasma, not within the blood cells. Infection with trypanosomes is known as *trypanosomiasis.* The types important to man are African trypanosomiasis or African sleeping sickness and South American trypanosomiasis or Chagas' disease.

Abastrin, a substance elaborated by *Trypanosoma,* has some antimicrobial activity.

■ **African trypanosomiasis.** African trypanosomiasis is seen in two forms: Gambian trypanosomiasis, agent *Trypanosoma gambiense,* and Rhodesian trypanosomiasis, agent *Trypanosoma rhodesiense.* Each is transmitted by a species of the tsetse fly. The fly becomes infected by ingesting the blood of a person with the disease, and the parasite undergoes a cycle of development in its body. When the parasites develop to a certain point, they invade the salivary glands of the fly, whence they are transferred to persons bitten. Cattle, swine, and wild animals, especially antelope, may harbor the parasites and be a source of human infection. Rhodesian trypanosomiasis is more virulent than the Gambian form. Early in the course of either form of trypanosomiasis there are acute episodes of fever and inflammation of lymph nodes as the trypanosomes multiply in the bloodstream. The Rhodesian form is usually fatal within a matter of months and rarely progresses to the chronic stages of the Gambian form. In the end stages of the disease, invasion of the brain and its coverings produces the celebrated and uncontrollable sleepiness.

■ **South American trypanosomiasis.** South American trypanosomiasis, caused by *Trypanosoma cruzi,* is transferred to man by small, bloodsucking, cricketlike insects from a reservoir in man and in domestic and wild animals such as dogs, cats, rats, armadillos, and opossums. Infection results from the contamination of the skin with insect feces and is not transferred by the actual insect bite. South American trypanosomiasis differs from African trypanosomiasis in that the parasites multiply in the tissues rather than in the blood. They reappear in the blood to be picked up by the vector. If the patient survives the acute stage, the disease becomes chronic, with the organisms localized in various organs.

An experimental vaccine has been prepared by killing the microorganisms of a culture by physical means—subjecting the trypanosomes to high-frequency sound waves, to pressure, or to mechanical forces evoked when the culture is shaken with glass beads. It has been used only in mice.

■ **Laboratory diagnosis.** During the fever, trypanosomes of the African disease may be demonstrated in Giemsa-stained films of peripheral blood. Concentration technics for peripheral blood facilitate the search for parasites. Smears and imprints of lymph nodes may contain them. *Trypanosoma cruzi* is identified in aspirated material from spleen, liver, lymph nodes, and bone marrow.

Leishmaniasis

Leishmaniasis is a protozoan disease caused by what is probably man's most ancient parasite. It exists in two forms: the visceral and the cutaneous.

■ **Visceral leishmaniasis (kala-azar, dumdum fever).** The visceral form of leishmaniasis is characterized by fever, enlargement of the spleen and liver, progressive emaciation, weakness, and, in untreated patients, death. The agent, *Leishmania donovani*, is transmitted from man to man by the bite of sand flies of the genus *Phlebotomus*. The disease is endemic among dogs, which may be a source of infection. It occurs in the countries bordering the Mediterranean Sea, in India, in the Middle East, in China, and in parts of Africa.

■ **Cutaneous leishmaniasis.** Cutaneous leishmaniasis is described by the presence of nodular and ulcerating lesions in the skin. There are two types. One, known as Oriental sore, Aleppo button, or Delhi boil, is caused by *Leishmania tropica*. The other, known as American leishmaniasis or espundia, is caused by *Leishmania braziliensis*. Cutaneous leishmaniasis is transmitted as is the visceral disease by sand flies. The individual lesions on the skin represent the bites of insects or the mechanical transfer of infection by scratching or some form of abrasion. This disease is seen in the same parts of the world as visceral leishmaniasis, but the two forms of leishmaniasis are said not to occur in exactly the same localities. Cutaneous leishmaniasis is occasionally seen in the United States in persons coming from endemic areas.

■ **Laboratory diagnosis.** In all forms of leishmaniasis the diagnosis is made by demonstrating the organisms in smears from lesions or in biopsies of involved tissues.

Trichomoniasis

Trichomoniasis is a widespread infection of the genitourinary tract caused by *Trichomonas vaginalis*. In women it is an intractable vaginitis with a profuse, cream-colored, foul-smelling discharge in which the trichomonads abound. In men the organisms are found in the prepuce and prostatic urethra, but symptoms seldom occur. The infection is transmitted by sexual intercourse. It is a venereal disease of generally unrecognized significance. The trichomonads are readily identified in vaginal discharges from the female and in urine and the prostatic discharges of the male.

Infections with intestinal flagellates

The most important intestinal flagellates found as cysts and trophozoites in stools are *Giardia lamblia*, *Trichomonas hominis*, and *Chilomastix mesnili* (Fig. 32-3). The latter two are not considered pathogenic by most protozoologists. The presence of *Giardia lamblia* in the upper part of the small intestine in man is usually associated with mild disturbances of the bowel. Giardiasis, infection with the organism, is related to a persisting diarrhea, malabsorption, and inflammatory changes in the lining of the small intestine (Fig. 32-8).

Malaria

Malaria is an acute febrile disease caused by the malarial parasite, a protozoon belonging to class Telosporea and genus *Plasmodium*. The word *malaria* is derived from Italian for "bad air," and the disease got its name in the eighteenth century because of its association with the ill-smelling vapors from the marshes around Rome.

Malaria, one of the most widely prevalent diseases in the world, remains the number one public health problem globally. More people have malaria than any other disease. It is a constant threat to more than a billion human beings. Prior to America's civil and military involvement in Southeast Asia, malaria was infrequently seen in the United States (60 cases reported in 1961). In 1970 there were over 3500 cases, mostly

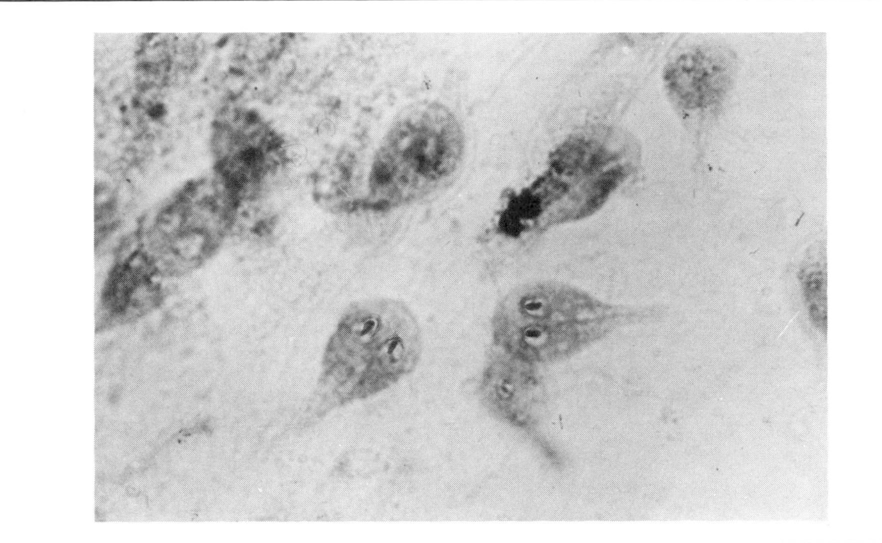

FIG. 32-8. *Giardia lamblia* as seen in gastric aspirate processed for cytologic examination, photomicrograph. Many are present.

in individuals returning from Southeast Asia. Malaria has been a scourge of war throughout the ages, and the Army Medical Corps states that this disease can put more men out of action than do battle casualties. In World War II there were over 490,000 cases of malaria with 8 million man-days lost.

■ **Types.** Malaria exists in three characteristic types,* each caused by its own distinct species of *Plasmodium,* as follows:

1. *Tertian*—a paroxysm of chill and fever every 48 hours; cause, *Plasmodium vivax* (the most common)
2. *Quartan*—a paroxysm of chill and fever every 72 hours; cause, *Plasmodium malariae* (the least common)
3. *Estivoautumnal* or *malignant*—irregular paroxysms; cause, *Plasmodium falciparum*†

The first two types are known as *regular intermittent types.* After the paroxysm of chill and fever, temperature returns to normal, and the patient is fairly comfortable until the next paroxysm. In the third or *remittent* type the fever varies in intensity, but the patient does not become completely afebrile. Estivoautumnal (pernicious) malaria is the most severe in its consequences and is the treacherous form. The clinical picture is diverse, sometimes obscure, oftentimes dramatic, and this type can be rapidly fatal. Death, it is said, may come within a matter of hours. If there is mixed infection, falciparum is the dominant type.

---

*A fourth parasite is *Plasmodium ovale,* whose appearance and life cycle are very much like those of *Plasmodium vivax.* Infection with this parasite, tertian malaria, is usually mild. It is not widely distributed over the world.

†Ninety percent of malaria in Southeast Asia is caused by *Plasmodium falciparum.* Although this is true, 85% of malaria in returnees is caused by *Plasmodium vivax.* In the Korean conflict malaria was almost entirely vivax. This is less severe clinically but is more likely to lie dormant only to recur many months after the initial episodes. There is little tendency for relapses with falciparum malaria.

■ **Modes of infection.** The different species of malarial parasites are closely related and are transmitted in the same way—by the bite of a female mosquito of the genus *Anopheles.* Of this genus, nearly 100 species may transmit the infection naturally. Man is the main reservoir of infection. In malaria-infected countries many of the inhabitants become asymptomatic carriers. They harbor the parasites in their blood (parasitemia) and tissues without manifesting the disease. Repeated attacks seem to give the patient some immunity. Unrecognized infections and insufficient treatment lead to the carrier state, so important in the spread of malaria.

The disease is occasionally transferred by the use of contaminated hypodermic syringes, as is common among heroin addicts, or by blood transfusion. There is an increased awareness of this hazard in blood banking partly because of the likelihood of the source of infection being a blood donor who is a drug addict.

A reservoir of malaria exists in monkeys, from which the infection is transmitted to man and other primates by certain forest species of *Anopheles.*

■ **Life story of the parasite.** There are two major events in the complicated life story of the malarial parasite (see Fig. 32-9).

FIG. 32-9. The malarial merry-go-round—life cycle of malarial parasite. *Preerythrocytic phase:* **1,** Sporozoite entering liver cell. *Erythrocytic phase:* **2,** Merozoite entering red blood cell; **3,** trophozoite or ring form; **4,** growing trophozoite; **5,** preschizont; **6,** schizont (segmentation stage); **7,** liberated merozoites; **8,** microgametocyte; **9,** macrogametocyte; **10,** exflagellation of microgametocyte with microgametes attached; **11,** fertilization of a macrogamete by a microgamete; **12,** ookinete; **13,** immature oocyst; **14,** mature oocyst discharging sporozoites.

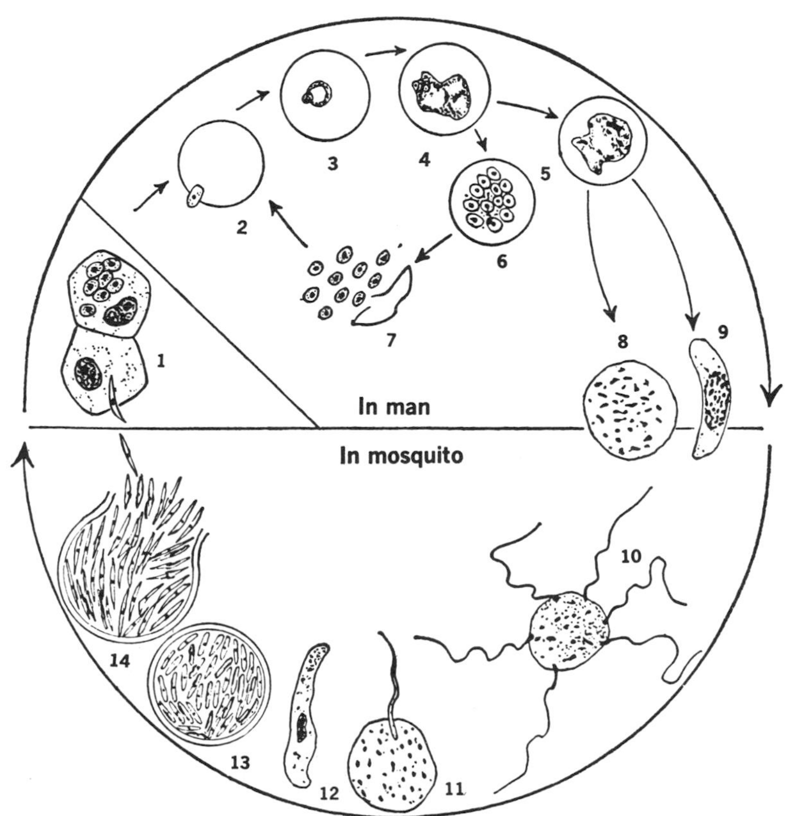

ASEXUAL DEVELOPMENT IN MAN. When young malarial parasites *(sporozoites)* are introduced into the bloodstream by the mosquito bite, they localize in the cells of the liver where they multiply. This is the *preerythrocytic phase.* (Some 500,000 parasites must be injected by the mosquito for a human being to be infected.) After 6 to 9 days young parasites *(merozoites)* are released into the bloodstream. The *erythrocytic phase* begins when each one bores into a red blood cell on which it feeds and develops. The parasite does not fully utilize the hemoglobin of the red cell. Because of this, granules of pigment (an iron porphyrin hematin) accumulate within its cytoplasm. This residual product is the malarial pigment; it is not a normal breakdown product of hemoglobin. When the parasite reaches maturity, it is known as a *schizont.* Within the red cell the mature schizont arranges itself into a number of segments. Suddenly the segments separate, release another generation of merozoites and destroy the red cell. This is *segmentation.* Some of the merozoites are destroyed by the white blood cells, but the majority bore into red blood cells to repeat the process. In estivoautumnal malaria two or even three or four parasites invade a single red cell (see Figs. 32-10 to 32-12).

About 2 weeks (sometimes longer) after the infecting mosquito bite, enough parasites are present for the red blood cell destruction to cause trouble. The *incubation period* covers the initial preerythrocytic phase and the first 2 weeks or so of the erythrocytic phase. The time elasping between the entrance of a parasite into a red blood cell and its segmentation is the *periodicity* of the parasite. For *Plasmodium vivax* it is 48 hours and for *Plasmodium malariae,* 72 hours. For *Plasmodium falciparum* it is usually 48 hours, but not regularly so.

The paroxysms of chills and fever in malaria stem from the liberation of metabolic by-products of the parasite and toxic breakdown products from the disrupted blood cell. The paroxysms are sharp in tertian and quartan malaria because all the parasitized cells rupture at about the same time. In estivoautumnal infections some of the para-

FIG. 32-10. *Plasmodium vivax* (parasite of tertian malaria), development in red blood cell: **1,** Young ring form (trophozoite); **2** to **5,** further stages in development of parasite; **6,** mature schizont (made up of many segments or merozoites); **7,** microgametocyte (male sexual parasite); **8,** macrogametocyte (female sexual parasite).

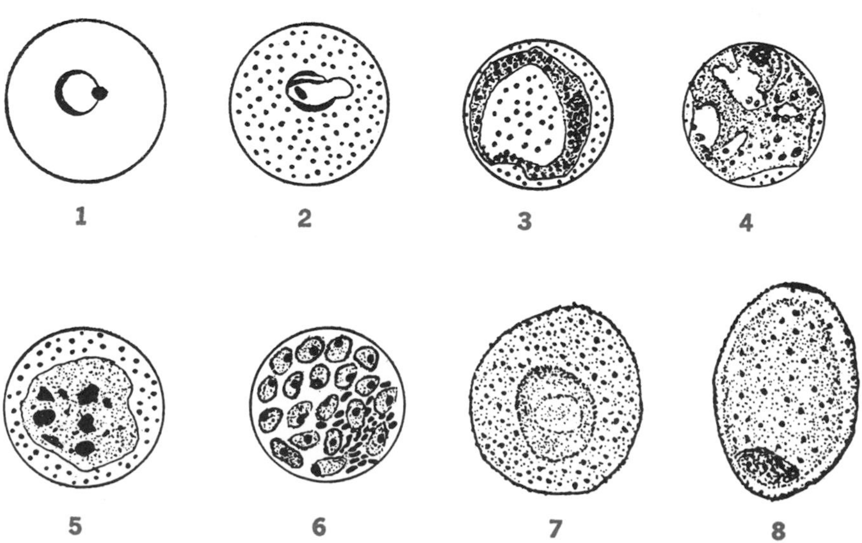

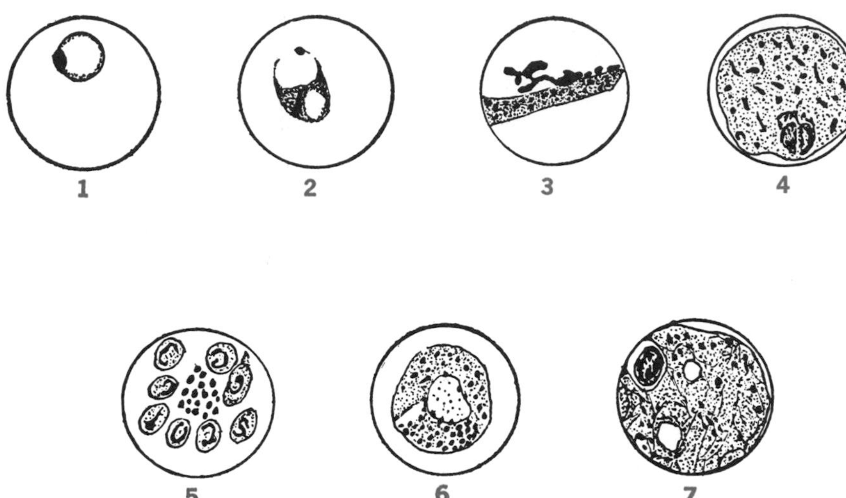

FIG. 32-11. *Plasmodium malariae* (parasite of quartan malaria), development in red blood cell: **1,** Young ring form (trophozoite); **2** to **4,** further stages in development of parasite; **5,** mature schizont (made up of merozoites); **6,** microgametocyte; **7,** macrogametocyte.

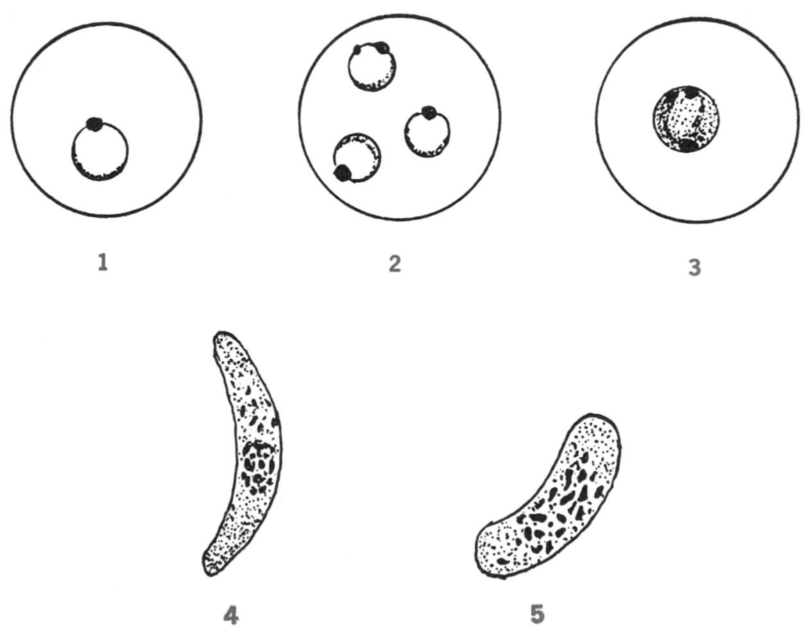

FIG. 32-12. *Plasmodium falciparum* (parasite of estivoautumnal malaria), development in red blood cell: **1,** Trophozoite; **2,** three trophozoites (multiple infection characteristic of *Plasmodium falciparum*) in red blood cell; **3,** schizont, early formation and production of pigment (this form seldom found in peripheral blood); **4,** macrogametocyte; **5,** microgametocyte.

sitized cells rupture ahead of time and some rupture behind time, so that several hours are required for the whole brood of parasites to be released. This explains why chills are usually absent and the fever may be continuous.

Some of the parasites do not repeat the asexual phase of development but produce male and female sexual forms or gametocytes, the sexual development of which is completed within the stomach of an *Anopheles* mosquito. Sexual forms do not appear in the blood until the infection is of 2 or 3 weeks' duration.

SEXUAL DEVELOPMENT IN THE MOSQUITO. When a female *Anopheles* mosquito ingests male and female sexual parasites from the blood of an infected person, a rather complicated sexual cycle begins in its stomach (Fig. 32-9). First, the female parasite or *macrogamete* is fertilized by flagellar structures, *microgametes*, that break away from the male parasite or *microgametocyte* by a process of exflagellation. These correspond to spermatozoa in the higher forms of life. The fertilized parasite or *ookinete* bores into the stomach wall, becomes encysted (the *oocyst*), and divides into many small spindle-shaped parasites or *sporozoites*. The cyst ruptures, and the sporozoites are carried by the lymphatic system to the salivary gland of the mosquito, which is so constructed that the parasites are ejected in the saliva when the mosquito bites.

■ **Pathology.** The chief pathologic changes in malaria are related to destruction of red blood cells (hemolysis). Hemolysis leads to different degrees of anemia and jaundice. The parasitic infection seems to make the blood more viscous, and the sticky parasitized cells plug and obstruct small blood vessels. This is prone to occur with falciparum malaria and accounts for its worst complication, cerebral malaria. The liver and spleen enlarge (hepatosplenomegaly), partly because the presence of the malarial pigment stimulates the reticuloendothelial system to activity. Reticuloendothelial cells ingest the pigment and deposit it in liver, spleen, and bone marrow. In acute malaria the spleen is moderately enlarged, soft, and friable. In chronic malaria it is enlarged and markedly fibrotic. Such a spleen easily ruptures as the result of a blow or fall or even spontaneously.

A dreaded complication is blackwater fever, which develops in certain cases with acute and massive hemolysis. Large amounts of hemoglobin are released into the plasma (hemoglobinemia), spilling over into the urine (hemoglobinuria). There is a severe kind of acute renal failure associated with the passage of reddish black urine.

On a worldwide basis malaria results in a greater morbidity and mortality than any other infectious disease.

■ **Laboratory diagnosis.** Malarial parasites may be easily seen in the red cells of properly prepared (and stained) thin smears of peripheral blood taken just before or at the peak of the paroxysm. Wright's and Giemsa's stains are most often used. The young parasites, seen as blue rings with a red chromatin dot attached, have a signet-ring appearance. The quartan ring is thicker than the tertian, and the estivoautumnal ring is thin and hairlike. The latter often has two chromatin dots. Full-grown malarial parasites almost completely fill the red cell and contain numerous red granules. Just before the red cell is ruptured, the parasites assume a segmented or rosette pattern. So-called malarial crescents are sausage-shaped gametocytes with a chromatin mass near their center and are the sexual parasites of estivoautumnal malaria. They are frequently seen in the blood, whereas the fully developed parasite or schizont is seldom so in this form of malaria.

At times, when organisms may be too few to be seen in thin smear, a sample of blood may be smeared thickly on a glass slide. This thick smear is treated to remove hemoglobin from the erythrocytes and then stained. The disadvantage of the thick smear is that the shape and appearance of the parasites are altered in preparation. Its advantage is that a much greater volume of blood may be examined within a given length of time.*

Fluorescent antibody technics are also used to stain the parasite specifically. A soluble antigen fluorescent antibody (SAFA) diagnostic test fills a need in the blood bank in the screening of donors.

A small dose of quinine or other antimalarial drug will drive the parasites out of the peripheral blood; therefore it is practically useless to examine the blood for malaria right after these drugs are given.

The laboratory diagnosis of malaria *does not consist of merely finding the parasites but includes the determination of the species.*

■ **Mosquitoes transmitting malaria.** Malaria is transmitted by various species of *Anopheles* mosquitoes. It is transmitted only by the female because the male lives on fruits and vegetables. Since the common house mosquito *(Culex)* is not a vector, it is important to distinguish it from *Anopheles.* The *Culex* mosquito bites during the daytime; the *Anopheles* at night or about dusk. The wings of *Anopheles* are spotted, whereas those of *Culex* are not. When *Culex* is resting on a wall, its body is almost parallel to the wall; the body of *Anopheles* stands at an acute angle (see Fig. 10-2, p. 150).

■ **Prevention.** Prevention of malaria depends upon blocking the transfer of the infection from person to person by mosquitoes. Recommended for this purpose are (1) screening of houses, (2) draining and oiling of ponds of water to prevent mosquito breeding and using minnows to destroy the larvae, (3) proper treatment of patients with antimalarial drugs, and (4) the detection and cure of carriers. The presence of an animal reservoir greatly complicates the problem of malarial control in those countries where the jungles swarm with monkeys and other primates.

Development of a vaccine for malaria is still experimental. Partially purified material from the malarial plasmodium is used.

Balantidiasis

Balantidium coli is the most important intestinal ciliate and the largest protozoon to invade man (Fig. 32-3, *D*). It is seen in two life stages: the cyst and the motile trophozoite. In some cases it seems to be a harmless inhabitant of the large intestine, but usually its presence is associated with diarrhea. It may invade the intestinal wall and produce ulcerations or abscesses with intense dysentery that may even cause death (Fig. 32-13).

Balantidium coli is a normal inhabitant of the large intestine of the domestic hog. Man is probably infected by ingesting cysts passed by the hog. The laboratory identification of cysts and trophozoites in the stools or in the exudate from intestinal ulcers makes the diagnosis.

Toxoplasmosis

Toxoplasma gondii, the cause of toxoplasmosis, is a delicate, boat-shaped, obligate intracellular parasite, somewhat similar to *Leishmania.* It is easily killed by physical agents. It is a cosmopolitan sporozoan, being found in animals and birds all over the world.

---

*In falciparum malaria thick smears may not show parasites for several days after the onset.

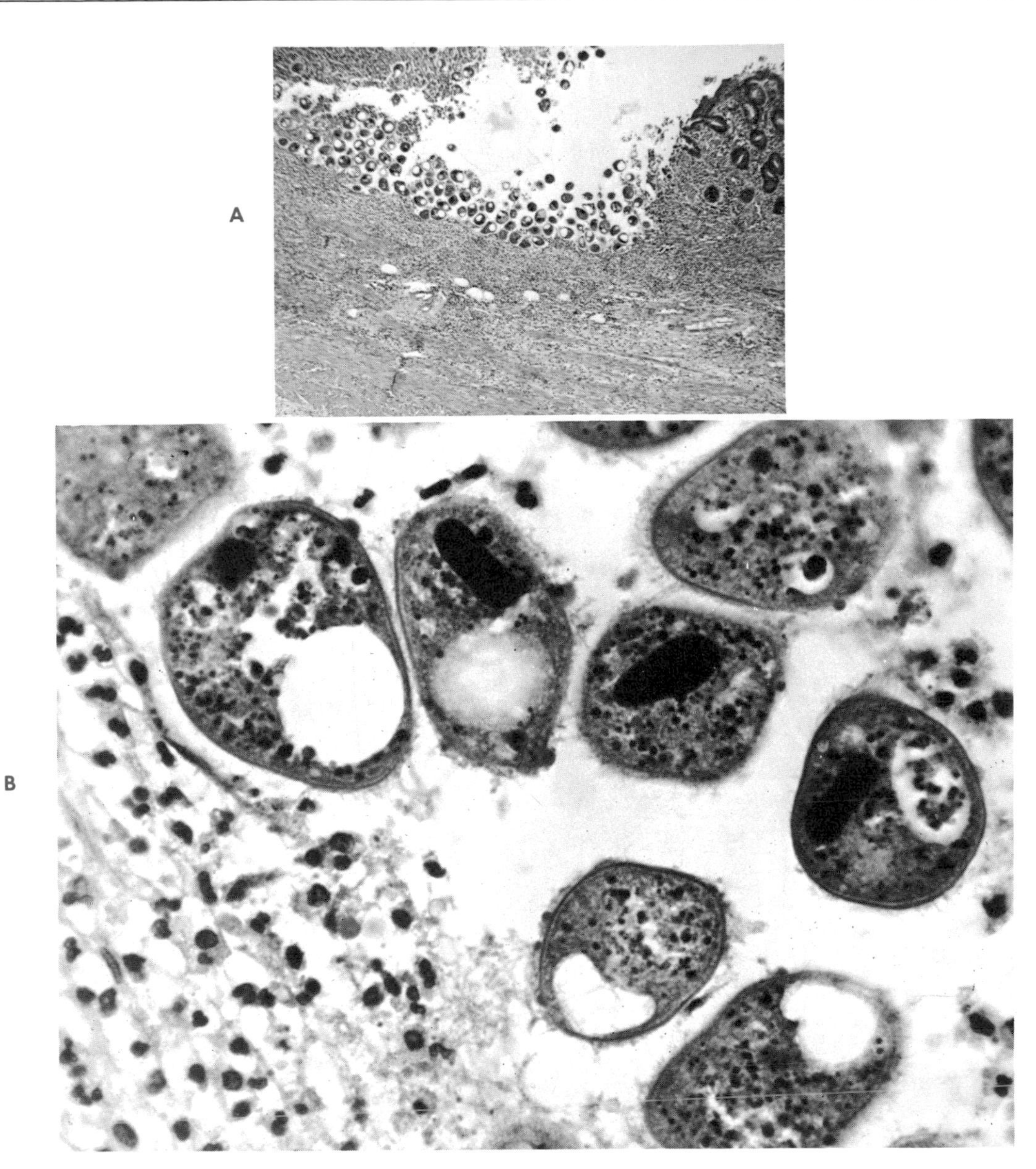

FIG. 32-13. *Balantidium coli* in stained tissue section. (H & E stain.) **A,** Balantidiasis of appendix. In crater of an ulcer of appendiceal wall are numerous organisms cut in cross section (×35). **B,** Cross-sectional areas of *Balantidium coli* as seen with higher microscopic magnification (×430). (From Anderson, W. A. D., editor: Pathology, ed. 6, St. Louis, 1971, The C. V. Mosby Co., vol. 1.)

■ **The disease.** The organism can invade practically any tissue cell and is especially prone to affect cells of the reticuloendothelial organs, including the lining cells of the blood vessels. Within a cell the organisms rapidly proliferate in typical fashion to form a rosette that may become a cyst containing some 3000 parasites (Fig. 32-14). Inflammation may be present in the lungs, lymph nodes, eyes, and brain. Characteristic are the focal deposits of calcium in nervous tissue (especially in utero). In man, the disease occurs in two forms: acquired and congenital. The acquired or adult form

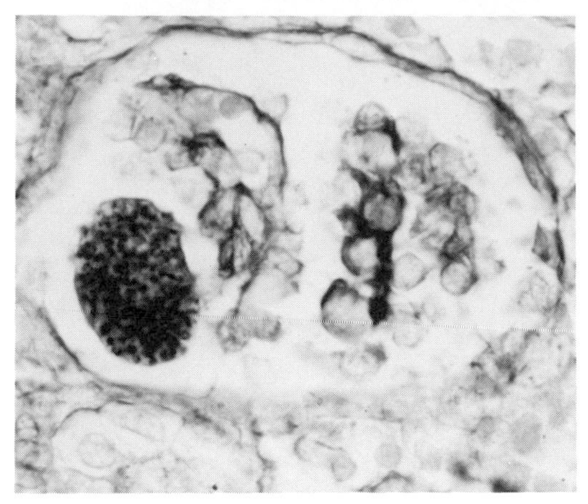

FIG. 32-14. Toxoplasma cyst (on left), photomicrograph of microsection of kidney from infant dead of disease.

often passes unnoticed. One third of adults in the United States have been infected at some time. The congenital form is acquired in utero from a mother most probably with no history of previous infection. The newborn baby becomes very ill, develops a skin rash, turns yellow, and may have convulsions. If the baby survives the damage done in the brain (associated with the calcium deposits), he may be born with microcephaly, hydrocephalus, and mental retardation. *Toxoplasma gondii* produces changes in the eye that are designated as chorioretinitis, and the lesion is responsible for blindness in these infants. (No effective form of treatment is known for the disease.)

Toxoplasmosis is associated with the formation of tumors in birds; in man it is seen associated with neoplasms of the central nervous system.

■ **Sources and mode of infection.** Recent evidence indicates that this sporozoan has a life cycle in cats similar to that of the malarial parasite in mosquitoes. Cats pick up the parasites when they consume the intermediate hosts—infected birds and mice. In the intestine of the cat, which provides a peculiarly suitable habitat for the parasite, the organisms go through asexual and sexual stages of development, and the oocysts are passed in the feces. After being passed (into soil, sand, or litterbox) oocysts become infectious after 3 to 4 days in warm, moist surroundings. They can be recovered after many months from water or wet soil and are generally resistant to many chemical agents including ordinary disinfectants. However, drying and heat kill them. The common house cat is implicated as the primary host and a human reservoir.

The parasite is universal. Many animals harbor it, and it may persist in the raw flesh of the slaughtered animal until killed by heat, drying, or freezing. Eating raw or undercooked meat is a principal source of human infection. The National Livestock and Meat Board recommends that all meat be heated to at least 140° F. throughout to kill the toxoplasmas. (This is the "rare" reading on meat thermometers designed for home use.) Pregnant women should be very careful in handling cats, cat feces, or articles contaminated with cat feces, and should avoid altogether any contact with a strange cat or one newly brought into the household.

The organisms are taken into the human body by way of the mouth either because the individual has been in contact with an infected cat or because he has consumed infected meat. The toxoplasmas are released from the oocysts and migrate into body tissues and fluids.

For the congenital form to develop, the organisms must get into the bloodstream of the mother sometime after the first trimester of pregnancy. They establish a focus of infection in the placenta that enables them to penetrate the fetal circulation and so infect the fetus.

■ **Laboratory diagnosis.** These microorganisms, possessing a delicate crescent-shaped body with tapered ends, are easily stained and identified in smears made from body fluids, exudates, and diseased tissues. Toxoplasma can be cultured in tissue or in a fertile hen's egg. Serologic examinations include a complement-fixation test, a neutralization test, and an immunofluorescence test. The antibodies detected by the Sabin-Feldman dye test are those preventing parasites of a laboratory culture from taking up methylene blue dye. A toxoplasmin skin test is available. A mouse can be inoculated with suspicious material so that the organisms can be recovered and identified from that animal.

Toxoplasma may be detected in the feces of a cat. Direct smears are not adequate; the fecal flotation method must be used.

Pneumocystosis

Pneumocystis carinii causes pneumocystosis (diffuse interstitial pneumonitis, interstitial plasma cell pneumonia), an inflammatory process unique in the lungs. Little is known about the organism, and nothing is known as to the mode of infection. The organisms are studied in smears of material aspirated from the diseased lungs stained by the Papanicolaou technic or in tissue sections of lung (from autopsy) stained with a special silver stain (Fig. 32-15). In this disease the air sacs of the lungs are filled with

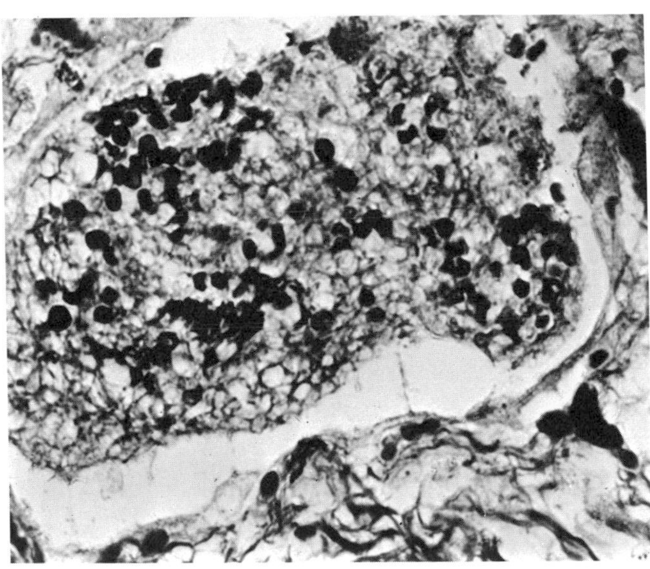

FIG. 32-15. *Pneumocystis carinii* in alveolus of lung. Tissue section stained with methenamine silver. Dark spheres are capsules of individual organisms. (×600) (From Anderson, W. A. D.: Pathology, ed. 6, St. Louis, 1971, The C. V. Mosby Co.)

a foamy, semiliquid, lightly staining material representing clustered masses of oval, minute organisms about 1μ in diameter, surrounded by a thin homogeneous capsule. The walls of the air sacs are permeated by inflammatory cells.

Pneumocystosis is another, not too rare, disorder of the "compromised" host. It was first encountered in debilitated infants. The vulnerability of much older patients stems from prolonged steroid hormone therapy, extended immunosuppression, or the presence of leukemia or other cancer of the lymphoid system. Cortisone (a steroid hormone) is actually believed to enhance the growth of these organisms.

QUESTIONS FOR REVIEW

1. Describe protozoa. What are their salient features?
2. Give the six classes of protozoa of medical interest. Characterize each.
3. Define: intermediate host, definitive host, organelle, pseudopod, cyst, flagella, cilia, trophozoite, parasite, endoparasite, ectoparasite.
4. How is amebiasis spread? How may it be prevented?
5. Outline the development of the malarial parasite in (a) man and (b) the mosquito.
6. Discuss the prevention of malaria.
7. Compare *Anopheles* mosquito with *Culex*.
8. Give the laboratory diagnosis for:
   a. Malaria
   b. Amebiasis
   c. Trypanosomiasis
   d. Leishmaniasis
   e. Toxoplasmosis
   f. Pneumocystosis
9. Characterize briefly: Chagas' disease, African trypanosomiasis, trichomoniasis, giardiasis, balantidiasis, toxoplasmosis, and pneumocystosis.
10. Compare the life cycle of *Plasmodium* with that of *Toxoplasma*.

REFERENCES. See at end of Chapter 26.

33 Metazoa: medical parasitology

METAZOA

Among multicellular animal parasites or *Metazoa,* three phyla are medically noteworthy as parasites of man: (1) *Platyhelminthes* or flatworms, including the two classes of *Trematoda* (flukes) and *Cestoidea* (tapeworms), (2) *Nematoda* (roundworms), and (3) *Arthropoda.* Included in the *Arthropoda* are mites, spiders, ticks, flies, lice, and fleas. As vectors, these are medically important in the transmission of disease* (Table 33-1).

Worms are elongated, invertebrate animals without appendages or bilateral symmetry. The laboratory diagnosis of metazoal, as with protozoal, infections depends most of the time upon the morphologic identification of the parasite or its ova (Fig. 33-1).

Trematodes (flukes)

The trematodes are flat, leaflike, nonsegmented parasites provided with suckers for attachment to the host. All species, except those that inhabit the bloodstream, are hermaphrodites and have operculate eggs. Some have the most complicated life histories in the animal kingdom.

■ **Life history.** The cycle of development of flukes is briefly as follows: The egg is passed from the body of the host, and the contained embryo develops into a ciliated organism, the *miracidium* (plural, *miracidia*). If water is present, the *miracidium* escapes from the egg and swims (about 5 or 6 hours) until it reaches an intermediate host (certain species of water snails) (Fig. 33-2). Hatching miracidia are phototrophic; that is, they swim toward light. This phenomenon can sometimes be demonstrated in fecal or urinary specimens. The miracidium penetrates the snail and forms a cyst in its lungs, in which many organisms develop. These wander to other parts of the snail and develop into minute worms called *cercariae* (Fig. 33-3).

One infested snail alone can release 169 million cercariae in 1 month's time. Being phototrophic the cercariae emerge during the hours of sunlight, and the greatest number (as the chance of getting the disease) is at high noon. (They seldom live more than 1 day.) They swim around until they attach themselves to blades of grass, where they encyst. Sometimes they enter other aquatic animals, such as certain fishes and crabs, to encyst. When the encysted organisms are swallowed by man, the definitive host, they develop into adult flukes in his tissues. The cercariae of the blood flukes gain access

*If the parasite is carried unchanged, the vector is a *mechanical* one. The parasite undergoes a series of developmental changes in the body of the *biologic* vector.

TABLE 33-1. OVERVIEW OF ARTHROPODA IN SPREAD OF DISEASE

Vector	Agents transmitted	Genus names of vectors	Examples of diseases spread
Mosquito	Metazoa Protozoa Viruses	*Anopheles* *Aedes* *Culex* *Mansonia*	Malaria, yellow fever, encephalitides, filariasis
Tick	Bacteria Rickettsia	*Dermacentor* *Rhipicephalus* *Amblyomma* *Ornithodorus*	Spotted fevers, tularemia, Q fever, relapsing fever
Mite	Rickettsia	*Trombicula* *Allodermanyssus*	Scrub typhus, rickettsialpox
Louse	Bacteria Rickettsia	*Pediculus*	Typhus fever, relapsing fever
Flea	Bacteria Rickettsia	*Xenopsylla* *Pulex*	Plague, murine typhus
Biting fly:			
Tsetse fly	Protozoa	*Glossina*	Sleeping sickness
Black fly	Metazoa	*Simulium*	River blindness
Sand fly	Protozoa Bacteria	*Phlebotomus*	Leishmaniasis, bartonellosis
Housefly	Metazoa Protozoa Bacteria Bedsoniae Viruses	*Musca*	Salmonellosis, bacillary dysentery, cholera, amebiasis, poliomyelitis, trachoma, trypanosomiasis, ascariasis
Triatomid bug	Protozoa	*Triatoma* *Rhodnius*	American trypanosomiasis

to the body of man through the skin. They make their way via the bloodstream to the portal venous system, where they mature into adult organisms.

■ **Classification.** Flukes are classified according to the area of the body in which their development into adult flukes is completed and their eggs are deposited. From the standpoint of habitat there are flukes that live in the intestine, the liver, the lungs, and the portal venous system with its tributaries. The last are called blood flukes. Of special note are three: *Schistosoma japonicum*, *Schistosoma mansoni*, and *Schistosoma haematobium*. The large intestinal fluke is *Fasciolopsis buski* (Fig. 33-4). *Clonorchis sinensis* and *Fasciola hepatica* are liver flukes, and *Paragonimus westermani* is the lung fluke.

■ **Pathogenicity.** The pathologic changes in the human body center about the eggs trapped in the tissues. Female blood flukes may migrate to the terminal vessels of the bladder and rectum to lay their eggs. (A single worm can deposit eggs in a given area for up to 20 years.) There the eggs set up an inflammatory reaction in the mucous membrane. This results in papillomatous thickenings. The eggs may escape into the lumen of the bladder or rectal canal to be passed in the urine or feces. The ova of flukes that

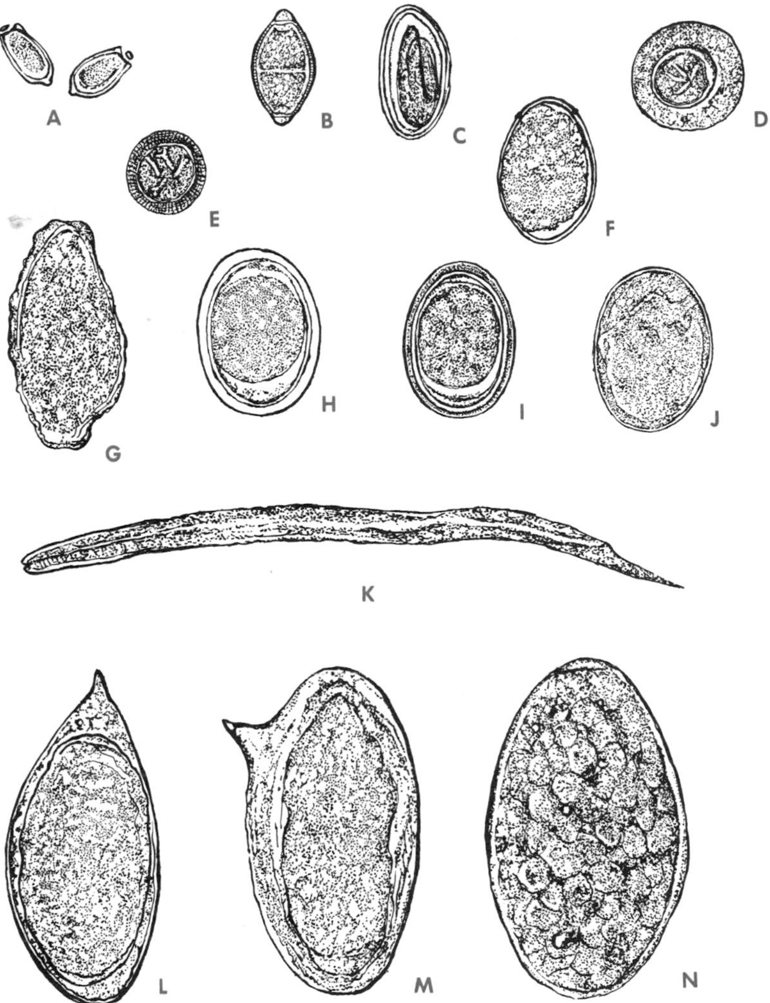

FIG. 33-1. Ova of metazoa in sketch. Note one larval form included. **A,** *Clonorchis sinensis.* **B,** *Trichuris trichiura.* **C,** *Enterobius vermicularis.* **D,** *Hymenolepis nana.* **E,** *Taenia solium.* **F,** *Diphyllobothrium latum.* **G,** *Ascaris lumbricoides,* unfertilized ovum. **H** and **I,** *Ascaris lumbricoides, fertilized ova.* **J,** *Schistosoma japonicum.* **K,** *Strongyloides stercoralis* larva. **L,** *Schistosoma haematobium.* **M,** *Schistosoma mansoni.* **N,** *Fasciolopsis buski.*

migrate to the lungs may be found either in sputum or with swallowed sputum in feces. Because of the anatomic relation of the liver to the intestines, we would expect to find the ova of flukes of this organ in the feces. It is important to look for ova in stool specimens, as practically all types may be found there, regardless of the habitat of a given fluke.

Trematode infections are prevalent in the Orient and in the tropics where contamination of fresh water by human feces is widespread. It is estimated that blood flukes or schistosomes infest 200 million persons. Infection with blood flukes is *schistosomiasis* or *bilharziasis.* It is one of man's oldest diseases; calcified ova have been

FIG. 33-2. *Australorbis glabratus,* snail vector for schistosomiasis in the Western Hemisphere. (From Med. World News **6:**35, Dec. 3, 1965; photograph by Pete Peters.)

found in Egyptian mummies. Also called snail fever, schistosomiasis is one of the world's most important medical problems. As a global disease, it is second only to malaria in the geographic extent of the incapacity and morbidity produced.

■ **Prevention.** Measures of control have been largely directed toward the elimination of the intermediate host, the snail.

Experimentally it has been found that if a small innocuous dose of *Klebsiella pneumoniae* is given to an animal infected with *Schistosoma mansoni,* the flukes die out and the host gets well. This, if feasible on a practical basis, would open a new era of biologic control of parasites.

Cestodes (tapeworms)

Tapeworms *(Taenia)* are intestinal parasites and produce digestive disturbances of variable degree.

■ **Anatomy.** Adult tapeworms typically have a small head, which buries itself in the intestinal mucosa and anchors the worm, and a nonsegmented neck (head and neck,

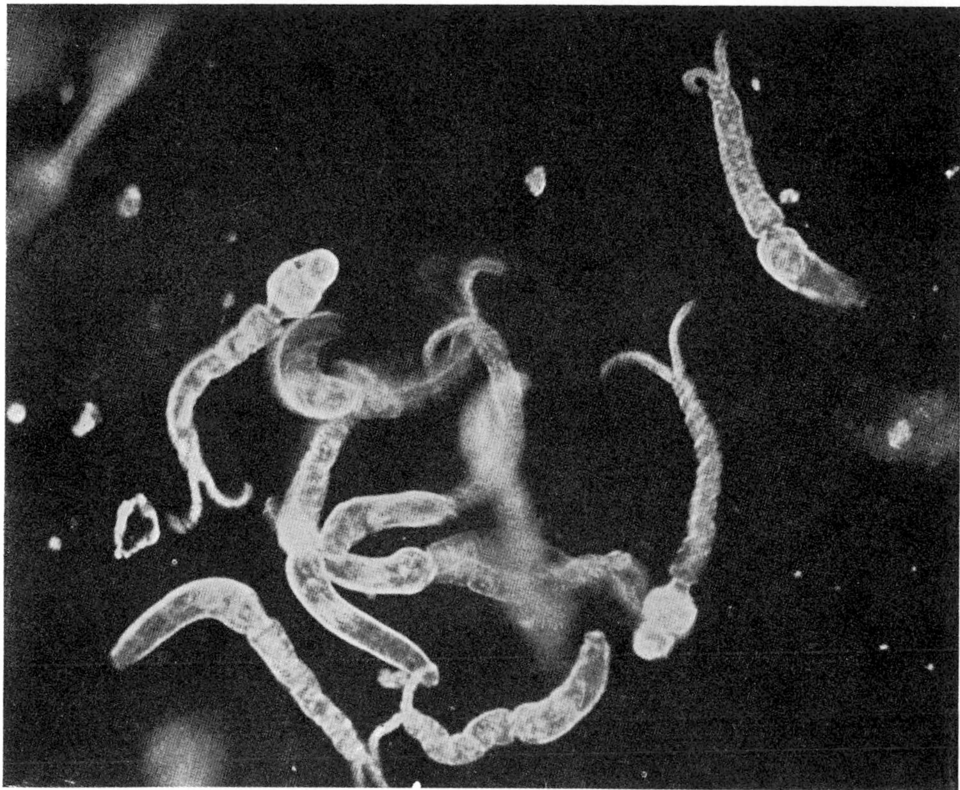

FIG. 33-3. Schistosome cercariae photographed by dark-field illumination. (Photograph by D. M. Blair; published by permission of F. Goodliffe, Southern Rhodesian Public Relations Department; from Gradwohl, R. B. H., Benitez Soto, L., and Felsenfeld, O.: Clinical tropical medicine, St. Louis, 1951, The C. V. Mosby Co.)

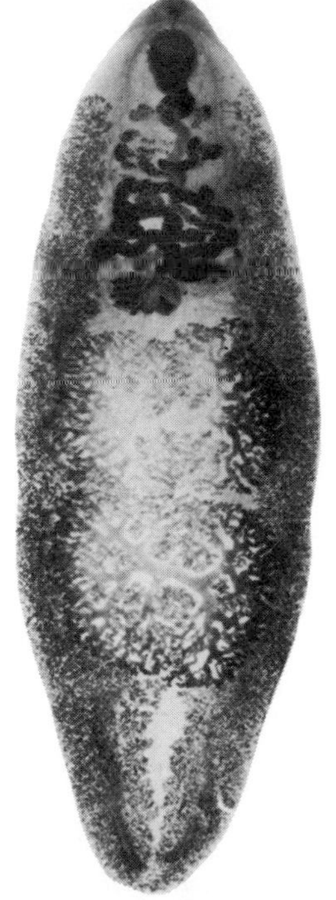

FIG. 33-4. *Fasciolopsis buski.* (From Frankel, S., Reitman, S., and Sonnenwirth, A. C.: Gradwohl's clinical laboratory methods and diagnosis, ed. 7, St. Louis, 1970, The C. V. Mosby Co.)

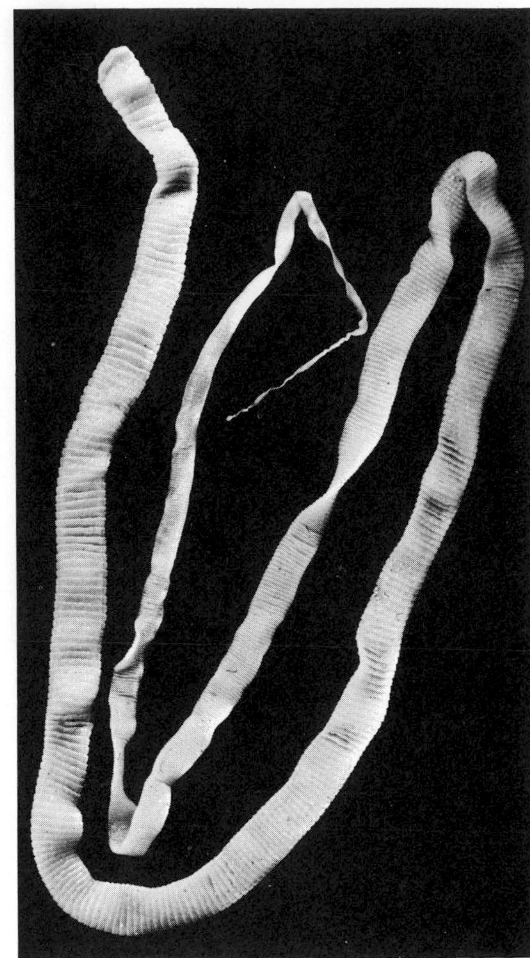

FIG. 33-5. Tapeworm. Note very small head and successively larger segments. Young proglottids are budded from scolex and neck (center); oldest (gravid) proglottids shown at upper left. (From Hickman, C. P.: Integrated principles of zoology, ed. 4, St. Louis, 1970, The C. V. Mosby Co.)

FIG. 33-6. Scolex of pork tapeworm, **A,** compared with that of beef tapeworm, **C. A,** Scolex of *Taenia solium* with apical hooks and suckers. **B,** Hooks of *Taenia solium.* **C,** Scolex of *Taenia saginata* with suckers only. (From Hickman, C. P.: Integrated principles of zoology, ed. 4, St. Louis, 1970, The C. V. Mosby Co.)

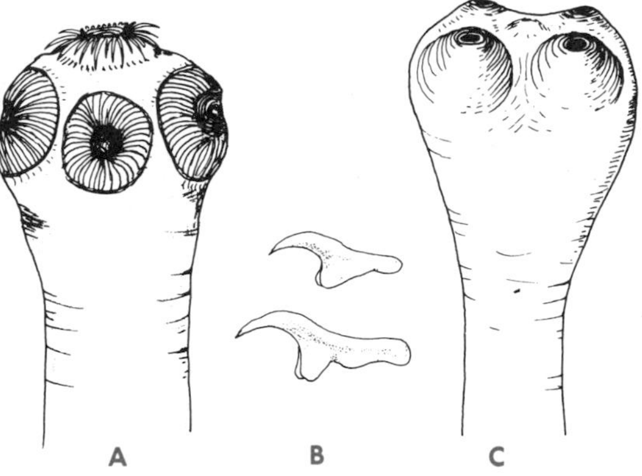

collectively spoken of as the *scolex*) to which are attached in line a variable number of segments *(proglottids).* New ones are formed by a process of segmentation from the scolex—the youngest segment is the one joining the scolex; the last segment is the oldest. Up to a certain point the farther the segment from the scolex, the larger it is. There is no alimentary canal. Each segment obtains its nourishment from the host's intestinal juices by osmosis.

The head is extremely small by comparison with the remainder of the body, often the size of a pinhead (Fig. 33-5). It is provided with hooklets or suckers, or both, for attachment to the intestinal wall (Fig. 33-6). The hooklets are arranged in one or more rows around a small prominence *(rostellum)* situated on the head. In at least one species, attachment is accomplished by suctorial grooves on the sides of the head. Treatment that fails to recover the head, regardless of the number of segments removed,

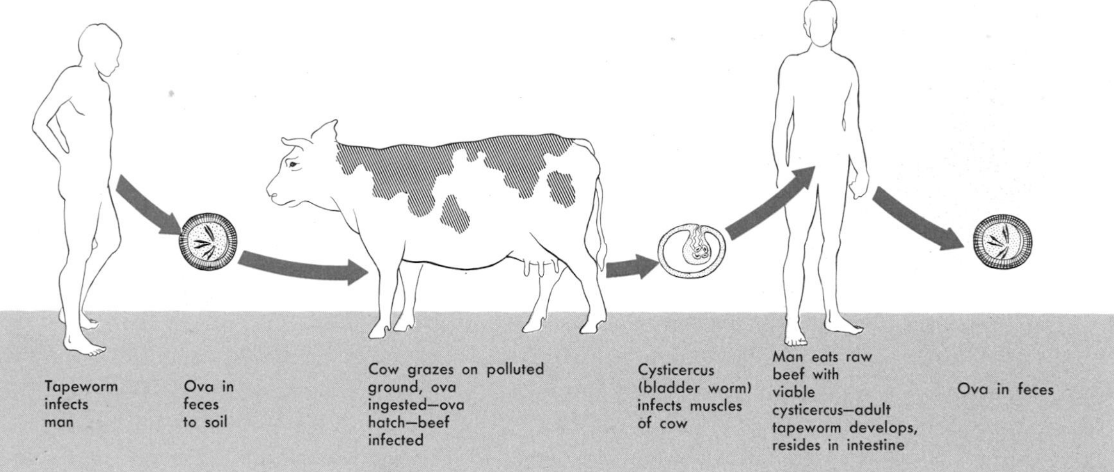

FIG. 33-7. Life cycle of *Taenia saginata.*

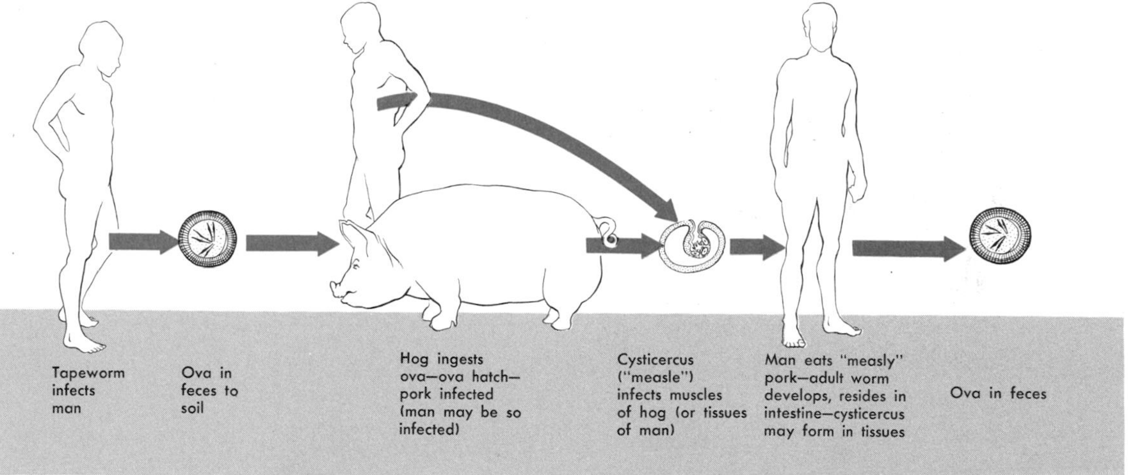

FIG. 33-8. Life cycle of *Taenia solium.*

is valueless because the head immediately replaces the lost segments. The peculiar shape, arrangement, and deeply embedded position of the hooklets often make removal of the head from the intestine extremely difficult.

Each fully developed segment is a sexually complete hermaphrodite. From the scolex to the other end of the worm, there are the following:

1. Undeveloped segments: *immature proglottids*
2. Segments with both male and female elements: well-developed, *mature proglottids*
3. Segments filled by the egg-laden uterus: *gravid proglottids*
4. Degenerating gravid proglottids

In a few species the ova are extruded from the segment through a birth pore. In most species, however, no birth pore is present, and the ova escape from the proglottid through a longitudinal slit. The gravid proglottids toward the end of the tapeworm separate and may be passed by the feces. A person can harbor a tapeworm without ova appearing in the stool because the segments may be expelled before the ova are liberated.

■ **Life cycle.** Tapeworms have a larval and an adult cycle of existence (Figs. 33-7 and 33-8). As a rule, the cycles take place in different species of animals. The adult cycle occurs within the intestinal canal, and the larval cycle within the tissues of the host.

The egg develops into an adult tapeworm as follows: Through a series of changes an embryo is formed within it. After the eggs are swallowed by a susceptible intermediate host, larvae are set free. By means of their hooklets the larvae penetrate the intestinal wall and pass into the tissues where the hooklets are lost. They then reach the bloodstream to be carried to different parts of the body to lodge and each to develop into a scolex. The irritation caused by formation of the scolex sets up a tissue reaction, and a cyst wall is formed around it. The scolex encased in the cyst is termed a *cysticercus.* When the raw or insufficiently cooked flesh of the animal containing the cyst is eaten by a susceptible host, the cyst wall is digested, the scolex attaches itself to the intestinal wall of the new host, and an adult tapeworm develops.

In worms (example, *Diphyllobothrium*) in which the eggs escape by a birth pore, ciliated embryos escape from the egg after it is passed. They take up an aquatic existence and swim until they gain access to certain species of freshwater fish. They parasitize the fish with the help of certain species of *Cyclops,* or water fleas, which act as transferring hosts. Consumption of the raw or poorly cooked fish transfers the parasite to man.

FIG. 33-9. Gravid proglottid of *Taenia saginata,* **A,** compared with that of *Taenia solium,* **B.** The uterus of *Taenia saginata* has more than 14 primary lateral branches; that of *Taenia solium* has less than 12.

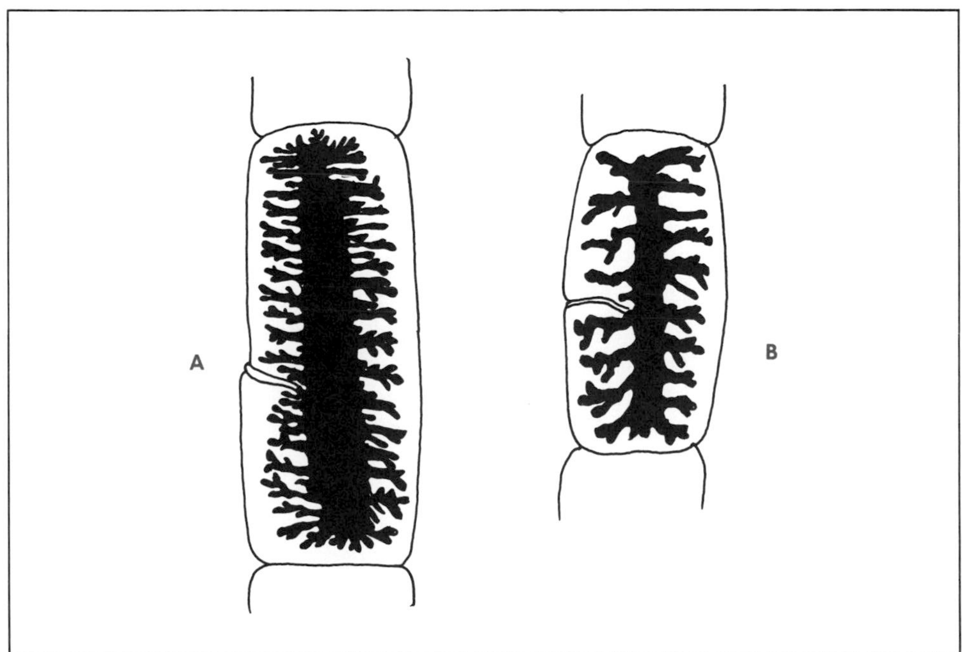

Taenia saginata (beef tapeworm, unarmed tapeworm)

Infection by *Taenia saginata* is quite common. The cow and giraffe are the intermediate hosts. Typically only one worm is present in man. The segments have independent motility and may escape through the anal canal.

■ **Anatomy.** The organism *Taenia saginata* ranges in length from 4 to 10 meters. The head is small (1.5 mm. in diameter), pear shaped, and somewhat quadrangular. It has four suckers but no hooklets. The absence of hooklets gives it the name "unarmed" tapeworm. The neck is rather long and slender. The mature segments measure 5 to 7 mm. by 18 to 20 mm. The uterus extends along the midline and gives off 20 to 30 delicate branches on each side. The eggs are spherical or ovoid in shape and yellow or brown in color. They measure 20μ to 30μ by 30μ to 40μ.

Taenia solium (pork tapeworm)

Infection with *Taenia solium* is rare in America and is acquired by eating measly pork. As a rule only one worm is present, but occasionally two or more are found.

■ **Biologic features.** Of the two worms, *Taenia solium* is shorter than *Taenia saginata,* measuring 2 to 8 meters in length. The head is very dark in color and globular or quadrangular in shape. It is provided with four suckers and two rows of hooklets projecting from a rostellum. The neck is threadlike. The mature segments measure 5 to 6 mm. by 10 to 12 mm.

The ova resemble those of *Taenia saginata.* Practically, it is impossible to distinguish the two. The differentiation of the worms, therefore, depends upon the characteristics of the uterus in the terminal proglottids. There are more branches coming from the sides of the uterus in the beef tapeworm than in the pork tapeworm (Fig. 33-9). Man may also become infected by swallowing the eggs of *Taenia solium* because it is possible for both cycles of development to occur in the human being.

Hymenolepis nana (dwarf tapeworm)

The smallest and the one most frequently found in man, the dwarf tapeworm, measures from 1 to 4 cm. in length. The head is round and provided with four suckers and a single row of 24 to 30 hooklets. The ova are characteristic. As a rule, many worms and ova are present, but the worms are so degenerated as to be unrecognizable.

Infection is spread directly from one person to another; there is no intermediate host. Eggs containing a fully developed embryo are released from a disintegrating end segment and passed in the feces. The eggs hatch in the stomach or small intestine of the new host and develop in the mucous membrane of the small intestine. The larval forms pass into the lumen and attach to the intestinal wall lower down. In about 2 weeks adult worms appear.

Dipylidium caninum (dog tapeworm)

Dipylidium caninum is common in cats and dogs. It is 15 to 30 cm. in length. The head is small with four suckers and about 60 hooklets arranged in four rows. The ova occur in groups and are usually passed within the segments.

Diphyllobothrium latum (fish tapeworm or broad Russian tapeworm)

Diphyllobothrium latum usually measures 3 to 6 meters in length, but occasionally a length of 12 meters is reached. The head, flattened and almond shaped, is provided

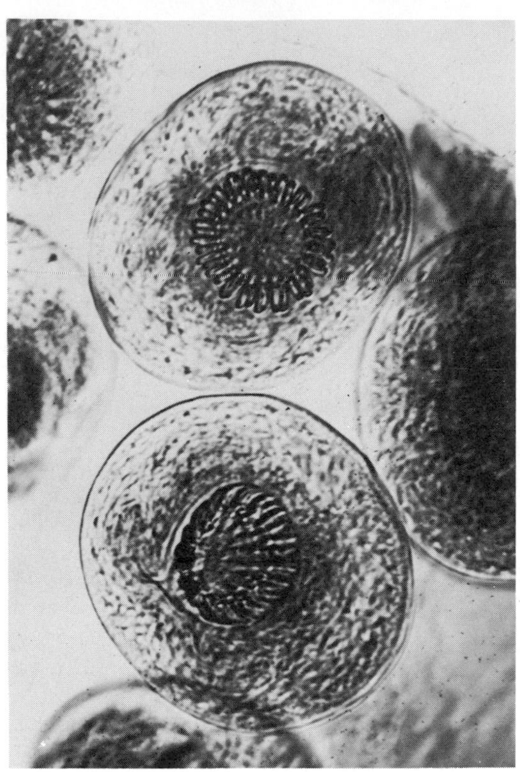

FIG. 33-10. Hydatid sand *(Echinococcus granulosus).* (From Frankel, S., Reitman, S., and Sonnenwirth, A. C.: Gradwohl's clinical laboratory methods and diagnosis, ed. 7, St. Louis, 1970, The C. V. Mosby Co.)

with two lateral grooves for attachment to the intestinal mucosa of the host. The segments show a characteristic brown or black rosette formation (uterus filled with ova).

The presence of this worm gives rise to irregular fever, digestive disturbances, and a blood picture identical to that of pernicious anemia. Manifestations promptly subside after its removal.

Echinococcus granulosus

Echinococcus granulosus is a tiny worm 3 to 6 mm. long with a pear-shaped head. It has four suckers, 30 to 36 hooklets, but only one each of the immature, mature, and gravid proglottids. The adult host is the dog and other canines. By ingesting the eggs man can become the larval or intermediate host and the unique development of the scolex within the cystic cavity can occur in his organs. The liver is a favored site. The result is termed the *hydatid cyst.* Free-floating scolices and brood capsules, structures within which scolices are formed, in cyst fluid are known as *hydatid sand* (Fig. 33-10). The gravity of the infection in man is determined by the location and the proportions that the cyst takes on with time.

Nematodes (roundworms)

Nematodes are nonsegmented worms with a flattened cylindrical body tapering toward both ends. The mouth is frequently surrounded by thick lips or papillae, and there is a complete digestive tract. The sexes are distinct. The male is shorter and more

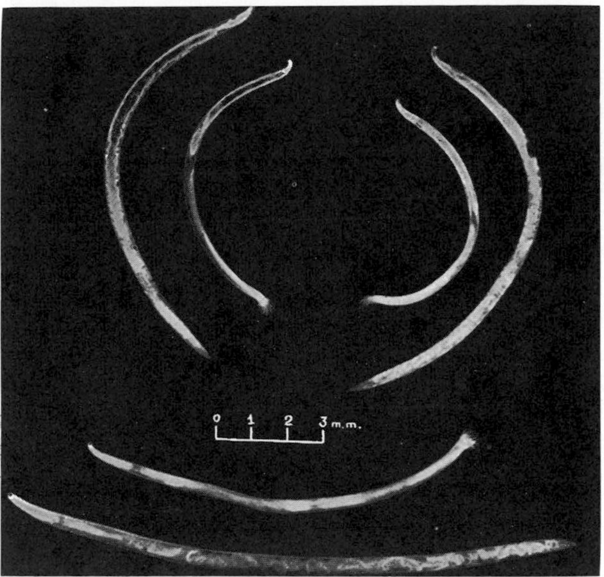

FIG. 33-11. Hookworms. Above, *Nector americanus* (male and female), two pairs; below, *Ancylostoma duodenale* (male and female), one pair. Note millimeter scale. (Original figure of P. Kourí.) (From Frankel, S., Reitman, S., and Sonnenwirth, A. C., editors: Gradwohl's clinical laboratory methods and diagnosis, ed. 7, St. Louis, 1970, The C. V. Mosby Co., vol. 12.)

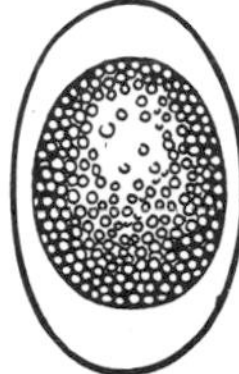
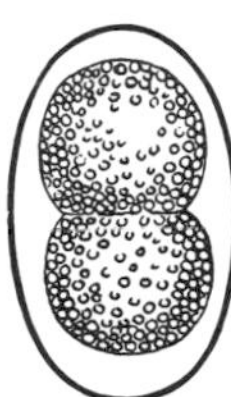
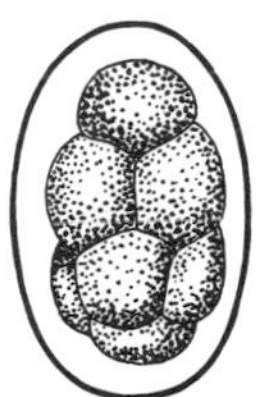

FIG. 33-12. Hookworm eggs in sketch to show development from earliest stage to the formation of the embryo.

slender than is the female. The adults inhabit the intestinal tract of man, and as a rule, there is no intermediate host. The typical nematode life cycle passes through a series of stages from the larval forms to the adult worm.

The severity of the digestive disorders related to nematode infection depends upon the load of worms carried.

Ancyclostoma duodenale and Necator americanus

Ancylostoma duodenale and *Necator americanus* are, respectively, the Old World and New World hookworms. They resemble each other closely but differ in several important details (Fig. 33-11).

■ **Anatomy.** Both species are pale red and pointed at both ends. As a rule, the adult *Ancylostoma duodenale* is larger than *Necator americanus.* Its mouth has a pair of ventral hooks on each side of the midline and a pair of dorsal hooks. The mouth of *Necator americanus* is provided with plates instead of ventral hooks and has a distinct dorsal conical toothlike structure.

The ova of the species are practically identical, except that those of *Necator americanus* are the larger. They are oval or oblong in shape but in certain positions appear spherical. They have three distinct parts—the shell, the yolk, and a clear space between the yolk and shell (see Fig. 33-12). The thin smooth shell appears as a distinct line. Eggs that have been passed for 24 hours or more show well-developed embryos. The ova tend to adhere noticeably to glass or other surfaces. Advantage is taken of this trait in certain diagnostic procedures, but it makes the thorough washing of laboratory glassware imperative.

FIG. 33-13. Life cycle of *Ancylostoma duodenale.*

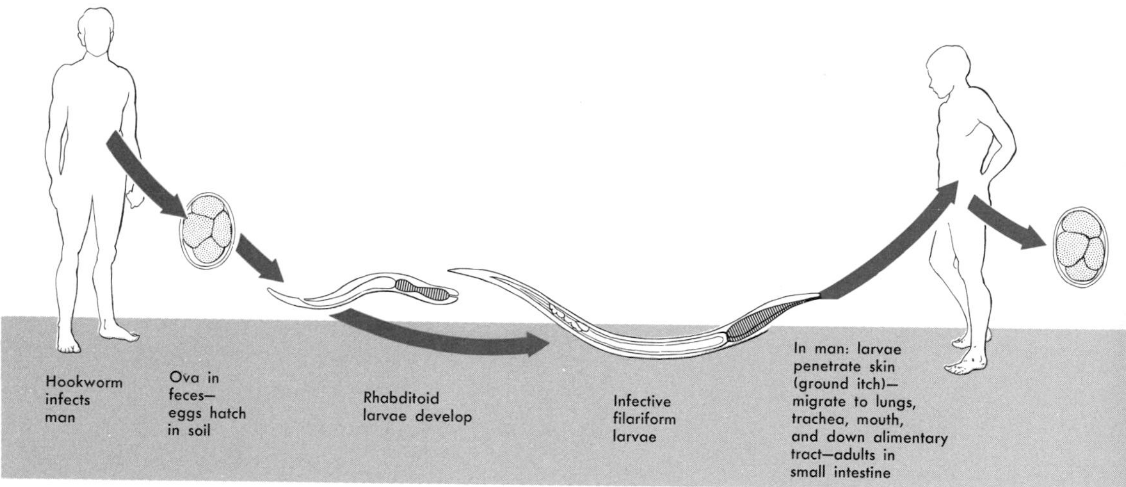
Hookworm infects man
Ova in feces—eggs hatch in soil
Rhabditoid larvae develop
Infective filariform larvae
In man: larvae penetrate skin (ground itch)—migrate to lungs, trachea, mouth, and down alimentary tract—adults in small intestine

FIG. 33-14. Life cycle of *Strongyloides stercoralis.*

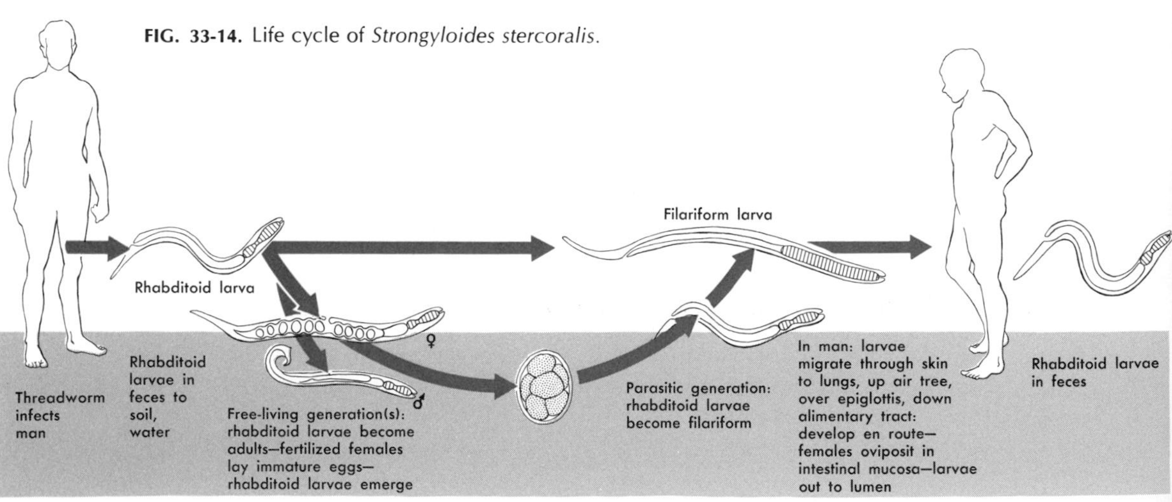
Filariform larva
Rhabditoid larva
♀
♂
Threadworm infects man
Rhabditoid larvae in feces to soil, water
Free-living generation(s): rhabditoid larvae become adults—fertilized females lay immature eggs—rhabditoid larvae emerge
Parasitic generation: rhabditoid larvae become filariform
In man: larvae migrate through skin to lungs, up air tree, over epiglottis, down alimentary tract: develop en route—females oviposit in intestinal mucosa—larvae out to lumen
Rhabditoid larvae in feces

■ **Habitat.** The adult worms live in the small intestine attached to the mucous membrane, where their presence produces a characteristic train of events. Great numbers are usually present. They tear the tissues to get to the small blood vessels, from which blood is pumped into their intestines. However, they wastefully extravasate much of the blood into the lumen of the intestinal tract. A profound anemia is secondary to their bloodsucking activities.

■ **Life history.** The life history of the hookworm is as follows (see Fig. 33-13): After the ova are passed, development begins with the proper temperature and moisture. The larvae* hatch and undergo certain developmental changes whereby they can infect a new host. When the larvae contact the skin of man, they penetrate it, producing a dermatitis (ground itch). They pass by the lymph and bloodstream to the lungs. In the lungs they gain access to the bronchi and are carried by the bronchial secretions to the pharynx, where they are swallowed. After they reach the small intestine, they develop into adult worms. The adult worm is seldom found in the feces if there is no anthelmintic treatment. Hookworms parasitize an estimated 456 million persons.

Strongyloides stercoralis

The adult *Strongyloides stercoralis* is only about 2 mm. long. It has a four-lipped mouth and an esophagus that extends through the anterior fourth of the body. Male worms have not been found in man. The adult females live deep in the intestinal mucosa, where the ova are deposited. The larvae hatch in the intestines and are passed in the stool (see Fig. 33-14). Neither adult worms nor ova appear in the feces unless active purgation is present. The larvae are 250μ to 500μ in length. They are actively motile, and, when fresh, are constantly wiggling and bending but have little progressive motion. The disturbance produced in the laboratory preparation is often noticed under the microscope before the larvae are seen. Infection is acquired when the larvae penetrate the skin or are accidentally swallowed.

***Rhabditoid* larvae are emerging first-stage larvae of typical shape. This is the free-living stage. Rhabditoid larvae can metamorphose into postfeeding, mature *filariform* larvae—long delicate threadlike organisms. This is the infective stage.

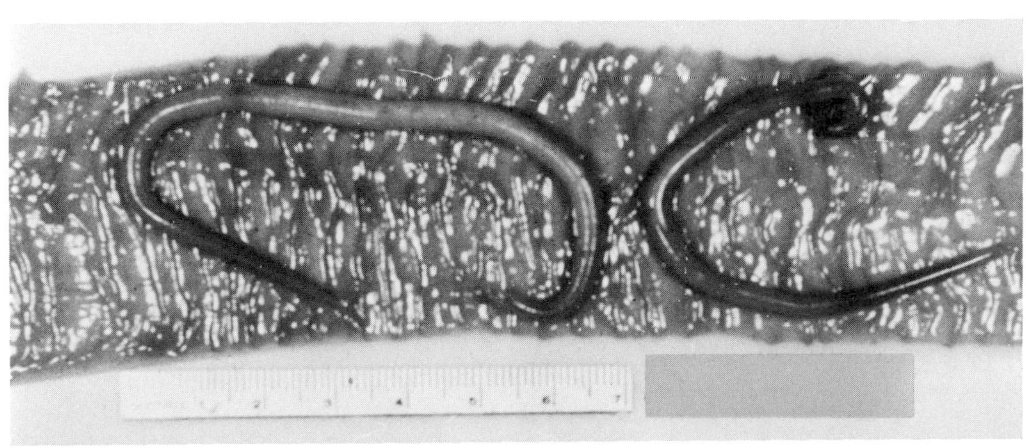

FIG. 33-15. *Ascaris lumbricoides* in lumen of intestine.

Ascaris lumbricoides (eelworm or roundworm)

Ascaris lumbricoides is the largest intestinal nematode and is harbored by approximately 644 million persons. It is fusiform in shape and yellow or reddish in color. The male measures 15 to 20 cm. in length, the female 20 to 40 cm. The head is relatively small, and the oral cavity has three serrated lips. This worm looks like the ordinary earthworm but is not so red (Fig. 33-15).

The habitat of *Ascaris lumbricoides* is the upper end of the small intestine, but it may be found in any part of the intestinal tract, free in the peritoneal cavity, or in the trachea and bronchi. Several are usually present, clinging together to form palpable masses or even causing intestinal obstruction. Infection is most frequently seen in children under 10 years of age.

The fertilized eggs are oval in shape and average 48μ in diameter and 62μ in

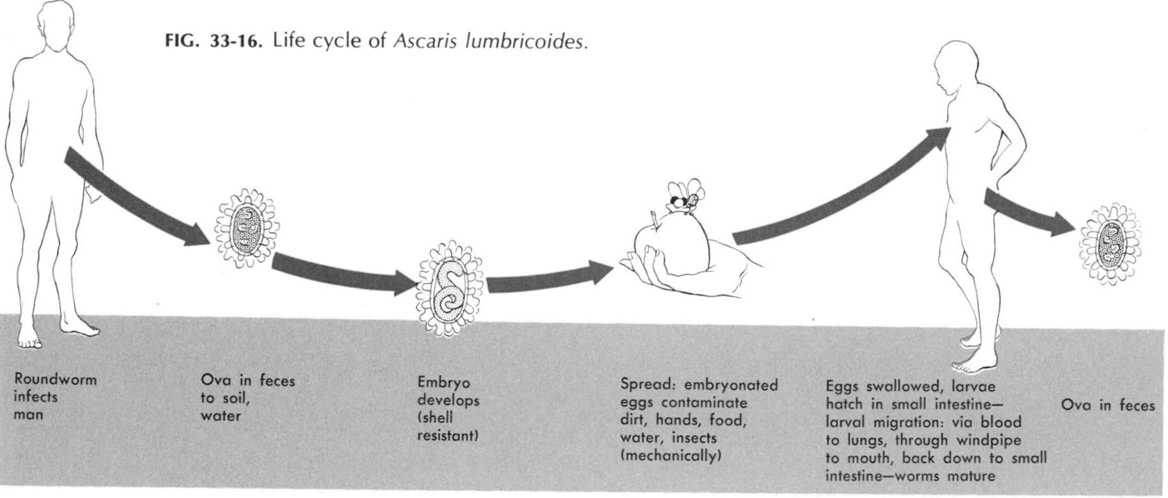
Roundworm infects man
Ova in feces to soil, water
Embryo develops (shell resistant)
Spread: embryonated eggs contaminate dirt, hands, food, water, insects (mechanically)
Eggs swallowed, larvae hatch in small intestine—larval migration: via blood to lungs, through windpipe to mouth, back down to small intestine—worms mature
Ova in feces

FIG. 33-16. Life cycle of *Ascaris lumbricoides*.

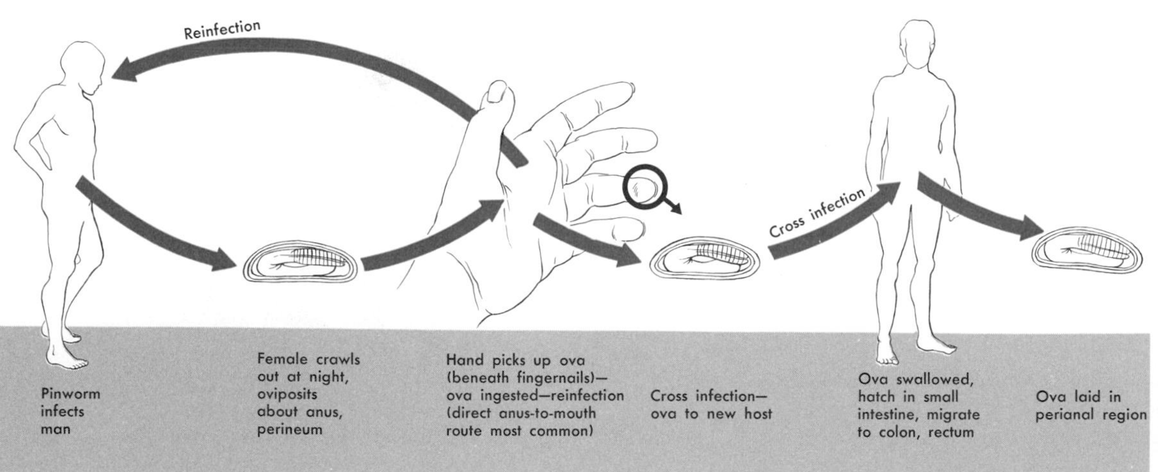
Reinfection
Cross infection
Pinworm infects man
Female crawls out at night, oviposits about anus, perineum
Hand picks up ova (beneath fingernails)—ova ingested—reinfection (direct anus-to-mouth route most common)
Cross infection—ova to new host
Ova swallowed, hatch in small intestine, migrate to colon, rectum
Ova laid in perianal region

FIG. 33-17. Life cycle of *Enterobius vermicularis*.

length. If only female worms are present, unfertilized eggs will be found. They are elongated, irregular in shape, and bear little resemblance to the fertilized egg. They frequently escape detection.

Adult worms appear in the feces only with active purgation. After the eggs are passed, segmentation takes place, and an embryo develops. If these mature eggs are swallowed, the embryos escape from the eggs and pass to the lungs by the bloodstream. They reach the intestines in the same manner as do hookworm larvae (see Fig. 33-16).

Enterobius vermicularis (Oxyuris vermicularis, threadworm, pinworm, or seatworm)

Infection with *Enterobius vermicularis* is the most prevalent worm infection of children and adults in the United States. The males of this species measure 3 to 5 mm. in length; the females measure about 10 mm. The adult female worms migrate through the anus and deposit their eggs on the perianal region, most frequently at night. Because of the peculiar laying habits of the female, the eggs seldom occur in the feces but are present around the anal region. They are best found by scraping this region and examining the scrapings. Eggs may be picked up from the skin of the perianal region by means of swabs made of cellulose adhesive tape, the sticky side applied to the skin. The eggs are removed from the tape by toluene and identified under the microscope. An enema may be given, and the adult worms can be identified in the stool that is passed.

Deposition of eggs on the perianal skin results in intense itching. In children, pinworms should be suspected from this finding alone. The small child may maintain the infection from ova collected under his fingernails when he scratches himself.

Infection comes from swallowing the eggs (see Fig. 33-17), after which male and female parasites hatch out at the lower end of the small intestine. After fertilizing the females, the males die, and the females migrate to the colon and rectum. The patient may continually reinfect himself, and parents may acquire the infection from their children. Eggs may be widely disseminated in a household or in an institution, in the dust, in the clothing, bedding, on furniture, on doorknobs, and so on. To eliminate them is an exasperating and almost hopeless task.

Trichuris trichiura (Trichocephalus trichiurus, Trichocephalus dispar, whipworm)

Trichuris trichiura is characterized by a long threadlike neck that makes up about one half the length of the body. The male is 30 to 45 mm. in length; the female is somewhat longer, 45 to 50 mm. The worms live in the cecum and large intestine, with the slender end of the worm embedded in the mucosa. The worms themselves are rare in the feces, and the eggs are not abundant. Generally, symptoms are related to the number of worms in the bowel (Fig. 33-18).

Trichinella spiralis

Trichinella spiralis is the cause of trichinosis. It is the smallest worm, with the exception of *Strongyloides stercoralis,* found in the intestinal canal. It is barely visible with the unaided eye. The males are about 1.5 mm. in length and the females 3 to 4 mm. in length. The posterior end of the male is bifid and has two tonguelike appendages.

The infection is primarily one of rats, propagated because rats eat their dead. Hogs acquire the infection from rats, and man becomes infected by eating insufficiently cooked pork. Pork is not the only source. Man has acquired the infection from eating

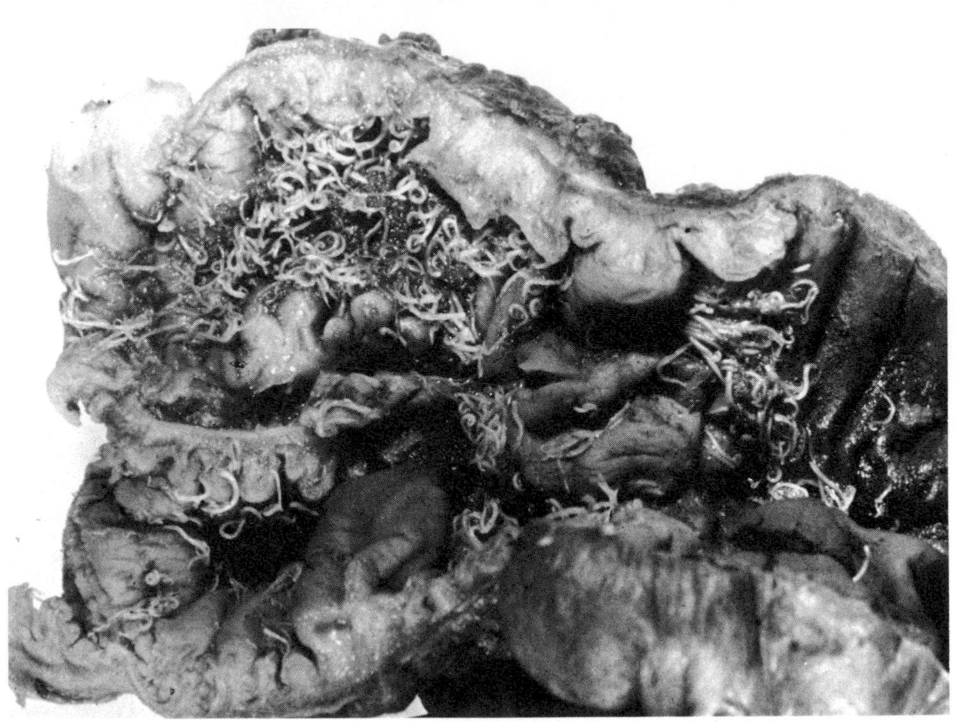

FIG. 33-18. *Trichuris trichiura,* massive infestation in a young child producing severe hemorrhagic diarrhea and death. The segment of bowel is opened to show the many whipworms on the mucosal surface. (From Anderson, W. A. D.: Pathology, ed. 6, St. Louis, 1971, The C. V. Mosby Co.)

bear meat. Polar bears are said to be heavily infected. The fact that it is found in the arctic region indicates that the parasite possesses a unique tolerance for cold.

■ **Life history.** The cycle of development of *Trichinella spiralis* (Fig. 33-19) is practically the same in man as in other animals. Pork containing the encysted larvae is eaten and the cyst capsule digested away. The larvae pass to the small intestine, where they mature. After copulation the males die, and the females embed in the mucous membrane, where they give birth to as many as 1000 to 1500 larvae. These larvae migrate by the lymph and bloodstream to the skeletal muscles, where they encyst, become encapsulated, and subsequently calcify (Fig. 33-20).

The free larvae measure 90μ to 100μ in length and 6μ in diameter. They may be found in the blood and spinal fluid during the period of migration (6 to 22 days after infection).

After encystment occurs, the coiled embryos may be found with the low power of the microscope in a teased portion of muscle and the diagnosis made. The cysts are most frequently found at the tendinous insertions of the muscle, and the muscles most frequently infected are the pectoralis major, the outer head of the gastrocnemius, the deltoid, and the lower portion of the biceps. The cysts appear as white specks, measuring 250μ by 400μ. The long axis of the cyst extends in the same general direction as the fibers of the muscle. Muscle biopsy is the surest method in diagnosis.

FIG. 33-19. Life cycle of *Trichinella spiralis.*

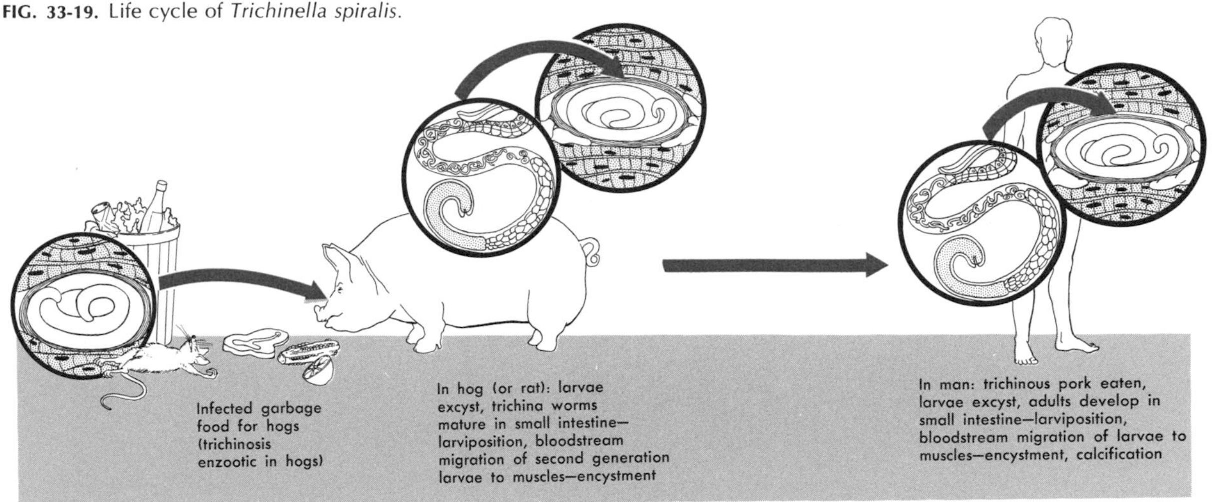

Infected garbage food for hogs (trichinosis enzootic in hogs)
In hog (or rat): larvae excyst, trichina worms mature in small intestine—larviposition, bloodstream migration of second generation larvae to muscles—encystment
In man: trichinous pork eaten, larvae excyst, adults develop in small intestine—larviposition, bloodstream migration of larvae to muscles—encystment, calcification

■ **Trichinosis.** When the parasites are developing in the intestines, gastrointestinal disturbances are prominent. These appear 2 to 3 days after ingestion of the contaminated pork. During this time the adult worms may be found in the feces. When the larvae migrate, fever, delirium, rheumatic pains, and labored respiration are present. This period begins at the end of 1 week after infection and lasts 1 or 2 weeks. When encystment begins, edema and skin eruptions appear. This period lasts about 1 week. After the disease becomes chronic, muscular pains of a rheumatic character may be present for months.

■ **Laboratory diagnosis.** Intense eosinophilia, commonly over 500 eosinophilic leukocytes per cubic millimeter of blood, is a feature of all stages of the disease. A skin test is available. A flocculation test, the Sussenguth-Kline test, becomes positive 2 to 3 weeks after infection and remains so for 10 months or longer. There are other serologic tests including complement fixation, latex agglutination, hemagglutination, and fluorescent antibody technics.

Identification of the worms in feces or larvae in other clinical specimens is usually impractical because of the course of the disease. Biopsy of a tender muscle may not be done. Generally the serologic tests must be used for indicating recent or current infection. The skin test is consistent with more remote infection.

■ **Prevention.** There is no simple inspection method at a slaughterhouse for the detection of trichinas in a carcass of meat. The elimination of the disease rests practically with adequate cooking of pork. For example, one should cook a pork roast at an oven temperature of at least 350° F., allowing 35 to 50 minutes per pound. Smoking, pickling, heavy seasoning, or spicing, does not make uncooked pork products safe. Freezing meat at −15° C. for 30 days or at −28.9° C. for 6 to 12 days eliminates the larvae.

Wuchereria bancrofti (Bancroft's filaria worm)

Filariasis is infection with the filarial worms, which incorporate certain unique features in their life cycle. The best known is *Wuchereria bancrofti.* Man is its reservoir of infection and the mosquito its vector. Of historical note is the fact that the first

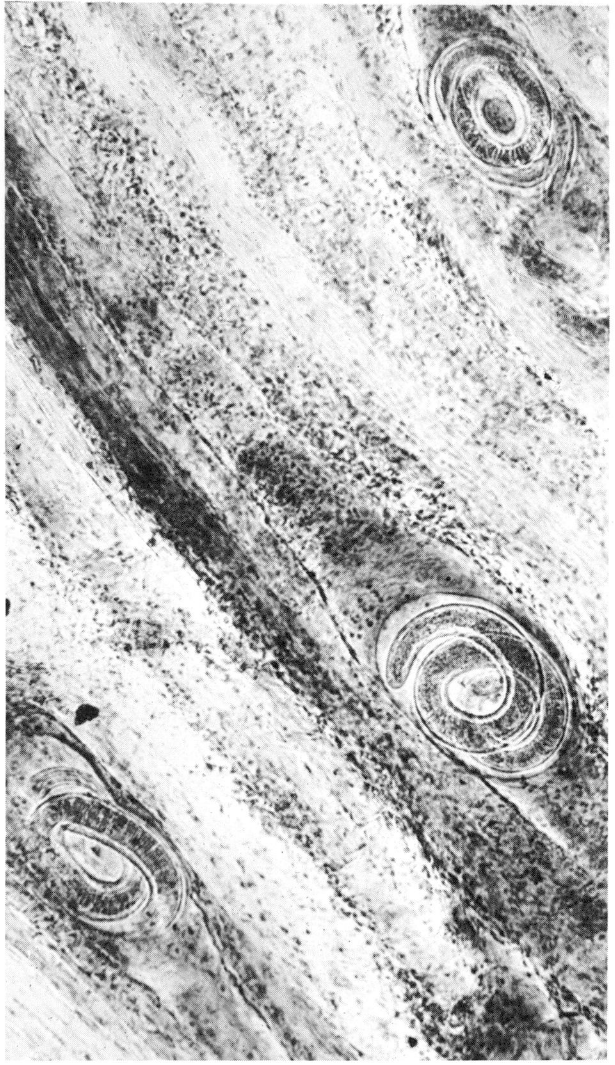

FIG. 33-20. *Trichinella spiralis* encysted in muscle, microscopic section. Larvae may live 10 to 20 years in these cysts. (From Hickman, C. P.: Integrated principles of zoology, ed. 4, St. Louis, 1970, The C. V. Mosby Co.)

demonstration of the mosquito as vector of disease was in connection with Bancroft's filaria worm, and the work suggested to Sir Ronald Ross the possibility that malaria might be similarly transmitted.*

■ **Life cycle.** The life cycle of *Wuchereria bancrofti* is sketched as follows (Fig. 33-21): When man, the definitive host, is infected, slender white adult male and female worms

*In 1878, Sir Patrick Manson, in Amoy, China, showed *Culex* mosquitoes to be the natural transmitters of filarial worms. In Great Britain, Manson is known as the father of tropical medicine.

FIG. 33-21. Life cycle of *Wuchereria bancrofti.*

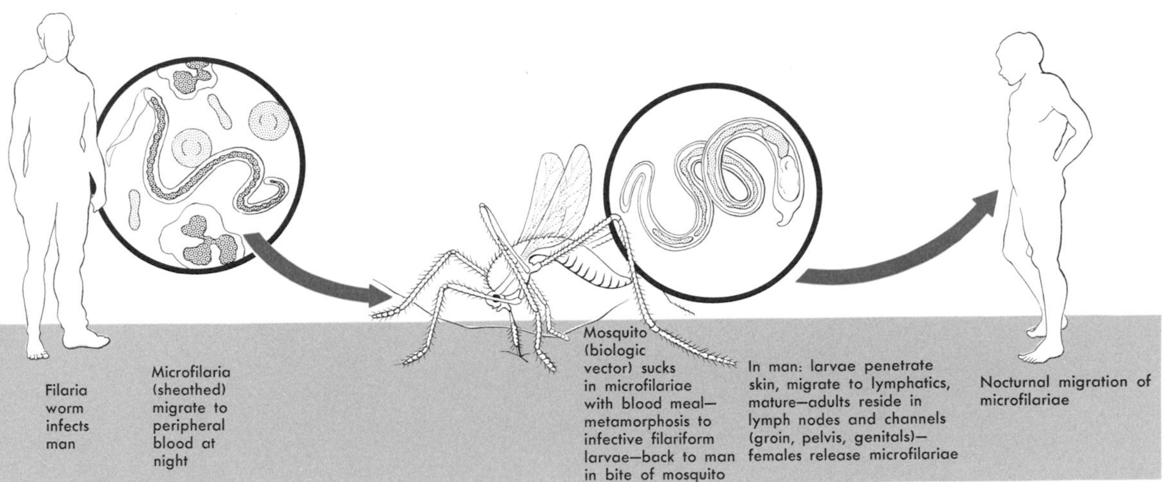
Filaria worm infects man
Microfilaria (sheathed) migrate to peripheral blood at night
Mosquito (biologic vector) sucks in microfilariae with blood meal—metamorphosis to infective filariform larvae—back to man in bite of mosquito
In man: larvae penetrate skin, migrate to lymphatics, mature—adults reside in lymph nodes and channels (groin, pelvis, genitals)—females release microfilariae
Nocturnal migration of microfilariae

reside in the lymphatic system. The females are 80 to 100 mm. in length and 0.24 to 0.3 mm. in diameter; the males are 40 mm. long and 0.1 mm. in diameter. In the uterus of the adult female, embryos develop as tightly coiled threads within eggshells. About the time an egg is laid, the embryo uncoils into a tiny, delicate, eel-like form. The eggshell remains applied about the elongated embryo and it is said to be sheathed. In some species of filarial worms, a naked or unsheathed embryo is discharged. The embryos in the lymph and bloodstream become *microfilariae* (Fig. 33-22) with remarkable habits. One is their migration into the peripheral blood at night at a time that coincides with the feeding time of the vector mosquito. During the day, they are hidden away in an undetermined site.

Taken into the body of the mosquito, the microfilariae lose their sheath and develop in a larval series. In time, the infective larvae escape from the mosquito as it takes a blood meal, returning to the definitive host in whose lymphatic channels they continue their growth. Adolescent worms gather within sinuses of lymph nodes of the groin and pelvis, where they mature and mate to repeat the cycle.

■ **The disease.** In localizing within the human lymphatic system the adult worms start an inflammatory process that progresses from acute to chronic stages. It is associated with tissue changes causing obstruction of lymphatic vessels, stagnation of lymph flow, and proliferation of connective tissue. The skin of the lower extremities and external genitalia becomes thickened, coarse, and redundant. The tremendous enlargement of the affected part becomes a great burden to the victim. The results of long years of infection, the end stage lesion, is referred to as *elephantiasis.* This disorder was well known to the ancient Hindu physicians around 600 B.C.

Filariasis is a widely distributed disease in tropical areas of the world and extends into some of the subtropical areas. Man is the only known definitive host. The appropriate mosquitoes breeding close by the dwelling places of human beings with microfilariae circulating in their bloodstream are two factors that favor the continuous

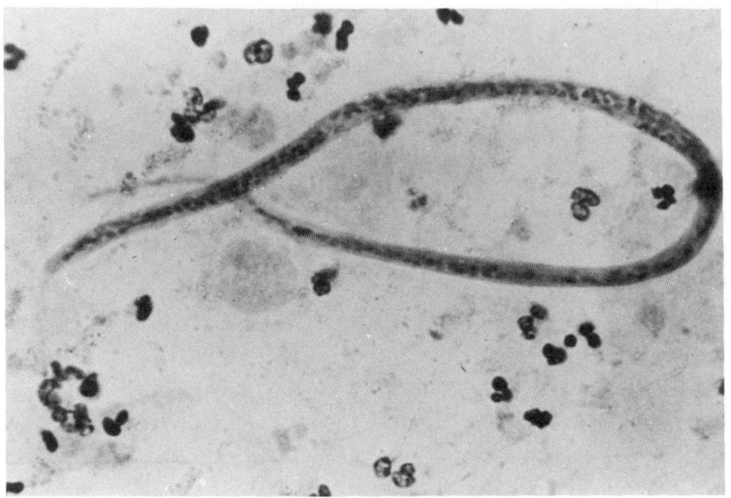

FIG. 33-22. Microfilaria in blood smear, photomicrograph.

spread of this disease. Therefore, methods of prevention and control must be designed for treatment of the human carrier and elimination of the insect vector.

■ **Laboratory diagnosis.** The diagnosis of filariasis lies in the identification of the microfilariae found in blood films. In a wet mount of blood, they are seen to move gracefully, pushing the red blood cells gently aside. They may be seen also in a dried, fixed blood smear stained with Wright's stain. For best results with *Wuchereria bancrofti,* peripheral blood films are made between the hours of 10 P.M. and 2 A.M.

Onchocerca volvulus

Onchocerciasis or river blindness is a disease afflicting an estimated 300 million persons in Africa and Latin America. It is produced by the filarial worm, *Onchocerca volvulus.* A hardy black gnat of the genus *Simulium,* which breeds in fast-flowing mountain streams, transmits the infection as intermediate host. In the bite of the fly, larvae of the threadlike worm are deposited beneath the skin, the site of residence of the adult worms. The female worm gives birth to unsheathed microfilariae, which pass into the lymphatic system. They are rarely found in peripheral blood. The microfilariae migrate throughout the body via the lymphatics. They die upon reaching the eyes, but not until they have induced serious changes in the tissues of the eye, which can in time lead to complete blindness.

Macroscopic examination of feces for parasites

Sometimes the nurse or medical attendant must examine feces macroscopically for parasites. In such a case, the procedure is as follows:

Improvise a sieve by removing both ends of a can and tying one or two layers of gauze over one opening. Allow water to run slowly over the feces and break up the masses with a glass rod or wooden applicator. Note the presence of worms. If necessary, remove worms and examine with a hand lens or the eyepiece of the microscope. Small worms may be mounted in a drop of water for macroscopic examination. If a tapeworm is present, be sure to search carefully for the head. After tapeworm treatment, the entire quantity of feces passed should be saved in order that the search for the head may be thorough. If it is suspected that the patient is suffering from typhoid fever or any other condition in which pathogenic bacteria appear in the feces, the washings should be received in a vessel containing a disinfectant. Needless to state, the sieve, glassware, and utensils used should be properly sterilized or carefully discarded.

■ **Objects likely to be mistaken for intestinal worms.** Segmented strands of mucus may be mistaken for tapeworms. The same is true of banana fibers because of their segmented structure and oval cells. Fibers of celery and green vegetables may be mistaken for roundworms and orange fibers for pinworms.

QUESTIONS FOR REVIEW

1. Define microfilariae, hydatid sand, mechanical vector, biologic vector, ova, scolex, proglottid, ground itch, definitive host, cysticercus, hermaphrodite, operculate, phototrophic.
2. List five arthropod vectors in disease transmission. Give the diseases spread by them.
3. Briefly describe the three phyla of metazoa concerned in disease production in man.
4. Outline the life cycle of the following:
   a. Blood flukes
   b. Beef tapeworm
   c. Pork tapeworm
   d. Fish tapeworm
   e. Hookworm
   f. Pinworm
   g. Roundworm
   h. Filarial worms
5. Discuss the pathogenicity of flukes.
6. What is elephantiasis? River blindness?
7. Give the laboratory diagnosis for
   a. Schistosomiasis
   b. Trichinosis
   c. Filariasis
   d. Onchocerciasis
   e. Enterobiasis
8. List objects mistaken for worms in a *macro*scopic stool examination.
9. Comment on the control and prevention of metazoan diseases.

REFERENCES FOR CHAPTERS 32 AND 33

Aziz, E. M.: *Strongyloides stercoralis* infestation, Southern Med. J. **62:**806, 1969.

Babb, R. R., Peck, O. C., and Vescia, F. G.: Giardiasis, J.A.M.A. **217:**1359, 1971.

Barrett-Connor, E.: Amebiasis, today in the United States, Calif. Med. **114:**1, (March) 1971.

Barrett-Connor, E.: Human fluke infections, Southern Med. J. **65:**86, 1972.

Beck, J. W., and Barrett-Connor, E.: Medical parasitology, St. Louis, 1971, The C. V. Mosby Co.

Belding, D. L.: Textbook of parasitology, New York, 1968, Appleton-Century-Crofts.

Bowman, J. E., et al.: Published malaria standards misleading and dangerous, Lab. Med. **2:**19, (Jan.) 1971.

Cahill, K. M.: Tropical diseases in a tropical climate, Hosp. Pract. **3:**56, (April) 1968.

Carter, J. P.: Nutrition and parasitism, Southern Med. Bull. **59:**31, (Oct.) 1971.

Cutting, R. T.: The army malaria program—prelude to triumph, Southern Med. Bull. **57:**29, (June) 1969.

Dangers of eating bear meat, editorial, J.A.M.A. **220:**274, 1972.

Expert Committee on Malaria, Fourteenth report, WHO Techn. Rep. Ser. no. 382, 1968.

The facts about toxoplasmosis: Today's Health **50:**64, (July) 1972.

Faust, E. C., Beaver, P. C., and Jung, R. C.: Animal agents and vectors of human diseases, Philadelphia, 1968, Lea & Febiger.

Faust, E. C., et al.: Craig and Faust's Clinical parasitology, Philadelphia, 1970, Lea & Febiger.

Federman, S.: Underwater weapon in war on mosquitoes, Today's Health **46:**14, (June) 1968.

Feldman, H. A.: Toxoplasma and toxoplasmosis, Hosp. Pract. **4:**64, (March) 1969.

Felsenfeld, O.: Synopsis of clinical tropical medicine—pathogenesis, clinical picture, diagnosis, prognosis, and therapy, St. Louis, 1965, The C. V. Mosby Co.
Fletcher, J. R., et al.: Acute *Plasmodium falciparum* malaria, Arch. Intern. Med. **129:**617, 1972.
Glor, B. A. K.: Falciparum malaria in Vietnam: Clinical manifestations and nursing care requirements, Milit. Med. **134:**181, (March) 1969.
Gould, S. E.: The story of trichinosis, Amer. J. Clin. Path. **55:**2, 1971.
Gould, S. E.: Trichinosis in man and animals, Springfield, Ill., 1970, Charles C Thomas, Publisher.
Greenberg, J. H.: Public health problems relating to the Vietnam returnee, J.A.M.A. **207:**697, 1969.
Hagan, A. D.: Malaria in Vietnam—1969, Southern Med. Bull. **58:**19, (April) 1970.
The immunology of malaria, Canad. Med. Ass. J. **106:**852, 1972.
Jones, J. C.: The sexual life of a mosquito, Sci. Amer. **218:**108, (April) 1968.
Jose, D. G., Gatti, R. A., and Good, R. A.: Eosinophilia with *Pneumocystis carinii* pneumonia and immune deficiency syndromes, J. Pediat., **79:**748, 1971.
Juniper, K., Jr., et al.: Serologic diagnosis of amebiasis, Amer. J. Trop. Med. **21:**157, 1972.
Marcial-Rojas, R. A.: Pathology of protozoal and helminthic disease, Baltimore, 1971, The Williams & Wilkins Co.
Medical News: Tracking mosquitoes with jar and paddle, J.A.M.A. **203:**37, (Feb. 12) 1968.
Miller, J. H., and Abadie, S. H.: Common intestinal parasites of the United States, Southern Med. Bull. **59:**11, (Oct.) 1971.
Mining an abandoned vein, editorial, J.A.M.A. **211:**1536, 1970.
New criteria for histolytica, Med. World News **11:**27, (March 20) 1970.
Quinn, R. W.: The epidemiology of intestinal parasites of importance in the United States, Southern Med. Bull. **59:**20, (Oct.) 1971.
Ramos-Morales, F., Sotomayor, Z. R., Diaz-Rivera, R., et al.: Manson's schistosomiasis in Puerto Rico, Bull. N. Y. Acad. Med. **44:**317, 1968.
Robson, V. L.: Malaria alert, J. Amer. Med. Wom. Ass. **22:**321, (May) 1967.
Rodgerson, E. B.: The diagnosis of *Trichomonas vaginalis,* Woman Physician **25:**576, 1970.
Seah, S. K. K., and Flegel, K. M.: African trypanosomiasis in Canada, Canad. Med. Ass. J. **106:**902, 1972.
Smith, H. A., Jones, T. C., and Hunt, R. D.: Veterinary pathology. Philadelphia, 1972, Lea & Febiger.
Sonnenwirth, A. C., Keating, J. P., and Waltman, S.: Malaria (1969), J.A.M.A. **209:**687, 1969.
Sundman, P. O.: The flight of the eagle, New York, 1970, Pantheon Books, Inc.
Teschan, P. E., editor: Panel on Malaria, Ann. Intern. Med. **70:**127, (Jan.) 1969.
Thompson, J. H., Jr.: How to detect blood and tissue parasites, Lab. Med. **2:**42, (April) 1971.
Thompson, J. H., Jr.: How to detect in intestinal parasites, Lab. Med. **1:**31, (April) 1970.
Thompson, J. H.: Useful laboratory procedures: Examination of blood smears for malaria and other organisms, III, Bull. Path. **10:**6, 1969.
Wand, M., and Lyman, D.: Trichinosis from bear meat, J.A.M.A. **220:**245, 1972.
Wilkinson, J. F.: The deadly animal that adopts people, Today's Health **45:**8, (Jan.) 1967.
Winslow, D. J., and Connor, D. H.: Malaria, Res. Phys. **14:**67, (Nov.) 1968.

LABORATORY SURVEY OF UNIT FIVE*

PROJECT

The pyogenic cocci

Part A—Study of morphology of pyogenic cocci

1. Examine microscopically prepared gram-stained smears of the following:
   a. *Staphylococcus aureus*
   b. *Streptococcus* species
   c. *Diplococcus pneumoniae*
   d. *Neisseria* species
2. Examine microscopically the capsule stain of *Diplococcus pneumoniae.* (A suitable specimen of sputum may be used if available.)
3. Note details of morphologic arrangement, gram-staining reactions. Make comparisons.

Part B—Study of cultural characteristics

1. Make cultures of a hemolytic *Staphylococcus aureus* on the following:
   a. Nutrient agar—note pigment.
   b. Blood agar—note hemolysis.
   c. Mannitol salt agar—note growth on medium with a high concentration of salt and fermentation of mannitol.
2. Make culture of *Diplococcus pneumoniae* on blood agar. Note alpha hemolysis and characteristic colonies.
3. Make culture of *Neisseria catarrhalis* on chocolate agar.
4. Examine a blood agar plate showing colonies of *Streptococcus pyogenes.* Note beta hemolysis.

Part C—Identification of colonies of cocci

1. Make a culture of a nasal swab from the nose.
   a. Inoculate a blood agar plate and mannitol salt agar. Incubate.
   b. Note characteristics of colonies. Use gram-stained bacterial smears to study morphology of bacteria in colonies present.
   Note: The differential characteristics of the organisms growing on blood agar should be discussed by the instructor.
2. Make a culture of a throat swab rubbed over your tonsils (if present) and the back part of your throat.
   a. Inoculate a blood agar plate.
   b. Note colonies. Make gram-stained smears.

*Since it would be impossible to investigate thoroughly all of the organisms in this group, we will present only a few preliminary tests indicating the nature of each organism and the detection of its infection.

Part D—Typing of pneumococci with demonstration by instructor and discussion

1. Discuss methods of typing pneumococci.
2. Compare pneumococci from cultures and from body fluids as sputum or peritoneal fluid of mouse. What important structure is not so well developed in pneumococci from cultures as in organisms from the body?
3. Demonstrate method of taking cultures to detect meningococcal carriers.
4. Distribute prepared smears.
   a. Pus containing gonococci stained with methylene blue and by Gram's method
   b. Purulent cerebrospinal fluid containing meningococci stained with methylene blue and by Gram's method
   Note the similarity of meningococci and gonococci.

PROJECT
The enteric bacilli

Part A—Study of morphology of gram-negative bacilli

1. Examine microscopically gram-stained smears of the following:
   a. *Escherichia coli*
   b. *Salmonella typhi* (or other *Salmonella* species)
   c. *Shigella flexneri*
   d. *Vibrio cholerae*
2. Note detail and gram-staining reaction.

Part B—Motility in hanging-drop preparation with demonstration by instructor and discussion

1. Emphasis on technic of making hanging-drop and on the precautions in handling a preparation of living infectious organisms.
2. Demonstration of motility of a pathogen, such as one of the salmonellae

Part C—Study of cultural characteristics

1. Make cultures of *Escherichia coli* on the following:
   a. Blood agar
   b. MacConkey agar
   c. *Salmonella-Shigella* agar
   d. Endo's medium
   e. Eosin-methylene blue agar
   f. Russell's double sugar agar
   g. Bismuth sulfite agar
   h. Lead acetate agar
2. Select a pathogen such as one of the salmonellae and make corresponding cultures on the media just mentioned for culture of *Escherichia coli.*
3. Compare the growth obtained on these media for the pathogen with that of the nonpathogen.

Note: The instructor should discuss the varied appearance of the colonies obtained and the reasons. The student should identify the lactose and nonlactose fermentors. The importance of this biochemical test should be stressed.

4. Make culture of *Pseudomonas aeruginosa* on blood agar. Note pigment produced. Examine smears of *Pseudomonas aeruginosa.*

Part D – Demonstration of a serologic reaction by instructor with discussion

1. The macroscopic slide agglutination test using growth of *Salmonella typhi* (or other *Salmonella)* and *Salmonella* polyvalent diagnostic serum
2. Discussion of bacterial agglutination in type-specific serum
3. Review of the significance of the agglutination test as a serologic reaction and diagnostic role of the Widal test in typhoid fever

---

PROJECT
Brucella and Pasteurella

1. Demonstration by the instructor
   a. Cultures of *Brucella*
   Note: Because of danger of infection, cultures of *Pasteurella* should be omitted.
   b. Prepared stained smears of these gram-negative rods
   c. The slide agglutination test for undulant fever
   d. The preserved liver of a guinea pig that has died of tularemia if available.
2. Discussion by instructor: diagnostic tests for tularemia and brucellosis

---

PROJECT
The hemophilic bacteria

Demonstration by instructor with brief discussion of the cough plate method of detecting the presence of *Bordetella pertussis*

PROJECT
Gram-positive bacilli

Part A – Study of morphology of the gram-positive bacilli – both aerobic and anaerobic

1. Examine microscopically prepared smears.
   a. *Corynebacterium diphtheriae* – in smears stained with:
      (1) Methylene blue
      (2) Albert's stain
   b. *Clostridium tetani* – in gram-stained smears
   c. *Bacillus anthracis* – in gram-stained smears
2. Note morphology of the bacteria, gram-staining reaction, arrangements, presence and position of spores. Make comparisons.
3. Compare prepared slides of the diphtheria bacillus with those of diphtheroid bacilli. What is the value of differential stains?

Part B – Study of cultural characteristics

1. Study prepared cultures of *Corynebacterium diphtheriae.*
   a. Examine a 24-hour culture on Löffler's serum medium.
   b. Examine a culture on blood-tellurite agar.
2. Study prepared cultures of diphtheroid bacilli.
3. Note morphology of colonies, especially on the differential media.
4. Compare organisms grown on one medium when examined in stained smears with those grown on the other.
5. Compare the growth of the diphtheria bacilli with that of the diphtheroid bacilli.

Part C—Study of a biochemical reaction

1. Demonstrate the stormy fermentation of milk.
   a. Boil a tube of skim milk for 10 minutes to drive off oxygen.
   b. Use sterile technic to inoculate heavily the tube of milk from an anaerobic culture of *Clostridium perfringens.* Allow the milk to cool to around 50° C. before the inoculation is made.
   c. Obtain a tube of sterile mineral oil.
   d. Pour the oil over the surface of the milk to form a surface layer about ½ inch in width.
   e. Incubate the tube of milk at 37° C.
2. Observe nature of bacterial growth in the milk under anaerobic conditions.

Part D—Demonstration of animal inoculation by instructor with discussion

1. Determination of the toxigenicity of a strain of suspected *Corynebacterium diphtheriae* in the guinea pig by the intracutaneous virulence test
2. Emphasis on applications of animal virulence tests
3. Comparison of a positive and a negative test result

Part E—Demonstration of the Schick test by instructor with discussion

Part F—Demonstration of the technic of throat culture by instructor with brief discussion

1. Emphasis on proper technic, precautions, and reasons for taking throat cultures
2. Use of differential stains for throat smears
   Students may stain suitable smears with:
   a. Albert's stain
   b. Löffler's alkaline methylene blue stain
   c. Gram I stain with Gram II stain as a mordant
3. Discussion of differential features noted microscopically

PROJECT

Acid-fast bacteria

Part A—Study of morphology of acid-fast bacteria

1. Use a specimen of tuberculous sputum that has been autoclaved. Why is this necessary?
2. Prepare smears of the sputum and make acid-fast stains. Make a drawing in color (red and blue) of the organisms present.
3. Make gram-stained smears of sputum. Compare the two stained preparations.
4. Examine microscopically prepared smears of:
   a. *Mycobacterium tuberculosis*
   b. *Mycobacterium leprae*
   c. Atypical mycobacteria
   d. *Nocardia asteroides*

Part B—Study of cultural characteristics

1. Examine prepared cultures of *Mycobacterium tuberculosis* on Lowenstein-Jensen medium.
2. Examine prepared cultures of atypical mycobacteria.

3. Compare the cultural characteristics of the atypical mycobacteria with those of the mycobacteria of tuberculosis.
4. Why do we not examine cultures of *Mycobacterium leprae?*

Part C—Animal inoculation in the laboratory diagnosis of tuberculosis with demonstration by instructor and discussion

1. Study of organs of a guinea pig that has died from tuberculosis

   The instructor can carefully perform an autopsy on a tuberculous guinea pig. The effect of the bacilli on different organs should be demonstrated and discussed. How do the intestinal ulcers of tuberculosis differ from those of typhoid fever?

Note: Organs must be fixed and noninfectious.

2. Demonstration of fixed tuberculous specimens from a human being, for example, a tuberculous lung with cavity formation
3. Presentation of microscopic tissue sections of tubercles
4. Demonstration of technic of inoculating a guinea pig to detect the presence of *Mycobacterium tuberculosis*

Part D—Skin testing in tuberculosis with demonstration by instructor and discussion

1. Technic of doing a Mantoux test
2. Technic of the tine (multiple puncture) test
3. Interpretation of test results, where positive and where negative

---

PROJECT
Spirochetes

Part A—Study of morphology of spirochetes

1. Examine microscopically prepared smears of:
   a. *Borrelia recurrentis*—Wright's stain of peripheral blood film
   b. *Treponema pallidum*—silver stain of infected tissue
   c. Fusospirochetal organisms of Vincent's angina—stained with Gram stains I and II
2. Compare.

Part B—Study of spirochetes from the mouth

1. Take scrapings from between the teeth with a toothpick.
2. Make a thin smear. Let dry and fix in flame.
3. Stain with carbolfuchsin for 1 minute.
4. Note morphology of organisms present.

PROJECT
Fungi

Part A—Study of unicellular fungi

1. Observe the morphology of yeasts in a wet mount.
   a. Rub up a small portion of yeast cake in about 5 to 10 ml. of distilled water.
   b. Place a drop of the suspension on a microslide.

c. Cover with a cover glass carefully so as to avoid bubbles.
d. Examine with the low- and high-power objectives of the microscope.
2. Observe the morphology of yeasts in a wet mount admixed with Gram's iodine.
Note: The starch granules from the yeast cake are stained a dark blue. By contrast the yeast cells appear yellowish.

Part B — Study of the morphology of multicellular fungi

1. Observe the morphology of the common bread mold *Rhizopus.*
   a. Obtain a piece of bread containing a black moldy growth.
      (1) Examine the natural growth macroscopically, that is, with the naked eye.
      (2) Examine the growth using a strong hand lens.
      Note: Observe the color, consistency, texture, and general appearance of the black mold growing on the bread.
   b. Observe the black mold microscopically.
      (1) Carefully tease off some of the mold and transfer to microslide. Place in a drop of water. Carefully position a cover glass.
      (2) Use the low- and high-power objectives of the microscope to examine the mycelial fragments of this mold. Reduce the opening of the iris diaphragm to enhance contrast in the specimen.
      Note: Observe the appearance of the hyphae and note whether they possess partitions. Observe the nuclei and the appearance of the fruiting bodies.
2. Observe the morphology of the common powdery green mold on citrus fruit. Proceed as above.
3. Make drawings. Note the presence of hyphae in test specimens of mold. Are they septate or nonseptate? Are nuclei present?

Part C — Study of morphology and cultural characteristics of *Candida albicans*

1. Observe the growth of *Candida albicans* on:
   a. Blood agar (Incubate at 37° C.)
   b. Sabouraud agar (Incubate one plate at 37° C. and the other at room temperature — 20° C.)
2. Prepare smears from colonies on blood and Sabouraud agar. Stain by Gram's method.
3. Prepare wet mounts of fungous growth. Examine microscopically.

Part D — Study of a biochemical reaction of yeast

1. Place about 20 ml. of 2% lactose in one large test tube (1 by 8 inches).
2. Place about 20 ml. of 2% dextrose in another.
3. Rub up a cake of yeast in a few milliliters of water and add one half the substance to each of the above tubes.
4. Place either test tube at an angle of about 45° in a vessel of water warmed to 42° C. What happens?

Note: This test is used to determine whether sugar found in the urine of a nursing mother is lactose or dextrose.

PROJECT

Protozoa

Part A—Study of morphology and biologic functions of protozoa in hay infusion

Note: Look for a protozoon of the genus *Paramecium.* Note its motility, feeding pattern, and expulsion of waste material from the body of the parasite. Observe cell detail.

1. Procure a few drops of hay infusion containing paramecia; place on a clean slide and add a few grains of sand to hold cover glass away from the organisms.
2. Gently apply cover glass.
3. Observe microscopically. Note movement of paramecia. Do they seem to have a purposeful or aimless movement?

Note: In the microscopic study of the wet mount, use the low-power objective of the scope first. Then study the organisms under the high-power objective. Reduce the amount of light coming through the microscope by closing the iris diaphragm. What is the effect of this?

4. Sketch a quiescent organism?
5. Place a drop of india ink on a microslide and mix with a drop of hay infusion (containing *Paramecium*).
6. Examine microscopically. Note activities of protozoa in regard to particles of carbon in the ink. Note currents in oral groove. Note ingestion of particles, attempt at digestion (india ink is indigestible), and expulsion of particles.

Part B—Study of morphology of an intestinal ciliate

1. Prepare a wet mount of a specimen of fecal material from a guinea pig.
   a. Mix a small portion of the contents of the cecum of the animal with a drop of isotonic saline solution on a microslide.
   b. Superimpose a cover glass.
2. Examine microscopically using both low-power and high-power objectives.
3. Look especially for the large ciliate *(Balantidium)* that often inhabits the cecum.
4. Note the presence of various flagellates.
5. Observe morphology and movements of protozoa present.
6. Note comparative size of protozoa and bacteria. Make a drawing of each type of protozoon found.

Part C—Study of morphology of important pathogenic protozoa

1. Obtain prepared slides for study of the following:
   a. Malarial parasites in stained blood films
      (1) Thick films
      (2) Thin films
   b. Trypanosomes in stained blood film
   c. *Entamoeba histolytica* in iron hematoxylin–stained fecal smear
   d. *Giardia lamblia* in iron hematoxylin—stained fecal smear
   e. *Trichomonas vaginalis* in a Papanicolaou–stained smear from cervix of uterus
   f. *Toxoplasma gondii* in hematoxylin and eosin–stained tissue section
2. Examine the protozoa microscopically. Note morphologic features and location in the body where found. Make drawings.
3. Compare the vegetative and cyst forms of *Entamoeba histolytica* in stained fecal smears.

Note: The vegetative forms of *Entamoeba histolytica* may be demonstrated by the instructor in a fresh specimen of feces, if such is available.

EVALUATION FOR UNIT FIVE

Part I

In the following exercises, indicate the number on the right that completes the statement or answers the question.

1. Which of the following are gram-positive cocci occurring typically in pairs?

(a) Pneumococci	1. all of these
(b) Gonococci	2. a, c, and e
(c) Streptococci	3. a, b, and d
(d) Meningococci	4. a
(e) Staphylococci	5. a and e

2. Which of the following diseases may be caused by *Streptococcus pyogenes?*

(a) Puerperal sepsis	1. a and b
(b) Osteomyelitis	2. a, b, and d
(c) Scarlet fever	3. c
(d) Erysipelas	4. none of these
(e) Septic sore throat	5. all of these

3. *Coxiella burnetii* is different from the rickettsiae causing typhus fever and Rocky Mountain spotted fever in that:

(a) It can produce a skin rash in the patient.	1. a
(b) It does not require an insect vector.	2. b
(c) It is found only in the United States.	3. b and d
(d) It is more resistant to heat.	4. none of these
	5. c

4. Which of the following statements apply to the enteric bacilli?

(a) They are all gram-negative.	1. b
(b) They are all airborne infections.	2. e
(c) They are all strictly anaerobic.	3. d and e
(d) Their mode of invasion is through the alimentary tract.	4. a and d
(e) They all invade the bloodstream during the course of disease.	5. all but c

5. Which one of the following facts about typhoid fever is used as the basis for a diagnostic laboratory test?

(a) Bacilli enter the lymphatics of the intestines.	1. e
(b) Symptoms are caused by the release of endotoxins when the organisms disintegrate.	2. d
(c) The bacilli often lodge in the bone marrow.	3. c
(d) The antibodies formed against the bacilli are agglutinins.	4. b
(e) Bacilli often lodge in "rose spots" on the skin.	5. a

6. What characteristic do members of the genus *Haemophilus* have in common?
   (a) They destroy red blood cells by dissolving them.
   (b) They are found only in the red blood cells of an infected person.
   (c) They cannot be grown in the laboratory except in living animals.
   (d) They grow in the laboratory only in a medium that contains hemoglobin.
   (e) They all cause diseases of the central nervous system only.

   1. a
   2. b
   3. c
   4. d
   5. d and e

7. Which of the following answer this description: anaerobic, sporeforming organisms that produce exotoxins?
   (a) *Corynebacterium diphtheriae*
   (b) *Clostridium tetani*
   (c) *Clostridium perfringens*
   (d) *Clostridium botulinum*
   (e) *Streptococcus pyogenes*

   1. a
   2. all but e
   3. all but a
   4. b and d
   5. all but a and e

8. Which of the following statements concerning diphtheria are true?
   (a) There is a method of testing a person's immunity to the disease.
   (b) Most children from 1 to 6 years of age are susceptible to the disease.
   (c) Diphtheria is an airborne disease.
   (d) A patient is proved to have diphtheria if his throat smears show an acid-fast organism.
   (e) A susceptible child exposed to the disease should be given diphtheria antitoxin as soon as possible.

   1. all but d
   2. all but b
   3. all but d and e
   4. all but e
   5. all of these

9. The following types of tuberculosis seldom, or never, occur in man:
   (a) Human
   (b) Bovine
   (c) Avian
   (d) Cold blooded

   1. c
   2. c and d
   3. a
   4. b
   5. b, c, and d

10. In testing sputum from a patient suspected of having tuberculosis, the bacteriologist can do the following for the laboratory identification of the organism:
   (a) Culture the specimen in the absence of oxygen
   (b) Use a special medium for the culture of the specimen
   (c) Attempt to identify the causative organism in smears by means of the Gram stain
   (d) Inoculate a guinea pig with the suspected material
   (e) Attempt to identify the causative organism in smears by means of the acid-fast stain

   1. a
   2. b
   3. c and d
   4. b and d
   5. b, d, and e

11. Which of the following diseases confer little, if any, immunity?
   (a) Pneumonia
   (b) Meningococcal meningitis
   (c) Gonorrhea
   (d) Syphilis
   (e) Osteomyelitis

   1. a, b, and d
   2. a, b, and e
   3. c and d
   4. a
   5. all but b

12. We are aware of the following facts concerning pathogenic rickettsiae:
   (a) They grow only within living cells.
   (b) They may be grown within the embryonated hen's egg.
   (c) They can be separated from bacteria by filtration.
   (d) They are always found within the nucleus of the cell.
   (e) They are transmitted by insects.

   1. a
   2. a, b, and e
   3. c
   4. d
   5. none of these

13. We are aware of the following facts concerning viruses:
   (a) They cannot be seen with an ordinary light microscope.
   (b) They grow only within living cells.
   (c) They cannot be separated from media by filtration.
   (d) They cause only one important communicable disease—smallpox.
   (e) They confer no immunity with disease.

   1. a
   2. a and b
   3. a, b, and c
   4. c
   5. none of these

14. The rickettsial diseases are spread primarily by the following route:
   (a) Droplet infection
   (b) Contaminated water
   (c) Contaminated food
   (d) Insects and rodents
   (e) Direct contact

   1. a
   2. b
   3. c
   4. d
   5. e

15. The presence of an infection caused by *Treponema pallidum* can be demonstrated by three laboratory procedures as follows:
   (a) Gram's stain
   (b) Complement-fixation test
   (c) Dark-field examination
   (d) Growth on special culture media
   (e) Precipitation tests

   1. b, c, and d
   2. b, c, and e
   3. c, d, and e
   4. a, d, and e
   5. a, b, and c

16. Mr. W. is diagnosed as having malaria. His clinical course is characterized by chills and fever that occur regularly once every 3 days. Which of the following statements apply to this patient?
   (a) His disease is the estivoautumnal type.
   (b) He is infected with *Plasmodium vivax.*
   (c) He is suffering from the tertian type of malaria.
   (d) His malaria is the quartan type.
   (e) He is infected with *Plasmodium malariae.*

   1. a and e
   2. a and b
   3. c and e
   4. b and d
   5. d and e

17. What sort of specimen should be collected from a patient if the doctor wants to ascertain the presence of *Entamoeba histolytica?*
   (a) Early morning sputum placed in refrigerator until examined
   (b) Catheterized urine specimen kept at room temperature until examined
   (c) Stool specimen kept warm until examined
   (d) Voided urine specimen incubated until examined
   (e) Stool specimen refrigerated until examined

   1. c
   2. b and c
   3. d and c
   4. b
   5. e

18. In meningococcal meningitis the following are true:
   (a) Credé's technic is useful in prophylaxis.
   (b) An exotoxin is prominent.
   (c) The Waterhouse-Friderichsen syndrome may be a feature.
   (d) The causative organism is found in the cerebrospinal fluid.
   (e) There are never skin lesions.

   1. a and b
   2. a and e
   3. c and d
   4. c, d, and e
   5. none of these

19. Rubella (German measles) is an important disease because
   (a) The virus is related to the arboviruses.
   (b) There is a mild skin rash.
   (c) The virus is teratogenic.
   (d) The virus is oncogenic.
   (e) The virus persists.

   1. a
   2. c and d
   3. b, c, and d
   4. c and e
   5. e

20. Hospital strains of coagulase-positive staphylococci:
   (a) Are a threat to newborn infants
   (b) Are maintained by carriers among personnel
   (c) Are distinguished by phage-typing patterns
   (d) Never become resistant to antibiotics
   (e) Do not cause infections of surgical wounds

   1. a
   2. a and b
   3. b and c
   4. a, b, and e
   5. d and e

21. The substance elaborated by streptococci that initiates the dissolution of fibrin clots is:
   (a) Spreading factor
   (b) Leucocidin
   (c) Streptodornase
   (h) Streptolysin
   (e) Streptokinase

   1. a
   2. b
   3. c
   4. d
   5. e

22. Which of the following is not typical for cholera:
   (a) Explosive outbreaks
   (b) Severe dehydration
   (c) Transmitted by water
   (d) Caused by a vibrio
   (e) Exo-endotoxin important

   1. a, b, c, and d
   2. all of these
   3. a and c
   4. a, b, and c
   5. none of these

23. Pulmonary tuberculosis:
   (a) Invariably develops when tubercle bacilli enter the respiratory tract
   (b) Is usually caused by the bovine bacillus in the United States
   (c) Results in immediate-type allergy
   (d) Is more likely to develop in a tuberculin-positive than in a tuberculin-negative person
   (e) Involves the lymphatics in the primary complex

   1. e
   2. d
   3. c
   4. b
   5. a

24. Which of the following is most likely to constitute a source of streptococcal upper respiratory tract infection?
   (a) A healthy throat carrier
   (b) A healthy nasal carrier
   (c) The skin of a patient with scarlet fever rash
   (d) Dust contaminated with streptococci
   (e) The hands of a throat carrier

   1. a
   2. b
   3. c
   4. d
   5. e

25. *Mycobacterium leprae:*
   (a) Tends to appear in large numbers in nasal mucosa
   (b) Causes a highly contagious disease
   (c) Infections exist in United States
   (d) Grows well on media used for the tubercle bacillus
   (e) Causes leprosy

   1. a and b
   2. c and d
   3. b, c, and d
   4. a, c, and e
   5. e

26. Urinary tract infection is rarely, if ever, caused by
   (a) *Proteus vulgaris*
   (b) *Escherichia coli*
   (c) *Pseudomonas aeruginosa*
   (d) *Mycobacterium tuberculosis*
   (e) *Mycobacterium smegmatis*

   1. a
   2. b
   3. c
   4. d
   5. e

27. Compared to most pathogenic bacteria, most pathogenic fungi require:
   (a) A higher temperature for growth
   (b) More complicated nutrients for growth
   (c) More time to grow
   (d) Less oxygen for growth
   (e) A higher moisture content for growth

   1. a, b, and c
   2. b
   3. c and d
   4. c
   5. d and e

28. A diagnosis of tuberculous meningitis is confirmed by:
   (a) X-ray examination of lung
   (b) Subcutaneous injection of tuberculin
   (c) Sputum examination
   (d) Blood culture
   (e) Culture of cerebrospinal fluid

   1. e
   2. d and e
   3. d
   4. b
   5. a, b, c, and d

29. All of the following are notable features of *Mycobacterium tuberculosis except:*
   (a) Grows very slowly
   (b) Highly resistant to drying
   (c) Spore former
   (d) Strict aerobe
   (e) Rich in lipids

   1. a
   2. b and c
   3. c
   4. d and e
   5. e

30. Which one of the following parasites most commonly causes blockage of the lumen of the intestinal tract?
   (a) *Taenia solium*
   (b) *Taenia saginata*
   (c) *Giardia lamblia*
   (d) *Ascaris lumbricoides*
   (e) *Necator americanus*

   1. a
   2. b
   3. c
   4. d
   5. e

31. A blood sample is often useful in the diagnosis of all of the following *except:*
   (a) Malaria
   (b) Brucellosis
   (c) Meningococcal meningitis
   (d) Bacillary dysentery
   (e) Cholera

   1. a
   2. b and d
   3. c and d
   4. d and e
   5. e

32. The diagnosis of hookworm disease is based mainly on:
   (a) An increase in antibody titer
   (b) Identification of skin lesions
   (c) Characteristic eggs in stool
   (d) Inclusion bodies in cells
   (e) Complement fixation test

   1. a, b, and c
   2. b, c, and d
   3. c
   4. c, d, and e
   5. d

33. A parasite seen often in the urine, especially in men is:
   (a) *Trichomonas vaginalis*
   (b) *Enterobius vermicularis*
   (c) *Giardia lamblia*
   (d) *Chilomastix mesnili*
   (e) *Trichuris trichiura*

   1. a
   2. b
   3. c
   4. d
   5. e

34. A parasite prevalent in the Arctic is:

(a) *Loa loa*	1. b
(b) *Fasciolopsis buski*	2. c
(c) *Onchocerca volvulus*	3. a
(d) *Trichinella spiralis*	4. e
(e) *Clonorchis sinensis*	5. d

35. The majority of respiratory tract infections are caused by:

(a) *Staphylococcus aureus*	1. a and b
(b) *Staphylococcus albus*	2. c
(c) Hemolytic streptococcus	3. d and e
(d) Friedländer's bacillus	4. c, d, and e
(e) A virus	5. e

36. Pneumonia in a meat packer may be caused by

(a) Q fever	1. a
(b) Tularemia	2. b
(c) Brill's disease	3. c
(d) Rickettsialpox	4. d
(e) Psittacosis	5. e

37. In tropical Africa 50% to 70% of the malignant tumors occurring in children are represented by:

(a) Burkitt's lymphoma	1. a
(b) Acute leukemia	2. b
(c) Kaposi sarcoma	3. c
(d) Hodgkin's disease	4. d
(e) Neuroblastoma	5. e

38. Echovirus is easily isolated from all of these except:

(a) Cerebrospinal fluid	1. a
(b) Nose	2. a and b
(c) Stools	3. c
(d) Throat	4. d and e
(e) Fingernail scrapings	5. e

39. A 20-year-old man has a hard, indurated, ulcerated penile papule. He believes it is healing with applications of vaseline and denies that it was ever painful. The most useful diagnostic procedure would be to take scrapings for:

(a) Gram stain	1. a and b
(b) Acid-fast stain	2. c
(c) Tissue culture	3. d
(d) Dark-field microscopy	4. d and e
(e) Fluorescent antibody study	5. e

40. Nosocomial infections:

(a) Must be reported to local health department	1. a, b, c, and d
	2. a, b, c, d, and e
(b) Are acquired in the hospital	3. b
(c) Usually are viral	4. c
(d) Are invariably airborne	5. b and e
(e) Are caused in great number by enteric bacilli	

41. The presence of "sulfur granules" in an exudate from a draining sinus is diagnostic for:

(a) Nocardiosis	1. a
(b) Actinomycosis	2. a and b
(c) Coccidioidomycosis	3. b
(d) Dermatophytosis	4. c and d
(e) Histoplasmosis	5. e

42. Disease in the central nervous system may result from

(a) Epidemic typhus	1. none of these
(b) Rabies immunization	2. all of these
(c) Mumps	3. a
(d) Cryptococcosis	4. a and b
(e) Torulosis	5. a, b, c, and d

43. The agent that causes ornithosis is classified as:

(a) Virus	1. a
(b) Protozoa	2. b
(c) Fungus	3. c
(d) Mycoplasma	4. d
(e) Rickettsia	5. none of these

44. Mycoplasmas differ from bacteria in that:

(a) Their colonial morphology is different.	1. all of these
(b) They lack cell walls.	2. a, b, c, and d
(c) They do not grow on cell-free media.	3. a, b, and c
(d) They produce disease only in animals other than man.	4. a and b
	5. e
(e) Their colonies are usually quite large.	

45. Diagnostic procedures of value in the laboratory identification of the fungi include:

(a) Gram stain	1. all of these
(b) Slide cultures	2. all of these except a
(c) Phenomenon of fluorescence	3. none of these
(d) Culture on Sabouraud agar	4. a
(e) Fermentation reactions	5. d and e

46. Spontaneous mutation of the virus is now known to be a problem in the epidemiology and control of:

(a) Influenza	1. all of these
(b) Rabies	2. none of these
(c) Rubeola	3. a, b, and c
(d) Rubella	4. a
(e) Yellow fever	5. d and e

47. The patient with leukemia or bone marrow failure is vulnerable to infection with which of the following fungi:

(a) *Histoplasma*	1. a
(b) *Coccidioides*	2. b
(c) *Candida*	3. c
(d) *Blastomyces*	4. d and e
(e) *Actinomyces*	5. d

48. The following are important fungi and ordinarily benign but they may be serious opportunist pathogens:

(a) *Candida*	1. all of these
(b) *Aspergillus*	2. none of these
(c) *Mucor*	3. all of these except b
(d) *Rhizopus*	4. all of these except e
(e) *Cryptococcus*	5. c

49. The three most important genera of dermatophytes are:

(a) Tinea	1. a, b, and c
(b) *Trichophyton*	2. b, c, and d
(c) *Epidermophyton*	3. c, d, and e
(d) *Microsporum*	4. b, d, and e
(e) *Candida*	5. a, c, and d

50. Although its true nature is as yet undefined, the following observations have been made in regard to the Australia antigen:

(a) It does not grow in tissue culture.
(b) It has neither DNA nor RNA backbone.
(c) It is of no consequence in blood banking.
(d) It is significantly related to serum hepatitis.
(e) It is prevalent in drug addicts.

1. a, b, d, and e
2. b, c, d, and e
3. a, b, and c
4. d and e
5. none of these

Part II

Comparisons: Match the item in Column B to the phrase (or word) in Column A best describing it.

STAPHYLOCOCCUS AUREUS WITH STREPTOCOCCUS PYOGENES:

COLUMN A

_________ 1. Can produce hyaluronidase
_________ 2. Phage typing in epidemiologic study
_________ 3. Serologic typing in epidemiologic study
_________ 4. Can produce greenish discoloration on blood agar
_________ 5. Healthy carrier state possible
_________ 6. Nonpathogenic in man
_________ 7. Septicemia a dangerous event
_________ 8. Produces localized abscesses
_________ 9. Produces cellulitis

COLUMN B

(a) *Staphylococcus aureus*
(b) *Streptococcus pyogenes*
(c) Both
(d) Neither

TYPHOID FEVER WITH BACILLARY DYSENTERY:

COLUMN A

_________ 10. Ulceration of bowel
_________ 11. Man only source of infection
_________ 12. Blood culture important
_________ 13. May be transmitted by water
_________ 14. Penicillin effective in infection
_________ 15. Organisms nonmotile
_________ 16. Organisms ferment lactose

COLUMN B

(a) Typhoid fever
(b) Bacillary dysentery
(c) Both
(d) Neither

TETANUS WITH GAS GANGRENE:

COLUMN A

_________ 17. Antibiotic therapy highly effective
_________ 18. Polyvalent antiserum useful
_________ 19. Active immunity advised
_________ 20. Organisms found in feces of man and animals
_________ 21. Dirt of wound enhances growth of organisms
_________ 22. Organisms produce potent exotoxin
_________ 23. Disease result of wound infection
_________ 24. Blood culture positive early in disease

COLUMN B

(a) Tetanus
(b) Gas gangrene
(c) Both
(d) Neither

ARBOVIRUSES WITH PICORNAVIRUSES:

	COLUMN A	COLUMN B
________	25. Transmitted by arthropods	(a) Arboviruses
________	26. Includes polioviruses	(b) Picornaviruses
________	27. Includes viruses of equine encephalomyelitis	(c) Both
________	28. Several immunologically distinct viruses	(d) Neither
________	29. Birds are important in cycle in nature	
________	30. Includes rabies virus	
________	31. Includes herpesvirus	
________	32. Man accidental host usually	
________	33. Important diseases in man	
________	34. Ecologic hodgepodge	
________	35. Many member viruses in the group	

COCCIDIOIDOMYCOSIS WITH HISTOPLASMOSIS:

	COLUMN A	COLUMN B
________	36. Disease occurs in primary and progressive forms	(a) Coccidioidomycosis
________	37. Organisms reproduce by endosporulation	(b) Histoplasmosis
________	38. Found in southwestern part of United States	(c) Both
________	39. Found in central Mississippi Valley	(d) Neither
________	40. Causes paracoccidioidal granuloma	
________	41. Organisms parasitize cells of reticuloendothelial system	
________	42. Also known as speleonosis	
________	43. Disease contracted from inhalation of spore-bearing dust	
________	44. Causes disease in lungs	
________	45. Organism grows in the laboratory as a mold	

Part III

1. Place the letter indicating the organism from the column on the right in front of the statement regarding it in the column on the left.

________	1. The organism infecting cattle and causing undulant fever in man	(a) *Klebisella pneumoniae*
________	2. The organism resembling the enteric group but causing respiratory infection in man	(b) *Bacillus anthracis*
________	3. The aerobic, gram-negative organism that caused epidemics of the Black Death during the Middle Ages	(c) *Actinobacillus mallei*
________	4. An organism attacking persons who handle infected animal products such as hides, wool, or hair	(d) *Vibrio cholerae*
________	5. The organism causing rabbit fever	(e) *Brucella abortus*
________	6. An organism losing its importance as a pathogen in man since the advent of the automobile	(f) *Pasteurella pestis*
		(g) *Pasteurella tularensis*

2. Match the given disease in the following list with its causative agent and a specific diagnostic test. From Column 1, select the appropriate causative agent and place its corresponding number opposite the disease in the first blank to the right. From Column 2, select the appropriate diagnostic test and place its corresponding number opposite the disease in the second blank to the right.

DISEASE	PATHOGENIC AGENT	DIAGNOSTIC TEST
Malaria	______	______
Lobar pneumonia	______	______
Diphtheria	______	______
Infectious mononucleosis	______	______
Typhoid fever	______	______
Typhus fever	______	______
Enterobiasis	______	______
Lymphopathia venereum	______	______
Herpes simplex	______	______
Hookworm disease	______	______
Epidemic cerebrospinal meningitis	______	______
Ringworm of scalp	______	______
Rabies	______	______
Primary atypical pneumonia	______	______
Whooping cough	______	______
Boils	______	______
Syphilis	______	______
Tuberculosis	______	______
Scarlet fever	______	______

COLUMN 1 (Pathogenic agents)

(1) *Staphylococcus aureus*
(2) *Streptococcus pyogenes*
(3) *Salmonella typhi*
(4) *Bordetella pertussis*
(5) *Mycobacterium tuberculosis*
(6) *Treponema pallidum*
(7) *Neisseria meningitidis*
(8) *Diplococcus pneumoniae*
(9) *Plasmodium*
(10) Neurotropic virus
(11) *Rickettsia prowazekii*
(12) *Bedsonia*
(13) *Microsporum*
(14) *Corynebacterium diphtheriae*
(15) Epstein-Barr virus
(16) *Mycoplasma pneumoniae*
(17) *Herpesvirus hominis*, type 2
(18) *Necator americanus*
(19) Pinworms
(20) *Herpesvirus hominis*, type 1.

COLUMN 2 (Diagnostic tests)

(1) Widal agglutination test
(2) Weil-Felix reaction
(3) Phage typing
(4) Thick blood smears
(5) Demonstration of Negri bodies
(6) Neufeld typing method
(7) Cold agglutination test
(8) Wassermann test
(9) Mantoux test
(10) Cough plate culture method
(11) Cerebrospinal fluid culture
(12) Heterophil antibody test
(13) Schultz-Charlton phenomenon
(14) Inoculation of Löffler's blood serum culture medium
(15) Use of 10% potassium hydroxide in wet mount
(16) Intranuclear inclusion bodies in skin scrapings
(17) Stool examination for ova and parasites
(18) Anal swabs taken at night
(19) Frei test

3. True-False. Circle either *T* or *F* accordingly.

T F (1) Madura foot is one manifestation of the dermatomycosis also known as "athlete's foot."

T F (2) The gonococcus is especially dangerous as it is able to live for a long time outside of the human body.

T F (3) The tubercle bacillus is able to live but for a brief period after being expelled from the body.
T F (4) A most important characteristic of yeasts is their ability to produce fermentation.
T F (5) Abscesses are circumscribed or localized areas of inflammation.
T F (6) A substance, aflatoxin, produced by a mold, is also known to induce cancers in animals.
T F (7) Cellulitis is a limited type of inflammation.
T F (8) Pneumococci cause pseudomembranes to be formed in the body.
T F (9) Leprosy is infectious but not contagious.
T F (10) Chancre and chancroid are identical terms and mean the same.
T F (11) Chancres always manifest themselves within 48 hours after contact.
T F (12) A father cannot give syphilis to his unborn child unless he gives it first to the mother and she in turn transmits it to the fetus.
T F (13) Enlarged buboes are large ulcers on the skin of patients with tuberculosis.
T F (14) Mucous patches always occur during the primary stage of syphilis.
T F (15) Serologic examinations for syphilis are most accurate if run immediately after the chancre occurs.
T F (16) Congenital syphilis always appears right after birth.
T F (17) The Wassermann test is very specific in the diagnosis of syphilis; that is, it will be positive only in those instances in which the patient has syphilis and no other disease.
T F (18) Hansen's disease is tuberculosis.
T F (19) Leprosy is also known as Hansen's disease.
T F (20) From the infectious standpoint, it is safer to live with an active case of tuberculosis than with a case of leprosy.
T F (21) Tuberculin testing is more important in adults than in children.
T F (22) Tuberculosis usually produces lesions in the lungs, but it may occur in other areas of the body without pulmonary changes.
T F (23) Only a few fungi are normal inhabitants of the human body.
T F (24) The main insect vector for yellow fever is *Aedes aegypti*, which can survive and transmit the infection in the southern parts of the United States.
T F (25) Viruses are commonly grown in the laboratory in semisolid media containing egg.
T F (26) Latent syphilis is caused by a weak type of spirochete.
T F (27) *Actinomyces* is often called ray fungus.
T F (28) Pyogenic bacteria are so called because they produce pus in tissues.
T F (29) *Candida* occurs and causes disease on mucous membranes.
T F (30) Coccidioidomycosis produces a chronic lesion in the lung that simulates tuberculosis.
T F (31) Meningitis may be caused by meningococci, staphylococci, and pneumococci.
T F (32) Sulfur granules are found in the lesions of blastomycosis.
T F (33) Sporotrichosis involves the skin and regional lymph vessels.
T F (34) Actinomycosis is acquired by eating infected pork.
T F (35) Bronchopneumonia is usually a bilateral inflammation of the lung.

UNIT SIX

MICROBES: public welfare

34 The microbe everywhere

35 The microbiology of water

36 The microbiology of food

37 Biologic products for immunization

38 Recommended immunizations

Laboratory survey of unit six

Evaluation for unit six

34 The microbe everywhere

Persons in daily contact with diseases caused by bacteria are likely to look upon microbes as agents of harm only, but such is not the case. *The majority of bacteria and other microbes are helpful to man, animals, and plants, all of which depend upon microbes for their very existence.* In a broad sense microorganisms producing disease form but a small and inconspicuous group. Nothing gives us a better idea of the broad scope of microbial activity than does a consideration of how bacteria and other microbes affect our daily lives.

MICROBES IN THE PROCESSES OF NATURE

The *microbiology of nature* defines the role that bacteria and other microbes play in the various processes of nature. Bacteria and other microbes have been said to be nature's garbage disposal system and fertilizer factory for they are the active agents in the decomposition of dead organic matter of animal origin, releasing the elements needed for the growth of plants and returning them to the soil. Bacteria purify sewage by living on the impurities in it and converting these to inoffensive substances that serve as food material for plants. As water trickles through the soil, bacteria help to filter it. Though contaminated when it enters the earth, water may trickle out pure and clean.

■ **Participation in the cycle of an element.** The chief elements entering the bodies of animals and plants are nitrogen, carbon, hydrogen, and oxygen, which are combined with other less common elements to form complex proteins. Animals depend upon plants for these elements. Plants receive their hydrogen and oxygen from water and must obtain their carbon and nitrogen from an inorganic compound containing these elements. The assimilation of an element from inorganic compounds, its conversion into organic compounds to form the bodies of plants and animals, and its subsequent reappearance in the inorganic state to be used again, constitute the *cycle* of the element.

CARBON CYCLE. Plants can assimilate carbon only from carbon dioxide. Carbon dioxide is present in the air in small quantities, but the supply must be constantly renewed. This comes about as carbon dioxide, a waste product of metabolism, is constantly eliminated in the breath of all animals (see Fig. 34-1).

If animals were sufficiently numerous to eat all the plants, converting them into carbon dioxide and excreting the gas, the supply of plant food for animals and of carbon dioxide for plants would be perfectly balanced. However, tissues of certain plants, such as wood and cellulose, which contain much carbon, when eaten by animals, pass through the intestinal canal unchanged. Therefore, the carbon in them is not converted into carbon dioxide and released. Other plants die with much carbon bound

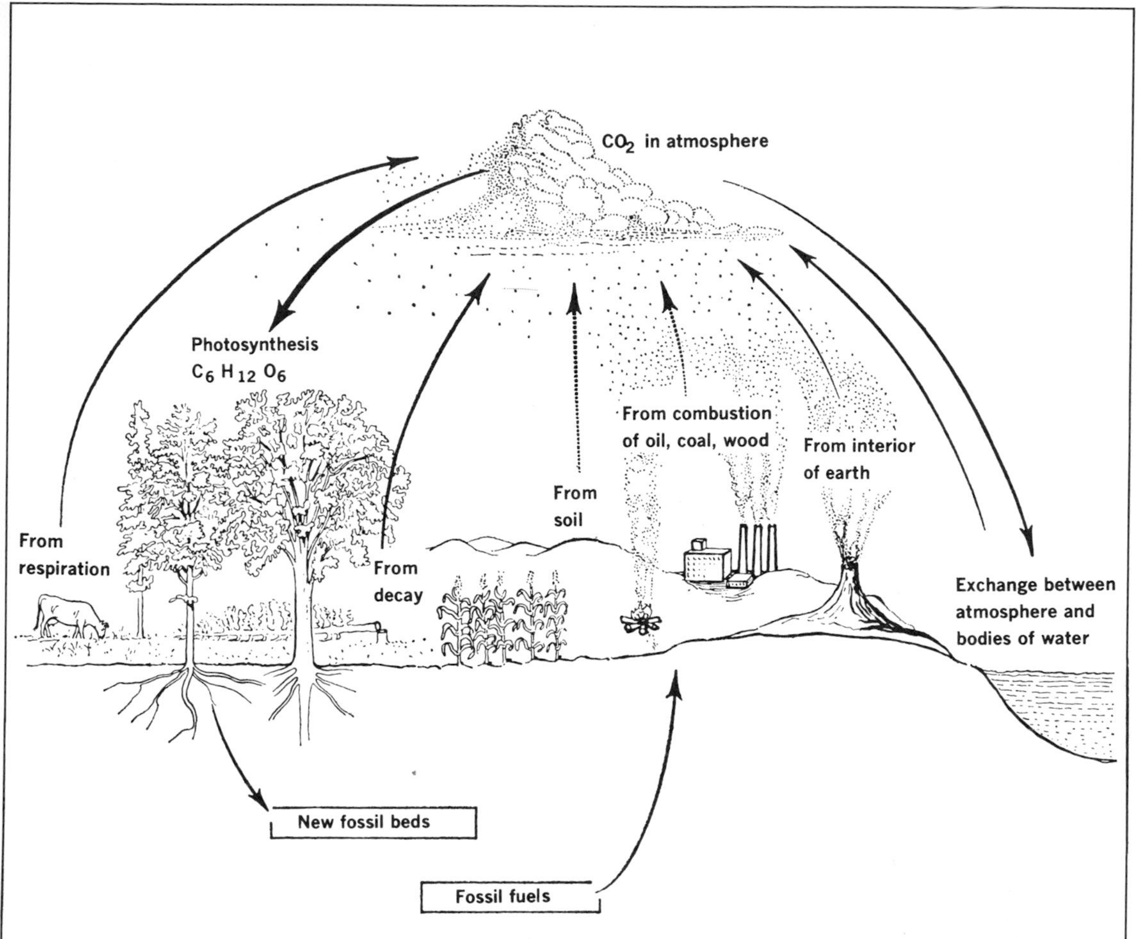

FIG. 34-1. Carbon cycle. The sources of carbon dioxide from the earth's surface and within its depths as well as assimilation of carbon dioxide from the atmosphere by plant life through photosynthesis are depicted. (From Arnett, R. H., Jr., and Braungart, D. C.: An introduction to plant biology, ed. 3, St. Louis, 1970, The C. V. Mosby Co.)

in their bodies. Carbon is also present in the body tissues of all animals when they die. If there were not some method of recovering this carbon in a form available for plant growth (CO_2), plant life would soon cease and animal life would shortly thereafter. To recover the carbon, fungi and bacteria attack animal excreta, dead plants, and animal bodies, and the process of decay and decomposition begins. The complex carbon compounds are broken down by microbes into carbon dioxide, which floats away in the air to become plant food.

NITROGEN CYCLE. Nitrogen is one of the most important elements in the composition of the plant body. It is abundant in the air but not in a form suitable for plant use. To be so suitable, it must be in the form of nitrates. Nitrates present in the soil in very small amounts must constantly be renewed. Microbes participate by (1) the decomposition of organic matter and (2) the conversion of the nitrogen of the air into nitrates (see Fig. 34-2).

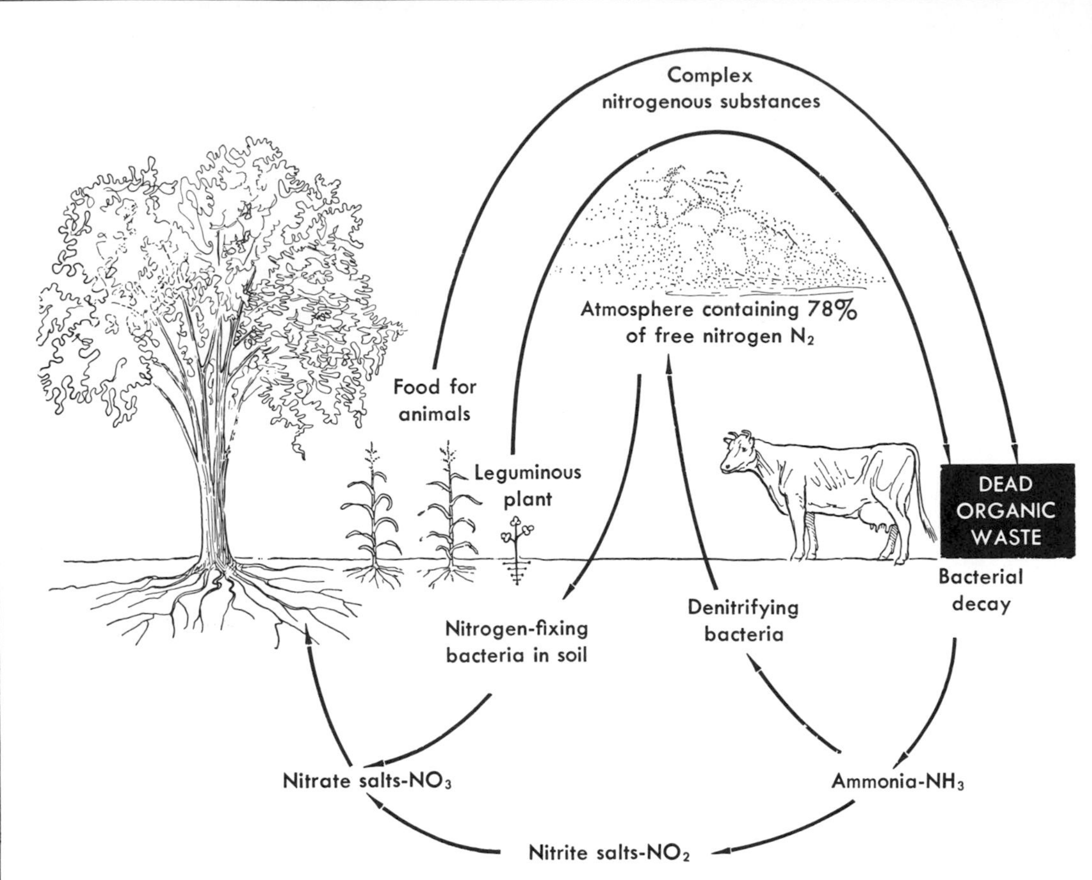

FIG. 34-2. Nitrogen cycle. Various sources are indicated for the nitrogen needed by plant and animals for protein production. Note that microbes of the soil convert a part of the nitrogen stepwise into ammonia, nitrite, and nitrate. Nitrate is also derived from nitrogen of the air by bacterial action and through the oxidizing action of lightning. Nitrate from these sources is taken up by roots of plants. (From Arnett, R. H., Jr., and Braungart, D. C.: An introduction to plant biology, ed. 3, St. Louis, 1970, The C. V. Mosby Co.)

Organic matter is broken down by different microorganisms with various actions to form simpler compounds, among which is ammonia. Ammonia is attacked by *nitrifying* bacteria (genus *Nitrosomonas*) which convert it into nitrites. Other nitrifying bacteria (genus *Nitrobacter*) convert the nitrites into nitrates, which are absorbed by the plant roots and built into complex nitrogen compounds like those of the original organic matter.

The term *nitrogen fixation* applies to the recovery of free nitrogen from the air and its synthesis into nitrates for use by plants. The first step in nitrogen fixation is the formation of ammonia. This is accomplished by two groups of microorganisms. One (genus *Azotobacter*) lives in the soil. Another group (genus *Rhizobium*) penetrating the root hairs, develops symbiotically in the cells of certain legumes, like clover. It receives its nourishment from the plant and forms nodules filled with the rapidly multiplying bacteria. They bind the nitrogen of the air into a compound incorporated into the bacterial body, which is removed and used by the plant. Animals eat the plants. When the plant dies and decays, the nitrogenous compounds are set free in the soil. Soils in which such products of decay are present are fertile. *Microbiology of the soil* deals with the many aspects of the population of microorganisms, upon which fertility is dependent.

OTHER CYCLES. Other elemental cycles are those of phosphorus and sulfur.

MICROBES IN ANIMAL NUTRITION

Many years ago Pasteur expressed the idea that animal life would not be possible were it not for microbial inhabitants of the intestinal tract.* Bacteria take an active part in the digestive processes of both lower animals and man. Many of the vitamins essential to animal and human nutrition are formed in the large bowel by the microbial inhabitants. Large amounts of vitamins in the B complex group are so produced. The elaboration of a certain vitamin by bacteria may constitute a major source of it, such as the production of vitamin K. Vitamin K is important in the prevention of hemorrhage in the body and is primarily supplied to the body from the lumen of the intestine, where it is manufactured by bacteria, especially by *Escherichia coli.* This organism produces about 1000 units of vitamin K per gram weight of dry bacteria.

Administration of the modern broad-spectrum antibiotics (for example, the tetracyclines) may so sharply reduce the microbial population of the intestinal tract as to cause a sizable reduction in the production of vitamin substances there. Depletion of vitamin K is especially serious prior to surgery, and the vitamin may have to be administered to the patient as a drug to prevent hemorrhage. Microbes may have even much greater importance in the physical well-being of lower animals and man than is apparent with our present state of knowledge.

MICROBES IN INDUSTRY

Microbiology of industry involves the roles that microbes play in processes of commerce and industry, such as the following.

■ **Manufacture of dairy products.** One of the aspects of *dairy microbiology* is the manufacture and processing of certain milk products. Broadly speaking, this subject may be included with *microbiology of agriculture,* which encompasses the relation of microbes to domestic animals and to the soil.

When milk is allowed to stand, it sours and curdles because of the conversion of lactose (milk sugar) into lactic acid by bacterial action. This is known as *ripening.* When milk is churned, the fat globules coalesce and take up a small amount of milk solids and water to form butter. The residue is *buttermilk.* The bacteria that play the key role in ripening are the ordinary lactic acid bacteria,† normal inhabitants of milk. Ripening is often hastened by adding cultures of these lactic acid-producers known

*"If microscopic beings were to disappear from our globe, the surface of the earth would be encumbered with dead organic matter and corpses of all kinds, animal and vegetable. . . . Without them, life would become impossible because death would be incomplete." (Louis Pasteur, 1861.)

†Principal genera are *Streptococcus, Leuconostoc, Pediococcus,* and *Lactobacillus.*

as "starters." Originally the purpose of ripening was to preserve the milk until enough to churn had accumulated. The acidity of sour milk inhibits growth of other microbes. Now ripening is done to give flavor to the butter. Butter is usually prepared from cream instead of whole milk. Cultured buttermilk is prepared by adding ordinary butter starters to fresh sweet milk.

Cheese consists of the curd of milk separated from the liquid portion either by the bacterial fermentation of the lactose in the milk to form lactic acid or by the action of the milk-curdling enzyme, *rennin* (from the stomach of a calf.) Cheese prepared by the action of lactic acid is known as cottage cheese. When renin is the curdling agent, the water is drained away, and the curd is pressed into a block and partly dried. This is green cheese. When green cheese is set aside for a time, it undergoes changes (ripens) that give it color, flavor, and odor. Ripening is induced by enzymes in the curd, as well as by molds, yeasts, and bacteria. The kind of cheese manufactured depends, to a great extent, upon the kind of organisms that cause it to ripen. For instance, Swiss cheese is ripened by gas-forming bacteria, and the liberated gas is responsible for the many holes. Cheese is usually named after the locality in which it is produced, and a certain type of cheese may have several different names, depending on where it is manufactured. For instance, Cheddar (after Cheddar, England). Wisconsin, American, and brick cheeses are basically the same.

■ **Manufacture of alcohol and alcoholic beverages.** Alcoholic beverages are of two classes: those that are manufactured by fermentation alone and those in which the alcohol is distilled from the fermented mixture. The distillate has a higher alcoholic content than does the beverage manufactured by fermentation alone. Among the beverages manufactured by fermentation alone are wine and beer. Among those that are manufactured by distillation are commercial alcohol, whiskey, rum, and brandy. The manufacture of both fermented and distilled beverages depends basically upon the conversion of sugar into alcohol by yeasts. Commercial fermentations are brought about mainly by the enzymes of yeasts (*Saccharomyces* species).

Wines are made directly by allowing grape juice sugars to ferment. Beer is made from grains, most often barley. Grains do not contain sugar at first, but when the grain is soaked in water, the seeds germinate and the starch-splitting enzyme known as *diastase* (amylase) is produced. It converts the starch of the grain into the sugars maltose and glucose. Grain starch that has been converted into sugar is known as *malt.* The maltose and glucose in the malt are converted into alcohol by yeasts. If foreign organisms grow in beer, they give it an undesirable taste. This is "disease" of beer.

Purification of alcoholic beverages by distillation depends upon the fact that alcohol has a much lower boiling point than does water. When a fermented mixture is heated to a temperature considerably below the boiling point of water, the alcohol distills over, leaving the water behind. Commercial alcohol and whiskey are usually made by distilling fermented grains. Brandy is made by distilling fermented fruits. Rum is made by distilling fermented cane juice. The mixture of fermenting grains used in the manufacture of alcohol and alcoholic beverages is known as *mash.* The mash from which rye whiskey is made contains 51% or more rye, whereas that used for the manufacture of bourbon contains 51% or more corn. The mash used in the production of alcohol is usually a mixture of corn (88.5%), barley (9.75%), and rye (1.75%), known as *spirit mash.*

Industrial alcohol (ethyl) has been for years manufactured from molasses. Alcohol may be produced by the action of bacteria on the cellulose of plants. In this method, by-products of the lumber industry such as sawdust, trimmings, and remains of trees are used. A large amount of alcohol is obtained as a by-product when petroleum is cracked to form gasoline.

■ **Baking.** An essential step in baking is the same as that for the manufacture of alcoholic beverages, the conversion of sugar into alcohol and carbon dioxide (by yeast). It is the carbon dioxide and not the alcohol that plays the important role in baking. When yeasts are mixed with dough, their enzymes, notably zymase, ferment the sugar present; the carbon dioxide thus produced riddles the dough with small holes, causing it to rise. When the dough is baked, the heat volatilizes the alcohol and causes the carbon dioxide to expand and to be driven off, leaving the bread light and spongy. Some sugar must be present or fermentation will not take place. The dough may contain a small amount of diastase, which converts some of the starch of the dough into sugar. If white flour is used in which the amylase has been destroyed in the refining process, sugar must be added.

■ **Manufacture of vinegar.** There are several methods of making vinegar. It may be made from wine or cider or from grains that have undergone alcoholic fermentation. In most methods the first step is the production of alcohol by fermentation, and the second step is the conversion of the alcohol into acetic acid. The first step differs in no manner from the production of alcohol for other purposes. When the alcoholic liquor is exposed to the air, a film known as mother of vinegar and containing acetic acid bacteria (gram-negative rods in genera *Acetomonas* and *Acetobacter*) forms on the surface. These bacteria convert alcohol into acetic acid according to the following formula:

$$C_2H_5OH + O_2 = CH_3COOH + H_2O$$

If the process continues too long, the bacteria decompose the acetic acid into carbon dioxide and water, and the vinegar loses some of its strength.

■ **Production of sauerkraut.** Sauerkraut is finely shredded cabbage that has been allowed to ferment in brine formed by salt and cabbage juice. The finely shredded cabbage is placed in layers in a cask, and salt is sprinkled over each layer. The layers are packed closely together by a weight placed upon them. The salt extracts juice from the cabbage, and the bacteria present convert the sugar of the juice into lactic acid, which helps to give sauerkraut its peculiar flavor.

■ **Tanning.** The recently removed hide or skin of an animal must be treated in some way to preserve it, or its microbial population will destroy it. The leather is soaked and softened in numerous changes of cool water to remove dirt, blood, and other debris. Cool water retards bacterial growth. The excess flesh is trimmed from the hide or skin and the hair is loosened. One method of loosening the hair is to soak the hide in a solution of lime. It has been found that old solutions, which contain many more bacteria, remove the hair more effectively than do fresh ones, which contain few. During the process the bacteria partially decompose and soften the leather.

After the loosened hair is scraped away, minerals are removed by washing or by chemical action. Before they are tanned, the hides of some animals are pickled. Pickling is carried out by treatment with sulfuric acid and salt. Tanning is accomplished by the use of vegetable tannins or chemicals such as alum.

■ **Curing of tobacco.** When tobacco leaves are cured, they change in texture and flavor and take on a brown color. These changes are accompanied by a decrease in sugar, nicotine, and water. The process is apparently one of oxidation induced by fermentation.

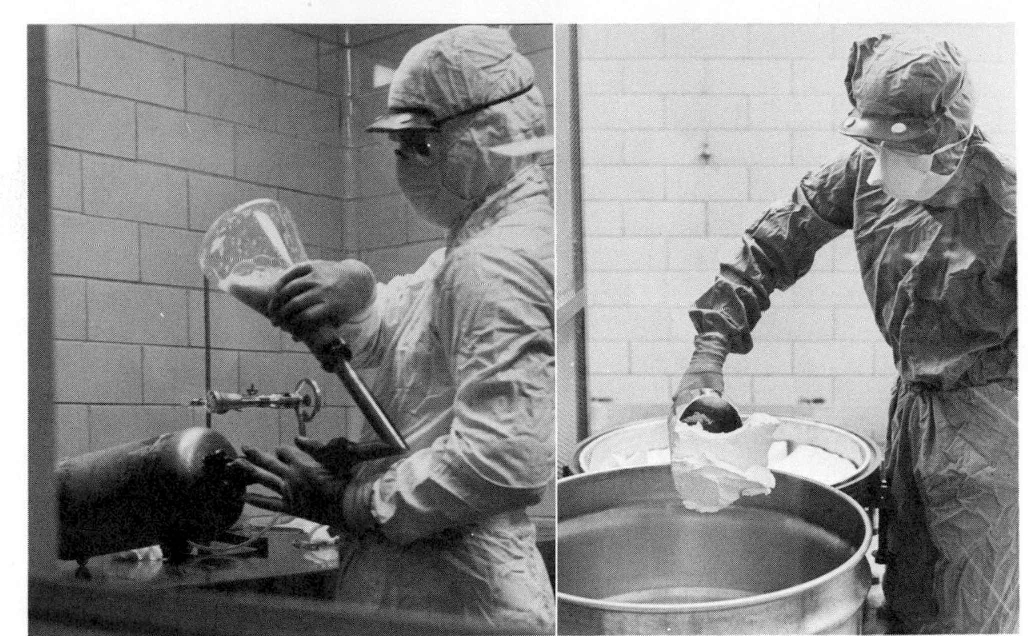

FIG. 34-3. Production of antibiotics. The manufacture of antibiotics is a complex process involving fermentation, filtration, extraction, and other intricate operations, many of which must be performed under sterile conditions. The antibiotic is grown in large fermentation tanks under controlled conditions, is filtered to remove impurities, and is finally recovered in crystalline form. **A,** Recharging seed tank with microbial culture in the fermentation process. **B,** Handling of bulk antibiotic under sterile conditions. (Courtesy Eli Lilly & Co., Indianapolis, Ind.)

■ **Retting of flax and hemp.** In flax and hemp plants, the fibers are closely bound to the wood and bark of the plant by a gluelike pectin. When these fibers are separated from the bark and wood, the commercial products linen and hemp are obtained. The dissolution of the pectin to effect separation is accomplished by the action of certain aerobes and the anaerobic butyric acid bacteria (organisms in genus *Clostridium*).

■ **Manufacture of antibiotics.** Today the manufacture of antibiotics is a major pharmaceutical activity. Over 50 such are available today for plants, animals and man. The basic materials used are cultures of molds or soil bacteria. Penicillin is produced by molds of the genus *Penicillium,* and antibiotics, such as streptomycin and the tetracyclines, are obtained from higher bacteria of the genus *Streptomyces.* The development of penicillin and streptomycin fermentations into industrial processes was the beginning of the field of bioengineering. (Fig. 34-3 and 34-4). At present, the annual output of one antibiotic (penicillin) alone is easily more than 1 million pounds. Within recent years the sales of internal anti-infectives in the United States have amounted to more than $500 million, and antibiotics form a large part of such sales. Prescriptions written for such pharmaceutical compounds have numbered more than 133 million.

■ **Oil prospecting.** It has been found that certain types of bacteria are extremely plentiful in the soil overlying deposits of oil. This finding has been utilized to locate such deposits. The growth of bacteria is supported by gases that seep from the oil to the surface.

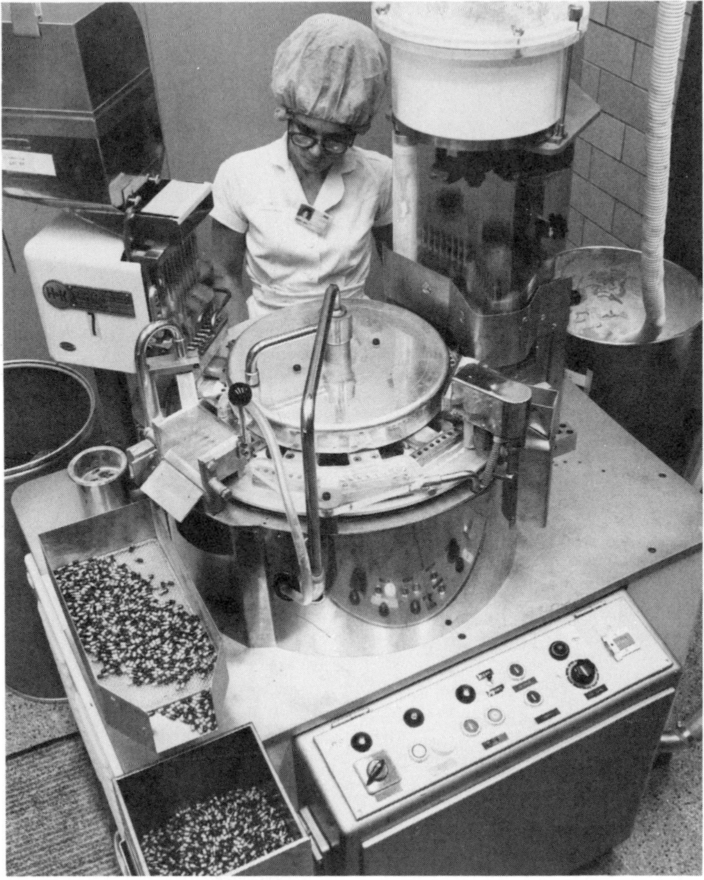

FIG. 34-4. Production of antibiotics. Highly automated, high-speed equipment fills several capsules at a time with precise amount of antibiotic. (Courtesy The Upjohn Company, Kalamazoo, Mich.)

■ **Other industrial processes.*** In addition to the practical applications of microbes just given, the following may be mentioned:

1. Manufacture of acetone (fermentation of sugars—*Clostridium acetobutylicum*)
2. Manufacture of lactic acid (fermentation of molasses—lactic acid bacteria)
3. Manufacture of dextran, plasma expander (fermentation of cane sugar—*Leuconostoc mesenteroides*)
4. Manufacture of butyl alcohol (fermentation of sugars—butyric acid bacteria)
5. Manufacture of vitamins (for example, cyanocobalamine or vitamin B_{12}, riboflavin or vitamin B_2, and vitamin C)
6. Microbiological assay—determination of the amount of amino acids and vitamins in tissues and body fluids (for example, assay of vitamin B_{12})
7. Manufacture of steroid hormones, cortisone and hydrocortisone (action of certain molds on plant steroids)

*The theoretical study of fermentation has been constantly stimulated by the industrial values of the many end products obtained.

8. Manufacture of citric acid used in lemon flavoring (oxidation of sugar—*Aspergillus niger*)
9. Manufacture of enzyme detergents (subtilisins, alkaline proteases from *Bacillus subtilis,* added to final detergent product)
10. Manufacture of insecticides (protein of *Bacillus thuringensis* lethal to larvae [caterpillars] of serious food crop pests in genus *Lepidoptera*)
11. Production of food in seas and oceans (photosynthesis carried on by certain marine microbes, role similar to that of plants on land)
12. Production of ensilage for animal feed (fermentation of sugars in shredded green plants by lactic acid bacteria)
13. Manufacture of plant growth factors (example, gibberellins)
14. Manufacture of enzymes (for example, amylases, proteases, pectinases)
15. Manufacture of amino acids (for example, glutamate, lysine)
16. Manufacture of flavor nucleotides (for example, inosinate, guanylate)
17. Manufacture of polysaccharides (example, xanthan polymer)
18. Sewage disposal (see Chapter 35)

QUESTIONS FOR REVIEW

1. What is meant by the cycle of an element?
2. Briefly outline the nitrogen cycle and the carbon cycle.
3. What is nitrogen fixation? Its importance?
4. What determines the fertility of the soil?
5. What causes the root nodules on clover plants? What purpose do they serve?
6. What organism is important in the synthesis of vitamin K in the intestine?
7. List 10 practical applications made of microorganisms.
8. Explain how cheese is made. What is dairy microbiology?
9. What is diastase? What is its role in fermentation?
10. Give specific examples of microbes (by name) used in industry.

REFERENCES. See at end of Chapter 38.

35 The microbiology of water

Practically all waters under natural conditions contain microbes, including protozoa, bacteria, fungi, and viruses. Some contain many microbes, others contain few. The number and kind of microbes present depend upon the source of water, the addition of the excreta from man and animals, and the addition of other contaminated material.

Since sewage contains the pooled excreta from both the sick and well, it necessarily contains pathogenic organisms, especially those that leave the body by the feces or urine. Sewage must be properly disposed of to avoid contamination of water supplies and the spread of disease by flies or other agents.

Sanitary microbiology (the microbiology of sanitation) pertains to drinking water supplies and sewage disposal.

■ **Sanitary classification.** From a sanitary standpoint waters may be classified as potable, contaminated, and polluted. A *potable* water is one free of injurious agents and pleasant to the taste. In other words, it is a satisfactory drinking water. A *contaminated* water is one that contains dangerous microbial or chemical agents. A contaminated water may be of pleasing taste, odor, and appearance. A *polluted* water is one of unpleasant appearance, taste, or odor. Because of its content it is unclear and unfit for use. It may or may not be contaminated with disease-producing agents.

The pollution of water supplies is a major health problem today. Not only may sanitary sewage be discharged into water, but complex wastes from industry and agriculture may be also. In increasing volume such substances as plastics, detergents, insecticides, oils, animal and vegetable matter, chemical fertilizers, and even radioactive materials are released from factories, canneries, poultry-processing plants, oil fields, and farms. The persistence of synthetic detergents in waste water and sewage reflects a failure of microbial action. Bacteria do not necessarily possess the enzymes to degrade them chemically.

Fish reflect a measure of the purity of a water supply since they cannot live and thrive in the water of a river or lake that is heavily polluted.

■ **Sources of water.*** A water supply may come from (1) rain or snow water, (2) surface water (shallow wells, rivers, ponds, lakes, and waste water), and (3) ground water (deep wells and springs) (Fig. 35-1). Generally, surface water contains more microbes than does either ground or rain water, and ground water contains more than rain water. Surface water contains many harmless microbes from the soil, and in the vicinity of cities it is often contaminated with sewage bacteria. (The majority of soil microorganisms are found in the upper 6 inches of the earth's crust.) Surface water in

---

*Water is the most abundant of substances on the earth and is almost the only inorganic fluid found simultaneously in nature as gas, liquid, or solid. It is surprisingly difficult to obtain or keep as a pure substance. It is contaminated easily (or recontaminated after purification).

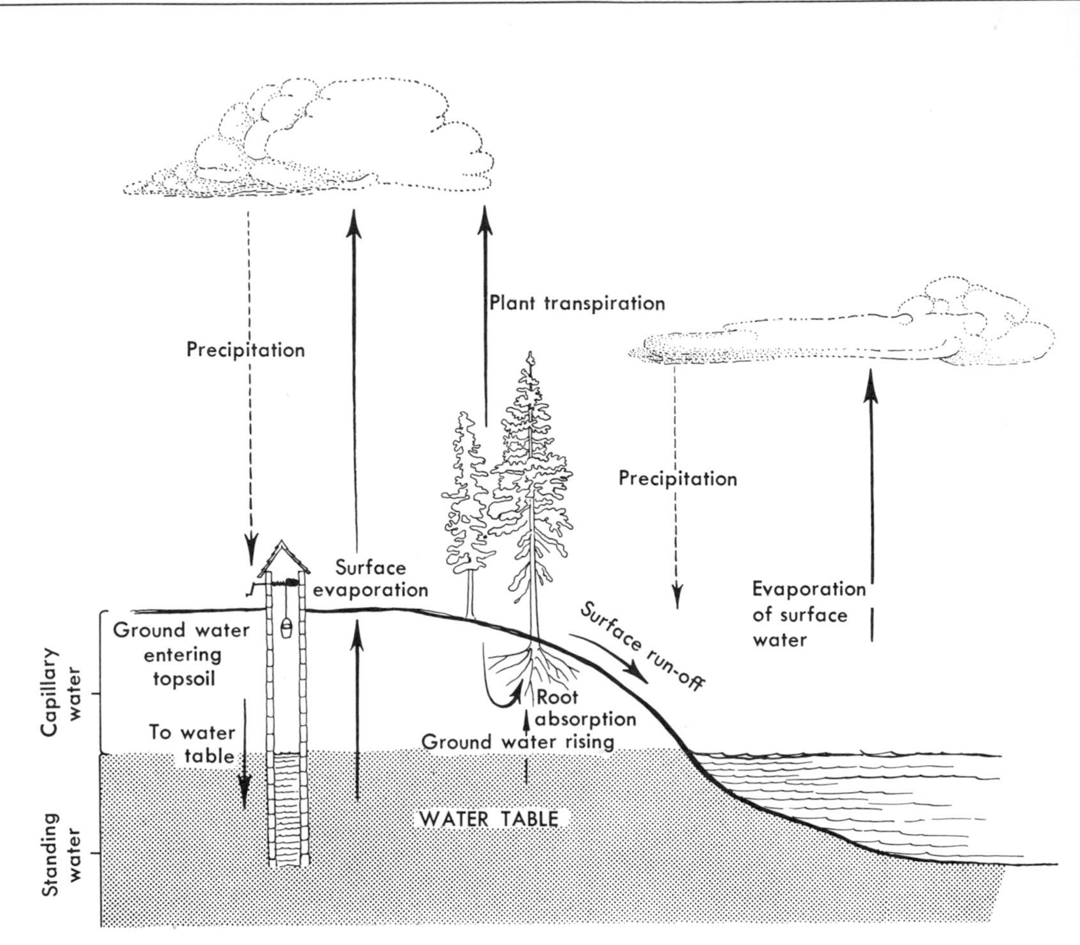

FIG. 35-1. Water cycle. This physical phenomenon is almost independent of life. However, water is necessary for metabolism in all organisms. (From Arnett, R. H., Jr., and Braungart, D. C.: An introduction to plant biology, ed. 3, St. Louis, 1970, The C. V. Mosby Co.)

sparsely settled localities may be comparatively safe, but the only safe rule is *not* to use it without purification.

Unless shallow wells are properly constructed, they may become contaminated with the drainage from outhouses, stables, and other outbuildings. Improperly constructed shallow wells have been responsible for many outbreaks of typhoid fever and dysentery in rural communities. For a shallow well to be safe, its upper portion must be lined with an impervious material so that water from the surface will not seep into it, and it should be located so that outhouses drain away from it. Of course, it must be kept tightly closed. Shallow springs may be just as dangerous as shallow wells.

Deep well water and deep spring water usually contain few microorganisms because these are filtered out as the water trickles through the layers of the earth, but like shallow wells, deep wells must be protected against pollution by surface water. Microbial contamination may occur when a well is situated within 200 feet of the source of contamination, and chemical contamination may occur for a distance of 400 feet. Contamination is more likely to occur during wet weather.

■ **Waterborne diseases.** The sources for bacteria in water are many—soil, air, the water itself, and the decaying bodies and excreta of man and animals. As a result many different kinds may be found there, including well-known pathogens as well as the harmless water bacteria. There is increased concern in this country as to the presence of viruses in water since most human ones multiply in the alimentary canal. Human enteric viruses have been recovered from a third of surface water samples examined. Ground waters may also transmit virus.

The diseases spread by water are notably those whose causative agents leave the body by way of the alimentary or urinary tracts, and water-borne infections are mostly contracted by drinking contaminated water. They may be caused by a variety of microorganisms. When a waterborne epidemic occurs, most cases appear within a few days, indicating that all were infected at about the same time. Most significant of the waterborne diseases are typhoid fever, bacillary dysentery, amebic dysentery, salmonellosis, viral hepatitis, and cholera. Epidemics of acute diarrhea often follow heavy contamination of water with sewage. Diseases that on occasion are transmitted by water are tularemia, anthrax, leptospirosis, schistosomiasis, ascariasis, poliomyelitis, and other enterovirus (coxsackievirus and echovirus) infections. Pathogenic bacteria do not live long in water, and they do not multiply there; but they may persist for a time in water that is cool and contains considerable organic matter.

Certain opportunist pathogens, including many species of *Pseudomonas,* are referred to as water organisms (water bugs) since they can survive and propagate in a wet, stagnant environment. They can grow in sterile water on traces of phosphorus and sulfur and are generally resistant to antibiotics and disinfectants. In hospitals they lurk as an unsuspected source of infection in sinks, drains, faucets, and the oxygen therapy apparatus. They can gain a foothold in the air ducts, mist generators, and humidifying pans in incubators of newborn and premature nurseries. Most susceptible to these water organisms are newborn infants, patients with surgical wounds, and individuals with chronic illness. Control of this source of infection lies in heat sterilization, absolute cleanliness, and removal of the moist, stagnant areas.

Paris and numerous European cities have dual water supplies, that is, water supplies in which drinking and household water is carried in one set of mains while water for industrial purposes and fire fighting is carried in another set. This is a very dangerous arrangement since accidental cross connections may occur between the two.

■ **Bacteriologic examination of water.** We know from experience that it is almost impossible to isolate from water the organisms responsible for the most important waterborne diseases because few are present to begin with and they do not multiply in water. Sanitarians and public health workers, therefore, have concluded that the only safe method to prevent waterborne diseases is to condemn any fecally polluted water as being unfit for human use. It *might* contain harmful organisms, since these practically always gain access to water in the feces or urine of man. Whether fecal pollution exists would be determined by examining the water for colon bacilli *(Escherichia coli),* abundant in feces. Certain other bacteria that resemble colon bacilli may or may not be of fecal origin. Also found in water, these bacteria ferment lactose with the formation of gas. In practical water analysis, therefore, testing includes the group spoken of as the *coliform* bacteria (The presence of coliforms increases the likelihood of viruses being present. Note that viruses have been identified in sewage as indicated

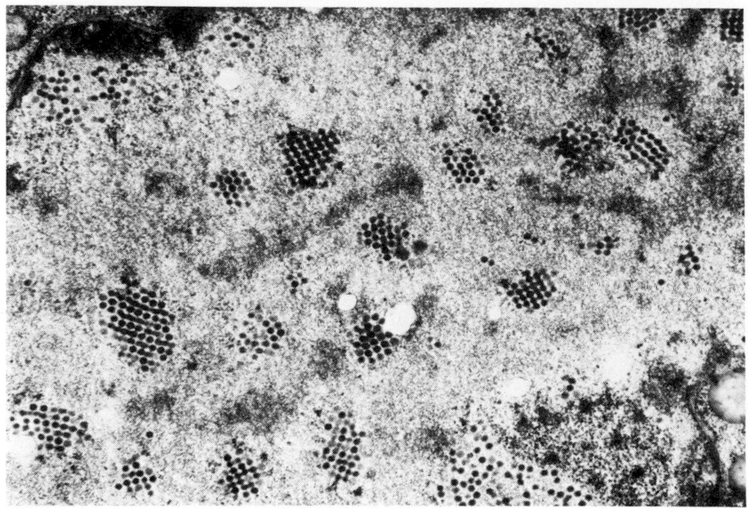

FIG. 35-2. Adenovirus crystals identified in sewage. (Courtesy Division of Microbiology and Infectious Diseases, Southwest Foundation for Research and Education, San Antonio, Texas.)

in Fig. 35-2). The method of procedure recommended by the American Public Health Association and the American Water Works Association consists of three tests: presumptive, confirmed, and completed. The following are the tests as outlined in *Standard Methods for the Examination of Water and Wastewater.*

PRESUMPTIVE TEST. The presumptive test is based upon the fact that coliform bacteria ferment lactose with the production of gas. Lactose broth fermentation tubes are inoculated with 0.1, 1, and 10 ml. portions of the test water and incubated for 24 hours ± 2 hours at 35° ± 0.5° C. If no gas has formed, the tubes are incubated for another 24 ± 3 hours. The presence of gas is *presumptive* evidence that members of the coliform group of bacteria are present and suggests fecal contamination. The smallest amount in which the fermentation of lactose occurs is an index of the number of organisms present.

CONFIRMED TEST. In the confirmed test, fermentation tubes of brilliant green lactose bile broth or solid differential media, such as Endo's medium or eosin–methylene blue agar, are inoculated with material from the tubes showing gas. If upon incubation at 35° ± 0.5° C., the brilliant green lactose bile broth shows gas at the end of 48 ± 3 hours or if colonies with the characteristic appearance of members of the coliform group appear on the solid medium within 24 ± 2 hours, the *confirmed* test is positive.

COMPLETED TEST. If brilliant green lactose bile broth was used in the confirmed test, the completed test is done by streaking Endo's medium or eosin–methylene blue agar plates from the tubes showing gas and incubating them at 35° ± 0.5° C. for 24 ± 2 hours. From these plates (or the original plates if Endo's medium or eosin–methylene blue agar was used in the confirmed test) colonies typical for or similar to those of the coliform bacteria are transferred to a lactose broth fermentation tube and an agar slant. These are incubated at 35° ± 0.5° C. and inspected at the end of 24 ± 2 hours and 48 ± 3 hours. After incubation, if microscopic examination of the growth on the agar slants reveals small gram-negative organisms and if the lactose fermentation is positive, the identification of coliform bacteria is said to be *completed.* At one time it was believed that when a positive test was obtained a determination of the fecal or nonfecal origin of the bacteria should be made, but most sanitarians in the United States feel that water to be satisfactory for drinking should be free of both fecal and nonfecal organisms.

Included in *Standard Methods for the Examination of Water and Wastewater* is the *membrane filter procedure* of detecting coliform bacteria. In this method, developed in Germany during World War II, the water is filtered through a thin disk of bacteria-retaining material. After filtration, the disk is inverted over an absorbent pad saturated with modified Endo's medium and incubated for 20 ± 2 hours at $35° \pm 0.5° C$. All organisms that produce a dark purplish green colony with a metallic sheen are considered to be coliforms. Colonies of coliforms are counted, and an estimation of the number per 100 ml. of water sample is made.

European bacteriologists have devised systems whereby the water is also examined for *Streptococcus faecalis* and *Clostridium perfringens,* both of which are just as constantly present, but not so plentiful, in the feces as are coliforms. Some American workers feel that fecal streptococci are more practical and reliable indicators of fecal contamination of water than are coliforms. Streptococci in the viridans group are suggested because, unlike the fecal streptococci, they strictly parasitize the alimentary tract of man and are not found in other animals or on plants. It has been suggested also that phages could be used as accurate indicators of fecal pollution of water.

■ **Drinking water standards.** The United States Public Health Service has adopted certain standards for drinking water used in interstate commerce. Originally designed to protect the traveling public, these, to a great extent, have served as standards throughout the country. Included are specifications relating to (1) source and protection of water, (2) physical and chemical properties, (3) bacteriologic content, and (4) limits on concentration of radioactivity.

Drinking water must contain no impurity that would offend sight, taste, or smell, and substances that might have deleterious physiologic effects must not be introduced.

Investigation of the problem of viruses in water is just beginning. Since minimal amounts of virus can produce infection, total removal of viruses from a water supply for human consumption appears indicated.

■ **Water purification.** Water may be purified by natural means. However natural methods are usually slow and uncertain. As water trickles through the earth, microbes are filtered out. Water standing in lakes and ponds undergoes some degree of purification because of the combined action of sunlight, sedimentation, dilution of impurities, and destruction of bacteria by protozoa. Streams may be purified as they flow along, but the possibility of contaminated material entering the stream along its course is so great that one cannot depend upon self-purification of the water. *Therefore, all surface waters must be regarded as potentially dangerous unless subjected to artificial purification.**

*In unusual circumstances small quantities of water can be made relatively safe for drinking by the addition of two drops of chlorine bleach, or two drops of tincture of iodine, or Halazone, a commercial tablet, to one pint of water, provided that the mixture is allowed to stand for 30 minutes.

FIG. 35-3. East Side water purification plant, Dallas, Texas. From the chemical plant in the foreground where ferrous sulfate and lime are added to the water to the pump house in the background from where water is pumped to the reservoir, steps in the purification process are indicated. Between these points there are rapid mixers, flocculators, settling basins, and sand filters. Water is chlorinated several times at points along the way. (Courtesy Dallas Water Utilities, Dallas, Texas.)

Generally, artificial water purification (Fig. 35-3) is carried out by a combination of physical and chemical methods—sedimentation and filtration combined with the action of chemicals to soften and clarify the water, and chlorination (see p. 247). *Sedimentation* is the process whereby solids in the water, such as mud and organic matter, are removed. Water is held in settling basins for a number of hours to allow the large particles to settle out. A chemical such as alum or ferric sulfate is added to coagulate the suspended organic matter and microbes present, and the precipitated coagulum is removed mechanically. Some clarification of water occurs with sedimentation. *Filtration* is the process that puts the final sparkle in the water.

The average rapid filter that is in use has the following units from the floor to the surface: (1) watertight floor with grooves and tile for draining the filtered water away, (2) layer of gravel from 12 to 18 inches thick, and (3) layer of sand. The gravel at the bottom is of such a size as to pass through a 2-inch screen mesh. The size of the gravel gradually decreases as the surface of the layer where the gravel will pass through

FIG. 35-4. Wastewater treatment plant, Dallas, Texas, aerial survey. Note layout of settling basins, trickling filters, and microbial digestors. Each of the 29 circular trickling filters shown in this photograph is a half-acre rock bed receiving the effluent from primary settling basins. Here organic wastes are oxidized by bacteria and algae growing on the rocks. Sludge from primary settling basins goes to a separate group of heated tanks called digestors where by a process of anaerobic decomposition the solids are broken down by microbes. (Courtesy Dallas Water Utilities, Dallas, Texas.)

a $1/16$-inch mesh is approached. The sand layer is about 2.5 feet thick. Rapid sand filters can filter over 200 million gallons per day per acre of filter area. By rapid filtration 90% to 95% of all bacteria are removed more or less mechanically by the layer of sand at the top of the filter. Efficient filtration removes all pathogenic and most saprophytic organisms. Filters must be cleaned often. This is done by reversing the flow of water and agitating the upper layer of sand. In some cases air is forced through the filter from the bottom to the top. A rapid filter is made in units so that just a part of the filter is closed for cleaning at any one time.

After filtration, water is universally chlorinated. There are three methods in common use:

1. *Simple chlorination.* This method is the addition to water of a standard effective amount of chlorine. Chlorine is usually added from cylinders connected with regulating equipment to control the exact amount.
2. *Ammonia-chlorine treatment.* This treatment results in compounds called chloramines, formed when chlorine is added to water containing ammonia. Chloramines are less active as germicidal agents but are more stable than free chlorine.
3. *Superchlorination.* This process involves the addition of a much larger dose of chlorine than with simple chlorination and the removal subsequently of the excess.

FIG. 35-5. Water reclamation. An "activated sludge" pilot plant in Dallas, Texas, of a type used by many cities over the nation. This is a 1 million-gallons-a-day facility. In an activated sludge process microbes are mixed with sewage to consume its organic products. In this photograph one large tank is maintained in an aerobic state to facilitate growth of sewage microbes. It is seeded by sludge (primarily microorganisms) settled from a second large tank, the secondary clarifying tank. The input of raw sewage or sewage already settled into the aeration tank provides substrate for microbial action. (Courtesy Dallas Water Utilities, Dallas, Texas.)

Certain chemicals in hard water soften the water by removing dissolved limestone. When mixed with raw, untreated water, activated carbon absorbs odors and tastes resulting from the decaying vegetation and organic matter present.

■ **Fluoride content of water and tooth decay.** Although a discussion of the fluoride content of water may not lie within the scope of a book devoted strictly to microbiology, fluoridation of water is mentioned because it is so important in public health. It is well known that the fluoride content of the drinking water of infants and children profoundly influences the development of their teeth. If too much fluoride is present (4 to 5 ppm water), the condition known as fluorosis or mottled enamel develops; if too little, the incidence of dental caries is greatly increased. Adult teeth do not seem to be affected. The optimum ratio of fluoride to water for drinking by infants and young children is 1 ppm. When this is present, there is sufficient fluoride to retard tooth decay significantly but not enough to stain or mottle the teeth, that is, to cause fluorosis. In nature the fluoride content of water varies from much less to much more than 1 ppm water, but the proper level may be maintained by adding fluoride to water whose content is low or by chemically removing fluoride from water whose content is too high.

■ **Ice.** When pathogenic bacteria gain access to ice, the majority die, but a few survive. It is estimated that about 10% survive for more than a few hours. Ice that has been stored for 6 months is practically bacteria free. Few instances of the transmission of disease in ice are known. The use of water of high sanitary quality in the manufacture of ice has lessened the possibility that ice itself will act as a conveyer of disease. However, the handling of ice in the home, eating places, and cold drink stands may be a source of infection.

■ **Purification of sewage.** Today sewage is processed in treatment plants to render it harmless before it is discharged into any body of water (Fig. 35-4). Many methods of sewage purification are in use. The procedures involved are screening and sedimentation to remove the larger particles, chemical treatment to remove small particles, microbial action to digest organic matter, aeration to induce oxidation, filtration through sand, and chlorination. There is no known way available to remove all viruses from sewage (see Fig. 35-2).

The breakdown of human waste in sewage by microbes is essentially the same as decay of organic material in other situations. The sanitary engineer is much concerned with this fact. He designs the sewage disposal plant to facilitate and speed up microbial activity and to ensure its completion. (See Fig. 35-5.)

■ **Swimming pool sanitation.** Swimming pools may convey conjunctivitis, ear infections, skin diseases, and intestinal infections, unless kept in a sanitary condition. The source of infection is a person using the pool. Swimming pool water should be kept in the same state of purity as drinking water. This is done with frequent changes of water and the use of disinfectants. Chlorine is the most widely used one. The highest possible concentration of chlorine to water should be maintained, about 1 ppm. A higher concentration is irritating to the eyes. Before entering the pool, each person should be required to take a shower bath using soap. Persons who have infections of any kind should not be allowed to use the pool.

Some authorities prefer iodine to chlorine as a disinfectant for swimming pools. Its disinfectant action is not as inhibited by the presence of organic matter, and there is less eye and skin irritation than with chlorine. Bromine has also been used as a swimming pool disinfectant.

QUESTIONS FOR REVIEW

1. Name the diseases that may be spread by water. Give the causal agent.
2. Discuss the bacteriologic examination of water.
3. Describe the purification of water by (a) natural means, (b) artificial means.
4. List the sources of water. From a public health standpoint how do waters from these sources differ? State the sources of the microbes in water.
5. Why is sewage of such great sanitary significance?
6. What specifications are made for drinking water standards?
7. What is the relation of the fluoride content of water to tooth decay?
8. Discuss water pollution today as a public health problem. (Consult references.)
9. Give the sanitary classification of water.

REFERENCES. See at end of Chapter 38.

36 The microbiology of food

MILK

If obtained in a pure state and kept pure, milk is our best single food. If improperly handled, it is our most dangerous one. Milk is an important conveyer of infection because it is an excellent culture medium and is consumed uncooked. *Microbiology of dairy products* deals with (a) the microbes that make milk and milk products unfit for human consumption, (b) the prevention of disease in cattle, and (c) the manufacture of milk products.

■ **Bacteria in milk.** Bacteria gain access to milk by many routes. Milk as secreted by the mammary glands of perfectly healthy cows is usually sterile, but as the cow is milked, bacteria from the teats and milk ducts get in, and by the time the milk enters the receptacle, contamination has taken place. Bacteria may be present within the udders of diseased cows. Unclean and unsanitary milking utensils, pasteurizing tanks, and milk containers as well as dust and manure are important sources of contamination.

Under the best conditions the bacterial content of milk is relatively high. Some bacteria, though they come from external sources, are so commonly present as to be regarded as normal milk bacteria; and pathogenic bacteria may also be present. Normal milk bacteria aid in the souring of milk; they also may destroy its value as a food; pathogens cause disease.

The two most important sources of pathogenic bacteria found in milk are the people who handle it and the infected cows. The organisms most often found in milk are staphylococci, streptococci, lactic acid bacteria, and enteric organisms.

Although it is not the only criterion, the number of bacteria in milk is the best index of its sanitary quality. The number present depends upon the number originally introduced and the temperature at which the milk is kept. Many species of bacteria begin to multiply as soon as they gain access to milk. This can be largely prevented if the milk is rapidly chilled to 10° C. (50° F.) as soon as it is obtained and kept cold. If milk is kept at that temperature, no great increase in the number of bacteria will occur, but if the milk is allowed to warm up, there is a rapid rise in numbers.

Milk of high quality may contain a few hundred bacteria per milliliter. Bad milk contains millions of bacteria per milliliter. All authorities agree that a single high bacterial count does not necessarily mean that the milk is of poor quality and that to be significant a high bacterial count must occur day after day. The presence of coliform bacteria points to fecal contamination.

■ **Milkborne diseases.** The diseases spread by milk may be classified as (1) diseases of human origin so transmitted and (2) diseases of milk-producing animals transmitted to man by the milk of the infected animal. Most notable of the diseases that primarily affect man and may be transmitted by milk are typhoid fever, salmonelloses, bacillary and amebic dysentery, scarlet fever, infantile diarrhea, diphtheria, poliomyelitis, infectious hepatitis, and septic sore throat. The diseases that primarily affect milk-producing animals and that may be transmitted to man by the milk of the infected animal are undulant fever, Q fever, and bovine tuberculosis. Undulant fever may be contracted from the milk of infected cows or goats.

The spread of human diseases by milk usually results from the contamination of the milk by the discharges of a handler who is ill of the disease or is a carrier. Milk-borne epidemics typically appear among the patrons of a given dairy, and the source of infection usually is traced to a food handler in that dairy.

Bovine tuberculosis, especially in children under 5 years of age, usually comes from the ingestion of the milk of tuberculous cows. In cattle, tuberculosis most often affects the lungs, but the sputum is swallowed and excreted in the feces. As a rule, the bacilli get into the milk by fecal contamination, but in some cases the udder becomes infected, and the bacilli are excreted directly into the milk. The milk of a single cow excreting *Mycobacterium tuberculosis* in her milk may render the mixed milk of the whole herd infectious. The danger of contracting tuberculosis through milk is completely eliminated by pasteurization of milk and the tuberculin testing of cows.

A disease not of bacterial origin but of great historic interest is *milk sickness* (called milksick by the pioneers). This disease cost the life of many Midwestern pioneers, including the mother of Abraham Lincoln. Milk sickness is caused by the ingestion of milk from cows that have been poisoned by eating certain species of goldenrod and the white snakeroot. The poisonous principles in these plants are excreted in the milk of the affected cow.

■ **Pasteurized milk.** *Pasteurization* is the process of heating milk to a temperature high enough to kill all *nonspore-bearing pathogenic* bacteria but not high enough to affect the chemical composition of the milk. There are two methods of pasteurization. One is known as the holding method; the other as the continuous-flow, high-temperature, short-time, or flash method. In the former the milk is held in tanks or vats, where it is subjected to a comparatively low temperature for a comparatively long period of time while being constantly agitated. In the latter the milk is subjected to a higher temperature for a short period of time as it flows through the pasteurizing equipment (see Fig. 36-1). After both methods the milk should be transferred through pipes, where it is chilled, to the bottling machines. All milk should be bottled or received into clean sanitary containers (glass, cardboard, plastic).

The *Grade "A" Pasteurized Milk Ordinance—1965 Recommendations of the Public Health Service* are that in the holding method the milk must be subjected to a temperature of *at least* 145° F. (63° C.) for *at least* 30 minutes and that in the continuous-flow method the milk must be subjected to a temperature of *at least* 161° F. (72° C.) for *at least* 15 seconds. Note carefully that pasteurization does *not* completely sterilize milk and that pasteurization should never be used to cover gross negligence in sanitation and the sanitary handling of milk.

■ **Pasteurized versus raw milk.** Pasteurization is the single most crucial item in the maintenance of a safe milk supply. Objections to pasteurization might be that (1) it destroys vitamins, (2) dairymen may not be so careful when they know that the milk is to be pasteurized, and (3) pasteurization may be carelessly and ineffectively done. The latter two objections can be eliminated by thorough inspection systems. Vitamin C, which is destroyed, can easily be replaced in the diet or given as ascorbic acid.

FIG. 36-1. High-temperature short-time (HTST) method of pasteurization. Milk is received from the dairy farm in tank trucks and placed in holding tanks. Milk is pasteurized at 175° F. for 16 seconds through a plate heat-exchanger with holding tube and timing pump. It is also clarified, standardized, homogenized, and steam heated to 190°-194° for 30 seconds, vacuum-cooled back to 175°, and cooled at 33° through a plate heat-exchanger containing cooling medium. All is done automatically from a central control panel, utilizing sanitary air-operated valves for routing the milk. Cleaning and sanitizing is also done by automatic circulation of cleaning and sanitizing compounds. Note the Vac-Heat unit in the foreground. This unit ensures the same milk flavor year-round by removing various volatile feed-off flavors, which vary during the year's four seasons. (Courtesy CP Division, St. Regis, Dallas, Texas.)

The temperature of pasteurization is not high enough to affect milk and is carefully controlled during the process. Heating milk to high temperatures can drastically change it. Boiling milk causes decomposition of its proteins, changes in its phosphorus content, precipitation of calcium and magnesium, expulsion of carbon dioxide, combustion of sugar, and destruction of enzymes.

■ **Requirements for a safe milk supply.** To ensure a safe milk supply, the following requirements must be met:

1. The cows must be healthy, well fed, and free of diseases such as tuberculosis, brucellosis, and mastitis.
2. All individuals handling milk must be free of infectious organisms, and their hands and person must be clean.
3. The premises must be kept clean.
4. The udders and flanks of the cows must be washed before milking.

5. The milking utensils and machinery that contact the milk must be kept sterile and should be so constructed as to keep out dust and flies. (Practically all milking today in commercial dairies is done mechanically.)
6. The milk must be chilled immediately to 10° C. (50° F.) and kept cold.
7. It must be pasteurized under carefully controlled conditions and again chilled.
8. Directly after pasteurization, it must be placed into the container reaching the customer.
9. It must be delivered cold (10° C.) in refrigerated trucks and immediately refrigerated at its destination.
10. A uniform statewide sanitary milk code must provide for proper inspection of dairies and pasteurization plants.
11. Local laboratory services must check on the purity, cleanliness, and safety of the milk.

■ **Milk grading.** Different states have different systems of grading their milk supply, but all are based generally upon the sanitation of the dairy, the health of the cows, the methods of handling the milk, and the bacterial content. In addition to the bacterial count* of milk, testing for coliforms is done to detect contamination after pasteurization. Milk is also checked for possible adulteration.

The phosphatase test is used in grading to check that milk has been adequately pasteurized. The enzyme phosphatase is mostly destroyed at the temperature of pasteurization. A significant amount detectable in milk indicates incomplete pasteurization or the presence of raw milk mixed in with a sample of pasteurized milk.

Several grades of milk are recognized by the *Milk Ordinance and Code—1953 Recommendations of the Public Health Service,* but only one, Grade A pasteurized milk, is allowed in interstate commerce for retail sale. This is milk that has been pasteurized, cooled, and packaged in accord with precise Public Health Service specifications. Its bacterial content has been reduced by pasteurization from no more than 300,000 per ml. to no more than 20,000 per ml., with coliforms no more than 10 per ml.

The general improvement in milk sanitation has been such that most authorities feel that there is no longer any practical necessity for special grades of milk. Therefore the present tendency is to stress the production of one standard, high-quality, safe, pure milk readily available to people in all walks of life at a reasonable cost.

■ **Milk formulas for hospital nurseries.** The preparation of infant milk formulas in hospitals should be in accord with standards set by the hospital medical staff or appropriate health agency. For proper microbiologic control, the American Academy of Pediatrics in its Standards and Recommendations for the Hospital Care of Newborn Infants recommends that technics of formula preparation be checked at least once a week. As part of their surveillance plan, random samples of the ready-to-use formula are sent to the laboratory to be cultured. The limit for the bacterial plate count of a formula sample is 25 organisms per milliliter. In the identification of organisms present there should be none other than sporeformers. Otherwise a break in technic is indicated, and responsible authorities should be notified at once.

---

*Bacterial count is obtained by either the plate count (p. 94) or the direct microscopic clump count. In the direct microscopic clump count, an accurately measured amount of milk (0.01 ml.) is placed on a slide and spread into a thin smear of uniform thickness. The dried smear is stained. With a microscope calibrated so that the area of the field is known, the bacteria in a number of representative fields are counted and the number of bacteria in the whole specimen calculated. In this clump count, both individual bacteria and unseparated groups of bacteria are counted as units.

■ **Bacterial action on milk.** Milk may be decomposed by the action of bacteria on its carbohydrates or proteins. The former is spoken of as fermentation of milk, the latter as putrefaction of milk. Souring comes from fermentation and is referred to as normal milk decomposition.

The souring of milk (lactic acid fermentation) is the result of the formation of insoluble casein by the action of lactic acid on caseinogen. Lactic acid is formed by the action of bacteria on lactose (milk sugar). *Streptococcus lactis, Lactobacillus acidophilus, L. bifidus, L. casei, L. bulgaricus, Escherichia coli,* and others are organisms responsible.

Putrefaction of milk seldom occurs, but when it does, it is a dangerous event. Putrefactive changes come from the action of sporebearing and anaerobic bacilli on the proteins of the milk. The milk is finally converted into a bitter liquid, having little resemblance to fresh milk.

Slimy or ropy milk usually comes from the action of *Alcaligenes viscolactis* or similar organisms, but it may be from other organisms such as streptococci and the lactic acid bacteria. In Norway, ropy milk is considered a delicacy. Americans consider it unfit for use.

Spontaneous alcoholic fermentation may occur rarely, but it is a feature of manufacturing processes in which sugar and yeasts or bacteria are added to milk. Among the products are kumyss or koumiss, kefir, leben, and yoghurt or matzoon. Yoghurt is the milk to which Metchnikoff referred in his book on the prolonged life of the Bulgarian tribes. Yoghurt is the forerunner of the present-day Bulgarian buttermilk.

The chief characteristic of yoghurt is its high acidity, produced by the growth of *Lactobacillus bulgaricus.* Because of this, it is considered to be one of the safest of all foods. Archaeologists working in underdeveloped areas feel that natural yoghurt, a major food item in the diet of many peoples, is always nutritious and safe to eat. Yoghurt and buttermilk are at times used in the nursing care of patients with advanced cancerous growths, secondarily infected and malodorous. The alkaline medium of the growth allows bacterial growth and fermentation activities to flourish. The application of yoghurt or buttermilk changes the tissue environment of the cancerous area to an acid one, thus inhibiting the growth of the odor-forming microbes.

Bacterial growth may color milk red, yellow, or blue.

■ **Ice cream.** Although ice cream is frozen, not all bacteria are destroyed. In fact, some pathogenic bacteria live longer in ice cream than in milk. Typhoid bacilli have been known to live as long as 2 years in ice cream. Outbreaks of typhoid fever, septic sore throat, diphtheria, food poisoning, and scarlet fever have been traced to ice cream. To prevent it from spreading disease, it must be made with pasteurized milk and handled properly thereafter. Proper handling means cleanliness of utensils and factory and proper health supervision of employees.

■ **Butter.** Although derived from milk, butter only poorly supports microbial growth because it is made up chiefly of fat and water and is deficient in protein and sugar. However, certain bacteria do grow in it and render it rancid. The ordinary saprophytic organisms found in butter are the various types of lactic acid bacteria, especially streptococci. Improperly prepared butters may contain *Escherichia coli* and various yeasts and molds. The presence of an excessive quantity of yeasts in butter indicates that proper sanitary precautions were not exercised in preparation. *Mycobacterium tuberculosis* and the atypical mycobacteria have been found in butter. Typhoid bacilli have been found, but they tend to die therein. Contamination of butter with pathogenic organisms may be eliminated by pasteurization of the cream used in making the butter, proper sterilization of utensils, and strict supervision of personnel.

FOOD

The *microbiology of food* is studied from many standpoints. The soil and agricultural microbiologist thinks of the microbes (free living and symbiotic) in the soil. These are utilized by food plants in growth and development. Microbes aid in the return of waste products of food utilization to the soil, where the waste products are used again by microbes and food plants. The food microbiologist studies the part that microbes play in the manufacture of bread, butter, cheese, and many other foods and the part that microbes play in the spoilage of foods and how this may be prevented. Public health microbiologists, primarily interested in the health of the community, study food as to what diseases it may convey, the harmful products of its spoilage, and how both foodborne disease and food spoilage may be prevented. The diseases that food conveys are discussed here.

Food poisoning

When foods are related to disease, they are considered first as carriers of infection, as indicated on p. 149 (examples, poliomyelitis, hepatitis).

Next come a group of illnesses spoken of as food poisoning. By definition, food poisoning is an acute illness resulting from the ingestion of some injurious agent in food. It is classified as follows:

1. Nonmicrobial type
   a. Individual idiosyncrasy
   b. Chemical factor from:
      (1) Foods naturally poisonous (toadstools and the like)
      (2) Poisons accidentally or intentionally added (plant sprays and the like)
2. Microbial type
   a. Food intoxications—effect of ingested toxin (for example, botulism)
   b. Food infections—multiplication of ingested microorganisms
      (1) Bacterial (for example, salmonellosis)
      (2) Parasitic (for example, trichinosis)

A characteristic of food infections and intoxications is that many persons eat the offending food and most develop the disease. In food infection, the diagnosis is made if the offending organism is cultured from the feces of the patient and from the food eaten. In food intoxication, diagnostic efforts are directed toward finding the offending toxin in the food eaten. Examination of the feces of the patient is of little value.

Staphylococci,* *Clostridium botulinum*, and, less often, certain other organisms produce food intoxication. *Salmonella* species and rarely *Streptococcus faecalis* (enterococcus) causes food infection. Table 36-1 lists bacterial types of food poisoning and the prominent features of each.

*Food poisoning, especially infection and intoxication caused by staphylococci, was formerly known as "ptomaine" poisoning. This usage is incorrect because ptomaines play no part in food poisoning.

TABLE 36-1. TYPES OF BACTERIAL FOOD POISONING

Organism	Foods	Onset (after food eaten)	Clinical features	Toxin	Outcome
Food intoxication					
Staphylococcus	Potato salad Chicken salad Cream fillings Milk Meats Cheese	1 to 6 hours	Vomiting Diarrhea Abdominal cramps Prostration	Enterotoxin	Severe symptoms of short duration (death rare)
Clostridium botulinum	Canned string beans, corn, beets, ripe olives	2 hours to 8 days	Vomiting Diarrhea Double vision Difficulty in swallowing, speaking Respiratory paralysis	Neurotoxin	Mortality high
Food infection					
Streptococcus	Sausage Poultry dressing Custard	2 to 18 hours	Nausea Pain Diarrhea	None	Recovery 1 to 2 days
Salmonella	Eggs Poultry Meats	7 to 72 hours	Diarrhea Severe abdominal pain Fever Prostration	None	Recovery—few days (death rare)

■ **Food intoxication.** Staphylococcal food intoxication is caused by the action of an enterotoxin liberated by some strains of staphylococci. The staphylococci multiply in the offending food before it is eaten and elaborate their toxin there. (They do not multiply in the intestinal tract.) Manifestations come from the ingested toxin. Staphylococci require a period of not less than 8 hours to elaborate enough toxin in food to cause symptoms. After the food is eaten, the disorder appears within 6 hours. Recovery occurs in 24 to 48 hours. Staphylococcal food intoxication usually follows the ingestion of starchy foods, especially potato salad, custards, and pies. When the offending food is meat, it is usually ham. Many outbreaks have been traced to chicken salad. This is probably the most common type of food intoxication.

Botulism is a very serious type of true food intoxication produced by the ingestion of food containing the potent exotoxin of *Clostridium botulinum* (p. 337). The major source of botulism is improperly processed home-canned vegetables of low acid content. *Before they are eaten,* such food should always be cooked *at the temperature of boiling water for at least 15 minutes* and thoroughly stirred and mixed. Contaminated foods may have unpleasant odors somewhat characteristic for a given food, but these off-odors are not easily recognized and do not suggest spoiled food. The cans or containers holding the food do not necessarily bulge, a conventional sign of food spoilage.

■ **Food infections.** Food infection is caused by the multiplication of bacteria that have been taken into the intestinal canal. The responsible bacteria are most often salmonellae. Of special consequence in food infection are *Salmonella enteritidis* (Gärtner's bacillus), *Salmonella typhimurium*, and *Salmonella choleraesuis*. Once inside the intestine, they multiply rapidly and induce widespread inflammation, with severe manifestations—nausea, vomiting, diarrhea, and fever.

The disease usually occurs as an explosive outbreak following a meal attended by a large number of persons. Food cooked in large quantities is more often the source of infection than is food cooked in small quantities because heat is not so likely to penetrate and destroy the organisms in a large quantity as in a small quantity of food. Most outbreaks occur during the warm months. The incubation period for *Salmonella* food infection is 7 to 72 hours, and recovery occurs usually within 3 or 4 days; but in severe infections death may occur within 24 hours.

The insufficiently cooked meat of infected animals may convey the disease to man, but usually salmonellae reach the food (often, but not necessarily, meat) from outside sources. Such sources are the intestinal contents of the slaughtered animals, the intestinal contents of animals (especially rats and mice) that have contacted the food incriminated, and (probably most important) human carriers of the bacteria. *Salmonella* infections are common in rats, and these animals often become carriers. Cockroaches have been shown to convey *Salmonella* food poisoning. Salmonellae are abundant in the intestinal tract of poultry, and may be present on the shell of an egg. Contaminated eggs broken commercially may be a source of organisms in egg products. Foods, such as the meringue on pies, contaminated with *Salmonella* may have no abnormal odor or taste.

Prevention of *Salmonella* food poisoning depends upon cleanliness in handling food, the proper cooking of food, the proper refrigeration of food that has been cooked, and the detection of carriers. Frequent and careful handwashing by food handlers is very important. Food that remains on their skin for a period of time provides nourishment for microbial contaminants, some foodborne, and thus contributes to their survival in the environment.

Under the provisions of the Egg Products Inspection Act of 1970, all commercial eggs broken out of the shell for manufacturing use must be pasteurized. The egg or egg product is heated to 60° to 61° C. (140 to 143° F.) and held there for 3½ to 4 minutes. The final product must be free of *Salmonella.*

Rarely streptococci cause food poisoning of the infection type. At times the addition of large numbers to the already resident enterococci in the bowel may trigger gastroenteritis. The food usually responsible contains meat that has stood at warm temperatures for several hours, and the organism usually recovered from the feces is *Streptococcus faecalis.* The incubation period is 2 to 18 hours, and the illness lasts a very short time.

Other organisms sometimes infecting food are *Clostridium perfringens, Bacillus cereus, Proteus* bacilli, and *Escherichia coli.* In outbreaks traced to these, it has been believed that the microbes were able to multiply in food, especially meat that had stood at room temperature overnight or longer.

Measures to safeguard food

The following points are advised to safeguard a food supply:

1. Appearance, taste, and smell are not always reliable indicators that a food is safe for consumption.

2. Any unusual change in color, consistency, or odor, or the production of gas in food means that the food should be discarded *without being eaten.*
   Note: If suspect food has been tasted or eaten, it should not be discarded. It should be kept for 48 hours in the event that it might have to be examined for the presence of *Clostridium botulinum* exotoxin.
3. Dishes, cutlery, can openers, utensils, equipment of all kinds used to prepare, serve, and store food should be clean and sanitary.
4. Hands that prepare, serve, and store food should be clean. Preferably food is handled as little as possible.
5. Foods served raw should be washed carefully and thoroughly.
6. Bacteria grow fastest in the temperature range 40° to 140° F.
   a. Hot food should be kept hot (above 140° F.).
   b. Cold food should be kept cold (below 40° F.). All dairy foods should be refrigerated.
   c. Cooked food that is to be refrigerated should be cooled quickly and refrigerated promptly. If cooked food is to be kept longer than a few days it should be frozen.
   d. Perishable foods should be kept chilled if taken on a trip or picnic. Foods may spoil easily if exposed to a warmer temperature for no longer than a half hour before they are eaten.
   e. Extra care should be taken with foods easily contaminated by microbial growth—meats, poultry, stuffing, gravy, salads, eggs, custards, and cream pies.

Food preservation

There are numerous methods of preserving food. The following are a few.

■ **Refrigerating.** Refrigeration is one of the best and most universally used methods of food preservation. It has several distinct advantages: (1) There is little change in the composition of food, therefore taste, odor, and appearance are preserved; (2) the nutritive value of the food is not reduced; (3) there is no decrease in digestibility; and (4) there is little effect upon vitamin content. During refrigeration pathogenic bacteria are checked and many destroyed. When refrigeration is discontinued, those that remain viable begin to multiply. Some saprophytic organisms are able to multiply at refrigeration temperatures.

■ **Freezing.** Freezing as a way of preserving food is in widespread use today because of its application in the preparation of "convenience" food items. Quick freezing is carried out at very low temperatures (−35° F. with a holding temperature of −10° F.). A preliminary step to freezing is blanching. The food is heated quickly to inactivate enzymes that would break down proteins and change the texture, flavor, and appearance of the food. Blanching is followed by quick cooling in ice water. Food that is frozen retains its nutrients and is palatable. Freezing of meat kills encysted larvae therein. Since cold storage foods decompose rapidly when thawed, they should be consumed right away.

■ **Drying.** Microbes are unable to multiply for lack of water.

■ **Smoking.** The food is dried and preservatives from the smoke added to it.

■ **Pickling.** The high acid content of the medium prevents microbial multiplication.

■ **Salting and preserving in brine.** The osmotic pressure of the medium is so changed that microbes will not multiply. Sometimes water is extracted from the microbial cell.

■ **Canning.*** Heat destroys microorganisms. The container with its content of food is hermetically sealed to keep out contaminants. The sealing also excludes free oxygen, which would aid the growth of molds, yeasts, and most species of pathogenic bacteria.

Commercial canning is a very safe method of food preservation. It is done scientifically and under supervision by the indicated health agency. Home canning on the other hand may be done under rough and ready conditions by unskilled hands. Since there is always a possibility for food contamination, home canned foods, especially meats and vegetables, should always be heated prior to consumption.

■ **Preserving.** The food is heated and sugar added. A large amount of sugar retards bacterial multiplication in the same manner as does salt. However, it does not retard the multiplication of fungi. Heating and sealing have the same effect as in canning.

■ **Cooking.** Microbes are killed by heat, the composition of the food is changed, and water may be removed.

■ **Chemicals.** Benzoic acid is the only chemical preservative permitted by law. The amount must be stated on the label.

■ **Ionizing radiation.** Ultraviolet irradiation has already been used successfully by food industries to treat the air in storage and processing rooms, to prevent the growth of fungi on shelves and walls of preparation rooms, and to destroy parasites in meat.

Cold sterilization of foods by means of ionizing radiation is under investigation by the United States Armed Forces and the Atomic Energy Commission, and one of the major obstacles encountered has been the terrific resistance of the spores of *Clostridium botulinum* to the effects of radiation (a problem with seafoods).

■ **Food in vending machines.** Vending machines dispense today many foods that support the rapid growth of infectious or toxigenic microbes. The food to be safe for consumption must be prepared under sanitary conditions and be transported in properly refrigerated vehicles. In the vending machine they must be maintained at proper temperatures and replaced often.

---

*In 1810, Nicholas Appert (1752-1841), encouraged by Napoleon Bonaparte, laid the foundation of modern canning and preserving industries by his discovery of the process of heating foods to boiling temperatures and sealing them hermetically in suitable containers. About the same time, the tin can was patented.

QUESTIONS FOR REVIEW

1. Name diseases spread by contaminated milk and food.
2. Give differential characteristics of milkborne and waterborne epidemics.
3. What is meant by Grade A pasteurized milk?
4. What is flash pasteurization?
5. Give the requirements for a safe milk supply.
6. Make an outline showing the different kinds of food poisoning. Name causal agents.
7. Differentiate food intoxication from food infection. Give examples.
8. List the different methods of food preservation.
9. List practical measures to safeguard food.
10. Explain briefly: lactic acid bacteria, normal milk bacteria, yoghurt, buttermilk, holding method of pasteurization, milk sickness, the phosphatase test, souring of milk, food idiosyncrasy, low acid foods, high acid foods, blanching of foods.
11. Indicate how a bacterial count of a milk sample is done.

REFERENCES. See at end of Chapter 38.

37 Biologic products for immunization*

The two terms "vaccine" and "immune serum" are often confused, any product of this nature being spoken of as a "serum." Vaccines and immune serums differ in fundamental properties, method of production, and resultant type of immunity.

A *vaccine* is the causative agent of a disease (bacterium, toxin, rickettsia, virus, or other microbe) so modified as to be incapable of producing the disease yet at the same time so little changed that it is able, when introduced into the body, to elicit the production of specific antibodies against the disease. Vaccines are always antigens; therefore, they always produce active immunity. They find their greatest usefulness in the *prevention* of disease. Important vaccines are those for the prevention of typhoid fever, smallpox, poliomyelitis, diphtheria, tetanus, and rabies.

An *immune serum* is the serum of an animal (or man) that has been immunized against a given infectious disease. Its salient feature pertains to the antibodies it contains. Immune serums confer passive immunity; this immunity results from their antibodies, which do not stimulate further antibody production. Both vaccines and immune serums are specific in their action; that is, they induce immunity to no disease other than the one for which they are prepared. Immune serums are of two types: antitoxic and antibacterial.

IMMUNE SERUMS (PASSIVE IMMUNIZATION)

■ **Antitoxins.** Antitoxins are immune serums that neutralize toxins. They may be prepared artificially, and they also develop in the body as a result of repeated slight infections. This is why some adults are immune to diphtheria. Antitoxins have no action on the bacteria that produce the toxins. For example, diphtheria antitoxin neutralizes diphtheria toxin in the tissues and body fluids but has no effect on the diphtheria bacilli growing in the throat and producing the toxin. Neutralization of the circulating toxin in the body does favor the defense mechanisms operating to eliminate the membrane formed in the throat.

PREPARATION. Antitoxins can be successfully prepared only against exotoxins. The antitoxins that have been used longest and have saved most lives are those against diphtheria and tetanus toxins. Both are prepared in the same general way. The bac-

*In this chapter are presented the biologic products that have been standardized and are available. The administration of certain of these for the production of passive immunity is given. In the next chapter are recommended schedules for the production of active immunity.

teria are grown in a liquid culture until the medium contains a large amount of exotoxin. The bacilli are then filtered from the medium and its toxin content. It is necessary to determine the strength of each lot of toxin or antitoxin because two lots prepared exactly alike seldom have the same strength.

The unit of toxin is measured as the *minimum lethal dose* (MLD). It is the amount of toxin that when injected into a test animal will kill it in a prescribed time. For diphtheria toxin it is the amount that will kill a guinea pig weighing 250 grams in 4 days, and for tetanus toxin it is the amount that will kill a 350-gram guinea pig in the same length of time.* To determine the MLD, one begins with a very small dose and gives increasing amounts of toxin to a series of guinea pigs. The animals receiving small doses may not be affected or else become but slightly ill and recover; those receiving larger doses become ill and some may die, but it is more than 4 days before death occurs. Finally, an animal receiving a still larger dose dies on the fourth day. The amount of toxin given this animal contains one MLD.

When the toxin is found to be of suitable strength, horses† are immunized with it. First, a small dose is used; it is increased at each of several successive injections. The first dose of toxin may be preceded by an injection of antitoxin. At the time that experience has shown antitoxin production to be at its height, the horse is bled, and the antitoxin strength of its serum tested. If the serum is found to contain sufficient antitoxin, it is further refined and purified for use. Refining serves (1) to concentrate the antitoxin and (2) to eliminate horse serum protein. Horse proteins can sensitize the recipient of the antitoxin. If an immune serum is given the second time to a sensitized person, anaphylactic shock may result. It is *not* the antibodies in antitoxins and other immune serums that lead to allergic manifestations but the *serum protein* of the animal used in preparing the immune serum.

STANDARDIZATION. There are three methods of standardizing antitoxins‡: (1) animal protection tests, (2) skin reactions, and (3) flocculation tests. We shall briefly describe how diphtheria antitoxin is standardized by determining its protective action against diphtheria toxin in a susceptible animal. First, there are certain definitions to be understood.

1. *Standard antitoxin*—an antitoxin of known strength prepared, stored, and distributed to antitoxin manufacturers by a controlling agency
2. *Unit of antitoxin*—an amount of antitoxin equivalent to one unit of standard antitoxin§
3. *L+ dose of diphtheria toxin*—an amount of toxin that, when combined with one unit of antitoxin, causes the death of a 250-gram guinea pig on the fourth day

*Tetanus toxin is so potent a poison that 1 ml. of a broth culture of tetanus bacilli may contain enough toxin to kill 75,000 guinea pigs.

†In a few instances, cattle instead of horses are used.

‡In 1896, Paul Ehrlich introduced methods of standardizing toxins and antitoxins. To him goes the credit for the concept of MLD.

§This definition is used instead of the original definition of Ehrlich—the amount of antitoxin required to neutralize 100 MLD of toxin—since toxin also contains variable quantities of toxoid. Although toxoid has no disease-producing capacity, it can combine with antitoxin. Different lots of antitoxin tested against different lots of toxin therefore have different strengths. The standard antitoxin unit used in this country contains sufficient antitoxin to neutralize 100 MLD of the particular toxin that Ehrlich used originally in establishing his unit of antitoxin.

Different amounts of diphtheria toxin are mixed with one unit of standard diphtheria antitoxin, and the mixtures are tested in 250-gram guinea pigs. The L+ dose of toxin as thus determined is mixed with different amounts of the antitoxin being tested. These are injected into 250-gram guinea pigs, and the mixture that causes the death of a guinea pig on the fourth day obviously contains one unit of antitoxin. Tetanus antitoxin is standardized in much the same manner as diphtheria antitoxin, but different amounts of antitoxin and toxin are used. Botulism and gas gangrene antitoxins are standardized in a similar manner, but the mouse is the test animal.

Standardization of antitoxins by skin reactions is based upon the same underlying principles as standardization by animal protection tests. The toxin and antitoxin are injected into the skin of rabbits or guinea pigs. The production of a skin reaction by this method has the same significance as the death of a guinea pig in the guinea pig antitoxin-toxin injection method.

The flocculation method depends upon the occurrence of flocculation in a test tube when antitoxin and toxin are brought together in certain proportions.

Since antitoxin can be successfully prepared only against the few organisms producing extracellular toxins, the number of antitoxins is limited. Generally speaking, all antitoxins should be given *early* in the disease and *in sufficient amount* because they cannot repair injury already done.

LISTING OF ANTITOXINS. *Botulinus antitoxin* (an equine antitoxin) is given to prevent damage from toxin not already taken up by the central nervous system. Therefore, it is of value when given before manifestations have appeared but is of little effect after that time. Antitoxin against one type of *Clostridium botulinum* is ineffective against the toxin of the other types. To be therapeutically expedient, antitoxin must be available against toxins of the A, B, and E types of bacilli, the ones that cause disease in man. Types A and B antitoxins are marketed in the United States but not type E antiserum, which is produced in Denmark and Canada. Recently supplies of type E antitoxin have been stockpiled by the Center for Disease Control in Atlanta, Georgia, and made available on a 24-hour emergency basis.*

Diphtheria antitoxin is a therapeutic agent that has given possibly more brilliant results than any other. By its use much suffering has been prevented and many lives saved. In the production of diphtheria antitoxin in horses, toxoid has replaced diphtheria toxin as the immunizing agent. Refinements in manufacture give a diphtheria antitoxin that contains more than 5000 units per milliliter and is less likely to trigger serum reactions than antitoxins previously manufactured, because in production a very high proportion of the horse serum protein is removed.

Diphtheria antitoxin is *not* effective against toxin that has combined with the body cells, and no amount of antitoxin given late in the disease can repair injury already incurred. For best results, the disease must be recognized early, and sufficient antitoxin must be given at once. After subcutaneous administration, the antitoxin content of the blood does not reach its maximum for 72 hours. Injections should, therefore, be made *intramuscularly* or *intravenously*.

The Committee on Infectious Diseases of the American Academy of Pediatrics *insists* that antitoxin be given as soon as the clinical diagnosis of diphtheria is made, with *no delay* even in waiting for bacteriologic results. The site of the membrane, the degree of toxicity, and the length of illness are much more reliable guidelines for deter-

*See also p. 339.

mining the dose of diphtheria antitoxin than the weight and age of the patient. The Committee's schedule for the administration of diphtheria antitoxin is given in Table 37-1.

When diphtheria antitoxin is given, remember a *serum reaction* is a possibility, especially when the antitoxin is given *intravenously*. Preliminary testing for serum hypersensitivity is advised.

Although antibiotics have no effect upon the toxin of *Corynebacterium diphtheriae*, certain ones attack the organism. The administration of these antibiotics is a valuable adjunct, never a substitute. It reduces the number of secondary invaders, decreases the severity and length of the illness, and helps to eliminate carriers.

TABLE 37-1. ADMINISTRATION OF ANTITOXIN IN DIPHTHERIA*

Duration of illness	48 hours		Over 48 hours		
Lesions	Throat	Membrane in larynx	Membrane in nasopharynx	Brawny swelling of neck	Extensive disease (3+ days)
Dose of antitoxin (units)	20,000 to 40,000	20,000 to 40,000	40,000 to 60,000	80,000 to 120,000	80,000 to 120,000
Route of administration†	1. Intravenously or 2. Intravenously up to one-half; rest intramuscularly	1. Intravenously or 2. Intravenously up to one-half; rest intramuscularly	Intravenously	Intravenously	Intravenously

*Recommendations of the Committee on Infectious Diseases (1970) of the American Academy of Pediatrics.
†Preferred route is intravenous (after eye and skin tests for hypersensitivity) to neutralize toxin rapidly. If patient reacts to antitoxin, desensitization is indicated.

Tetanus antitoxin (TAT) or *antitetanic serum* (ATS) is a product usually obtained from horses, but it may also be produced in cattle. Although patients have undoubtedly benefited by the use of tetanus antitoxin, in the person allergic to horse serum there is a real hazard in its use. *Tetanus immune globulin (human)* or *antitetanus globulin* (ATG) is prepared from the blood of persons actively immunized with tetanus toxoid. As a gamma globulin fraction of a hyperimmunized human being, it contains antitoxin without any foreign protein. Table 37-2 compares the human antitoxin with that from the animal.

If the tetanus-susceptible wound is seen immediately after injury and if it can be adequately cared for by standard surgical technics and antibiotics, most authorities believe that the risk involved in giving equine or bovine tetanus antitoxin is not justified regardless of the immune status of the patient. On the other hand, if the patient delays seeking medical attention for a day or more or if the wound is one in which adequate surgical care is impossible because of its extent and nature, then protection must be given to nonimmunized persons and to those patients who have failed to take their tetanus booster in the preceding 5 years.

Gas gangrene antitoxin is prepared against the organisms important in causing gas gangrene. Its value is questioned. Many surgeons do not give it.

TABLE 37-2. EVALUATION OF BIOLOGIC PRODUCTS FOR PASSIVE IMMUNIZATION IN TETANUS

	Tetanus antitoxin (TAT)	
Point of comparison	**Equine or bovine***	**Human**
Comparative effectiveness	Less effective	10:1 (more effective)
Dosage	Usual—3000 to 5000 units†	Usual adult dose—250 units
Duration of protective levels of antitoxin in recipient	5 days	Up to 30 days (detectable levels reported to 14 weeks)
Danger of allergic reaction	Present; estimated 6% to 7% or more	Remote
Cost	Less expensive	More expensive
Evaluation	Not advocated if human product available	Recommended in allergic persons and generally

*The majority of persons sensitive to horse serum are also sensitive to bovine serum.
†Recommendations as to administration of tetanus antitoxin are variable.

■ **Antivenins.** Antivenins have been prepared against the venoms of snakes and the black widow spider. They are prepared in the same general way as antitoxins, that is, by immunization of a horse with serial doses of the venom of the snake or the spider. Antivenins may be prepared against a single species of snake or against a group of closely related species. Whether snake antivenin is prepared against a species or a group of species depends upon the geographical location in which it is to be used. North American antisnakebite serum (also known as Antivenin [Crotalidae] and *Crotalus* antitoxin) is effective against rattlesnakes, copperheads, and cottonmouth moccasins, the most common poisonous snakes of North America. It is not effective against the venom of the coral snake. To combat the potent poison of the North American coral snake, licensed antivenin is distributed to state or local health departments through the Center for Disease Control in Atlanta, Georgia.

The use of antivenin in snakebites or black widow spider bite should in no way replace first aid and supportive measures.

■ **Antibacterial serums.** Before the discovery of the antimicrobial compounds, antibacterial serums played an important part in the treatment of various infections. Among such serums were those against meningococci (antimeningococcal serum) and pneumococci (antipneumococcal serum). Most of these were prepared by injecting horses with the given bacteria. Rabbits were employed in the manufacture of antipneumococcal serum, and a specific serum for each known type of pneumococcus was prepared. A serum against *Haemophilus influenzae,* type b, has been prepared by injecting rabbits with the bacilli and was formerly used in the treatment of meningitis caused by *Haemophilus influenzae,* type b. Antipertussis serum has been prepared by injecting rabbits with *Bordetella pertussis* and its products. This serum has been used with success in very young children to prevent the disease in those exposed, when given early enough, and to decrease the severity of the established disease.

Antibacterial serums act to destroy bacteria. It is thought that they combine with a surface antigen, thereby rendering the bacteria more susceptible to leukocytes.

A serum that acts upon several strains of bacteria is *polyvalent;* one that acts upon only one strain, *univalent.* Today the treatment of disease processes with antibacterial serums has almost completely been supplanted by sulfonamide and antibiotic therapy.

■ **Antiviral serums.** The best known antiviral serum is *antirabies hyperimmune serum.* It is prepared by injecting horses with rabies virus. (In any serum of equine origin there is always the danger of an anaphylactic reaction or severe serum sickness.) Its purpose is to establish an immunity directly after exposure and to provide a passive immunity until an active one can be established by vaccination. See Table 38-2, p. 622, for the recommendations of the World Health Organization.

For passive immunization in rabies, a human antiserum, rabies immune globulin of human orgin, has been developed but as yet is not widely used.

■ **Convalescent serum.** Convalescent serum therapy consists of the injection of the whole blood or serum of a person *recently* recovered from a disease into one ill of the disease (as a therapeutic measure) or into one exposed to the disease (as a preventive measure). By so doing, the blood of the convalescent patient containing antibodies confers a passive immunity upon the recipient. This type of therapy has been used with good results in measles, whooping cough, and scarlet fever.

Adult immune serum is the serum from a person who has had the disease (example, measles) many years previously. The serum of an adult recently vaccinated against the disease is of value in preventing infection in those exposed and in reducing the severity of disease in those already ill. *Hyperimmune human antipertussis serum* (prepared by giving pertussis vaccine to a person who has recovered from the disease) has been used successfully in the treatment of whooping cough.

Caution should be observed in giving convalescent serum or any human serum because it may transmit the agent of serum hepatitis!

■ **Gamma globulin.** Gamma globulin is the one of several fractions of the globulin of blood plasma with which antibodies are associated. It is prepared commercially by a series of precipitations in pooled normal adult plasma, in venous blood, or in pooled extracts of human placentas, with varying concentrations of alcohol at a low temperature. The gamma globulin fraction removed is more than 40 times as rich in antibodies as the original plasma from which it is taken. Five hundred milliliters of blood yields an average dose.

Gamma globulin, usually dispensed as *immune serum globulin (human)* (ISG), is emphasized as indicated in Table 37-3. *Gamma globulin is always given intramuscularly —never intravenously** (see Fig. 37-1). As ordinarily given, gamma globulin is one of the benign injectables, being associated with few side reactions.

VACCINES

■ **Bacterial vaccines.** Bacterial vaccines are suspensions of killed bacteria in isotonic sodium chloride solution. After culture for 24 to 48 hours, the bacterial growth is emulsified in sterile saline solution and the number of bacteria in each milliliter of the washings is determined. Isotonic sodium chloride solution is added until the desired number per milliliter is obtained. The bacteria are then killed at the lowest possible lethal temperature in the shortest possible time. They may be killed without

*The consequence can be a severe shocklike state.

TABLE 37-3. CLINICAL VALUE OF IMMUNE SERUM GLOBULIN (GAMMA GLOBULIN) (ISG)*

Disease	Indications	Preparation
Measles†	Modification of symptoms after vaccination; use only with Edmonston live virus vaccine	Measles immune globulin (human) *or* ISG, *or* poliomyelitis immune globulin (human) assayed for measles antibody (Cutter, Lederle, Parke-Davis)
	Prevention:	
	1. Children with normal immune mechanisms	Immune serum globulin (human)
	2. Children with immune system disorder	Immune serum globulin (human)
	Modification	Immune serum globulin (human)
Rubella‡	Prevention only in female in first trimester of pregnancy	Immune serum globulin (human)
Varicella (chickenpox)	Modification	Immune serum globulin (human)
Immunologic deficiency syndromes (agammaglobulinemia, hypogammaglobulinemia, dysgammaglobulinemia)§	Replacement therapy	Immune serum globulin
Rh hemolytic disease‖	Prevention	Rh_0(D) immune globulin (human) (Ortho)
Mumps	Prevention	Mumps immune globulin (human) (Breon, Cutter)
Poliomyelitis¶	Prevention	Immune serum globulin (human)
Serum hepatitis	Prevention (value undetermined)	Immune serum globulin (human)

*Compiled from the 1972 Physicians' Desk Reference, Oradell, N.J., 1971, Medical Economics, Inc., and the 1971 Red book of the American Academy of Pediatrics. Diphtheria immune (human) globulin, zoster immune globulin (ZIG), and rabies immune globulin of human origin (RIGH) have also been prepared.
†See also p. 626 for schedule of measles vaccination.
‡See also Chapter 38.
§See also p. 177 for discussion.
‖Rh immune globulin (Rh_0GAM) is discussed on p. 185.
¶See also Chapter 38.

Dose	Schedule	Route	Comments
0.02 ml./kg. body weight	One dose	Intramuscular	Inject measles vaccine in one arm, this in other, with separate syringe
0.22 ml./kg. body weight			Inject within 6 days of exposure; give live virus vaccine in 8 or more weeks
20 to 30 ml.			
0.05 ml./kg. body weight			Give within 6 days, if possible (rarely indicated)
0.25 to 0.44 ml./kg. body weight (20 ml. total advised by some)	One dose	Intramuscular	Subject controversial
0.55 to 1.3 ml./kg. body weight (20 to 30 ml. total sometimes advised)		Intramuscular	Give within 3 days of exposure; may decrease severity of illness
0.60 ml./kg. every 2 to 4 weeks Double dose at onset of therapy	One dose a month	Intramuscular	Maximum dose 20 to 30 ml. at any one time; gamma globulin deficiency here; adequate levels must be supplied to protect against infections
One vial	One dose	Intramuscular	Give to nonsensitized Rh-negative mothers after delivery of Rh-positive infant, or after abortion
2.5 to 7.5 ml.	One dose	Intramuscular	Efficiency not known
0.3 ml./kg. body weight	One dose (second dose can be given 5 weeks after first)	Intramuscular	Possible protection for 2 to 6 weeks if given before exposure or development of symptoms
10 ml.	One dose within 1 week after transfusion; one dose (10 ml. 1 month later	Intramuscular	For adults — not recommended for children; recommended for high-risk adults only; give right after transfusion

Continued.

TABLE 37-3. CLINICAL VALUE OF IMMUNE SERUM GLOBULIN (GAMMA GLOBULIN) (ISG)*—cont'd

Disease	Indications	Preparation
Infectious hepatitis	Prevention or modification in children or adults	
	1. Short-term, moderate risk	Immune serum globulin (human)
	2. Long-term, intense exposure	
	Prophylaxis (in institutions)	
Pertussis‡	Prevention in infants under 2 years	Pertussis immune serum globulin (human) (Breon, Cutter)
	Treatment	
Tetanus‡	Prevention—children and adults	Tetanus immune globulin (human) (Parke-Davis)
	Treatment	
Vaccinia-smallpox‡	Prevention	Vaccinia immune globulin (human) (VIG) (American Red Cross)
	Protection of children	
	Treatment	

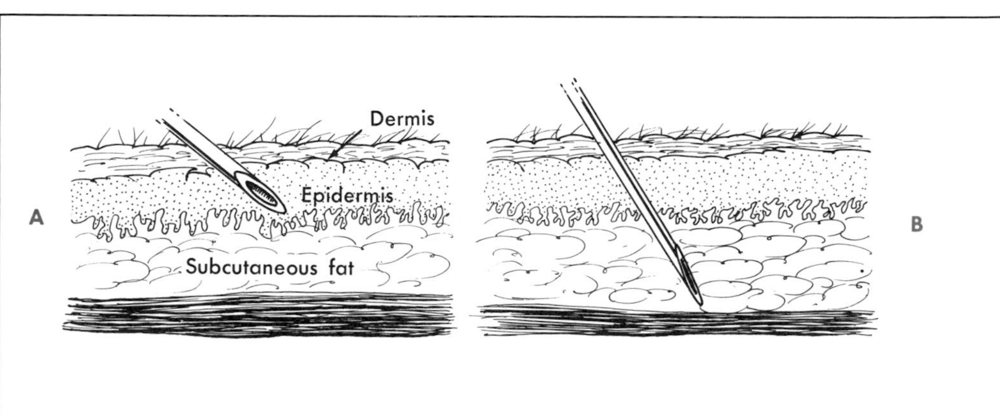

FIG. 37-1. Routes of injection, diagrammatic sketches. Note position and depth of needle. **A,** Intradermal. **B,** Subcutaneous. **C,** Intramuscular. **D,** Intravenous.

Dose	Schedule	Route	Comments
0.02 to 0.04 ml./kg. body weight	One (repeat in 3 to 5 months if necessary)	Intramuscular	Give within 1 week of exposure
0.06 ml./kg. body weight	One (repeat in 5 to 6 months if necessary)		
0.06 to 0.10 ml./kg. body weight	One (repeat in 5 months)		Protection for 1 year
1.5 ml. (1.25 to 2.5 ml.)	One dose (repeat in 1 week)	Intramuscular	Product from donors hyperimmunized with pertussis vaccine; protection not reliable
1.5 ml.	One dose every 1 to 2 days for three to five doses; *or* 4 to 6 ml. at once		Use larger doses with severe disease
250 to 1500 units	One dose	Intramuscular	Product from person hyperimmunized with tetanus toxoid
5000 to 20,000 units	May be repeated 1 month later		Active immunization should be started with tetanus toxoid
0.3 ml./kg. body weight	One dose	Intramuscular	Product from blood collected from military recruits 3 to 4 weeks after reaction to smallpox vaccination; released free of charge
0.6 ml./kg. body weight			

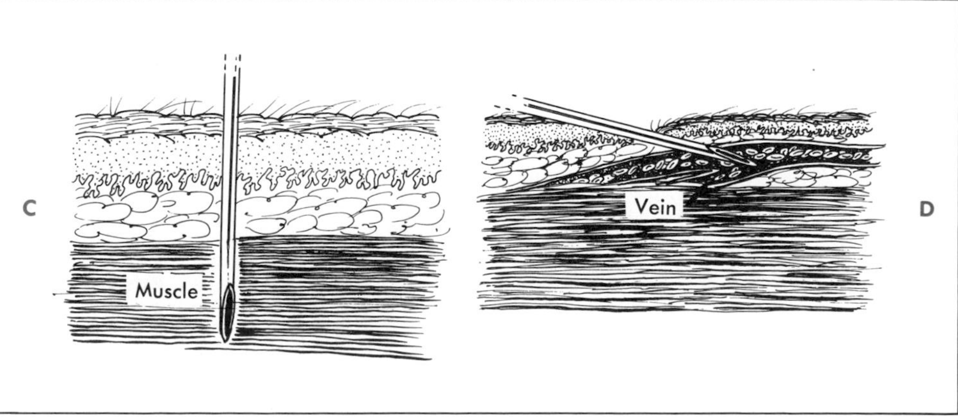

heating by formalin, cresol, or Merthiolate. In some vaccines the bacteria are killed by ultraviolet light. The finished product is cultured to ensure its sterility. Vaccines are usually prepared so that the number of dead microorganisms given at a single injection ranges from 100 million to 1 billion.

Bacterial vaccines are of two types: *stock vaccines,* made from stock cultures maintained in the laboratory, and *autogenous vaccines,* made from the organisms of a specific lesion in the patient for whom the vaccine is being prepared. A *mixed* vaccine is one containing bacteria belonging to two or more species. A *polyvalent vaccine* contains several strains of organisms belonging to the same species, for instance, one containing several strains of *Pseudomonas aeruginosa.* Bacterial vaccines show their best results in the prevention of typhoid fever, whooping cough, and plague.

LISTING OF BACTERIAL VACCINES. *BCG vaccine (bacille Calmette-Guérin)* is made from a strain of bovine tubercle bacilli cultivated on artificial media (containing bile) so long that they have completely lost their virulence for man. In the United States freeze-dried BCG vaccine is available in two concentrations, the higher for use in the multiple puncture technic, and the lower for intradermal inoculation. Administered orally or intradermally, BCG vaccine is given *only to persons with a negative tuberculin test.* The immunity induced lasts over a considerable number of years.

Being tested experimentally is a BCG vaccine aerosol. Persons to be vaccinated breathe in vaccine droplets suspended in a manmade mist for some 45 minutes or so.

BCG vaccine has been used extensively in European countries to immunize children. The World Health Organization in recent years has vaccinated over 250 million persons with BCG. This has been an especially worthwhile program in parts of the world where the incidence of tuberculosis is high. In the United States the prevalence rate is low. BCG vaccination, a controversial issue for 50 years, has not been used widely. The Public Health Service recommends it only for those tuberculin-negative individuals in close contact with known untreated cases of tuberculosis.

Cholera vaccine given to the military personnel of the United States Armed Forces sent to duty in cholera-endemic areas contains 8 billion killed vibrios per milliliter. Cholera vaccine is toxic and must be injected at frequent intervals to maintain immunity. A new oral vaccine that contains killed vibrios in water is being tested. It can be given daily and induces antibody formation in the intestinal tract, the actual site of involvement in the disease. Another new vaccine, a cholera toxoid (inactivated cholera toxin), will soon be available.

Pertussis vaccine is a saline suspension of the killed organisms. Pertussis vaccine is commonly prepared and used in the alum-precipitated and aluminum hydroxide– and aluminum phosphate–adsorbed forms in combination with diphtheria and tetanus toxoids. Severe neurologic reactions rarely complicate administration of pertussis vaccine.

Plague vaccine of the United States Armed Forces contains 2 billion killed plague bacilli per milliliter.

Tularemia vaccine has been prepared by the United States Army as an inhalant type of vaccine. Large numbers of persons can be vaccinated against this disease simply by marching through a room containing vaccine droplets in the atmosphere. The interest of the military in this disease stems from its possibilities in bacteriologic warfare. Airborne immunization is being evaluated experimentally in other diseases transmitted by droplet infection. *Foshay's tularemia vaccine* is made up of killed organisms and produces an immunity that lasts about a year.

Typhoid vaccine for the U. S. Armed Forces is an acetone-inactivated dried strain of *Salmonella typhi* (the AKD vaccine). An oral vaccine containing live attenuated typhoid bacilli is being tested. The U. S. Public Health Service discredits paratyphoid A and B vaccines as immunizing agents and has implicated them in reactions succeeding administration of mixed vaccines.

■ **Toxin-antitoxin.** A mixture of diphtheria toxin and antitoxin with a small excess of the former, *diphtheria toxin-antitoxin* has been used to produce permanent active (not passive) immunity to diphtheria. Toxin-antitoxin is unstable and may be toxic. Manufacturers have discontinued it.

■ **Toxoids.** *Botulinum toxoid,* an effective pentavalent preparation available for active immunization against botulism, incorporates toxoids (see below) for the antitoxins of *Clostridium botulinum,* types A, B, C, D, and E. It is given in three spaced injections followed by a booster.

Diphtheria toxoid is diphtheria toxin detoxified so that it cannot cause diphtheria but can induce the formation of specific antitoxin. Formalin 0.2% to 0.4% is added to diphtheria toxin and the mixture incubated at 37° C. until detoxication is complete (several weeks). This treatment, used with other bacterial toxins as well, reduces toxicity but preserves the antigenic properties of toxin. After purification to remove inert protein the preparation is available as *plain* or *fluid toxoid.* Fluid toxoids are used in the United Kingdom and Canada.

Alum added to diphtheria toxoid precipitates the antigenic portion. The precipitate, after being washed and suspended in sterile physiologic saline solution, is *diphtheria toxoid, alum-precipitated.* Its advantage is that the alum is not absorbed but remains at the site of injection. The toxoid slowly separates from it to give a prolonged antigenic stimulation. If aluminum hydroxide or aluminum phosphate is added to the liquid toxoid, the antigenic portion adheres to the particles of aluminum compound. This process is known as *adsorption,* and diphtheria toxoid so treated is known as *aluminum hydroxide-* or *aluminum phosphate-adsorbed diphtheria toxoid* (depot toxoid or antigen). Its properties are similar to those of the alum-precipitated toxoid. Diphtheria toxoid should not be given to a person more than 10 to 12 years of age without preliminary sensitivity tests.

Tetanus toxoid establishes permanent immunity to tetanus. It is manufactured in the liquid (or fluid), alum-precipitated, aluminum hydroxide–adsorbed, and aluminum phosphate–adsorbed forms.

Most manufacturers market mixtures of diphtheria and tetanus toxoids (DT) or mixtures of diphtheria and tetanus toxoids and pertussis vaccine (DTP). Combinations are available in the unconcentrated, alum-precipitated, aluminum hydroxide–adsorbed and aluminum phosphate–adsorbed forms. They have been proved to be very satisfactory.

■ **Viral vaccines.** With the exception of the antibiotics, nothing has done more to protect us against infection than the unborn chick. Viral vaccines have been possible largely because of the development of modern technics for cultivating viruses in embryonated hen's eggs and also in tissue cultures, thus furnishing the large supply of virus essential to the production of vaccines. In chick embryos or in tissue cultures, cultivation of a virus involving multiple transfers from one medium to another (serial passage) can alter the organism's pathogenicity. A normally virulent virus can thus be *attenuated* (weakened) or domesticated. Fortunately with loss of virulence there is no loss of antigenicity. This makes possible the production of effective vaccines.

In a killed or *inactivated* virus vaccine, the infectivity of the virus and its ability to reproduce have been destroyed by physical or chemical means, but its capacity to induce antibody formation has been preserved. Formalin is a standard inactivating agent. Killed virus vaccines must be injected in several doses, but live virus vaccines can be taken orally or inhaled (as an aerosol).

Virologists have long preferred living agents to killed or inactivated viruses for vaccines. A living agent continues to multiply in the body of the animal or man to which it is given and thus exerts a prolonged and increasingly strong stimulation to the host to make antibodies. The immunity so produced is stronger and longer lasting because the presence of the live virus simulates actual infection. The possibility that the virus will revert to its original level of pathogenicity in the vaccinated person is an obvious disadvantage to the use of the living agent. Live viral vaccines are well known in veterinary medicine. The two best known in human medicine are the measles vaccine and the live poliovirus vaccine, Sabin-type strains.

LISTING OF VIRAL VACCINES. *Adenovirus vaccine* has been prepared by the formalin and ultraviolet inactivation of viral types grown on monkey kidney cells. Because of the discovery that some of the *inactivated* adenovirus serotypes produced tumors in experimental animals, active immunization with these agents has been suspended. Trials with live adenovirus types 4 and 7 vaccine given orally in an enteric-coated tablet are in operation, since live adenoviruses presumably have no capacity to cause cancers.

Influenza virus vaccine is prepared from virus grown in the fertile hen's egg, inactivated with formalin, and concentrated (Figs. 37-2 and 37-3).

FIG. 37-2. Production of influenza vaccine. Influenza virus injected into fertile hen eggs along an assembly line. (Courtesy Eli Lilly & Co., Indianapolis, Ind.)

The Division of Biologics Standards regularly reviews the formulation of influenza vaccines, suggesting changes as needed to take in the strains expected to cause trouble during the next flu season. Usually the strains of the recent epidemics are known, and the most practical arrangement from year to year seems to be one wherein two strains recently implicated are placed in a bivalent vaccine. It is believed that a bivalent vaccine with greater amounts of antigen is preferable to a polyvalent one containing smaller amounts of four or five strains. In the event that a large-scale epidemic is anticipated, a monovalent vaccine might be indicated.

The peculiar ability of influenza viruses to change their antigenic structure from time to time poses a problem to the manufacturer of flu vaccines. A strain against which there is no protection in an existing polyvalent vaccine can easily emerge. Great care must be taken that strains containing a wide pattern of antigenic substances are selected for vaccine production.

FIG. 37-3. Harvest of virus-containing fluid from fertile hen eggs in production of influenza vaccine. (Courtesy Eli Lilly & Co., Indianapolis, Ind.)

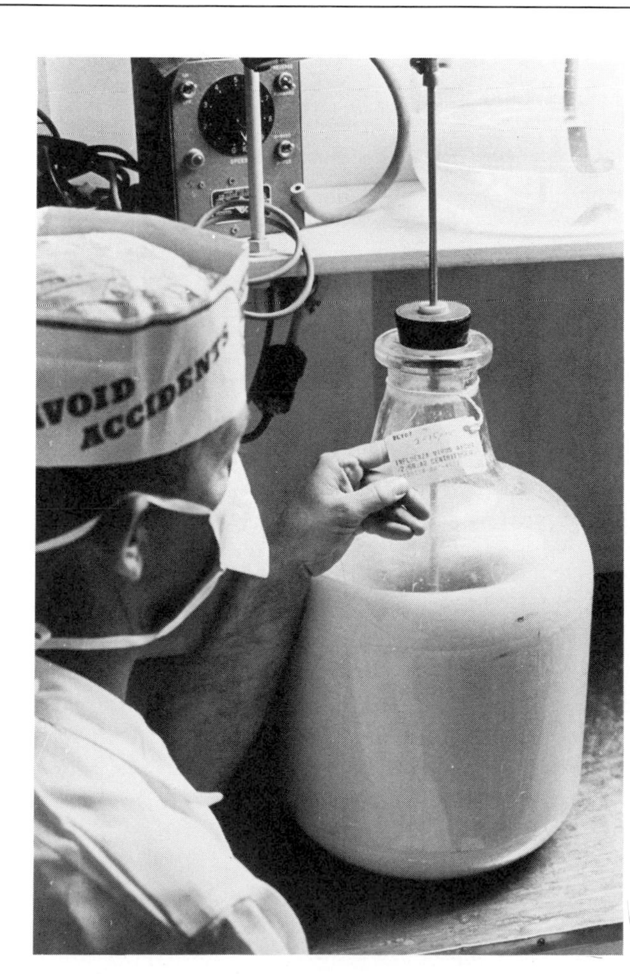

The production of active immunity by influenza virus vaccine has met with considerable success. However, the immunity is short lived, the vaccine sometimes gives fairly severe reactions, and the subject can become sensitized to egg protein as well as react to a previously existing sensitivity. Highly purified vaccines with most of the egg protein eliminated are being made available. A live virus vaccine still in the experimental stage is an aerosol vaccine.

Measles vaccines are of several types (and must be carefully refrigerated): (1) *Attenuated live virus vaccine* (Edmonston B vaccine) is prepared from chick embryo or renal cell cultures of the Edmonston B strain of the measles virus, developed by John F. Enders and associates. Live virus vaccines give permanent protection. (2) *Live virus vaccines* (Edmonston strains) have been further attenuated by additional chick embryo passages. These are the *further attenuated vaccines* of Schwarz and Moraten. (3) *Inactivated virus vaccines* are derived from the Edmonston strain virus propagated in chick embryo or monkey tissue cultures, inactivated with formaldehyde, concentrated, and precipitated with alum. Their use is contraindicated.

Mumps virus vaccine is an attenuated live virus vaccine adapted to the chick embryo and given in a single injection. The vaccine contains egg proteins and a small quantity of neomycin. Routine administration of mumps vaccine is contraindicated before puberty. At the present time, it is believed best to let the younger child develop his own immunity. The duration of protection afforded by the mumps vaccine is at least 4 years.

Salk poliomyelitis vaccine is prepared by growing all three types of poliomyelitis virus separately in tissue cultures on the kidney cells of rhesus monkeys. The viral suspensions are filtered to remove cells and other particles. Formaldehyde is added or ultraviolet irradiation is used to inactivate the virus. In the last step of production the three suspensions are mixed. Salk vaccine is a highly effective vaccine that is judged to be 90% preventive.

Sabin poliomyelitis vaccine is prepared with the three types of attenuated live poliovirus grown in human (not monkey) cell culture. After the vaccine has been fed to an individual, the virus of the vaccine multiplies in the lining and in the lymphoid tissue of the alimentary tract. A satisfactory immune response from the vaccinated person occurs usually within 7 to 10 days. The implantation of the virus in the bowel blocks further infection by the same type of a wild poliovirus. This materially cuts down on the number of carriers. The fear that competition from other enteroviruses might check growth of the poliovirus in the intestinal tract has been largely obliterated by studies of children in the tropics. Their alimentary tracts literally swarm with viruses and yet they show antibody responses indicating growth of poliovirus. To minimize possible interference from other enteric viruses, it is recommended that the oral vaccine be given during the spring and winter months in temperate climates.

The chief disadvantage of the Sabin vaccine is that it is difficult to preserve. It can be kept frozen for years but only *for 7 days in an ordinary refrigerator* and *3 days at room temperature!* Tap water cannot be used to dilute the vaccine since *the contained chlorine is destructive to the poliovirus.* Like other vaccines manufactured on monkey kidney cells, the final preparation contains traces of penicillin and streptomycin from the tissue culture medium. The overall effectiveness of the Sabin vaccine is 94% to 100%.

Rabies vaccines are of two kinds. One is prepared from the brain and spinal cord of rabbits suffering with rabies induced by the inoculation of fixed virus (p. 440).

FIG. 37-4. Virus-laden duck embryo used to produce rabies vaccine. Frozen, finely ground embryonic tissue is inactivated, filtered, and dried under sterile conditions. (Courtesy Eli Lilly & Co., Indianapolis, Ind.)

Just before death the brain and cord are removed under the strictest aseptic precautions and virus is inactivated. The phenol-inactivated vaccine is known as *Semple vaccine*. An emulsion is prepared by grinding the virus-containing material with water or isotonic saline solution either during the process of inactivation or after it is complete. The fact that this vaccine contains brain and spinal cord tissue is thought to be responsible for the severe complication of encephalomyelitis that sometimes follows its administration. Probably on an allergic basis, this disease may lead to paralysis and even death.

The other rabies vaccine is a recently developed one free of nervous tissue. It is prepared by growing the fixed rabies virus in the tissues of the embryonated duck egg (Fig. 37-4). To the suspension of embryonic duck tissue, a virucidal agent is added to inactivate the virus. The duck embryo vaccine (DEV) is dried and administered after dilution with distilled water. Because of the absence of central nervous system effects, this vaccine is less hazardous than the one prepared from the nervous tissues of the rabbit. However, it is not quite as immunogenic.

For dogs a vaccine is grown on a series of chick embryos to reduce its virulence. Such a live virus vaccine, given in a single dose, will produce immunity of over 3 years' duration in the dog. In areas where the vampire and other bats exist, it may be necessary to immunize livestock.

Currently an experimental rabies vaccine has been grown in nonneural tissue culture. There is also work being done to minimize the amount of fat in nervous tissue used to grow the virus.

Rubella vaccine, live, attenuated, exists as several preparations, three licensed and one experimental. One vaccine is a live virus strain originally attenuated by 77 passages in green monkey kidney and then prepared in duck embryo culture. A second is the same strain prepared in dog kidney culture. The third vaccine, the Cendehill strain, was isolated in green monkey kidney and then passed 51 times in rabbit kidney culture. These three vaccines are effective. A fourth vaccine (unlicensed) is being developed in human fibroblast tissue culture.

Episodes of arthritic joint pain, found especially in the older age groups, complicate the use of the licensed vaccines, especially the dog kidney one. Although arthritis was a disturbing event when first recognized as a consequence to vaccination, it has usually turned out to be mild and self-limited, without sequelae. Hypersensitivity reactions to egg protein and neomycin may occur.

Smallpox vaccine is prepared by the inoculation of female calves 6 months to 1 year old with cowpox virus. After the skin of the abdomen is shaved and disinfected, it is scratched with a needle (scarified) in many places. The scratches are just deep enough to bring a little blood. The virus is rubbed into these scratches. At the end of 6 days the skin of the abdomen is thickly broken out with vesicles (blisters) that contain the modified virus used for vaccination. With the most careful aseptic technic, the tops of the vesicles are opened, and the sticky exudate is removed. The exudate is mixed with four times its weight of glycerin and water (equal parts) containing 1% phenol for preservation. The vaccine is purified and if found to be of sufficient potency and sterile, it is ready for use. Some manufacturers add brilliant green to the vaccine to inhibit bacterial growth. They also paint the vaccinated area on the calf with this dye. This vaccine, made from the dermal lymph of the calf, is the one most widely accepted at present.

A smallpox vaccine as effective as the calf vaccine is produced by growing the vaccinia virus in the bacteria-free tissues of the chick embryo. Vaccine may also be prepared by growing the virus in tissue cultures.

If smallpox vaccine is not shipped and stored at freezing temperatures, it rapidly loses its potency. There is a lyophilized (freeze-dried) vaccine that can be stored at temperatures ranging from 35° to 50° F. without losing its potency for a period of 18 months. Once this vaccine is reconstituted, it must be refrigerated.

Smallpox vaccination (see also p. 619) means infecting a person with a strain of vaccinia or cowpox virus. The viruses of cowpox and of smallpox are quite similar in makeup and serologic properties. Some authorities believe that cowpox originated by the transfer of smallpox from man to the cow, with an adaptation of the virus to this animal to produce a milder disease.

Yellow fever vaccine is prepared by growing a strain of the virus in chick embryos. It was developed at a time when little thought was given to the avian leukosis viruses. A new strain of the virus is being tested with technics that filter out any such contaminants. It will ultimately replace the old one.

■ **Rickettsial vaccines.** Rickettsiae for vaccine production have been grown in lice, in rodent lung, and in tissue cultures. The vaccines in use at the present time are prepared by growing the rickettsiae in the yolk sac of the developing chick embryo. A vaccine so prepared is referred to as a *Cox vaccine.* Cox vaccines containing the respective rickettsiae have been prepared against epidemic typhus fever, murine typhus fever, Rocky Mountain spotted fever, and Q fever. In addition, a Rocky Mountain spotted fever vaccine, found to be effective, has been made by grinding up the bodies of infected ticks.

Vaccination against the rickettsial diseases is advised for persons likely to contact the infectious agent because of living conditions or occupation. Protection against epidemic typhus is strongly urged for those who have been in contact with a patient. Although typhus immunization is not required by any country in the world as a condition of entry, it is nevertheless recommended to travelers when travel plans include a geographic area that is not only infected but one in which living conditions are generally poor.

■ **Summary.** Table 37-4 is a summary for material just presented and indicates prospects for the future.

TABLE 37-4. CURRENT STATUS OF COMMON VACCINES*

1. Vaccines satisfactory in:
   - Diphtheria
   - Measles
   - Mumps
   - Poliomyelitis
   - Rocky Mountain spotted fever
   - Rubella
   - Tetanus
2. Vaccines being improved for:
   - Cholera
   - Influenza
   - Smallpox
   - Tuberculosis
   - Yellow fever
3. Vaccines being developed for:
   - Meningococcal meningitis
   - Mycoplasma pneumonia
   - Pneumococcal pneumonia
   - Rabies
   - Respiratory syncytial virus disease
   - Shigellosis
   - Streptococcal infections (group A)
4. Vaccines with improved product desired (no work in progress) for:
   - Pertussis
5. Vaccines needed for:
   - Varicella
   - Cytomegalic inclusion disease
   - Gonorrhea
   - Hepatitis

*After Medical World News **12:**31 (Feb 5), 1971.

PRECAUTIONS IN THE ADMINISTRATION OF BIOLOGIC PRODUCTS

To ensure the development of the desired immunity and to prevent insofar as possible certain untoward side effects or complications, the following precautions in the administration of vaccines, serums, and other biologic products are strongly advised:

1. *Read carefully the label on the package and the accompanying leaflet when any biologic product is to be given!*
2. Disinfect properly the skin at the site of injection. (Note also proper disinfection of the surface of the stoppered vial of medication.)
3. Use a sterile syringe and needle for each injection, preferably a sterile *disposable* unit. (Using several needles with the same syringe can be a way to transmit serum hepatitis.)
4. When the needle is placed into the subcutaneous or intramuscular area, pull the plunger of the syringe outward to check that the needle has not inadvertently entered a vein. If it has, blood wells up in the syringe.

5. Immunize only well children. Delay any such procedure for sick children until their recovery.
6. *Anaphylactic and allergic accidents in biotherapy are unpredictable—the next one may be yours!* Carefully question the recipient of the injection as to reactions to previous doses of the same or comparable biologic products. Inquire carefully as to previous injection of horse serum, any known allergy to horse serum protein, and any history of allergic conditions such as asthma. Did the child have reactions to a previous dose, such as fever or sleeplessness? Was there an area of redness and tenderness about the injection site? If the child had only a mild reaction, the injections may be continued. In some instances the amount of the material must be reduced and the overall number of injections increased. If the child had a severe reaction, do not repeat the injection. Consult the physician.
7. Ascertain the presence of allergy to egg or chicken dander if certain viral vaccines prepared in chick embryos (influenza, yellow fever, and measles) are to be given. Practically speaking if the recipient can eat eggs without event, it is safe to give the vaccine. Some of the newer vaccines may contain neomycin to which an allergic state may also exist.
8. Note these special situations:
   a. Do not immunize persons with altered or deficient immunologic mechanisms.
   b. Do not immunize persons receiving steroid hormones.
   c. Do not vaccinate children for smallpox who have a generalized dermatitis such as eczema because of the danger of a generalized vaccinia. Children with skin diseases should be segregated from children being vaccinated against smallpox.
   d. Unless there is an emergency, do not immunize during a poliomyelitis outbreak.

■ **Tests for hypersensitivity.** Preliminary testing to detect hypersensitivity by all means is mandatory.

The following tests for hypersensitivity must be used discreetly, and a syringe containing 1 ml. of epinephrine 1:1000 should be handy (see also p. 206).

1. In the *scratch* test, a 1:10 dilution of the biologic product in isotonic saline solution (tetanus antitoxin, for example) is applied to the abraded skin.
2. In the *eye* test, one drop of a 1:10 dilution is placed in the conjunctival sac. If this is negative, the eye test may be repeated with undiluted material.
3. The *intradermal* skin test is usually carried out with 0.02 to 0.03 ml. of a 1:10 dilution. *(An intradermal test with 0.1 ml. of undiluted TAT may be fatal.)* If indicated, skin tests may be carried out serially with 0.02 to 0.03 ml. of 1:10,000, 1:1000, 1:100, and 1:10 dilutions. The positive reaction is a hivelike wheal with redness.
4. In the *intravenous* test, the patient's blood pressure is recorded before the 5-minute injection of 0.1 ml. of biologic product in 10 ml. of isotonic saline solution and every 5 minutes afterward for 30 minutes. If the blood pressure falls 20 points or more, a hypersensitive state is indicated.

■ **Injection site for biologic products.** *The optimal site for injection* is the outer lateral aspect of the thigh (Fig. 37-5). By directing the needle downward as it enters the tissues at the junction of the upper third with the lower two thirds, one can deliver serum or

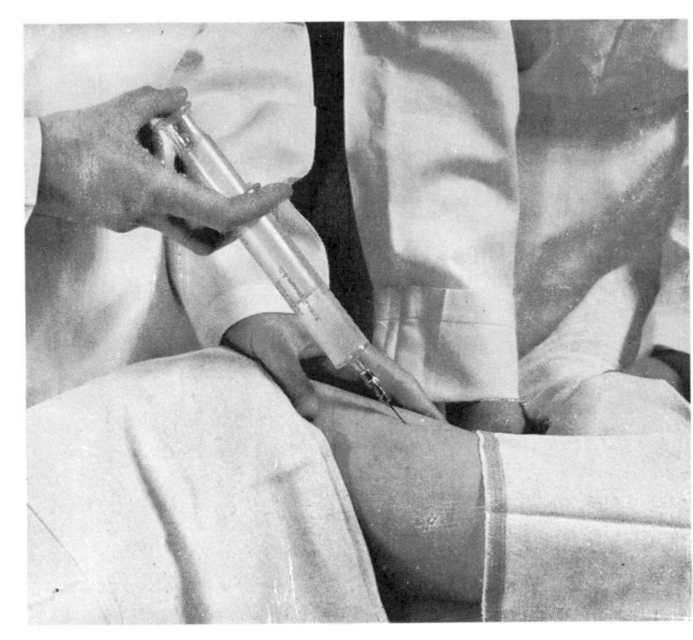

FIG. 37-5. Administration of serum into outer lateral aspect of thigh at junction of upper third with lower two thirds. The needle points downward to deliver serum into the middle third of the thigh. (From Top F. H., Sr.: Communicable and infectious diseases, ed. 5, St. Louis, 1964, The C. V. Mosby Co.)

other biologic product into the middle third of the thigh. This area is preferred because (1) in the event of an impending reaction a tourniquet can be placed on the thigh above the injection site, and the absorption of the biologic product greatly delayed, and (2) here there are no important anatomic structures to be injured by pressure of injected material. If speed of absorption is desired, the fascia lata, a tense band stretched across this area, facilitates it.

STANDARDS FOR BIOLOGIC PRODUCTS

■ **Manufacture.** It is so important that serums, vaccines, and other biologic products, including human blood and its derivatives, meet acceptable standards of safety, purity, and potency that, when offered for sale, import, or export in interstate commerce, they must be manufactured under the license and regulations of the federal government. Authority for this is delegated to the Food and Drug Administration of the United States Department of Health, Education, and Welfare.

■ **Label.** Each package must be properly labeled. The name, address, and license number of the manufacturer, the lot number, and expiration date must be shown.

■ **Date of expiration.** On the *date of issue* the product is placed on the market. This must be within a certain time after manufacture, depending upon the kind of product and the temperature of storage. The *expiration date* means the date beyond which the product cannot be expected to exert its full potency or usefulness. For instance, if a package of diphtheria antitoxin has a stated potency of 10,000 units and an expiration date of Sept. 1, 1976, it means that properly stored the package will contain 10,000 units of diphtheria antitoxin on Sept. 1, 1976. An excess of antitoxin in the package makes allowance for this. The expiration date of most biologic products is 1 year after manufacture or issue.

QUESTIONS FOR REVIEW

1. Define or briefly explain: vaccine, immune serum, antitoxin, antivenin, antibacterial serum, convalescent serum, MLD, date of issue (of vaccine), expiration date (of vaccine), autogenous vaccine, lyophilized vaccine, immune serum globulin.
2. Cite at least four differences between vaccines and immune serums.
3. What is the term used to designate the strength of tetanus and diphtheria antitoxins?
4. Briefly discuss the three methods of standardizing diphtheria antitoxin.
5. Name three antitoxins of importance. Indicate the manufacture of one.
6. List eight diseases for which immune serum globulin may be of benefit. Indicate how gamma globulin is used to modify or prevent disease.
7. How is toxoid prepared? Name three important toxoids.
8. What is BCG? Comment on its usefulness.
9. Compare live virus vaccines with inactivated virus vaccines.
10. List and briefly describe the measles vaccines.
11. Name immunizing agents used in combination.
12. How are rabies vaccines prepared?
13. List the rubella virus vaccines.
14. Give the optimal site for injection of biologic products.
15. List the tests for hypersensitivity.
16. Give precautions in the administration of biologic products.

REFERENCES. See at end of Chapter 38.

38 Recommended immunizations

In this chapter methods of immunization for the prevention of the more important infectious diseases are given. Since opinions from public health physicians and authorizing agencies concerning the most efficacious methods of immunization are not uniform, the student will note variations in procedures endorsed in standard references.

RECOMMENDATIONS OF THE COMMITTEE ON INFECTIOUS DISEASES OF THE AMERICAN ACADEMY OF PEDIATRICS*

The Red Book of the American Academy of Pediatrics is widely accepted as a source of the best immunization procedures in both children and adults. Table 38-1 schedules the protection recommended by the Committee on Infectious Diseases for normal infants and children from 2 months of age through the sixteenth year of life (after which the immunizing procedures are those for adults).

■ **Combined active immunization.** In the schedule of Table 38-1, combined diphtheria and tetanus alum (depot) toxoids and *Bordetella pertussis* vaccine are preferred as primary immunizing agents over the nonadsorbed (fluid or plain) mixtures. A dose,† is injected deep into the deltoid or midlateral muscles of the thigh or given by intradermal jet injection (see Fig. 37-1).

The standard preparations of *combined diphtheria-tetanus toxoids (DT)* used in infants and young children contain 7 to 25 Lf. (flocculating units) diphtheria toxoid per dose. The *adult type of combined tetanus-diphtheria toxoids (Td)* with adjuvant used in teenagers and adults contains not more than 2 Lf. diphtheria toxoid per dose. Td contains less antigen than DT, since fairly severe reactions in older children and adults may be associated with the increase of diphtheria antigen. Adult type toxoids are specially purified preparations.

Live oral poliovirus vaccine is recommended over the killed vaccine because it is more effective as an immunizing agent, the antibodies induced persist longer, and it is easier to give. Trivalent OPV is the one preferred. (If monovalent virus-type feeding is carried out, the order is type 1 followed by type 3 and then type 2.)

The immunity for each disease produced by the combined method of immunization is as great as would be obtained by separate immunization against the given disease.

---

*A copy of the Red Book is obtained from the American Academy of Pediatrics, Inc., Evanston, Ill. The Academy has developed a personal immunization card, billfold size and plastic, to provide a permanent record.

†The concentration of antigens varies in various products. The package insert supplied with the vaccine should be carefully studied as to volume of dose.

TABLE 38-1. SCHEDULE FOR ACTIVE IMMUNIZATION AND TUBERCULIN TESTING IN NORMAL INFANTS AND CHILDREN*

Age		Administration of immunizing agents†		
Months	**Years**	**Combined antigens**	**Others**	**Tuberculin testing‡**
2		DTP (diphtheria-tetanus toxoids, adsorbed, and pertussis vaccine)	Trivalent OPV (oral poliovirus vaccine)	
4		DTP	Trivalent OPV	
6		DTP	Trivalent OPV	
12		Measles-Rubella or Measles-Mumps-Rubella§	Live measles vaccine (2 to 3 days after tuberculin test)	X
18		DTP	Trivalent OPV	
	4 to 6	DTP	Trivalent OPV	
	14 to 16	Td (adult-type combined tetanus-diphtheria toxoids)		
	Thereafter	Td every 10 years		

*See also text.
†Routine smallpox vaccination no longer recommended.
‡Risk of exposure indicates frequency of tuberculin testing.
§Combined viral vaccines (single injection) can be given from ages 1 to 12.

■ **Tetanus.** For individuals to whom the combined primary immunizations of early childhood do not apply, the Red Book recommends administration of alum tetanus toxoid. The total recommended dose of the manufacturer may be given in fractional doses of 0.05 or 0.1 ml. A booster (recall) injection of alum tetanus toxoid should be given 1 year later.

When the child or older person sustains an injury and the wound remains clean, the Committee on Infectious Diseases feels that in a fully immunized child no booster dose is needed unless more than 10 years have elapsed since the last dose. When the child or older person incurs a contaminated wound likely to be complicated by tetanus, such as a deep puncture wound, a dog bite, certain crusted or suppurating lesions, and wounds containing soil or manure, T (tetanus toxoid with adjuvant) or Td should be given regardless of the immunization status of the patient *unless* the patient is known to have received Td or T within the past five years. If he has, the inoculation at this time need not be given.* It has been found that the immunity conferred by tetanus toxoid is extremely long lasting† and that frequent recall doses seem to provoke reactions.

■ **Pertussis.** Combined immunization as indicated in Table 38-1 is preferable, utilizing the antigens combined with adjuvant, especially when immunization is begun under the age of 6 months. Pertussis vaccine is not given beyond the age of 6 years.

*Nonimmunized individuals also require passive immunization. See p. 599.
†The U.S. Public Health Service recommends that tetanus boosters of adsorbed toxoid be given at 10-year intervals.

When the young child has been intimately exposed to whooping cough or there is an epidemic and protective immunity must be developed as soon as possible, pertussis vaccine adsorbed should be given intramuscularly in three doses of 4 protective units each, at least a month apart. Routine recall injections are given 12 months later and at school age with adsorbed vaccine. Exposure recall injections indicated for children up to the age of 6 years are given with adsorbed vaccine.

■ **Diphtheria.** In infants and young children, diphtheria toxoid is given alone *only* when there is definite reason for not using the combined immunization schedule, such as the occurrence in the child of a severe reaction following the injection of the multiple vaccines or the exposure of the child to a possible or known case of diphtheria. The toxoid with adjuvant is given in three fractional (0.05 to 0.1 ml.) doses.

TOXOID SENSITIVITY TEST (ZOELLER, MOLONEY). Since even small doses of diphtheria toxoid can cause a reaction in a highly sensitive person, when the adult type toxoids are not available, a toxoid sensitivity test should be done beforehand. One-tenth milliliter of *fluid* diphtheria toxoid diluted 1:50 or 1:100 in saline is injected intracutaneously. The test site is read after 18 to 24 hours. A positive reaction, indicating hypersensitivity, is a red indurated area. In that person the positive toxoid sensitivity test is presumptive evidence of immunity from natural exposure, and further toxoid should *not* under any circumstances be given. Those with no reaction, a negative test, should receive diphtheria toxoid adsorbed as a single antigen, or diphtheria and tetanus toxoids, childhood type (DT).

■ **Smallpox.** The preferred method of smallpox vaccination is either the multiple pressure method or intradermal jet injection with calf lymph virus.* In preparation of the arm, acetone or ether is suitable; alcohol, especially if medicated, should *not* be used because it destroys the virus. However, no preparation at all is better than vigorous cleansing that might abrade the skin. The vaccination site† should be kept dry, preferably uncovered, until the scab falls off. The site of a primary vaccination should be inspected between the sixth and the eighth days, the revaccination site between the fourth and the seventh days.

After vaccination *primary vaccinia* (Fig. 38-1) occurs in persons with no immunity. On the fourth day a small circumscribed solid elevation of the skin *(papule)* appears; on the eighth day the lesion contains fluid *(vesicle)* and has a pink areola around it that averages several centimeters in diameter. On the ninth day it is filled with pus *(pustule)*

---

*There are three methods for smallpox vaccination:

1. In the *multiple pressure* technic, a sterile sharp needle is held tangentially to the skin. Pressure is applied to an area about one-eighth inch in diameter. For primary vaccination in a smallpox-free area, 6 to 10 pressures within 5 or 6 seconds suffice. For revaccination 20 to 30 pressures within 30 seconds are given. A trace of blood welling up within half a minute indicates the right amount of force exerted against the drop of vaccine on the skin.
2. In the *multiple puncture* method, a forked needle dipped in vaccine is touched to the surface of the skin. The needle is held perpendicular to the skin, and (if the geographic area is free of smallpox) two to three punctures are made in an up-and-down manner within a small area about one-eighth inch in diameter.
3. The *jet injection* method requires a specially constructed injector that delivers about 0.1 ml. of a specially purified vaccine into the skin (Fig. 38-2).

†"Primary vaccination and revaccination are best performed on the outer aspects of the upper arm, over the insertion of the deltoid muscle, or behind the midline. Reactions are less likely to be severe on the upper arm than on the lower extremity or other parts of the body. With proper technic, resultant scars are small and unobtrusive." (Kempe, C. H., and Benenson, A. S.: Smallpox immunization in the United States, J.A.M.A. **194:**161, 1965.)

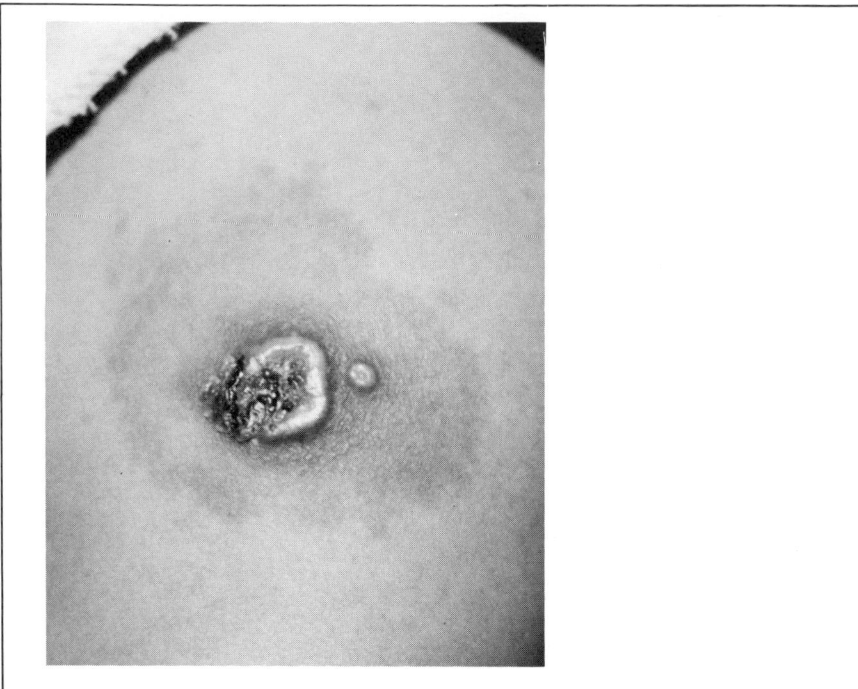

FIG. 38-1. Smallpox vaccination. Primary vaccinia on arm of year-old child. Note red areola surrounding an umbilicated partly encrusted pustule with small satellite.

and has begun to dry up by the end of 14 days. The formation of a scar at the site is evidence of a successful vaccination.

After revaccination, the following may be seen:

1. *Major reaction**—a vesicular or pustular lesion, an area of hardening, or a reddened zone about a central crust or ulcer—indicates virus has multiplied at the site.
2. *Equivocal reaction*—any other reaction—indicates vaccination did not "take." Equivocal reactions may come as a consequence of immunity adequate to suppress the virus or may represent only an allergic reaction to an inactive vaccine.

No reaction means inactive vaccine or faulty technic. In such a case vaccination should be repeated. Vaccine virus is very susceptible to even moderate warmth and should be kept in the freezing compartment of the refrigerator. *The vaccine virus must be kept frozen.*

INDICATIONS. The U. S. Public Health Service currently recommends that primary smallpox vaccination be discontinued on a routine basis in the United States. The risk from the disease is small, and on occasions severe reactions present to the vaccine. Accordingly the American Academy of Pediatrics no longer advocates this immunization routinely. It stresses vaccination in the event of exposure and in preparation for

*The World Health Organization's Expert Committee on Smallpox defines only two responses to smallpox vaccination—"major" and "equivocal."

foreign travel. The U.S. Public Health Service advises personnel in medical and allied health fields to be vaccinated every 3 years since there is a potential hazard to these persons from unsuspected cases accidentally imported into the country.

All persons, including those who have had smallpox, should be vaccinated if exposed to the disease. Those who live and work under conditions of constant exposure should be vaccinated every year. This includes laboratory workers and residents in areas of the world where smallpox is endemic and occasionally epidemic.

CONTRAINDICATIONS. Children with eczema and skin lesions such as poison ivy or impetigo, extensive wounds, or burns should not be vaccinated against smallpox because generalized vaccinia, a condition resembling smallpox, may develop; nor should they be allowed to contact a person who has an active vaccination lesion. It would be dangerous to vaccinate a patient with depressed immunologic responses, either because of a disease such as dysgammaglobulinemia, lymphoma, leukemia, or blood dyscrasia, or because of therapy being given with steroids, antimetabolites, alkylating agents, or ionizing radiation. Pregnant women must *not* be vaccinated.

■ **Mumps.** Live attenuated mumps vaccine is given in a 1-ml. subcutaneous injection, followed by a second dose in 2 weeks. Yearly boosters are recommended by some authorities. Fever reactions have not been noted. The vaccine is advocated for susceptible children approaching puberty, adolescents, and adults, particularly males with no history of the disease. In special circumstances vaccination may be considered for younger children in closed populations, such as special schools, camps, institutions, and so on. It should not be given to pregnant women or to children less than 12 months of age.

■ **Rubella.** Rubella immunization is carried out with a single subcutaneous injection of the vaccine. It is advocated for boys and girls between the ages of 1 year and puberty and especially for those of kindergarten age. Vaccinees will shed virus from their throats for 2 or more weeks afterward, but they are not thought to be infectious. It is felt that vaccination gives long-range protection.

Immunization is contraindicated for infants less than 1 year of age, for pregnant women, and for individuals with altered immune states (leukemia, lymphoid malignancies, recipients of immunosuppressive therapy). It should not be given routinely to adolescent girls or adult women because of the danger that it might complicate a pregnancy. It is not given in severe febrile illness.

Some 32 million children have received rubella vaccine in the all-out effort to block the circulation of rubella virus in population groups and thus avoid the exposure of pregnant women to the virus.

■ **Rabies.** With rabies we have an exception to the rule that there is not enough time between exposure and onset of disease to establish an *active* immunity to protect the person exposed. Since the incubation period of rabies may be much longer than that of most diseases, an active immunity can usually be established before the disease begins.* The onset of the disease *must* be prevented because there is no definitive treatment; the mortality is close to 100%.

Table 38-2 outlines the recommendations of the WHO Expert Committee on

---

*Smallpox and rabies are the only diseases in which the exposed person can be protected against development of the disease by the production of an active immunity. In smallpox an active immunity develops very quickly. In rabies the period of incubation is so long that an active immunity can be produced before the disease takes effect.

TABLE 38-2. RECOMMENDATIONS FOR THE PREVENTION OF RABIES*

Exposure	Condition of vaccinated or unvaccinated animal inflicting injury: At exposure	Condition of vaccinated or unvaccinated animal inflicting injury: During 10 days' observation	Recommended treatment†
1. Indirect contact	Healthy or rabid	Healthy or rabid	None
2. Licks			
a. Intact skin	Rabid	–	None
b. Skin lesions, scratches; mucous membrane intact or abraded	1. Healthy	Clinical signs of rabies or laboratory diagnosis	Serum now; start vaccine at first signs of rabies in the animal
	2. Signs veterinarian judges to suggest rabies	Healthy	Serum now; start vaccine; stop at 5 days if animal is normal
	3. Rabid, escaped, killed, or unknown	Rabid	Serum now; start vaccine
3. Bites			
a. Mild exposure	1. Healthy	Clinical signs of rabies or laboratory diagnosis	Serum immediately; start vaccine at first sign of rabies in animal
	2. Signs veterinarian judges to suggest rabies	Healthy	Serum and vaccine at once; stop at 5 days if animal is normal
	3. Rabid, escaped, killed, or unknown	–	Serum and vaccine at once
	4. Wild wolf, fox, bat, skunk, others	–	Serum and vaccine at once
b. Severe exposure such as multiple wounds or face, finger, and neck wounds	1. Healthy	Clinical signs of rabies or laboratory diagnosis	Serum at once; start vaccine at first sign of rabies in animal
	2. Signs veterinarian judges to suggest rabies	Healthy	Serum and vaccine at once; stop at 5 days if animal is normal
	3. Rabid, escaped, killed, or unknown	Rabid	Serum and vaccine at once
	4. Wild wolf, fox, bat, skunk, others	–	Serum and vaccine at once

*Modified from Technical Report Series No. 321 (1966), WHO Expert Committee on Rabies, Fifth Report, Geneva, 1966, World Health Organization.
†See text for the dosage and administration of vaccine and serum.

Rabies for rabies prevention. Active immunization, as indicated in this table, is carried out in a series of 14 to 21 daily doses of rabies vaccine (duck embryo or brain tissue types). A booster dose is given 10 days and 20 days after the last dose.

The best prophylaxis against rabies after exposure is the use of hyperimmune serum in combination with vaccine. The hyperimmune serum (horse serum*) is given

*Rabies immune globulin human (RIGH) has been developed.

as soon as possible. Within 24 hours after the bite 1000 units per 40 pounds of body weight of antirabies serum is given intramuscularly. If more than 24 (up to 72) hours has elapsed, two to three times this amount of serum is given. Part of the dose of rabies serum may be injected around the wound inflicted by the animal for local antiviral effect. If reexposure of an ordinary nature occurs, one booster dose of duck embryo vaccine is given. If reexposure is severe, five daily doses of vaccine are given followed by a booster 20 days later.

Prophylactic vaccination is now advised for laboratory workers, veterinarians, spelunkers, and others likely to contract the disease. Two subcutaneous inoculations of 1 ml. of duck embryo rabies vaccine each into the deltoid area, 1 month apart, are followed by a booster dose 6 months later. Boosters are recommended every 2 to 3 years.

The amount of protection needed from hyperimmune serum and the vaccine depends upon many factors (see Table 38-2) such as the type of animal, the location of the bite, the extent of injury, the protection afforded by clothing, whether rabies is present in the community, and whether observation of the animal is possible. *It is very important that the animal inflicting the bite be apprehended and kept for a 10-day period of observation.* (Rats and mice do not transmit rabies.)

Although the method of preparing antirabic vaccine differs greatly from the original method of Pasteur, the administration of antirabic vaccine is still referred to as the *Pasteur treatment.* It is befitting that his name be thus remembered.

■ **Tuberculosis.** BCG vaccination against tuberculosis is done as superficially as possible into the skin of the upper arm over the deltoid or triceps muscle. To the newborn infant 0.05 ml. is given; to the older tuberculin-negative individual 0.1 ml. Tuberculin testing must be done before vaccination except in infants less than 2 months of age. A preliminary radiograph of the chest may be indicated. Two to three months after vaccination, tuberculin testing is repeated, and if negative, vaccination is repeated.

BCG vaccination is advised for children living or traveling in areas of the world where tuberculosis is a major health problem and for those likely to be exposed to adults with the disease. It should be carried out 2 months before exposure.

RECOMMENDATIONS OF THE U. S. PUBLIC HEALTH SERVICE

■ **Influenza.** The U. S. Public Health Service recommends annual influenza vaccination for persons in whom the onset of flu would represent an added health risk, such as individuals of all ages with chronic debilitating diseases (diabetes, cardiovascular ailments, pulmonary disease, and others). There is some indication for immunization of persons responsible for furnishing essential public services, such as law enforcement officers, firemen, and many others.

The primary series usually consists of two doses given *subcutaneously* (beneath the skin) 6 to 8 weeks apart. (In the package insert supplied with his vaccine, the manufacturer gives the recommended dose for adults and children and the schedule for administration.) If an individual received one or more doses of the current vaccine containing the epidemic strains in the previous year, he may need only a single subcutaneous booster. Otherwise he should be given a primary series. Once a cycle of annual immunization is begun, only single boosters annually are required. Vaccination should be completed by the middle of November.

The status of influenza immunization (both vaccine and schedule) can change

TABLE 38-3. ADMINISTRATION OF POLIOVIRUS VACCINES

Group	Doses	Poliovirus type(s)	Time interval	Comments
I. Administration of trivalent oral poliovirus vaccine (OPV)				
Infants	First	Trivalent OPV (I, II, III)		Primary immunization begun at 6 to 12 weeks of age; may be given with DPT
	Second	Trivalent OPV	8 weeks after first dose (no less than 6 weeks)	
	Third	Trivalent OPV	8 to 12 months after second dose	
	Booster	Trivalent OPV		Given at time child enters school; boosters with OPV otherwise not indicated
Children and adolescents	First	Trivalent OPV		
	Second	Trivalent OPV	6 to 8 weeks after first dose	
	Third	Trivalent OPV	8 to 12 months after second dose	Third dose given as early as 6 weeks after second dose in unusual circumstances
Adults*	First	Trivalent OPV		Recommended for adults at high risk; pregnancy no indication or contraindication for immunization
	Second	Trivalent OPV	6 to 8 weeks after first dose	
	Third	Trivalent OPV	8 to 12 months after second dose	
II. Administration of monovalent oral poliovirus vaccine (OPV)				
All ages	First	Monovalent OPV (I)		
	Second	Monovalent OPV (III)	6 to 8 weeks after first dose	
	Third	Monovalent OPV (II)	6 to 8 weeks after second dose	
	Fourth	Trivalent OPV (I, II, III)	8 to 12 months after third dose of monovalent vaccine	

*Routine immunization is not indicated for most adults residing in continental United States.

Continued.

TABLE 38-3. ADMINISTRATION OF POLIOVIRUS VACCINES— cont'd

Group	Doses	Poliovirus type(s)	Time interval	Comments
III. Administration of inactivated poliovirus vaccine (IPV)				
All ages	First	IPV		Parenteral administration; immunization series four doses; for infant 6 to 12 weeks of age, IPV may be given with DPT
	Second	IPV	4 weeks after first dose	
	Third	IPV	4 weeks after second dose	
	Fourth	IPV	6 to 12 months after third dose	
	Booster	IPV	2- to 3-year interval	

rapidly. For this reason the recommendations of its Committee on Immunization Practices are regularly published by the U.S. Public Health Service and should be consulted before any procedure is carried out.

■ **Poliomyelitis.** The U. S. Public Health Service is urging regular immunization against poliomyelitis for all children from early infancy. Routine immunization of infants should begin at 6 to 12 weeks of age and be completed against all three types of the virus. The schedule for oral vaccination is given in Table 38-3. Oral poliovirus vaccine (OPV) is more widely used in the United States than is inactivated poliovirus vaccine (IPV), and the trivalent OPV has largely replaced the monovalent form in immunization programs.

Vaccination is contraindicated in patients with severe underlying diseases such as leukemia or generalized malignancy or those with lowered resistance because of therapy.

■ **Measles.** The U. S. Public Health Service urges immunization for all susceptible children (those not having had vaccine or natural measles). The prime target groups are children at least 1 year of age and susceptible children entering nursery school, kindergarten, or elementary school (Fig. 38-2). Protection against measles is especially needed for children in the high risk groups, that is, children in institutions and those with chronic heart or pulmonary disease and cystic fibrosis. Vaccination of adults is not stressed since most individuals are serologically immune by the age of 15 years.

Table 38-4 gives the currently acceptable schedules for measles immunization.

Live attenuated vaccine should not be given to pregnant women or to patients being treated with steroids, irradiation, antimetabolites, or those agents that depress an individual's immunologic capacities. Leukemia, lymphoma, and other generalized malignancies are also contraindications, as is untreated tuberculosis.

■ **Typhoid and paratyphoid fevers.** The U. S. Public Health Service no longer recommends routine typhoid immunization in the United States even for individuals in flood disaster areas. Circumstances where vaccination would be indicated include an intimate exposure to a known carrier, an outbreak of the disease in a community,

TABLE 38-4. SCHEDULES FOR MEASLES IMMUNIZATION

Schedule	Type of vaccine*	Age	Dose (Subcutaneous)†	Remarks
1.	Live attenuated (Edmonston B strain) measles virus vaccine	12 months and older	One	Although live attenuated viral vaccine may be given safely, many physicians may wish to add Measles Immune Globulin because of fewer clinical reactions
2.	Live attenuated (Edmonston B strain) plus Measles Immune Globulin	12 months and older	One, plus 0.01 ml. per pound of body weight Measles Immune Globulin at different site with different syringe	
3.	Live, further attenuated (Schwarz and Moraten strains)	12 months and older	One	Clinical reactions those of schedule 2; Measles Immune Globulin *not* indicated

*Only live attenuated virus vaccines should be used! As soon as possible children who may have received inactivated vaccine should be revaccinated with live virus vaccine. It has been found that children having received inactivated vaccine develop a sinister illness when they are exposed to natural measles several years later.
†No booster needed.

or the event of travel to an endemic area. The Public Health Service does not advise paratyphoid immunization because there is good evidence that paratyphoid A and B vaccines are worthless. (Schedules for typhoid immunization may be found in the Red Book of the American Academy of Pediatrics.)

REQUIREMENTS FOR UNITED STATES ARMED FORCES

Immunization procedures for the United States Armed Forces are regularly evaluated and updated. So that the requirements may be precisely related to the geographic area of military duty, the world is divided into the following:

Area I (routine)—United States (includes the 50 states, District of Columbia, Virgin Islands, Puerto Rico, Wake Island, Midway Island), Canada, Greenland, Iceland, Marshall Islands, Guam, Pacific islands east of the 180th meridian, North Pole, South Pole, Bermuda, Bahama Islands, Baja California, and a strip of Mexico 50 miles south of the U. S. border.

Area II—All areas outside Area I

Area IIC (cholera)—Arabian Peninsula, Afghanistan, Burma, Ceylon, Republic of China, Hong Kong, India, Indonesia, Korea, Communist China, Macao,

Malaysia, Pakistan, Philippines, Thailand, Iran, Iraq, Syria, Turkey, Lebanon, Israel, Jordan, Kuwait, African continent north of the Sahara, and Madagascar.

Area IICP (cholera and plague) — Laos, Cambodia, and Vietnam.

Area IIY (yellow fever) — Central America southeast of the Isthmus of Tehuantepec, Panama, and South America.

Area IIYC (yellow fever and cholera) — Africa south of the Sahara

Table 38-5 summarizes the immunization for United States military personnel.

TABLE 38-5. IMMUNIZATIONS REQUIRED FOR UNITED STATES ARMED FORCES

Immunizing agent	Basic series*	Required Area I (excluding alert forces)	Required Area II	Reimmunization Area I	Reimmunization Area II
1. Smallpox vaccine	Vaccination to be read after 6 to 8 days; successful reaction required	✓	✓	3 years	3 years
2. Typhoid vaccine (acetone killed, dried)	Single 1.0 ml infection — Area I Two 1.0 ml. injections 4 weeks apart — alert forces	✓	✓	None	0.5 ml. — 3 years
3. Tetanus and diphtheria toxoids	Two 0.5 ml. injections 1 to 2 months apart with third dose of 0.1 ml. 12 months later	✓	✓	0.1 ml. — 6 years; 0.5 ml. after injury or burn	0.1 ml. — 6 years; 0.5 ml. after injury or burn
4. Poliovirus vaccine, live oral (OPV) trivalent, types I, II, and III	Three oral doses 8+ weeks apart	✓	✓	None	None
5. Yellow fever vaccine, USP†	One 0.5 ml. injection of concentrated vaccine diluted 1:10		✓ (Areas IIY, IIYC)		0.5 ml. 1:10 dilution — 10 years
6. Influenza virus vaccine, USP†	One 1.0 ml. injection	✓	✓	1.0 ml. — 1 year	1.0 ml. — 1 year
7. Cholera vaccine, USP†	0.5 ml. injection followed by 1.0 ml. in 1+weeks		✓ (Areas IIC, IIYC, IICP)		0.5 ml. — 6 months
8. Plague vaccine, USP†	Two *intramuscular* injections — 1.0 ml. followed in 3 months by 0.2 ml.		✓ (Area IICP)		0.2 ml. (IM) — 6 months

*Smallpox vaccination is given either by multiple puncture technic or by intradermal jet. Plague vaccine must be given intramuscularly. Other injections are subcutaneous or intramuscular. (See Fig. 37-1.)

†The vaccine listed in the United States Pharmacopeia.

Military dependents and civilians in the employ of the armed forces or organizations serving the armed forces are not required to take immunizations while within the United States or Canada. Outside the United States they are required to take essentially the same ones as members of the armed forces.

INTERNATIONAL TRAVEL

■ **Immunizations.** For those persons planning to travel, the Division of Foreign Quarantine of the U. S. Public Health Service has prepared a booklet entitled *Immunization Information for International Travel* (1973-1974). This booklet gives information on (1) immunization requirements for entrance into foreign countries, (2) immunization requirements for entrance into the United States, and (3) immunizations recommended as precautionary measures. This booklet may be obtained from the Superintendent of Documents, United States Government Printing Office, Washington, D.C. 20402.

The immunizations outlined in this booklet may be placed in two groups:

1. *Mandatory*
   a). *Smallpox.* Certificate of vaccination within the last 3 years required for entering some foreign countries and for reentering the United States, except from Canada, Mexico, and a few other countries.
   b). *Cholera.* For travel in certain countries, but not for reentry into the United States.
   c). *Yellow fever.* For travel in certain countries.
   Note: Yellow fever vaccination must be given at a designated Yellow Fever Vaccination Center, registered by the World Health Organization. Centers are located in 46 states, the District of Columbia, Puerto Rico, Canal Zone, and American Samoa. Yellow fever vaccination certificate is not valid until 10 days after primary immunization.
2. *Recommended**
   a). *Typhoid.*
   b). *Tetanus.*
   c). *Plague.* For travel in certain countries.
   d). *Epidemic typhus.* For travel in certain countries.
   e). *Poliomyelitis.*
   f). *Diphtheria.*
   g). *Influenza.*
   h). *Measles.*

The Red Book of the American Academy of Pediatrics also contains immunization information and recommendations for international travel. In Table 38-6, which presents vaccination schedules recommended by this organization, there is a breakdown of special requirements for travel by age groups.

■ **Where to find a doctor.** The International Association of Medical Assistance to Travelers, Inc., 745 Fifth Avenue, New York, N.Y. (or also Intermedic, 777 Third Avenue, New York, N.Y. 10017) supplies without charge a directory of English-speaking physicians available 24 hours a day in the different cities of the world. A long way from home one may also obtain medical aid from U.S. Embassies, Consulates, the Red Cross, travel agencies, the police, medical associations, hospitals, clinics, and the U.S. Armed Forces bases and installations.

*This booklet also advises suppressive medication for malaria (a major worldwide problem) prior to entry into an infected area, during sojourn there, and even for a number of weeks thereafter.

TABLE 38-6. SPECIAL IMMUNIZATION REQUIREMENTS BY AGE FOR INTERNATIONAL TRAVEL

Disease	Travel destination	Age	Doses of vaccine	Route of inoculation	Intervals	Recall needed
Cholera	Asian countries, Middle East	6 months to 4 years	(1) 0.1 ml. (2) 0.3 ml. (3) 0.3 ml.	Subcutaneously	7+ days between 1 and 2; 6+ months between 2 and 3	6 months
		5 to 9 years	(1) 0.3 ml. (2) 0.5 ml. (3) 0.5 ml.			
		10 years and older (adults also)	(1) 0.5 ml. (2) 1.0 ml. (3) 0.5 ml.			
Yellow fever	Central and South America, Africa	All ages	0.5 ml. of 1:10 dilution	Subcutaneously		10 years
Plague	High-risk area	Less than 1 year	(1) 0.1 ml. (2) 0.1 ml. (3) 0.04 ml.	Intramuscularly only	30 days between 1 and 2; 4 to 12 weeks between 2 and 3	3 months
		1 to 4 years	(1) 0.2 ml. (2) 0.2 ml. (3) 0.08 ml.			
		5 to 9 years	(1) 0.3 ml. (2) 0.3 ml. (3) 0.12 ml.			
		10 years and older (adults)	(1) 0.5 ml. (2) 0.5 ml. (3) 0.2 ml.			
Typhoid fever	Developing countries and those with low standards of sanitation	6 months to 9 years	(1) 0.25 ml. (2) 0.25 ml.	Subcutaneously	4+ weeks	1 to 3 years if exposed
		10 years and older (adults also)	(1) 0.5 ml. (2) 0.5 ml.			

■ **Traveler's diarrhea.** Traveler's diarrhea is known by the many colorful synonyms —Aztec two-step, Montezuma's revenge, Cromwell's curse, Delhi belly—that mark out a widespread geographic distribution, and preventive immunization does not exist for this unpleasant complication of the initial phase of the tourist's trip abroad (or out of the country). Traveler's diarrhea is described consistently by nausea and vomiting, abdominal cramps, chills, low-grade fever, and a diarrhea productive of loose and watery, foul-smelling stools. Manifestations, though distressing and temporarily incapacitating, disappear within a few days. Relapses or sequelae are rare. Microbes found in the stools and thereby implicated include *Shigella* species, *Salmonella* species, enteropathogenic *Escherichia coli,* and *Giardia lamblia.* However the cause is uncertain and a cure-all nonexistent.

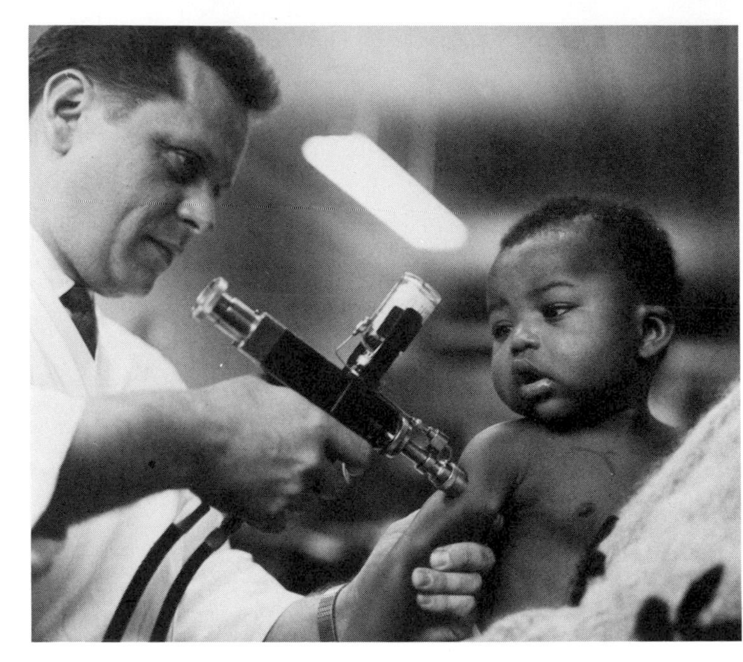

FIG. 38-2. Jet injector (air gun) delivery of measles vaccine to young Chicagoan—instant, pain free. (From Medical News, J.A.M.A. **196:**29, [May 23] 1966; courtesy American Medical Association.)

To minimize the hazard, especially where sanitary practices are substandard, certain precautions are stressed. The tourist can safely eat only foods thoroughly cooked or recently peeled. He can drink with impunity only boiled water, carbonated mineral water, or beverages boiled or carbonated in cans. Very hot water out of the tap is not likely to transmit viable enteric bacteria. After it has cooled, it can be used for oral hygiene and various other purposes.

■ **World Health Organization.** The World Health Organization, or WHO as it is usually called, is an agency of the United Nations set up to provide for the highest level of health for all peoples of the world. It works with the United Nations, governments, and special health groups to devise standards, to train personnel, to carry on research, and to improve public health in every way at the international level. To protect the sightseer of the jet age, it is concerned for uniform standards of hygiene in aviation. Guidelines are detailed for aspects of sanitation applied to aircraft and international airports, the design to provide every safety factor for the traveler in rapid transit over the globe.

REFERENCES FOR UNIT SIX

Abst, D. B., et al.: Time and cost factors to provide regular, periodic dental care for children in a fluoridated and nonfluoridated area: final report, J. Amer. Dent. Ass. **80:**770, 1970.

Balagtas, R. C., Nelson, K. E., Levin, S., et al.: Treatment of pertussis with pertussis immune globulin, J. Pediat. **79:**203, 1971.

Bauman, H. E.: Food microbiology, ASM News **38:**312, 1972.

Bell, J. A.: Viruses and water quality, editorial, J.A.M.A. **219:**1628, 1972.
Berg, G., editor: Transmission of viruses by the water route, New York, 1967, Interscience Publishers, Inc.
Brooks, G. F., and Buchanan, T. M.: Tetanus toxoid immunization of adults, Ann. Intern. Med. **73:**603, 1970.
Brown, H.: Tetanus, J.A.M.A. **204:**614, 1968.
Buynak, E. B., Weibel, R. E., Whitman, J. E., Jr., et al.: Combined live measles, mumps, and rubella virus vaccines, J.A.M.A. **207:**2259, 1969.
Byrne, E. B.: Prevention of viral hepatitis in hospital physicians, Res. Phys. **15:**62, (June) 1969.
Carper, J.: Disease prevention—tomorrow's best hope, Today's Health **47:**21, (March) 1969.
Collected recommendations of Public Health Service Advisory Committee on Immunization Practices, Morbid Mortal Weekly Rep. **21**(no. 25) (supp.):1, 1972.
Demain, A. L.: Application of the microbe to the benefit of mankind: Challenges and opportunities, ASM News **38:**237, 1972.
Diefenbach, V. L.: The fluoridation story: Putting the smile on young faces, Today's Health **45:**60, (Feb.) 1967.
Di Palma, J. R.: The problem of food additives, RN **33:**53, (June) 1970.
Dunham, C. L.: Food preservation with ionizing radiation, Texas J. Med. **63:**54, (Sept.) 1967.
Edsall, G.: The current status of tetanus immunization, Hosp. Pract. **6:**57, (July) 1971.
Egeberg, R. O.: Fluoridation for all: a national priority, Today's Health **48:**30, (June) 1970.
Eickhoff, T. C.: Committee on Immunization (Infectious Diseases Society of America): Immunization against influenza: Rationale and recommendations, J. Infect. Dis. **123:**446, 1971.
Fireman, P., Friday, G., and Kumate, J.: Effect of measles vaccine on immunologic responses, Pediatrics **43:**264, 1969.
Frazier, W. C.: Food microbiology, New York, 1967, McGraw-Hill Book Co.
Freese, A. S.: Salmonella: Food poison plus, Today's Health **47:**34, (April) 1969.
Frye, J.: Secondhand water? Why not? Today's Health **44:**56, (March) 1966.
Goodheart, B. A.: A second look at the German measles vaccine, Today's Health **49:**46, (March) 1971.
Gordon-Smith, C. E.: Prospects for the control of infectious disease, Proc. Roy. Soc. Med. **63:**1181, 1970.
Grade "A" Pasteurized Milk Ordinance 1965 Recommendations of the United States Public Health Service, U.S. Public Health Bulletin No. 229 (revision), Washington, D.C., 1967, U.S. Government Printing Office.
Grand, M. G., et al.: Clinical reactions following rubella vaccination, J.A.M.A. **220:**1569, 1972.
Hayflick, L.: Human virus vaccines: Why monkey cells? Science **176:**813, 1972.
Houran, J. B.: Is your lab water up to standards? Lab Management **10:**52, (Jan.) 1972.
Hungate, R. E.: Status of microbiological research in relation to waste disposal, ASM News **38:**364, 1972.
Immunization Department of the Army Technical Bulletin, Department of the Navy Publication, Department of the Air Force Pamphlet, TB MED 114, NAVMED P-5052-15A, AFP 161-9, Washington, D.C., May 25, 1970, U.S. Government Printing Office.
Impure ice, editorial: J.A.M.A. **17:**118, 1891.
Insalata, N. F., et al.: Industrial advantages of FA testing for salmonellae, Food Techn. **26:**124, (April) 1972.
Insalata, N. F., et al.: A 24-hour method for the detection of coagulase-positive staphylococci in fish and shrimp, Food Techn. **26:**78, (May) 1972.
Karchmer, A. W., Friedman, J. P., Casey, H. L., et al.: Simultaneous administration of live virus vaccines, Amer. J. Dis. Child **121:**382, 1971.
Kempe, C. H.: To vaccinate or not, Hosp. Pract. **3:**28, (Sept.) 1968.
Kermode, G. O.: Food additives, Sci. Amer. **226:**15, (March) 1972.
Kilbourne, E. D.: Influenza: The vaccines, Hosp. Pract. **6:**103, (Oct.) 1971.
Kilbourne, E. D., and Smillie, W. G., editors: Human ecology and public health, New York, 1969, The Macmillan Co.
Koenig, M. G.: Trivalent botulinus antitoxin, Ann. Intern. Med. **70:**643, 1969.
Knutson, J. W.: Water fluoridation after 25 years, J. Amer. Dent. Ass. **80:**765, 1970.
Kretzer, M. P., and Engley, F. B., Jr.: Preventing food poisoning, RN **33:**50, (June) 1970.
Lane, J. M., Ruben, F. L., Abrutyn, E., et al.: Deaths attributable to smallpox vaccination, 1959 to 1966, and 1968, J.A.M.A. **212:**441, 1970.
Lawson, R. B.: Current status of immunizations, Postgrad. Med. **49:**111, (June) 1971.

Leininger, H. V., et al.: Microbiology of frozen cream-type pies, frozen cooked-peeled shrimp, and dry food-grade gelatin, Food Techn. **25:**28, (March) 1971.

Lerman, S. J., and Gold, E.: Measles in children previously vaccinated against measles, J.A.M.A. **216:**1311, 1971.

Linnemann, C. C., Jr., et al.: Measles antibody in previously immunized children, Amer. J. Dis. Child. **124:**53, (July) 1972.

Loofbourow, J. C. Cabasso, V. J., Roby, R. E., et al.: Rabies immune globulin (human), clinical trials and dose determination, J.A.M.A. **217:**1825, 1971.

Measles vaccines, editorial: J.A.M.A. **202:**1098, 1967.

Milk ordinance and code—1953 Recommendations of Public Health Service, U.S.P.H.S. Bull. no. 229, Washington, D. C., 1953, U. S. Government Printing Office.

Miller, L. W., et al.: Diphtheria immunization, Amer. J. Dis. Child. **123:**197, 1972.

Mitchell, R., editor: Water pollution microbiology, New York, 1972, John Wiley & Sons, Inc.

Morse, L. J., et al.: The Holy Cross College football team hepatitis outbreak, J.A.M.A. **219:**706, 1972.

Neumann, H. H.: Foreign travel immunization manual, Springfield, Ill., 1971, Charles C Thomas, Publisher.

Nickerson, W. J.: Microbial degradation and transformation of wastes, ASM News **38:**367, 1972.

Ogra, P. L., and Herd, J. K.: Arthritis associated with induced rubella infection, J. Immunol. **107:**810, 1971.

Pace, W. E.: Microbiological aspects of food irradiation, Milit. Med. **134:**215, 1969.

Paton, B. C.: Bites—human, dog, spider and snake, Surg. Clin. N. Amer. **43:**537, (April) 1963.

Paul, R. A.: An environmental model for swimming pool bacteriology, Amer. J. Public Health **62:**770, 1972.

Physician's desk reference to pharmaceutical specialities and biologicals, Oradell, N.J., 1973, Medical Economics, Inc., subsidiary of Litton Publications, Inc.

Pitel, M.: The subcutaneous injection, Amer. J. Nurs. **71:**76, 1971.

Plorde, J. J.: Advice to foreign travelers, Postgrad. Med. **51:**179, (Jan.) 1972.

Plotkin, S. A.: Rubella vaccination, editorial, J.A.M.A. **215:**1492, 1971.

Plotkin, S. A., and Clark, H. F.: Committee on Immunization (Infectious Disease Society of America): Prevention of rabies in man, J. Infect. Dis. **123:**227, 1971.

Price, J. F., and Schweigert, B. S., editors: The science of meat and meat products, San Francisco, 1971, W. H. Freeman & Co.

Public Health Service Drinking Water Standards 1962, U.S.P.H.S. publication no. 956, Washington, D. C., 1962, U. S. Government Printing Office.

Purdom, P. W., editor: Environmental health, New York, 1971, Academic Press Inc.

Rauch, P.: Avoiding injuries from injections, Nursing '72 **2:**12, (May) 1972.

Recommendation of the Public Health Service Advisory Committee on Immunization Practices: Combination live virus vaccines; measles and rubella; and measles, mumps, and rubella, Morbid Mortal Weekly Rep. **20**(16):145, 1971.

Reimann, H., editor: Food borne infections and intoxications, New York, 1969, Academic Press Inc.

Report of the Committee on Infectious Diseases, 1971, Evanston, Ill., 1971, American Academy of Pediatrics, Inc.

Rose, N. J., Schnurrenberger, P. R., and Martin, R. J.: Rabies prophylaxis, the physician's dilemma, Arch. Environ. Health **23:**57, 1971.

Rosenberg, F. A., Dondero, N., and Heukelekian, H.: Indicators of household well water pollution, Amer. J. Public Health **58:**452, (March) 1968.

Sanford, J. P.: Personal communication, 1972.

Schiff, G. M., et al.: Two-year follow-up of rubella vaccines in public school system, Amer. J. Dis. Child. **123:**193, 1972.

Schultz, M. G.: Entero-Vioform for preventing travelers' diarrhea, editorial, J.A.M.A. **220:**273, 1972.

Schwartz, W. F.: Communities strike back, Amer. J. Nurs. **71:**724, 1971.

Shea, K. P.: Infectious cure, Environment **13:**43, (Jan.-Feb.) 1971.

Sikes, R. K.: Rabies vaccines, Arch. Environ. Health **19:**862, 1969.

Snyder, J.: About the meat you are buying, Today's Health **49:**38, (Dec.) 1971.

Standard methods for the examination of dairy products, Washington, D. C., 1967, American Public Health Association, Inc.

Standard methods for the examination of water and wastewater, Washington, D. C., 1971, American Public Health Association, Inc.

Stanfield, J. P., et al.: Diphtheria-tetanus-pertussis immunization by intradermal jet infection, Brit. Med. J. **2:**197, 1972.

Stokes, J., Jr.: Pediatric immunization and the new Academy schedule, Hosp. Pract. **7:**127, (May) 1972.

Stokes, J., Jr., et al.: Trivalent combined measles-mumps-rubella vaccine, J.A.M.A. **218:**57, 1971.

Suggested ordinances and regulations covering public swimming pools, Washington, D. C., 1964, American Public Health Association, Inc.

Taylor, S. L., Fina, L. R., and Lambert, J. L.: New water disinfectant, Appl. Microbiol. **20:**720, 1970.

Thomas, W. C., Jr., Black, A. P., Freund, G., et al.: Iodine disinfection of water, Arch. Environ. Health **19:**124, 1969.

Toshach, S., and Thorsteinson, S.: Detection of staphylococcal enterotoxin by the gel diffusion test, Canad. J. Public Health **63:**58, 1972.

United States designated yellow fever vaccination centers, Morbid Mortal Weekly Rep. **20**(9)(supp.):1, 1971.

Vaccination certificate requirements for international travel, Morbid Mortal Weekly Rep. **21**(11)(supp.):1, 1972.

Vaheri, A., et al.: Isolation of attenuated rubella vaccine virus from human products of conception and uterine cervix, New Eng. J. Med. **286:**1071, 1972.

Villarejos, V. M., et al.: Combined live measles-rubella virus vaccine, J. Pediat. **79:**599, 1971.

Wallace, R. B., et al.: Joint symptoms following an area-wide rubella immunization campaign—report of a survey, Amer. J. Public Health **62:**658, 1972.

When is infectious waste not infectious waste? Hospitals, J.A.H.A. **46:**56, (May 1) 1972.

Wilkins, J., et al.: Live further attenuated rubeola vaccine, Amer. J. Dis. Child. **123:**190, 1972.

Williams, E. W.: Frozen foods: Biography of an industry, Boston, 1970, Cahners Book Division.

Wilson, G. S.: The hazards of immunization, New York, 1967, Oxford University Press, Inc.

Wolford, E. R.: Negative air pressure conveying, Food Techn. **26:**37, (Feb.) 1972.

Woodson, R. D., and Clinton, J. J.: Hepatitis prophylaxis abroad: Effectiveness of immune serum globulin in protecting Peace Corps volunteers, J.A.M.A. **209:**1053, 1969.

Wyll, S. A., and Grand, M. G.: Rubella in adolescents, J.A.M.A. **220:**1573, 1972.

Zollar, L. M., and Mufson, M. A.: Parotitis of non-mumps etiology, Hosp. Pract. **5:**93, (Aug.) 1970.

LABORATORY SURVEY OF UNIT SIX

PROJECT

Bacteriology of water

Part A – Bacterial plate count of water

1. Obtain from the instructor the following:
   a. Sample of water – may be water of low bacterial count (tap water) or water of high bacterial count (river water)
   b. Sterile pipettes
   c. Three tubes, each of which contains 9 ml. of sterile water
   d. Three sterile Petri dishes
   e. Three tubes of nutrient agar
2. Prepare the dilutions of the water sample
   a. To one tube of sterile water, add 1 ml. of the water sample and mix.
   b. Transfer 1 ml. from this first tube with a new sterile pipette to a second tube of 9 ml. of sterile water and mix.
   c. Repeat the procedure by taking a new pipette, transferring 1 ml. from the second to the third tube and mixing.
3. Calculate the dilution of water in each tube. Why is a new pipette used for each procedure?
4. Mix the diluted water samples with a suitable culture medium.
   a. With a new pipette, transfer 1 ml. from each tube to a Petri dish, beginning with the highest dilution and proceeding to the lowest. Why do you go from the highest to the lowest dilution?
   b. Melt tubes of agar in water bath, let cool to 42°-45 C., and add one tube to each Petri dish.
   c. Mix the agar with water by rotating, let the agar solidify, invert the plates, and incubate at 37° C. for 24 hours.
5. Select a plate on which the colonies are distinctly separated and count the colonies.
6. Calculate from the dilution the bacterial content of the water per milliliter. Record your results below.

Plate no.	*Dilution*	*No. of bacterial colonies present*	*Estimated no. of bacteria per ml. in water sample*
1			
2			
3			

Part B—Presumptive test for coliforms in water

1. Obtain from the instructor the following:
   a. Sample of contaminated water
   b. Sterile pipettes
   c. Nine fermentation tubes containing lactose (or lauryl tryptose) broth
2. Inoculate a sample of contaminated water as follows:
   a. To each of two fermentation tubes, add 0.1 ml. of the water sample.
   b. To two more tubes, add 1 ml. of the water sample.
   c. To the remaining five fermentation tubes, add 10 ml. portions of the water sample.
3. Mix the water sample with the contents of each tube by gently rolling the tube between the palms of the hands.

Note: Do not let the contents of a tube come in contact with the cotton plug.

4. Place tubes in incubator at 37° C. and examine at end of 24 and 48 hours. For each period, note the percentage of gas formed in each tube as indicated by the portion of the collecting tube that is filled with gas.
5. Record your findings below:

Tube no.	*Amount of water sample*	*Percentage gas formed in fermentation tube in*		*Presumptive test*	
		24 hrs.	*48 hrs.*	*Positive*	*Negative*
1	0.1 ml.				
2	0.1 ml.				
3	1.0 ml.				
4	1.0 ml.				
5	10.0 ml.				
6	10.0 ml.				
7	10.0 ml.				
8	10.0 ml.				
9	10.0 ml.				

Part C—Demonstration of a positive confirmed test by the instructor with discussion

1. Selection of one of the fermentation tubes from the presumptive test in which gas formation is nicely shown
2. Demonstration of plates (Endo's medium or eosin–methylene blue agar) made from the tube selected
3. Observation of characteristics that differentiate coliform colonies from those of other organisms on the medium

PROJECT

Bacteriology of milk

Part A — Bacterial plate count of milk

1. Use the same method here that was used in obtaining the bacterial plate count of water (p. 634). Use milk samples instead of water. Adapt the technic so that the dilution bottles contain 99 ml. instead of 9 ml. of sterile water.
2. Plate the milk sample dilutions and incubate at 37° C. for 48 hours.
3. Count the colonies and calculate the number of bacteria per milliliter of the original milk sample. Are the colonies on the plates all alike, or are there several varieties?
4. Chart results.

Plate no.	*Dilution*	*No. of bacterial colonies present*	*Estimated no. of bacteria per ml. of milk*
1			
2			
3			

Part B — Effect of pasteurization on milk

1. Obtain from the instructor the following:
   a. Two tubes of raw milk
   b. Two 99 ml. dilution bottles
   c. Two tubes of nutrient agar
   d. Pipettes
2. Pasteurize one raw milk sample as follows:
   a. Place one tube of milk in a water bath and raise the temperature of the water to 65° C.
   b. Hold at this temperature for 30 minutes.
3. Prepare dilutions of both milk samples.
   a. Make a 1:100 dilution of this milk and the unheated milk.
   b. With separate sterile pipettes, place 1 ml. of each dilution in a Petri dish and mix with melted agar as outlined in steps (b) and (c) in the experiment on counting bacteria in water (p. 634).
   c. Observe. Which plate contains more bacteria? Why?
   d. Chart results.

Sample	*Bacterial plate count*	*No. of bacteria per ml. in milk sample*
Pasteurized milk (heated to 65° C. for 30 min.)		
Raw milk (not heated)		

Part C—Phosphatase test*

Note: This test depends on the fact that raw milk contains an enzyme known as phosphatase, which is more resistant to pasteurization than is any known pathogenic bacterium. Its inactivation therefore is a reliable index that the milk has been properly pasteurized and is safe.

1. Test the two samples of milk with which you have previously worked—one a sample of raw milk and one a sample of raw milk that you have pasteurized. Proceed as follows for a given sample.
   a. Pipette 5 ml. of buffered substrate into a test tube.
   b. Add 0.5 ml. of milk sample and shake. Label tube.
   c. Place the test tube in a 37° C. incubator or water bath for 30 minutes. Remove.
   d. Add 6 drops of CQC reagent and 2 drops of catalyst to the tube. Stopper test tube, mix contents, and incubate for 5 minutes.

   Note: If the milk sample has not been properly pasteurized, a blue color will appear after about 5 minutes. If the milk has been properly pasteurized, the color will be gray or light brown. No attempt is made in this exercise to extract the pigment and quantitate it.
2. Compare the results obtained for the two samples of milk—one pasteurized and one not.

Part D—Methylene blue reduction test

1. Work in pairs.
2. Obtain two milk samples from the instructor—one of low bacterial count and one that is heavily contaminated.
3. On each sample, carry out the test as follows:
   a. Place 10 ml. of each sample of milk in a test tube. Label tube.
   b. Add 1 ml. of methylene blue thiocyanate solution† to each test tube.
   c. Place rubber stoppers in the two tubes and mix thoroughly by inverting.
   d. Place in water bath at 37° C.
   e. Observe tubes occasionally to determine decolorization.

   Note: The more bacteria present, the more rapid the methylene blue is decolorized. On this basis the test milk may be classified as follows:

   Class 1. Excellent quality milk, not decolorized in 8 hours.
   Class 2. Milk of good quality, decolorized in less than 8 hours, but not less than 6 hours.
   Class 3. Milk of fair quality, decolorized in less than 6 hours, but not less than 2 hours.
   Class 4. Poor quality milk, decolorized in less than 2 hours.
4. Classify the two milk samples that you and your partner have tested. Record findings.

---

*The method of preparing the reagents needed in the test may be found in *Standard Methods for the Examination of Dairy Products*, published by the American Public Health Association.

†Prepared from methylene blue thiocyanate tablets (about 8.8 gm.) certified by the Commission on Standardization of Biological Stains. Tablets may be obtained from laboratory supply houses. Dissolve one tablet in 200 ml. hot sterile distilled water and allow solution to stand overnight. Store in amber glass bottles away from light.

REFERENCES

Standard methods for the examination of dairy products, microbiological and chemical, Washington, D. C., 1967, American Public Health Association, Inc.

Standard methods for the examination of water and wastewater, Washington, D. C., 1971, American Public Health Association, Inc.

EVALUATION FOR UNIT SIX

Part I

Check the number from the column on the right that indicates the correct answer to the question or that correctly completes the statement.

1. Why does the bacteriologist measure the safety of a water supply by testing for the comparatively nonpathogenic colon bacillus?

(a) Although the colon bacillus is normal in the intestines, if ingested, it usually causes disease.	1. a
(b) The colon bacillus indicates that the water supply is probably contaminated with fecal material.	2. all of these
(c) Pathogenic organisms from infected fecal material are hard to isolate in water.	3. b and c
(d) Although fecal material from some sources may be safe, that from other sources may carry disease germs.	4. c, d, and e
(e) Presence of the colon bacillus means that the typhoid bacillus is also present.	5. b, c, and d

2. Measles vaccines are licensed by:

(a) American Medical Association	1. a
(b) National Institutes of Health	2. b
(c) Children's Bureau	3. c
(d) Federal Food and Drug Administration	4. d
(e) American Public Health Association	5. e

3. Authorities recommend that all children be actively immunized against the following diseases:

(a) Pertussis	1. a and c
(b) Diphtheria	2. b and d
(c) Smallpox	3. c, d, and e
(d) Rabies	4. a, b, and c
(e) Scarlet fever	5. all of them

4. In which of the following diseases can toxoid be used for active immunization?

(a) Smallpox	1. c and d
(b) Typhoid fever	2. a, c, and d
(c) Diphtheria	3. c and e
(d) Tetanus	4. a and e
(e) Whooping cough	5. all of these

5. In which of the following is active immunization of doubtful, uncertain, or not universally accepted value?

(a) Smallpox	1. e
(b) Tuberculosis	2. d
(c) Influenza	3. c
(d) Typhoid fever	4. b
(e) Diphtheria	5. a

6. A woman, 8 weeks pregnant, is exposed for the first time to rubella. What is recommended of the following?

(a) Nothing	1. a
(b) Administration of live attenuated rubella vaccine	2. b
(c) Administration of killed vaccine	3. c
(d) Administration of gamma globulin from pooled normal adult blood	4. d
(e) A course of antibiotic therapy	5. e

7. Which of the following tests for the pasteurization of milk is most practicable in the detection of raw milk mixed with pasteurized milk or in indicating milk incompletely heated in the pasteurization process?

(a) Acidity test	1. a and b
(b) Phosphatase test	2. a, b, and c
(c) Catalase test	3. b and c
(d) Test for coliforms	4. c and e
(e) Methylene blue test	5. b

8. An unprotected family of five—two parents and children aged 2, 4, and 6 years—is exposed to a case of poliomyelitis. (Members of the family have not been immunized against the disease.) Subsequently, the six year old becomes ill, and the clinical diagnosis of paralytic poliomyelitis is made. Which of the following is effective in preventing infection in the other members of the family?

(a) Administration of gamma globulin from pooled normal adult blood	1. a
(b) Administration of Sabin vaccine	2. b
(c) Administration of Salk vaccine	3. c
(d) A course of antimicrobial therapy	4. d and e
(e) Administration of an antiviral compound	5. none of these

9. A smallpox vaccination certificate is valid for:

(a) Six months	1. a
(b) One year	2. b
(c) Three years	3. c
(d) Five years	4. d
(e) Ten years	5. e

10. The presence of 1 part per million of fluoride in drinking water may lead to:

(a) Mottled tooth enamel	1. a and b
(b) Decreased number of dental caries	2. b
(c) Impaired renal function	3. a and e
(d) Softening of the bones	4. b and c
(e) Fluorosis	5. d and e

Part II

Most of the people who ate the potato salad at the picnic became ill with nausea and vomiting in a period of time 2 to 6 hours after the meal was served. Of those who did *not* eat the potato salad, none became ill.

1. The most likely organism causing the illness would be: (Circle one)
   (a) Group A streptococcus
   (b) Enterococcus
   (c) Coagulase-positive staphylococcus
   (d) Coagulase-negative staphylococcus
   (e) none of above
2. The symptoms were probably caused by:
   (a) An infection of the alimentary canal
   (b) A toxin produced by the organisms growing in the food before it was eaten
   (c) A toxin produced by organisms growing in the intestinal tract
   (d) A substance resulting from bacterial decomposition of the food
   (e) None of these

3. The organism associated with the illness would most likely be identified by
   (a) Culture of the potato salad
   (b) Culture of the vomitus from a patient
   (c) Culture of stool specimens
   (d) Microscopic examination of the remaining salad
   (e) None of these
4. Epidemiologic investigation of the episode of food poisoning revealed that the food handler who prepared the salad had an infection of the hand. An organism was cultured from the sample of the salad dressing used in preparation of the potato salad. To determine the source of the food contamination, one should compare this organism with any organisms recovered from cultures taken. If organisms are found to be of the same species, they are compared by means of:
   (a) A precipitin test to determine serologic type
   (b) The fluorescent antibody technic
   (c) A test to determine bacteriophage type
   (d) A slide agglutination test
   (e) A complement fixation test

Part III

1. Using the letter in front of the appropriate term from column B, indicate the nature of the immunizing substance in column A.

COLUMN A	COLUMN B
________ Antirabies hyperimmune serum	(a) Dead bacteria
________ BCG Vaccine	(b) Attenuated bacteria
________ Botulinum toxoid	(c) Inactivated virus
________ Cholera vaccine	(d) Live attenuated virus
________ Diphtheria toxoid	(e) Modified exotoxin
________ Gamma globulin	(f) Antibodies (immunoglobulins)
________ Immune serum globulin (human)	
________ Influenza vaccine	
________ Measles immune globulin	
________ Measles vaccine (Edmonston B strain)	
________ Mumps immune globulin	
________ Mumps vaccine	
________ Pertussis vaccine	
________ Poliomyelitis immune globulin	
________ Plague vaccine	
________ Rabies duck embryo vaccine	
________ Rabies chick embryo vaccine	
________ Rubella vaccine	
________ Smallpox vaccine	
________ Salk poliomyelitis vaccine	
________ Sabin poliomyelitis vaccine	
________ Semple vaccine	
________ Tetanus antitoxin	
________ Tetanus immune globulin	
________ Tetanus toxoid	
________ Vaccinia immune globulin	
________ Yellow fever vaccine	

2. The following is a list of diseases that are spread by ingestion of either contaminated milk or water. If the disease is spread by contaminated water, place a *W* in the blank space in front of the disease; if it is spread by contaminated milk, then place an *M* in the blank space. Some of

these diseases may be spread by both milk and water, in which case both an *M* and a *W* are to be placed in the blank space. If the disease is not spread by either medium, the space is to be left blank.

________ Typhoid fever
________ Bacillary dysentery
________ Septic sore throat
________ Syphilis
________ Bovine tuberculosis
________ Actinomycosis
________ Scarlet fever
________ Infantile diarrhea
________ Rabies
________ Amebic dysentery
________ Q fever
________ Relapsing fever
________ Infectious jaundice
________ Salmonelloses
________ Infectious hepatitis
________ Cholera
________ Rocky Mountain spotted fever
________ Brucellosis
________ Diphtheria
________ Tetanus

3. True-False. Circle either the T or the F.

T F 1. Bacteria are able to obtain carbon and nitrogen from the air because they contain chlorophyl.
T F 2. Pathogenic bacteria that grow in the soil never form spores.
T F 3. Bacteria are seldom found below 6 feet in the soil.
T F 4. Diseases may be transmitted by means of an intermediate host.
T F 5. Drinking water contains a large number of bacteria always.
T F 6. A silo depends upon bacteria for its action.
T F 7. The causative organism of a disease must be isolated before a vaccine may be made.
T F 8. In food poisoning from *Salmonella,* digestive disturbances are prominent.
T F 9. *Salmonella* food poisoning has been chiefly associated with canned vegetables.
T F 10. Toxin preformed in food may cause food poisoning.
T F 11. Toxoids are not immunogenic because the part of the molecule responsible for toxicity has been masked or destroyed.
T F 12. A person with a negative Schick test is always completely immune to diphtheria.
T F 13. Material for active immunization against diphtheria is best produced by boiling the diphtheria toxin.
T F 14. Ultraviolet radiation is bactericidal because the heat that results from the ultra-violet radiation coagulates the bacterial proteins.
T F 15. Bacterial counts of pasteurized milk cannot be expected to yield any information useful in preventing disease because pasteurization kills all pathogenic organisms except spore forms.
T F 16. Most cases of pulmonary tuberculosis are acquired by the drinking of infected milk.
T F 17. Food poisoning is often associated with picnics because the food served has not been properly refrigerated for several hours.
T F 18. Food is an important vehicle in the transmission of syphilis.
T F 19. Most microbes in nature are not harmful to man.
T F 20. Ammonia is converted by nitrifying bacteria into nitrites.
T F 21. The term "nitrogen fixation" refers to the recovery of free nitrogen from the air by unicellular animals.
T F 22. Lactic acid bacteria are normal milk inhabitants.
T F 23. The chief milk curdling enzyme is diastase.
T F 24. *Saccharomyces* species are important in commercial fermentations.
T F 25. *Bacillus subtilis* provides alkaline proteases for the manufacture of enzyme detergents.

GLOSSARY

abscess circumscribed collection of pus.

acid-fast bacteria bacteria that do not lose their stain when treated with an acid after being stained with an aniline dye.

acquired immunity immunity that a person acquires after birth.

active carrier person or animal who becomes a carrier after recovery from a disease.

active immunity immunity brought about by the activity of certain body cells of the person becoming immune on direct exposure to antigen.

acuminate pointed, tapering.

acute short in time.

acute disease disease that runs a rapid course with more or less severe manifestations.

adjuvant substance mixed with antigen to enhance antigenicity and antibody response.

aerate to charge with air or other gas.

aerobe organism whose growth requires the presence of oxygen.

aerosols finely divided antiseptic solutions sprayed into the air.

afferent bringing to or into.

affinity attraction.

agammaglobulinemia deficiency or absence of gamma globulin in the blood usually associated with increased susceptibility to infection.

agar gelatinous substance prepared from Japanese seaweed used as a base for solid culture media.

agglutination visible clumping of cells suspended in a fluid.

agglutinins antibodies that cause agglutination.

agglutinogen any substance that, acting as an antigen, stimulates the production of agglutinins.

algid cold; an algid fever is one in which the patient goes into a state of collapse.

alimentary tract digestive tract.

alkaloid basic substance found in plants that is usually the part of the plant that has medicinal properties.

allergen substance capable of bringing about an allergic state when introduced into the body of a susceptible person.

allergy (hypersensitivity) state in which the affected person exhibits unusual manifestations upon contacting an allergen.

alum toxoid toxoid treated with an aluminum compound.

amboceptor substance that combines with cells and complement to dissolve cells (example, hemolysin).

ameba, amoeba (pl., **amebas, amoebae)** protozoon that moves by extruding fingerlike processes (pseudopods).

amebiasis infection with pathogenic amebas; acute amebiasis is known as amebic dysentery.

ameboid resembling an ameba.

amino acids organic chemical compounds containing an amino (NH_2) group and a carboxyl (COOH) group that form the chief structure of proteins.

anaerobe organism that grows only or best in the absence of atmospheric oxygen.

anaphylactoid resembling anaphylaxis.

anaphylaxis state of hypersusceptibility to a protein resulting from a previous introduction of the protein into the body.

angstrom (Å) one tenth of a millimicron or one 254 millionth of an inch.

antagonism mutual resistance.

antibacterial serum antiserum that destroys or prevents the growth of bacteria.

antibiotic agent produced by one organism that will destroy or inhibit another organism.

antibody agent in the body that destroys or inactivates certain foreign substances that gain access to the body, particularly microbes and their products.

anticoagulant agent that prevents coagulation.

antigen substance that, when introduced into the body, causes the production of antibodies.

antiluetic antisyphilitic.

antiseptic substance that prevents the growth of bacteria.

antiserum immune serum.

antitoxin immune serum that neutralizes the action of a toxin.

antivenin antitoxic serum for snake venom.

aphthous characterized by the presence of small ulcers.

arboviruses viruses borne by and isolated from arthropods.

ascites abnormal collection of fluid in the peritoneal cavity.

aseptic free from living microorganisms.

aspirate to draw by suction as when fluid is removed with a syringe or material is drawn into the lungs during inspiration.
ataxia failure of muscular coordination.
atopy human allergy with hereditary background.
attenuated weakened.
autoantibody antibody formed against autoantigens.
autoantigen antigens present in the same individual as the antibody-producing cells and not "foreign" to the body.
autoclave apparatus for sterilizing by steam under pressure; pressure steam sterilizer.
autogenous vaccine vaccine made from a culture of bacteria obtained from the patient himself.
autoimmune disease disease wherein autoimmunization against certain body proteins is the pathogenetic mechanism.
autoimmunity unusual state resulting from the production *(autoimmunization)* by the body of antibodies against its own proteins.
autoinfection infection of one part of the body by bacteria derived from some other part of the body.
autopsy examination of the internal organs of the dead body.
autotrophic organisms capable of forming their proteins and carbohydrates out of inorganic salts and carbon dioxide.

bacillus (pl., **bacilli)** rod-shaped bacterium.
bacteremia condition in which bacteria are in the bloodstream but do not multiply there.
bacteria (sing., **bacterium)** unicellular organism of the plant kingdom; germs.
bactericide (bactericidal) agent lethal to bacteria.
bacteriology science that treats of bacteria.
bacteriolysins antibodies that cause the solution of bacteria in the presence of complement.
bacteriophages (phages) bacterial viruses.
bacteriostasis inhibition of bacterial growth; bacteria, however, not directly killed.
Bang's disease contagious abortion of cattle.
BCG (bacillus of Calmette-Guérin) vaccine against tuberculosis made from a bovine strain of tubercle bacilli attenuated through long culturing; name derived from the two French scientists developing the strain.
benign mild in character; not malignant.
biologic transfer of infection mode of transfer of infection from host to host by an animal or insect in which the agent causing the disease undergoes a cycle of development.
biology science that treats of living things, both animals and plants.
biotherapy treatment of disease with a living agent or its products.
blastomycosis infection with blastomyces.
blood serum (see **serum).**
boil abscess of the skin and subcutaneous tissue.
botany science that treats of plant life.
broad-spectrum term used to indicate that an antibiotic is effective against a large array of micoorganisms.
bronchiolitis inflammation of the bronchioles.
bronchopneumonia small focal areas of inflammatory consolidation in the lungs.
bubo inflammatory enlargement of a lymph node often with the formation of pus.
buffer substance or system of substances in body fluids that lessens the effect of addition of acids or alkalies.

capnophilic growing best in the presence of carbon dioxide.
capsule envelope that surrounds certain bacteria.
carbohydrates class of organic chemical compounds composed of carbon, hydrogen, and oxygen, the latter two in the proportion to form water; to this class belong the sugars, starches, and cellulose.
cardiolipin lipid extract of fresh beef heart used as the antigen in serologic tests for syphilis.
carditis inflammation of the heart.
caries decay of bone; caries in the teeth (dental caries) are layman's cavities.
carrier person in apparent health who harbors a pathogenic agent in his body.
caseous cheeselike.
catabolism breaking down of complex bodies into products of simpler composition.
catalyst substance that speeds up a chemical reaction but is not decreased in amount by the reaction.
catarrhal characterized by an outpouring of mucus.
cell minute protoplasmic structure, the anatomic and physiologic unit of all animals and plants; that is, all animals and plants are made up of one or more cells, and their activities depend on the combined activities of those cells.
cellulitis diffuse inflammation of connective tissues.
Celsius inventor of the temperature scale setting 100 degrees between the freezing point at zero and the boiling point of water.
centigrade thermometer temperature scale with 100 degrees between the melting point of ice at zero and the boiling point of water at 100.
centrifugalize, centrifuge to subject to centrifugal force.
centrifuge machine by which centrifugation is effected.
centrioles two or more granules contained in the centrosome and prominent during cell division.

centrosome (cell center) condensed portion of the cytoplasm of certain cells containing the centrioles; it plays an important part in cell division.

chain reaction series of successive reactions in which each succeeding reaction depends on the preceding one and which, when once begun, continues until one or more of the chemicals taking part in the reaction are exhausted.

chancre initial lesion of syphilis.

chemotaxis reaction to a chemical whereby cells are attracted (positive chemotaxis) or repelled (negative chemotaxis) by the chemical.

chemotherapeutic agent agent used in chemotherapy.

chemotherapy treatment of disease by the administration of drugs that destroy the causative organism of the disease but do not injure the patient.

chlorophyl coloring matter of green plants.

cholesterol fatlike alcohol ($C_{27}H_{45}OH$) occurring as crystals with notched corners; found in blood, bile, egg yolk, seeds of plants and animal fats.

chromatin stainable part of the nucleus of a cell forming a network of fibrils; it is deoxyribonucleic acid attached to a protein base and is the carrier of the genes in inheritance.

chromatin granules karyosomes.

chromidia chromatin material scattered throughout the protoplasm of cells, with organized nuclei.

chromogenic pigment producing.

chromosomes rod-shaped masses of chromatin that appear in the cell nucleus during mitosis; they play an important part in cell division and transmit the hereditary characteristics of the cell.

chronic long, continued.

cilia (sing., cilium) hairlike processes that spring from certain cells and by their action create currents in lipids; if the cells are fixed, the lipid is made to flow, but if the cells are those of unicellular organisms suspended in the liquid, the cells move.

ciliates unicellular organisms that move by means of cilia.

clinical founded on actual observation.

clinical case person ill and showing signs of a disease.

clone progeny of a single cell.

coagulase enzyme that hastens coagulation.

coagulation formation of a blood clot.

coccus (pl., cocci) a sphere-shaped bacterium.

coliform bacteria group of bacteria consisting of *Escherichia coli* and related intestinal inhabitants.

colon bacillus *Escherichia coli.*

colony visible growth of bacteria on a culture medium; all the progeny of a single preexisting bacterium.

commensualism symbiosis in which one individual is benefited and the other is not affected.

communicable capable of being transmitted from one person to another.

complement lytic substance found in normal blood capable of destroying bacteria or other cells when combined with the antigen-antibody complex.

complement fixation destruction or inactivation of complement brought about by the combination of antigen, antibody, and complement; this is the basis of the complement fixation tests for syphilis and certain other diseases.

complication disease state concurrent with another disease.

condyloma wartlike growth.

congenital existing at the time of birth or shortly thereafter.

consolidation solidification, as of the lung in pneumonia.

constitutional relating to the makeup of the body as a whole.

consumption wasting away; a lay term meaning pulmonary tuberculosis.

contagious highly communicable; in common parlance, a disease that is easily "caught."

contamination soiling with infectious material.

convalescent carrier carrier who harbors the organisms of a disease during recovery from the disease.

convalescent serum serum of a person recently recovered from a disease; in a few cases the injection of convalescent serum seems to be of value in treatment or in prevention of the disease in others.

Coombs' test test to detect the presence of globulin antibodies on the surface of red blood cells; used to detect sensitized red cells in hemolytic diseases.

coproantibodies antibodies formed in the colon.

coprophilic affinity for feces and filth (bacteria).

counterstain second stain of a different color applied to a smear to make effects of second stain more distinct.

Credé's method instillation of a 2% silver nitrate solution into each eye of the newborn infant to prevent ophthalmia neonatorum.

crisis sudden change in the course of a disease; diseases that terminate by a sudden change for the better are said to end by crisis.

culture growth of microorganisms on a nutrient

medium; to grow microorganisms on such a medium.

culture media (sing., **medium**) artificial food material upon which microbes are grown (cultured).

cutaneous pertaining to the skin; cutaneous inoculation is done by rubbing the infectious material on the abraded skin.

cycle series of changes considered to lead back to the starting point.

cyst abnormal saccular structure containing liquid, air, or solid material; a stage in the history of certain protozoa during which time the encysted organism is protected by a surrounding wall.

cysticercus larval form of the tapeworm in the tissues of the intermediate host; it develops upon entry into the intestinal canal of the definitive host into an adult worm.

cytology science treating of the study of cells, their origin, structure, and function.

cytolysin antibody that can dissolve cells.

cytopathogenic (cytopathic) effect pathologic changes in cells of a given tissue culture referable to the action of some injurious agent, especially those from viruses grown in tissue culture.

cytoplasm protoplasm of a cell other than that of the nucleus.

dactylitis inflammation of a finger or toe.

Dakin's solution neutral solution of sodium hypochlorite once used in the disinfection of wounds.

defibrinate to remove the fibrin of blood to prevent clotting.

definitive final, ending.

degenerate to undergo progressive deterioration.

degeneration deterioration; change from a higher to a lower form.

degerm to remove bacteria from skin by mechanical cleansing or application of antiseptics.

deodorant substance that destroys unpleasant odors.

deoxyribonucleic acid (DNA) one of the two nucleic acids that have been identified; essential for biologic inheritance.

dermatitis inflammation of the skin.

dermatomycoses superficial fungous infections of the skin and appendages.

dermatophyte fungus parasitic for the skin.

dermotropic affinity for the skin.

desensitization condition wherein an organism does not react to a specific antigen to which it previously did; process of bringing about this state.

desquamation shedding of the superficial layer of the skin in scales or shreds.

detergent cleansing agent

DEV Duck embryo vaccine for rabies.

diarrhea abnormal frequency or fluidity of bowel movements.

diatomaceous earth earth made up of petrified bodies of diatoms or unicellular algae.

Dick test skin test to determine susceptibility to scarlet fever.

differential stain stain distinguishing between different groups of organisms.

diplococci cocci occurring in pairs.

direct contact spread of a disease more or less directly from person to person.

disinfectant substance that disinfects.

disinfection destruction of all disease-producing organisms and their products.

dissociation separation; dissolution of relations.

DNA (see **deoxyribonucleic acid**).

DTP diphtheria-tetanus-pertussis vaccine.

droplet infection infection conveyed by the spray thrown off from the mouth and nose during talking, coughing, etc.

drug-fast resistance to the action of drugs; microbes able to withstand action of a given drug.

dysentery diarrhea plus blood and mucus in the stool; associated with inflammation of alimentary tract.

ecology science of organisms as affected by factors of their environment.

ectoenzyme enzyme excreted through the plasma membrane of the cell forming it into the surrounding medium.

ectoplasm outer clear zone of the cytoplasm of a unicellular organism.

efferent bearing away.

electrophoresis application of an electric current to separate substances that move faster in an electrical field from those that move more slowly.

elementary bodies virus particles.

encephalitis inflammation of the brain.

encephalomyelitis inflammation of the brain and spinal cord.

endemic more or less continuously present in a community.

endocarditis inflammation of the endocardium or lining membrane of the heart including that of the heart valves.

endoenzyme enzyme liberated only when the cell that produces it disintegrates.

endogenous originating within the organism.

endoplasm zone of granular cytoplasm found near the nucleus of many unicellular organisms.

endoplasmic reticulum membrane-limited, canalicular organelle in the cytoplasm of cells.

endothelium flattened cell lining of heart, blood vessels, and lymph channels.

endotoxin toxin liberated only when the cell producing it distintegrates.
enteric bacteria bacteria isolated from the gastrointestinal tract.
enterotoxin toxin bringing about diarrhea and vomiting usually succeeding ingestion of contaminated food.
enteroviruses viruses isolated from gastrointestinal tract.
enzyme catalytic substance secreted by a living cell capable of changing other substances without undergoing any change itself.
eosinophil granular leukocyte whose granules stain with eosin.
epidemic disease that attacks a large number of persons in a community at the same time.
epidemiology science that treats of epidemics.
epithelium cells that cover body surfaces and mucous membranes and line glands.
essential (disease) of unknown cause; idiopathic.
etiology cause.
exacerbation increase in the severity of a disease.
exanthem febrile disease accompanied by a skin eruption.
exanthematous relating to exanthema or generalized disease accompanied by skin eruption.
excrete to cast out waste products.
exfoliate to come off in strips or sheets, particularly the stripping of the skin after certain exanthematous diseases.
exfoliative cytology study of cells cast off from a body surface.
exogenous coming from the outside of the body.
exotoxin toxin secreted by a microorganism into the surrounding medium.
extracellular outside of cells.
extraneous outside the organism and not belonging to it.
extrinsic from without.
exudate fluid and formed elements of the blood extravasated into the tissues or cavities of the body as part of the inflammatory reaction.

facultative having the power to do a thing although not ordinarily doing it.
familial affecting several members of the same family.
feedback return of part of output of a system as input.
ferment enzyme.
fermentation breaking down of complex organic compounds, particularly carbohydrates by enzymes.
fever abnormally high body temperature usually related to a disease process.
fibrinolysin substance that dissolves or destroys fibrin.
filaria long, threadlike round worm that lives in the circulatory or lymphatic system.
filariform resembling filariae.
filtrate liquid passed through a filter.
filtration process of passing a liquid through a filter.
fix to make firmly attached or set.
fixed virus (rabies) street or wild virus adapted to rabbit and less virulent for man and the dog.
flagella (sing., **flagellum**) long, hairlike processes that by their lashing activity cause the organism to move; one or more flagella may be attached to one or both ends of the organism or completely around it.
flagellates organisms that move by means of flagella.
flocculation test test dependent upon the coalescence of finely divided or colloidal particles into larger visible particles.
fluid toxoid toxoid not further treated with aluminum compounds.
fluorescence property of emitting light after exposure to light.
fluorescent antibodies antibodies that fluoresce.
focal infection localized site of more or less chronic infection from which bacteria or their products are spread to other parts of the body.
fomites substances other than food that may transmit infectious organisms.
fractional sterilization heating of a material to be sterilized at a low temperature for a given time, on three or four successive days, with storage in the interval under conditions suitable for bacterial growth; the heating kills vegetative bacteria and any spores developing into vegetative bacteria during the incubation periods.
fulminating sudden, severe, and overwhelming.
fumigation exposure to the fumes of a gas that destroys bacteria, vermin, etc.
Fungi Imperfecti fungi lacking sexual spores.
fungicide agent destructive to fungi.
fungus (pl., **fungi**) unicellular and multicellular vegetable organisms that feed on organic matter; molds, mushrooms, and toadstools.
furuncle a boil.
furunculosis presence of a number of boils.

gamete cell, either male or female, that undergoes sexual reproduction.
gamma globulin one of the fractions of the globulin of blood and the one with which antibodies are associated.
gene biologic unit of heredity, self-reproducing and located in a definite position (locus) on a particular chromosome.

generalized infection infection involving the whole body.
germ microbe.
germ cell cell specialized for reproduction.
germicide agent that destroys germs.
Giemsa stain stain (azure and eosin dyes) used to demonstrate protozoa, viral inclusion bodies, and rickettsiae.
globulin class of proteins characterized by being insoluble in water but soluble in weak solutions of various salts; one of the important proteins of the plasma of the blood.
glucose dextrose, grape sugar.
gnotobiosis science of rearing and keeping animals either born germ free or with a limited known microbial flora.
Golgi apparatus organelle of cell cytoplasm made up of an irregular network of canals or solid strands.
gram-negative bacteria decolorized by Gram's method but stain with the counterstain (red or brown).
gram-positive bacteria not decolorized by Gram's method; they retain the original violet color of the Gram stain and are not stained by the counterstain.
Gram stain, Gram's method method of differential staining devised by Hans Christian Gram, a Danish bacteriologist.
granulocyte cell containing granules within its cytoplasm; granular leukocyte (polymorphonuclear neutrophil, eosinophil, or basophil).
granuloma circumscribed collection of reticuloendothelial cells surrounding a point of irritation.
granulomatous inflammation characterized by pronounced response of the reticuloendothelial system.
grouping classification (see also **typing**).
gumma granuloma of late stages of syphilis.

habitat place in which a plant or animal is found in nature.
halophile salt loving
hectic fever daily recurring fever characterized by chills, sweating, and flushed countenance.
helix a spiral.
hematogenous originating in the blood or borne by the blood.
hemoglobin oxygen-carrying red pigment of the red blood cell.
hemoglobinophilic pertaining to organisms that grow especially well in culture media containing hemoglobin.
hemolysin antibody that in the presence of complement dissolves red blood cells.
hemolysis lysis of red blood cells.
hemolytic having the power to cause hemolysis.
hemophilic blood loving; hemoglobinophilic.
hepatitis inflammation of the liver.
hepatogenous originating in the liver.
hereditary transmitted through the members of a family from generation to generation.
herd immunity resistance to disease related to immunity of high proportion of members of a group.
heterologous derived from an animal of another species.
heterophil having affinity for antigens or antibodies other than the one for which it is specific.
heterotrophic organisms requiring a simple form of carbon for metabolism.
Hinton test precipitation test for syphilis.
homeostasis tendency to stability within the internal environment or fluid matrix of an organism.
homologous serum jaundice (see serum hepatitis).
host animal or plant upon which a parasite lives.
hydrolysis decomposition from the incorporation and splitting of water.
hyperimmune quality of possessing a degree of immunity greater than that found under similar circumstances.
hyperplasia increase in size related to increase in number of component units.
hyperpyrexia very high fever, usually more than 106°F.
hypersensitivity (see allergy).
hypertonic solution having a higher osmotic pressure than that of a reference solution.
hypha (pl., **hyphae)** one of the filaments composing a fungus.
hypotonic solution having a lower osmotic pressure than that of a reference solution.

icterus jaundice.
idiopathic of unknown cause.
idiosyncrasy individual and peculiar susceptibility or sensitivity to a drug, protein, food, or other agent.
immune exempt from a given infection.
immune bodies antibodies.
immune globulin sterile preparation of globulin from blood; it is with the gamma globulin fraction of the plasma proteins that antibodies are normally associated.
immune serum serum containing immune bodies.
immunity natural or acquired resistance to a disease.
immunohematology branch of hematology treating of immune bodies in the blood.
immunologist one versed in immunity.
immunology science that deals with immunity.

impetigo contagiosa infectious vesicular and pustular eruption most often seen on the faces and other body areas of children.

inclusion bodies round, oval, or irregularly shaped particles in cytoplasm or nucleus of cells parasitized by viruses; colonies of viruses.

incompatible not capable of being mixed without undergoing destructive chemical changes or acting antagonistically.

incubate to promote the growth of microorganisms by placing them in an incubator.

incubation period period intervening between the time of infection and the appearance of the manifestations of a disease.

incubator cabinet in which a constant temperature is maintained for the purpose of growing cultures of bacteria.

indicator something that renders visible the completion of a reaction.

indirect contact transfer of infection by means of inanimate objects, contaminated fingers, water, food, and the like.

infection invasion of the body by pathogenic agents with their subsequent multiplication and the production of disease; inflammation from living agents.

infectious having qualities that may transmit disease.

infectious granuloma granuloma from a specific microbe.

infestation invasion of the body by macroscopic parasites such as insects; refers particularly to parasites on the surface of the body.

inflammation protective reaction on the part of the tissue of the body brought about by the presence of an irritant—reaction to injury.

inhibition diminution or arrest of function.

inoculate to implant microbes or infectious material onto culture media; to introduce artifically a biologic product or disease-producing agent into the body.

inoculation process of inoculating.

insecticide substance that destroys insects.

intercellular between cells.

intercurrent infection infection that attacks a person already ill of another disease.

interferon protein produced by cells infected with certain viruses; viral reproduction and spread is inhibited thereby.

intermittent characterized by periods of activity separated by periods of quietude.

intermittent sterilization (see **fractional sterilization**).

intoxication poisoning.

intracellular within a cell.

intracutaneous (see **intradermal**).

intradermal within the substance of the skin.

intraperitoneal within the peritoneal cavity.

intraspinal within the vertebral canal.

intravenous within a vein.

intrinsic from within.

involution forms abnormal forms assumed by microorganisms growing under unfavorable conditions.

iodophors disinfectants, the germ-killing iodine of which is carried by a surface-active solvent.

isoantigens antigens from an individual of the same species.

isolate to close all avenues by which a person may spread infection to others; to separate from others.

isologous pertaining to the same species.

isotonic solution having the same osmotic pressure as that of the standard reference solution.

jaundice deposition of bilirubin in skin and tissues with resultant yellowish discoloration in a patient with hyperbilirubinemia.

Kahn test precipitation test for syphilis.

karyosome chromatin mass or knot on the network of fibrils in the cell nucleus.

karyotype chromosomal makeup of a cell, individual, or species.

keratitis inflammation of cornea.

keratoconjunctivitis inflammation of the cornea and conjunctiva.

kernicterus cerebral manifestations of severe jaundice in the newborn infant; degeneration of nerve cells caused by accumulation of bilirubin in brain.

Kline test precipitation test for syphilis.

Koch's postulates certain requirements that must be met before a given microorganism can be considered the cause of a certain disease.

lactose milk sugar.

lag phase period of time elapsing between a stimulus and the resultant reaction.

larva (pl., **larvae**) young of any animal differing in form from its parent.

latent seemingly inactive, potential, concealed.

leptospirosis disease produced by members of genus *Leptospira*.

lesion specific pathologic structural or functional (or both) change brought about by disease.

leukocidin substance that destroys leukocytes.

leukocyte white blood cell.

leukocytosis transient protective increase in leukocytes in the blood in response to an injury.

leukopenia abnormal decrease in leukocytes in the blood.

lipase fat-splitting enzyme.

lobule subdivision of an organ or anatomic part bounded by some kind of structural demarcation.
local infection one that is confined to a restricted area.
locus position.
lues venerea syphilis.
lumbar pertaining to that part of the back between the ribs and pelvis.
lumen space inside a tubular or hollow structure.
lumpy jaw actinomycosis in cattle.
lymphadenitis inflammation of a lymph node.
lymphadenopathy disease of a lymph node (or nodes).
lymphocyte nongranular white blood cell of lymphoid origin.
lymphocytosis increase in lymphocytes in the blood.
lymphoid tissue connective tissue support and lymphocytes.
lymphoma neoplasm of lymphoid tissues.
lyophilization creation of a stable product by rapid freezing and drying of frozen product in a vacuum.

macrophage large mononuclear wandering phagocytic cell originating in the reticuloendothelial system.
macroscopic visible to the naked eye.
macular consisting of small flat reddish spots in the skin.
malignant virulent; going from bad to worse.
Mantoux test tuberculin skin test.
Mazzini test precipitation test for syphilis.
meatus opening.
mechanical transfer (of infection) referring to the transfer of infection by insects in which the infectious agent is transferred mechanically and undergoes no cycle of development in the body of the particular insect.
medical technologist specialist in technics of tests vitally related to medicine.
meiosis special type of cell division during the maturation of the sex cells by which the normal number of chromosomes is halved.
membranous croup lay term indicating diphtheria.
metabolism sum total of the chemical changes whereby nutrition and functional acitivies of the body are maintained.
metachromatic granules granules of deeply staining material found in certain bacteria.
metazoa multicellular animals.
microaerophilic term applied to microorganisms that require free oxygen for their growth but in an amount less than that of the oxygen of the atmosphere.
microbe microscopic unicellular organism, either animal or plant.
microbiology science that treats of microbes.
microgram one ten-thousandth of a milligram or 1/1,000,000 of a gram.
micromillimeter one thousandth part of a millimeter or 1/25,000 of an inch; a micron.
micron (see micromillimeter).
microorganism animal or plant of microscopic size.
microscopy study of objects by means of the microscope.
miliary small, resembling a millet seed in size.
minimum lethal dose (MLD) smallest dose that will cause death.
mitochondrion (pl., **mitochondria)** small spherical or rod-shaped cytoplasmic organelle(s).
mitosis indirect cell division.
mixed culture culture containing two or more kinds of organisms.
mixed infection infection with two or more kinds of organisms.
mixed vaccine vaccine containing two or more kinds of organisms.
molds multicellular fungi.
morbid pertaining to or affected with disease.
mordant chemical added to a dye to make it stain more intensely.
morphologic pertaining to shape or form.
multicellular composed of many cells.
mutation change or alteration in form or qualities; a permanent transmissible change in the characteristics of an offspring from those of its parents.
mycelium vegetative part of a fungus, consisting of many hyphae.
mycobacteriosis disease produced by certain unclassified members of the genus *Mycobacterium.*
mycology science that deals with fungi.
mycosis disease caused by fungi.

nasopharynx portion of the pharynx above the palate.
natural immunity immunity with which a person or animal is born.
necrosis death of a mass of tissue while yet a part of the living body.
negative staining staining the background, not the organism.
Negri bodies inclusion bodies found in certain cells of the brain of an animal with rabies.
neoplasm new growth of abnormal cells of autonomous nature (scientific term for layman's "cancer").
neuritis inflammation of a nerve.
neurosyphilis syphilis affecting the central nervous system.
neurotropic affinity for the central nervous system or nervous tissue.
NIH National Institutes of Health.

nitrogenous relating to or containing nitrogen.

nocturnal occurring during the night.

nodule small circumscribed mass of tissue; a small node.

nonpathogenic not productive of disease.

normal flora bacterial content of a given area remaining reasonably constant as to quantity and proportions of certain types during health.

nosocomial pertaining to the hospital or infirmary.

nosology science of classification of disease.

nucleic acids complex chemical substances closely associated with the transmission of genetic characteristics of cells; the two identified are ribonucleic acid and deoxyribonucleic acid; cells of bacteria and higher organisms contain both RNA and DNA; viruses contain one or the other but not both.

nucleolus (pl., **nucleoli**) body within the nucleus of a cell that takes part in the metabolic processes of the cell and plays a part in its multiplication.

nucleoprotein simple basic protein combined with a nucleic acid.

nucleus (pl., **nuclei**) central, compact portion of a cell that is the functional center of the cell.

old tuberculin (OT) special type of tuberculin.

oncogenic tumor producing.

opalescent resembling an opal in the display of colors.

opportunists microbes that produce infection only under especially favorable conditions.

opsonins substances in the blood that render microorganisms more susceptible to the action of phagocytes.

optimum temperature temperature at which bacteria grow best (as applied to bacterial growth).

OPV oral polio vaccine.

organ part of the body that performs a specific function or functions.

organelle specialized part of a protozoon that performs a special function; specific particles of organized living substance present in almost all cells.

organism living being, either animal or vegetable.

osmosis passage of fluids or other substances through a membrane.

osteomyelitis inflammation of the bone marrow.

otitis media inflammation of the middle ear.

pandemic very widespread epidemic, even of worldwide extent.

parasite animal or vegetable organism that lives on another organism.

parasitology science that treats of parasites and their effects on other living organisms.

parenchyma specialized functioning tissue or cells of an organ.

parenterally in some manner other than by the intestinal tract.

parietal relating to the wall of a cavity.

paroxysm sudden attack of a disease or acceleration of the manifestations of existing disease.

passive carrier carrier who harbors the causative agent of a disease without having had the disease.

passive immunity immunity produced without the body of the person or animal that becomes immune taking any part in its production.

pasteurization heating of milk for a short time to a temperature that will destroy pathogenic bacteria but will not affect the food properties and flavor.

pathogenic capable of causing disease.

pathogenicity disease-producing quality.

pathognomonic characteristic or indicative of a disease.

Paul-Bunnell test heterophil antibody test on serum; important in the diagnosis of infectious mononucleosis.

pepsin digestive enzyme, occurring in the gastric contents and acting on proteins.

peritoneum serous membrane that lines the abdominal walls and invests the abdominal viscera.

peritonitis inflammation of the peritoneum.

permanent carrier carrier who harbors a disease-producing agent for months or years.

permeability state or quality of being permeable, that is, may be passed through, as through a porous membrane.

petechia (pl., **petechiae**) pin-point hemorrhage.

Petri dish round glass dish with cover used for growing bacterial cultures.

Peyer's patches collection of lymphoid nodules packed together to form oblong elevations of the mucous membrane of the small intestine, their long axis corresponding to that of the intestine.

pH measure of alkalinity or acidity.

phage typing use of bacteriophages and their lytic properties to classify bacteria.

phagocyte cell capable of ingesting bacteria or other foreign particles.

phagocytic related to phagocytes or phagocytosis.

phagocytosis process of ingestion by phagocytes.

phenol coefficient disinfecting property of a chemical as compared with that of phenol (carbolic acid).

photosynthesis construction of glucose from carbon dioxide and water by sunlight in the presence of chlorophyl.
Pirquet's test (von Pirquet) tuberculin skin test.
plaques visible areas of cellular damage caused by a virus inoculated into susceptible tissue culture cells, analogous to colonies of bacteria on an agar plate.
plasma fluid portion of the circulating blood; the fluid portion of clotted blood is known as *serum*.
plasma membrane outer membrane encasing the protoplasm of a cell.
plasmolysis shrinking of a cell when suspended in a hypertonic solution.
plasmoptysis swelling and bursting of a cell when suspended in a hypotonic solution.
plastids small bodies found in the cytoplasm of cells; they have to do with cell nutrition and contain the chlorophyl of green plants.
pleomorphism existence of different forms in the same species.
pleurisy inflammation of the pleura.
pleuropneumonia infectious pneumonia and pleurisy of cattle.
pneumonia inflammatory consolidation or solidification of lung tissue because of the presence of an exudate blotting out the air-containing spaces.
pneumonitis inflammation of the supporting framework of the lung.
pneumotropic affinity for the lungs.
polar bodies deeply staining bodies found in one or both ends of certain species of bacteria.
pollution state of being unclean; as used in bacteriology, containing harmful substances other than bacteria.
precipitation process of clumping of proteins in solution by the addition of a specific precipitin.
precipitins antibodies that cause precipitation.
prehension act of taking hold; grasping.
preservative substance that is added to a product to prevent bacterial growth and consequent spoiling.
primary first (first focus of disease).
primary infection first of two infections, one occurring during the course of the other.
prognosis forecast of the outcome of a disease.
prophylaxis prevention of disease.
proprietary referring to the fact that a given item is a commercial one.
protein one of a group of complex organic nitrogenous compounds, widely distributed in plants and animals, forming the principal constituents of cell protoplasm; they are essentially combinations of amino acids and their derivatives.
proteolytic bringing about the digestion or liquefaction of proteins.
protoplasm living material of which cells are composed.
protozoology science that treats of protozoa.
protozoon (pl., **protozoa;** adj., **protozoan)** unicellular animal organism.
provocative dose dose stimulating the appearance of a given effect.
pseudomembrane fibrinous exudate forming a tough membranous structure on the surface of the skin or mucous membrane.
pseudopod temporary protoplasmic process put forth by a protozoon for purposes of locomotion or obtaining food.
ptomaines basic substances resembling alkaloids formed during the decomposition of dead organic matter.
pure culture culture containing only one species of organism.
purulent containing pus.
pus fluid product of inflammation consisting of leukocytes, bacteria, dead tissue cells, foreign elements, and fluid from the blood.
pustule circumscribed elevation on the skin containing pus.
putrefaction decomposition of proteins.
pyelonephritis inflammation of the kidney parenchyma and pelvis.
pyemia form of septicemia in which the organisms in the bloodsteam lodge in the organs and tissues and set up secondary abscesses.
pyogenic pus forming.

quiescent not active.
quinsy peritonsillar abscess.

racial immunity immunity peculiar to a race.
radiant energy energy from a radioactive source.
raw not pasteurized (as applied to milk).
reaction of immunity reaction indicating that a person is immune to a test substance, particularly applied to the reaction of a smallpox-immune person following smallpox vaccination.
receptors term used by Ehrlich in his side-chain theory of immunity to denote specialized portions of the cell that combine with foreign substances (foods, toxins, etc.).
recrudescent recurrence of symptoms of a disease after a period of days or weeks.
recurrent reappearance of symptoms after an intermission.
reduction removal of oxyg[illegible] from a compound or the addition of hydr[illegible] to [illegible] compound.
remission temporary cessa[illegible] [illegible]anifestations of a disease.
remittent characterized by re[illegible]

rennin milk-curdling enzyme.

replication process by which genetic determinants are duplicated during cell multiplication so that identical genetic characters are passed onto the next generation.

resident bacteria bacteria normally occuring at a given anatomical site.

resistance inherent power of the body to ward off disease.

reticuloendothelial system system of phagocytic cells scattered through various organs and tissues, particularly the spleen, liver, bone marrow, and lymph nodes, playing an important part in immunity.

Rh factor blood factor, agglutinogen, found on the red blood cells of most Caucasians; factor so-named because of its occurrence on the red blood cells of rhesus monkeys.

rhinitis inflammation of the nose.

ribonucleic acid (RNA) one of two nucleic acids (see also **DNA**); when of viral origin, RNA is infectious for susceptible cells.

ribosomes ribonucleoprotein granules in cell cytoplasm.

rickettsia minute microorganisms that must live within cells; animal reservoirs of infection are important, and organisms are transmitted to man by arthropod vectors.

ringworm fungous disease of the skin.

rodent gnawing mammal (rats, mice, etc.).

rose spots characteristic spots that occur over the lower portion of the trunk and abdomen in typhoid fever.

salmonellosis infection with an organism of the genus *Salmonella;* manifestations may be varied.

sanitary conducive to health.

sanitize to reduce the number of bacteria to a safe level as judged by public health standards.

sapremia condition in which the products of the action of saprophytic bateria on dead tissues are absorbed into the body and produce disease.

saprophyte organism that normally grows on dead matter.

Schick test skin test to detect susceptibility to diphtheria.

scrofula tuberculosis of the lymph nodes, particularly those of the neck.

secondary infection infection occurring in a host already suffering from an infection.

sedimentation settling of solid matter to the bottom of a liquid.

selective action tendency on the part of disease-producing agents to attack certain parts of the body.

Semple vaccine rabies vaccine prepared from rabbit brain treated with phenol.

sepsis poisoning by microbes or their products.

septic relating to or caused by the presence of pathogenic organisms or their poisonous products.

septicemia systemic disease caused by the invasion of the bloodstream by pathogenic organisms, with their subsequent multiplication therein.

serology branch of science that deals with serums, especially immune serums.

serous surface surface of the smooth membrane lining one of the closed cavities of the body.

serum fluid that exudes when the blood coagulates; plasma of the blood from which the plasma protein fibrinogen has been removed.

serum hepatitis type of viral jaundice succeeding administration of human serum.

signs objective disturbances produced by disease observed by the physician, nurse, or person attending the patient.

simian viruses viral contaminants in tissue cultures of normal monkey cells.

simple stain stain using only one dye.

sinus tract leading from an area of disease to the body surface.

skin test dose (STD) unit of measurement of scarlet fever toxin, the amount required to produce a positive reaction on the skin of a person susceptible to scarlet fever.

smear very thin layer of material spread on a glass slide.

species immunity immunity peculiar to a species.

spirilla (sing., **spirillum**) bacteria of a corkscrew shape.

spontaneous occurring without external stimulation.

spontaneous generation theory that held that organisms arose spontaneously.

sporadic disease disease that occurs in neither an endemic nor epidemic.

spore highly resistant form assumed by certain species of bacteria when grown under adverse influences; the reproductive cells of certain types of organisms.

sporulation production of spores or division into spores.

stain dye used in bacteriologic and histologic technics.

staining coloring cells or tissues with dyes.

staphylococci (sing., **staphylococcus**) cocci that occur in irregular grapelike clusters.

sterile free of living microorganisms and their products.

sterilization process of making sterile.

stock vaccine vaccine made from cultures other than those from the patient who is going to receive the vaccine.
strain subdivision of species.
streptococci cocci that divide in such a manner as to form chains.
streptolysin O and S hemolysins produced by streptococci.
stroma supporting framework of an organ or gland.
subacute between acute and chronic in time.
subcutaneous under the skin.
substrate substance upon which an enzyme acts.
sulfur granules small yellow granules present in the pus from the lesions of actinomycosis.
suppuration formation of pus.
suppurative forming pus.
susceptibility state or quality of being vulnerable.
sycosis barbae folliculitis of beard.
symbiosis mutually advantageous association of two or more organisms.
symbiotic relating to symbiosis.
symptoms subjective disturbances of disease, felt or experienced by the patient but not directly measurable; for example, pain—the patient feels it definitely but it cannot be seen, heard, touched, etc.
syndrome set of symptoms and signs occurring together in a complex; etiologic agents variable.
system group of organs concerned in performing the same general function; the entire organism.
systemic relating to a system; relating to the entire organism instead of a part.

taxonomy branch of biology treating of arrangement and classification of animals and plants.
technic method of performing an operation, test, or other procedure.
template pattern, guide, or blueprint.
terminal disinfection disinfection of a room after it has been vacated by a patient.
terminal infection infection with streptococci or other pathogenic bacteria that occurs during the course of a chronic disease and causes death.
thallophytes division of the plant kingdom to which fungi and bacteria belong.
tinea a dermatomycosis.
tissue collection of cells forming a definite structure.
tissue culture cultivation of tissue cells away from the human or animal body.
titer concentration of infective microbes in a medium; amount of one substance to correspond with a given amount of another substance.
tolerance ability to endure without ill effect.
toxemia presence of toxins in the blood.
toxin poisonous substance elaborated during the growth of pathogenic bacteria.
toxoid toxin treated in such a manner that its toxic properties are destroyed without affecting its antibody-producing properties.
transduction transmission of a genetic factor from one bacterial cell to another by a viral agent.
transformation artificial conversion of bacterial types within a species by transfer of DNA from one bacterium to another.
trophozoite active, motile, feeding stage of a protozoan organism.
tubercle granuloma that forms the unit specific lesion of tuberculosis.
tuberculin toxic protein extract obtained from tubercle bacilli.
tuberculous affected with tuberculosis.
tumor mass.
typing (classification) determination of the category to which an individual, object, microbe, and the like, belongs with respect to a given standard of reference.

ulcer circumscribed area of inflammatory necrosis of the epithelial lining of a surface.
ulceration process of ulcer formation.
unicellular composed of but a single cell.
unilateral confined to one side.
unit standard of measurement.

vaccinate to introduce a vaccine into the body.
vaccination introduction of a vaccine into the body.
vaccine causative agent of a disease so modified that it is incapable of producing disease while retaining its power to cause antibody formation.
vaccinia cowpox.
vaccinoid modified vaccinia.
variation deviation from the parent form of an organism.
varioloid mild form of smallpox occurring in a person who has been vaccinated or previously had the disease.
VDRL (Venereal Disease Research Laboratory) test precipitation test for syphilis.
vector carrier of disease-producing agents from one host to another, especially an arthropod (fly, mosquito, flea, louse, or other insect).
vegetative bacteria nonspore-forming bacteria or spore-forming bacteria in their nonsporulating state.
venereal transmission of disease by intimate sexual contact.

vesicle (blister) small circumscribed elevation of the skin containing a thin nonpurulent fluid.
viremia presence of virus in the bloodstream.
virology science that treats of viruses and viral diseases.
virucide agent destroying or inactivating viruses.
virulence ability of an organism to produce disease.
virus submicroscopic agent of infectious disease; requires living cells for its proliferation.
virus-neutralizing antibodies antibodies that inactivate viruses.
viscerotropic affinity for the internal organs of the chest or abdomen.
viscus (pl., **viscera)** internal organ, especially one of the abdominal organs.
vital functions functions necesary for the maintenance of life.
vitamins certain little-understood food substances whose presence in very small amounts is necessary for the normal functioning of the body cells.

Wassermann test complement fixation test for syphilis devised by August von Wassermann.
Weil-Felix reaction nonspecific but highly valuable agglutination test for typhus fever in which the organism agglutinated is a member of the *Proteus* group.
wheal circumscribed elevation of the skin caused by edema of the underlying connective tissue.
Widal test agglutination test for typhoid fever.
wild virus virus found in nature.
Wright's stain mixture of eosin and methylene blue used to demonstrate blood cells and malarial parasites.

X rays (roentgen rays) highly penetrating form of ionizing radiation produced in special high-voltage equipment.

yeasts unicellular fungi.

zoology science that treats of animal life.
zoonosis (pl., **zoonoses)** disease of animals that may be secondarily transmitted to man.

INDEX

A

ABO blood group system, 180, 182
Abortion, contagious, in animals, 361
Abscess, 292
Absorption, bacterial, 100
Acetic acid production, 573
Acid, as disinfectant, 248
Acid-fast bacteria, 470-483
  project for, 552-553
Acid-fast stains, 61-63
Actinobacillus mallei, 368-369
Actinomyces, 490, 503
Actinomycosis, 503-504
Adaptation of bacteria, 50
Adenoviruses, 403, 436-437
  diagnosis of, 410
  pneumonia from, inclusion bodies in, 401
  vaccine for, 608
Adjuvants, 170
Aerobes, 80
  gram-positive bacilli, 324-331
  obligate, 80
Aerobic respiration, 101
Aerosols
  BCG vaccine, 606
  for disinfection, 276
Aflatoxins, 492
Agammaglobulinemia, 177
  gamma globulin in, 602-603
Agar, 84-85
Agar slant, 85
Agglutination tests, 192-195, 200
  hemagglutination, 195
  lymphoagglutination, 180
  project for, 218
Agglutinins, 192
  of blood, 182-183
  cold, 377
  group, 192
  immune, 192
  nonspecific, 194
  against typhoid bacilli, 345
Agglutinogens, 192
Agriculture, microbiology of, 571
Air
  disinfection of, 276
  spread of infection in, 149
Alastrim, 424
Alcohol
  ethyl, 244-245, 251
  isopropyl, 245, 251
  production of, 572
Alcoholic beverages, manufacture of, 572
Aleppo button, 516
Alexin; *see* Complement
Alementary tract; *see* Gastrointestinal tract
Allergens, 202
Allergy, 202-213; *see also* Hypersensitivity
  and anaphylaxis, 205-207
  and Arthus phenomenon, 209-210
  and asthma, 207
  atopic, 207
  characteristics of, 202
  and contact dermatitis, 212
  cross allergenicity to penicillins, 258
  delayed type, 203, 205, 210-212
  desensitization in, 212
  to drugs, 210-211, 255-256
  and hay fever, 207
  immediate type, 203, 205-210
  infectious, 211
  mechanisms in, 203-205
  project for, 222
  and serum sickness, 209
  skin tests in, 200
  tests for detection of, 200, 212, 614, 619
  types of, 203
  urticaria and skin eruptions in, 207-208
Allografts, 180
Amantadine hydrochloride, 263
Amebase, 510
Amebiasis, 511-515
Amebicides, 239
Amphotericin B, 262
Amphyl, 246
Ampicillin, 258
Anaerobes, 80
  culture of, 94-95
  gram-positive bacilli, 332-339
  obligate, 80
Anaerobic respiration, 101
Anaphylaxis, 205-207
  agents causing, 207, 208
  in immunization procedures, 614
  in man, 206-207
  passive, 205

Anaphylaxis—cont'd
role of smooth muscle in, 206
Anastomoses, vascular, 156
Anatoxins, 103
Ancylostoma duodenale, 536-538
Angstrom, 29
Animals
bites of; *see* Bites
inoculation of, 110-112
project for, 132
nutrition of, microbes in, 571
as source of infection, 152
Anopheles mosquito, 521, 522
Antagonism, in bacteria, 82-83
Anthrax, 152, 329-331
Antibiotics, 252-263
broad-spectrum, 254, 255
manufacture of, 574
microbial sources of, 253
mode of action of, 254-255
narrow-spectrum, 254, 255
selective toxicity of, 255
sensitivity testing of bacteria, 95-96
project for, 281
side effects of, 255-256
spectrum of activity in, 254, 255
Antibodies, 169, 170-171
antigen-antibody reaction, 171-173
autoantibodies, 178
classification of, 170, 172
complete, 172
Coombs test for, 186-187
detection of, 189, 190
fluorescent, 195, 200
formation of, 170-171
heterophil, 194
incomplete or blocking, 173
nature of, 170
reaginic, 207
in syphilis, 384
virus-neutralizing, 195, 200, 407-408
Antigen(s), 169-170, 176-177
allergens, 202
Australia, 184, 449
of blood groups, 182
H antigens, 345
hepatitis-associated, 449
heterophil, 194
Milan, 449
O antigens, 345
partial, 169
transplantation, 180
tumor, 181
Vi antigens, 345
Antigen-antibody reaction, 171-173
detection of, 189-201
project for, 218-221
Antigenic determinant or marker, 171
Antiglobulin test, 186
Antilymphocyte globulin, 181
Antiseptics, 239, 242-250
action of, 240-241
Antistreptolysin O titer, 300
Antitoxins, 103, 596-600
botulinus, 339, 598
diphtheria, 598-599
gas gangrene, 599
listing of, 598
preparation of, 596-597
standardization of, 597-598
tetanus, 599, 600
toxin-antitoxin
in diphtheria, 607
flocculation, 191
Antivenins, 600
Antiviral agents, 263
Arboviruses, 403-404, 441-442
Argyrol, 244
Arizona, 353
Arthropod-borne diseases, 528
Arthus phenomenon, 209-210
Ascaris lumbricoides, 539
Aschoff bodies, 304
Ascomycetes, 487
Asepsis, 239
Aspergillosis, 492
Aspergillus, 487, 489, 492
Asthma, 207
Athlete's foot, 491
Atopic allergy, 207
Attenuation
of bacteria, 50
of viruses, 607
Australia antigen, 184, 449
Autoclave, 232-234, 266
operation of, 233-234
project for, 280
Autograft, 179
Autoimmune diseases, 178-179
Autoimmunization, 178
Autotrophic organisms, 79
Avery, Oswald T., 22

B

Bacilli
Actinobacillus mallei, 368-369
anthrax, 329-331
Bacteroides, 374
Bartonella, 378
Battey, 480
Bordet-Gengou, 358-360
botulism, 337-339
Brucella, 361-364
Calymmatobacterium granulomatis, 370
clostridia, 332-339

Bacilli—cont'd
coliform, 349-353
detection in water, 579-581
comma, 390
diphtheria, 324-329
diphtheroid, 328
Ducrey, 360
dysentery, 347-349
enteric, 340-356
project for, 550-551
Escherichia coli, 349-353
Gärtner's, 593
gram-negative
enteric, 340-356
Mycoplasma, 376-378
Pseudomonas, 371-373
small bacilli, 357-370
gram-positive
aerobic, 324-331
anaerobic, 332-339
project for, 551-552
Hansen's, 481
hemophilic, 357-361
project for, 551
influenza, 357-358
Koch-Weeks, 360
lactobacilli, 374-375
Listeria monocytogenes, 375
Mima, 376
Morax-Axenfeld, 361
Mycoplasma, 376-378
paracolon, 340, 352-353
paratyphoid, 346-347
Pasteurella, 364-368
Proteus, 353
Pseudomonas, 371-374
Salmonella, 342-347
Shigella, 347-349
smegma, 480
tetanus, 332-335
tubercle, 470
typhoid, 343-346
Bacitracin, 260
disk test, 300
Bacteremia, 146
in endotoxin shock, 354-355
Bacteria, 40-52
acid-fast, 470-483
project for, 552-553
adaptation of, 50
antibacterial serums, 600-601
attenuation of, 50
biologic activities of, 97-106
project for, 127-129
biologic needs of, 78-83
capsule formation in, 43-44
cell division in, 48
cell wall of, 42-43
Bacteria—cont'd
chemical composition of, 44-45
classification of, 5-11
compared to other microbes, 398
culture media for, 78, 84-88
direct microscopic clump count of, 589
dissociation of, 49
distribution of, 40
project for, 126
drug fastness of; *see* Resistance to drugs
electricity affecting, 81
environmental factors affecting, 78-82
project for, 126-127
examination of; *see* Visualization of microbes
genetic substance of, 50-52
growth by-products of, 81
identification of, 112-113
intermicrobial transfer in, 51-52
interrelations of, 82-83
kinship to plants, 40
light affecting, 81
medically related activities of, 102-105
metabolic activities of, 97-102
moisture affecting, 79
morphology of, 41-48
variants in, 50
motility of, 45-46
mutations in, 50
nonpathogenic, 40
nutrition of, 78-79
osmotic pressure affecting, 81-82
oxygen affecting, 80
pathogenic, 40
pH affecting, 80
plate count of, 93-94
project for, 132
production of L-forms, 48-49
R-type colonies of, 49
reactions in culture media, 80
reproduction of, 48-49
rod-shaped; *see* Bacilli
S-type colonies of, 49
secondary effects of, 105
sensitivity tests of, 95-96
shape of, 41-42
size of, 42
spherical; *see* Cocci
spiral-shaped; *see Spirillum;* Spirochetes; Vibrio
sporulating, 47, 48
staining of, 59-63
project for, 72
structure of, 42-44
temperature affecting, 79-80
variations in, 49-52
vegetative, 47
Bactericides, 239
Bacteriocins, 105
Bacteriology, definition of, 2

Bacteriolysins, 192
Bacteriophages, 293-294, 398, 411-413
  characteristics of, 411
  diagnosis of, 413
  life cycle of, 411-413
Bacteriostasis, 239
Bacteroides, 374
Baking, yeast enzymes in, 573
Balantidiasis, 522
Balantidium coli, 510, 522
Bancroft's filaria worm, 544-546
Bang's disease, 361
Barber's itch, 491
Bard-Parker germicide, 245, 251
Bartonella, 378
Basidiomycetes, 487
Basophils, and allergic reactions, 203, 204
Bathroom fixtures, disinfection of, 275
Battey bacilli, 480
BCG vaccine, 606, 623
  aerosol, 606
  and leprosy prevention, 483
Bedsoniae, 463
  compared to other microbes, 398
  and inclusion bodies, 401
  infections with, 463-467
    in animals, 467
    cat scratch disease, 467
    diagnosis of, 410
    lymphopathia venereum, 467
    ornithosis and psittacosis, 463-465
    skin tests in, 409
    trachoma, 465-467
Beer, manufacture of, 572
Beijerinck, Martinus Willem, 19
Benzalkonium chloride, 243, 251
Benzoic acid, 248
Betadine, 247
Bilharziasis, 529-530
Biochemical reactions, 107-110
Biologic activities of bacteria, 97-106
Biologic needs of bacteria, 78-83
Biologic products for immunization, 596-616
Biology, definition of, 2
Bites
  antivenin for, 600
  of cats and dogs, microbiology of, 368
  and rabies prevention, 440
  rat-bite fever, 391, 505
Blackwater fever, 521
Blastomycin test, 494
Blastomycosis, 492-494
  North American, 492-494
  South American, 494
Bleaching powder, 250
Bliss, Eleanor, 20
Blood
  agglutinins of, 182-183
Blood—cont'd
  collection of specimens
    for cultures, 115
    for serologic examinations, 116
  cross-matching of, 186
    project for, 218-219
  pathogens found in, 121
  as portal of exit in infections, 145
  Rh factor in, 184
  transfusions of
    and Coombs test, 186-187
    hepatitis from, 184
    reactions from, 183-184
Blood groups, 185-186
Blumberg, Baruch, 24
Boas-Oppler bacillus, 374
Boiling, for sterilization, 231
  cold boiling, 238
  project for, 280
Bordet, Jules Jean Baptiste, 19
Bordet-Gengou bacillus, 358-360
Bordetella
  parapertussis, 358
  pertussis, 358-360
    characteristics of, 357, 358-359
    diagnosis of infection, 359
    immunity to, 359-360
    modes of infection, 359
    pathogenicity of, 359
    prevention of infection, 360
Boric acid, 248
Borrelia
  buccalis, 389
  recurrentis, 388
  refringens, 389
  vincentii, 389
Botulinum toxoid, 607
Botulinus antitoxin, 339, 598
Botulism, 592
  nature of, 337
  prevention of, 338-339
Brieger, Ludwig, 17
Brill-Zinsser disease, 458
Bronchial secretions, collection of specimens, 118
Bronchopneumonia, 306
Brucella, 361, 364
  abortus, 361, 362
  characteristics of, 361
  diagnosis of infection, 363-364
  immunity to, 364
  melitensis, 361, 362
  pathogenicity of, 361-362
  prevention of infection, 364
  project for, 551
  sources and modes of infection, 362-363
  suis, 361, 362
Brucellergen, 364
Brushes, sterilization of, 270

Bubo
climatic, 467
in plague, 365
Bubonic plague, 152
Buchner, Hans, 17
Budd, William, 16
Budding, of fungi, 486
Burkitt, Denis, 25
Burkitt's lymphoma, 418
Burning, for sterilization, 236-237
Butter
bacteria in, 590
production of, 571
Buttermilk, 571, 590

C

C-reactive protein, 192
Calamine lotion, 244
Calcium hypochlorite, 250
Calymmatobacterium granulomatis, 370
Cancer
immunity in, 181-182
and viruses, 414-418
Candidiasis, 494-495
Canning of food, 595
Capnophiles, 80
Capsid, of virus, 398
Capsomeres, of virus, 398
Capsule formation, in bacteria, 43
Carbenicillin, 258
Carbolic acid, 245
Carbomycin, 260
Carbon cycle, microbes in, 568-569
Caries, dental, 374-375
and fluoride in water, 584
Carriers
active, 150
animal, 152, 153
in cholera, 390
convalescent, 150
in diphtheria, 327
human, 150-151
insect, 150
intermittent, 151
intestinal and urinary, 150
of meningococci, 320
oral, 150
passive, 150
in plague, 366
in typhoid fever, 344-345
Carrión's disease, 378
Caseation, in tuberculosis, 474
Cat scratch disease, 467
Cats
distemper in, 454
microbiology of bites, 368
Cavity formation, in tuberculosis, 474
Ceepryn chloride, 243
Cell(s), 26-39
anatomy of, 26-38
antimicrobial agents affecting, 254-255
bacterial, 40-52
cytoplasm of, 29, 31-33
division of, 39
bacterial, 48
hyperplasia of, 402
immunologically competent, 171
nucleus of, 29, 33-38
project for study of, 69-71
protoplasm of, 26
shape of, 26
size of, 27-29
structure of, 29-39
Cell-mediated immunity, 168, 176-177
Cell wall, 32
bacterial, 42-43
Centrioles, 32
Centrosome, 32
Cephalosporins, 258-259
Cercariae, 527
Cerebrospinal fluid
collection of specimens, 119
pathogens in, 122
Cestodes, 530-535; *see also* Tapeworms
Chagas' disease, 515
Chain, Ernst Boris, 21
Chancre, in syphilis, 381-382
Chancroid, 360
Cheese, production of, 572
Chemical agents
affecting bacteria, 81
for sterilization, 239-263
Chemical composition, of bacteria, 44-45
Chemosynthesis, 100-101
Chemotaxis, 81
Chemotherapeutic agents, 250-263
anaphylaxis from, 208
drug idiosyncrasy, 210
drug intolerance, 210
resistance to, 52, 256
and sensitivity testing of bacteria, 95-96
sensitivity to, 210-211
testing for, 95-96
Chemotherapeutic index, 250
Chickenpox, 425-427
gamma globulin in, 602-603
inclusion bodies in varicella-zoster, 401
Chilomastix, 510
mesnili, 516
Chlamydiae; *see* Bedsoniae
Chloramines, 248
Chloramphenicol, 262
Chlorinated lime, 250
Chlorination of water, 581, 583, 585
Chlorine, 247-248

Chlorophyll, antimicrobial effects of, 250
Chlortetracycline, 262
Cholera, 390
  vaccine for, 606, 626-627, 628, 629
Cholera red reaction, 390
Chromatin granules, 33
Chromoblastomycosis, 502
Chromosomes, 33-37
  recombinant, 51
  of virus, 398
Cidex, 245, 251
Cilia, of protozoa, 509
Ciliatea, 510
Circulation
  collateral, 156
  lymphatic, 163
Cisternal puncture, 119
Citrobacter organisms, 353
Classification
  of antibodies, 170, 172
  of bacteria, 5-11
  biologic, 3
  of enzymes, 99-100
  of fungi, 487-488
  of viruses, 402-405
Clindamycin, 260
Clonorchis sinensis, 528
Clorox, 248
Clostridium, 332-339
  botulinum, 332, 337-339, 592
  characteristics of, 332
  novyi, 332, 335-336
  perfringens, 332, 335-336
    detection in water, 581
  septicum, 332, 335-336
  tetani, 332-335
Clothing, soiled, disinfection of, 274
Coagulase, 104, 290
Coagulase test, 104, 290
Cocci, 41
  Diplococcus pneumoniae, 304-308
  Gaffkya tetragena, 294
  gonococcus, 310-322
  gram-negative, 310-322
  gram-positive, 288-308
  meningococcus, 317-322
  pneumococcus, 304-308
  pyogenic, 288-308
    project for, 549-550
  staphylococcus, 288-294
  streptococcus, 294-304
Coccidioidin test, 498-499
Coccidioidomycosis, 495-499
Coccobacilli, 42
Coenzymes, 98
Cold viruses, 438
Colebrook, Leonard, 21
Colicins, 105
Coliform bacilli, 349-353
  Arizona-Edwardsiella-Citrobacter, 352-353
  characteristics of, 349
  detection in water, 579-581
  Escherichia coli, 349-351
  Klebsiella-Enterobacter-Serratia, 351-352
  pathogenicity of, 349
  "Providence" group, 353
Colistin, 260
Collateral circulation, 156
Collection of specimens, 114-124
Colorado tick fever, 448
  diagnosis of, 410
Comma bacillus, 390
Commensalism, 82
Communicable disease, 142
  transmission of, 148-150
Complement, 173-175
  fixation of, 175, 189-191, 200
    project for, 217
    in syphilis, 385
Condyloma acuminatum, 453
Conjugation, in bacteria, 51
Conjunctiva, smears and cultures from, 119
  pathogens found in, 122
Contamination, 140
  of water, 577
Convalescence, 146
  carriers in, 150
Convalescent serum, 601
Coombs test, 186-187
Coons, Albert, 22
Copper sulfate, 244
Coronaviruses, 404, 438
Corynebacterium diphtheriae, 324-329
Coryza, acute, 438
Councilman bodies, 401
Count, bacterial, 93-94, 589
Cox, Herald A., 23
Cox vaccine, 612
Coxiella burnetii, 462
Coxsackieviruses, 404, 435-436, 446
Credé's method, for prevention of ophthalmia neonatorum, 316
Cresol, 245
Crick, F. H. C., 22, 23
Croup, membranous, 324
Cryptococcosis, 499
Culex mosquito, 522
Cultural methods, 88-95
  for anaerobic bacteria, 94-95
  and bacterial plate count, 93-94
  in brucellosis, 363
  in embryonated hen's egg, 95, 405
  end result of, 89-91
  for fungi, 488
  in gonorrhea diagnosis, 313
  incubator, 88-89

Cultural methods—cont'd
inoculation, 88
for meningococci, 320
mixed culture, 91
in Petri dish, 85, 91
pour plate method, 91
projects for, 130-132, 221-222
pure cultures, 91-92
slide cultures, 95
streak plates in, 93
tissue cultures, 95, 405
for viruses, 405-407
Culture media, 78, 84-88
agar, 84-85
differential, 86
nonsynthetic, 88
project for, 129-130
Sabouraud, 488
selective, 86
sterilization of, 85-86
synthetic, 88
Cultures
from conjunctiva, 119
from nose and throat, 118
Cyst(s)
in amebiasis, 511
formation by protozoa, 509
hydatid, 535
and tapeworms, 533
Cysticercus, 533
Cytochrome oxidase, 110
Cytology, definition of, 26
Cytolysis, 192
Cytomegalic inclusion, 401
Cytomegalic inclusion disease, 401, 430
Cytomegalovirus, as teratogen, 414
Cytopathic effect, of viruses, 399
Cytoplasm, 29, 31-33
in protozoa, 508

D

Dairy microbiology, 571-572, 586-591
Dakin's solution, 248
Dark-field illumination, 58-59
Darling's disease, 499
Davaine, Casimir Joseph, 14
Death point, thermal, for bacteria, 231
Decontamination, 240
Defense mechanisms, 155-164
anatomic barriers in, 155
chemical factors in, 156
fever in, 156-157
interferon in, 156
lymphoid system in, 162-164
lysozyme in, 156
phagocytosis in, 157-159
physiologic reserves in, 156
reticuloendothelial system in, 157-161
Defense mechanisms—cont'd
spleen in, 163
thymus in, 163
Defervescence period, in infections, 146
Degenerative forms of bacteria, 50
Degerming, 240
Dehydrogenases, 101
Delhi boil, 516
Demethylchlortetracycline, 262
Dengue fever, 448
Dental caries, 374-375
and fluoride in water, 584
Deodorants, 239
Deoxyribonuclease
formation by bacteria, 109
streptococcal, 297
Deoxyribonucleic acid (DNA), 34-37
Deoxyriboviruses, 402
Dermatitis
blastomycetic, 492
contact, 212
Dermatomycoses, 489, 491-492
Dermotropic viruses, 405
Desenex, 248
Desensitization, in allergies, 212
Detergents, 241
enzyme, 242
Deuteromycetes, 488
Dhobie itch, 491
Diabetes, mycoses in, 495, 501
Diaparene chloride, 243
Diarrhea
infantile, 349-351
traveler's, 629-630
Dick, George, 19, 20
Dick, Gladys, 19, 20
Dick test, 200, 301, 302
Diffusion immunoassay, 197-199
Digestion, bacterial, 100
Dihydrostreptomycin, 259
Diphtheria antitoxin, 598-599
Diphtheria bacilli, 324-329
carriers of, 327
characteristics of, 324-325
diagnosis of infection with, 326, 327-328
extracellular toxin of, 325
immunity to, 328
pathogenicity of, 325-326
prevention and control of infection, 329
sources and modes of infection, 326-327
Diphtheria toxin-antitoxin, 607
Diphtheria toxoid, 607, 619, 627, 628
sensitivity test for, 619
with tetanus toxoid, and pertussis vaccine, 617, 618
Diphtheroid bacilli, 328
Diphyllobothrium latum, 535
Diplobacilli, 42

Diplococcus, 42
  pneumoniae, 304-308
Diplornavirus, 404
Dipylidium caninum, 535
Dishes, sterilization of, 275
Disinfectants, 239, 242-250, 266
  acid, 248
  action of, 240-241
  alcohols and aldehydes, 244-245
  chlorophyll, 250
  dyes, 249
  ethylene oxide, 249
  ferrous sulfate, 250
  halogen compounds, 247-248
  heavy metal compounds, 243-244
  lime preparations, 249-250
  oxidizing agents, 248-249
  phenol and derivatives, 245-247
  project for, 280-281
  qualities of, 240
  sulfur dioxide, 250
  surface-active compounds, 242-243
Disinfection
  of air, 276
  of articles for public use, 275
  concurrent, 273
  of feces, 274
  of hands, 270-271
    for nurses, 273-274
    project for, 282
  in infectious diseases, 273-275
  of mucous membranes, 273
  of shoes, 274
  of skin, 271-273
  of sputum, 274-275
  terminal, 270, 273, 275
  of urine, 274
  of wound, 273
Disposable items, use of, 268, 269, 613
Dissociation, bacterial, 49
Distemper of dogs and cats, 454
Distillation, and alcohol manufacture, 572
Division of cells, 39
  bacterial, 48
Döderlein's bacilli, 374
Dogs
  distemper in, 454
  microbiology of bites, 368
Domagk, Gerhard, 21
Donovan bodies, 370, 401
Doxycycline, 262
Dressings and linens, sterilization of, 269
Droplet infection, 148, 149, 408
Drug therapy; *see* Chemotherapeutic agents
Dubos, René Jules, 21
Ducrey's bacillus, 360
Dumdum fever, 516
Durham, Herbert Edward, 18
Dyes, as disinfectants, 249
Dysentery, amebic, 511
Dysentery bacilli, 347-349
  characteristics of, 347-348
  diagnosis of infection, 348
  immunity to, 348
  mode of infection, 348
  pathogenicity of, 348
  prevention of infection, 348-349
Dysgammaglobulinemia, 177
  gamma globulin in, 602-603

E

Ear, external, normal flora in, 142
Eating utensils, sterilization of, 275
Eaton agent, 377
Echinococcus granulosus, 535
Echoviruses, 404, 438, 447
Ectoparasite, 508
Ectoplasm, in protozoa, 508
Edwardsiella, 353
Ehrlich, Paul, 18
Electricity, affecting bacteria, 81
Electron microscope, 28-29
  for virus studies, 408
Electrophoresis, 197
  immunoelectrophoresis, 201
Elephantiasis, 545
Embden-Meyerhof glycolytic pathway, 102
Embryonated hen's egg, cultures in, 95, 405
Encephalitides, 441-443
  diagnosis of, 410
  epidemic, 442-443
  equine, 441-442
  postinfection, 443
Endemic infections, 148
Enders, John F., 23, 24
Endocarditis, streptococcal, 303
Endoenzymes, 99
Endoparasite, 508
Endoplasm, in protozoa, 508
Endoplasmic reticulum, 32-33
Endotoxins, 102, 103, 144
  shock from, 354-355
Energy storage, in bacteria, 101
Entamoeba histolytica, 511-515
Enteric bacilli
  coliform, 349-353
  endotoxin shock from, 354-355
  gram-negative, 340-356
  in hospital-acquired infections, 355-356
  paratyphoid, 346-347
  project for, 550-551
  Proteus, 353
  Salmonella, 342-347
  Shigella, 347-349
Enterobacter organisms, 352
Enterobius vermicularis, 540-541

Enterococci, 296
Enterotoxin, 290
Enteroviruses, 404, 435
Environmental factors
  affecting bacteria, 78-82
  affecting fungi, 486-487
Enzyme(s), 97-100
  activation of, 98
  characteristics of, 97-99
  classification of, 99-100
  demonstration of, 110
  in detergents, 242
  feedback control of, 98
  proteolytic, 109
  in yeasts, uses of, 572, 573
Enzyme system, 98
Epidemics, 143-144, 148
Epidemiology, 148
Epidermophyton, 491
Epstein-Barr virus, 418, 452-453
Ergotism, 503
Erysipelas, 302-303
Erythrasma, 502
Erythrocytes
  immunology of, 182-187
  rouleau formation in, 219
Erythrogenic toxin, 297
  in scarlet fever, 301
Erythromycin, 260
Escherichia coli, 349-351
Espundia, 516
Ethambutol, 252
Ethylene oxide, 249
  for sterilization, 265
Evans, Alice C., 19, 20
Exoenzymes, 99
Exotoxins, 102, 103
  necrotizing, 144, 290
Eye
  normal flora in, 142
  ophthalmia neonatorum, 311-312, 316
  smears and cultures from conjunctiva, 119, 122
  in toxoplasmosis, 524
  in trachoma, 465-467

F

Face, dangerous triangle in, 292
Facultative organisms, 79, 80
Farcy, 368-369
Fasciolopsis buski, 528
Favus, 491
Feces
  collection of specimens, 118-119
  disinfection of, 274
  examination of, for parasites, 546
  pathogens in, 122
  as portal of exit in infections, 145
Feedback control, negative, of enzymes, 98
Fermentation, 101
  and alcohol manufacture, 572
  of milk, 590
  project for, 127-128
  and sauerkraut production, 573
  of sugars, 107-109
Ferrous sulfate, 250
Fever, 156-157
Fibrils, cellular, 33
Fibrinolysin, 104, 297
Fièvre boutonneuse, 461
Filaments, cellular, 33
Filariasis, 544-546
Filth, and spread of infection, 149
Filtration
  for sterilization, 230-231
    project for, 279
  for water purification, 582-583
Fishing, in cultures, 91
Fixing, of bacteria in smear, 59
Flagella
  in bacteria, 45
  in protozoa, 509
Flagellates, 510
  intestinal, 516
Flaming, of bacteria in smear, 59
Flax and hemp, retting of, 574
Flea-borne diseases, 528
  typhus, 458
Fleming, Alexander, 20-21
Flocculation
  tests, 200
    in syphilis, 385
  toxin-antitoxin, 191
Flora of body, normal, 140, 141-142
Florey, Howard, 21
Flukes, 527-530
Fluorescent antibodies, 195-196, 200
Fluorescent antibody test
  treponemal antibody absorption test in syphilis, 386
  in virus infections, 408
Fluorescent dyes, for sterilization, 238
Fluoride content of water, and tooth decay, 584
Fly-borne diseases, 528
Fomites, 149
Food
  infections from, 149, 592-593
  intoxication, 592
  microbiology of, 591-595
  poisoning, 591-593
  preservation of, 594-595
  safeguards for, 593-594
  and traveler's diarrhea, 630
Foot
  athlete's, 491
  Madura, 504
Foot-and-mouth disease, 454

Formaldehyde, 245
Formalin, 245, 251
Foshay's tularemia vaccine, 606
Fracastoro, Girolamo, 12
Frambesia, 388
Freezing of food, 594
Friedländer bacillus, 351
Fumagillin, 262
Fumigation, 239, 276
Fungi, 484-505
  actinomycetes, 490
  classification of, 487-488
  conditions affecting growth of, 486-487
  ergot, 503
  Imperfecti, 485, 488
  importance of, 489
  infections from, 488-505; *see also* Mycoses
  laboratory study of, 488
  pathology from, 488-489
  project for, 553-554
  ray, 503
  reproduction of, 485-486
  structure of, 484
Fungicides, 239
Furazolidone, 252
Fusobacterium fusiforme, 374
Fusospirochetal disease, 389

G

Gaffkya tetragena, 294
Gametocytes, 521
Gamma globulin, 601, 602-605
Gamna-Favre bodies, 401
Gangrene, gas, 335-336
  antitoxin for, 599
Gärtner's bacillus, 593
Gas gangrene, 335-336
  antitoxin for, 599
Gas sterilization, 265
Gastrointestinal tract
  normal flora in, 141
  as portal of entry for infection, 143
  tuberculosis of, 476
Gel diffusion, 197-199
Gelatin, liquefaction of, 109
Genetic code, 36
Genetic substance, of bacteria, 50-52
Gengou, Octave, 19
Genitourinary tract
  normal flora in, 141
  as portal of entry for infection, 143
Gentamicin sulfate, 260
Geotrichosis, 502
Germicides, 239
Giardia, 510
Giardiasis, 516
Gilchrist's disease, 492
Glanders, 368-369
Glandular fever, 452-453
Gloves, reusable, sterilization of, 269-270
Glutaraldehyde, 245, 251
Golgi apparatus, 33
Gonococcus, 310-322
  characteristics of, 310-311
  compared to meningococcus, 318
  diagnosis of infection, 313-314
  pathogenicity of, 311-312
  prevention of infection, 316-317
  smears for, 119
  social importance of gonorrhea, 315-316
  sources and modes of infections, 313
  and VD pandemic, 317
Gooseflesh, 156
Graft-versus-host reaction, 181
Grafts, 179-181; *see also* Transplantation of tissue
Gram-negative organisms
  cocci, 310-322
  enteric bacilli, 340-356
  Mycoplasma, 376-378
  Pseudomonas, 371-373
  small bacilli, 357-370
Gram-positive organisms
  aerobic bacilli, 324-331
  anaerobic bacilli, 332-339
  cocci, 288-308
  project for, 551-552
Gram stain, 60-61
  negative, 61
  positive, 61
  variable, 61
Gramicidin, 257
Granules
  chromatin, 33
  metachromatic, 44
  sulfur, in actinomycosis, 503
Granuloma
  coccidioidal, 498
  inguinale, 370
    inclusion bodies in, 401
  paracoccidioidal, 494
Granulomatosis infantiseptica, 375
Granulomatous inflammation, 160
Gray syndrome, from chloramphenicol, 262
Griffith, Frederick, 22
Griseofulvin, 262
Gruber, Max von, 18
Guarnieri bodies, 401
Gumma, syphilitic, 383

H

H antigens, 345
Haemophilus
  aegyptius, 360
  ducreyi, 360
  influenzae, 357-358, 431
    characteristics of, 357, 358

Haemophilus—cont'd
influenzae—cont'd
serum against, 600
parainfluenzae, 361
pertussis; see Bordetella pertussis
suis, 360, 433
Halazone, for water purification, 581
Halogen compounds, 247-248
Halophiles, 82
Hand brushes, sterilization of, 270
Hands
disinfecting of, 270-271
for nurses, 273-274
project for, 282
infections from contaminated fingers, 149
Hanging-drop preparations, 57-58
Hansen's bacillus, 481
Haplosporea, 510
Haptens, 169
Haverhill fever, 505
Hay fever, 207
Heat
bacterial, 105
for sterilization
dry, 236-237
moist, 231-236
project for, 279
Hemagglutination, 195
inhibition of, 195
as test for rubella, 421
in test for virus infections, 408
viral, 195, 200
Hemolysins, 103-104, 144, 290, 297
alpha, 104
beta, 104
Hemolysis
and complement fixation, 190
in malaria, 521
project for, 219-221
Hemolytic disease, Rh, gamma globulin in, 602-603
Hemophilic bacteria, 357-361
project for, 551
Hemopoiesis, 157
Hemorrhagic septicemia, pasteurellas of, 367-368
Hemp and flax, retting of, 574
Hepatitis
infectious, 451
gamma globulin in, 604-605
posttransfusion, 184
serum, 451
gamma globulin in, 602-603
viral, 448-451
Hepatitis-associated antigen, 449
Herpes
simplex, 427-429
inclusion bodies in, 401
zoster, 425-427
Herpesviruses, 403
as teratogens, 414
Herrelea species, 376
Heterotrophic organisms, 79
Hexachlorophene, 246-247
Hexagerm, 247
Hi-Sine, 247
Histamine, and allergic reactions, 203
Histocompatibility, 170
Histoplasmin test, 501
Histoplasmosis, 499-501
Historical aspects of microbiology, 12-25
HL-A system, 180
Holmes, Oliver W., 14
Homograft, 180
Homostatic graft, 179
Hooke, Robert, 14
Hookworms, 536-538
Hospital-acquired infections
enteric bacilli in, 355-356
Pseudomonas, 372
staphylococcal, 292-293
Host, 79, 508
compromised, 181
defenses of, 155-164
Humoral immunity, 168, 169-175
Hutchinson's teeth, 387
Hyaluronidase, 105, 297
Hydatid cyst, 535
Hydrocephalus, acute, in tuberculosis, 476
Hydrogen acceptor or carrier, 101
Hydrogen peroxide, 248
Hydrogen sulfide production, by bacteria, 109
Hydrolases, 100
Hydrolysis, 100
of starch, 109
Hydrophobia, 438-441
Hymenolepis nana, 535
Hyperplasia, virus-induced, 402
Hypersensitivity, 202
to biologic products, testing for, 614, 619
to infections, 211
Hypertonic solutions, 81
Hyphae, 485
Hypogammaglobulinemia, 177
gamma globulin in, 602-603
Hypotonic solutions, 81

I

Iatrogenic reactions, 207
Ice, bacteria in, 584-585
Ice cream, bacteria in, 590
Idoxuridine, 263
Immune agglutinins, 192
Immune bodies; *see* Antibodies
Immune serum, 596-601
Immune system, 165, 167-168
Immunity
acquired, 166
active, 166, 167
and antibody formation, 170-171

Immunity—cont'd
and antigen activity, 169-170, 176-177
and antigen-antibody reaction, 171-173
artificially acquired, 166, 167
autoimmunity, 178
in brucellosis, 364
in cancer, 181-182
cell-mediated, 168, 176-177
compared to humoral immunity, 176
and complement activity, 173-175
definition of, 165
in diphtheria, 328
disturbances in, 177-179
in dysentery, 348
humoral, 168, 169-175
compared to cell-mediated immunity, 176
and infection, 165-167
in influenza, 433
level of, 167
lymphoid tissue role in, 168
to meningococcus, 320
natural, 166
nonspecific or innate, 167
passive, 166-167
in plague, 366
to pneumococci, 308
in poliomyelitis, 445
in rickettsial disease, 456
in rubella, 423
in scarlet fever, 302
specific or adaptive, 168
to staphylococci, 294
to streptococci, 300
in tuberculosis, 479
in tularemia, 367
in typhoid, 343-346
in typhus fever, 459
in viral diseases, 409-411
in whooping cough, 359-360
Immunization procedures, 166
adenovirus vaccine, 608
for anthrax, 331
antibacterial serums, 600-601
antitoxins, 596-600
antivenins, 600
antiviral serums, 601
autogenous vaccines, 606
BCG vaccine, 606, 623
biologic products for, 596-616
botulinum toxoid, 607
botulinus antitoxin, 339
for brucellosis, 364
cholera toxoid, 606
cholera vaccine, 606, 626-627, 628, 629
combined diphtheria-tetanus toxoids, 617
convalescent serum, 601
Cox vaccine, 612
current status of common vaccines, 613
for diphtheria, 329
Immunization procedures—cont'd
diphtheria toxoid, 607, 619, 627, 628
sensitivity test for, 619
for distemper, 454
encephalomyelitis after, 443
for equine encephalomyelitis, 442
gamma globulin, 601, 602-605
influenza virus vaccine, 608-610, 623, 627, 628
injection routes in, 604-605, 614-615
for international travel, 628-630
measles vaccine, 610, 625, 626, 628
mumps virus vaccine, 610, 621
for paratyphoid fever, 625-626
passive, 596-601
pertussis vaccine, 360, 606, 618-619
plague vaccine, 606, 627, 628, 629
poliomyelitis vaccines, 610, 617, 624-625, 627, 628
precautions in administration of products in, 613-615
rabies vaccines, 610-611, 621-623
recommendations for
by American Academy of Pediatrics, 617-623
by U. S. Public Health Service, 623-626
requirements for U. S. Armed Forces, 626-628
rickettsial vaccines, 612-613
rubella vaccine, 611-612, 621
smallpox vaccine, 425, 612, 619-621, 627
and standards for biologic products, 615-616
stock vaccines, 606
and tests for hypersensitivity, 614, 619
for tetanus, 335
tetanus toxoid, 607, 618, 627
toxin-antitoxin in, 607
toxoids, 103, 607
tularemia vaccine, 606
typhoid vaccine, 346, 607, 625-626, 627, 628, 629
for typhus, epidemic, 628
vaccines, 601-613
bacterial, 601-607
viral, 607-612
yellow fever vaccine, 612, 627, 628, 629
Immunoassay, diffusion, 197-199
Immunobiology, 165
Immunochemistry, 168
Immunoelectrophoresis, 201
Immunofluorescence, 195-197
Immunoglobulins, 170
absence of, 177
classes of, 172-173
detection of, 189
IgE and allergic reactions, 204
immunologic activities of, 174-175
Immunologic deficiency syndromes, 177
gamma globulin in, 602-603
Immunologic reactions, 189-201
agglutination, 192-195
application of, 200
C-reactive protein, 192
complement fixation, 189-191

Immunologic reactions—cont'd
  cytolysis, 192
  diffusion immunoassay, 197-199
  electrophoresis, 197
  in fungous diseases, 488
  hemagglutination, 195
  immunoelectrophoresis, 201
  immunofluorescence, 195-197
  neutralization tests, 195
  nitroblue tetrazolium test, 195
  opsonocytophagic tests, 195
  precipitation, 191-192
  toxin-antitoxin flocculation, 191
  in Wassermann test, 191
Immunologic tolerance, 177
Immunologically competent cell, 171
Immunology, 165-188
  definition of, 2
  and lymphoid system, 162-164
  of red blood cells, 182-187
  and tissue transplantation, 179-181
Immunopathology, 202
Immunosuppression, 177
  and transplantation of tissues, 180-181
Incineration, for sterilization, 236-237
Inclusions, cellular, 31, 33, 399, 401
  cytomegalic inclusion disease, 401, 430
  in trachoma, 465
Incubation period, in infections, 145-146, 147-148
Incubator, bacteriologic, 88-89
Indole production, by bacteria, 109
Industry, microbes in, 571-576
Infection, 140-154
  communicable, 142
  course of, 145-146
  defense against; *see* Defense mechanisms
  defervescence period in, 146
  definition of, 140
  from direct contact, 148
  disinfection procedures in, 273-275
  droplet, 148, 149, 408
  endemic, 148
  epidemic, 143-144, 148
  exogenous, 142
  factors influencing occurrence of, 143-145
  focal, 146
  from food, 149, 592-593
  general effects in, 145
  generalized, 146
  hospital-acquired
    enteric bacilli in, 355-356
    Pseudomonas, 372
    staphylococcal, 292-293
  and how microbes cause disease, 144
  hypersensitivity to, 211
  and immunity, 165-167
  and immunology, 165-188
  incubation period in, 145-146, 147-148
  from indirect contact, 148-149
Infection—cont'd
  inoculation, 146
  from insects, 149-150
  invasion of body in, 140-143, 146
  and Koch's postulates, 154
  latent, 146
  local effects in, 145
  localized, 146
  lymphatics in, 163-164
  mixed, 146
  noncommunicable, 142
  and normal flora of body, 140, 141-142
  pandemic, 148
  pattern of, 145-148
  portals of entry in, 143
  portals of exit in, 145
  primary, 146
  prodromal symptoms in, 146
  reservoir of, 150
  responses to, 164
  secondary, 146
  self-limited, 146
  sources of, project for, 216-217
  sporadic, 148
  spread of, 148-153
  subclinical, 146
  superinfections, 256, 489
  susceptibility to, 165
  terminal, 148
  transmission of, 148-150
  in travelers, 152
  types of, 146-148
Influenza, 430-433
  diagnosis of, 433
  epidemiology of, 432-433
  immunity to, 433
  prevention of, 433
  recent epidemics of, 433
  swine, 433
  vaccine for, 608-610, 623, 627, 628
Influenza bacillus, 357-358
Influenza viruses, 431-432
  diagnosis of, 410
Injection routes, in immunization procedures, 604-605, 614-615
Inoculation
  of cultures, 88
  infection from, 146
  of laboratory animals, 110-112
    project for, 132
Insects, infections from, 149-150
Interference
  between staphylococci strains, 294
  between virus strains, 409
Interferon, 156, 263, 409
  inducers of, 411
International Association of Medical Assistance to Travelers, 628
Intestinal tract; *see* Gastrointestinal tract

Intoxication food, 592
Invasion period, in infections, 146
Involution, of bacteria, 50
Iodine, 247, 251
  for water purification, 581, 585
Iodophors, 247, 251
Iosan, 247
Isoantibodies, 169
Isoantigens, 169
Isoenzyme, 99
Isograft, 179
Isomerases, 100
Isoniazid, 252

J

Jenner, Edward, 14

K

Kala-azar, 516
Kanamycin, 259-260
Kendrick, Pearl, 20
Kenny, Méave, 21
Kinases, bacterial, 104, 144
Kircher, Athanasius, 12
Kitasato, Shibasaburo, 17
Klebsiella, 351
  pneumoniae, affecting flukes in schistosomiasis, 530
Knapp, R. E., 19
Koch postulates, 154
Koch, Robert, 16
Koch-Weeks bacillus, 360
Koprowski, Hilary, 23
Kuru, 443

L

L-forms in bacteria, production of, 48
Lactic acid bacteria, 571, 590
  cultures of, 571-572
Lactobacilli, 374-375
Lancefield, Rebecca, 19, 20
Landsteiner, Karl, 19
Laveran, Alphonse, 16
Leeuwenhoek, Antonj van, 13
Leishmania, 510
  braziliensis, 516
  donovani, 516
  tropica, 516
Leishmaniasis, 515-516
  cutaneous, 516
  visceral, 516
Lenses, of microscope, 55
Leprosy, 481-483
Leptospirosis, 388-389
Lethal dose, minimum, of toxins, 597
Lethal factor, 144, 290
Leukemia, 414
Leukocidins, 104, 144, 290, 297
Leukocytes, 160
  polymorphonuclear, 160
Leukocytosis, 160
Leukoproteases, 160
Leukovirus, 404
Ligases, 100
Ligatures, sterlization of, 270
Light
  affecting bacteria, 81
  bacteria producing, 105
Lime preparations, 249-250
Lincomycin, 260
Linens, sterilization of, 269, 274
Lipschütz bodies, 401
Liquefaction, of gelatin, 109
Lister, Joseph, 16
Listeria monocytogenes, 375
Liver, viral diseases of, 448-451
Localization, elective, of microbes, 144
Lockjaw, 332
Louse-borne diseases, 528
  typhus, 457
Lumbar puncture, 119
Lungs, tuberculosis of, 474-476
Lupus erythematosus, 179
Lyases, 100
Lymph node, 163
Lymphadenopathy, 164
Lymphatic circulation, 163
Lymphoagglutination, 180
Lymphocytes, 163
  antilymphocyte globulin, 181
  immunologic competence of, 171
  thymus-dependent, 168, 176
  thymus-independent, 168
Lymphocytotoxicity, 180
Lymphoid follicles, 168
Lymphoid system, 162-164
Lymphoid tissue, 163
  aplasia of, 177
  role in immunity, 168
Lymphoma, Burkitt, 418
Lymphopathia venereum, 467
  diagnosis of, 410
  inclusion bodies in, 401
Lysergic acid diethylamide (LSD), 503
Lysine decarboxylase, 110
Lysis, 413
Lysogeny, 413
Lysol, 247
Lysosomes, 33
Lysozyme, 156

M

M protein, 296
MacLeod, Colin, 22
Macrogametes, 521
Macromolecules, 26
  synthesis of, 102

Macrophages, 160
Madura foot, 504
Malaria, 516-522
  diagnosis of, 521-522
  estivoautumnal or malignant, 517
  life story of parasite in, 518-521
  modes of infection in, 518
  mosquitoes transmitting, 522
  prevention of, 522
  quartan, 517
  regular intermittent, 517
  remittent, 517
  tertian, 517
  types of, 517
Mallein, 369
Malt, 572
Malta fever, 362
Mantoux test, 477
Martin, Louis, 17
Mast cells, and allergic reactions, 203, 204
McCarty, Maclyn, 22
Measles, 420
  diagnosis of, 410
  gamma globulin in, 602-603
  German, 421-423; *see also* Rubella
  and giant cell pneumonia, inclusion bodies in, 401
  vaccine for, 610, 625, 626, 628
Media for growth of microbes, 87-88; *see also* Culture media
Melanin, 33
Melioidosis, 373-374
Membrane
  nuclear, 29, 38
  plasma, 31-32
    of bacteria, 43
  semipermeable, of bacterial cell, 81
  undulating, in protozoa, 509
Meningitis
  organisms in, 321, 322
  tuberculous, 476
Meningococcus, 317-322
  antimeningococcal serum, 600
  characteristics of, 317-318
  compared to gonococci, 318
  diagnosis of infections, 320
  groups of, 318
  immunity to, 320
  pathogenicity of, 318-319
  prevention of infection, 322
  sources and mode of infection, 319-320
Merbromin, 243
Mercresin, 244
Mercurochrome, 243
Merozoites, 519
Merthiolate, 244, 251
Mesophiles, 79
Mesosome, 44
Metabolic activities of bacteria, 97-102
Metachromatic granules, 44
Metal compounds, as disinfectants, 243-244
Metaphen, 243
Metazoa, 527-547
  cestodes, 530-535
  nematodes, 536-546
  trematodes, 527-530
Metchnikoff, Élie, 16, 18
Methisazone, 263
Microaerophiles, 80
Microbes
  in animal nutrition, 571
  classification of, 5-11
  definition of, 2
  in industry, 571-576
  naming of, 3
  in nature, 568-571
Microbiology
  of agriculture, 571
  dairy, 571, 586-591
  definition of, 2
  of food, 591-595
  historical aspects of, 12-25
  of industry, 571-576
  of nature, 568
  of sanitation, 577
  scope of, 4
  of water, 577-585
Microfilariae, 545
Microgametes, 521
Microgametocyte, 521
Micrometer, ocular, 55-57
Micron, 27
Microscope(s), 53-55
  adjustment of, 55
  compound, 54-55
    care of, 67
    project for, 67-68
  electron, 28-29
    for virus studies, 408
  equivalent focal distance of, 55
  fluorescent, 196
  for fungi studies, 488
  immersion objective for, 55
  lenses of, 55
  project for bacterial studies with, 71
  simple, 54
Microsporum, 491
Microtubules, 33
Microvilli, 32
Milan antigen, 449
Milk, 586-591
  bacteria in, 586
    project for, 636-637
  bacterial action on, 571, 590
  butter from, 571, 590
  digestion by bacteria, 110
  formulas for hospital nurseries, 589

Milk—cont'd
grading of, 589
ice cream from, 590
infections from, 149, 472, 586-587
pasteurized, 236, 587
processing of, 571-572
requirements for safe supply of, 588-589
Milk sickness, 587
Mima species, 376
Miracidium, 527
Mite-borne diseases, 528
Mitochondria, 32
Mitosis, 39
Moisture, affecting bacteria, 79
Molds, 484
importance of, 489
plant diseases from, 503
Molluscum bodies, 401
Molluscum contagiosum, 425
inclusion bodies in, 401
Moniliasis, 494-495
Monocytes, 160
Mononucleosis, infectious, 452-453
Morax-Axenfeld bacillus, 361
Moraxella lacunata, 361
Morphology of bacteria, 41-48
variants of, 50
Mosquito-borne diseases, 447-448, 528
malaria, 516-522
Motility of bacteria, 45-46
Mouth, Vincent's angina of, 328, 374, 389
Mucor, 487, 489, 501
Mucormycosis, 501
Mucous membranes, disinfection of, 273
Mumps, 451-452
gamma globulin in, 602-603
vaccine for, 610, 621
Murine typhus, 458
Mutations in bacteria, 50
Mycelium, 484-485
Mycetoma, 504
Mycobacteria, unclassified, 480-481
Mycobacteriosis, 480
Mycobacterium
avium, 471
bovis, 471
leprae, 481-483
tuberculosis, 470-479, 490
in butter, 590
characteristics of, 471
importance of, 470-471
in milk, 587
pathogenicity of, 473
species of, 471-472
toxic products of, 472
Mycology, 484-505
definition of, 2
medical, 491
Mycoplasma, 376-378
characteristics of, 376
compared to other microbes, 398
diagnosis of infection, 378
pathogenicity of, 376-377
T-strain, 377
Mycoses, 488-505
actinomycosis, 503-504
in animals, 502
aspergillosis, 492
blastomycosis, 492-494
candidiasis, 494-495
coccidioidomycosis, 495-499
cryptococcosis, 499
dermatomycoses, 489, 491-492
histoplasmosis, 499-501
nocardiosis, 504
phycomycosis, 501
in plants, 502-503
primary, 489
secondary, 489
sporotrichosis, 501
superficial, 488, 491-492
systemic, 489, 492-501
Mycotoxins, 503
Myxoviruses, 403

N

Nature, microbiology of, 568-571
Necator americanus, 536-538
Needham, John, 13
Needles and syringes, disposable, 268, 613
Negri bodies, 401, 440
Neisseria
catarrhalis, 315, 322
flava, 315, 322
gonorrhoeae, 310-317
meningitidis, 315, 317-322
sicca, 322
Nematodes
Ancylostoma duodenale, 536-538
Ascaris lumbricoides, 539
Enterobius vermicularis, 540-541
Necator americanus, 536-538
Onchocerca volvulus, 546
Strongyloides stercoralis, 539
Trichinella spiralis, 542-544
Trichuris trichiura, 541
Wuchereria bancrofti, 544-546
Neomycin, 259
Neo-Silvol, 244
Nervous system, central, viral diseases of, 438-445
Neurotoxins, 334
Neurotropic viruses, 405
Neutralization tests, 195
Neutrophils, 160
Niacin test, 110
Nitrate reduction, by bacteria, 109

Nitric acid, fuming, 248
Nitroblue tetrazolium test, 195
Nitrofurans, 252
Nitrogen cycle, microbes in, 569-571
Nitromersol, 243
Nocardia, 490, 504
Nocardiosis, 504
Nose, cultures from, 118
Novy, Frederick George, 19
Nuclear membrane, 29, 38
Nucleolus, 38
Nucleoplasm, 29
Nucleus, 29, 33-38
  of bacteria, 44
Nutrition of bacteria, 78-79
Nystatin, 262

O

O antigens, 345
Odors, bacterial, 105
Oidiomycin, 495
Oil prospecting, and bacteria in soil, 574
Oleandomycin, 260
Onchocerca volvulus, 546
Oncogenesis, viruses in, 414-418
Oocyst, 521
Ookinete, 521
Operating room, sterilization of, 270
Ophthalmia neonatorum, 311-312
  prevention of, 316
Ophthalmic test, for allergy detection, 212
Opsonocytophagic tests, 195
Organelles, 31
  of protozoa, 508
Oriental sore, 516
Ornithosis, 463-465
Oroya fever, 378
Osmophiles, 82
Osmotic pressure, affecting bacteria, 81-82
O-syl, 246
Otomycosis, 502
Oxidase, 101
Oxidase reaction, 110, 310
Oxidation, biologic, 101
  classes of, 101-102
Oxidizing agents, as disinfectants, 248-249
Oxidoreductases, 100
Oxygen, affecting bacteria, 80
Oxytetracycline, 262
Oxyuris vermicularis, 540

P

Papovaviruses, 403
Para-aminosalicylic acid, 252
Paracolon bacilli, 340, 352-353
Paragonimus westermani, 528
Parainfluenza viruses, 435
Paramyxoviruses, 403, 435
Parasites, 79
Parasitology
  definition of, 2
  examination of feces for parasites, 546
  facultative parasites, 79
  metazoa, 527-547
  protozoa, 508-526
Paratyphoid bacilli, 346-347
  vaccine for, 625-626
Paresis, general, 384
Park, William Hallock, 17
Paromomycin, 263
Parotitis, epidemic, 451-452
Parrot fever, 463-465
Parvoviruses, 403
Pasteur, Louis, 15-16
Pasteurella, 364-368
  of hemorrhagic septicemia, 367-367
  multocida, 368
  pestis, 364-366
  project for, 551
  tularensis, 366-367
    characteristics of, 366
Pasteurization, 236, 587
Patch test, for allergy detection, 212
Pathogenic organisms, 40, 143
  in specimen collections, 121-123
Paul-Bunnell test, 194, 200
Penicillin, 257-258
  allergic reactions to, 211
  cross allergenicity of, 258
  semisynthetic, 258
  susceptibility to, project for, 281
Penicillinase, 256, 258
Penicilliosis, 502
Penicillium, 487
Peritoneal fluid
  collection of specimens, 119
  pathogens in, 123
Peroxisome, 33
Pertussis, 358-360
  antipertussis serum, 600
    hyperimmune, 601
  gamma globulin in, 604-605
  vaccine for, 606, 618-619
    with diphtheria-tetanus toxoids, 617, 618
Peruvian verruca, 378
Petri dish, 85
  cultures in, 91
Pets, as sources of infection, 152, 153
Pfeiffer's bacillus, 357
pH, affecting bacteria, 80
Phages; *see* Bacteriophage
Phagocytosis, 157-159
  and opsonocytophagic tests, 195
Phagolysosome, 157
Phagosome, 157

Phenol, 245, 251
coefficient test, 241
Phenylalanine deaminase, 110
pHisodern, 247
pHisoHex, 247
Phosphorylation, 101
Photochromogens, 480
Phycomycetes, 487
Phycomycosis, 501
Picodnaviruses, 403
Picornaviruses, 404, 435
Piedra, 502
Pigment production by bacteria, 105
project for, 128-129
Pili, in bacteria, 46
Pinworms, 540-541
Placenta
as portal of entry for infection, 143
transfer of immunity through, 166
Plague
bubonic, 152, 365
pneumonic, 365
septicemic, 365
vaccine for, 606, 627, 628, 629
Plague bacillus, 364-366
carriers of, 366
and clinical types of plague, 365
diagnosis of infection, 366
immunity to, 366
prevention of infection, 366
Plaques, phage, 413
Plasma cells, 163
Plasma membrane, 31-32
of bacteria, 43
Plasmodium, 510, 516
falciparum, 517
life cycle of, 518-521
malariae, 517
ovale, 517
vivax, 517
Plasmolysis, 81
Plasmoptysis, 81
Plate count, bacterial, 93-94
project for, 132
Plating, of cultures, 91
Pleomorphism, 50
Pleural fluid
collection of specimens, 119
pathogens in, 123
Pleurisy, tuberculous, 476
Pleuropneumonia-like organisms, 376-378
Pneumobacillus, 351
Pneumococcus, 304-308
antipneumococcal serum, 600
characteristics of, 304
diagnosis of infection, 307-308
differentiation from streptococci, 308
immunity to, 308
Pneumococcus—cont'd
pathogenicity of, 305-306
prevention of infections, 308
sources and modes of infection, 306-307
toxic products of, 305
types of, 304-305
Pneumocystis carinii, 510, 525
Pneumocystosis, 525-526
Pneumonia, 305-306
adenovirus, inclusion bodies in, 401
alba, 387
bronchopneumonia, 306
hypostatic, 306
interstitial plasma cell, 525
lobar, 306
measles giant cell, inclusion bodies in, 401
primary atypical, 377
tuberculous, 476
viral, 438
Pneumotropic viruses, 405
Poisoning, food, 591-593
Polar bodies, 44, 324
Poliomyelitis, 443-445
epidemiology of, 445
gamma globulin in, 602-603
immunity to, 445
prevention of, 445
transmission of, 444-445
vaccines for, 610, 617, 624-625, 627, 628
Polioviruses, 404, 444
diagnosis of, 410
Polluted water, 577
Polymyxin, 260
Portals of entry, in infections, 143
Portals of exit, in infections, 145
Potassium permanganate, 249
Pour plate method, for cultures, 91
Poxviruses, 402
Precipitation tests, 200
in syphilis, 385
in virus infections, 408
Precipitin reaction, 191-192
Preservation of food, 594-595
Preservatives, 239
Pressure, osmotic, affecting bacteria, 81-82
Prodromal symptoms, in infections, 146
Proglottids, of tapeworm, 530
Prophage, 413
Protargol, 244
Protein
C-reactive, 192
M protein, 296
Proteolytic enzymes, 109
Proteus bacilli, 353
Protophyta, 4
Protoplasm, 26
Protoplasts, 42

Protozoa, 508-511
  characteristics of, 508-510
  classification of, 510
  cyst formation by, 509
  diseases from, 511-526
  locomotion of, 509
  project for, 555
  reproduction of, 509-510
  structure of, 508-509
Protozoan diseases, 511-526
  amebiasis, 511-515
  balantidiasis, 522
  giardiasis, 516
  leishmaniasis, 515-516
  malaria, 516-522
  toxoplasmosis, 522-525
  trichomoniasis, 516
  trypanosomiasis, 515
"Providence" organisms, 353
Pseudomembrane, in diphtheria, 325
Pseudomonas
  aeruginosa, 371-372
  pseudomallei, 373-374
Pseudopods, of protozoa, 509
Psittacosis, 152, 463-465
  inclusion bodies in, 401
Psychrophiles, 79
Ptomaine poisoning, 591
Public welfare
  disinfection for articles used by public, 275
  immunizations in; *see* Immunization procedures
  and microbes everywhere, 568-576
  and water microbiology, 577-585
Puerperal sepsis, 304
Pulse rate, and temperature, 157
Puncture
  cisternal, 119
  lumbar, 119
Purification
  of sewage, 585
  of water, 581-584
Pus
  collection of specimens, 119
  pathogens in, 123
Pustule, malignant, 329
Putrefaction, 109
  of milk, 590
Pyemia, 148, 292
Pyogenic cocci, 288-308
  project for, 549-550

Q

Q fever, 462
  vaccine for, 612

R

Rabies, 152, 438-441
  antirabies hyperimmune serum, 601

Rabies—cont'd
  diagnosis of, 410, 440
  immune globulin in, 601
  inclusion bodies in, 401, 440
  incubation period in, 440
  prevention of, 440-441
  vaccines for, 610-611, 621-623
Radiant energy, affecting bacteria, 81
Radiation, ionizing
  food treated with, 595
  for sterilization, 238
Ramon, Gaston Leon, 17
Rat-bite fever, 391, 505
Reactions
  biochemical, 107-110
  immunologic, 189-201
Reagins, 207
  in syphilis, 384
Recombinant chromosomes, 51
Rectal swabs, for collection of feces, 118
Refrigeration of food, 594
Rejection of grafts, 180
Relapsing fever, 388
Rennin, 572
Reoviruses, 404, 437
Reproduction of bacteria, 48-49
Reservoir of infection, 150
Resident population of microbes, normal, 140, 141-142
Resistance to drugs by bacteria, 52, 256
  and sensitivity testing, 95-96
Resistance factor, 256
Respiration
  aerobic, 101
  anaerobic, 101
Respiratory tract
  discharges from
    disinfection of, 274
    as portal of exit in infections, 145
  normal flora in, 141
  as portal of entry for infection, 143
  protective mechanisms in, 155
  virus diseases of, 430-438
    accute syndromes, 434-437
Respiratory viruses, 434, 435-437
Reticuloendothelial system, 157-161
  anatomy of, 159-161
  and phagocytosis, 157-159
Reticulum, endoplasmic, 32-33
Retting of flax and hemp, 574
Rh factor, 184
Rh hemolytic disease, gamma globulin in, 602-603
Rhabdoviruses, 403
Rhagades, syphilitic, 387
Rheumatic fever, streptococcal, 303-304
Rhinosporidiosis, 502
Rhinoviruses, 404, 438
  of cattle, 454

Rhizopodea, 510
Rhizopus, 487, 489, 501
Ribonucleic acid (RNA), 37-38
  messenger RNA, 37
Ribosomes, 33
Riboviruses, 402
Rickettsia, 456-457
  akari, 461
  collection of specimens, 119-120
  compared to other microbes, 398
  filtrable, 231
  pathogenicity of, 456
  prowazekii, 457
  quintana, 462
  ricketsii, 460
  special properties of, 456
  tsutsugamushi, 461
  typhi, 458
Rickettsial diseases, 457-462
  in animals, 462
  diagnosis of, 457
  Q fever, 462
  scrub typhus, 461-462
  spotted fever group, 460-461
  trench fever, 462
  typhus fever group, 457-460
  vaccines for, 612-613
Rickettsialpox, 461
Rifampin, 261
Ringworm, 491
Ristocetin, 260-261
Risus sardonicus, 332
Rocky Mountain spotted fever, 456, 460-461
  vaccine for, 612
Ross, Ronald, 16
Rouleau formation, in red cells, 219
Roundworms, 539
Rous, Peyton, 24
Roux, Émile, 16
Rubella, 421-423
  congenital, 421
  diagnosis of, 410, 421
  gamma globulin in, 602-603
  immunity to, 423
  vaccines for, 611-612, 621
  virus as teratogen, 414
Rubeola, 420; *see also* Measles

S

Sabin, Albert, 23-24
Sabin vaccine, 610
Sabouraud culture medium, 488
Saccharomyces, 487
Safety, rules for, 66
Salicylic acid, 248
Saline solutions, sterilization of, 270
Salisbury viruses, 438
Saliva, as portal of exit in infections, 145
Salk, Jonas, 23
Salk vaccine, 610
Salmonella, 342-347
  characteristics of, 342
  food poisoning from, 593
  gallinarum, 347
  paratyphi, 346-347
  pathogenicity of, 343
  sources and modes of infection, 342-343
  typhi, 343-346
  typhimurium, 347
Sanitation, microbiology of, 577
Sanitizing, 240
Sapremia, 148
Saprophytes, 79
  facultative, 79
Sarcinae, 42
Satellite phenomenon, 358
Sauerkraut, production of, 573
Scarlet fever, 301-302
  erythrogenic toxin in, 301
  immunity to, 302
  prevention of, 302
Schick test, 200, 328
Schistosoma, 528
Schistosomiasis, 529-530
Schizonts, 519
Schultz-Charlton phenomenon, 302
Schultz-Dale reaction, 206
Schwann, Theodor, 14
Scolex, of tapeworm, 530
Scotochromogens, 480
Scrapie, 443
Scrofula, 473
Scrub regimen, surgical, 271
Scrubbing, for sterilization, 230
Scutula, in favus, 491
Seatworms, 540-541
Sedimentation
  for sterilization, 231
  for water purification, 582
Semipermeable membrane, of bacterial cell, 81
Semmelweis, Ignaz Philipp, 14
Sensitivity testing, of bacteria, 95-96
Sepsis, puerperal, 304
Septicemia, 148
  hemorrhagic, pasteurellas of, 367-368
  puerperal, 304
  staphylococcal, 292
Serology, 189
  project for, 217
Serratia organisms, 352
Serum hepatitis, 451
Serum, immune, 596-601
Serum sickness, 209
Sewage, purification of, 585
Shape
  of bacteria, 41-42

Shape—cont'd
  of cells, 26
Shigella, 347-349
Shingles, 425-427
Shock
  anaphylactic, 205
  endotoxin, 354-355
Shoes, disinfection of, 274
Side effects of antibiotics, 255-256
Silver nitrate, 244
Silvol, 244
Size
  of bacteria, 42
  of cells, 27-29
Skin
  allergic reactions in, 207-208
  contact dermatitis, 212
  fungous infections of, 489, 491-492
  leishmaniasis of, 516
  normal flora in, 141
  as portal of entry for infection, 143
  preoperative preparation of, 271-273
  protective mechanisms in, 155
  viral diseases of, 420-430
Skin tests, 200, 212
  in blastomycosis, 494
  in candidiasis, 495
  in coccidioidomycosis, 498-499
  cross reactions in, 494, 499
  in histoplasmosis, 501
  for hypersensitivity to biologic products, 614
  in toxoplasmosis, 525
  in tuberculosis, 211, 477
  in viral and bedsonial diseases, 409
Sleeping sickness, 442
  African, 515
Slide cultures, 95
Slow virus infections, 443
Smallpox, 424-425
  diagnosis of, 410, 425
  gamma globulin in, 604-605
  prevention of, 425
  transmission of, 424-425
  vaccine for, 612, 619-621, 627, 628
    contraindications to, 621
    indications for, 620-621
  and vaccinia, 425
    inclusion bodies in, 401
Smears, 59
  from conjunctiva, 119
  for gonococci, 119, 313
Smegma bacilli, 480
Snails, and life cycle of flukes, 527
Snake bites, antivenin for, 600
Snuffles, in congenital syphilis, 387
Soaps, 242-243
Sodium hypochlorite, 248
Sodium perborate, 248
Sodoku, 391
Spallanzani, Abbé Lazzaro, 13
Specimen collection, 114-124
  blood
    for cultures, 115
    pathogens found in, 121
    for serologic examinations, 116
  bronchial secretions, 118
  cerebrospinal fluid, 119
    pathogens found in, 122
  and cultures from nose and throat, 118
  feces, 118-119
    pathogens found in, 122
  general rules for, 114-115
  nose and throat secretions, 118
    pathogens found in, 121
  pathogens found in, 121-123
  peritoneal fluid, 119
    pathogens found in, 123
  pleural fluid, 119
    pathogens found in, 123
  pus, 119
    pathogens found in, 123
  and shipment of specimens, 120, 123
  smears and cultures from conjunctiva, 119
    pathogens found in, 122
  smears for gonococci, 119, 313
  sputum, 116-117
    pathogens found in, 122
  urine, 115
    pathogens found in, 121
  for viral and rickettsial laboratory, 119-120
Spectinomycin, 260
Speleonosis, 499
Spider bites, antivenin for, 600
Spirillum, 41
  minus, 391, 505
Spirochetes, 41, 379-392
  Borrelia recurrentis, 388
  classification of, 380
  Leptospira, 388-389
  project for, 553
  Treponema pallidum, 380-388
Spleen, 163
Splenomegaly, 164
Spores, 46-48
  formation in fungi, 485-486
Sporotrichosis, 501
Sporozoites, 519, 521
Sporulating bacteria, 47, 48
Spotted fever, Rocky Mountain, 460-461
  vaccine for, 612
Sputum
  collection of specimens, 116-117
  disinfection of, 274-275
  pathogens found in, 122
Staining procedures, 59-63
  acid-fast, 61-63

Staining procedures—cont'd
  counterstains, 61
  differential stains, 60-63
  Gram, 60-61
  negative (relief) staining, 63
  project for, 72
  simple stains, 60
  special stains, 63
Staphene, 246
Staphylococcus, 42, 288-294
  albus, 289
  aureus, 289
  bacterial interference between strains of, 294
  bacteriologic diagnosis of, 293
  characteristics of, 288-290
  citreus, 289
  food intoxication from, 591, 592
  hospital-acquired infection, 292-293
  immunity to, 294
  pathogenicity of, 291-292
  pathologic conditions from, 292
  prevention and control of infections, 294
  sources and modes of infection with, 293
  toxic products of, 290
Staphylokinase, 290
Staphylolysin, 104, 290
Starch, hydrolysis of, 109
Steam, for sterilization, 232-235
  compared to gas sterilization, 265
Steinhardt, Edna, 19, 20
Sterilization, 230
  by boiling, 231
  of brushes, 270
  chemical agents in, 239-263
  by chemotherapeutic agents, 250-263
  of culture media, 85-86
  of dressings, 269
  by dry heat, 236-237
  by filtration, 230-231
    project for, 279
  by fluorescent dyes, 238
  fractional or intermittent, 235-236
  by gas, 265
    compared to steam sterilization, 265
  of gloves, 269-270
  of linens, 269, 274
  by moist heat, 231-236
  of operating room, 270
  of operation site, 271-273
  by pasteurization, 236, 587
  physical agents in, 230-238
  by scrubbing, 230, 271
  by sedimentation, 231
  by steam, 232-235
    compared to gas sterilization, 265
  by sunlight, 81, 237
  of surgical instruments and supplies, 265-270
  of sutures and ligatures, 270
Sterilization—cont'd
  testing efficiency of, 282
  of thermometers, 267, 275
    project for, 282
  of tubing, 267, 269
  by ultrasonics, 238
  by ultraviolet radiation, 237-238
  of vaccines, 276
  of water and saline solutions, 270
  by x-rays, 238
Sternberg, George M., 17-18
Stewart, Sarah, 19, 20
Stool specimens
  collection of, 118-119
  pathogens found in, 122
Storage of energy, in bacteria, 101
Straus reaction, 369
Streak plates, 93
Streaking, of cultures, 91
Streptobacillus, 42
  moniliformis, 505
Streptococcus, 42, 294-304
  alpha-hemolytic, 295, 296
  in animal infections, 299
  beta-hemolytic, 295-296
  biologic properties of, 298
  characteristics of, 295
  classification of, 295-296
  diagnosis of infection, 300
  differentiation from pneumococci, 308
  endocarditis from, 303
  enterococcus, 296
  erysipelas from, 302-303
  faecalis, detection in water, 581
  food poisoning from, 593
  gamma, 295, 296
  in human diseases, 299
  immunity to, 300
  lactic, 296
  pathogenicity of, 297-299
  prevention of infections, 300
  and puerperal sepsis, 304
  pyogenes, 296, 299
  rheumatic fever from, 303-304
  scarlet fever from, 301-302
  serologic groups of, 298
  sources and modes of infection, 299-300
  throat inflammation from, 303
  toxic products of, 296-297
  viridans, 296
Streptodornase, 297
Streptokinase, 104, 297
Streptolysin, 104, 297
Streptolysin O, antibodies to, 300
Streptomyces, 490
Streptomycin, 259
Strongyloides stercoralis, 539

Structure
  of bacteria, 42
  of cells, 29-39
Sugars, fermentation of, 107-109
Sulfonamides, 251-252
Sulfur dioxide, 250
Sulfur granules, in actinomycosis, 503
Sunlight, affecting bacteria, 81, 237
Superinfection, 256, 489
Surface tension, in disinfection, 241
Surgical instruments and supplies, sterlization of, 265-270
Susceptibility, 165
Sussenguth-Kline test, 543
Sutures, sterilization of, 270
Swabs, rectal, for collection of feces, 118
Swimming pool sanitation, 585
Symbiosis, 82
Syncytial virus, respiratory, 435
Synergism, 82
Syphilis, 380-388
  acquired, 381-386
  cardiovascular, 384
  congenital 386-388
  diagnosis of, 385
    false negative and false positive reactions in, 385-386
  immunity to, 384-385
  incubation stage of, 381
  latent stage of, 383
  modes of infection, 381
  neurosyphilis, 384
  prevention of, 386
  primary stage of, 381-382
  secondary stage of, 383
  tertiary stage of, 383
  Wassermann test for, 191
  and yaws, 388
Syringes and needles, disposable, 268, 613

T

Tabes dorsalis, 384
Taenia, 530; *see also* Tapeworms
  saginata, 534, 535
  solium, 534-535
Tanning, 573
Tapeworms, 530-535
  anatomy of, 530-532
  beef, 534, 535
  dog, 535
  dwarf, 535
  fish, 535
  life cycle of, 533-534
  pork, 534-535
Telosporea, 510
Temperature
  affecting bacteria, 79-80
  of body, 157
    fever, 156-157
Temperature—cont'd
  of body—cont'd
    and pulse rate, 157
Teratogenesis, and viruses, 413-414
Tetanus, 152
  antitoxin for, 599, 600
  gamma globulin in, 604-605
  toxoid for, 607, 618, 627
    with diphtheria toxoid and pertussis vaccine, 617, 618
Tetanus bacillus, 332-335
  characteristics of, 332, 333
  distribution of, 333
  extracellular toxin of, 333-334
  pathogenicity of, 334
  pathology from, 334
  prevention of infection, 335
  sources and modes of infections, 334
Tetracycline, 261
Thallophyta, 4
Thermal death point, for bacteria, 231
Thermometers, sterilization of, 267, 275
  project for, 282
Thermophiles, 79
Thimerosal, 244
Threadworms, 540-541
Throat
  cultures from, 118
  inflammation from streptococci, 303
Thrush, 494-495
  vaginal, 495
Thymus, 163
  dysplasia of, 177
Thymus-dependent lymphocytes, 168
Tick-borne diseases, 528
  Colorado tick fever, 448
  diagnosis of, 410
  Rocky Mountain spotted fever, 460
Tine tests
  histoplasmin, 501
  tuberculin, 477
Tinea
  barbae, 491
  capitis, 491
  pedis, 491
Tissue cultures, 95
  for viruses, 405
Tissue typing, for transplantation, 180
Titers, 189
Tobacco, curing of, 573
Togavirus, 404
Tolerance, immunologic, 177
Tooth conditions
  decay, 374-375
    and fluoride content of water, 584
  Hutchinson teeth, 387
Torulosis, 499
Toxemia, 148
Toxicity, 144

Toxigenicity, 102
Toxin(s), 102-103
  fungal, 492
  minimum lethal dose of, 597
  mycotoxins, 503
Toxin-antitoxin
  in diphtheria, 607
  flocculation, 191
Toxoids, 103, 607; *see also* Immunization procedures
Toxoplasma, 510
  gondii, 510, 522
Toxoplasmin skin test, 525
Toxoplasmosis, 522-525
  acquired, 523-524
  congenital, 524
  diagnosis of, 525
  sources of infection in, 524-525
Trachoma, 465-467
  inclusion bodies in, 401
Transduction, in bacteria, 52
Transfer, intermicrobial, 51-52
Transferases, 100
Transformation, in bacteria, 52
Transfusions
  and Coombs test, 186-187
  cross matching of blood in, 186
  hepatitis from, 184
  reactions from, 183-184
Transmission of communicable diseases, 148-150
Transplantation of tissue, 179-181
  antigens in, 180
  graft-versus-host reaction, 181
  histocompatibility in, 180
  immunosuppression in, 180-181
  and rejection of grafts, 180
  tissue typing for, 180
Transport, active, mechanisms in, 100
Travelers
  diarrhea in, 629-630
  immunization of, 628-630
  infections in, 152
Trematodes, 527-530
Trench fever, 462
Trench mouth, 389
Treponema
  microdentium, 389
  pallidum, 380-388
    immobilization test, 385, 386
  pertenue, 388
Triatomid, diseases from, 528
Trichinella spiralis, 542-544
Trichinosis, 542-544
Trichocephalus, 541
Trichomonas, 510
  hominis, 516
  vaginalis, 516
Trichomoniasis, 516
Trichophyton, 491
Trichuris trichiura, 541
Trophozoites, vegetative, 511
Trypanosoma, 510
  cruzi, 515
  rhodesiense, 515
Trypanosomiasis, 515
  African, 515
  Gambian, 515
  Rhodesian, 515
  South American, 515
Tsutsugamushi disease, 461
Tubercle bacillus, 470
Tubercles, in tuberculosis, 473-474
Tuberculin, 472
Tuberculin tests, 477-479
Tuberculosis, 470-479
  animals as source of infection in, 152
  BCG vaccine in, 606, 626
  chronic, 473
  diagnosis of, 476-477
  immunity to, 479
  of intestines, 476
  lesions in, 473-474
  meningeal, 476
  miliary, 474
  from milk ingestion, 472, 587
  prevention of, 479
  primary, 473
  pulmonary, 474-476
  skin test in, 211
  sources of infection in, 472-473
  spread of, 474
  tuberculin tests in, 477-479
Tubing, sterilization of, 267, 269
Tularemia, 152, 366-367
  clinical types of, 367
  diagnosis of, 367
  modes of infection in, 367
  vaccine for, 606
Tumors
  and immunity, 181-182
  virus-induced, 414-418
Tunnicliff, Ruth, 20
Turgidity, 81
Tyndallization, 236
Typhoid bacillus, 343-346
  characteristics of, 343
  diagnosis of infection, 345
  immunity to, 345-346
  pathogenicity of, 344
  portal of entry for, 344
  portal of exit for, 344
  prevention of infection, 346
  spread of, 344-345
  vaccine for, 607, 625-626, 627, 628, 629
Typhus fevers, 152, 457-460
  diagnosis of, 458-459
  epidemic, 457
    vaccine for, 612, 628
  immunity to, 459

Typhus fevers—cont'd
  murine, 458
    vaccine for, 612
  prevention of, 459-460
  recrudescent, 458
  scrub, 461-462
Tyrocidine, 257
Tyrothricin, 257

U

Ulcer, tuberculous, 474
Ultramicroscopic objects, 29
Ultrasonics, for sterilization, 238
Ultraviolet radiation, for sterilization, 237-238
Undecylenic acid, 248
Undulant fever, 152, 362
Undulating membrane, in protozoa, 509
Urea, splitting of, by bacteria, 109
Urease, 109
Urine
  collection of specimens, 115
  disinfection of, 274
  pathogens found in, 121
  as portal of exit in infections, 145
Urticaria, in allergy, 207-208

V

Vaccines, 166, 596, 601-613; *see also* Immunization procedures
  bacterial, 601-607
  sterilization of, 276
  viral, 607-612
Vaccinia, 425, 619-620
  gamma globulin in, 604-605
  inclusion bodies in, 401
Vagina, normal flora in, 142
Vancomycin, 260
Variations in bacteria, 49-52
Varicella, 425-427
  gamma globulin in, 602-603
  inclusion bodies in, 401
Variola, 424-425; *see also* Smallpox
VDRL tests, 385
Vegetations and endocarditis, 303
Vegetative bacteria, 47
Venereal diseases, 317
  chancroid, 360
  condyloma acuminatum, 453
  gonorrhea, 310-322
  granuloma inguinale, 370
  herpes progenitalis, 427
  lymphopathia venereum, 467
  syphilis, 380-388
  trichomoniasis, 516
Venoms, antivenin preparations for, 600
Verrucae
  in condyloma acuminatum, 453
  Peruvian, 378
Vi antigens, 345
Vibrio, 41
  cholerae, 390
  Eltor, 390-391
  fetus, 391
  parahemolyticus, 391
Villemin, Jean Antoine, 15
Vincent's angina, 374, 389
  diagnosis of, 328
Vinegar, manufacture of, 573
Viomycin, 259
Viricides, 239
Viroid, 398
Virology, definition of, 2
Viron, 398
Viropexis, 399
Virulence, 143
  transposal of, 144
Virulence tests, guinea pig
  for diphtheria, 325
  for tuberculosis, 477
Virus-neutralizing antibody, 195, 200, 407-408
Viruses, 396-419
  antiviral agents, 263
  antiviral serums, 601
  attenuated, 607
  bacteriophage, 293-294, 398, 411-413
  and cancer, 414-418
  classification of, 402-405
  collection of specimens, 119-120
  compared to other microbes, 398
  coxsackieviruses, 446
  cultivation of, 405-407
  cytopathic effect of, 399
  definition of, 396-397
  diseases from, 420-455
    arthropod-borne, 447-448
    of central nervous system, 438-445
    diagnosis of, 407-408, 410
    hepatitis, 448-451
    prevention of, 411
    of skin, 420-430
    skin tests in, 409
    spread of, 408
  echoviruses, 447
  filtrable, 231
  hemagglutination of, 195, 200
  hyperplasia from, 402
  immunity to, 409-411
  and inclusion bodies, 399, 401
  life cycle of, 399
  pathogenicity of, 399-402
  respiratory, 430-438
  in sewage, 579-580
  slow virus infections, 443
  structure of, 397-398
  and teratogenesis, 413-414
  vaccines for, 607-612

Viscerotropic viruses, 405
Visualization of microbes, 53-65
  dark-field illumination in, 58-59
  hanging-drop preparations in, 57-58
  with microscope, 53-55; *see also* Microscope
  with ocular micrometer, 55-57
  project for, 71-72
  stained bacteria in, 59-63
  unstained bacteria in, 57-59
  wet mount for, 58
Vollmer test, 477
von Behring, Emil, 17
von Economo's disease, 442-443
von Pirquet test, 477
von Plenciz, Marc Antony, 13
Vulvovaginitis, 495

W

Waksman, Selman Abraham, 22
Wassermann test, 191, 385
Water
  bacteriologic examination of, 579-581
    project for, 634-635
  fluoride content of, and tooth decay, 584
  and ice storage, 584-585
  polluted, 577
  potable, 577
  purification of, 581-584
  and saline solutions, sterilization of, 270
  sources of, 577-579
  standards for drinking water, 581
  and swimming pool sanitation, 585
  and traveler's diarrhea, 630
Waterborne diseases, 579
Waterhouse-Friderichsen syndrome, 319
Watson, J. D., 22, 23
Weil's disease, 388
Weil-Felix reaction, 200, 353, 458, 461
Wells, and water supply, 577-579
Wescodyne, 247
Wetting agents, 241
Whipworm, 541
Whitfield's ointment, 248
Whooping cough, 358-360; *see also* Pertussis
Widal reaction, 200, 345
Wiener, Alexander S., 19
Wines, manufacture of, 572
Woodruff, Alice, 19
Woodwork, disinfection of, 275
World Health Organization, 630
Worms, 527, 530-546; *see also* Metazoa
Wounds, disinfection of, 273
Wuchereria bancrofti, 544-546

X

X-rays, for sterilization, 238

Y

Yaws, 388
Yeasts, 484
  and alcohol manufacture, 572
  in baking, 573
  importance of, 489
Yellow fever, 447
  inclusion bodies in, 401
  vaccine for, 612, 627, 628, 629
Yersin, Alexandre, 17
Yoghurt, 590

Z

Zephiran chloride, 243
Ziehl-Neelsen stain, 61
Zinc peroxide, medicinal, 244
Zoomastigophorea, 510
Zygote, 510